CONVERSION FACTORS

Length

1 in. = 2.54 cm

1 m = 39.4 in. = 3.28 ft

1 mi = 5280 ft = 1609 m

1 km = 0.621 mi

1 angstrom (Å) = 10^{-10} m

1 light-year (ly) = 9.46×10^{15} m

Volume

1 liter = 1000 cm^3

1 gallon = 3.79 liters

Speed

1 mi/h = 1.61 km/h = 0.447 m/s

Mass

1 atomic mass unit (u) = 1.660×10^{-27} kg

(Earth exerts a 2.205-lb force on an object with 1 kg mass)

Force

1 lb = 4.45 N

Work and Energy

1 ft · lb = 1.356 N · m = 1.356 J

1 cal = 4.180 J

1 eV = 1.60×10^{-19} J

1 kWh = 3.60×10^6 J

Power

1 W = 1 J/s = 0.738 ft · lb/s

1 hp (U.S.) = 746 W = 550 ft · lb/s

1 hp (metric) = 750 W

Pressure

1 atm = 1.01×10^5 N/m^2 = 14.7 lb/in^2

    = 760 mm Hg

1 Pa = 1 N/m^2

PHYSICAL CONSTANTS

Gravitational coefficient on Earth g	9.81 N/kg
Gravitational constant G	6.67×10^{-11} N · m^2/kg^2
Mass of Earth	5.97×10^{24} kg
Average radius of Earth	6.38×10^6 m
Density of dry air (STP)	1.3 kg/m^3
Density of water (4 °C)	1000 kg/m^3
Avogadro's number N_A	6.02×10^{23} particles (g atom)
Boltzmann's constant k_B	1.38×10^{-23} J/K
Gas constant R	8.3 J/mol · K
Speed of sound in air (0°)	340 m/s
Coulomb's constant k_C	9.0×10^9 N · m^2/C^2
Speed of light c	3.00×10^8 m/s
Elementary charge e	1.60×10^{-19} C
Electron mass m_e	9.11×10^{-31} kg = 5.4858×10^{-4} u
Proton mass m_p	1.67×10^{-27} kg = 1.00727 u
Neutron mass m_n	1.67×10^{-27} kg = 1.00866 u
Planck's constant h	6.63×10^{-34} J · s

POWER OF TEN PREFIXES

Prefix	Abbreviation	Value
Tera	T	10^{12}
Giga	G	10^9
Mega	M	10^6
Kilo	k	10^3
Hecto	h	10^2
Deka	da	10^1
Deci	d	10^{-1}
Centi	c	10^{-2}
Milli	m	10^{-3}
Micro	μ	10^{-6}
Nano	n	10^{-9}
Pico	p	10^{-12}
Femto	f	10^{-15}

SOME USEFUL MATH

Area of circle (radius r) πr^2

Surface area of sphere $4\pi r^2$

Volume of sphere $\frac{4}{3}\pi r^3$

Trig definitions:

  sin θ = (opposite side)/(hypotenuse)

  cos θ = (adjacent side)/(hypotenuse)

  tan θ = (opposite side)/(adjacent side)

Quadratic equation:

  $0 = ax^2 + bx + c$,

  where $x = \dfrac{-b \pm \sqrt{b^2 - 4ac}}{2a}$

Brief Contents

PART 1 Mechanics

1 Introducing Physics 1
2 Kinematics: Motion in One Dimension 13
3 Newtonian Mechanics 51
4 Applying Newton's Laws 84
5 Circular Motion 118
6 Impulse and Linear Momentum 147
7 Work and Energy 176
8 Extended Bodies at Rest 217
9 Rotational Motion 251

PART 2 VIBRATIONS AND WAVES

10 Vibrational Motion 284
11 Mechanical Waves 315

PART 3 GASES AND LIQUIDS

12 Gases 352
13 Static Fluids 386
14 Fluids in Motion 415

PART 4 THERMODYNAMICS

15 First Law of Thermodynamics 441
16 Second Law of Thermodynamics 476

PART 5 ELECTRICITY AND MAGNETISM

17 Electric Charge, Force, and Energy 500
18 The Electric Field 535
19 DC Circuits 572
20 Magnetism 616
21 Electromagnetic Induction 649

PART 6 OPTICS

22 Reflection and Refraction 685
23 Mirrors and Lenses 712
24 Wave Optics 751
25 Electromagnetic Waves 784

PART 7 MODERN PHYSICS

26 Special Relativity 813
27 Quantum Optics 847
28 Atomic Physics 880
29 Nuclear Physics 921
30 Particle Physics 957

Help students learn physics by doing physics

Dear Colleague,

Welcome to the second edition of our textbook *College Physics: Explore and Apply* and its supporting materials (Mastering™ Physics, the *Active Learning Guide* (ALG), and our *Instructor's Guide*)—a coherent learning system that helps students **learn physics by doing physics**!

Experiments, experiments...   Instead of being presented physics as a static set of established concepts and mathematical relations, students develop their own ideas just as physicists do: they *explore* and analyze **observational experiments**, identify patterns in the data, and propose explanations for the patterns. They then design **testing experiments** whose outcomes either confirm or contradict their explanations. Once tested, students *apply* explanations and relations for practical purposes and to problem solving.

A physics tool kit   To build problem-solving skills and confidence, students master proven visual tools (representations such as motion diagrams and energy bar charts) that serve as bridges between words and abstract mathematics and that form the basis of our overarching problem-solving strategy. Our unique and varied problems and activities promote 21st-century competences such as evaluation and communication and reinforce our practical approach with photo, video, and data analysis and real-life situations.

A flexible learning system   Students can work collaboratively on ALG activities in class (lectures, labs, and problem-solving sessions) and then read the textbook at home and solve end-of-chapter problems, or they can read the text and do the activities using Mastering Physics at home, then come to class and discuss their ideas. However they study, students will see physics as a living thing, a process in which they can participate as equal partners.

Why a new edition?   With a wealth of feedback from users of the first edition, our own ongoing experience and that of a gifted new co-author, and changes in the world in general and in education in particular, we embarked on this second edition in order to refine and strengthen our experiential learning system. Experiments are more focused and effective, our multiple-representation approach is expanded, topics have been added or moved to provide more flexibility, the writing, layout, and design are streamlined, and all the support materials are more tightly correlated to our approach and topics.

Working on this new edition has been hard work, but has enriched our lives as we've explored new ideas and applications. We hope that using our textbook will enrich the lives of your students!

Eugenia Etkina

Gorazd Planinsic

Alan Van Heuvelen

"This book made me think deeper and understand better."

—student at *Horry Georgetown Technical College*

PEARSON

A unique and active learning approach promotes deep and lasting

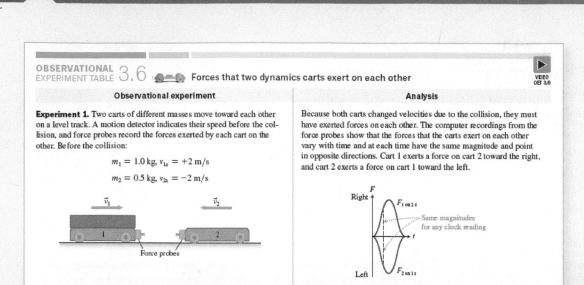

OBSERVATIONAL EXPERIMENT TABLE 3.6 — Forces that two dynamics carts exert on each other

VIDEO
OET 3.6

Observational experiment	Analysis
Experiment 1. Two carts of different masses move toward each other on a level track. A motion detector indicates their speed before the collision, and force probes record the forces exerted by each cart on the other. Before the collision: $m_1 = 1.0$ kg, $v_{1x} = +2$ m/s; $m_2 = 0.5$ kg, $v_{2x} = -2$ m/s $\vec{v}_1$ $\vec{v}_2$ Force probes	Because both carts changed velocities due to the collision, they must have exerted forces on each other. The computer recordings from the force probes show that the forces that the carts exert on each other vary with time and at each time have the same magnitude and point in opposite directions. Cart 1 exerts a force on cart 2 toward the right, and cart 2 exerts a force on cart 1 toward the left. $F_{1\ on\ 2\ x}$ — Same magnitudes for any clock reading $F_{2\ on\ 1\ x}$
Experiment 2. Cart masses and velocities before collision: $m_1 = 1.0$ kg, $v_{1x} = 0$ m/s (at rest); $m_2 = 0.5$ kg, $v_{2x} = -1$ m/s $\vec{v}_1 = 0$ $\vec{v}_2$	Although the forces that the carts exert on each other are smaller than in the first experiment, the magnitudes of the forces at each time are still the same. $F_{1\ on\ 2\ x}$ — Same magnitudes for any clock reading $F_{2\ on\ 1\ x}$
Experiment 3. Cart masses and velocities before collision: $m_1 = 1.0$ kg, $v_{1x} = +2$ m/s; $m_2 = 0.5$ kg, $v_{2x} = -1$ m/s $\vec{v}_1$ $\vec{v}_2$	The same analysis applies. $F_{1\ on\ 2\ x}$ — Same magnitudes for any clock reading $F_{2\ on\ 1\ x}$

Pattern

In each experiment, independent of the masses and velocities, the force that cart 1 exerted on cart 2 $\vec{F}_{1\ on\ 2}$ had the same magnitude as the force that cart 2 exerted on cart 1 $\vec{F}_{2\ on\ 1}$.

UPDATED! Observational Experiment Tables and Testing Experiment Tables: Students must make observations, analyze data, identify patterns, test hypotheses, and predict outcomes. Redesigned for clarity in the second edition, these tables encourage students to explore science through active discovery and critical thinking, constructing robust conceptual understanding.

NEW! Digitally Enhanced Experiment Tables now include embedded videos in the Pearson eText for an interactive experience. Accompanying questions are available in Mastering Physics to build skills essential to success in physics.

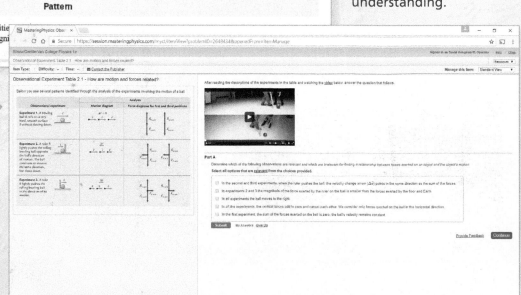

conceptual understanding of physics and the scientific process

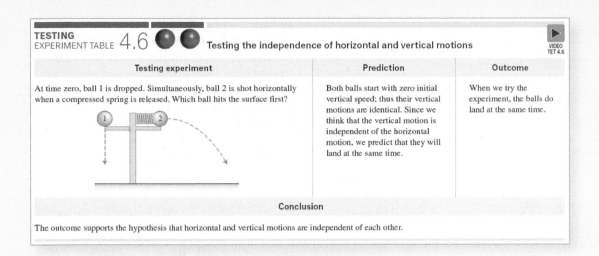

TESTING EXPERIMENT TABLE 4.6 ● ● Testing the independence of horizontal and vertical motions

VIDEO
TET 4.6

Testing experiment	Prediction	Outcome
At time zero, ball 1 is dropped. Simultaneously, ball 2 is shot horizontally when a compressed spring is released. Which ball hits the surface first?	Both balls start with zero initial vertical speed; thus their vertical motions are identical. Since we think that the vertical motion is independent of the horizontal motion, we predict that they will land at the same time.	When we try the experiment, the balls do land at the same time.

Conclusion

The outcome supports the hypothesis that horizontal and vertical motions are independent of each other.

"I like that the experiment tables... explain in detail why every step was important."

—student at *Mission College*

EXPANDED! Experiment videos and photos created by the authors enhance the active learning approach. Approximately 150 photos and 40 videos have been added to the textbook, as well as embedded in the Pearson eText, and scores more in the *Active Learning Guide* (ALG).

FIGURE 2.2 Long-exposure photographs of a moving cart with a blinking LED.

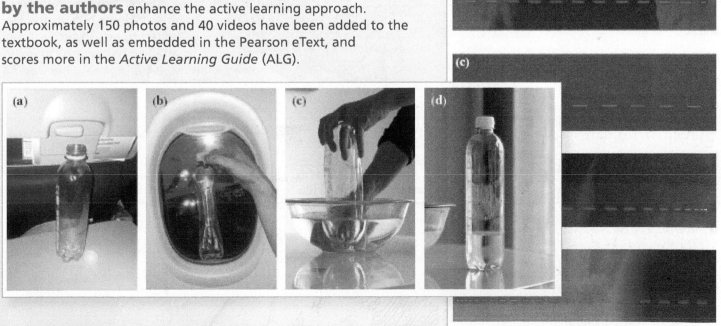

A wealth of practical and consistent guidance, examples, and opportunities

A four-step problem-solving approach in worked examples

consistently uses multiple representations to teach students how to solve complex physics problems. Students follow the steps of **Sketch & Translate, Simplify & Diagram, Represent Mathematically, Solve & Evaluate** to translate a problem statement into the language of physics, sketch and diagram the problem, represent it mathematically, solve the problem, and evaluate the result.

Physics Tool Boxes focus on a particular skill, such as drawing a motion diagram, force diagram, or work-energy bar chart, to help students master the key tools they will need to utilize throughout the course to analyze physics processes and solve problems, bridging real phenomena and mathematics.

"It made me excited to learn physics! It has a systematic and easy-to-understand method for solving problems."

—student at *State University of West Georgia*

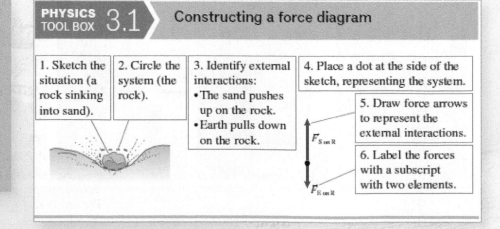

56. * A frog jumps at an angle 30° above the horizontal. The origin of the coordinate system is at the point where the frog leaves the ground. Complete Table P4.55 by drawing check marks in the cells that correctly connect the quantities in the first column that describe the motion of the frog and the descriptions of what is happening to these quantities while the frog is moving. Consider the frog as a point-like object and assume that the resistive force exerted by the air is negligible.

TABLE P4.55

Physical quantity	Remains constant	Is changing	Increases only	Decreases only	Increases, then decreases	Decreases, then increases
x-coordinate magnitude						
y-coordinate magnitude						
Direction of velocity						
Magnitude of velocity						
Direction of acceleration						
Magnitude of acceleration						

29. * Your friend Devin has to solve the following problem: "You have a spring with spring constant k. You compress it by distance x and use it to shoot a steel ball of mass m into a sponge of mass M. After the collision, the ball and the sponge move a distance s along a rough surface and stop (see Figure P7.29). The coefficient of friction between the sponge and the surface is μ. Derive an expression that shows how the distance s depends on relevant physical quantities."

FIGURE P7.29

Devin derived the following equation:

$$s = \frac{kmx^2}{2(m + M)^2 g\mu}$$

Without deriving it, evaluate the equation that [...] reasonable? How do you know?

NEW! Problem types include multiple choice with multiple correct answers, find-a-pattern in data presented in a video or a table, ranking tasks, evaluate statements/claims/explanations/measuring procedures, evaluate solutions, design a device or a procedure that meets given criteria, and linearization problems, promoting critical thinking and deeper understanding.

59. * Jeff and Natalie notice that a rubber balloon, which is first in a warm room, shrinks when they take it into the garden on a cold winter day. They propose two different explanations for the observed phenomenon: (a) the balloon is slowly leaking; (b) the balloon shrinks due to decreased temperature while the pressure in the balloon remains constant (isobaric compression). In order to test their proposed explanations, Jeff and Natalie perform three consecutive experiments: they measure the volume of the balloon and the temperature of the air near the balloon (1) in the room, (2) in the garden, and (3) again in the room. Their measurements, including uncertainties, are presented in the table below.

Exp. #	Location	Temperature	Volume of the balloon
1	Room	26.2 °C ± 0.1 °C	7500 cm³ ± 400 cm³
2	Garden	−15.3 °C ± 0.1 °C	6400 cm³ ± 400 cm³
3	Room	26.2 °C ± 0.1 °C	7300 cm³ ± 400 cm³

Based on the data, can Jeff and Natalie reject any of their hypotheses? Explain. Make sure you include uncertainties in your answer.

Pedagogically driven design and content changes

NEW! A fresh and modern design with a more transparent hierarchy of features and navigation structure, as well as an engaging chapter opener page and streamlined chapter summary, result in a more user-friendly resource, both for learning and for reference.

12

Gases

- Why does a plastic bottle left in a car overnight look crushed on a chilly morning?
- How hard is air pushing on your body?
- How long can the Sun shine?

When you inflate the tires of your bicycle in a warm basement in winter, they tend to look a bit flat when you take the bike outside. The same thing happens to a basketball—you need to pump it up before playing outside on a cold day. An empty plastic bottle left in a car looks crushed on a chilly morning. What do all those phenomena have in common, and how do we explain them?

BE SURE YOU KNOW HOW TO:
- Draw force diagrams (Section 3.1).
- Use Newton's second and third laws to analyze interactions of objects (Section 3.7 and 3.8).
- Use the impulse-momentum principle (Section 6.3).

IN CHAPTER 11, we learned that sound propagates due to the compression and decompression of air. But what exactly is being compressed? To answer this question and the ones above, we need to investigate what makes up a gas and how certain properties of gases can change.

352

---

308   **CHAPTER 10**   Vibrational Motion

Summary

Vibrational motion is the repetitive movement of an object back and forth about an **equilibrium position**. This vibration is due to the **restoring force** exerted by another object that tends to return the first object to its equilibrium position. An object's maximum displacement from equilibrium is the **amplitude** A of the vibration. **Period** T is the time interval for one complete vibration, and **frequency** f is the number of complete vibrations per second (in hertz). The frequency is the inverse of the period. (Section 10.1)

Object at end of spr

$F_{\text{Restoring}}$

$T = 2\pi$

Simple pendulum:

$F_{\text{Restoring}\,x} =$

$T = 2\pi\sqrt{\ }$

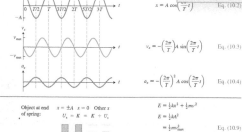

Simple harmonic motion is a mathematical model of vibrational motion when position x, velocity v, and acceleration a of the vibrating object change as sine or cosine functions with time. (Section 10.2)

$x = A\cos\left(\frac{2\pi}{T}t\right)$ Eq. (10.2)

$v_x = -\left(\frac{2\pi}{T}\right)A\sin\left(\frac{2\pi}{T}t\right)$ Eq. (10.3)

$a_x = -\left(\frac{2\pi}{T}\right)^2 A\cos\left(\frac{2\pi}{T}t\right)$ Eq. (10.4)

The **energy of a spring-object system** vibrating horizontally converts continuously from elastic potential energy when at the extreme positions to maximum kinetic energy when passing through the equilibrium position to a combination of energy types at other positions. (Section 10.3)

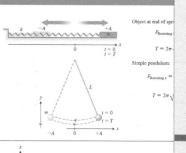

Object at end $x = \pm A$ $x = 0$ Other x
of spring: $U_s = K = K + U_s$

$E = \frac{1}{2}kx^2 + \frac{1}{2}mv^2$

$E = \frac{1}{2}kA^2$

$= \frac{1}{2}mv_{\max}^2$ Eq. (10.9)

The **energy of a pendulum-Earth system** converts continuously from gravitational potential energy when it is at the maximum height of a swing to kinetic energy when it is passing through the lowest point in the swing to a combination of energy types at other positions. (Section 10.5)

Simple $\pm A$ 0 At other places
pendulum: $U_g = K = K + U_g$

$E = mgy + \frac{1}{2}mv^2$

$E = mgy_{\max}$

$= \frac{1}{2}mv_{\max}^2$

Resonant energy transfer occurs when the frequency of the variable external force driving the oscillations is close to the natural frequency f_0 of the vibrating system. (Section 10.8)

REVISED! Streamlined text, layout, and figures throughout the book enhance the focus on central themes and topics, eliminating extraneous detail, resulting in **over 150 fewer pages** than the first edition and allowing students to study more efficiently.

enhance ease of use for students and instructors alike

FIGURE 19.14 A green LED. The electric circuit in (b) is used to collect the I-versus-ΔV data plotted in (c).

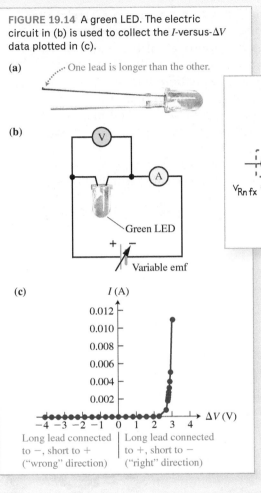

(a)

⋯ One lead is longer than the other.

(b)

V

A

Green LED

$+$ $-$

Variable emf

(c)

I (A)

0.012
0.010
0.008
0.006
0.004
0.002

-4 -3 -2 -1 0 1 2 3 4 ΔV (V)

Long lead connected to $-$, short to $+$ ("wrong" direction)

Long lead connected to $+$, short to $-$ ("right" direction)

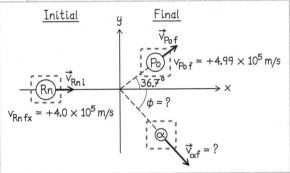

Initial   Final

$\vec{v}_{\text{Po f}}$

$\vec{v}_{\text{Rn i}}$

Rn

Po $v_{\text{Po f}} = +4.99 \times 10^5$ m/s

$36.7°$

$\phi = ?$

$v_{\text{Rn fx}} = +4.0 \times 10^5$ m/s

α $\vec{v}_{\alpha f} = ?$

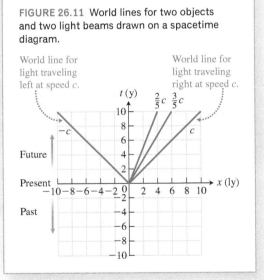

FIGURE 26.11 World lines for two objects and two light beams drawn on a spacetime diagram.

World line for light traveling left at speed c.

World line for light traveling right at speed c.

t (y) $\frac{2}{5}c$ $\frac{3}{5}c$

$-c$ c

10
8
6
4
2

Future

Present $-10\,-8\,-6\,-4\,-2$ 0 2 4 6 8 10 x (ly)

Past

-2
-4
-6
-8
-10

NEW, REVISED, and EXPANDED! Topics include capacitors, AC circuits, LEDs, friction, 2-D collisions, energy, bar charts for rotational momentum and nuclear energy, ideal gas processes, thermodynamic engines, semiconductors, velocity selectors, and spacetime diagrams in special relativity.

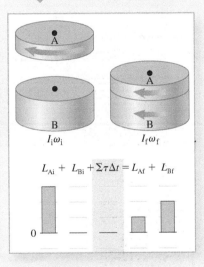

A

A

B   B

$I_i \omega_i$ $I_f \omega_f$

$$L_{\text{Ai}} + L_{\text{Bi}} + \Sigma \tau \Delta t = L_{\text{Af}} + L_{\text{Bf}}$$

0

NEW! Integration of vector arithmetic into early chapters helps students develop vector-related skills in the context of learning physics. **Earlier placement of waves and oscillations** allows instructors to teach these topics with mechanics if preferred. Coverage with optics is also possible.

A flexible learning system adapts to any method of instruction

2.9.9 Evaluate the solution

Class: Equipment per group: whiteboard and markers

Discuss with your group: Identify any errors in the proposed solution to the following problem and provide a corrected solution if there are errors.

Problem: Use the graphical representation of motion to determine how far the object travels until it stops.

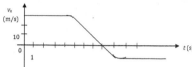

Proposed solution The object was at rest for about 5 seconds, then started moving in the negative direction and stopped after about 9 seconds. During this time its position changed from 30 m to – 10 m, so the total distance that it traveled was 40 m.

2.9.10 Observe and analyze

Class: Equipment per group: whiteboard and markers

Collaborate together with your group to figure this out: The figure below shows long exposure photos of two experiments with a blinking LED that was fixed on a moving cart. In both cases the cart was moving from right to left. The duration of the ON and OFF time for LED is 154 ms and the length of the cart is 17 cm. a) Specify the coordinate system and draw a qualitative velocity-time graph for the motion of the cart in both experiments; b) estimate the speed of the cart in the first experiment. Both photos were obtained from the same spot and with the same settings. Indicate any assumptions that you made.

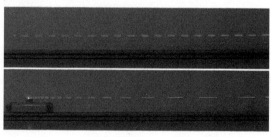

Etkina, Brookes, Planinsic, Van Heuvelen COLLEGE PHYSICS *Active Learning Guide*, 2/e © 2019 Pearson Education, Inc.

REVISED! The **Active Learning Guide** aligns with the textbook's chapters and supplements the knowledge-building approach of the textbook with activities that provide opportunities for further observation, testing, sketching, and analysis as well as collaboration, scientific reasoning, and argumentation. The *Active Learning Guide* can be used in class for individual or group work or assigned as homework and is now better integrated with the text. Now available via download in the Mastering Instructor Resource Center and customizable in print form via Pearson Collections.

> "It is much easier to understand a concept when you can see it in action, and not just read it."
>
> —student at *San Antonio College*

The **Instructor's Guide** provides key pedagogical principles of the textbook and elaborates on the implementation of the methodology used in the textbook, providing guidance on how to integrate the approach into your course.

2

Kinematics: Motion in One Dimension

In Chapter 2, students will learn to describe motion using sketches, motion diagrams, graphs, and algebraic equations. The chapter subject matter is broken into four parts:

I. *What is motion and how do we describe it qualitatively?*
II. *Some of the quantities used to describe motion and a graphical description of motion*
III. *Use of the above to describe constant velocity and constant acceleration motion*
IV. *Developing and using the skills needed to analyze motion in real processes*

For each part, we provide examples of activities that can be used in the classroom, brief discussions of why we introduce the content in a particular order and use of these activities to support the learning, and common student difficulties.

Chapter subject matter	Related textbook section	ALG activities	End-of-chapter questions and problems	Videos
What is motion and how do we describe it qualitatively?	2.1, 2.2	2.1.1–2.1.6, 2.2.1–2.2.4	Problems 1, 3	OET 2.1

Etkina/Planinsic/van Heuvelen 2e *Instructor's Guide* © 2019 Pearson Education, Inc.　　**2-1**

and provides tools for easy implementation

NEW! Ready-to-Go Teaching Modules created for and by instructors make use of teaching tools for before, during, and after class, including new ideas for in-class activities. The modules incorporate the best that the text, Mastering Physics, and Learning Catalytics have to offer and guide instructors through using these resources in the most effective way. The modules can be accessed through the Instructor Resources area of Mastering Physics and as pre-built, customizable assignments.

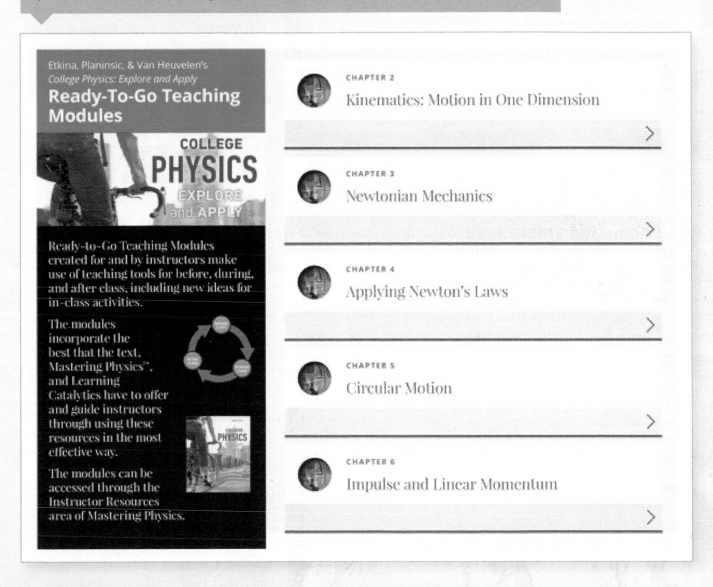

Etkina, Planinsic, & Van Heuvelen's
College Physics: Explore and Apply
Ready-To-Go Teaching Modules

COLLEGE
PHYSICS
EXPLORE
and APPLY

Ready-to-Go Teaching Modules created for and by instructors make use of teaching tools for before, during, and after class, including new ideas for in-class activities.

The modules incorporate the best that the text, Mastering Physics™, and Learning Catalytics have to offer and guide instructors through using these resources in the most effective way.

The modules can be accessed through the Instructor Resources area of Mastering Physics.

CHAPTER 2
Kinematics: Motion in One Dimension

CHAPTER 3
Newtonian Mechanics

CHAPTER 4
Applying Newton's Laws

CHAPTER 5
Circular Motion

CHAPTER 6
Impulse and Linear Momentum

Mastering Physics

Build a basic understanding of physics principles and math skills

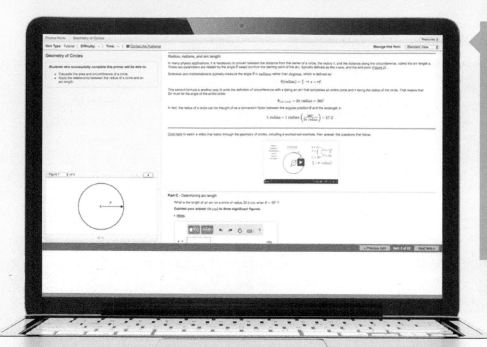

NEW! The Physics Primer relies on videos, hints, and feedback to refresh students' math skills in the context of physics and prepare them for success in the course. These tutorials can be assigned before the course begins as well as throughout the course as just-in-time remediation. The primer ensures students practice and maintain their math skills, while tying together mathematical operations and physics analysis.

Interactive Animated Videos provide an engaging overview of key topics with embedded assessment to help students check their understanding and to help professors identify areas of confusion. Note that these videos are not tied to the textbook and therefore do not use the language, symbols, and conceptual approaches of the book and ALG. The authors therefore recommend assigning these videos after class to expose students to different terminology and notation that they may come across from other sources.

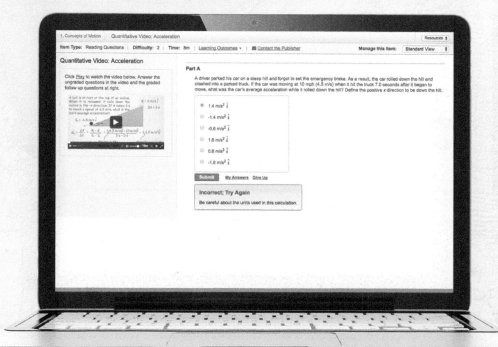

Dynamic Study Modules (DSMs) help students study effectively on their own by continuously assessing their activity and performance in real time and adapting to their level of understanding. The content focuses on definitions, units, and the key relationships for topics across all of mechanics and electricity and magnetism.

www.MasteringPhysics.com

Show connections between physics and the real world as students learn to apply physics concepts via enhanced media

NEW! Direct Measurement Videos are short videos that show real situations of physical phenomena. Grids, rulers, and frame counters appear as overlays, helping students to make precise measurements of quantities such as position and time. Students then apply these quantities along with physics concepts to solve problems and answer questions about the motion of the objects in the video.

NEW! End-of-chapter problem types and 15% new questions and problems include multiple choice with multiple correct answers, find-a-pattern in data presented in a video or a table, ranking tasks, evaluate statements/claims/ explanations/measuring procedures, evaluate solutions, design a device or a procedure that meets given criteria, and linearization problems. End-of-chapter problems have undergone careful analysis using Mastering Physics usage data to provide fine-tuned difficulty ratings and to produce a more varied, useful, and robust set of end-of-chapter problems.

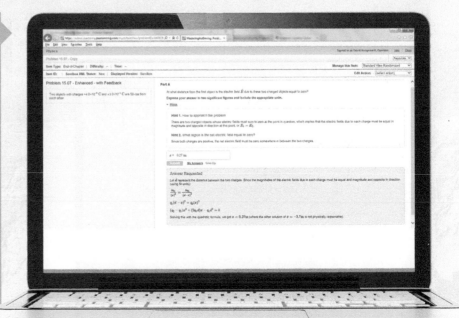

Give students fingertip access to interactive tools

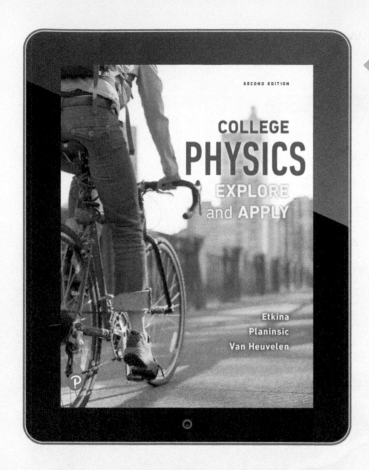

NEW! Pearson eText, optimized for mobile, seamlessly integrates videos such as the Observational Experiment Tables and other rich media with the text and gives students access to their textbook anytime, anywhere. Pearson eText is available with Mastering Physics when packaged with new books or as an upgrade students can purchase online.

Learning Catalytics™ helps generate class discussion, customize lectures, and promote peer-to-peer learning with real-time analytics. Learning Catalytics acts as a student response tool that uses students' smartphones, tablets, or laptops to engage them in more interactive tasks and thinking.

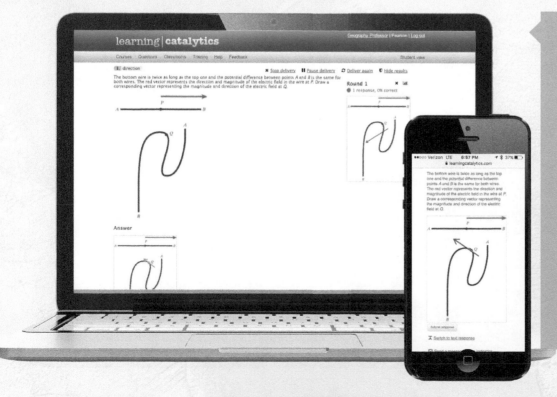

- **NEW!** Upload a full PowerPoint® deck for easy creation of slide questions.
- **NEW!** Team names are no longer case sensitive.
- Help your students develop critical thinking skills.
- Monitor responses to find out where your students are struggling.
- Rely on real-time data to adjust your teaching strategy.
- Automatically group students for discussion, teamwork, and peer-to-peer learning.

COLLEGE
PHYSICS

EXPLORE and APPLY

second edition

COLLEGE
PHYSICS
EXPLORE
and APPLY

VOLUME 2

Eugenia Etkina
RUTGERS UNIVERSITY

Gorazd Planinsic
UNIVERSITY OF LJUBLJANA

Alan Van Heuvelen
RUTGERS UNIVERSITY

New York, NY

Couseware Portfolio Management, Director: Jeanne Zalesky
Courseware Portfolio Manager: Darien Estes
Managing Producer: Kristen Flathman
Content Producer: Tiffany Mok
Courseware Director, Content Development: Jennifer Hart
Courseware Senior Analyst, Content Development: Alice Houston, Ph.D.
Senior Content Developer: David Hoogewerff
Courseware Editorial Assistant: Kristen Stephens and Leslie Lee
Rich Media Content Producer: Dustin Hennessey
Full-Service Vendor: Cenveo® Publisher Services
Full-Service Vendor Project Manager: Susan McNally, Cenveo® Publisher Services
Copyeditor: Joanna Dinsmore

Compositor: Cenveo® Publisher Services
Design Manager: Mark Ong, Side By Side Studios
Interior Designer: Lisa Buckley
Cover Designer: Lisa Buckley
Illustrators: Jim Atherton, Cenveo® Publisher Services
Rights & Permissions Project Manager: Kathleen Zander, Cenveo® Publisher Services
Rights & Permissions Management: Ben Ferrini
Photo Researcher: Karin Kipp, Cenveo® Publisher Services
Manufacturing Buyer: Stacey Weinberger
Director of Product Marketing: Allison Rona
Product Marketing Manager: Elizabeth Ellsworth Bell
Cover Photo Credit: Kari Medig/Aurora/Getty Images

Library of Congress Cataloging-in-Publication Data is on file with the Library of Congress.

www.pearson.com

ISBN 10: 0-134-86291-0 **ISBN 13:** 978-0134-86291-0 (Volume 2)
ISBN 10: 0-134-60182-3 **ISBN 13:** 978-0-134-60182-3 (Student Edition)
ISBN 10: 0-134-68330-7 **ISBN 13:** 978-0-134-68330-0 (NASTA)

1 2019

About the Authors

EUGENIA ETKINA is a Distinguished Professor at Rutgers, the State University of New Jersey. She holds a PhD in physics education from Moscow State Pedagogical University and has more than 35 years of experience teaching physics. She is a recipient of the 2014 Millikan Medal, awarded to educators who have made significant contributions to teaching physics, and is a fellow of the AAPT. Professor Etkina designed and now coordinates one of the largest programs in physics teacher preparation in the United States, conducts professional development for high school and university physics instructors, and participates in reforms to the undergraduate physics courses. In 1993 she developed a system in which students learn physics using processes that mirror scientific practice. That system, called *Investigative Science Learning Environment* (ISLE), serves as the basis for this textbook. Since 2000, Professor Etkina has conducted over 100 workshops for physics instructors, and she co-authored the first edition of *College Physics* and the *Active Learning Guide*. Professor Etkina is a dedicated teacher and an active researcher who has published over 60 peer-refereed articles.

GORAZD PLANINSIC is a Professor of Physics at the University of Ljubljana, Slovenia. He has a PhD in physics from the University of Ljubljana. Since 2000 he has led the Physics Education program, which prepares almost all high school physics teachers in the country of Slovenia. He started his career in MRI physics and later switched to physics education research. During the last 10 years, his work has mostly focused on the research of new experiments and how to use them more productively in teaching and learning physics. He is co-founder of the Slovenian hands-on science center House of Experiments. Professor Planinsic is co-author of more than 80 peer-refereed research articles and more than 20 popular science articles, and is the author of a university textbook for future physics teachers. In 2013 he received the Science Communicator of the Year award from the Slovenian Science Foundation.

ALAN VAN HEUVELEN holds a PhD in physics from the University of Colorado. He has been a pioneer in physics education research for several decades. He taught physics for 28 years at New Mexico State University, where he developed active learning materials including the *Active Learning Problem Sheets* (the *ALPS Kits*) and the *ActivPhysics* multimedia product. Materials such as these have improved student achievement on standardized qualitative and problem-solving tests. In 1993 he joined Ohio State University to help develop a physics education research group. He moved to Rutgers University in 2000 and retired in 2008. For his contributions to national physics education reform, he won the 1999 AAPT Millikan Medal and was selected a fellow of the American Physical Society. Over the span of his career he has led over 100 workshops on physics education reform. He worked with Professor Etkina in the development of the *Investigative Science Learning Environment* (ISLE) and co-authored the first edition of *College Physics* and the *Active Learning Guide*.

Preface

To the student

College Physics: Explore and Apply is more than just a book. It's a learning *companion*. As a companion, the book won't just tell you about physics; it will act as a guide to help you build physics ideas using methods similar to those that practicing scientists use to construct knowledge. The ideas that you build will be *yours*, not just a copy of someone else's ideas. As a result, the ideas of physics will be much easier for you to use when you need them: to succeed in your physics course, to obtain a good score on exams such as the MCAT, and to apply to everyday life.

Although few, if any, textbooks can honestly claim to be a pleasure to read, *College Physics: Explore and Apply* is designed to make the process interesting and engaging. The physics you learn in this book will help you understand many real-world phenomena, from why giant cruise ships are able to float to how telescopes work. The cover of the book communicates its spirit: you learn physics by exploring the natural world and applying it in your everyday life.

A great deal of research has been done over the past few decades on how students learn. We, as teachers and researchers, have been active participants in investigating the challenges students face in learning physics. We've developed unique strategies that have proven effective in helping students think like physicists. These strategies are grounded in *active learning with your peers*—deliberate, purposeful action on your part to learn something new. For learning to happen, one needs to talk to others, share ideas, listen, explain, and argue. It is in these deliberations that new knowledge is born. Learning is not passively memorizing so that you can repeat it later. When you learn actively, you engage with the material and—most importantly—share your ideas with others. You relate it to what you already know and benefit from the knowledge of your peers. You think about the material in as many different ways as you can. You ask yourself questions such as "Why does this make sense?" and "Under what circumstances does this not apply?" Skills developed during this process will be the most valuable in your future, no matter what profession you choose.

This book (your learning companion) includes many tools to support the active learning process: each problem-solving tool, worked example, observational experiment table, testing experiment table, review question, and end-of-chapter question and problem is designed to help you build your understanding of physics. To get the most out of these tools and the course, stay actively engaged in the process of developing ideas and applying them; form a learning group with your peers and try to work on the material together. When things get challenging, don't give up.

At this point you should turn to Chapter 1, Introducing Physics, and begin reading. That's where you'll learn the details of the approach that the book uses, what physics is, and how to be successful in the physics course you are taking.

To the instructor

Welcome to the second edition of *College Physics: Explore and Apply* and its supporting materials (Mastering^{TM} Physics, the *Active Learning Guide* (ALG), and the *Instructor's Guide*), a coherent learning system that helps our students learn physics as an ongoing process rather than a static set of established concepts and mathematical relations. It is based on a framework known as ISLE (the *Investigative Science Learning Environment*). This framework originated in the work of Eugenia Etkina in the early 1990s. She designed a logical progression of student learning of physics that mirrors the processes in which physicists engage while constructing and applying knowledge. This progression was enriched in the early 2000s when Alan Van Heuvelen added his multiple representation approach. While logical flow represents a path for thinking, multiple representations are thinking tools. Since 2001, when ISLE curriculum development began, tens of thousands of students have been exposed to it as hundreds of instructors used the materials produced by the authors and their collaborators. Research on students learning physics through ISLE has shown that these students not only master the content of physics, but also become expert problem solvers, can design and evaluate their own experiments, communicate, and most importantly see physics as a process based on evidence as opposed to a set of rules that come from the book.

Experiments, experiments… The main feature of this system is that students practice developing physics concepts by following steps similar to those physicists use when developing and applying knowledge. The first introduction to a concept or a relation happens when students observe simple experiments (called **observational experiments**). Students learn to analyze these experiments, find patterns (either qualitative or quantitative) in the data, and develop multiple explanations for the patterns or quantitative relations. They then learn how to test the explanations and relations in new **testing experiments**. Sometimes the outcomes of the experiments might cause us to reject the explanations; often, they help us keep them. Students see how scientific ideas develop from evidence and are tested by evidence, and how evidence sometimes causes us to reject the proposed explanations. Finally, students learn how tested explanations and relations are applied for practical purposes and in problem solving. This is the process behind the subtitle of the book.

Explore and apply To help students explore and apply physics, we introduce them to tools: physics-specific representations, such as motion and force diagrams, momentum and energy bar charts, ray diagrams, and so forth. These representations serve as bridges between words and abstract mathematics. Research shows that students who use representations other than mathematics to solve problems are much more successful than those who just look for equations. We use a representations-based problem-solving strategy that helps students approach problem solving without fear and eventually develop not only problem-solving skills, but also confidence. The textbook and ALG introduce a whole library of novel problems and activities that help students develop competencies necessary for success in the 21st century: argumentation, evaluation, estimation, and communication. We use photo and video analysis, real-time data, and real-life situations to pose problems.

A flexible learning system There are multiple ways to use our learning system. Students can work collaboratively on ALG activities in class (lectures, labs, and problem-solving sessions) and then read the textbook and solve end-of-chapter problems at home, or they can first read the text and do the activities using Mastering Physics at home, then come to class and discuss their ideas. However they study, students will see physics as a living thing, a process in which they can participate as equal partners.

The key pedagogical principles of this book are described in detail in the first chapter of the *Instructor's Guide* that accompanies *College Physics*—please read that chapter. It elaborates on the implementation of the methodology that we use in this book and provides guidance on how to integrate the approach into your course.

While our philosophy informs *College Physics*, you need not fully subscribe to it to use this textbook. We've organized the book to fit the structure of most algebra-based physics courses: we begin with kinematics and Newton's laws, then move on to conserved quantities, statics, vibrations and waves, gases, fluids, thermodynamics, electricity and magnetism, optics, and finally modern physics. The structure of each chapter will work with any method of instruction. You can assign all of the innovative experimental tables and end-of-chapter problems, or only a few. The text provides thorough treatment of fundamental principles, supplementing this coverage with experimental evidence, new representations, an effective approach to problem solving, and interesting and motivating examples.

New to this edition

There were three main reasons behind the revisions in this second edition. (1) Users provided lots of feedback and we wanted to respond to it. (2) We (the authors) grew and changed, and learned more about how to help students learn, and our team changed—we have a new co-author, who is an expert in educational physics experiments and in the development of physics problems. (3) Finally, we wanted to respond to changes in the world (new physics discoveries, new technology, new skills required in the workplace) and to changes in education (the Next Generation Science Standards, reforms in the AP and MCAT exams). Our

first edition was already well aligned with educational reforms, but the second edition strengthens this alignment even further.

We have therefore made the following global changes to the textbook, in addition to myriad smaller changes to individual chapters and elements:

- **An enhanced experiential approach**, with more experiment videos and photos (all created by the authors) and an updated and more focused and effective set of experiment tables, strengthens and improves the core foundation of the first edition. Approximately 150 photos and 40 videos have been added to the textbook, and even more to the ALG.

- **An expanded introductory chapter** (now Chapter 1) gives students a more detailed explanation of "How to use this book" to ensure they get the most out of the chapter features, use them actively, and learn how to think critically.

- **Integration of vector arithmetic in early chapters** allows students to develop vector-related skills in the context of learning physics, rather than its placement in an appendix in the first edition.

- **Earlier placement of waves and oscillations** allows instructors to teach these topics with mechanics if preferred. Coverage with optics is also still possible.

- **Significant new coverage of capacitors, AC circuits, and LEDs** (LEDs now permeate the whole book) expand the real-world and up-to-date applications of electricity.

- **Other new, revised, or expanded topics** include friction, 2-D collisions, energy, bar charts for rotational momentum and nuclear energy, ideal gas processes, thermodynamic engines, semiconductors, velocity selectors, and spacetime diagrams in special relativity.

- **Applications are integrated throughout each chapter**, rather than being grouped in the "Putting it all together" sections of the first edition, in order to optimize student engagement.

- **Problem-solving guidance is strengthened** by the careful revision of many Problem-Solving Strategy boxes and the review of each chapter's set of worked examples. The first edition Reasoning Skill boxes are renamed Physics Tool Boxes to better reflect their role; many have been significantly revised.

- **Streamlined text, layout, and figures** throughout the book enhance the focus on central themes and topics, eliminating extraneous detail. The second edition has over 150 fewer pages than the first edition, and the art program is updated with over 450 pieces of new or significantly revised art.

- **21st-century skills incorporated into many new worked examples and end-of-chapter problems** include data analysis, evaluation, and argumentation. Roughly 15% of all end-of-chapter questions and problems are new.

- **Careful analysis of Mastering Physics usage data** provides fine-tuned difficulty ratings and a more varied, useful, and robust set of end-of-chapter problems.

- **A fresh and modern design** provides a more transparent hierarchy of features and navigation structure, as well as an engaging chapter-opening page and streamlined chapter summary.

- **A significantly revised *Active Learning Guide*** is better integrated with the textbook, following the section sequence, and emphasizes collaboration, scientific reasoning, and argumentation.

All of the above sounds like a lot of work—and it was! But it was also lots of fun: we took photos of juice bottles sinking in the snow, we chased flying airplanes and running water striders, we drove cars with coffee cups on dashboards. Most exciting was our trip to a garbage plant to study and photograph the operation of an eddy current waste separator. Working on this new edition has enriched our lives, and we hope that using our textbook will enrich the lives of your students!

Instructor supplements

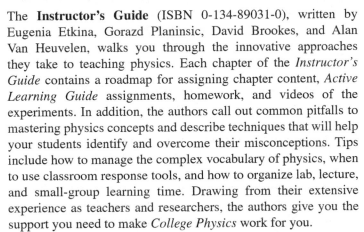

TIP | All of the following materials are available for download on the Mastering Physics Instructor Resources page.

The **Instructor's Guide** (ISBN 0-134-89031-0), written by Eugenia Etkina, Gorazd Planinsic, David Brookes, and Alan Van Heuvelen, walks you through the innovative approaches they take to teaching physics. Each chapter of the *Instructor's Guide* contains a roadmap for assigning chapter content, *Active Learning Guide* assignments, homework, and videos of the experiments. In addition, the authors call out common pitfalls to mastering physics concepts and describe techniques that will help your students identify and overcome their misconceptions. Tips include how to manage the complex vocabulary of physics, when to use classroom response tools, and how to organize lab, lecture, and small-group learning time. Drawing from their extensive experience as teachers and researchers, the authors give you the support you need to make *College Physics* work for you.

The **Active Learning Guide** workbook (ISBN 0-134-60549-7) by Eugenia Etkina, David Brookes, Gorazd Planinsic, and Alan Van Heuvelen consists of carefully crafted cycles of in-class activities that provide an opportunity for students to conduct observational experiments, find patterns, develop explanations, and conduct the testing experiments for those explanations described in the textbook before they read it. These learning cycles are interspersed with "pivotal" activities that serve different purposes: (a) to introduce and familiarize students with new representational techniques, (b) to give students practice with representational techniques, (c) to directly address ideas that we *know* students struggle with (the goal is to encourage that struggle so that students reach a resolution either through their own discussion or by the instructor giving a "time for telling" lecture at the end of the activity), and (d) to provide scaffolding for students to work through an example or a passage in the textbook. The ALG also contains multiple experiments that can be used in labs. Whether the activities are assigned or not, students can always use this workbook to reinforce the concepts they have read about in the text, to practice applying the concepts to real-world scenarios, or to work with sketches, diagrams, and graphs that help them visualize the physics. The ALG is downloadable to share with your class; you may also talk to your sales representative about printing a custom version for your students.

The **Instructor Resource Materials** (ISBN 0-134-87386-6) on the Mastering Physics Instructor Resources page provide invaluable and easy-to-use resources for your class, organized by textbook chapter. The contents include a comprehensive library of all figures, photos, tables, and summaries from the textbook in JPEG and PowerPoint formats. A set of editable **Lecture Outlines**, **Open-Ended Questions**, and **Classroom Response System "Clicker" Questions** in PowerPoint will engage your students in class. Also included among the Instructor Resource Materials are the **Test Bank**, **Instructor Solutions Manual**, **Active Learning Guide**, **Active Learning Guide Solutions Manual**, and **Instructor Guide**.

Mastering™ Physics is the leading online homework, tutorial, and assessment platform designed to improve results by engaging students with powerful content. All Mastering resources, content, and tools are easy for both students and instructors to access in one convenient location. Instructors ensure that students arrive ready to learn by assigning educationally effective content before class and encourage critical thinking and retention with in-class resources such as Learning Catalytics™. Students can master concepts after class through traditional and adaptive homework assignments that provide hints and answer-specific feedback. The Mastering gradebook records scores for all automatically graded assignments in one place, while diagnostic tools give instructors access to rich data to assess student understanding and misconceptions.

New for the second edition of this book, Mastering Physics includes activities for students to do before coming to class, as an alternative to working through the *Active Learning Guide* activities prior to reading the textbook. These activities focus students' attention on observational experiments, helping them learn to identify patterns in the data, and on testing experiments, helping them learn how to make a prediction of an outcome of an experiment using an idea being tested, not personal intuition. Both skills are very important in science, but are very difficult to develop.

The significantly revised **Instructor's Solutions Manual**, provided as PDFs and editable Word files, gives complete solutions to all end-of chapter questions and problems using the textbook's problem-solving strategy.

The **Test Bank**, which has also been significantly revised, contains more than 2000 high-quality problems, with a range of multiple-choice, true/false, short-answer, and regular homework-type questions. Test files are provided in TestGen® (an easy-to-use, fully networkable program for creating and editing quizzes and exams), as well as PDF and Word format.

Student supplements

Physics experiment videos, accessed via the eText, with a smartphone through this QR code, at https://goo.gl/s2MerO, or online in the Mastering Physics Study Area, accompany most of the Observational and Testing Experiment Tables, as well as other discussions and problems in the textbook and in the ALG. Students can observe the exact experiment described in the text.

VIDEO

The **Pearson eText,** optimized for mobile, seamlessly integrates videos and other rich media with the text and gives students access to their textbook anytime, anywhere.

- The Pearson eText mobile app offers offline access and can be downloaded for most iOS and Android phones/tablets from the Apple App Store or Google Play
- Accessible (screen-reader ready)
- Configurable reading settings, including resizable type and night reading mode
- Instructor and student note-taking, highlighting, bookmarking, and search

The **Student Solutions Manual** (ISBN 0-134-88014-5) gives complete solutions to select odd-numbered end-of-chapter questions and problems using the textbook's problem-solving strategy.

In addition to content assigned by the instructor and this text's accompanying experiment videos, **Mastering™ Physics** also provides a wealth of self-study resources:

- **Dynamic Study Modules** assess student performance and activity in real time. They use data and analytics that personalize content to target each student's particular strengths and weaknesses. DSMs can be accessed from any computer, tablet, or smart phone.
- **PhET simulations** from the University of Colorado, Boulder are provided in the Mastering Physics Study Area to allow students to explore key concepts by interacting with these research-based simulations.
- **24/7 access to online tutors*** enables students to work one-on-one, in real time, with a tutor using an interactive whiteboard. Tutors will guide them through solving their problems using a problem-solving-based teaching style to help them learn underlying concepts. In this way, students will be better prepared to handle future assignments on their own.

Acknowledgments

We wish to thank the many people who helped us create this textbook and its supporting materials. First and foremost, we want to thank our team at Pearson Higher Education, especially Jeanne Zalesky, who believed that the book deserved a second edition; Alice Houston, who provided careful, constructive, creative, enriching, and always positive feedback on every aspect of the book and the ALG; Darien Estes, who fearlessly made pivotal decisions that made the new edition much better; Susan McNally, who tirelessly shepherded the book through all stages of production; and David Hoogewerff, who oversaw the Mastering Physics component of the program. Tiffany Mok and Leslie Lee oversaw the new edition of the *Active Learning Guide* and other supplements. Special thanks to Jim Smith and Cathy Murphy who helped shape the first edition of the book. We also want to thank Adam Black for believing in the future of the project.

Although Michael Gentile is not a co-author on the second edition, this work would be impossible without him; he contributed a huge amount to the first edition and provided continuous support for us when we were working on the second edition. No words will

*Please note that tutoring is available in selected Mastering products, and in those products you are eligible for one tutoring session of up to 30 minutes duration with your course. Additional hours can be purchased at reasonable rates.

describe how grateful we are to have Paul Bunson on our team. Paul helped us with the end-of-chapter problem revisions and Mastering Physics and ALG activities, and provided many helpful suggestions, particularly on rotational mechanics, fluids, relativity, and quantum optics. In addition, he was the first to adopt the textbook even before the first edition was officially printed and since then has remained a vivid advocate and supporter of ISLE. We are indebted to Charlie Hibbard, who checked and rechecked every fact and calculation in the text. Brett Kraabel prepared detailed solutions for every end-of-chapter problem for the *Instructor's Solutions Manual*. We also want to thank all of the reviewers, in particular Jeremy Hohertz, who put their time and energy to providing thoughtful, constructive, and supportive feedback. We thank Matt Blackman for adding excellent problems to the Test Bank, Katerina Visnjic for her support of ISLE and the idea to expand energy bar charts to nuclear physics, and Mikhail Kagan for timely feedback. Our special thanks go to Lane Seeley for his thoughtful review of the energy chapter, which led to its deep revision. We thank Diane Jammula and Jay Pravin Kumar, who not only became avid supporters and users of ISLE but also helped create instructor resources for the second edition. We thank Ales Mohoric and Sergej Faletic for their suggestions on problems.

Our infinite thanks go to Xueli Zou, the first adopter of ISLE, and to Suzanne Brahmia, who came up with the Investigative Science Learning Environment acronym "ISLE" and was and is an effective user and tireless advocate of the ISLE learning strategy. Suzanne's ideas about relating physics and mathematics are reflected in many sections of the book. We are indebted to David Brookes, another tireless ISLE developer, whose research shaped the language we use. We thank all of Eugenia's students who are now physics teachers for providing feedback and ideas and using the book with their students.

We have been very lucky to belong to the physics teaching community. Ideas of many people in the field contributed to our understanding of how people learn physics and what approaches work best. These people include Arnold Arons, Fred Reif, Jill Larkin, Lillian McDermott, David Hestenes, Joe Redish, Stamatis Vokos, Jim Minstrell, David Maloney, Fred Goldberg, David Hammer, Andy Elby, Noah Finkelstein, David Meltzer, David Rosengrant, Anna Karelina, Sahana Murthy, Maria Ruibal-Villasenhor, Aaron Warren, Tom Okuma, Curt Hieggelke, and Paul D'Alessandris. We thank all of them and many others.

Personal notes from the authors

We wish to thank Valentin Etkin (Eugenia's father), an experimental physicist whose ideas gave rise to the ISLE philosophy many years ago, Inna Vishnyatskaya (Eugenia's mother), who never lost faith in the success of our book, and Dimitry and Alexander Gershenson (Eugenia's sons), who provided encouragement to Eugenia over the years. While teaching Alan how to play violin, Alan's uncle Harold Van Heuvelen provided an instructional system very different from that of traditional physics teaching. In Harold's system, many individual abilities (skills) were developed with instant feedback and combined over time to address the process of playing a complex piece of music. We tried to integrate this system into our ISLE physics learning system.

—Eugenia Etkina, Gorazd Planinsic, and Alan Van Heuvelen

Reviewers and classroom testers of the first and second editions

Ricardo Alarcon
Arizona State University

Eric Anderson
University of Maryland, Baltimore County

James Andrews
Youngstown State University

Aurelian Balan
Delta College

David Balogh
Fresno City College

Linda Barton
Rochester Institute of Technology

Ian Beatty
*University of North Carolina at
    Greensboro*

Robert Beichner
North Carolina State University, Raleigh

Aniket Bhattacharya
University of Central Florida

Luca Bombelli
University of Mississippi

Scott Bonham
Western Kentucky University

Gerald Brezina
San Antonio College

Paul Bunson
Lane Community College

Debra Burris
University of Central Arkansas

Hauke Busch
Georgia College & State University

Rebecca Butler
Pittsburg State University

Paul Camp
Spelman College

Amy Campbell
Georgia Gwinnett College

Kapila Castoldi
Oakland University

Juan Catala
Miami Dade College North

Colston Chandler
University of New Mexico

Soumitra Chattopadhyay
Georgia Highlands College

Betsy Chesnutt
Itawamba Community College

Chris Coffin
Oregon State University

Lawrence Coleman
University of California, Davis

Michael Crivello
San Diego Mesa College

Elain Dahl
Vincennes University

Jared Daily
Mid Plains Community College

Danielle Dalafave
The College of New Jersey

Charles De Leone
California State University, San Marcos

Carlos Delgado
Community College of Southern Nevada

Christos Deligkaris
Drury University

Dedra Demaree
Oregon State University

Karim Diff
Santa Fe Community College

Kathy Dimiduk
Cornell University

Diana Driscoll
Case Western Reserve University

Raymond Duplessis
Delgado Community College

Taner Edis
Truman State University

Bruce Emerson
Central Oregon Community College

Davene Eyres
North Seattle Community College

Xiaojuan Fan
Marshall University

Nail Fazleev
University of Texas at Arlington

Gerald Feldman
George Washington University

Frank Ferrone
Drexel University

Jane Flood
Muhlenberg College

Lewis Ford
Texas A&M University

Tom Foster
*Southern Illinois University,
    Edwardsville*

Joseph Ganem
Loyola University Maryland

Brian Geislinger
Gadsden State Community College

Richard Gelderman
Western Kentucky University

Lipika Ghosh
Virginia State University

Anne Gillis
Butler Community College

Martin Goldman
University of Colorado, Boulder

Greg Gowens
University of West Georgia

Michael Graf
Boston College

Alan Grafe
University of Michigan, Flint

Sigrid Greene
Alice Lloyd College

Elena Gregg
Oral Roberts University

Recine Gregg
Fordham University

John Gruber
San Jose State University

Arnold Guerra
Orange Coast College

Edwin Hach III
Rochester Institute of Technology

Steve Hagen
University of Florida, Gainesville

Thomas Hemmick
State University of New York, Stony Brook

Scott Hildreth
Chabot College

Zvonimir Hlousek
California State University, Long Beach

Jeremy Hohertz
Elon University

Mark Hollabaugh
Normandale Community College

Klaus Honscheid
Ohio State University

Kevin Hope
University of Montevallo

Joey Huston
Michigan State University

Richard Ignace
East Tennessee State University

Doug Ingram
Texas Christian University

George Irwin
Lamar University

Darrin Johnson
University of Minnesota

Adam Johnston
Weber State University

Mikhail Kagan
Pennsylvania State University

David Kaplan
Southern Illinois University, Edwardsville

James Kawamoto
Mission College

Julia Kennefick
University of Arkansas

Casey King
Horry-Georgetown Technical College

Patrick Koehn
Eastern Michigan University

Victor Kriss
Lewis-Clark State College

Peter Lanagan
College of Southern Nevada, Henderson

Albert Lee
California State University, Los Angeles

Todd Leif
Cloud County Community College

Eugene Levin
Iowa State University

Jenni Light
Lewis-Clark State College

Curtis Link
*Montana Tech of The University
    of Montana*

Donald Lofland
West Valley College

Susannah Lomant
Georgia Perimeter College

Rafael Lopez-Mobilia
University of Texas at San Antonio

Kingshuk Majumdar
Berea College

Gary Malek
Johnson County Community College

Eric Mandell
Bowling Green State University

Lyle Marschand
Northern Illinois University

Eric Martell
Millikin University

Donald Mathewson
Kwantlen Polytechnic University

Mark Matlin
Bryn Mawr College

Dave McGraw
University of Louisiana, Monroe

Ralph McGrew
Broome Community College

Timothy McKay
University of Michigan

David Meltzer
Arizona State University

William Miles
East Central Community College

Rabindra Mohapatra
University of Maryland, College Park

Enrique Moreno
Northeastern University

Joe Musser
Stephen F. Austin State University

Charles Nickles
University of Massachusetts, Dartmouth

Gregor Novak
United States Air Force Academy

Gonzalo Ordonez
Butler University

John Ostendorf
Vincennes University

Philip Patterson
Southern Polytechnic State University

Euguenia Peterson
Richard J. Daley College

Jeff Phillips
Loyola Marymount University

Francis Pichanick
University of Massachusetts, Amherst

Alberto Pinkas
New Jersey City College

Dmitri Popov
State University of New York, Brockport

Matthew Powell
West Virginia University

Stephanie Rafferty
Community College of Baltimore County, Essex

Roberto Ramos
Indiana Wesleyan University

Greg Recine
Fordham University

Edward Redish
University of Maryland, College Park

Lawrence Rees
Brigham Young University

Lou Reinisch
Jacksonville State University

Andrew Richter
Valparaiso University

Joshua Ridley
Murray State University

Melodi Rodrigue
University of Nevada, Reno

Charles Rogers
Southwestern Oklahoma State University

David Rosengrant
Kennesaw State University

Alvin Rosenthal
Western Michigan University

Lawrence Rowan
University of North Carolina at Chapel Hill

Roy Rubins
University of Texas at Arlington

Otto Sankey
Arizona State University

Rolf Schimmrigk
Indiana University, South Bend

Brian Schuft
North Carolina Agricultural and Technical State University

Sara Schultz
Montana State University, Moorhead

Bruce Schumm
University of California, Santa Cruz

David Schuster
Western Michigan State University

Lane Seeley
Seattle Pacific University

Bart Sheinberg
Houston Community College, Northwest College

Carmen Shepard
Southwestern Illinois College

Douglas Sherman
San Jose State University

Chandralekha Singh
University of Pittsburgh

David Snoke
University of Pittsburgh

David Sokoloff
University of Oregon

Nickolas Solomey
Wichita State University

Mark Stone
Northwest Vista College

Bernhard Stumpf
University of Idaho

Steven Sweeney
DeSales University

Tatsu Takeuchi
Virginia Polytechnic Institute and State University

Julie Talbot
University of West Georgia

Colin Terry
Ventura College

Beth Ann Thacker
Texas Tech University

James Thomas
University of New Mexico

John Thompson
University of Maine, Orono

Som Tyagi
Drexel University

David Ulrich
Portland Community College

Eswara P. Venugopal
University of Detroit, Mercy

James Vesenka
University of New England

Melissa Vigil
Marquette University

William Waggoner
San Antonio College

Kendra Wallis
Eastfield College

Jing Wang
Eastern Kentucky University

Tiffany Watkins
Boise State University

Laura Weinkauf
Jacksonville State University

William Weisberger
State University of New York, Stony Brook

John Wernegreen
Eastern Kentucky University

Daniel Whitmire
University of Louisiana at Lafayette

Luc Wille
Florida Atlantic University

Roy Wilson
Mississippi Gulf Coast Community College Perkinston

Brian Woodahl
Indiana University-Purdue University Indianapolis

Gary Wysin
Kansas State University, Manhattan

Chadwick Young
Nicholls State University

Jiang Yu
Fitchburg State College

Real-World Applications

Chapter 17

Everyday electrostatic phenomena	501, 503, 507, 533
Grounding	511
The Van de Graaff generator	524
Photocopiers	526
BIO Ventricular defibrillation	530
BIO Electrical forces across cell walls	530
BIO Bee pollination	530
BIO Calcium ion synapse transfer	533
Electrostatic exploration of underground materials	533

Chapter 18

BIO Electric field and potential due to the heart's dipole	540, 547, 569
Electron speed in an X-ray machine	548
Grounding	553
Shielding during lightning	555
BIO Sodium ions in the nervous system	557–8
BIO Body cells as capacitors	561
BIO Electrocardiography (ECG)	563–4, 569
BIO Electric field across body cell membranes	568, 569
BIO Energy used to charge nerve cells	568
BIO Electric field of a fish	569, 570
BIO Axon capacitance	569
BIO Capacitance of red blood cells	569
BIO Defibrillators	569
BIO Can sharks detect an axon $\vec{E}$ field ?	570
BIO Electrophoresis	570
BIO Energy stored in an axons electric field	570
Electrostatic air cleaning devices	571

Chapter 19

Light-emitting diodes (LEDs)	583, 591
Household wiring	589, 612
Circuit breakers and fuses	599, 602
Semiconductors	606–7
BIO Electrolytes	607
BIO Preventing electric shock	610
BIO Electric currents in the body	611
BIO Current produced by electric eels	612
Subway rail resistance	613
BIO Respirator detector	613
BIO Resistance of human nerve cells	613
Conductive textiles	613
Types of household switches	614
BIO Hands and arms as a conductor	614
BIO Current across the membrane wall of an axon	614
BIO Signals in nerve cells stimulate muscles	614
BIO Effects of electric current on the human body	615

Chapter 20

Magnetic and geographic poles of Earth	617
Cathode ray tubes	628
The auroras	632
Mass spectrometers	637, 646
BIO Intensity modulated radiation therapy (IMRT)	638
BIO Magnetic force exerted by Earth on ions in the body	645
Minesweepers	646
Magnetic field sensor on your mobile phone	646
BIO Magnetic resonance imaging (MRI)	647
BIO Power lines—do their magnetic fields pose a risk?	647

Chapter 21

BIO Transcranial magnetic stimulation (TMS)	653, 664, 681
Microphones and seismometers	654
Magnetic braking	659, 683
Eddy current waste separators	659
Electric generators	666–7
Transformers	674–6, 682s
BIO Breathing monitors	679, 681
Burglar alarm design	680
Induction cooktops	680
Metal detectors	680
BIO Magnetic field and brain cells	681
BIO Hammerhead sharks	683
BIO Magnetic resonance imaging	683
Magstripe readers	683
BIO Magnetic induction tomography (MIT)	684
BIO Measuring the motion of flying insects	684

Chapter 22

Pinhole cameras	689, 708
The red-eye effect	692
BIO Measuring concentration of blood glucose	695, 697
BIO Fiber optics endoscope	701, 702
Mirages	703
Color of the sky	703
BIO Vitreous humor	708
Rain sensors	709
Rainbows	710

Chapter 23

Improving visibility with convex mirrors	716
Digital projectors	731
Magnifying glasses	732, 739
Photography and cameras	734–5

Light field cameras | 736
BIO The human eye and corrective lenses | 736–8, 747–749
Telescopes and microscopes | 740–743, 748
BIO Seeing underwater | 745
BIO Blind spots | 745
Dentists' mirrors and lamps | 746, 747
BIO Fish eyes, unusual and usual | 746, 750
BIO Dissecting microscopes | 748
BIO Laser surgery for the eye (LASIK) | 749

Chapter 24

Chromatic aberration in optical instruments | 759
Colors on CDs and DVDs | 762
Spectrometers | 762
Soap bubble colors | 764, 767, 780
Lens coatings | 767
BIO Bird and butterfly colors | 764, 768
Resolving power of lenses | 772–4, 780
Memory capacity of a CD | 779
BIO Morpho butterfly wings | 781
BIO Ability of a bat to detect small objects | 781
BIO Diffraction-limited resolving power of the eye | 782
BIO What is 20/20 vision? | 782
Thin-film window coatings for energy conservation | 783

Chapter 25

Antennas | 791
Radar | 793–4, 810
The Global Positioning System (GPS) | 794
Microwave cooking | 795
Polarized sunglasses | 804, 805
Liquid crystal displays (LCDs) | 806
3-D movies | 807
BIO Ultraviolet A and B | 810
BIO X-rays used in medicine | 810
BIO Laser to treat cancer | 810

BIO Spider polarized light navigation | 811
BIO Human vision power sensitivity | 811
BIO Honeybee navigation | 812
Incandescent lightbulbs | 812

Chapter 26

Muons and cosmic rays | 821, 824, 829
Police radar speed detector | 835
Age of the universe | 837, 845
Gravitational waves and black holes | 839
GPS and relativity | 840
BIO Mass equivalent of metabolic energy | 845
Cherenkov radiation | 846
Tennis serve speed measurement | 846
Quasars | 846

Chapter 27

Measuring stellar temperatures | 849–850
X-rays | 868–872
Solar cells | 872, 878
Light-emitting diodes (LEDs) | 874
BIO Safe photon energy in tanning beds | 877
BIO Laser surgery | 877
BIO Pulsed laser replaces dental drills | 877
BIO Human vision sensitivity | 877
BIO Energy absorption during X-rays | 877
BIO Light detection by the human eye | 878
BIO Firefly light | 878
BIO Owl night vision | 878
BIO Photosynthesis | 878
BIO Radiation from our bodies | 879

Chapter 28

Spectral analysis to identify chemical composition | 892–5

Absorption spectrum of the Sun | 896
BIO Why are plants green? | 897
Lasers | 897–8
BIO Laser surgery | 899
Lidar mapping | 899
BIO Hydrogen bonds in DNA molecules | 913–914
BIO Welding the retina | 917
BIO Electron microscopes | 918
BIO Electron transport chains in photosynthesis and metabolism | 919

Chapter 29

Fusion energy in the Sun | 936
BIO Beta decay in our bodies | 942
BIO Determining the source of carbon in plants | 944
BIO Estimating blood volume | 946
Carbon dating and archeology | 947
BIO Relative biological effectiveness (RBE) of radiation | 950
Sources of ionizing radiation | 951
BIO Potassium decay in the body | 954, 955
BIO O$_2$ emitted by plants | 954
BIO Radiation therapy | 954
BIO Dating a Wisconsin glacier tree sample | 954
BIO Radiation dose | 955
BIO Dose during X-ray exam | 955
BIO Estimating the number of ants in a nest | 955
BIO Ion production due to ionizing radiation | 955
BIO Nuclear accidents and cancer risk | 955, 956
Neutron activation analysis | 956

Chapter 30

BIO Positron Emission Tomography (PET) | 957, 962

Contents

17 Electric Charge, Force, and Energy 500

17.1 Electrostatic interactions 501
17.2 Explanations for electrostatic interactions 504
17.3 Conductors and insulators (dielectrics) 507
17.4 Coulomb's force law 512
17.5 Electric potential energy 516
17.6 Skills for analyzing processes involving electric charges 521
17.7 Charge separation and photocopying 524
Summary 528 • Questions and Problems 529

18 The Electric Field 535

18.1 A model of the mechanism for electrostatic interactions 536
18.2 Skills for analyzing processes involving $\vec{E}$ fields 542
18.3 The V field: electric potential 546
18.4 Relating the $\vec{E}$ field and the V field 550
18.5 Conductors in electric fields 552
18.6 Dielectric materials in an electric field 555
18.7 Capacitors 558
18.8 Electrocardiography 563
Summary 565 • Questions and Problems 566

19 DC Circuits 572

19.1 Electric current 573
19.2 Batteries and emf 576
19.3 Making and representing simple circuits 578
19.4 Ohm's law 581
19.5 Qualitative analysis of circuits 586
19.6 Joule's law 589
19.7 Kirchhoff's rules 592
19.8 Resistor and capacitor circuits 596
19.9 Skills for solving circuit problems 600
19.10 Properties of resistors 602
Summary 608 • Questions and Problems 609

20 Magnetism 616

20.1 Magnetic interactions 617
20.2 Magnetic field 618
20.3 Magnetic force on a current-carrying wire 621

20.4 Magnetic force exerted on a single moving charged particle 628
20.5 Magnetic fields produced by electric currents 632
20.6 Skills for analyzing magnetic processes 634
20.7 Magnetic properties of materials 639
Summary 642 • Questions and Problems 643

21 Electromagnetic Induction 649

21.1 Inducing an electric current 650
21.2 Magnetic flux 654
21.3 Direction of the induced current 656
21.4 Faraday's law of electromagnetic induction 659
21.5 Skills for analyzing processes involving electromagnetic induction 662
21.6 AC circuits 668
21.7 Transformers 674
21.8 Mechanisms explaining electromagnetic induction 677
Summary 678 • Questions and Problems 679

22 Reflection and Refraction 685

22.1 Light sources, light propagation, and shadows 686
22.2 Reflection of light 689
22.3 Refraction of light 692
22.4 Total internal reflection 696
22.5 Skills for analyzing reflective and refractive processes 698
22.6 Fiber optics, prisms, mirages, and the color of the sky 701
22.7 Explanation of light phenomena: two models of light 704
Summary 706 • Questions and Problems 707

23 Mirrors and Lenses 712

23.1 Plane mirrors 713
23.2 Qualitative analysis of curved mirrors 715
23.3 The mirror equation 721
23.4 Qualitative analysis of lenses 725

23.5 Thin lens equation and quantitative analysis of lenses 730
23.6 Skills for analyzing processes involving mirrors and lenses 734
23.7 Single-lens optical systems 735
23.8 Angular magnification and magnifying glasses 739
23.9 Telescopes and microscopes 740
   Summary 744 • Questions and Problems 745

24 Wave Optics 751
24.1 Young's double-slit experiment 752
24.2 Refractive index, light speed, and wave coherence 757
24.3 Gratings: an application of interference 760
24.4 Thin-film interference 764
24.5 Diffraction of light 768
24.6 Resolving power 772
24.7 Skills for applying the wave model of light 774
   Summary 777 • Questions and Problems 778

25 Electromagnetic Waves 784
25.1 Polarization of waves 785
25.2 Discovery of electromagnetic waves 788
25.3 Applications of electromagnetic waves 793
25.4 Frequency, wavelength, and the electromagnetic spectrum 795
25.5 Mathematical description of EM waves and EM wave energy 797
25.6 Polarization and light reflection 802
   Summary 808 • Questions and Problems 809

26 Special Relativity 813
26.1 Ether or no ether? 814
26.2 Postulates of special relativity 817
26.3 Simultaneity 818
26.4 Time dilation 819
26.5 Length contraction 822
26.6 Spacetime diagrams 824
26.7 Velocity transformations 827
26.8 Relativistic momentum 828
26.9 Relativistic energy 830
26.10 Doppler effect for EM waves 834
26.11 General relativity 838
26.12 Global Positioning System (GPS) 840
   Summary 842 • Questions and Problems 843

27 Quantum Optics 847
27.1 Black body radiation 848
27.2 Photoelectric effect 853

27.3 Quantum model explanation of the photoelectric effect 859
27.4 Photons 864
27.5 X-rays 867
27.6 Photocells, solar cells, and LEDs 872
   Summary 875 • Questions and Problems 876

28 Atomic Physics 880
28.1 Early atomic models 881
28.2 Bohr's model of the atom: quantized orbits 885
28.3 Spectral analysis 892
28.4 Lasers 897
28.5 Quantum numbers and Pauli's exclusion principle 899
28.6 Particles are not just particles 903
28.7 Multi-electron atoms and the periodic table 907
28.8 The uncertainty principle 910
   Summary 915 • Questions and Problems 916

29 Nuclear Physics 921
29.1 Radioactivity and an early nuclear model 922
29.2 A new particle and a new nuclear model 924
29.3 Nuclear force and binding energy 928
29.4 Nuclear reactions 932
29.5 Nuclear sources of energy 935
29.6 Mechanisms of radioactive decay 939
29.7 Half-life, decay rate, and exponential decay 943
29.8 Radioactive dating 947
29.9 Ionizing radiation and its measurement 949
   Summary 952 • Questions and Problems 953

30 Particle Physics 957
30.1 Antiparticles 958
30.2 Fundamental interactions 962
30.3 Elementary particles and the Standard Model 966
30.4 Cosmology 972
30.5 Dark matter and dark energy 974
30.6 Is our pursuit of knowledge worthwhile? 978
   Summary 979 • Questions and Problems 979

Appendices
A Mathematics Review A-1
B Atomic and Nuclear Data A-11
C Answers to Select Odd-Numbered Problems A-15

Credits C-1

Index I-1

COLLEGE
PHYSICS
EXPLORE and APPLY

Electric Charge, Force, and Energy

- How does a photocopier work?
- Why is your hair attracted to a plastic comb after combing?
- How is a metal soda can different from a plastic bottle on a microscopic level?

When you make a photocopy, the toner inside the copier reproduces your original image on a new sheet of paper. How does the toner stick to the paper, and how does the toner "know" where to stick? To attract small particles of toner to its drum, a photocopier uses a fundamental mechanism that is similar to the one that causes a rubbed balloon to stick to your sweater. This mechanism involves a new type of force that we will learn about in this chapter.

BE SURE YOU KNOW HOW TO:

- Identify a system and construct a force diagram for it (Section 3.1).
- Identify a system and construct an energy bar chart for it (Section 7.2).
- Convert a force diagram and an energy bar chart into a mathematical statement (Sections 3.3 and Section 7.6).

IN CHAPTERS 12–16, we learned that all objects are made of tiny particles. What holds these particles together? For that matter, what holds the particles that make up the paper in this textbook together? The interaction responsible for holding particles together is the same interaction that makes toner stick to a copier drum and a balloon stick to your sweater, as well as many other phenomena that we observe in our everyday world.

17.1 Electrostatic interactions

If you rub a balloon with a wool sweater (**Figure 17.1a**) and then bring the sweater near the balloon, the sweater attracts the balloon (Figure 17.1b). If you bring a second balloon rubbed the same way near the first balloon, they repel each other (Figure 17.1c). Let's investigate this phenomenon in detail, in Observational Experiment **Table 17.1**.

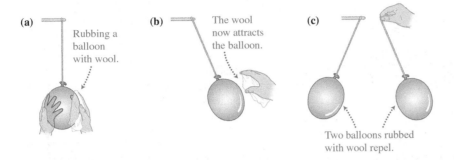

FIGURE 17.1 Rubbing different materials causes an interaction between them.

(a) Rubbing a balloon with wool.

(b) The wool now attracts the balloon.

(c) Two balloons rubbed with wool repel.

OBSERVATIONAL EXPERIMENT TABLE 17.1 Experimenting with rubbed objects

VIDEO
OET 17.1

Observational experiment	Analysis
Experiment 1. Hang two small balloons from two thin strings. Rub each balloon with felt. The balloons repel each other and settle into the position shown at right. Bringing the balloons closer to each other increases the angle between the strings. 	The force diagram shows that the balloons are in the equilibrium position. To settle in the position shown at left, they must exert a repulsive force on each other. To explain the increase in the angle between the strings, we need to assume that as the balloons are brought closer together, the forces that they exert on each other increase in magnitude. 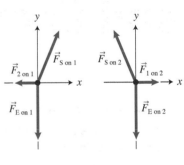
Experiment 2. Bring the felt used to rub the balloons close to one of the rubbed balloons. The balloon is attracted to the felt. The attraction strengthens as the felt is held closer. Felt used to rub balloon / Rubbed balloon	The felt exerts an unknown attractive force on the rubbed balloon. The force strengthens as the felt gets closer to the balloon. 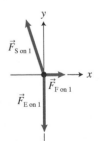
Experiment 3. Repeat Experiment 1, only this time rub the two small balloons with plastic wrap. The balloons repel. Bringing the balloons closer to each other increases the angle between the strings. 	Each balloon exerts an unknown repulsive force on the other balloon. This force increases in magnitude as the balloons get closer to each other. 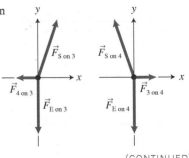

(CONTINUED)

Observational experiment	Analysis
Experiment 4. Bring the plastic wrap close to one of the balloons that was rubbed with plastic wrap in Experiment 3. The balloon is attracted to the plastic. The attraction gets stronger as the plastic wrap is held closer. Plastic wrap used to rub balloon	The plastic wrap exerts an unknown attractive force on the rubbed balloon. The force is stronger the closer the plastic wrap is to the balloon. 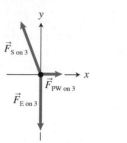
Experiment 5. Bring a balloon rubbed with felt near a balloon rubbed with plastic wrap. The balloons are attracted to each other. The attraction strengthens as the balloons are held closer. 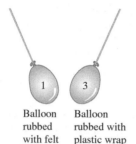 Balloon rubbed with felt Balloon rubbed with plastic wrap	An unknown force pulls each balloon toward the other. This force increases in magnitude as the balloons get closer to each other. 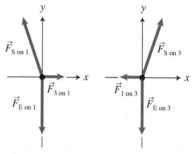

Patterns

- Balloons rubbed with the same material repel each other.
- A balloon rubbed with a particular material is attracted to that material.
- A balloon rubbed with felt attracts a balloon rubbed with plastic wrap.
- All of these effects are stronger the closer the objects are to each other.

TIP In figures, we will draw red pluses to note positive charges and blue minuses to note negative charges.

Many other objects exhibit similar effects. For example, you can use plastic rods instead of balloons and rub them with felt (like Experiment 1) or glass rods and rub them with silk. In all of these experiments, we find that more vigorous rubbing increases the magnitude of the forces that the objects exert on each other.

Ancient people observed similar effects with different materials. The Greeks noticed that amber, a fossilized resin-like material, attracted small, light objects after it was rubbed with wool. The word for amber in Greek is "electron"; thus this attraction was called electrical. Over time, the new property acquired by the materials due to rubbing came to be called **electric charge**. Physicists explained the observations by saying that the objects acquired **positive electric charge** or **negative electric charge**. Objects that interact with each other because they are charged are said to exert **electric forces** on each other. When the charged objects are at rest, the force is called an **electrostatic force**; the word *static* emphasizes that the objects are not moving with respect to the observer.

By convention, the charge that appears on a balloon rubbed with plastic wrap or on a glass rod rubbed with silk is called positive charge; the charge that appears on a balloon rubbed with wool or felt or on a plastic rod rubbed with felt is called negative charge. We can now reinterpret the patterns in Table 17.1:

1. Materials rubbed against each other acquire electric charge.
2. Two objects with the same type of charge repel each other.
3. Two objects with opposite types of charge attract each other.
4. Two objects made of different materials rubbed against each other acquire opposite charges.
5. The magnitude of the force that the charged objects exert on each other increases when the distance between the objects decreases.
6. Often more vigorous rubbing leads to a greater force exerted by the rubbed objects on each other.

Remember, we have *not* observed electric charge in any of these experiments. We *have* observed attraction and repulsion and used the concept of electric charge to explain these observations. The model of two electric charges is sufficient to explain all of the experiments that we have done so far.

CONCEPTUAL EXERCISE 17.1 **Mysterious Scotch tape**

Take two 9-inch-long pieces of Scotch tape and place them sticky side down on a plastic, glass, or wooden tabletop. Now pull on one end of each piece to remove them from the table. Bring the pieces of tape near each other. They repel, as shown at right. Can we explain the repulsion of the pieces of tape as an electric interaction?

The pieces of tape pulled from the table repel.

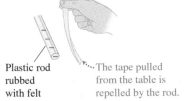
Glass rod rubbed with silk ....The same tape is attracted to this rod.
Plastic rod rubbed with felt ....The tape pulled from the table is repelled by the rod.

Sketch and translate The pieces of tape repel each other, exhibiting the behavior of like-charged objects. Thus, it is possible that the pieces of tape have electric charge. To test this hypothesis, we need to design an experiment whose outcome we can predict using the idea that the pieces are electrically charged. *If* they are, and we bring them one at a time next to a glass rod rubbed with silk (positively charged) and a plastic rod rubbed with felt (negatively charged), *then* they should be attracted to one and repelled by the other.

Simplify and diagram When we perform the experiment, we see that a piece of tape is repelled by the plastic rod rubbed with felt and attracted to the glass rod rubbed with silk. Does it mean that the pieces are electrically charged? We cannot say for sure, but our experiment does not rule out the hypothesis that the pieces of tape are electrically charged.

Try it yourself Take two pieces of tape and fold the end of each piece over about half a centimeter to make a "handle." Place one piece of tape on top of a table and the second piece on top of the first. Then pull them together off the table and separate them by pulling the "handles" away from each other. The pieces of tape attract each other. Design an experiment to determine the sign of each of their charges.

Answer According to our charge sign convention, a plastic rod rubbed with felt is negatively charged and a glass rod rubbed with silk is positively charged. Hold both rods near the two pieces of tape to observe repulsion or attraction—this will tell you the signs of the charges of the tape pieces.

Charged objects attract uncharged objects

So far, we have learned that objects rubbed with different materials can have opposite charges and attract each other electrically. However, when we bring a plastic comb that has been rubbed with felt near small bits of paper, the bits of paper are attracted to the comb (**Figure 17.2**). Small pieces of aluminum foil are attracted to the comb even more readily than paper. Why are uncharged objects attracted to charged objects? We explore this question in Observational Experiment **Table 17.2**.

FIGURE 17.2 Uncharged paper pieces are attracted to a charged comb.

OBSERVATIONAL EXPERIMENT TABLE 17.2  **Interactions of charged and uncharged objects**

 VIDEO OET 17.2

Observational experiment	**Analysis**
Experiment 1. Hang a balloon from a string next to a wall (neither the balloon nor the wall are charged). The balloon hangs straight down. Next, hang a negatively charged balloon near the same wall. The negatively charged balloon is attracted to the wall. Uncharged wall 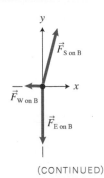	The uncharged wall exerts an attractive force on the negatively charged balloon. $\vec{F}_{\text{S on B}}$ $\vec{F}_{\text{W on B}}$ $\vec{F}_{\text{E on B}}$

(CONTINUED)

Observational experiment		Analysis	
Experiment 2. Hang a positively charged balloon near the uncharged wall. The balloon is attracted to the wall.	Uncharged wall	The uncharged wall exerts an attractive force on the positively charged balloon.	$\vec{F}_{\text{S on B}}$ $\vec{F}_{\text{W on B}}$ $\vec{F}_{\text{E on B}}$
Experiment 3. Repeat Experiment 1, only this time bring the negatively charged balloon near an uncharged metal plate. You may observe that the balloon has a stronger attraction to the plate than to the wall.	Uncharged metal plate	The magnitude of the force that the uncharged metal plate exerts on the negatively charged balloon is larger than the magnitude of the force that the wall in Experiment 1 exerts on the balloon.	$\vec{F}_{\text{S on B}}$ $\vec{F}_{\text{M on B}}$ $\vec{F}_{\text{E on B}}$
Experiment 4. Repeat Experiment 2, only this time bring the positively charged balloon near an uncharged metal plate. You may observe that the balloon has a stronger attraction to the plate than to the wall.	Uncharged metal plate	The magnitude of the force that the uncharged metal plate exerts on the positively charged balloon is larger than the magnitude of the force that the wall in Experiment 2 exerts on the balloon.	$\vec{F}_{\text{S on B}}$ $\vec{F}_{\text{M on B}}$ $\vec{F}_{\text{E on B}}$

Patterns

- The positively and negatively charged balloons were attracted to uncharged materials (the wall and the metal).
- The intensity of interaction was stronger with the metal plate than with the nonmetal wall.

Using the patterns in Table 17.2, we add a new statement to our conceptual model of the electric interaction: uncharged objects are attracted to both positively and negatively charged objects; uncharged metal objects are usually attracted more strongly than nonmetal objects.

In the next section, we will develop conceptual explanations for these and other observations.

REVIEW QUESTION 17.1 To decide whether an object is electrically charged, we need to observe its repulsion from some other object, not its attraction. Why is attraction insufficient?

17.2 Explanations for electrostatic interactions

Electrically charged objects interact with each other: they repel or attract. They also attract uncharged objects. What is the mechanism behind this interaction?

Is the electric interaction a magnetic interaction?

Do you recall observing any other interaction that involved repulsion? If you've ever played with magnets, you probably know that magnets have poles (north and south). Like poles repel, while unlike poles attract. Both poles attract objects that are not magnets but contain iron—nails, paper clips, etc. Electrically charged objects also attract and repel each other and attract uncharged objects. Perhaps the electric interaction is actually the magnetic interaction, just described using different terminology. We test this hypothesis in Testing Experiment Table 17.3.

TESTING EXPERIMENT TABLE 17.3 Testing the electric = magnetic interaction hypothesis

VIDEO TET 17.3

Testing experiment	Prediction	Outcome
Experiment 1. Bring a negatively charged plastic rod near the north pole of a magnet that is free to rotate about a pivot.	If the electric interaction is the same as the magnetic interaction, then the negatively charged rod should either attract or repel the north pole of the magnet.	The negatively charged rod attracts the north pole.
Experiment 2. Bring the negatively charged plastic rod near the south pole of a magnet that is free to rotate about a pivot.	If the electric interaction is the same as the magnetic interaction, then the negatively charged rod should repel the south pole of the magnet.	The negatively charged rod also attracts the south pole.

Conclusion

The outcome of the second experiment is inconsistent with the prediction. Thus we can reject the hypothesis that electric interaction is the same as magnetic. However, both experiments can be explained using our knowledge of electric interactions if we assume that a magnet *also* behaves like a piece of metal and thus attracts any charged object, the way all metals do.

The experiments in Table 17.3 show that the same charged rod attracts both the north and the south poles of a magnet, which is very different from the behavior of a magnet. This disproves our hypothesis that electric interactions are the same as magnetic interactions. We need a different mechanism to explain electrostatic interactions.

Fluid models of electric charge

Benjamin Franklin (1706–1790) proposed that a weightless electric fluid is present in all objects. According to Franklin, too much electric fluid in an object makes it positively charged and too little makes it negatively charged. If you rub two different types of material against each other, the electric fluid may move from one material to the other. The material that loses some electric fluid becomes negatively charged, and the material that gains some electric fluid becomes positively charged. The fluid can also move inside the uncharged object when another charged object is present, making its sides positive or negative. This movement of the fluid inside explains why an uncharged object is attracted to a charged object.

According to another model, there are two types of weightless electric fluids— positive and negative. In an object that is not rubbed, the two fluids are in balance.

After rubbing, one object loses some positive fluid and becomes negative, while the other one gains some positive fluid and becomes positive. Objects with an excess of the same fluid repel each other, and objects with an excess of different fluids attract. The fluids can move inside objects, making their sides charged even when the object overall is uncharged. Both of these fluid models account for all of our observations so far.

Particle model of electric charge

In 1897, J.J. Thomson, explaining the results of his experiments with cathode rays (you will learn more about cathode rays in Chapter 27), proposed that electric charge was not associated with a weightless fluid but was carried by the particles that comprise matter. In 1909, Robert Millikan and Harvey Fletcher tested this hypothesis by placing tiny droplets of oil between two horizontal metal plates that could be charged with opposite charges. Falling drops acquired some charge by interacting with ions in the surrounding air. When the plates were discharged, the drops fell down with a terminal speed due to viscous drag in air, which allowed the scientists to determine their size. The charge on the plates was then adjusted so that Earth's force, buoyant force and electric force added to zero and the only force left was drag. This force would eventually bring the droplets to a stop (**Figure 17.3**). They could determine the charge on the drop knowing its size and the charge on the plates.

Thousands of measurements demonstrated that the electric charge of each oil droplet changed only by multiples of some discrete unit of electric charge. This pattern indicated to Millikan and Fletcher that electric charge was carried by microscopic particles. Removing or adding these particles to an object changed the charge of the object. Each charged particle had a charge equal to some smallest indivisible amount of electric charge. These particles were called **electrons**. The electron also had a very small mass, several thousand times less than the mass of an average atom.

Contemporary model

Today we know that the explanation of charging by rubbing is actually very complicated. Atoms, the basic units of matter, consist of positively charged nuclei and negatively charged electrons. In normal conditions, the atoms are **neutral**, meaning that the total charge of each atom is zero because the positive charge of the nucleus is equal in magnitude to the negative charge of the electrons. Inside nuclei reside positively charged protons and neutral neutrons. You will learn about the structure of atoms and nuclei in Chapters 28 and 29, respectively (**Figure 17.4a**). It is possible for an electron to leave one object and move to a different object. Such a transfer occurs because the atoms of the material that gains electrons hold them more tightly than the atoms of the material that loses them. An atom that has lost one electron has a net charge of $+e$ (it has one more proton than it has electrons) and is called a **positive ion** (a positively charged sodium ion is shown in Figure 17.4b). An atom that gains an extra electron has a net charge of $-e$ and is called a **negative ion**.

We now understand why rubbed objects acquire opposite charges. Two objects start as neutral—the total electric charge of each is zero. During rubbing, one object gains electrons and becomes negatively charged. The other loses an equal number of electrons and with this deficiency of electrons becomes positively charged. Sometimes when you rub two objects against each other, no transfer of electrons occurs. When the electrons in both materials are bound equally strongly to their respective atoms, no transfer occurs during rubbing. Note that atomic nuclei or individual protons are never transferred during charging.

We can represent this process using integers. A neutral object can be represented with a zero. A zero can be made up of a sum of positive and negative numbers—for example, $+10 + (-10) = 0$. When we rub object A with a second object B that pulls an electron from object A, the negative number of object A becomes smaller while its positive number stays the same: $+10 + (-10) - (-1) = +10 + (-9) = +1$.

FIGURE 17.3 A simplified model of Millikan and Fletcher's oil drop experiment.

Three forces are exerted on the stationary drop D: gravitational force (by Earth, E), electric force (by the plates, P), and buoyant force (by air, A).

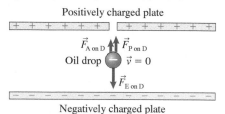

Positively charged plate

$\vec{F}_{A\ on\ D}$ $\vec{F}_{P\ on\ D}$

Oil drop $\vec{v} = 0$

$\vec{F}_{E\ on\ D}$

Negatively charged plate

FIGURE 17.4 A simplified model of (a) a carbon atom and (b) a sodium ion.

(a)

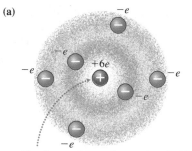

$-e$

$-e$

$+6e$

$-e$

$-e$

$-e$

$-e$

The charge of the carbon nucleus is $+6e$.

(b) The sodium ion Na$^+$ has 10 negative electrons and a nucleus with charge $+11e$.

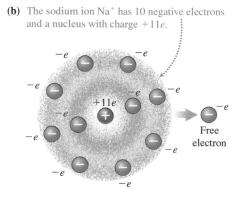

$-e$ $-e$

$-e$

$+11e$ $-e$ $-e$

$-e$ $-e$ Free electron

$-e$ $-e$

$-e$

Object A is positively charged. The model of charging by rubbing described above allows us to explain why balloons rubbed with different materials acquire different electric charges. A balloon rubbed with plastic wrap becomes positively charged because its electrons transfer to the wrap (which becomes negatively charged). The opposite effect occurs when the balloon is rubbed with felt. The plastic wrap must attract electrons more strongly than the balloon, and the felt must attract them less strongly.

CONCEPTUAL EXERCISE 17.2   **Attraction of shirt and sweater**

You pull your sweater and shirt off together and then pull them apart. You notice that they attract each other—a phenomenon called "static" in everyday life. Explain the mechanism behind this attraction and suggest an experiment to test your explanation.

The garments attract after they are pulled apart.

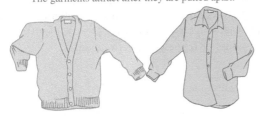

Sketch and translate Electric attraction between the sweater and shirt is a possible explanation. We need to explain how pulling the sweater and shirt apart leads to their opposite charges.

Simplify and diagram According to the particle model of electric charge, some of the electrons from one garment transfer to the other garment when they rub against each other. Before rubbing, both objects were neutral. If some negative electrons transfer from one object to the other, the former object now has a net positive charge while the latter has a net negative charge. If this hypothesis is correct, then a Scotch tape strip pulled off a table should be repelled by one of the garments and attracted to the other. When we perform the experiment, shown at top right, the outcome matches this prediction.

Negatively charged tape is attracted to the sweater . . .   . . . and repelled by the shirt.

Try it yourself You suspect that some objects after rubbing acquire a third type of charge. How can you test this hypothesis?

Answer

Imagine that rubbing a balloon with some new type of material (spandex, for example) gives the balloon a third type of charge. Then according to our pattern, two balloons rubbed with spandex should repel—and they do. But they should also attract both the balloon rubbed with felt and the balloon rubbed with plastic (if we assume that all differently charged objects attract each other). However, this never happens. If the balloon rubbed with spandex attracts the balloon rubbed with felt, it always repels the balloon rubbed with plastic, and vice versa. No experiment ever performed has required the existence of a third type of electric charge in order to explain its outcome.

REVIEW QUESTION 17.2 The model of charging by rubbing involved the transfer of negatively charged electrons. How then can a neutral object become positively charged?

17.3 Conductors and insulators (dielectrics)

We found in Table 17.2 that an electrically charged balloon was attracted to an uncharged wall and was slightly more strongly attracted to an uncharged sheet of metal. The former is a so-called insulator (dielectric) and the latter a conductor. The meaning of these terms will be apparent by the end of this section.

Conductors

Suppose that in metals, some electrons are not bound to their respective atoms but can move freely throughout the metal. When we bring a positively charged rod next to a metal bar (without touching it), these **free electrons** in the metal bar should move closer to the positively charged rod, leaving the other side of the metal bar with a deficiency of electrons and therefore positively charged (**Figure 17.5a**). The negatively charged side of the bar is closer to the positively charged rod and thus should be attracted to it more strongly than the more distant positively charged side of the bar is repelled by it. When a negatively charged object is brought near an uncharged metal bar, the free electrons are repelled, resulting in similar charge separation (but with opposite orientation) and attraction (Figure 17.5b). We can use the same model of the internal structure of a metal to explain the experiments in Table 17.2. The magnet, which is made of metal, was attracted to both positively and negatively charged objects. We now see that the attraction is electric in nature and has nothing to do with magnetism.

So far, we have explained the experiments that we already performed. Can we use the model of free electrons in metals to explain new experiments?

Consider the following experiment. An uncharged ball of aluminum foil hangs from a thin insulating string. When a negatively charged plastic rod is brought near the metal foil, the ball first moves toward the rod, touches it, and swings away. **Figure 17.6** helps explain this experiment using our model of free electrons in metals.

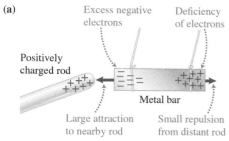

FIGURE 17.5 A metal bar is attracted to charged objects.

(a)

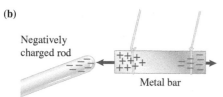

(b)

FIGURE 17.6 A neutral foil ball interacts with a negatively charged plastic rod.

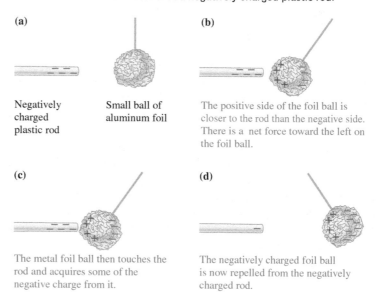

(a)

Negatively charged plastic rod

Small ball of aluminum foil

(b)

The positive side of the foil ball is closer to the rod than the negative side. There is a net force toward the left on the foil ball.

(c)

The metal foil ball then touches the rod and acquires some of the negative charge from it.

(d)

The negatively charged foil ball is now repelled from the negatively charged rod.

It looks as if our model of the freely moving charged particles in metals can explain a variety of experiments. It turns out that this model can be generalized to other materials. Metals are examples of electric **conductors**, materials in which some of the charged particles can move. In metals the moving charges are negatively charged free electrons. Some nonmetallic materials, such as graphite and silicon, can be conductors under certain conditions. In liquid conductors, such as nondistilled water or human blood, the moving charged particles can be positively or negatively charged ions. In hot gases, the moving charged particles can be both free electrons and ions.

The electroscope

Electroscopes, such as that shown in **Figure 17.7**, are useful tools for studying electrostatic interactions. They rely on the movement of free electrons in metal conductors. An electroscope consists of a metal ball attached to a metal rod that passes from the outside

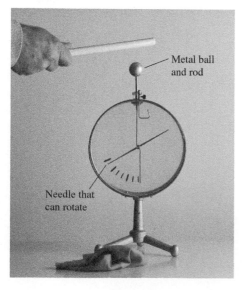

FIGURE 17.7 An electroscope.

Metal ball and rod

Needle that can rotate

through an insulating support into a glass-fronted metal enclosure. A very lightweight needle-like metal rod is connected on a pivot near the bottom of the larger rod.

Imagine that we bring a negatively charged object near the top of the electroscope ball without touching it. Then some of the free electrons in the ball and rod move away from the negatively charged object, leaving the ball with a positive charge. Electrons move to the lower part of the rod and to the needle, and as a result both become negatively charged. They repel each other, and the needle deflects away from the rod (**Figure 17.8a**). If we then take the negatively charged object away from the top of the electroscope, the needle returns to its original vertical position.

FIGURE 17.8 Charging an electroscope.

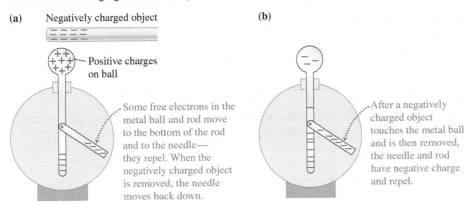

(a) Negatively charged object

Positive charges on ball

Some free electrons in the metal ball and rod move to the bottom of the rod and to the needle— they repel. When the negatively charged object is removed, the needle moves back down.

(b)

After a negatively charged object touches the metal ball and is then removed, the needle and rod have negative charge and repel.

Now, we repeat the experiment, only this time, we touch the top of the electroscope with the charged rod. We observe the same deflection; only this time when the rod is removed, the needle stays deflected (Figure 17.8b). Touching a negatively charged object to the ball on the top of the electroscope transfers electrons to the electroscope. If we then remove the negatively charged object from the electroscope, the electroscope ball, rod, and needle are now negatively charged throughout, and again the rod and needle repel each other. When we repeat the experiments with a positively charged object instead, we observe the same outcomes.

If instead of just touching the top of the electroscope with the charged rod, we carefully rub the charged rod against the top of the electroscope as if we are scrubbing the charges from the rod, we see that the angle of the deflection increases—the angle of deflection is related to the magnitude of the electric charge of the electroscope.

Dielectrics

The presence of electrically charged particles that can freely move inside conductors explains how a charged object attracts a neutral metal object. How can we explain the attraction of a neutral nonmetal object and a charged object, such as observed in Table 17.2?

Plastic, glass, and other nonmetal materials do not have free electrons or any other charged particles that are free to move inside. Such materials are called **insulators** or **dielectrics**. All the electrons are tightly bound to their atoms/molecules (**Figure 17.9a**). If a charged object (let's make it positive) is brought nearby, it exerts forces on the negatively charged electrons and on the positively charged nuclei inside the atoms of the dielectric material. Negatively charged electrons are pulled slightly closer to the charged bar, and the positively charged nuclei are pushed slightly away. This causes a slight atomic-scale separation of the negative and positive charges in the material, a process called **polarization**. When the atoms are in this polarized state, they are called **electric dipoles**. An electric dipole is any object that is overall electrically neutral but has its negative and positive charges separated. Polarization leads to

FIGURE 17.9 A simplified model of how a charged rod polarizes atoms in an insulating material.

(a)

Neutral atoms with charge uniformly distributed

(b) When a positively charged object is brought near the dielectric material, its atoms and molecules become polarized.

There is now a weak attraction between the charged object and the dielectric material.

(c) Polarization also occurs when we bring a negatively charged object near a nonconducting material.

FIGURE 17.10 The effect of a charged rod on metal and plastic cups.

(a)

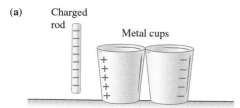

The rod polarizes the charge on the cups so that there are excess electrons on the right cup and a deficiency of electrons on the left cup.

(b)

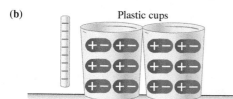

Atoms become polarized but there is no net charge separation.

(c)

When separated, the metal cups are oppositely charged.

(d)

When separated, the plastic cups are uncharged.

FIGURE 17.11 Free electrons in the body move toward a positively charged cup.

(a)   Initial

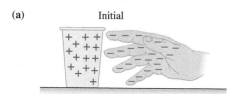

If your hand touches the cup, free electrons on the hand transfer to the cup until the net charge on the cup is zero.

(b)   Final

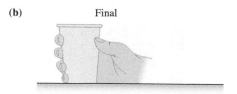

a small accumulation of charge on the surface of the object (depicted in Figure 17.9b). A similar polarization occurs if you bring a negatively charged object near an insulating material (Figure 17.9c).

Due to this polarization, electrically neutral pieces of paper are attracted to a charged comb that has been run through your hair, and an electrically neutral wall attracts a charged balloon.

Let's summarize what we have learned about the electric properties of materials.

> **Conductors and insulators** Materials can be divided into two groups: conductors and insulators (dielectrics). Both materials are composed of oppositely charged particles with a total charge of zero. In conductors, some of the charged particles can move freely. In dielectrics, charges can only be redistributed slightly a process called polarization.

We can test the difference between conductors and dielectrics. Imagine that you have two aluminum cups (conductors) and two plastic cups (dielectrics) placed on an insulating surface. The aluminum cups are touching each other, as are the plastic cups. You bring a negatively charged rod next to each pair of cups (**Figures 17.10a** and b). While holding the rod nearby, you separate the cups by pushing them apart with a plastic ruler. If our understanding of conductors and insulators is correct, the metal cups will be oppositely charged; the cup closer to the rod will be positively charged and the cup farther from the rod will be negatively charged (Figure 17.10c). The plastic cups should remain uncharged (Figure 17.10d). We can check this prediction using a strip of Scotch tape pulled off a table. The strip of tape is attracted to one of the metal cups and repelled from the other. However, the charged tape is attracted slightly to both plastic cups (each plastic cup becomes slightly polarized and, thus, attracts the negatively charged tape). Since the outcome of the experiment matched the prediction, we gain confidence in our model.

Experiments with metal cups give us another example of charging. We charged both cups without touching them with a charged object. We did not transfer any extra charge to both cups; we merely separated the existing charges. Such a process is called charging *by induction*. We are already familiar with charging *by rubbing* from the balloon experiments in Section 17.1, and with charging *by touching* from the electroscope experiments earlier in this section.

> TIP Electric charge is completely defined by the effects it produces. You cannot see electric charge.

Is the human body a conductor or a dielectric?

In the previous experiment you used a plastic ruler to separate the metal cups. If you had touched the metal cups with your hand instead of the ruler, the charged tape would have been attracted to both. This is because the human body is a conductor. As your hand approaches a positively charged cup, the free electrons on the surface of your hand and arm move toward it (**Figure 17.11a**). When you touch the positively charged metal cup, electrons from the surface of your hand travel to the cup until the cup has a net charge of approximately zero (Figure 17.11b). A similar process happens when you touch a negatively charged cup, only this time the electrons transfer from the cup to your skin. Sometimes when you touch a charged object, you feel a shock and might even see a spark. We will explain these phenomena later in the book.

Grounding

Instead of touching the charged metal cup with your hand, you can safely discharge it (make its net charge effectively zero) by connecting a wire to both the cup and a metal pipe that is driven into the ground (**Figure 17.12**). This process is called **grounding**. To briefly explain why a grounded metal cup discharges, let's assume that Earth is a large conductor. If the cup is charged positively, negative electrons in the ground are attracted toward the cup and travel from the ground to the cup, causing it to become neutral. For a negatively charged cup, electrons in the cup travel through the grounding wire into the ground. Grounding electric appliances is extremely important in order to avoid electric shocks. A more detailed explanation of grounding will be given in Chapter 18.

FIGURE 17.12 A wire that is "grounded" on one end to a pipe in the ground can discharge the positively charged cup.

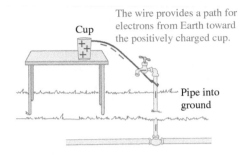

Discharging by moisture in the air

If you leave a negatively or positively charged object alone for a long time interval, the object can become neutral without being grounded. This can happen for two reasons. First, the air, which is made of neutral molecules and should ideally be a dielectric, in fact has a few electrically charged particles, which are attracted to any object that has an opposite charge. The second reason is the presence of water vapor in the air. The electric charge in water molecules is distributed in such a way that the molecule is a natural electric dipole even without the polarizing influence of an external charged object (**Figure 17.13a**). The hydrogen side of the water molecule is slightly positive, while the oxygen side is slightly negative, thus producing an electric dipole (Figure 17.13b). When water molecules are near a charged object (in this case negative), they will be attracted to it. On a humid day, a very thin film of water forms on all objects. This water layer works just like a wire in grounding. This is why it is a good idea to dry equipment for electrostatic experiments with a hair dryer before use.

FIGURE 17.13 A schematic sketch of a water molecule.

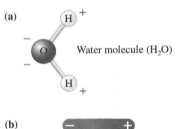

The net effect is an electric dipole.

Properties of electric charge

Think back to our explanation of charging by rubbing. When a plastic rod is rubbed with felt, the rod becomes negatively charged. The felt becomes positively charged. However, rubbing did not create the charges—it just transferred some electrons from one object to the other. If we assume that each transferred electron has the same electric charge independent of the presence of other electrons, then the total electric charge of an isolated system should be constant. The charge of a nonisolated system changes only if charged objects are transferred in or out of the system. Thus, similar to mass, momentum, and energy, *electric charge is a conserved quantity*; if it changes in one system, we can always find another system in which it is constant.

Another property of electric charge can be deduced from the Millikan-Fletcher experiments. Electric charge cannot be divided infinitely—the smallest amount of charge that an object can have is the charge of one electron. This property is called charge **quantization**.

Electric charge Electric charge (symbol q or Q) is a property of objects that participate in electrostatic interactions. Electric charge is quantized—you can only change an object's charge by increments, not continuously. Electric charge is conserved. The unit for electric charge is the coulomb (C). The smallest increment of charge is that of one electron $-e = -1.6 \times 10^{-19}$ C. The magnitude of the charge of one electron is sometimes called the *elementary charge*.

TIP ▸ Notice how small the charge of the electron is. Even if we place a billion electrons on an object, it will still have only a -1.6×10^{-10} C charge. Typical charges of objects through rubbing are about 10 times larger, but still are very small. I C is a *huge* charge!

Our next goal is to develop a quantitative description of the forces that charged objects exert on each other.

REVIEW QUESTION 17.3 One cannot charge a held metal object by rubbing. Why?

17.4 Coulomb's force law

We have observed in the experiments so far that the nearer rubbed objects are to each other, the greater the forces they exert on each other. The forces also increase when the magnitude of the charges of the interacting objects increases. So the force that a charged object exerts on another must depend on the distance r between them and on their electric charges q_1 and q_2.

Charles Coulomb determined the relationship between these quantities in 1785. The experimental apparatus he used is called a torsion balance (**Figure 17.14**). Coulomb hung a light glass rod from a thin wire. At the ends of the rod he attached identical small metal spheres. He then charged a third metal sphere identical in size to the first two and touched one of the spheres attached to the rod. Because the two touching spheres were identical, after touching they had the same electric charge, half of the original, and they repelled. The wire twisted until the torque exerted by the wire on the glass rod balanced the torque exerted by one charged sphere on the other. Coulomb measured the angle of this twist and used it to determine the electric force exerted by one sphere on the other. He measured the distance between the repelling spheres and varied it to see how the force of repulsion depended on the distance between them.

In Coulomb's time, scientists did not know how to directly measure electric charge, and the unit for electric charge did not exist. That is why Coulomb used relative charges rather than absolute charges. Imagine two identical metal spheres—one charged and one uncharged. If you bring them into contact, what would be the charge on each? Coulomb reasoned that since the spheres were identical, the charge would distribute equally between them. Using the same method described above to get the charge of one sphere equally distributed between two identical spheres and thus obtaining half of the original charge on both, he could achieve fractions of charge by touching a charged sphere to identical uncharged spheres many times. This way he could have charges of $q, \frac{1}{2}q, \frac{1}{4}q$, and so forth on the original sphere. He needed to have spheres of different charges to find out how the force exerted by one on the other depended on the magnitude of their charges. **Table 17.4** shows a simplified version of data that Coulomb might have collected. It indicates how the repulsive force that one metal sphere exerted on another depended on their separation and on their charges. (In reality, Coulomb's data did not follow such clear patterns and included experimental uncertainty.)

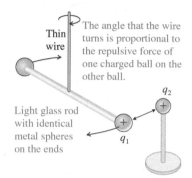

FIGURE 17.14 Coulomb's torsion balance, an apparatus for measuring the electric force between charged objects.

Thin wire

The angle that the wire turns is proportional to the repulsive force of one charged ball on the other ball.

q_2

Light glass rod with identical metal spheres on the ends

q_1

TABLE 17.4 Coulomb's data

Experiment	Charge q_1	Charge q_2	Distance r	Force $F_{q_1 \text{ on } q_2}$
1	1 (unit)	1 (unit)	1 (unit)	1 (unit)
2	$\frac{1}{2}$	1	1	$\frac{1}{2}$
3	$\frac{1}{4}$	1	1	$\frac{1}{4}$
4	1	$\frac{1}{2}$	1	$\frac{1}{2}$
5	1	$\frac{1}{4}$	1	$\frac{1}{4}$
6	$\frac{1}{2}$	$\frac{1}{2}$	1	$\frac{1}{4}$
7	$\frac{1}{4}$	$\frac{1}{4}$	1	$\frac{1}{16}$
8	1	1	2	$\frac{1}{4}$
9	1	1	3	$\frac{1}{9}$
10	1	1	4	$\frac{1}{16}$

To find patterns in the data presented in Table 17.4, let's focus on one variable at a time. In Experiments 1 and 2, the only quantity that changes is the magnitude of the charge q_1. When the charge is halved, the force exerted by q_1 on q_2 is also halved. Looking at Experiment 3, we see that dividing the charge by four reduced the force to one-fourth. The force seemed to be directly proportional to the magnitude of the charge q_1. Experiments 4 and 5 show the same pattern for the magnitude of the charge q_2. If a quantity is directly proportional to two independent quantities, it has to be proportional to their product. Thus, we infer from the table that $F_{q_1 \text{ on } q_2} \propto q_1 q_2$. All 10 experiments are consistent with this conclusion.

From the data we also see that doubling the distance between the objects decreased the force by a factor of 4, and tripling the distance decreased the force by a factor of 9. It appears that the force is inversely proportional to the square of the separation between the objects:

$$F_{q_1 \text{ on } q_2} \propto \frac{1}{r^2}$$

We can now combine these patterns into one mathematical expression, called **Coulomb's law**.

Coulomb's law   The magnitude of the electric force that point-like object 1 with electric charge q_1 exerts on point-like object 2 with electric charge q_2 when they are separated by a distance r is given by the expression

$$F_{q_1 \text{ on } q_2} = k_C \frac{|q_1||q_2|}{r^2} \tag{17.1}$$

where k_C is a proportionality constant called **Coulomb's constant** whose value using SI units is

$$k_C = 9.0 \times 10^9 \frac{\text{N} \cdot \text{m}^2}{\text{C}^2}$$

When using the above to determine the magnitude of the electric force, we do not include the signs of the electric charges q_1 and q_2.

The force that object 1 exerts on object 2 points away from object 1 if they have same sign charges and toward object 1 if they have opposite sign charges.

Notice that, mathematically, the expression for the electric force that two charged objects exert on each other,

$$F_{q_1 \text{ on } q_2} = k_C \frac{|q_1||q_2|}{r^2}$$

is analogous to the expression for the gravitational force that any two objects with mass exert on each other,

$$F_{m_1 \text{ on } m_2} = G \frac{m_1 m_2}{r^2}$$

The gravitational force depends on the masses of the objects, and the electric force depends on the charges of the objects. Also, the gravitational force is always an attractive force, whereas the electric force can be attractive or repulsive. The proportionality constants have very different values.

TIP   If the interacting objects are large conductors and near each other, the electrons will redistribute on their surfaces, making the effective distance smaller than the distance between the centers of the objects (and difficult to estimate). Thus, we only use Coulomb's law for point-like objects. These objects can be either conductors or dielectrics.

Comparing the magnitude of the electric force to the gravitational force

Let's compare the electric force to the gravitational force exerted by the nucleus, which consists of just one proton, on the electron in a hydrogen atom. A proton has a charge of $+1.6 \times 10^{-19}$ C and mass of 1.7×10^{-27} kg. An electron has a charge of -1.6×10^{-19} C and mass of 9.1×10^{-31} kg. They are separated in the atom by about 10^{-10} m. We call the proton charge 1 and the electron charge 2.

Electric force:

$$F_{q_1 \text{ on } q_2} = k_C \frac{q_1 q_2}{r^2} = (9.0 \times 10^9 \, \text{N} \cdot \text{m}^2/\text{C}^2) \frac{(1.6 \times 10^{-19} \, \text{C})^2}{(10^{-10} \, \text{m})^2}$$

$$= 2.3 \times 10^{-8} \, \text{N}$$

Gravitational force:

$$F_{m_1 \text{ on } m_2} = G \frac{m_1 m_2}{r^2} = (6.67 \times 10^{-11} \, \text{N} \cdot \text{m}^2/\text{kg}^2) \frac{(1.7 \times 10^{-27} \, \text{kg})(9.1 \times 10^{-31} \, \text{kg})}{(10^{-10} \, \text{m})^2}$$

$$= 1.0 \times 10^{-47} \, \text{N}$$

If we divide the electric force by the gravitational force, the result is about 2×10^{39}. The electric force that the proton exerts on the electron is about 2×10^{39} times greater than the gravitational force that the proton exerts on the electron! What about the gravitational force exerted by Earth on the electron?

$$F_{\text{E on } m_2} = G \frac{m_E m_2}{r^2} = (6.67 \times 10^{-11} \, \text{N} \cdot \text{m}^2/\text{kg}^2) \frac{(6.0 \times 10^{24} \, \text{kg})(9.1 \times 10^{-31} \, \text{kg})}{(6.4 \times 10^6 \, \text{m})^2}$$

$$= 8.9 \times 10^{-30} \, \text{N}$$

TIP Since Coulomb's law contains the product of the two charges, the electric force that object 1 exerts on object 2 is exactly the same in magnitude as the electric force that object 2 exerts on object 1. Coulomb's law is consistent with Newton's third law.

The gravitational force exerted by Earth on the electron is 18 orders of magnitude larger than the gravitational force exerted by the proton on the electron, but still about 22 orders of magnitude weaker than the electric force exerted by the proton on the electron. That is why physicists can confidently ignore gravitational forces when dealing with atomic size particles.

We have learned that the electric force that electrons and nuclei in atoms exert on each other is what holds atoms together. Although the atoms themselves are electrically neutral and should not interact with each other via electric forces, they do in fact interact. This interaction occurs because the electrons of one or more atoms can interact with the nuclei and electrons of other atoms. These forces are responsible for the existence of groups of bound atoms—molecules. In addition, the electric forces that the charged parts of electrically neutral molecules exert on each other are responsible for molecules forming liquids and solids at sufficiently low temperatures. Many forces that we encounter in everyday life—tension forces, friction forces, normal forces, buoyant forces, etc.—can be explained using Coulomb's law. You can use Coulomb's law to explain why a liquid exerts an upward buoyant force on a submerged object, or why carpet exerts a friction force on your feet.

CONCEPTUAL EXERCISE 17.3   **Interactions of charged objects**

Two equal-mass small aluminum foil balls A and B have electric charges $+q$ (on A) and $+3q$ (on B). (a) Compare the magnitude of the electric force that the foil ball with the smaller charge exerts on the ball with the larger charge ($F_{\text{A on B}}$) to the force that the larger charged ball exerts on the smaller charged ball ($F_{\text{B on A}}$). Justify your answer. (b) If the balls are suspended by equal-length strings from the same point, will one string hang at a greater angle from the vertical than the other? Justify your answer.

Sketch and translate A labeled sketch of the situation is shown at right. We choose each ball as a separate system of interest. Ball A

interacts with Earth, the string, and ball B. Ball B interacts with Earth, the string, and ball A.

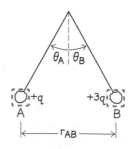

Simplify and diagram We model the balls as point-like objects (see the force diagrams below).

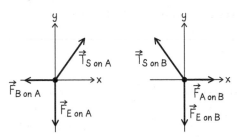

(a) The electric forces that the balls exert on each other have the same magnitude, according to Newton's third law and Coulomb's

law ($F_{\text{A on B}} = k_C q_A q_B / r_{AB}^2 = k_C q_B q_A / r_{AB}^2 = F_{\text{B on A}}$). (b) Since the balls have the same mass, Earth exerts the same force on each. Thus the forces exerted by the strings must also have the same magnitude since the net force exerted on each ball equals zero. Therefore, the angles of the strings relative to the vertical should also be the same.

Try it yourself Assume that the balls have twice the mass of the balls in the above exercise. By what factor should we increase the charges on the balls so that the angle made by the strings does not change?

Answer $\sqrt{2}$.

QUANTITATIVE EXERCISE 17.4   **Coulomb's law and proportional reasoning**

Three identical metal spheres A, B, and C are on separate insulating stands (Figure a below). You charge sphere A with charge $+q$. Then you touch sphere B to sphere A and separate them by distance d (Figure b). (a) Write an expression for the electric force that the spheres exert on each other. What assumptions did you make? (b) You now take

sphere C, touch it to sphere A (Figure c), and remove sphere C. You then separate spheres A and B by a distance $d/2$ (Figure d). Write an expression for the magnitude of the force that spheres A and B exert on each other.

Represent mathematically Assume that we can model the spheres as point-like objects. (a) Assume that after sphere A touches B, each of them has a charge of $+q/2$. (b) After C touches A, the charge on A halves again, so that A now has charge $+q/4$. The distance between A and B is $d/2$. We can use Coulomb's law to write an expression for the force that each charge exerts on the other for each part of the problem.

Solve and evaluate (a) The magnitude of the electric force that A and B exert on each other is

$$F_{\text{A on B}} = F_{\text{B on A}} = k_C \frac{(q/2)(q/2)}{d^2} = k_C \frac{q^2/4}{d^2} = k_C \frac{q^2}{4d^2}$$

(b) Now, the force that A and B exert on each other is

$$F'_{\text{A on B}} = F'_{\text{B on A}} = k_C \frac{(q/4)(q/2)}{(d/2)^2} = k_C \frac{q^2/8}{d^2/4} = k_C \frac{q^2}{2d^2}$$

Although one of the charges got smaller, the decrease in the distance between A and B more than compensated for this, and the force became twice as great as in the first case.

Try it yourself Repeat this exercise, starting with charge $+q$ on sphere A and no charge on spheres B and C. Touch B to A. Then touch C to B. Separate spheres B and C by a distance $d/2$. Write an expression for the force that B exerts on C.

Answer $F_{\text{B on C}} = F_{\text{C on B}} = k_C \frac{(q/4)^2}{(d/2)^2} = k_C \frac{q^2}{4d^2}$.

(a) $+q$

Charge A with $+q$.

A   B   C

(b) $+q/2$ $+q/2$

Touch B to A and separate them by distance d.

A   B

$\leftarrow d \rightarrow$

(c) $+q/4$ $+q/4$ $+q/2$

Touch C to A and remove C.

C

A   B

(d) $+q/4$ $+q/2$

Separate A and B by $d/2$.

A   B

$\leftarrow d/2 \rightarrow$

EXAMPLE 17.5 **Net electric force**

The metal spheres on the insulating stands from the previous exercise have the following electric charges: $q_A = +2.0 \times 10^{-9}$ C, $q_B = +2.0 \times 10^{-9}$ C, and $q_C = -4.0 \times 10^{-9}$ C. The spheres are placed at the corners of an equilateral triangle whose sides have length $d = 1.0$ m with C at the top of the triangle. What is the magnitude of the total (net) electric force that spheres A and B exert on C?

Sketch and translate A labeled sketch of the top view of the situation is shown below. All three angles of the triangle are 60°.

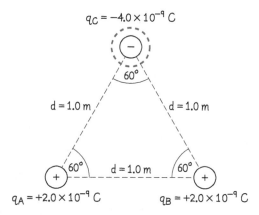

Simplify and diagram Assume that the distance between the spheres is much bigger than their radii so that they can be modeled as point-like objects. Choose sphere C as the system of interest and draw a force diagram for it. We are only interested in the magnitude of the net electric force exerted on C. Thus, we include in the diagram only the electric forces exerted on C by A and by B. We see that the vector sum of $\vec{F}_{A \text{ on } C}$ and $\vec{F}_{B \text{ on } C}$ points straight down (in the negative y-direction).

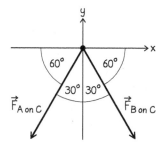

Represent mathematically Each of the spheres A and B exerts a force on C of magnitude

$$F_{A \text{ on } C} = F_{B \text{ on } C} = k_C \frac{|q_A||q_C|}{d^2}$$

where d is the length of a side of the triangle. Each force vector points along the line connecting the respective spheres. The x-components of the two force vectors cancel each other, and the resultant force has only a y-component. Using trigonometry, we find that the net y-component force that A and B exert on C is

$$F_{\text{Net on } C\,y} = (-F_{A \text{ on } C} \sin 60°) + (-F_{B \text{ on } C} \sin 60°)$$

$$= -2F_{A \text{ on } C} \sin 60° = -2\left(k_C \frac{|q_A||q_C|}{d^2}\right) \sin 60°$$

Solve and evaluate Inserting the appropriate values gives

$$F_{\text{Net on } C\,y} = -2\Big((9.0 \times 10^9 \text{ N} \cdot \text{m}^2/\text{C}^2)$$

$$\times \frac{(2.0 \times 10^{-9} \text{ C})(4.0 \times 10^{-9} \text{ C})}{1.0 \text{ m}^2} \Big) \sin 60°$$

$$= -1.2 \times 10^{-7} \text{ N}$$

The force has magnitude 1.2×10^{-7} N and points straight down. This is a very small force. If the spheres had relatively small masses (a few hundred grams) and were mounted on stands on which the surface exerted a regular friction force, we would probably not observe any effects of these electric forces.

Try it yourself What charges should you place at the corners of an equilateral triangle with a horizontal base so that the net force exerted on the top charge points horizontally to the left?

Answer

The bottom left could be $-q$, the bottom right $+q$, and the top center sphere could have any positive charge.

REVIEW QUESTION 17.4 Two charged objects (1 and 2) with charges $q_1 = q$ and $q_2 = 28q$ are placed r meters away from each other. What is the ratio of the electric force that object 1 exerts on 2 to the force that 2 exerts on 1? Explain your answer.

17.5 Electric potential energy

In the previous section we learned to describe the interactions of electric charges in terms of forces. In this section we will learn to describe these interactions in terms of energy.

When you have a system of two objects exerting gravitational forces on each other (an apple-Earth system, for example), the system possesses gravitational potential energy. Since Coulomb's law is mathematically very similar to the law of universal gravitation, it is reasonable to suggest that a system of electrically charged objects also possesses some sort of electric energy. Let's investigate this idea.

Electric potential energy: a qualitative analysis

Let's begin by looking at the electric potential energy of two like-charged objects that are part of the hypothetical "electric cannon" shown in **Figure 17.15a**. A positively charged cannonball is held near another fixed positively charged object in the barrel of the cannon. This situation is similar to that of an object pressed against a compressed spring. The cannonball, the other charged object, and Earth are the system. Since the cannonball is repelled from the fixed-charge object, the cannonball when released accelerates up the barrel and out of its end (Figure 17.15b). During this acceleration, the system's kinetic (K) and gravitational potential (U_g) energies both increase. If we assume that the system is isolated, then some other type of energy has to decrease so that these two can increase. This other type of energy must be the electric energy suggested above. We will call it **electric potential energy** U_q.

FIGURE 17.15 The electric potential energy of two like-charged objects.

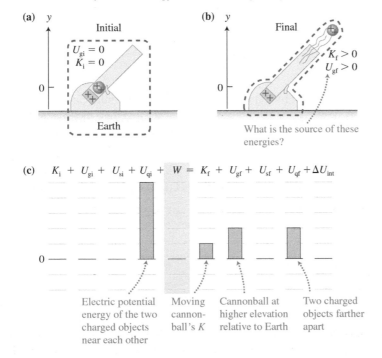

$$K_i + U_{gi} + U_{si} + U_{qi} + W = K_f + U_{gf} + U_{sf} + U_{qf} + \Delta U_{int}$$

The system comprising the two charged objects has electric potential energy. It seems that the electric potential energy of like-charged objects decreases as they get farther apart from each other. As the cannonball moves farther from the fixed-charge object, the electric potential energy of the system decreases as it is converted into kinetic and gravitational potential energy.

We can represent this process using a bar chart (Figure 17.15c). The initial state is the moment at which the cannonball is near the other fixed-charge object; the final state is when the cannonball is moving at the end of the barrel. For this process we assign the zero level for gravitational potential energy to be at the initial position of the cannonball. We must also make a zero-level assignment for electric potential energy. We say that when two charged objects are infinitely far apart, so that they essentially do not interact, they have zero electric potential energy.

Let us consider a situation in which two oppositely charged objects interact: a hypothetical "electrostatic nutcracker" (**Figure 17.16a**, on the next page). The system consists of two oppositely charged blocks, one of which can slide without friction. When the negatively charged block is released and moves nearer the nut, the kinetic energy of the system increases (Figure 17.16b). Thus electric potential energy must decrease. Assuming the zero level is when the two oppositely charged blocks are infinitely far apart, we conclude that the electric potential energy of a pair of unlike charges is less than zero.

FIGURE 17.16 The electric potential energy of two oppositely charged objects.

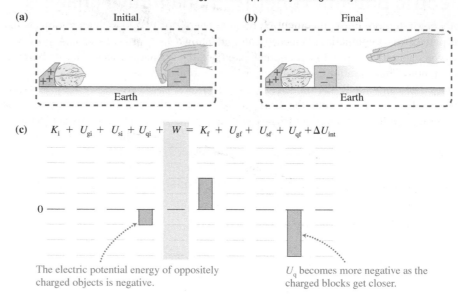

(a) Initial

(b) Final

Earth

Earth

(c) $K_i + U_{gi} + U_{si} + U_{qi} + W = K_f + U_{gf} + U_{sf} + U_{qf} + \Delta U_{int}$

0

The electric potential energy of oppositely charged objects is negative.

U_q becomes more negative as the charged blocks get closer.

The bar chart in Figure 17.16c represents this process. The initial electric potential energy of the system is negative. As the objects come closer together, the kinetic energy of the system increases and its electric potential energy decreases, becoming even more negative. Our next step is to devise a mathematical expression for electric potential energy.

Electric potential energy: a quantitative analysis

To derive an expression for the change in electric potential energy of two electric charges whose separation changes, we use the generalized work-energy principle from Chapter 7 (we assume that all objects are at the same temperature, thus there is no energy transferred by heating):

$$W = \Delta E_{system} \qquad (7.3)$$

where W is the work done on the system by objects in the environment and ΔE_{system} is the resulting change in the system's energy. To derive an expression for ΔU_q, we choose a hypothetical system that has the following feature: when work is done on this system, only its electric potential energy changes. The system shown in **Figure 17.17** consists only of the point-like charged objects 1 and 2 with like charges q_1 and q_2, initially separated by a distance r_i. The like charges exert a repulsive force on each other of magnitude

$$F_{q_1 \text{ on } q_2} = k_C \frac{|q_1||q_2|}{r^2} \qquad (17.1)$$

Imagine that we wish to prevent object 2 from flying away from 1 by exerting a force on 2 toward 1 of the same magnitude that the charged object 1 exerts on 2 (Figure 17.17a). If we push object 2 just a tiny bit harder, object 2 can be displaced a small distance Δr toward object 1 (Figure 17.17b). We use Eq. (7.1) to calculate the work ΔW_1 (here the symbol delta Δ indicates a small amount of work and not the change in work) done by the pushing force during this small displacement (the force and displacement are in the same direction; thus the cosine of the angle between them is 1):

$$\Delta W_1 = F_{P \text{ on } 2} \Delta r = \frac{k_C q_1 q_2}{r_i^2} \Delta r$$

Notice that there are no magnitude symbols for the charges because their signs are now important for the sign of work. Another important point is that this calculation is only

FIGURE 17.17 The work done in pushing two like-charged objects closer together.

(a)

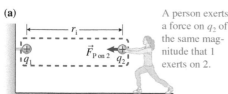

A person exerts a force on q_2 of the same magnitude that 1 exerts on 2.

(b)

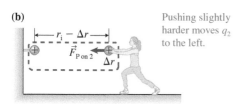

Pushing slightly harder moves q_2 to the left.

(c)

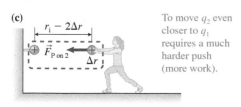

To move q_2 even closer to q_1 requires a much harder push (more work).

(d)

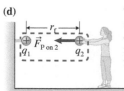

The total work is the sum of the work done in each tiny displacement from the initial to the final position.

approximate because the force needed to push object 2 increases as its separation from object 1 decreases. If Δr is small, the equation is a good approximation for the work done during that small displacement of object 2. If we wish to push object 2 through the next small displacement Δr (Figure 17.17c), we must exert a larger force since object 2 is now closer to object 1. For each step closer, the amount of work needed to move object 2 closer to object 1 increases.

The total work done in moving the charge from an initial separation r_i to a final separation r_f (Figure 17.17d) equals the sum of the work for each small step. This type of infinitesimal addition is done easily using calculus. The result of such a calculation is

$$W = \Delta W_1 + \Delta W_2 + \Delta W_3 + \cdots = \frac{k_C q_1 q_2}{r_f} - \frac{k_C q_1 q_2}{r_i} = k_C q_1 q_2 \left(\frac{1}{r_f} - \frac{1}{r_i} \right)$$

At each step of the process represented in Figure 17.17, there is no acceleration and therefore no change in kinetic energy or any other kind of energy. Thus we can reason that the only energy change of the system due to this work is the electric potential energy change. By substituting the above expression for W into Eq. (7.3) and zeros for all energy changes except the electric potential energy change, we find

$$k_C q_1 q_2 \left(\frac{1}{r_f} - \frac{1}{r_i} \right) = 0 + 0 + 0 + \Delta U_q + 0 + \cdots$$

Therefore,

$$\Delta U_q = k_C q_1 q_2 \left(\frac{1}{r_f} - \frac{1}{r_i} \right)$$

$$\Rightarrow U_{qf} - U_{qi} = \frac{k_C q_1 q_2}{r_f} - \frac{k_C q_1 q_2}{r_i}$$

Evidently, the electric potential energy of a system with two charged point-like objects separated by distance r is

$$U_q = \frac{k_C q_1 q_2}{r}$$

where q_1 and q_2 are the actual electric charges, not their magnitudes. This equation is consistent with the convention that we established earlier: the electric potential energy of interaction of two charged objects is zero when the distance between them is infinite. It is also consistent with our qualitative understanding of electric potential energy from the previous section. The energy of interaction of two like charges is positive, and that of two unlike charges is negative.

Electric potential energy The change in electric potential energy ΔU_q of a system of two charged objects q_1 and q_2 when they are moved from an initial separation r_i to a final separation r_f is

$$\Delta U_q = U_{qf} - U_{qi} = k_C q_1 q_2 \left(\frac{1}{r_f} - \frac{1}{r_i} \right) \tag{17.2}$$

Electric potential energy is measured in units of joules. Equation (17.2) is valid for both positively and negatively charged objects, provided the signs of the charges are included.

Two points are worth noting: (a) electric potential energy is proportional to $1/r$ and not to $1/r^2$ (as is the force in Coulomb's law), and (b) mathematically the expression for the electric potential energy is similar to the expression for the gravitational potential energy of a system with two objects with mass, $U_g = -G \dfrac{m_1 m_2}{r}$.

EXAMPLE 17.6 **Changes in electric potential energy**

Two oppositely charged objects (with positive charge $+q$ and negative charge $-q$) are separated by distance r_i. Will the electric potential energy of the system decrease or increase if you pull the objects farther apart? Explain.

Sketch and translate The initial state is the two oppositely charged objects (the system) a distance r_i apart. You pull the $-q$ charged object so that it is farther from the positively charged object—the final state—as shown below.

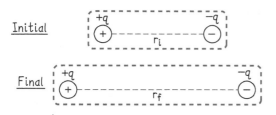

Because the direction of the force that you exert on the $-q$ charged object is the same as the direction of its displacement, positive work is being done on the system, increasing the electric potential energy of the system. If you pulled $-q$ infinitely far from $+q$, the final electric potential energy of the system would be zero. Thus, the initial electric potential energy must be negative. To understand this, imagine that you add a positive number to an unknown number and get zero: $x + 5 = 0$. What is the unknown number? $x = -5$ because $(-5) + 5 = 0$. The initial energy was negative.

Simplify and diagram We model the two objects as point-like particles, and we assume that only the electric potential energy changes. A bar chart representing the process is shown at top right.

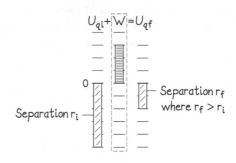

Represent mathematically We now use the bar chart and the generalized work-energy principle to represent the process mathematically:

$$E_i + W = E_f$$

$$\Rightarrow \frac{k_C(+q)(-q)}{r_i} + W = \frac{k_C(+q)(-q)}{r_f}$$

Solve and evaluate Notice that if r_i is less than r_f, the electric potential energy term on the left side is more negative than that on the right side—consistent with the bar chart. This means that the electric potential energy of the system is negative and changes from a larger negative value to a smaller negative value as the charges get farther apart—an increase in energy because positive work was done on the system. This outcome is consistent with our conceptual understanding of the process.

- -

Try it yourself In a hydrogen atom, a proton and an electron have charges of the same magnitude, 1.6×10^{-19} C, but opposite sign. The distance between them is 0.53×10^{-10} m. What is the change in electric potential energy if the electron is moved far away from the nucleus?

Answer 4.3×10^{-18} J.

FIGURE 17.18 The electric potential energy-versus-distance graphs for two charged objects.

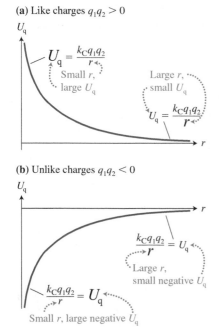

Graphing the electric potential energy versus distance

So far, we have represented electric potential energy with bar charts and with mathematics. Using a graph is another way to conceptualize this abstract physical quantity. Consider first how the electric potential energy changes as the separation r between two like-charged objects is varied. Because of the $1/r$ dependence, the electric potential energy approaches positive infinity when the separation approaches zero (see **Figure 17.18a**). The electric potential energy becomes less positive and approaches zero as the like charges are moved far apart. For a system with two unlike-charged objects, the electric potential energy approaches negative infinity when their separation approaches zero and becomes less negative (increases) and approaches zero as they are moved far apart (Figure 17.18b).

Electric potential energy of multiple charge systems

When a system has several interacting electrically charged objects, to determine the electric potential energy of such a system, we consider the electric potential energy of each pair of objects and then sum up all of the energies to find the total. Suppose that the system has three charged objects labeled 1, 2, and 3. Then the total electric potential energy is

$$U_q = U_{q12} + U_{q13} + U_{q23}$$

$$= \frac{k_C q_1 q_2}{r_{12}} + \frac{k_C q_1 q_3}{r_{13}} + \frac{k_C q_2 q_3}{r_{23}} \qquad (17.3)$$

where r_{12} is the distance between object 1 and 2, and similarly for the other terms. Remember that the signs of all charges must be included.

REVIEW QUESTION 17.5 How can we reduce the magnitude of the electric potential energy of a system of two electrically charged objects?

17.6 Skills for analyzing processes involving electric charges

While solving problems with electrically charged objects, you can use a problem-solving strategy similar to that used in the dynamics and energy chapters. The process is described on the left side of Problem-Solving Strategy 17.1 and illustrated on the right side for Example 17.7.

PROBLEM-SOLVING STRATEGY 17.1 **Analyzing processes involving electric charges**

EXAMPLE 17.7 **Where to put the charge?**

Two electrically charged objects with charges $q_1 = +1.0 \times 10^{-9}$ C and $q_2 = +2.0 \times 10^{-9}$ C are separated by 1.0 m. Where, with respect to charge 1, should you place a third electrically charged object so that the net electric force exerted on it by the first two objects is zero?

Sketch and translate

- Sketch the process described in the problem statement. Label the known physical quantities.
- Identify the unknowns.
- Choose an appropriate system.

We do not yet know where to place the third object. It could be in any of the three regions shown, but it has to be on the line connecting objects 1 and 2 so that the forces that 1 and 2 exert on this third object are parallel and pointing in opposite directions. We choose the object of unknown electric charge q as the system. The system interacts with objects 1 and 2. We need to determine the location r of the third charge.

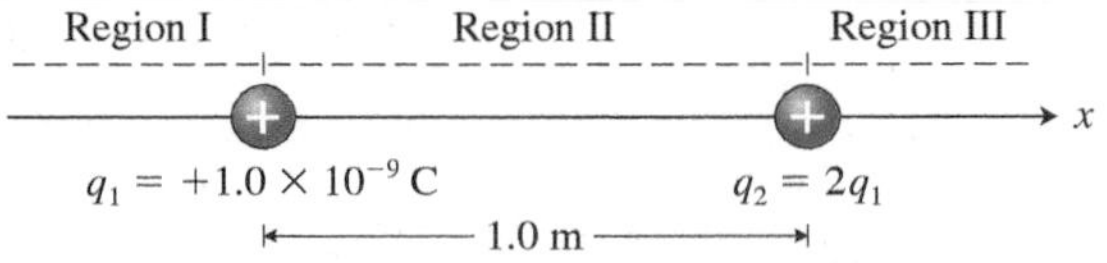

Region I   Region II   Region III

$q_1 = +1.0 \times 10^{-9}$ C $q_2 = 2q_1$

|← 1.0 m →|

(CONTINUED)

Simplify and diagram

- Decide whether you can consider the charged objects to be point-like.

- Decide what other interactions you will consider and what interactions you can ignore.

- Construct a force diagram for the system. Choose appropriate coordinate axes.

- If you are using the work-energy principle, construct an energy bar chart. Decide where the zeros for potential energies are.

Consider all objects to be smaller than the distances between them and, thus, point-like.

Draw force diagrams for the charged object q in the three possible regions. The net electric force exerted on q is the vector sum of the two forces exerted by q_1 and q_2.

(I) If q is positioned to the left of both charged objects, the forces due to q_1 and q_2 both point left and cannot add to zero.

(III) If q is positioned to the right of the two charges, both forces point right and cannot add to zero.

(II) If q is between the charges, the forces point in opposite directions and can add to zero. If q is closer to the smaller charge, the reduced distance will compensate for the smaller q_1.

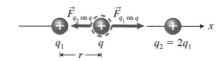

Represent mathematically

- Use the force diagram to apply the component form of Newton's second law to the process (or use the energy bar chart to apply the generalized work-energy equation).

- If necessary, use kinematics equations to describe the motion of the object.

Both forces are horizontal (their y-components are zero). Thus, we use only the horizontal x-component form of Newton's second law. With the positive x-direction toward the right and taking the signs of the components into account, we get

$$F_{q_1 \text{ on } q} + (-F_{q_2 \text{ on } q}) = 0$$

Using Coulomb's law for electric force, the fact that the total distance between the charges is 1.0 m, and that $q_2 = 2q_1$ and labeling the distance between q_1 and q as r, the above equation becomes

$$\frac{k_C q_1 q}{r^2} = \frac{k_C 2 q_1 q}{(1.0 \text{ m} - r)^2}$$

Divide both sides of the equation by $k_C q q_1$ to get

$$\frac{1}{r^2} = \frac{2}{(1.0 \text{ m} - r)^2}$$

Take the square root of both sides of the equation:

$$\frac{1}{r} = \frac{\sqrt{2}}{1.0 \text{ m} - r}$$

Solve and evaluate

- Rearrange the equation and solve for the unknown quantity.
- Verify that your answer is reasonable with respect to sign, unit, and magnitude.
- Also make sure the equation applies for limiting cases, such as objects having very small or very large charge.

Rearranging the last equation, we get

$$\sqrt{2}r = 1.0 \text{ m} - r \quad \text{or} \quad r = 0.41 \text{ m}$$

The location where a net electric force of zero will be exerted on q is a distance 0.41 m from charged object 1 in the direction of object 2. The result looks reasonable, as the unknown charge q should be closer to the smaller magnitude charge q_1 than to the larger charge q_2.

The distance that we found does not depend on the magnitude or sign of the charge q. The net force exerted on q by the two other objects will be zero regardless of the sign of q.

One limiting case is if one of the charges q_1 or q_2 is zero. If $q_1 = 0$, then only q_2 could exert a force on q and the net force could *not* be zero. The force equation becomes $0 + (-F_{q_2 \text{ on } q}) = 0$, which has no solution because the right and left sides are never equal to each other. So this limiting case is consistent with our mathematics.

Try it yourself Suppose we placed q slightly closer to q_1 than the position calculated in this example. What happens to q?

Answer

The magnitude of the force that q_1 exerts on q increases slightly, and the magnitude of the force that q_2 exerts on q decreases slightly. The net force exerted on q is no longer zero—if released, q will accelerate away from the desired position if it is negative and toward the desired position if it is positive.

In the next example we apply the steps of the problem-solving strategy to a process whose analysis requires an energy approach.

EXAMPLE 17.8   Radon decay in lungs

Suppose that a radon atom in the air in a home is inhaled into the lungs. While in the lungs, the nucleus of the radon atom undergoes radioactive decay, emitting an alpha (α) particle, which is composed of two protons and two neutrons. (These high-energy alpha particles can damage lung tissue—a reason for keeping radon concentration low in homes.) During this process, the radon nucleus turns into a polonium nucleus with charge $+84e$ and mass 3.6×10^{-25} kg. The alpha particle has charge $+2e$ and mass 6.6×10^{-27} kg. Suppose the two particles are initially separated by 1.0×10^{-15} m (the size of the nucleus) and are at rest. How fast is the alpha particle moving when it is very far from the polonium nucleus? (Note: 1 mm would be very far for such a process since even 1 mm is much larger than the size of a nucleus.)

Sketch and translate We draw a sketch showing the initial and final states of the system, choosing the product polonium nucleus and the

escaping alpha particle as the system. We need to determine the speed of the alpha particle when it is infinitely far from the polonium nucleus.

Simplify and diagram Model the nuclei and alpha particle as point-like objects and neglect the gravitational attraction between them. Assume that they are at rest at the start of the process and that we can ignore the final kinetic energy of the massive polonium nucleus. A work-energy bar chart for the process is shown at the right. In the initial state, the system has electric potential energy; in the final state, the alpha particle has kinetic energy, and since the particles are comparatively far apart we assume the system has zero electric potential energy.

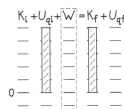

Represent mathematically Each nonzero bar in the bar chart turns into a nonzero term in the generalized work-energy equation.

$$\frac{k_C(+84e)(+2e)}{r_i} = \tfrac{1}{2}m_\alpha v_f^2$$

Rearranging the above, we get an expression for the final speed of the alpha particle:

$$v_f = \sqrt{\frac{2k_C(84e)(2e)}{m_\alpha r_i}}$$

(CONTINUED)

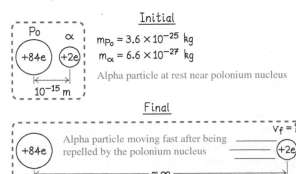

Initial

$m_{Po} = 3.6 \times 10^{-25}$ kg
$m_\alpha = 6.6 \times 10^{-27}$ kg

Alpha particle at rest near polonium nucleus

Final

Alpha particle moving fast after being repelled by the polonium nucleus

Solve and evaluate Substituting the appropriate values:

$$v_f = \sqrt{\frac{2k_C(84e)(2e)}{m_\alpha r_i}}$$

$$= \sqrt{\frac{2(9.0\times10^9\,\text{N}\cdot\text{m}^2/\text{C}^2)(84.0\times1.6\times10^{-19}\text{C})(2.0\times1.6\times10^{-19}\text{C})}{(6.6\times10^{-27}\,\text{kg})(1.0\times10^{-15}\,\text{m})}}$$

$$= 1.1 \times 10^8 \text{ m/s}$$

We estimate that the final alpha particle speed is about 10^8 m/s. (Note that when we study relativity in Chapter 26, we will find that we need a new expression for kinetic energy at these speeds, which reduces our answer by a factor of three [which is still a very high speed].)

Try it yourself In this example we assumed that the massive polonium nucleus remained at rest. Use impulse-momentum ideas to determine how fast the polonium nucleus was actually moving following the decay process. Compare the kinetic energies of the nucleus and the alpha particle.

Answer   in this example.

$v_{Po} = 0.0183v_\alpha$. The kinetic energies, which depend on the speed squared, are related by $K_{Po} = 0.0183K_\alpha$. We were justified in ignoring the final kinetic energy of the polonium

---

FIGURE 17.19 A charged Van de Graaff generator.

FIGURE 17.20 How a Van de Graaff generator works.

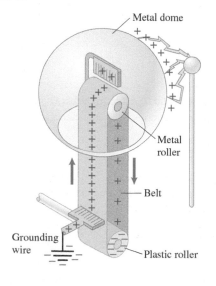

REVIEW QUESTION 17.6 How would our reasoning in Example 17.8 change if we chose the alpha particle alone as the system instead of both the nucleus and the escaping alpha particle?

17.7 Charge separation and photocopying

We can separate charge by rubbing together two objects that are made of different materials. However, the magnitude of the resulting charge is quite low. In humid weather the separated charges on these objects are quickly neutralized, making charge-related investigations difficult. To study the behavior of charged objects more easily, we need charges of large magnitudes. Specialized equipment such as Van de Graaff generators (**Figure 17.19**) and Wimshurst machines is commonly used for such studies in physics courses.

Van de Graaff generator

A Van de Graaff generator is made of a plastic cylinder with a motorized moving belt inside and a hollow metal dome at the top. When the belt is moving, you can hear a cracking sound and see sparks around the dome. Some generators have an attachment—a metal ball on an insulating stand. The sparks between the dome and this ball or any nearby metal objects can be large, implying large charge separation. How does this device function?

A Van de Graaff generator belt moves on two rollers (one on the top inside of the metal dome and the other on the bottom—see **Figure 17.20**). The lower plastic roller becomes negatively charged due to a repeated contact and then separation from the moving belt. The inner surface of the belt acquires an equal amount of positive charge. Because the positive charge on the belt is spread over a larger area than the negative charge on the plastic roller, the electrons in the lower metal comb are repelled more by the negative charge on the roller than they are attracted by the positive charge on the belt. Thus the electrons are pushed from the lower metal comb to the ground. Now, the positively charged metal comb attracts and removes electrons from the outer surface of the belt. As the positively charged outer surface of the belt reaches the dome, it attracts electrons from the upper metal comb, which is connected to the dome—the dome therefore becomes positively charged.

CONCEPTUAL EXERCISE 17.9   **Making your hair stand on end**

The woman in the photo has her hand on the dome of a Van de Graaff generator and her feet on an electrically insulated footstool. Why is her hair standing on end?

Sketch and translate Hair can stand up when the individual strands are repelled from each other and from the body. For this to happen, the hair and body need to have the same sign of electric charge.

Simplify and diagram If we assume that the human body is a fairly good conductor of electric charge, the woman's body is acting like an extension of the dome. Electrons flow from her to the dome, leaving her and each strand of her hair positively charged. This makes them repel from her body and from the other strands—they stand on end.

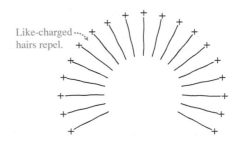

Like-charged hairs repel.

Try it yourself You place a stack of aluminum pie plates on top of an uncharged Van de Graaff generator and then turn on the generator. Predict what happens to the plates and explain your prediction.

Answer

If we assume that electrons transfer from the plates to the positively charged dome, the now positively charged plates repel one another and the dome. They fly up off the dome.

Now imagine that you are holding a grounded metal ball about 10 cm from the dome of a charged Van de Graaff generator. What is actually happening in the air that results in a flash and a sharp cracking sound? The explanation involves what are known as **cosmic rays**, high-energy particles that continually rain down on Earth's atmosphere from space. These cosmic rays ionize atoms, producing free electrons and positively charged ions. High-energy particles produced during naturally occurring radioactive decay can do the same. The positive charge on the dome of a Van de Graaff generator attracts the free electrons already present in the air, causing them to accelerate toward the dome, colliding with other atoms and causing some of them to ionize. When these free electrons recombine with atoms, light is produced and we see a spark.

EXAMPLE 17.10   **What causes the sparking from a Van de Graaff generator?**

The energy needed to remove an electron from a hydrogen atom is about 2×10^{-18} J (about the same for other atoms, too). The average distance a free electron in air will travel between collisions, called the mean free path, is about 10^{-6} m. The dome has a 0.15-m radius and a $+10^{-5}$-C charge. Could a free electron in the air gain enough kinetic energy to ionize an atom and cause a spark as it travels that short distance toward the charged dome?

Sketch and translate The situation is sketched below (not to scale). We need to compare the kinetic energy that the electron gains between collisions to the energy of ionization of the atom. Given that the mean free path is much smaller than the size of the dome, what values should we put for the initial and final location of the electron?

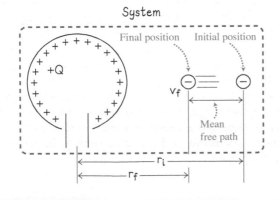

System

Final position   Initial position

$+Q$

v_f

Mean free path

r_i

r_f

Simplify and diagram Assume that the electron is near the surface of the dome. The system is the electron and the dome. We model both as point-like objects. This is reasonable for the electron and reasonable for the dome as well, but only because it is spherical in shape. (We will discuss that in more detail when we discuss the electric field in the next chapter.) An energy bar chart representing the process is shown at right. The zero level of electric potential energy is at infinity. The electron-dome system starts with negative electric potential energy and zero kinetic energy. As the electron accelerates toward the dome, its kinetic energy increases and electric potential energy decreases.

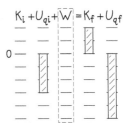

$K_i + U_{qi} + W = K_f + U_{qf}$

0 —

Represent mathematically Use the bar chart to help apply the generalized work-energy equation to the process:

$$\frac{k_C(Q)(-e)}{r_i} = \tfrac{1}{2}mv_f^2 + \frac{k_C(Q)(-e)}{r_f}$$

Solve and evaluate Our question concerns the kinetic energy $(1/2)mv_f^2$ of the electron—is the energy as large as or does it exceed the energy needed to ionize an atom $\Delta U \approx 2 \times 10^{-18}$ J?

$$\tfrac{1}{2}mv_f^2 = \frac{k_C(Q)(-e)}{r_i} - \frac{k_C(Q)(-e)}{r_f} = -k_C Qe\left(\frac{1}{r_i} - \frac{1}{r_f}\right)$$

$$= k_C Qe\left(\frac{r_i - r_f}{r_i r_f}\right)$$

(CONTINUED)

The value for $r_i - r_f$ is 10^{-6} m—the distance that the electron travels between the collisions. But what values should we choose for r_i and r_f to calculate their product? Given that the difference between the initial and final positions is tiny compared to the radius of the dome (0.15 m), we can assume that r_i and r_f are both equal to the radius of the dome, giving $r_i r_f = (0.15 \text{ m})^2$. Therefore,

$$\tfrac{1}{2}mv_f^2 = (9 \times 10^9 \text{ N} \cdot \text{m}^2/\text{C}^2)(10^{-5} \text{ C})(1.6 \times 10^{-19} \text{ C})\left(\frac{10^{-6} \text{ m}}{(0.15 \text{ m})^2}\right)$$

$$\approx (10^{10})(10^{-5})(10^{-19})\left(\frac{10^{-6}}{10^{-2}}\right) \text{ N} \cdot \text{m}$$

$$\approx 10^{-18} \text{ N} \cdot \text{m} = 10^{-18} \text{ J}$$

This is comparable to the energy needed to ionize an atom. Note that some electrons will travel farther than the average distance and therefore will have even more kinetic energy.

Try it yourself Why are we ignoring the positive ions in the air and considering only the acceleration of the free electrons due to the positive charge on the dome?

Answer The positive ions are much more massive than the electrons and do not accelerate to sufficient speeds to have the kinetic energy needed to ionize other atoms.

Wimshurst machine

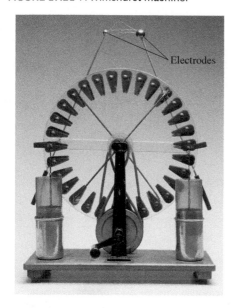

FIGURE 17.21 A Wimshurst machine.

The Wimshurst machine (**Figure 17.21**) is another device that can produce large charge separations. The device, invented in the 1880s, consists of two plastic or glass disks that rotate in opposite directions. Charge separation in the Wimshurst machine occurs by a complex process that we will not explain here. As a result of this process, one of the electrodes becomes positively charged and the other becomes negatively charged. When the two electrodes are brought near one another, we see sparks as long as 5–10 cm. The Wimshurst machine not only allows us to create large charges on the electrodes but also allows us to study the creation of the spark and its dependence on the distance between the electrodes—one goal of the next chapter.

A photocopier

The photocopying process depends on the electrical attraction of oppositely charged particles—like how a balloon rubbed on a wool sweater attracts tiny bits of paper or sugar off a tabletop. In a copy machine, the charged balloon is a drum that becomes electrically charged with an image of the page being copied, and the paper bits are dark toner particles that stick to the places where the drum is charged. The toner then transfers to paper in the exact location of the letters or pictures on the copied page.

Let's analyze this process in more detail. The drum is covered with a special photoconductive material. When we turn on the photocopier, the drum becomes positively charged beneath the photoconductive material and negatively charged on the outside of the photoconductive material (**Figure 17.22a**).

FIGURE 17.22 A schematic representation of the process of making a photocopy.

(a)

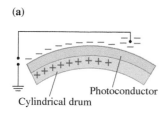

The photoconductor layer on the drum is charged negatively on the outside and positively on the inside.

(b)

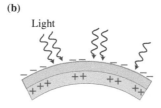

Light reflects from white parts of the page being copied and neutralizes the photoconductive layer. The remaining negative charge is an electrical image of the copied page.

(c)

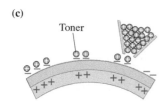

Positive toner particles stick only to the negatively charged part of the photoconductive layer.

(d)

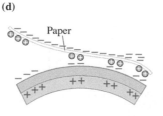

A negatively charged blank white paper attracts the dark toner particles, which are baked onto the page before its ejection from the machine.

During copying, a strong source of light moves across the page from under the glass cover. This light reflects off white regions of the page but does not reflect from dark regions. The reflected light reaches the photoconductive material where electrons on the surface of the photoconductive material absorb the light and move inside the photoconductive material to neutralize the positive charge. The drum surface is now neutral in places where light was reflected from the original page being copied. The drum remains negatively charged in areas below the dark text of the page that was being copied. Thus, the negative electrons on the top of the drum form an electrical image of the text being copied (Figure 17.22b).

The next step is covering the drum with the toner, which is positively charged. The toner sticks to the negatively charged "electrical image" on the drum (Figure 17.22c). Then a blank negatively charged paper wraps around the drum and pulls off the toner (Figure 17.22d). The drum and the paper are heated and pressed together to make this transfer more effective and to bake the toner on the surface of the paper. Finally, the new copy is ejected and a rubber material wipes the drum clean of remaining toner and illuminates it with light to remove all remaining charge.

REVIEW QUESTION 17.7 In a Van de Graaff generator, where does the energy emitted as light in a spark come from?

Summary

Electric charge is a physical quantity that characterizes how charged objects participate in electrostatic interactions. The unit of charge in the SI system is the coulomb (C).

- Charge comes in two types—positive and negative.
- Like charged objects repel and unlike charged objects attract.
- The smallest charge is the charge of an electron ($-e = -1.6 \times 10^{-19}$ C).
- Electric charge is constant in an isolated system. (Sections 17.1 and 17.2)

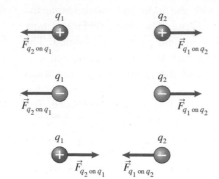

Conductors are materials in which electrically charged particles can move freely. (Section 17.3)

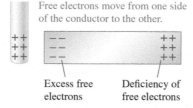

Free electrons move from one side of the conductor to the other.

Excess free electrons   Deficiency of free electrons

Insulators (dielectrics) are materials in which electrically charged particles cannot move freely. However, the electric charges in the atoms and molecules in the material can rearrange slightly (a process called polarization), allowing them to participate in electric interactions. (Section 17.3)

The charge in neutral atoms is slightly rearranged to create electric dipoles.

Coulomb's law is used to determine the magnitude of the electric force that point-like objects with charges q_1 and q_2 exert on each other when separated by a distance r. (Section 17.4)

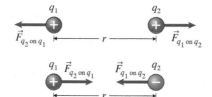

$$\left|F_{q_1 \text{ on } q_2}\right| = \left|F_{q_2 \text{ on } q_1}\right| = \frac{k_C |q_1||q_2|}{r^2} \quad \text{Eq. (17.1)}$$

$$k_C = 9.0 \times 10^9 \, \frac{\text{N} \cdot \text{m}^2}{\text{C}^2}$$

The **electric potential energy** U_q of a system of point-like objects with charges q_1 and q_2 separated by a distance r is positive for like charges and negative for unlike charges (include the signs of the charges when using this equation.) The zero level of electric potential energy is usually chosen to be when they are infinitely far apart. (Section 17.5)

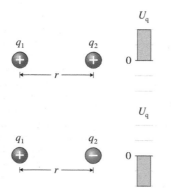

$$\Delta U_q = U_{qf} - U_{qi} = k_C q_1 q_2 \left(\frac{1}{r_f} - \frac{1}{r_i} \right) \quad \text{Eq. (17.2)}$$

Questions

Multiple Choice Questions

1. Which of the following occurs when two objects are rubbed against each other?
   (a) Both acquire the same sign charge.
   (b) They acquire opposite sign charge.
   (c) No new charges are created.
   (d) Both b and c can be correct.

2. With which statements do you disagree?
   (a) If two objects repel, they have like charges.
   (b) If two objects attract, they have opposite charges.
   (c) If two objects attract, one of them can be charged and the other one can be neutral.
   (d) None of them

3. Which explanation agrees with the contemporary model of electric charge?
   (a) If an object is charged negatively by rubbing, it has lost positively charged particles.
   (b) If an object is charged positively by rubbing, it has lost negatively charged particles.
   (c) If an object is charged positively by rubbing, it has acquired positively charged particles.
   (d) If an object is charged negatively by rubbing, it has acquired negatively charged particles.
   (e) a and c
   (f) b and d

4. When an object gets charged by rubbing, where does the electric charge originate?
   (a) It comes from the air surrounding the object.
   (b) It is created by the rubbing energy.
   (c) The process of rubbing leads to a redistribution of charge, not the creation of charge.

5. Choose all of the quantities that are constant in an isolated system.
   (a) Mass
   (b) Entropy
   (c) Electric charge
   (d) Momentum
   (e) Mechanical energy
   (f) Total energy
   (g) Temperature

6. Identically charged point-like objects A and B are separated by a distance d. You measure the force F that A exerts on B. If you now separate the objects by distance $1.4d$, what will be the new force that A exerts on B?
   (a) $1.4F$
   (b) $2.0F$
   (c) $\frac{1}{2}F$
   (d) $\frac{1}{1.4}F$
   (e) F

7. When separated by distance d, identically charged point-like objects A and B exert a force of magnitude F on each other. If you reduce the charge of A to one-half its original value, and the charge of B to one-fourth, and reduce the distance between the objects by half, what will be the new force that they exert on each other?
   (a) $2F$
   (b) $\frac{1}{2}F$
   (c) $4F$
   (d) $\frac{1}{4}F$
   (e) F

8. Balloon A has charge q, and identical mass balloon B has charge $10q$. You hang them from threads near each other. Choose all of the statements with which you agree.
   (a) The force that A exerts on B is $1/10$ the force that B exerts on A.
   (b) The force that A exerts on B is 10 times the force that B exerts on A.
   (c) A and B exert the same magnitude forces on each other.
   (d) The angle between the thread supporting A and the vertical is less than the angle between the thread supporting B and the vertical.

9. Imagine that two charged objects are the system of interest. When the objects are infinitely far from each other, the electric potential energy of the system is zero. When the objects are close to each other, the electric potential energy is positive. Which of the following statements is(are) incorrect?
   (a) Both objects are positively charged.
   (b) Both objects are negatively charged.
   (c) One object is negatively charged and the other one is positively charged.

10. Two objects with charges $+q$ and $-2q$ are separated by a distance r. You slowly move one of the objects closer to the other so that the distance between them decreases by half. Considering the two objects as the system, which statements are *incorrect*?
    (a) The electric potential energy of the system doubles.
    (b) The electric potential energy of the system decreases by half.
    (c) The magnitude of the electric potential energy doubles.
    (d) The magnitude of the electric potential energy decreases by half.

11. Charged point-like objects A and B are separated by a distance d. Object A is fixed in place, while object B can be moved. Choose object A alone as the system. You slowly push object B closer to A, decreasing the distance between them to $0.3d$. Based on the given information, choose all of the correct statements.
    (a) The electric potential energy of the system increases.
    (b) The electric potential energy of the system decreases.
    (c) The electric potential energy of the system does not change.
    (d) You do positive work on the system.
    (e) You do negative work on the system.
    (f) You do zero work on the system.

12. If you move a negatively charged balloon toward a neutral dielectric wall, the electric potential energy of the balloon-wall system
    (a) increases.
    (b) decreases.
    (c) is constant but nonzero.
    (d) is constant and zero.

Conceptual Questions

13. Describe the differences between the electric force and the gravitational force and some experiments that can be explained through these differences.

14. Describe the difference between charged objects and magnets and some experiments that can be explained through these differences.

15. At one time it was thought that electric charge was a weightless fluid. An excess of this fluid resulted in a positive charge; a deficiency resulted in a negative charge. Describe an experiment for which this hypothesis provides a satisfactory explanation and another experiment for which it does not.

16. What experiments can you do to show that there are only two kinds of charge?

17. An object becomes positively charged due to rubbing. Does its mass increase, decrease, or remain the same? Explain. How is its mass affected if it becomes negatively charged? Explain.

18. List everything that you know about electric charges.

19. What experimental evidence supports the idea that conducting materials have freely moving electrically charged particles inside them?

20. You have an aluminum pie pan with pieces of aluminum foil attached to it. Predict what will happen if you touch the pan with a plastic rod rubbed with wool.

21. You have a charged metal ball. How can you reduce its charge by half?

22. You have a foam rod rubbed with felt and a small aluminum foil ball attached to a thread. Describe what happens when you slowly approach the ball with the rod and then touch the ball. Explain why this happens.

23. A positively charged metal ball A is placed near metal ball B. Measurements demonstrated that the force between them is zero. Is ball B charged? Explain.

24. Show that if the charge on B in the previous question is positive but of small magnitude, the balls will be attracted to each other.

25. Two metal balls of the same radius are placed a small distance apart. Will they interact with the same magnitude force when they have like charges as when they have opposite charges? Explain your answer. Include a sketch showing the charge distribution.

26. Describe the experiments that were first used to determine a quantitative expression for the magnitude of the force that one electrically charged object exerts on another electrically charged object. What are the limitations of this expression?

27. The electrical force that one electric charge exerts on another is proportional to the product of their charges—that is, $F_{1 \text{ on } 2} \propto q_1 q_2$. How would Coulomb's observations be different if the force were proportional to the sum of their charges $(q_1 + q_2)$?

28. Why isn't Coulomb's law valid for large conducting or dielectric objects, even if they are spherically symmetrical?

29. How is electric potential energy similar to spring potential energy? How is it different?

Problems

Below, **BIO** indicates a problem with a biological or medical focus. Problems labeled **EST** ask you to estimate the answer to a quantitative problem rather than derive a specific answer. Asterisks indicate the level of difficulty of the problem. Problems with no * are considered to be the least difficult. A single * marks moderately difficult problems. Two ** indicate more difficult problems.

   In some of these problems you may need to know that the mass of an electron is 9.1×10^{-31} kg and the charge is -1.6×10^{-19} C.

17.1–17.3 Electrostatic interactions, Explanations for electrostatic interactions, and Conductors and insulators (dielectrics)

1. **BIO** **Ventricular defibrillation** During ventricular fibrillation, the muscle fibers of the heart's ventricles undergo uncoordinated rapid contractions, resulting in little or no blood circulation. To restore the heart's normal rhythm, a defibrillator sends an abrupt jolt of about -0.20 C of electric charge through the chest into the heart. How many electrons pass through the body during this defibrillation?

2. * You rub two 2.0-g balloons with a wool sweater. The balloons hang from 0.50-m-long very light strings. When you attach the strings together at the top, the balloons hang away from each other each string making an angle of 37° with the vertical. (a) Represent the situation with the force diagram for each balloon and determine the magnitudes of the forces on the diagram. (b) What can you say about the magnitudes of the forces that the balloons exert on each other? Explain. (c) Will the ratio of the forces that the balloons exert on each other change if the charge on one balloon is two times larger than on the other? How do you know?

3. * Two balloons of different mass hang from strings near each other. You charge them about the same amount by rubbing each balloon with wool. Draw a force diagram for each of the balloons. Compare the angles of the threads with the vertical. How do your answers depend on whether the balloons have the same or different magnitude charge?

4. * **Lightning** A cloud has a large positive charge. Assume that this is the only cloud in the sky over Earth and that Earth is a good electrical conductor. Draw a sketch showing electric charge distribution on Earth due to the cloud's electric charge. Explain why a person's hair might stand on end before a lightning strike.

5. Sodium chloride (table salt) consists of sodium ions of charge $+e$ arranged in a crystal lattice with an equal number of chlorine ions of charge $-e$. The mass of each sodium ion is 3.82×10^{-26} kg and that of each chlorine ion 5.89×10^{-26} kg. Suppose that the sodium ions could be separated into one pile and the chlorine ions into another. What mass of salt would be needed to get 1.00 C of charge into the sodium ion pile and -1.00 C into the chlorine ion pile?

17.4 Coulomb's force law

6. * **EST** (a) Earth has an excess of 6×10^5 electrons on each square centimeter of surface. Determine the electric charge of Earth in coulombs. (b) If, as you walk across a rug, about 10^{-22} kg of electrons transfer to your body, estimate the number of excess electrons and the total charge in coulombs on your body.

7. Determine the electrical force that two protons in the nucleus of a helium atom exert on each other when separated by 2.0×10^{-15} m.

8. * Determine the number of electrons that must be transferred from Earth to the Moon so that the electrical attraction between them is equal in magnitude to their present gravitational attraction. What is the mass of this number of electrons?

9. **BIO** **Ions on cell walls** The membrane of a body cell has a positive ion of charge $+e$ on the outside wall and a negative ion of charge $-e$ on the inside wall. Determine the magnitude of the electrical force between these ions if the membrane thickness is 0.80×10^{-9} m. Ignore the effect of the material in which the ions are located.

10. * **Hydrogen atom** In a simplified model of a hydrogen atom, the electron moves around the proton nucleus in a circular orbit of radius 0.53×10^{-10} m. Determine at least four physical quantities related to this information.

11. * Three 100 nC charged objects are equally spaced on a straight line. The separation of each object from its neighbor is 0.3 m. Find the force exerted on the center object if (a) all charges are positive, (b) all charges are negative, and (c) the rightmost charge is negative and the other two are positive.

12. ** Two objects with charges q and $4q$ are separated by 1.0 m. (a) Determine the charge (both sign and magnitude) and position of a third object that causes all three objects to remain in equilibrium (all three objects are on the same line). (b) Is the equilibrium stable or unstable? How do you know? (Hint: An object is in a stable equilibrium if a small displacement of the object from the equilibrium results in a force exerted on the object that points opposite to the direction of displacement.)

13. * **Salt crystal** Four ions (Na^+, Cl^-, Na^+, and Cl^-) in a row are each separated from their nearest neighbor by 3.0×10^{-10} m. The charge of a sodium ion is $+e$ and that of a chlorine ion is $-e$. Determine the electric force exerted on the chlorine ion at the right end of the row due to the other three ions.

14. * A $+10^{-6}$-C charged object and a $+2 \times 10^{-6}$-C charged object are separated by 100 m. Where should a -1.0×10^{-6}-C charged object be located on a line between the positively charged objects so that the net electrical force exerted on the negatively charged object is zero?

15. ** **BIO** **Bee pollination** Bees acquire an electric charge in flight from friction with the air, which causes pollen to cling to them. The pollen is then attracted to the stigma of the next flower (Figure P17.15). Suppose the bee's body has a charge of -10^{-9} C and is about 3×10^{-3} m from the front edge of a spherical granule of pollen of diameter 5×10^{-5} m. Charged particles in the pollen become polarized with $+10^{-11}$ C on the front edge and -10^{-11} C on the backside of the pollen ($3 \times 10^{-3} + 5 \times 10^{-5}$) m from the bee. What useful physical quantities can you determine?

FIGURE P17.15

16. * A triangle with equal sides of length 10 cm has -2.0-nC charged objects at each corner. Determine the electrical force (magnitude and direction) exerted on the object at the top corner due to the two objects at the base of the triangle.

17. You have a small metal sphere fixed on an insulating stand and another sphere attached to a plastic handle. You charge one of the spheres and then touch it with the second sphere. Draw a force-versus-time graph for the force that one sphere exerts on the other as you move the sphere on the handle slowly and steadily away from the sphere on the stand. What information do you need in order to determine the magnitudes of the quantities on the graph?

18. * After the experiment in Problem 17.17, you touch the sphere on the stand with your hand. Then you touch that sphere with the sphere on the handle. Draw a force-versus-time graph for the force that one sphere exerts on the other as you move the sphere on the handle slowly and steadily away from the sphere on the stand. How is this graph different from the graph in the previous problem?

19. ** Coulomb's law is formulated for point-like charged objects. Imagine that you have two point-like charged objects q_1 and q_2 of unspecified sign separated by a distance d and you also have two large spheres of radius R with the same charges q_1 and q_2. The distance between the centers of the spheres is d. Compare the electric force that the point-like objects exert on each other to the force that the two spheres exert on each other. Consider all possibilities.

17.5 Electric potential energy

20. (a) Determine the change in electric potential energy of a system of two charged objects when a -15.0-nC charged object and a -40.0-nC charged object move from an initial separation of 50.0 cm to a final separation of 10.0 cm. (b) What other quantities can you calculate using this information?

21. You have a system of two positively charged objects separated by some arbitrary finite distance. (a) What is the sign of their potential energy? (Remember that charges that are infinitely far from each other have zero potential energy.) (b) What can you do to decrease this energy? (c) Draw an energy bar chart for this process of decreasing the energy.

22. You have a system of two negatively charged objects separated by some arbitrary finite distance. (a) What is the sign of their potential energy? (Remember that charges that are infinitely far from each other have zero potential energy.) (b) What can you do to decrease this energy? (c) Draw an energy bar chart for this process of decreasing the energy.

23. Repeat (a)–(c) of Problem 17.22 for a system with a negatively charged object and a positively charged object separated by some arbitrary finite distance.

24. The metal sphere on the top of a Van de Graaff generator has a relatively large positive charge. In which direction must a positively charged ion in the air move relative to the sphere in order for the electrical energy of the ion-generator system to decrease? Justify your answer.

25. * EST An electron is 0.10 cm from an object with electric charge of $+3.0 \times 10^{-6}$ C. (a) Determine the magnitude of the electrical force $F_{O \text{ on } e}$ that the object exerts on the electron. (b) The electron is pulled so that it moves to a distance of 0.11 cm from the charged object. Determine the magnitude of the electrical force $F'_{O \text{ on } e}$ exerted by the charged object on the electron when at this distance. (c) Estimate the work done by the average force pulling the electron

$$\left(\frac{F_{O \text{ on } e} + F'_{O \text{ on } e}}{2} \right) \Delta x$$

(d) Compare this number to the change in electric potential energy of the electron–charged-object system as the electron moves away from the object. Why should the numbers be approximately equal?

26. * (a) An object with charge $q_4 = +3.0 \times 10^{-9}$ C is moved to position C from infinity (Figure P17.26). $q_1 = q_2 = q_3 = +10.0 \times 10^{-8}$ C. Determine as many work-energy quantities as you can that characterize this process. Make sure you specify the system. (b) Repeat your calculations, but for $q_1 = q_3 = -10.0 \times 10^{-9}$ C and $q_2 = +10.0 \times 10^{-9}$ C.

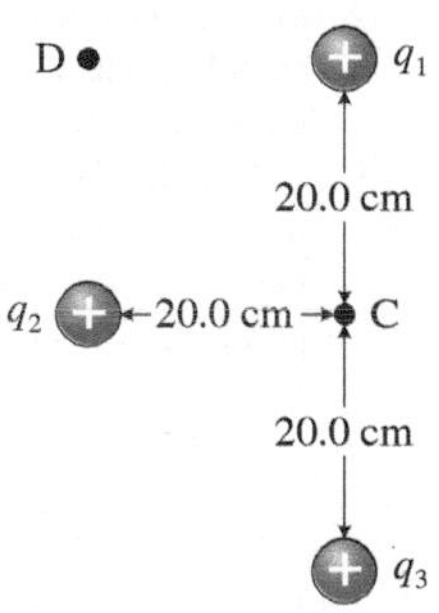

27. * An object with charge $+2.0 \times 10^{-9}$ C is moved from position C to position D in Figure P17.26. $q_1 = q_3 = +10.0 \times 10^{-9}$ C and $q_2 = -20.0 \times 10^{-9}$ C. All four charged objects are the system. What work-energy-related quantities can you determine for the process?

28. ** Two small spheres with charges $q_1 = +8$ nC and $q_2 = -4$ nC are placed at marks $x_1 = -2$ cm and $x_2 = 13$ cm. Where along the x-axis should you place a third sphere with charge q_3 so that (a) the sum of the forces exerted by the first two spheres on the third one is zero, and (b) the total electric potential energy of the three spheres does not depend on q_3?

29. * Two small objects with charges $+Q$ and $-Q$ are fixed a distance a apart. (a) Show that if a third small object with an arbitrary charge q is placed halfway between the first two objects, the total electric potential energy does not change even though we have added a new object to the system. (b) While solving this problem, your friend came to the following conclusion: "If adding the third charged object does not change the total electric potential energy of the system (the system comprises all three objects), then there must be a way to remove the third object from the system without doing any work." Do you agree or disagree with your friend? If you disagree, explain why. If you agree, describe a path along which you can remove the third object from the system so that the total work done on the third object is zero.

30. * A stationary block has a charge of $+6.0 \times 10^{-8}$ C. A 0.80-kg cart with a charge of $+4.0 \times 10^{-8}$ C is initially at rest and separated from the block by 0.4 m. The cart is released and moves along a frictionless surface to a distance of 1.0 m from the block. Determine as many values of the physical quantities describing the motion of the cart as you can.

FIGURE P17.26

D •

q_1 +

20.0 cm

q_2 + ←—20.0 cm—→• C

20.0 cm

q_3 +

17.6 Skills for analyzing processes involving electric charges

31. Figure P17.31 shows four different configurations with equal-magnitude charges that are placed at the corners of equal-size squares. Rank the configurations A to D in order of increasing electric potential energy.

FIGURE P17.31

32. * **Equation Jeopardy 1** The solution to a problem is represented by the following equation:

$(9.0 \times 10^9 \text{ N} \cdot \text{m}^2/\text{C}^2)(+20.0 \times 10^{-9} \text{ C})(+50.0 \times 10^{-9} \text{ C})/(0.1 \text{ m})^2$
$+ (9.0 \times 10^9 \text{ N} \cdot \text{m}^2/\text{C}^2)(+10.0 \times 10^{-9} \text{ C})(+50.0 \times 10^{-9} \text{ C})/(0.5 \text{ m})^2 = (10 \text{ kg})a_x$

Sketch a situation that the equation might represent and formulate a problem for which it is a solution (there are multiple possibilities).

33. * **Equation Jeopardy 2** The solution to a problem is represented by the following equation:

$(9.0 \times 10^9 \text{ N} \cdot \text{m}^2/\text{C}^2)(+20.0 \times 10^{-9} \text{ C})(+50.0 \times 10^{-9} \text{ C})/(5 \text{ cm}) =$
$\frac{1}{2}(10 \text{ g})v_x^2 + (9.0 \times 10^9 \text{ N} \cdot \text{m}^2/\text{C}^2)(+20.0 \times 10^{-9} \text{ C})(50.0 \times 10^{-9} \text{ C})/(20 \text{ cm})$

Sketch a situation that the equation might represent and formulate a problem for which it is a solution (there are multiple possibilities).

34. * **Equation Jeopardy 3** The solution to a problem is represented by the following equation:

$(9.0 \times 10^9 \text{ N} \cdot \text{m}^2/\text{C}^2)(+20.0 \times 10^{-9} \text{ C})(+50.0 \times 10^{-9} \text{ C})/(10 \text{ cm})^2$
$- (9.0 \times 10^9 \text{ N} \cdot \text{m}^2/\text{C}^2)(+10.0 \times 10^{-9} \text{ C})(+50.0 \times 10^{-9} \text{ C})/(5 \text{ cm})^2 = (10 \text{ g})a_x$

Sketch a situation that the equation might represent and formulate a problem for which it is a solution (there are multiple possibilities).

35. * **Equation Jeopardy 4** The solution to a problem is represented by the following equation:

$(9.0 \times 10^9 \text{ N} \cdot \text{m}^2/\text{C}^2)(-20.0 \times 10^{-9} \text{ C})(+50.0 \times 10^{-9} \text{ C})/(20 \text{ cm})$
$= \frac{1}{2}(10 \text{ g})v_x^2 + (9.0 \times 10^9 \text{ N} \cdot \text{m}^2/\text{C}^2)(-20.0 \times 10^{-9} \text{ C})(+50.0 \times 10^{-9} \text{ C})/(5 \text{ cm})$

Sketch a situation that the equation might represent and formulate a problem for which it is a solution (there are multiple possibilities).

36. * **Evaluate the solution** Metal sphere 1 has charge q and is attached to an insulating stand that is fixed on a table. Metal sphere 2 has charge $3q$ and is attached to an insulating stand that can be moved along the table. The distance between the centers of the spheres is initially $a = 35$ cm. If we slowly move sphere 1 toward sphere 2 until the distance between their centers is equal to $b = 25$ cm, we do 10^{-3} J work on the system consisting of the two spheres. (a) Determine the magnitudes and the signs of the charges on the spheres. (b) Your friend provides the following solution for (a): The signs of the charges are either both positive or both negative because work is done on the system as they are moved closer together. The magnitudes of the charges can be determined using the generalized work-energy principle:

$$k_C \frac{q \cdot 3q}{a^2} + W = k_C \frac{q \cdot 3q}{b^2}$$

$$\Rightarrow q = \sqrt{\frac{W}{3k_C}\left(\frac{1}{b^2} - \frac{1}{a^2}\right)^{-1}} = \sqrt{\frac{10^{-3} \text{ J} \cdot \text{C}^2}{3(9 \times 10^9 \text{ N} \cdot \text{m}^2)}(16.0 \text{ m}^{-2} - 8.2 \text{ m}^{-2})^{-1}}$$

$$= 6.9 \times 10^{-8} \text{ C}$$

Evaluate the solution and correct any errors you find.

37. * Construct separate force diagrams for each charged object shown in Figure P17.37. Use two-letter subscripts identifying each force.

FIGURE P17.37

$q_1 = -2q$ $q_2 = +q$ $q_3 = +q$

|←—— a ——→|←—— a ——→|

38. ** The six objects shown in **Figure P17.38** have equal-magnitude electric charge. Adjacent objects are separated by distance a. Write an expression in terms of q and a for the force that the five objects on the right exert on the positive charge on the left.

FIGURE P17.38

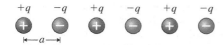

39. * A small metal ball with positive charge $+q$ and mass m is attached to a very light string, as shown in **Figure P17.39**. A larger metal ball with negative charge $-Q$ is securely held on a plastic rod to the ceiling. Write an expression for the magnitude of the force T that the string exerts on the ball. Define any other quantities used in your expression.

FIGURE P17.39

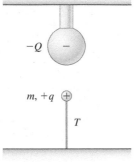

40. * Four objects each with charge $+1.0 \times 10^{-7}$ C are located at the corners of a square whose sides are 2.0 m long. Determine the values of two physical quantities describing the situation but not provided in the givens.

41. * Two 5.0-g aluminum foil balls hang from 1.0-m-long threads that are suspended from the same point at the top. The charge on each ball is $+5.0 \times 10^{-9}$ C. Make a list of the physical quantities that you can determine using this information. Determine the values of two of those physical quantities.

42. * A 6.0-g ball with charge $+3.0 \times 10^{-8}$ C hangs from a 0.50-m-long string at an angle of 87° below the horizontal. The string is attached at the top on the wall above a second charged object, as shown in **Figure P17.42**. (a) Determine the charge on the second object. (b) Make a list of other physical quantities that you can determine using the information in the problem. Describe how you can determine two of those quantities.

FIGURE P17.42

43. * A 0.40-kg cart with charge $+4.0 \times 10^{-8}$ C starts at rest on a horizontal frictionless surface 0.50 m from a fixed object with charge $+2.0 \times 10^{-8}$ C. When the cart is released, it moves away from the fixed object. (a) How fast is the cart moving when very far (infinity) from the fixed charge? (b) When 2.0 m from the fixed object?

44. A dust particle has an excess charge of 4×10^6 electrons and a mass of 5×10^{-8} g. It is initially held at $d = 10^{-3}$ m directly above a fixed object of the same size that is charged with the same number of electrons. Both objects are in an evacuated container. (a) Determine the maximum height of the dust particle after its release. (b) Show how you can determine two other physical quantities relevant to the process.

45. **Electric accelerator** A micro-transporter moves from one side of an evacuated chamber to another. It is powered by equal-magnitude, opposite-sign charges Q and $-Q$ on top of two station points (see **Figure P17.45**). If a 7×10^{-3}-N friction force opposes the transporter's motion, what must be the magnitude of Q to make the transporter move with an initial acceleration of 1.0 m/s²? The transporter's mass is 1.0 g and it is carrying a positive charge of 8×10^{-10} C.

FIGURE P17.45

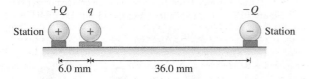

46. * You are holding at rest a small sphere A with electric charge $+q$ and mass m at a distance d from another small sphere B with charge Q fixed on top of an insulating support. What are the largest and smallest magnitudes of the force that your hand exerts on sphere A? Make sure you analyze different possibilities.

17.7 Charge separation and photocopying

47. ** A Van de Graaff generator is placed in rarefied air at 0.04 times the density of air at atmospheric pressure. The average distance that free electrons move between collisions (the mean free path) in that air is about 6×10^{-6} m. Determine the positive charge needed on the generator dome so that a free electron located 0.20 m from the center of the dome will gain at the end of the mean free path length the 2.0×10^{-18} J of kinetic energy needed to ionize a hydrogen atom during a collision.

48. * Two protons each of mass 1.67×10^{-27} kg and charge $+e$ are initially at rest and separated by 1.0×10^{-14} m (approximately the radius of the largest nucleus). When released, the protons fly apart. (a) Determine the change in their electric potential energy when they are 1.0×10^{-10} m apart (approximately the radius of an atom). (b) If the electric potential energy change is converted entirely into the kinetic energy of the protons (shared equally), what is the speed of one proton when 1.0×10^{-10} m from the other proton?

49. * Two protons, initially separated by a very large distance (r_i is infinity), move directly toward each other with the same initial speed v_i. (a) Determine their initial speeds if the distance of closest approach when their speeds are zero is 4.0×10^{-14} m. (b) Determine some other physical quantity relevant to the process.

50. * An alpha particle consists of two protons and two neutrons together in one nucleus with a mass of 6.64×10^{-27} kg and charge $+2e$. The alpha particle flies at 3.0×10^7 m/s from a large distance toward the nucleus of a stationary gold atom (charge $+79e$). (a) Make a list of physical quantities that you can determine using this information and explain how you will calculate one of them. (b) Determine the distance of the alpha particle from the gold nucleus when it stops.

51. * Determine the speed that the proton shown in **Figure P17.51** must be moving in order to get within 1.0×10^{-15} m of the helium-3 nucleus that has two protons and one neutron. Assume that the helium nucleus is a point-like object and does not move.

FIGURE P17.51

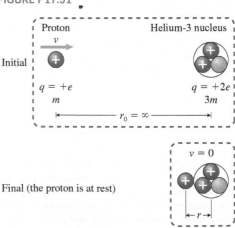

General Problems

52. ** Suppose that Earth and the Moon initially have zero charge. Then 1000 kg of electrons are transferred from Earth to the Moon. Determine the radius of a stable Moon orbit when both the electrical and gravitational forces of attraction are exerted on the Moon and it completes one rotation about Earth in 27.5 days.

53. * **BIO** **Calcium ion synapse transfer** Children have about 10^{16} synapses that can transfer signals between neurons in the brain and between neurons and muscle cells. Suppose that these synapses simultaneously transmit a signal, sending 1000 calcium ions (Ca^{2+}) across the membrane at each synaptic ending. Determine the total electric charge transfer in coulombs during that short time interval. By comparison, a lightning flash involves about 5 C of charge transfer. (Note: This is a fictional scenario. All human neurons do not simultaneously produce signals.)

54. A small ball D has a charge of -10 nC and cannot move. You move another small ball E that has a charge of -5 nC from point 1 to point 4 as shown in **Figure P17.54**. Determine the work done by you on ball E when you move it (a) from 1 to 4 along the path shown via 2 and 3 (determine the work for each step separately), and (b) along a straight line from 1 to 4. (c) Compare your two answers for parts (a) and (b) and explain the comparison.

FIGURE P17.54

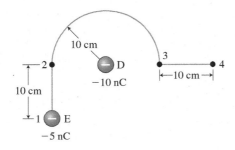

55. * Two small balls A and B with equal charges $+q$ are fixed at distances a and $2a$ from the origin of a coordinate system as shown in **Figure P17.55**. You move a third equally charged ball C along the z-axis from negative infinity to positive infinity. Draw a qualitative graph that shows how the electric potential energy of the three-ball system depends on the position of ball C along the z-axis.

FIGURE P17.55

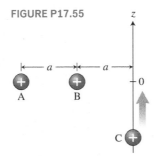

56. ** **EST** **Shocking your friend**
(a) You shuffle across the rug and then place your finger near your friend's nose, causing a small spark that transfers about 10^{-9} C of charge from you to your friend. Determine the number of electrons transferred. (b) Estimate the fraction of electrons in your body that were transferred to the friend. Note that the electron mass is about $1/20{,}000$ the mass of the atom—the nuclei are much more massive—and the mass of an average atom in the body is about 2×10^{-26} kg. (Note: Don't get hung up on minutia—this is a rough estimate.)

Reading Passage Problems

Static cling You pull your clothes from the dryer and find that they stick together. You take off your coat and find that your pants stick to your legs. Static cling like this can occur for two reasons. First, different types of atoms have greater or lesser affinity for additional electrons. When two different materials are rubbed together, the atoms with a greater electron affinity attract electrons from the material with a lesser electron affinity. Because of this, the substance that gains electrons becomes negatively charged and the substance that loses electrons becomes positively charged. The differently charged materials attract and stick together.

Second, static cling can occur between charged and uncharged objects. For instance, you may notice that a sock removed from the dryer is attracted to an uncharged sweater you are wearing. Or sometimes your skirt sticks to your legs. This happens because the molecules in a charged piece of clothing cause the electric charge inside the molecules of the nearby uncharged objects to slightly redistribute (to become polarized) so that the unlike charge of the molecule moves closer to the charged object and is attracted more than the same molecular charge of the same sign, which is slightly farther away (see Figures 17.9b and c).

Some people use fabric softener to prevent static cling. This product coats cloth fibers with a thin layer of electrically conductive molecules, thus preventing buildup of static electricity. You can also use a metal clothes hanger to remove electric charge from already clean charged clothes. Brush the hanger on the inside of the garment from top to bottom.

Shoes scuffing on different surfaces can also cause electric charge transfer. For that reason, hospital personnel wear special shoes in hospital operating rooms to avoid sparking that might ignite flammable gases in the room.

57. You rub a balloon against your wool sweater and then place the balloon on the wall—it sticks. Why?
(a) The balloon and wall have opposite electric charges.
(b) The molecules on the wall redistribute their charge so that the charge opposite that on the balloon is nearest the balloon.
(c) Electric charge in Earth is pulled to the part of the wall nearest the balloon.
(d) a and c are correct.
(e) a, b, and c are correct.

58. As you unload the clothes dryer, you find a sock clinging to a shirt. As you pull the sock off the shirt, you do positive work on the sock. What is the main form of energy increase?
(a) Gravitational   (b) Elastic
(c) Electric potential   (d) Thermal
(e) c and d

59. Table salt, Na^+Cl^-, is made of ionized sodium and chlorine atoms. Which atom is most attractive for an excess electron?
(a) Na   (b) Cl
(c) They are equally attractive.

60. You add fabric softener to your next load of wash. Your clothes do not cling after they emerge from the dryer. What do the water softener molecules do?
(a) Carry away all the excess electrons.
(b) Cause sparks in the dryer that discharge the excess charge on the clothes.
(c) Form a protective coating that prevents the charge from joining the clothes.
(d) Make the cloth fluffy so the charge comes off naturally.
(e) Join the cloth molecules and make a conductive layer that prevents excess charge from accumulating on the clothes.

61. You put strips of aluminum foil in the dryer along with your clothes. Which answer below best represents the condition of the clothes after leaving the dryer?
(a) The clothes are uncharged because the excess charge is on the aluminum strips.
(b) The clothes are uncharged because they transfer excess charge to the strips, which transfer it to the metal dryer walls.
(c) The clothes are charged but the strips are not because they are conductive.
(d) The clothes are charged because the strips are not connected to anything.
(e) None of these answers is reasonable.

62. You remove electric charge from your clean pants by rubbing a metal clothes hanger down the inside of the pants. Which answer below represents the best explanation for why this works?
(a) The charge travels from the cloth to the metal to your hand through your body to the ground.
(b) The metal hanger absorbs all the charge.
(c) The metal causes tiny sparks that send the charge into the air.
(d) The metal provides charge to the cloth that neutralizes its charge.
(e) None of these answers is reasonable.

Electrostatic exploration Geologists sometimes analyze the distribution of materials under Earth's surface, materials such as iron, water, oil, or dry soil. The process they use is called electrostatic exploration. Electrodes are placed in the ground about 800 m apart. An electric generator is connected to the electrodes and causes them to become oppositely charged. The opposite-sign charges on the electrodes cause electrically charged ions in the matter below the surface to move. The moving ions are detected by equipment on the surface. This helps the geologist decide what type of matter is below the surface. What causes this motion? To help answer this question, determine the net electric force exerted on ions at different places below the surface.

63. Which arrow in **Figure P17.63** best represents the force exerted on a positive ion at position A?
    (a) I   (b) II   (c) III
    (d) IV   (e) V   (f) VI

FIGURE P17.63

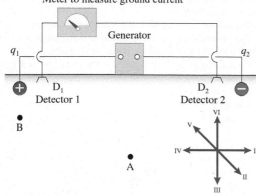

64. Which arrow in Figure P17.63 best represents the force exerted on a negative ion at position A?
    (a) I   (b) II   (c) III
    (d) IV   (e) V   (f) VI

65. Which arrow in Figure P17.63 best represents the force exerted on a positive ion at position B?
    (a) I   (b) II   (c) III
    (d) IV   (e) V   (f) VI

66. Which arrow in Figure P17.63 best represents the force exerted on a negative ion at position B?
    (a) I   (b) II   (c) III
    (d) IV   (e) V   (f) VI

67. Based on the analysis in the first four questions, which charge detector in Figure P17.63 will be collecting negative ions?
    (a) D_1 (b) D_2
    (c) Both D_1 and D_2 (d) Neither D_1 nor D_2

68. The resistance to the motion of electrons through different types of materials is, in decreasing order, dry soil, moist soil, underground water, and iron ore. How can this knowledge and the measurement of the charge reaching the detector per unit time help identify what type of material is under Earth's surface?
    (a) More electrons will reach the detector from iron than from dry soil.
    (b) The electrons reaching the detector will carry water with them if water is under the surface.
    (c) The electrons reaching the detector are affected very little by what's under the surface.
    (d) The detector will get more electric charge if the resistance to flow is less, and resistance is related to the type of material.
    (e) a and d

The Electric Field

Imagine that you are in a car during a thunderstorm. Lightning strikes the road just in front of you. Are you in danger? Many people think that cars are safe because the rubber tires isolate them from the ground. We will learn in this chapter why it is not the rubber tires but the metal body of the car that protects passengers from the effects of the lightning.

- Why is it safe to sit in a car during a lightning storm?
- Why do we need Earth or a large conductor for grounding?
- How does electrocardiography work?

PREVIOUSLY (IN CHAPTER 17) we learned how to describe electrostatic interactions in two ways: (1) with a force exerted by one charged object on another and (2) with the electric potential energy possessed by systems of charged objects. We also discovered that charged objects can exert forces on each other without being in direct contact. This is only the second interaction we have encountered with this property, the gravitational interaction being the first. How does one charged object "know" about the presence of another when they are not in direct contact? This chapter helps us answer this question and develop a new approach for analyzing electric processes.

BE SURE YOU KNOW HOW TO:

- Find the force that one charged object exerts on another charged object (Section 17.4).
- Determine the electric potential energy of a system (Section 17.5).
- Explain the differences in the internal structure of electric conductors and dielectrics (Section 17.3).

FIGURE 18.1 The electric field model.

(a)

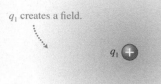

q_1 creates a field.

The field is
stronger near q_1
than farther
from it.

(b)

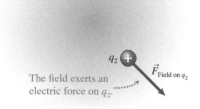

q_2

$\vec{F}_{\text{Field on } q_2}$

The field exerts an
electric force on q_2.

FIGURE 18.2 Visualizing an electric field using
an elastic sheet.

(a)

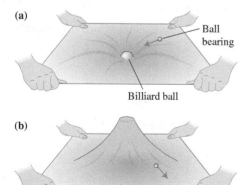

Ball
bearing

Billiard ball

(b)

18.1 A model of the mechanism for electrostatic interactions

When we pull two pieces of Scotch tape off a table and place them near each other, they repel. How does an electrically charged object, such as a piece of tape, affect the other piece of tape without touching it? You could ask the same question about gravitational interactions—how does the Sun pull on Earth even though they are millions of kilometers apart?

Historically, there have been two answers to this question. The first model was called *interaction at a distance*. Isaac Newton supported this model for the gravitational interaction, and Charles Coulomb supported it for the electrostatic interaction. In this model, both interactions just happen. For example, if you move one of the two interacting charges, the other charge instantaneously "senses" that movement and responds accordingly.

The second model for electric interactions, suggested by Michael Faraday, involved an agent that acts as an intermediary between the charges. It was some sort of *electric disturbance* in the region surrounding them. Physicists call this electric disturbance an **electric field**. According to this **electric field model**, a charged object q_1 electrically disturbs the region surrounding itself (**Figure 18.1a**). If you place a second charged object q_2 in this region, the electric field due to object 1 exerts a force on object 2 (Figure 18.1b). The same is true for object 2—it creates its own field that exerts a force on object 1.

We can apply the field model to the gravitational interaction. For example, how does the Sun pull on Earth, located about 150 million km away? According to the field model, the Sun gravitationally disturbs the region around it in an invisible way. This disturbance is called a **gravitational field**. The Sun's gravitational field exerts a force on Earth that causes it to travel in an almost circular orbit around the Sun.

An analogy may help you visualize the idea of a field. Imagine a very large horizontal elastic sheet that has been pulled tight. The sheet represents space with no charges (no field). Now imagine a billiard ball resting on the sheet. The ball causes the sheet to bend, more so closer to the ball than farther away from it (see **Figure 18.2a**). The presence of the ball causes a disturbance of space (the sheet), and the bending of the sheet represents the field due to the billiard ball. Now suppose a much smaller object such as a ball bearing is placed on the sheet, near the billiard ball. The ball bearing is so small that it does not disturb the sheet field. But the billiard ball's field disturbs the ball bearing—less when it is farther away and more when it is closer to the billiard ball. If the ball bearing is released, it will move toward the billiard ball as if they attract each other. An analogous model for visualizing repulsion can be obtained by stretching the elastic sheet over a pole and again placing the ball bearing on the bent surface (see Figure 18.2b). The billiard ball's or the pole's distortion of the elastic sheet acts as an intermediary, allowing the ball bearing and the billiard ball or pole to interact with each other, even though they are not in contact. Similarly, the electric field acts as an intermediary between charged objects, allowing the objects to interact without direct contact with each other.

Making analogies serves two purposes: (1) to understand a new phenomenon using a phenomenon that we already understand and (2) to predict new phenomena. The elastic sheet analogy allows us to make a prediction based on the electric field model: the interaction between the objects changes at a finite, nonzero speed. When you move one of the two interacting charges (object 1), the other one does not "sense" that movement instantly—it takes time for the change in the field to reach object 2. According to the model of interaction at a distance, the prediction is different: any change in the position of object 1 leads to the *immediate* change in the force exerted on object 2. Which prediction matches the outcome of an experiment testing these models? We will learn the answer to this question when we study electromagnetic waves in Chapter 25.

It is important to remember that every analogy we make has limitations. The two most important limitations of the elastic sheet analogy are the following: (1) In the analogy, there would be no visible interaction between the billiard ball and the ball bearing if Earth did not exert gravitational forces on both balls, while the interaction between charged objects does not need a "third agent." (2) In the analogy, the ball bearing can move only on the two-dimensional surface of the sheet, while charged objects can move freely in three-dimensional space.

Gravitational field due to a single object with mass

Let's develop a gravitational field approach for the gravitational force that one object with mass exerts on another. In the region near Earth, Earth's contribution to the gravitational field is the dominant one. Imagine Earth and one of three small objects A, B, or C. We call these objects *test objects*, because we use them as detectors to probe the field (**Figure 18.3a**). We place the test objects one at a time at the same location near Earth. The masses of these test objects are $m_A = m$, $m_B = 2m$, and $m_C = 3m$. Consider the gravitational force that Earth exerts on each test object (Figure 18.3b).

$$F_{\text{E on A}} = G\frac{m_E}{r^2}m_A = m_A\left(G\frac{m_E}{r^2}\right)$$

$$F_{\text{E on B}} = G\frac{m_E}{r^2}m_B = m_B\left(G\frac{m_E}{r^2}\right) = 2F_{\text{E on A}}$$

$$F_{\text{E on C}} = G\frac{m_E}{r^2}m_C = m_C\left(G\frac{m_E}{r^2}\right) = 3F_{\text{E on A}}$$

The directions of these forces are all toward the center of Earth. However, the magnitudes of the forces differ: they are proportional to the masses of the test objects. Despite the differences in magnitude, however, the ratio of the magnitude of the force exerted on each object and the mass of that object is identical for all three objects:

$$\frac{F_{\text{E on A}}}{m_A} = \frac{F_{\text{E on B}}}{m_B} = \frac{F_{\text{E on C}}}{m_C} = G\frac{m_E}{r^2}$$

Consider the objects to be near Earth's surface. When we substitute the values of G, m_E, and Earth's radius r_E, we find that

$$G\frac{m_E}{r_E^2} = \left(6.67 \times 10^{-11}\,\frac{\text{N} \cdot \text{m}^2}{\text{kg}^2}\right)\frac{(5.97 \times 10^{24}\,\text{kg})}{(6.37 \times 10^6\,\text{m})^2} = 9.8\,\text{N/kg}$$

Since this value does not depend on the mass of any test object, we speculate that this value might be a mathematical description of the "strength" of Earth's gravitational field at a particular location. Since the gravitational force has direction, we say that the gravitational field close to Earth's surface at a particular location has a magnitude of 9.8 N/kg and points directly toward the center of Earth. Until now, we have called this quantity free-fall acceleration. Now we characterize the gravitational field using the quantity $\vec{g}$ field. We define the $\vec{g}$ **field** at any location as the gravitational force exerted by the field on a test object at that location, divided by the mass of that object:

$$\vec{g} = \frac{\vec{F}_{\text{Field on Object}}}{m_{\text{Object}}} \tag{18.1}$$

The magnitude of the field at any location does not depend on the test object; it depends on the mass of the object(s) creating the field (in this case, Earth) and the location where the field is measured: $g = G(m_E/r_E^2)$.

Electric field due to a single point-like charged object

Let's use a similar approach to construct a physical quantity for the "strength" of the electric field. Imagine an object with positive electric charge Q (we use capital Q to denote the object whose field we are investigating), and one of three test objects K,

FIGURE 18.3 Earth exerts a gravitational force on each of the three test objects placed at the same location.

(a)

We place test objects A, B, or C one at a time at the same location relative to Earth.

(b)

$\vec{F}_{\text{E on test object}}$ ⬅●

The gravitational force that Earth exerts on a test object

TIP Equation (18.1) is an operational definition for the $\vec{g}$ field; the expression $g = G(m_E/r^2)$ is the cause-effect relationship for g.

FIGURE 18.4 An object with charge Q exerts an electric force on one of three charged objects placed at the same location.

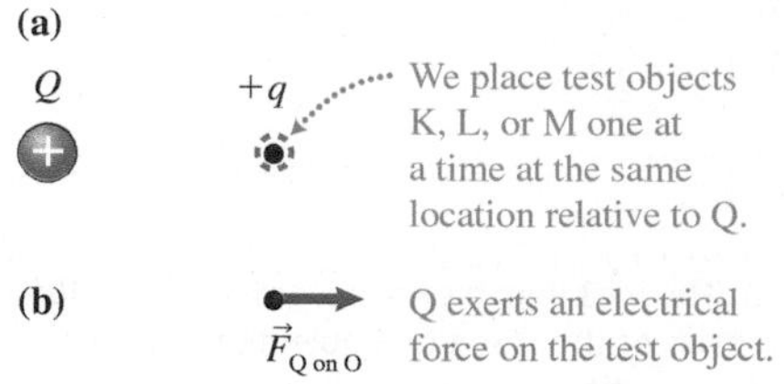

(a)

Q $+q$

We place test objects K, L, or M one at a time at the same location relative to Q.

(b)

$\vec{F}_{Q \text{ on } O}$

Q exerts an electrical force on the test object.

L, or M placed at a distance r from the center of object Q (**Figure 18.4a**). Objects K, L, and M have small positive charges $q_K = q$, $q_L = 2q$, and $q_M = 3q$ (here we use lowercase q to denote that these are test objects). Use Coulomb's law to compare the magnitudes of the electric forces that Q exerts on K, L, or M (Figure 18.4b):

$$F_{Q \text{ on } K} = k_C \frac{Qq_K}{r^2} = q_K \left(k_C \frac{Q}{r^2} \right)$$

$$F_{Q \text{ on } L} = k_C \frac{Qq_L}{r^2} = q_L \left(k_C \frac{Q}{r^2} \right) = 2F_{Q \text{ on } K}$$

$$F_{Q \text{ on } M} = k_C \frac{Qq_M}{r^2} = q_M \left(k_C \frac{Q}{r^2} \right) = 3F_{Q \text{ on } K}$$

Notice that Q exerts electric forces proportional to the electric charge of the test objects K, L, or M. However, the ratio of the magnitude of the electric force exerted on objects K, L, or M and the electric charge of each object equals the same value:

$$\frac{F_{Q \text{ on } K}}{q_K} = \frac{F_{Q \text{ on } L}}{q_L} = \frac{F_{Q \text{ on } M}}{q_M} = k_C \frac{Q}{r^2}$$

Since these ratios all have the same value, the ratio $(F_{Q \text{ on } K}/q_K)$ could be a mathematical description of the strength of the electric field produced by the charge Q at that location.

We can now think of the objects K, L, and M as probes of the electric field—called **test charges**. We use these test charges to examine the electric field produced by some object Q—the **source charge**. If we agree that a test charge is always positive and relatively small in magnitude so that it has little effect on the source charge, we can define the physical quantity that characterizes the electric field at a location as a vector with magnitude $(F_{Q \text{ on } q_{\text{test}}}/q_{\text{test}})$ that points in the direction of the electric force that is exerted on the positive test charge placed at that location. This physical quantity is called the $\vec{E}$ **field**.

The $\vec{E}$ field is a physical quantity that characterizes properties of space around charged objects. To determine the $\vec{E}$ field at a specific location, place an object with a small positive test charge q_{test} at that location and measure the electric force exerted on that object. The $\vec{E}$ field at that location equals the ratio

$$\vec{E} = \frac{\vec{F}_{Q \text{ on } q_{\text{test}}}}{q_{\text{test}}} \tag{18.2}$$

and points in the direction of the electric force exerted on the positive test charge. The $\vec{E}$ field is independent of the test charge used to determine the field. The unit of the electric field is newtons per coulomb (N/C).

TIP Equation (18.2) is an operational definition for the $\vec{E}$ field; the expression $E = k_C Q/r^2$ is the cause-effect relationship.

Using the above operational definition of the $\vec{E}$ field at a point, we found that the magnitude of the $\vec{E}$ field produced by a point-like source object with charge Q was

$$E = \frac{F_{Q \text{ on } q_{\text{test}}}}{q_{\text{test}}} = k_C \frac{|Q|q_{\text{test}}}{r^2 q_{\text{test}}} = k_C \frac{|Q|}{r^2}$$

We can interpret this field as follows:

$$E = k_C \frac{|Q_{\text{Source}}|}{r^2} \tag{18.3}$$

Effect   Cause

The object with electric charge Q is the cause, and the $\vec{E}$ field produced by it is the effect. Note also that the $\vec{E}$ field vector at any location points *away* from the object creating the field if Q is positive and *toward* it if Q is negative.

$\vec{E}$ field due to multiple charged objects

To determine the magnitude and direction of the $\vec{E}$ field at a particular point caused by more than one charged object, we start by investigating the field created by two charged objects in Observational Experiment **Table 18.1**.

OBSERVATIONAL
EXPERIMENT TABLE 18.1 $\vec{E}$ field due to two charged objects

VIDEO
OET 18.1

Observational experiment	Analysis
Experiment 1. Two identical metal spheres are on insulating stands with equal positive charge. You charge a small aluminum foil ball positively and hang it from an insulated string between the two spheres. When hanging closest to the left sphere, the foil ball is repelled away, toward the right sphere. When hanging nearer the right sphere, it is repelled away, toward the left sphere. At the exact middle between the spheres, the foil ball hangs straight down. (Note that the figure shows the outcomes of three independent experiments.)	Regard the foil ball as a test charge to determine the direction of the $\vec{E}$ field at different places between the spheres. The $\vec{E}$ field points right when nearest the left sphere, is zero in the exact middle between the spheres, and points left when nearest the right sphere.
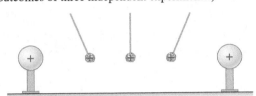	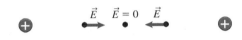
Experiment 2. Repeat Experiment 1, only this time with the left sphere charged negatively and the right sphere charged positively. When the foil ball hangs at any place between the spheres, the ball is attracted toward the left negative sphere and repelled from the right sphere. (Note that the figure shows the outcomes of three independent experiments.)	The foil ball indicates that the $\vec{E}$ field points left at all points between the spheres. The different tilts of the strings indicate slightly different magnitudes of the $\vec{E}$ field vector.

Patterns

- In Experiment 1, the zero field at the exact middle must have been due to the vector addition of equal-magnitude but oppositely directed $\vec{E}$ fields caused by the two spheres.
- In Experiment 1, when on either side of the middle, the nearer sphere caused a greater repulsion than the farther sphere. The opposing $\vec{E}$ fields only partially canceled each other.
- In Experiment 2, both spheres caused an $\vec{E}$ field to the left, which added to form a stronger $\vec{E}$ field to the left.

It appears that when multiple charged objects are present, each object creates its own contribution to the $\vec{E}$ field. At a particular location, the field is the *vector* sum of the $\vec{E}$ field contributions due to all of these charges. This summation effect is an example of the superposition principle. The fact that we add the $\vec{E}$ field contributions means that they do not affect each other, but they all affect a test charge for $\vec{E}$ fields placed at a particular location.

Superposition principle for $\vec{E}$ fields The $\vec{E}$ field at a point of interest is the *vector* sum of the individual contributions to the $\vec{E}$ field of each charged object, and it is called $\vec{E}_{net}$:

$$\vec{E}_{net} = \vec{E}_1 + \vec{E}_2 + \vec{E}_3 + \cdots \qquad (18.4)$$

Physics Tool Box 18.1 illustrates step by step how to use the superposition principle and graphical vector addition to estimate the direction of the $\vec{E}$ field at a point of interest.

PHYSICS TOOL BOX 18.1 ▶ **Estimating the $\vec{E}$ field at a position of interest**

1. Place an imaginary positive test charge at the point of interest, A. Use this test charge to estimate the direction of the $\vec{E}$ field contribution of each charged object creating the field at A, which is the same as the direction of the force that each object exerts on the test charge.

2. Draw an arrow in the direction of the $\vec{E}$ field contribution of each charged object q_1, q_2, and q_3.

3. Estimate the approximate relative lengths of the $\vec{E}$ fields due to each charged object.

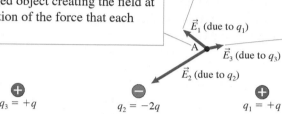

4. Graphically add $\vec{E}_1$, $\vec{E}_2$, and $\vec{E}_3$ to estimate the net $\vec{E}$ field at point A.

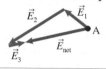

CONCEPTUAL EXERCISE 18.1 ▶ **The $\vec{E}$ field in body tissue due to the heart dipole charge**

The muscles of the heart continually contract and relax, making the heart an electric dipole with equal-magnitude positive and negative electric charges, such as shown in the figure at right. Estimate the direction of the $\vec{E}$ field at position A in the body tissue to the left side of the midpoint between the dipole charges.

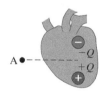

Sketch and translate Imagine placing a positive test charge at position A to estimate the direction of the forces exerted on it by each dipole charge. We use the directions of those forces to find the direction of the corresponding $\vec{E}$ fields.

Simplify and diagram We draw arrows representing the $\vec{E}$ field at position A due to each dipole charge (see figure at right). Then we graphically add the $\vec{E}$ fields to find the direction of the net $\vec{E}$ field.

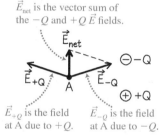

$\vec{E}_{net}$ is the vector sum of the $-Q$ and $+Q$ $\vec{E}$ fields.

$\vec{E}_{+Q}$ is the field at A due to $+Q$.
$\vec{E}_{-Q}$ is the field at A due to $-Q$.

Try it yourself Repeat this exercise for a position that is in the body tissue several centimeters to the right side of the center of the heart dipole.

Answer The net $\vec{E}$ field on the right side also points up in the same direction as the $\vec{E}$ field on the left side of the heart dipole.

$\vec{E}$ field lines

We have just learned how to estimate the $\vec{E}$ field at a single location in the vicinity of several charged objects. To represent the overall shape of the $\vec{E}$ field, we use electric field lines, or $\vec{E}$ *field lines*. Imagine two source charges ($+Q$ and $-Q$) arranged as shown in **Figure 18.5a**. We start by using the superposition principle and graphical vector addition to find the net electric field $\vec{E}_{net}$ at position A due to both charges (see Figure 18.5a). Next, we use the same method to draw another $\vec{E}_{net}$ vector for a position near the tip of our first $\vec{E}_{net}$ vector (Figure 18.5b). The field at this location points toward the right and downward a little, since the negative charge $-Q$ is slightly closer and more attractive to a positive test charge at that point than the slightly more distant positive charge $+Q$ is repulsive to it.

Next, we draw a third $\vec{E}$ field vector for a position near the tip of the second $\vec{E}$ field vector. We continue drawing the $\vec{E}_{net}$ vectors for adjacent positions along the direction of the previous $\vec{E}_{net}$ vector, eventually getting a series of $\vec{E}_{net}$ vectors that seem to follow one after the other from the positive charge $+Q$ to the negative charge $-Q$, as shown in Figure 18.5c. The $\vec{E}_{net}$ vectors from the positive charge to A have a similar shape and are also shown in Figure 18.5c. Now, draw a single line that passes through each position tangent to these vectors. The line looks as shown in Figure 18.5d. If you repeat the process for a series of positions farther and closer to a line between the two charges, you get a series of lines such as shown in Figure 18.5e—the $\vec{E}$ field lines for the source charges shown in Figure 18.5a.

TIP In this book all figures are two-dimensional, but in real life the situations are three-dimensional. The $\vec{E}$ field vectors and $\vec{E}$ field lines are really in a three-dimensional space.

Using a similar process, you could construct the $\vec{E}$ field lines for the single positively charged object shown in **Figure 18.6a**, for the single negatively charged object shown in Figure 18.6b, and for the two equal-magnitude positively charged objects shown in Figure 18.6c. Note that at locations closer to the charged objects, the lines are closer together than when the locations are farther away. This means that the number of lines per unit area (the density of the lines) is larger where the $\vec{E}$ field is stronger. Similarly, if you have two source charges, then at the same relative location with respect to the charges, the magnitude of the $\vec{E}$ field is larger next to a bigger charge. Thus the number of lines emanating from the charge represents the magnitude of that charge. If you have two charged objects next to each other with the magnitude of one charge twice the magnitude of the other, and you draw six lines coming out of the first object, then the second object should have 12 lines coming out of it.

Summary—$\vec{E}$ field lines

- $\vec{E}$ field lines are drawn so that the $\vec{E}$ field vectors at positions on those lines are tangent to the lines.
- $\vec{E}$ field lines begin on positively charged objects and end on negatively charged objects (these charged objects do not necessarily have to be present on the sketches).
- The number of $\vec{E}$ field lines beginning or ending on a charged object is proportional to the magnitude of that object's charge. Therefore, on an electric field diagram you will see twice as many $\vec{E}$ field lines leaving a charge of $+2q$ as leaving a charge of $+1q$.
- The density of the $\vec{E}$ field lines in a region is proportional to the magnitude of the $\vec{E}$ field in that region.
- The lines do not cross; if they did, the force exerted on a test charge placed in that location would be undetermined.

FIGURE 18.5 Constructing $\vec{E}$ field lines.

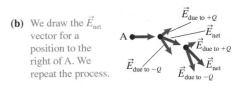

(a) The net field at A due to the source charges

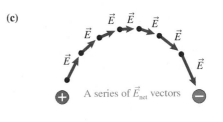

(b) We draw the $\vec{E}_{net}$ vector for a position to the right of A. We repeat the process.

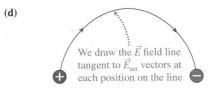

(c) A series of $\vec{E}_{net}$ vectors

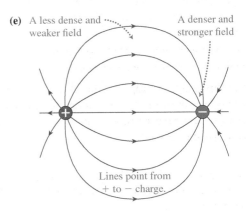

(d) We draw the $\vec{E}$ field line tangent to $\vec{E}_{net}$ vectors at each position on the line

(e) A less dense and weaker field   A denser and stronger field

Lines point from $+$ to $-$ charge.

FIGURE 18.6 $\vec{E}$ field lines for (a) a single positive charge, (b) a single negative charge, and (c) two positive charges.

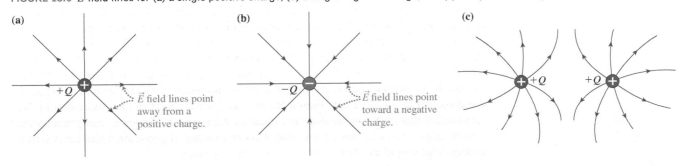

(a) $+Q$ $\vec{E}$ field lines point away from a positive charge.

(b) $-Q$ $\vec{E}$ field lines point toward a negative charge.

(c) $+Q$ $+Q$

TIP The $\vec{E}$ field lines give information about the force that the electric field exerts on a test charge placed at any location, and the resulting acceleration of the test charge, but not the direction of motion of the test charge.

CONCEPTUAL EXERCISE 18.2   Uniform $\vec{E}$ field

Draw $\vec{E}$ field lines for a large, uniformly charged plate of glass.

Sketch and translate A uniformly charged plate of glass is shown in the figure at right. "Uniformly charged" means that the amount of electric charge located on each unit area of the surface is constant throughout the plate.

Simplify and diagram Assume that the plate of glass is infinitely large in both perpendicular directions in the plane of its surface. Choose a position of interest to the right of the plate. Think of the $\vec{E}$ field at this position as caused by each small segment of positive source charge on the plate. Notice that the y-component of the $\vec{E}$ field from one of the charge segments on the plate (for example, the position 1 segment in the figure below) is canceled by the y-component of the $\vec{E}$ field from a charge segment at some other location on the plate (position 2). This occurs for every charge segment. Therefore, the y-component of the $\vec{E}$ field is zero at every position to the right of the plate.

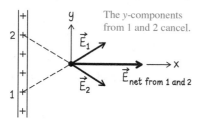

The y-components from 1 and 2 cancel.

Consequently, the $\vec{E}$ field is perpendicular to the plate (see the first figure in the next column) at every point to the right of the plate (close or far away). The $\vec{E}$ field lines are parallel at every position. Therefore,

the $\vec{E}$ field lines must point away from the plate and be perpendicular to it (see the second figure below).

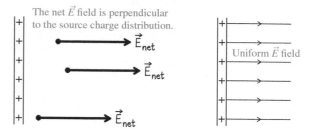

The net $\vec{E}$ field is perpendicular to the source charge distribution.

Uniform $\vec{E}$ field

The plate is infinitely large, so the $\vec{E}$ lines should look the same at any point on the right side of the plate. If the $\vec{E}$ field decreased in magnitude as you moved farther from the plate, the lines would have to spread farther apart. But they don't. The field lines must have a uniform density—the same separation between the $\vec{E}$ field lines everywhere. This means the $\vec{E}$ field has the same magnitude everywhere in this region. The uniform charge on the plate produces a uniform $\vec{E}$ field.

Try it yourself Imagine that an infinitely large, uniformly positively charged plate oriented in the plane perpendicular to the page produces an $\vec{E}$ field of magnitude E. You add another plate, negatively charged, at a distance d from the first one and parallel to it. What is the magnitude of the $\vec{E}$ field to the left of the plates, between them, and to the right? Note that the fields created by the two plates are independent of each other.

Answer   0, 2E, 0.

The analysis of the uniform $\vec{E}$ field in the last Conceptual Exercise will be very useful later in the chapter, especially when we study capacitors.

REVIEW QUESTION 18.1 How do you estimate the direction of the $\vec{E}$ field at a point located near two point-like charged objects?

18.2 Skills for analyzing processes involving $\vec{E}$ fields

You will encounter two main types of problems involving $\vec{E}$ fields. In the first type, you will know the charge distribution and must determine the $\vec{E}$ field at one or more points caused by these source charges. In the second type, you will need to analyze a process involving a charged object that is stationary or moving in a given $\vec{E}$ field. Let's start by describing a strategy for solving the first type of problem.

Determining the $\vec{E}$ field produced by given source charges

When you have several objects (with charges Q_1, Q_2, etc.) contributing to the electric field at a particular location, to determine the direction and magnitude of the $\vec{E}_{net}$ field vector at that location, you first need to follow the steps in Physics Tool Box 18.1 to qualitatively determine the direction of $\vec{E}_{net}$. You can then start the calculations. Choose coordinate axes and write expressions for the magnitudes of the $\vec{E}$ field vector contributions of each charged object:

$$E_1 = k_C \frac{|Q_1|}{r_1^2}; \quad E_2 = k_C \frac{|Q_2|}{r_2^2}; \quad \text{etc.}$$

Then find the x- and y-components of each contribution and add the respective components algebraically to find $E_{x\,net}$ and $E_{y\,net}$, which are the components of $\vec{E}_{net}$. Knowing these components, you can use the Pythagorean theorem to determine the magnitude of $\vec{E}_{net}$:

$$E_{net} = \sqrt{E_{x\,net}^2 + E_{y\,net}^2}$$

and next use trigonometry to determine as needed the angles that $\vec{E}_{net}$ makes with the coordinate axes.

This approach is implemented in Example 18.3. Follow the example closely and focus on each step in turn.

EXAMPLE 18.3 $\vec{E}$ **field due to multiple charges**

Two small metal spheres attached to insulating stands reside on a table a distance d apart. The left sphere has positive charge $+q$ and the right has negative charge $-q$. Determine the magnitude and direction for the net $\vec{E}$ field at a distance d above the center of the line connecting the spheres.

Sketch and translate The sketch shows the source charges $+q$ and $-q$ that are producing the $\vec{E}$ field and the place (the point of interest) where we want to determine the net $\vec{E}$ field—both its magnitude and direction. To estimate the direction of the vectors, imagine placing a positively charged test object q_T at that point. (Remember, the fields exist without the test charge.)

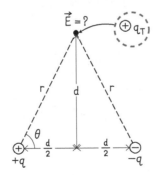

Simplify and diagram Next, construct an $\vec{E}$ field diagram for the point of interest, showing an $\vec{E}$ field arrow for each source charge that contributes to the $\vec{E}$ field at that point (see figure at right). Include coordinate axes in the diagram. Notice that the $\vec{E}$ field vectors due to each source charge point in the direction of the electric force that each of the source charges would exert on the imaginary positive charge placed at the location of interest.

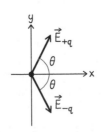

Represent mathematically Use Eq. (18.3) to determine the magnitude of the $\vec{E}$ field contributions from both $+q$ and $-q$: $E_{+q} = E_{-q} = k_C(q/r^2)$.

Use the Pythagorean theorem to determine the distance r from $+q$ and $-q$ to the point of interest:

$$r^2 = \left(\frac{d}{2}\right)^2 + d^2 = \frac{5d^2}{4}$$

Thus, the magnitude of the $\vec{E}$ field contributions from both $+q$ and $-q$ is

$$E_{+q} = E_{-q} = \frac{4k_C q}{5d^2}$$

The angle θ that each $\vec{E}$ field vector makes relative to the positive x-axis can be determined using the triangle in the first figure:

$$\tan\theta = \frac{\text{opposite side}}{\text{adjacent side}} = \frac{d}{d/2} = 2.0 \quad \text{or} \quad \theta = 63.4°$$

Now determine the x- and y-components of the $\vec{E}$ field.

$$E_{x\,net} = E_{+qx} + E_{-qx} = E_{+q}\cos\theta + E_{-q}\cos\theta$$

$$= 2E_{+q}\cos\theta = 2\left(\frac{4k_C q}{5d^2}\right)\cos 63.4° = 0.716\frac{k_C q}{d^2}$$

$$E_{y\,net} = E_{+qy} + E_{-qy}$$

$$= +E_{+q}\sin 63.4° + (-E_{-q})\sin 63.4° = 0$$

The y-components of the two $\vec{E}$ field contributions add to zero.

Solve and evaluate The magnitude of the net $\vec{E}$ field is

$$E_{net} = \sqrt{E_{x\,net}^2 + E_{y\,net}^2} = \sqrt{\left(\frac{0.716\,k_C q}{d^2}\right)^2 + 0^2} = 0.72\frac{k_C q}{d^2}$$

and it points in the positive x-direction.

(CONTINUED)

Try it yourself Determine the $\vec{E}$ field at point I in the figure at right. (a) Determine the magnitudes of the $\vec{E}$ field contributions at point I from charges $+q$ and $-q$. (b) Determine the x- and y-components of those $\vec{E}$ field contributions. (c) Determine the net x- and y-components of the $\vec{E}$ field at point I. (d) Finally, determine the magnitude and the direction of the $\vec{E}$ field at point I.

Answer

(a) 225 N/C, 144 N/C; (b) (225 N/C, 0) and (−115 N/C, +87 N/C); (c) (+110 N/C, +87 N/C); (d) 140 N/C, 38° above +x-axis (toward the right).

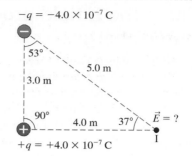

Analyzing a process involving a known $\vec{E}$ field

If the $\vec{E}$ field at some location is known, we can use Eq. (18.2) to determine the electric force that the field exerts on an object with charge q if at that location:

$$\vec{F}_{E \text{ on } q} = q\vec{E} \tag{18.5}$$

We can now combine our understanding of electric field and force with our knowledge of Newton's laws to solve problems.

PROBLEM-SOLVING STRATEGY 18.1 Electric field and Newton's second law

---

EXAMPLE 18.4 **A charged object in a known $\vec{E}$ field**

A spring made of a dielectric material with spring constant 5.0 N/m hangs from a large, uniformly charged horizontal plate. The uniform $\vec{E}$ field produced by the plate has magnitude 1.0×10^6 N/C and points down. A 0.020-kg ball with charge $+4.0 \times 10^{-8}$ C hangs at the end of the spring. Determine the distance the spring is stretched from its equilibrium length. Assume that $g = 10$ N/kg.

Sketch and translate

- Draw a labeled sketch of the situation. Include the symbols for the known and unknown quantities that you plan to use.
- Choose the system.

The situation is sketched at the right. The spring has stretched distance Δy from its equilibrium length. The charged ball (B) is the system.

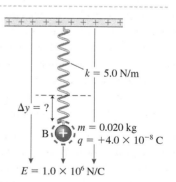

Simplify and diagram

- Determine the $\vec{E}$ field produced by the environment. Is it produced by point-like charges (making it nonuniform) or by large charged plates (making it uniform)?
- Construct a force diagram for the system if necessary. Do not forget the axes.
- State any assumptions you have made.

The field is uniform. The upward direction is positive.
Assume that the spring has no mass.
A force diagram for the ball is shown at the right. S stands for the spring and P for the plate.

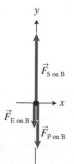

Represent mathematically

- Use the force diagram to help apply Newton's second law in component form. Use component addition to determine the net force along each coordinate axis. Determine the acceleration of the system if needed.

- If necessary, use kinematics equations to describe the motion of the system.

Solve and evaluate

- Combine the above equations and complete the solution.

- Check the direction, magnitude, and units and decide whether the result makes sense in limiting cases.

The ball is not accelerating, so the forces exerted on the ball add to zero. None of the forces have x-components:

$$F_{\text{S on B } y} + F_{\text{E on B } y} + F_{\text{P on B } y} = 0$$
$$\Rightarrow k\,\Delta y + (-mg) + (-qE) = 0$$
$$\Rightarrow \Delta y = \frac{mg + qE}{k}$$

$$\Delta y = \frac{(0.02\ \text{kg})(10\ \text{N/kg}) + (4.0 \times 10^{-8}\ \text{C})(1.0 \times 10^6\ \text{N/C})}{5.0\ \text{N/m}}$$

$$= 0.048\ \text{m} = 48\ \text{mm}$$

The units for the stretch distance are the units of length, as they should be. Check using limiting cases. As both the gravitational force and the electric force point in the same direction, eliminating either of them should reduce the distance that the spring stretches. Note in the equation above that Δy decreases if $m = 0$ or if $q = 0$. Also, a stiff spring (larger k) should stretch less if we have a ball of the same mass and the same electric charge—also consistent with our result.

Try it yourself If the plate were negatively charged, producing an $\vec{E}$ field with the same magnitude but now pointing up, what would the spring stretch or compression be?

Answer   +32 mm (the spring is stretched).

EXAMPLE 18.5	Electric field deflects a tiny ink ball

Inside an inkjet printer, a tiny ball of black ink of mass 1.1×10^{-11} kg with charge -6.7×10^{-12} C moves horizontally at 40 m/s. The ink ball enters an upward-pointing uniform $\vec{E}$ field of magnitude 1.0×10^4 N/C produced by a negatively charged plate above and a positively charged plate below. The plates are used to deflect the ink ball so that it lands at a particular spot on a piece of paper. Determine the deflection of the ink ball after it travels 0.010 m in the $\vec{E}$ field. Assume that $g = 10$ N/kg.

Sketch and translate After sketching the situation (see figure below), we choose the charged ink ball as the system. We break the problem into two parts: first we determine the acceleration of the ball due to the forces being exerted on it, and then we use kinematics to determine its vertical displacement as it passes through the region with the electric field.

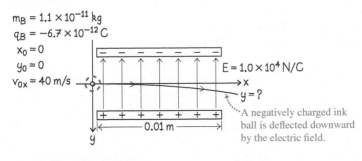

$m_B = 1.1 \times 10^{-11}$ kg
$q_B = -6.7 \times 10^{-12}$ C
$x_0 = 0$
$y_0 = 0$
$v_{0x} = 40$ m/s
$E = 1.0 \times 10^4$ N/C
$y = ?$
0.01 m

A negatively charged ink ball is deflected downward by the electric field.

Simplify and diagram Assume that the ink ball is a point-like object and that the plates produce a uniform $\vec{E}$ field. A force diagram for the ball is shown at right. Earth and the electric field created by the plates (P) exert downward forces on the negatively charged ball (B).

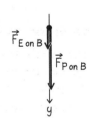

Represent mathematically The forces exerted on the ink ball do not have x-components, so we use only the y-component form of Newton's second law to determine the y-component of the acceleration (we choose the positive y-direction pointing down):

$$a_y = \frac{1}{m_B}\sum F_y = \frac{1}{m_B}(F_{\text{P on B } y} + F_{\text{E on B } y})$$

$$= \frac{1}{m_B}(q_B E_y + m_B g_y) = \frac{q_B E_y}{m_B} + g_y$$

The ink ball does not accelerate in the x-direction, so the x-component of its velocity remains constant at 40 m/s during the time interval it takes to traverse the region with nonzero $\vec{E}$ field. During this time interval, $\Delta t = \Delta x/v_{0x}$, the ball's displacement in the y-direction is

$$y - y_0 = v_{0y}\,\Delta t + \frac{1}{2}a_y\,\Delta t^2$$

(CONTINUED)

Solve and evaluate Substituting for a_y, Δt, and $v_{0y} = 0$ gives

$$y - y_0 = 0 + \frac{1}{2}\left(\frac{q_B E_y}{m_B} + g_y\right)\left(\frac{\Delta x}{v_{0x}}\right)^2$$

We can now insert the known information ($E_y = -1.0 \times 10^4\,\text{N/C}$, $q_B = -6.7 \times 10^{-12}\,\text{C}$, $m_B = 1.1 \times 10^{-11}\,\text{kg}$, and $g_y = +10\,\text{N/kg}$) into the above to determine the ink ball deflection:

$$y - y_0 = \frac{1}{2}\left(\frac{(-6.7 \times 10^{-12}\,\text{C})(-1.0 \times 10^4\,\text{N/C})}{1.1 \times 10^{-11}\,\text{kg}} + 10\,\text{N/kg}\right)$$

$$\times \left(\frac{0.010\,\text{m}}{40\,\text{m/s}}\right)^2 = 1.9 \times 10^{-4}\,\text{m} = 0.19\,\text{mm}$$

The distance unit is correct, and the value seems reasonable given the size of a letter on a printed page. To get a bigger deflection, we can use a larger electric field.

- -

Try it yourself What should the mass of the ink ball be so that if the direction of the $\vec{E}$ field is reversed, the ball will have zero acceleration?

Answer $6.7 \times 10^{-9}\,\text{kg}.$

REVIEW QUESTION 18.2 You have a point-like object with charge $+Q$ on the left and a second object with charge $-2Q$ at a distance d on the right. Sketch the $\vec{E}$ field vector at the same distance d directly above the midpoint between the charged objects.

18.3 The V field: electric potential

So far, we have been studying electric fields using a force approach. However, in many problems in mechanics and electrostatics, an energy approach allows us to focus on the initial and final stages of a process without knowing what happened in between. Can we describe properties of space around charged objects using the concepts of work and energy? In order to do this, we need to describe space around charged objects not as the force-related $\vec{E}$ field, but instead as an energy-related field.

Electric potential due to a single charged object

FIGURE 18.7 Consider the electric potential energy of a system with source charge Q and test charge $q = q_K$, q_L, or q_M.

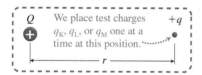

To construct this new way of describing space around charged objects, we will think of the electric interaction in terms of the electric potential energy. The source charge Q creates an electric field around it, and charged objects K, L, or M with different small electric charges q_K, q_L, or q_M (test charges) are placed one at a time at a distance r from Q (**Figure 18.7**). Recall (from Chapter 17) that if we consider the electric potential energy of two charged objects to be zero when they are infinitely far from each other, then the electric potential energy of two point-like charged objects 1 and 2 separated by a distance r is

$$U_{q_1 q_2} = \frac{k_C q_1 q_2}{r}$$

Thus, the electric potential energy of charges Q and q_K, q_L, or q_M is

$$U_{Q q_K} = \frac{k_C Q q_K}{r} = q_K\left(\frac{k_C Q}{r}\right)$$

$$U_{Q q_L} = \frac{k_C Q q_L}{r} = q_L\left(\frac{k_C Q}{r}\right)$$

$$U_{Q q_M} = \frac{k_C Q q_M}{r} = q_M\left(\frac{k_C Q}{r}\right)$$

If we divide the electric potential energy of the Q-q_K, Q-q_L, or Q-q_M charge pair by the charges q_K, q_L, or q_M (that is, U_{qQ}/q), we find that all three ratios equal the same quantity:

$$\frac{U_{Q q_K}}{q_K} = \frac{U_{Q q_L}}{q_L} = \frac{U_{Q q_M}}{q_M} = \frac{k_C Q}{r}$$

Since these ratios all have the same value (kQ/r), this ratio could be a mathematical representation of another physical quantity characterizing space around charge Q at a distance r from Q. We call this physical quantity the *V* **field** or **electric potential** at a particular location that is a distance r from the charge Q creating the field.

V **field (or electric potential) due to a single charge** To determine the *V* field due to a single source charge at a specific location, place a test charge q_{test} at that location and determine the electric potential energy $U_{Qq_{test}}$ of a system consisting of the test charge and the source charge that creates the field. The *V* field at that location equals the ratio

$$V = \frac{U_{Qq_{test}}}{q_{test}} = \frac{k_C Q}{r} \tag{18.6a}$$

The unit of electric potential is joule/coulomb (J/C) and is called the **volt** (V).

    We can also think of the *V* field at a specific location (let's say location A) as the ratio of the work that an external force needs to do on a positive test charge q_{test} to bring it to the location A from infinity and that charge:

$$V_A = \frac{W_{from\ \infty\ to\ A}}{q_{test}} \tag{18.6b}$$

The superposition principle and the *V* field due to multiple charges

Now that we can determine mathematically the *V* field produced by a single point-like charged object [Eq. (18.6)], we can use the superposition principle to determine the *V* field at a specific location produced by several charged objects. Using the same superposition principle idea we used for the $\vec{E}$ field, we have

$$V = V_1 + V_2 + V_3 + \cdots = \frac{k_C Q_1}{r_1} + \frac{k_C Q_2}{r_2} + \frac{k_C Q_3}{r_3} + \cdots \tag{18.7}$$

where $Q_1, Q_2, Q_3, \ldots$ are the source charges (including their signs) creating the field and $r_1, r_2, r_3, \ldots$ are the distances between the source charges and the location where we are determining the *V* field. Because the *V* field is a scalar field rather than a vector field (like the $\vec{E}$ field is), it is much easier to apply the superposition principle.

QUANTITATIVE EXERCISE 18.6   **Electric potential due to heart dipole**

Suppose that the heart dipole charges $-Q$ and $+Q$ at a particular instant (the heart does change during the cardiac cycle) are separated by distance d as in the figure. Write an expression for the *V* field due to both charges at point A, a distance d to the right of the $+Q$ charge.

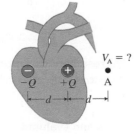

$V_A = ?$

Represent mathematically Use Eq. (18.7) to write an expression for the *V* field:

$$V = V_{+Q} + V_{-Q} = \frac{k_C(+Q)}{r_{+Q}} + \frac{k_C(-Q)}{r_{-Q}}$$

Solve and evaluate Inserting the distances, we get

$$V = \frac{k_C(+Q)}{d} + \frac{k_C(-Q)}{2d} = +\frac{k_C Q}{2d}$$

- -

Try it yourself Write an expression for the *V* field at a distance d to the left of $-Q$.

Answer $V = \frac{k_C(-Q)}{d} + \frac{k_C(+Q)}{2d} = -\frac{k_C Q}{2d}$.

Finding the electric potential energy when the V field is known

If we know the electric potential at a specific location, we can rearrange the definition of the V field [Eq. (18.6)] to determine the electric potential energy of a system that includes the source charges and another charge q that is in the vicinity of the source charges:

$$U_{\text{Source } Q \text{ and } q} = qV_{\text{Due to source } Q} \tag{18.8}$$

For example, we could use this equation to determine the electric potential energy of the heart dipole in Quantitative Exercise 18.6 and a sodium ion in the tissue near the dipole.

Since the value of the electric potential depends on the choice of zero level (as does electric potential energy), we often use the difference in electric potential between two points, called the **potential difference**, to analyze processes using electric potential.

> **Potential difference** ΔV between two points A and B is equal to the difference in the values of electric potential at those points: $\Delta V = V_B - V_A$. Potential difference is also called **voltage**. This term for potential difference is often used in electric circuits, the subject of the next chapter.

Potential difference is a very useful quantity—if we know the potential difference between two points A and B, we can predict the direction of acceleration of a charged object. A positively charged object accelerates from regions of higher electric potential toward regions of lower potential (like an object falling to lower elevation in Earth's gravitational field). A negatively charged particle tends to do the opposite, accelerating from regions of lower potential toward regions of higher potential. Consider the next example.

EXAMPLE 18.7   X-ray machine

Inside an X-ray machine is a wire (called a filament) that, when hot, ejects electrons. Imagine one of those electrons, now located outside the wire. It starts at rest and accelerates through a region where the V field increases by 40,000 V. The electron stops abruptly when it hits a piece of tungsten at the other side of the region, producing X-rays (we will learn more about that in Chapter 27). How fast is the electron moving just before it reaches the tungsten?

Sketch and translate The situation is sketched in the figure. The system consists of the electron and the electric field of the tube. We choose the zero level of the V field to be at the initial position of the electron, just outside the filament.

Simplify and diagram Neglect the gravitational force that Earth exerts on the electron given that the time of flight is very short. A work-energy bar chart for the process is shown at right. In the initial state, the system has neither electric potential energy (the electric

potential is zero there) nor kinetic energy (the electron is not moving). In the final state, the system has positive kinetic energy and negative electric potential energy. The electric potential energy is determined using $U_q = qV$ with $q < 0$ and $V_f > 0$.

Represent mathematically Each nonzero bar in the bar chart turns into a term in the generalized work-energy equation:

$$0 = \tfrac{1}{2} m_e v_f^2 + (-e)V_f$$

Solve and evaluate

$$v_f = \sqrt{-\frac{2(-e)V_f}{m_e}}$$

$$= \sqrt{\frac{-2(-1.6 \times 10^{-19}\,\text{C})(4.0 \times 10^4\,\text{V})}{9.11 \times 10^{-31}\,\text{kg}}} = 1.2 \times 10^8\,\text{m/s}$$

Limiting case check: If the electric charge of the electron were zero, it would not participate in the electric interaction at all and its final speed should be zero; it is. If the electric potential at the final location were the same as at the initial location, the final speed of the electron should be zero; and it is. Additionally, the speed of the electron is very high, similar to Example 17.8; the result obtained with the corrections for higher speeds that we will learn in Chapter 26 gives us an estimate that is three times smaller but still very high.

Try it yourself A 0.10-kg cart with a charge of $+6.0 \times 10^{-5}$ C rests on a 37° incline. Through what potential difference must the cart move along the incline so that it travels 2.0 m up the incline before it stops? The cart starts at rest. Assume that $g = 10$ N/kg.

Answer   The potential difference needed is $-20{,}000$ V.

Equipotential surfaces—representing the *V* field

$\vec{E}$ field lines help us visualize the $\vec{E}$ field. We will now develop a new way to visualize the *V* field at different points surrounding one or more charged objects. We start by considering a positively charged point-like charged object $+Q$ (shown in **Figure 18.8**). The values of the *V* field at points A, B, and C are all the same. The values of the *V* field at D, E, and F are all the same but different than the values at A, B, and C (in this case, they are smaller than the values for A, B, and C). If we imagine the surface to which all the points of equal *V* field value belong (for example, points A, B, and C), that surface will be a sphere surrounding the source charge $+Q$. These surfaces of constant electric potential *V* are called **equipotential surfaces**. The surfaces are spheres (they look like circles on a two-dimensional page) because the *V* field produced by a point-like charged object has the same value at all points equidistant from the charged object. The electric potential energy of a system with the source charge $+Q$ and a test charge $+q$ is constant if the test charge remains on an equipotential surface. No work is needed to move the test charge $+q$ to some other place on that equipotential surface.

The $\vec{E}$ field lines for a single positively charged source charge point outward along radial lines. The spherical equipotential surfaces are perpendicular to the $\vec{E}$ field lines at every point (**Figure 18.9a**). This is a general feature of every charge distribution; the $\vec{E}$ field lines and the equipotential surfaces are perpendicular to each other at every point in the region surrounding the charges (see the $\vec{E}$ field lines and the *V* field surfaces for the dipole and two equal-magnitude positive charge distributions in Figures 18.9b and c). Note also that the values of the *V* field of the surfaces decrease in the direction of the $\vec{E}$ field lines.

FIGURE 18.8 Equipotential surfaces produced by an object with charge $+Q$.

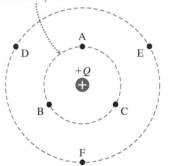

An equipotential surface has the same *V* field value at each point on the surface.

FIGURE 18.9 The equipotential surfaces and $\vec{E}$ field lines produced by (a) a positively charged point object, (b) an electric dipole, and (c) two equal-magnitude positively charged objects.

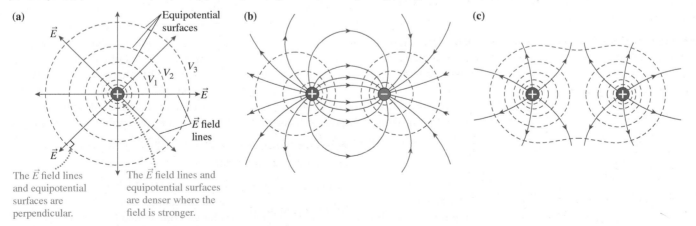

(a)

The $\vec{E}$ field lines and equipotential surfaces are perpendicular.

The $\vec{E}$ field lines and equipotential surfaces are denser where the field is stronger.

(b)

(c)

The relative distances between neighboring equipotential surfaces that vary by the same amount (for example, 10 V) are smaller where the $\vec{E}$ field is greater—the *V* field changes faster where the $\vec{E}$ field is stronger. Consider the equipotential surfaces and the

FIGURE 18.10 Changes in the *V* field created by point-like charge *Q* shown (a) on a *V*-versus-*r* graph and (b) using equipotential surfaces.

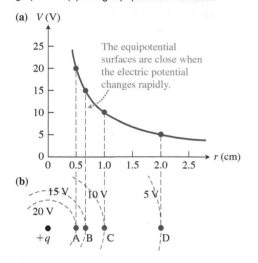

(a)

The equipotential surfaces are close when the electric potential changes rapidly.

(b)

FIGURE 18.11 Visualizing equipotential surfaces using an elastic sheet.

FIGURE 18.12 An experiment to derive a relationship between the $\vec{E}$ field and the *V* field.

(a) A stick pushes on a positively charged object in a uniform $\vec{E}$ field.

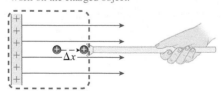

(b) The charged object is not accelerating, so the forces add to zero.

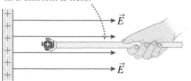

(c) The stick (not in the system) does negative work on the charged object.

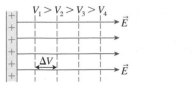

(d) ΔV is constant in this case because *E* is constant.

$V_1 > V_2 > V_3 > V_4$

graph of *V* field-versus-distance from $+Q$ (*V*-versus-*r* graph) shown in **Figure 18.10**. The charge creating the field is $+1.1 \times 10^{-11}$ C. The closer to $+Q$, the more rapidly the *V* field changes with changing distance *r*. The adjacent equipotential surfaces vary by the same amount of potential from one surface to the next and are closer together when nearer $+Q$ and farther apart when farther away from $+Q$. The density of the $\vec{E}$ field lines is greatest near the $+Q$ charge and less dense when far from it, as shown in Figures 18.9b and c.

The elastic sheet analogy that we used in Section 18.1 to visualize an electric field can also help us to visualize *V* fields and equipotential surfaces.

The disturbance of the elastic sheet due to the billiard ball can also be described by assigning to each point on the sheet a number that tells us the energy of the billiard ball–ball bearing system. In our case, the analog of the electric potential energy is gravitational potential energy. We choose this energy to be zero when the ball bearing is very far away from the billiard ball, where the sheet is undisturbed (horizontal). As the ball bearing approaches the billiard ball, the sheet is deformed and the energy of the system is no longer zero. If we divide this energy by the mass of the ball bearing, the value will depend only on the location, not on the mass, of the ball bearing. In this way, each point on the sheet can be assigned a number (a scalar). These numbers represent the analogous *V* field on the elastic sheet. In our case, points that have the same analogous *V* value form concentric circles around the billiard ball (see **Figure 18.11**). When the ball bearing is moved around one of these circles, the energy of the system remains the same regardless of where on the circle the ball bearing is. A different circle indicates a different value of analogous *V*. Thus these circles are analogous to equipotential surfaces in the electric field.

REVIEW QUESTION 18.3 Compare the work needed to move an object with charge *q* from point A to point B to the work needed to move the same object from point C to point D in the electric potential region depicted in Figure 18.10a. Explain.

18.4 Relating the $\vec{E}$ field and the *V* field

We know qualitatively that the *V* field varies most rapidly with position where the $\vec{E}$ field is strongest. We can also examine this idea quantitatively.

Deriving a relation between the $\vec{E}$ field and ΔV

Consider the uniform $\vec{E}$ field produced by an electrically charged infinitely large glass plate. We attach a small object with charge $+q$ to the end of a wooden stick and place the charged object and stick in the electric field produced by the plate (**Figure 18.12a**). The electric field exerts a force on the charged object in the positive *x*-direction $F_{\vec{E} \text{ on O} x} = +qE_x$. The stick exerts a force on the charged object with an *x*-component $F_{\text{S on O} x} = -F_{\text{S on O}}$ pointing toward the plate (Figure 18.12b) in the negative *x*-direction. If the charged object does not move or moves slowly with zero acceleration, the *x*-component form of Newton's second law applied to the charged object is $\sum F_x = F_{\vec{E} \text{ on O} x} + F_{\text{S on O} x} = +qE_x - F_{\text{S on O}} = 0$, or

$$F_{\text{S on O}} = qE_x$$

Now, let's do a work-energy analysis for a process in which the charged object (still attached to the stick) is moved slowly a small distance Δx farther from the plate (Figure 18.12c). For our system we choose the charged object and the electric field, but not the stick, which is part of the environment and does work on the system. The stick exerts a force on the charged object opposite its displacement and does negative work on the system:

$$W = F_{\text{S on O}} \Delta x \cos(180°) = -F_{\text{S on O}} \Delta x = -qE_x \Delta x$$

The only energy change is the system's electric potential energy because the positively charged object moves farther away from the positively charged plate. Applying the generalized work-energy principle, we get $U_{qi} + W = U_{qf}$, or

$$W = \Delta U_q = q\,\Delta V$$

Setting these two expressions for work equal to each other and canceling the common q, we get

$$\Delta V = -E_x\,\Delta x \qquad (18.9)$$

Equivalently, the component of the $\vec{E}$ field along the line connecting two points on the x-axis is the negative change of the V field divided by the distance between those two points:

$$E_x = -\frac{\Delta V}{\Delta x} \qquad (18.10)$$

TIP Equation (18.10) suggests that another unit for the $\vec{E}$ field is V/m.

The magnitude of the $\vec{E}$ field component in a particular direction indicates how fast the V field (electric potential) changes in that direction (Figure 18.12d). The $\vec{E}$ field vector points in the direction in which the V field decreases fastest with position, hence the minus sign in Eq. (18.10). Similar equations apply for other directions if the situation is in two or three dimensions. Although we derived Eqs. (18.9) and (18.10) using the example of a uniform $\vec{E}$ field, the equations represent a general result that relates the component of the $\vec{E}$ field in the chosen direction to the rate of change of the V field in that direction. The relation between the $\vec{E}$ field and V field tells us two things: (1) in a region where the V field is constant, the $\vec{E}$ field is zero, and (2) if you have two points at different potentials, the closer those points are, the stronger the $\vec{E}$ field between them will be.

QUANTITATIVE EXERCISE 18.8   **Connecting electric potential to force and energy**

The figure below shows a region of space with an electric field, represented using vertical lines for equipotential surfaces with the corresponding values of the potential. (a) What are the forces exerted by the electric field on a particle with charge -20 nC placed at locations A and B? (b) The particle is released from rest at point A. What is its kinetic energy at point B? The mass of the particle is very small, so you can ignore any changes in gravitational potential energy.

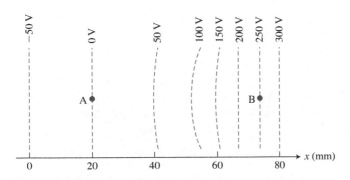

Represent mathematically (a) Because the problem situation is one-dimensional, we can use Eq. (18.10) to determine the component of the $\vec{E}$ field in the x-direction and then use this information to determine the forces exerted on the particle at two locations.

$$E_x = -\frac{\Delta V}{\Delta x}$$

To determine the forces, we use Eq. (18.2) in component form:

$$E_x = \frac{F}{q_{test}} \Rightarrow F = E_x q_{test}$$

In our case, this means $F_x = qE_x$.

(b) Using the work-energy principle, we can write $U_{qi} + K_i = U_{qf} + K_f$ for the system consisting of the charged particle and the electric field. Given that the initial kinetic energy is zero, the final kinetic energy is equal to the change in electric potential energy:

$$K_f = U_{qi} - U_{qf} = q(V_i - V_f)$$

Solve and evaluate (a) From the figure we see that in the region near point A, the V field changes by 50 V over the distance of 20 mm. From these data, we get

$$E_{Ax} = -\frac{\Delta V}{\Delta x} = -2500 \text{ V/m}$$

Near B, the V field changes by 50 V over the distance of $20/3$ mm; therefore, we get

$$E_{Bx} = -7500 \text{ V/m}$$

The minus signs indicate that the $\vec{E}$ field points in the negative x-direction.

(CONTINUED)

Next, we determine the forces exerted on the particle at two locations:

$$F_{Ax} = qE_{Ax} = (-20 \times 10^{-9}\,\text{C})(-2500\,\text{N/C}) = 5 \times 10^{-5}\,\text{N}$$
$$F_{Bx} = qE_{Bx} = (-20 \times 10^{-9}\,\text{C})(-7500\,\text{N/C}) = 15 \times 10^{-5}\,\text{N}$$

Both force components are positive; this means that the forces point in the positive x-direction. Note that when calculating the force it is more convenient to use units N/C than V/m for the $\vec{E}$ field.

(b) $K_f = q(V_i - V_f) = (-20 \times 10^{-9}\,\text{C})(0\,\text{V} - 250\,\text{V}) = 5 \times 10^{-6}\,\text{J}$.

The components of the forces exerted on the particle are positive: does this make sense? Since the V field is decreasing in the negative

x-direction, the $\vec{E}$ field points in this direction. Therefore, the force exerted on the negative charge at A and B must point in the positive x-direction. Thus our answer about forces makes sense. Given that the negatively charged particle accelerates in the direction of decreasing potential, the increase in kinetic energy also makes sense.

Try it yourself What is the maximum kinetic energy of an electron launched from point B toward A such that the electron does not pass the 50-V equipotential line?

Answer $K > 2.4 \times 10^{-17}\,\text{J}.$

REVIEW QUESTION 18.4 Imagine that you have an electric field in some large region of space. The electric potential at location P is 10 V and at another faraway location N is 3 V. Your friend says that the $\vec{E}$ field magnitude at point P is greater than that at N. Do you agree with your friend? Justify your opinion.

18.5 Conductors in electric fields

We can use the concepts of the $\vec{E}$ field and the V field to understand more precisely the behavior of electrically conducting materials and applications such as shielding and grounding.

Electric field of a charged conductor

Imagine a metal sphere of radius R that we touch with a negatively charged plastic rod (**Figure 18.13a**). Some of the excess electrons on the rod move to the metal sphere (Figure 18.13b). The electrons transferred to the sphere create an electric field in the metal that causes free electrons that are already in the metal to accelerate away from that spot. As a result of repulsion, the free electrons in the conductor quickly redistribute until equilibrium is reached, at which point the $\vec{E}$ field inside the conductor becomes zero (Figure 18.13c). If it were not zero, the electrons would continue to move. Given that the electrons stop moving, the component of the $\vec{E}$ field at the sphere's surface that is parallel to the surface is zero as well. Hence, outside the sphere, the $\vec{E}$ field vector is

FIGURE 18.13 Charging a metal sphere.

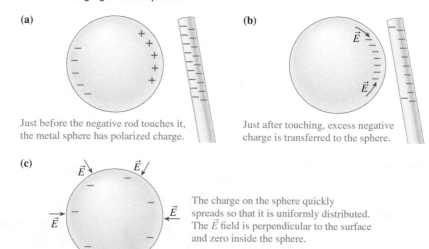

(a) Just before the negative rod touches it, the metal sphere has polarized charge.

(b) Just after touching, excess negative charge is transferred to the sphere.

(c) The charge on the sphere quickly spreads so that it is uniformly distributed. The $\vec{E}$ field is perpendicular to the surface and zero inside the sphere.

perpendicular to the surface. Since the $\vec{E}$ field is zero within and parallel to the surface of the sphere, this also means that the V field has the same value at all points on the sphere. If there were an electric field within or parallel to the surface of the sphere, it would exert a force on free electrons inside that would cause them to move.

Outside the charged conductor the field is not zero. **Figures 18.14a** and **b** illustrate the $\vec{E}$ field lines and equipotential surfaces for a point-like charged object with charge $+q$ and for a metal sphere of radius R with the same charge $+q$ on its surface. We won't do the calculation here since it requires calculus, but the magnitude of the $\vec{E}$ field and the value of the V field outside the sphere are the same as that produced by a point-like charged object:

$$E = k_C \frac{|q|}{r^2} \text{ and } V = k_C \frac{q}{r}$$

where r is the distance from the center of the sphere $(r \geq R)$. **Figures 18.15a** and **b** show graphs of $\vec{E}$ field and V field magnitudes as functions of the distance from the center of the sphere with charge q. Inside the sphere, $E = 0$ and $V = k_C (q/R)$, the same value as on the surface.

How do we know the potential is constant throughout the sphere and on the surface? When $\vec{E} = 0$ inside the sphere, no work is needed to bring a test charge to the surface. Therefore, the potential inside the sphere should be equal to the potential on the surface. What we have found here for one case is true in general:

1. When any conductor is charged, all electric charges reside on the surface.
2. The $\vec{E}$ field inside the conductor is zero.
3. The interior and the surface of the conductor are at the same potential; the surface is always an equipotential surface.

The above summary helps explain an important application—*grounding*.

Grounding

Grounding discharges an object made of conducting material by connecting it to Earth. Here we will explain how grounding works by applying our new understanding of conductors and electric potential.

Imagine that we have two conducting metal spheres of radii R_1 and R_2 with charges q_1 and q_2, respectively (**Figure 18.16a**). If we connect them with a long metal wire, electrons will move between and within the spheres until the V field on the surfaces

FIGURE 18.16 (a) When you connect the different size spheres by a metal wire, they are at the same electric potential. (b) The charge/area and the $\vec{E}$ field are greater on the small sphere.

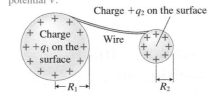

(a) The conducting spheres are at the same electric potential V.

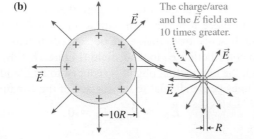

FIGURE 18.14 The $\vec{E}$ field lines and equipotential surfaces for (a) a point charge and (b) a charged metal sphere.

(a)
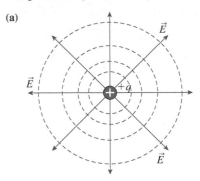

(b)
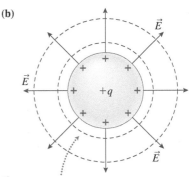

$\vec{E}$ field lines and equipotential surfaces outside the metal sphere are the same as for the point-like charge.

FIGURE 18.15 Graphs showing how E and V change with the distance r from the center of a charged metal sphere of radius R.

(a)
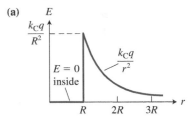

(b)
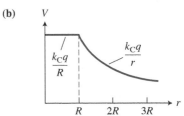

of and within both spheres achieves the same value. (As long as there is potential difference between the spheres, there will be a net force exerted on an electron and it will continue to accelerate; the process stops when the electrons are in equilibrium.) Because

$$V_1 = \frac{k_C q_1}{R_1} \text{ and } V_2 = \frac{k_C q_2}{R_2}$$

and $V_1 = V_2$, the ratio of the charges on the surface of each sphere is

$$\frac{q_1}{q_2} = \frac{R_1}{R_2}$$

This result tells us that when two spheres are connected, the bigger sphere has a greater electric charge than the smaller sphere. For example, suppose that sphere 1 has a radius that is 10 times greater than sphere 2 ($R_1 = 10R_2$). Large sphere 1 is originally uncharged and small sphere 2 has charge q. If we connect the two spheres with a wire, we find that $10/11$ of small sphere 2's original charge q will be transferred to large sphere 1 $[Q_1 = (10/11)q]$ and $1/11$ will remain on small sphere 2 $[Q_2 = (1/11)q]$. This result follows from two conditions:

$$\frac{Q_1}{Q_2} = \frac{R_1}{R_2} \text{ and } Q_1 + Q_2 = q$$

When we ground a conducting object, we are effectively connecting it with a wire to a sphere of radius $R = 6400$ km $= 6{,}400{,}000$ m, the radius of Earth. The object, the wire, and Earth become a single large equipotential surface, and the electric charge on the conducting object approaches zero. Therefore, you can safely touch the grounded conducting object (which might be a toaster or a high-definition television) while standing on the ground without experiencing an electric shock.

Another important consequence of connecting different size conductors is that the $\vec{E}$ field lines will be denser on the surface of the small sphere and the $\vec{E}$ field will be stronger. In the above example with spheres of radii $R_1 = 10R_2$, the large sphere had a greater charge $Q_1 = (10/11)q$ and the small sphere had charge $Q_2 = (1/11)q$. The *charge per unit surface area* (or surface charge density) for the large sphere is

$$\frac{Q_1}{A_1} = \frac{10q}{11 \cdot 4\pi(10R_2)^2} = \frac{q}{11 \cdot 4\pi \cdot 10R_2^2}$$

while for the smaller sphere, the charge per unit surface area is

$$\frac{Q_2}{A_2} = \frac{q}{11 \cdot 4\pi R_2^2}$$

The surface charge density is 10 times greater on the small sphere (Figure 18.16b). The $\vec{E}$ field lines are 10 times denser on the surface of the small sphere, and the electric field is 10 times stronger. As a result, on a pointed surface, such as a lightning rod, the electric field is strongest at the tip.

Uncharged conductor in an electric field—shielding

In our chapter-opening story we asked if you would be safe in a car during a lightning storm. Now we can answer that question. Imagine that we take a noncharged hollow conducting object and place it in a region with a uniform $\vec{E}$ field whose value is $\vec{E}_0$ (**Figure 18.17a**). The free electrons inside the object redistribute due to electric forces, producing their own contribution $\vec{E}_1$ to the $\vec{E}$ field (Figure 18.17b). This continues until the net $\vec{E}$ field within the conducting object is reduced to zero and the field outside the conductor has field lines perpendicular to the surface of the conductor (Figure 18.17c):

$$\vec{E}_{net} = \vec{E}_0 + \vec{E}_1 = 0$$

FIGURE 18.17 A hollow conducting object in an external $\vec{E}_0$ field.

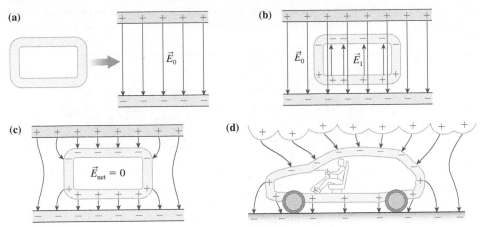

(a)

(b) $\vec{E}_0$ $\vec{E}_1$

(c) $\vec{E}_{net} = 0$

(d)

The $\vec{E}$ field produced by the environment $\vec{E}_0$ is cancelled by the contribution produced by the conductor. The net $\vec{E}$ field outside the conducting object is now a superposition of two $\vec{E}$ fields. Because the conductor effectively "destroys" the external $\vec{E}$ field inside itself by creating the oppositely directed $\vec{E}_1$ field, the interior is protected from the external field. This effect is called **shielding**.

The hollow conducting object we just described might well be a car. Note that because the $\vec{E}$ field inside the conductor is zero, people inside the car during a lightning storm are completely unaffected by the outside electric processes as long as they do not touch the surface of the car (Figure 18.17d).

REVIEW QUESTION 18.5 In this section you read that the $\vec{E}$ field inside the conductor is always zero. Why is this true?

18.6 Dielectric materials in an electric field

In this section we will investigate the behavior of dielectric (insulating) materials. Recall from Chapter 17 that dielectric materials are attracted to both positively and negatively charged objects. How does our concept of the electric field contribute to the explanation of this phenomenon?

All atoms are composed of positively charged nuclei surrounded by negatively charged electrons (**Figure 18.18a**). If an atom in a dielectric material resides in a region with an external electric field $\vec{E}_0$, the field exerts a force on the atom's positive nucleus in the direction of the field $\vec{E}_0$ and a force on the atom's negatively charged electrons in the opposite direction. The nucleus and the electrons are displaced slightly in opposite directions until the force that the field exerts on each of them is balanced by the force they exert on each other due to their attraction (Figure 18.18b). Now the centers of the positive and negative charges are spatially separated. Such a system is familiar to us as an electric dipole (Figure 18.18c). We can represent an electric dipole using a vector $\vec{p}$ that points from the negative charge center of the system to its positive charge center.

However, some molecules, such as water, are natural electric dipoles even when the external $\vec{E}$ field is zero (**Figure 18.19a**, on the next page). Without an external electric

FIGURE 18.18 The electric field polarizes an atom.

(a)

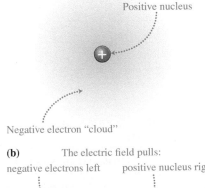

Positive nucleus

Negative electron "cloud"

(b)   The electric field pulls:
negative electrons left   positive nucleus right

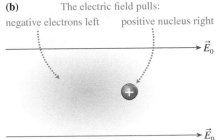

$\vec{E}_0$

$\vec{E}_0$

The charge displacement is greatly exaggerated.

(c) The atom becomes an electric dipole.

$\vec{p}$

FIGURE 18.19 Polar water molecules in an external electric field.

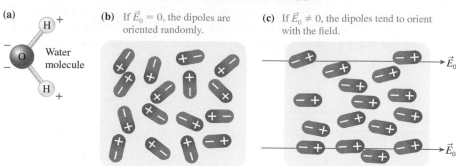

(a)   Water molecule

(b) If $\vec{E}_0 = 0$, the dipoles are oriented randomly.

(c) If $\vec{E}_0 \neq 0$, the dipoles tend to orient with the field.

FIGURE 18.20 The field $\vec{E}_0$ causes polarization and an internal field $\vec{E}_1$.

(a) The field $\vec{E}_0$ causes the polarization of molecules.

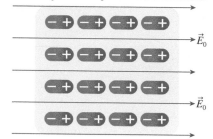

(b) The net effect of the polarization is a net negative charge on the left, a positive charge on the right, and an internal field $\vec{E}_1$.

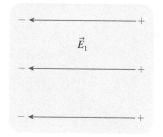

TABLE 18.2 Dielectric constants for different materials[1]

Material	Dielectric constant (κ)
Vacuum	1.0000
Dry air	1.0006
Wax	2.25
Glass	4–7
Paper	3–6
Axon membrane	8
Body tissue	8
Ethanol	26
Water	81

[1] At 20° C.

field, water molecules bump into each other and the dipoles are oriented randomly (Figure 18.19b). When the external $\vec{E}$ field is nonzero, the electric field exerts forces in opposite directions on the ends of the molecules, producing a torque that tends to rotate the molecules so that their electric dipoles align with the $\vec{E}$ field (Figure 18.19c).

When a dielectric material is placed in a nonzero $\vec{E}$ field, the atomic and molecular electric dipoles that result form a layer of positive electric charge on one surface of the material and a negative layer on the other surface (**Figure 18.20a**). These layers produce an additional contribution to the $\vec{E}$ field, an $\vec{E}_1$ field inside the dielectric material that points in the opposite direction to the external field $\vec{E}_0$ (Figure 18.20b). The net electric field inside is $\vec{E} = \vec{E}_0 + \vec{E}_1$. Since $\vec{E}_1$ points opposite $\vec{E}_0$, $E < E_0$. Thus, dielectric materials reduce the magnitude of the $\vec{E}$ field inside the material, similar to what happens in conductors. However, unlike conductors, the $\vec{E}$ field within does not decrease to zero. Thus, a dielectric material cannot completely shield its interior from an external electric field.

The ability of a dielectric material to decrease the $\vec{E}$ field varies from material to material. Physicists use a physical quantity to characterize this ability—the **dielectric constant** κ (Greek letter kappa). The dielectric constant of a material determines how much the material reduces the external $\vec{E}$ field (the field produced by the environment) within itself. The larger the value of κ, the more the $\vec{E}_0$ external field is reduced in the material.

The dielectric constant of a material is defined as follows:

$$\kappa = \frac{E_0}{E} \tag{18.11}$$

where E_0 is the magnitude of the electric field produced by the environment and $E = E_0 - E_1$ is the magnitude of the electric field within the dielectric material. You can see from Eq. (18.11) that the dielectric constant is dimensionless. **Table 18.2** lists sample dielectric constants for various materials.

Using Eq. (18.11), we can calculate the magnitude of the electric force that one charged object (object 1) exerts on another charged object (object 2) when both are inside a material with a dielectric constant κ:

$$F_{1 \text{ on } 2 \text{ in dielectric}} = q_2 E_{1 \text{ in dielectric}} = q_2 \frac{E_{1 \text{ in vacuum}}}{\kappa} = \frac{F_{1 \text{ on } 2 \text{ in vacuum}}}{\kappa}$$

The force that object 1 exerts on object 2 is reduced by κ compared with the force it would exert in a vacuum. Inside the dielectric material, Coulomb's law is now written as

$$F_{1 \text{ on } 2} = k_C \frac{|q_1||q_2|}{\kappa r^2} \tag{18.12}$$

Note in **Figure 18.21** that the positive ends of the dipole water molecules group around a negative ion and in effect reduce its net charge. The same thing happens to a positive ion in water. There is a reduction in the force that charged objects exert on each other when in that dielectric material. We can interpret Eq. (18.12) in this way:

$$F_{1 \text{ on } 2} = k_C \frac{\left(\dfrac{q_1 q_2}{\kappa}\right)}{r^2}$$

The charge product $q_1 q_2$ in Coulomb's law has been reduced by a factor of $1/\kappa$.

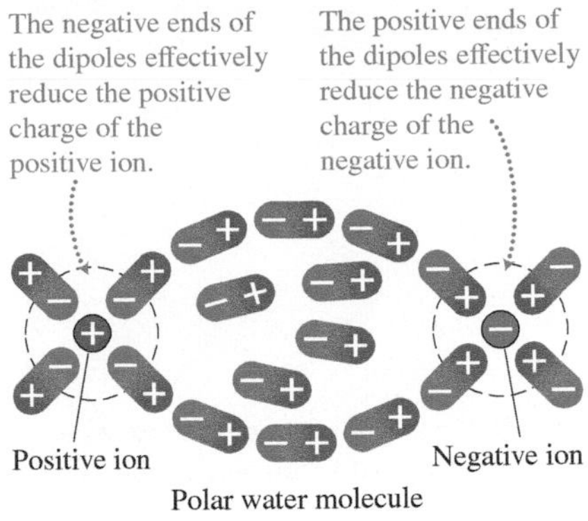

Salt dissolves in blood but not in air

The presence of dielectric materials that have polar molecules (like water) has dramatic effects on processes occurring inside the human body. One important function of water is to dissolve compounds, making their elements available for physiological use. For instance, the nervous system uses sodium ions in the transmission of information. Solid salt (NaCl) dissolves into sodium (Na^+) and chlorine (Cl^-) ions in the blood, which is made up primarily of water.

Table salt comes in the form of an *ionic crystal*. Because a chlorine atom is more attractive to electrons than a sodium atom, an electron is transferred from a sodium atom (leaving it with electric charge $+1.6 \times 10^{-19}$ C) to the chlorine atom (giving it electric charge -1.6×10^{-19} C). The electric force that the oppositely charged sodium and chlorine ions exert on each other holds the salt crystal together. The distance between the two ions is about the same as the characteristic size of a molecule (about 10^{-10} m). The magnitude of the electrical force that the ions exert on each other is then

$$F_{\text{Na on Cl}} = F_{\text{Cl on Na}} = k_C \frac{|q_{\text{Na}}||q_{\text{Cl}}|}{r^2}$$

$$= (9 \times 10^9 \text{ N} \cdot \text{m}^2/\text{C}^2)\frac{(1.6 \times 10^{-19} \text{ C})^2}{(10^{-10} \text{ m})^2} \approx 2 \times 10^{-8} \text{ N}$$

The electric potential energy of a sodium-chlorine ion pair is about

$$U_q = k_C \frac{q_{\text{Na}} q_{\text{Cl}}}{r}$$

$$= (9 \times 10^9 \text{ N} \cdot \text{m}^2/\text{C}^2)\frac{(+1.6 \times 10^{-19} \text{ C})(-1.6 \times 10^{-19} \text{ C})}{(10^{-10} \text{ m})} \approx -2 \times 10^{-18} \text{ J}$$

The energy is negative because of the opposite charges of the ions. To separate the ions, approximately $+2 \times 10^{-18}$ J must be added to the system.

Atoms, molecules, and ions also have positive kinetic energy due to their random thermal motion. A rough estimate of this energy is $K = (3/2)k_B T$, an expression we learned (in Chapter 12) for the kinetic energy of the particles in an ideal gas, where k_B is Boltzmann's constant (do not confuse it with the k_C used in Coulomb's law). The average positive kinetic energy at room temperature (300 K) is approximately

$$K = \frac{3}{2}k_B T = \frac{3}{2}(1.4 \times 10^{-23} \text{ J/K})(300 \text{ K}) \approx 6 \times 10^{-21} \text{ J}$$

This positive kinetic energy of the individual ions is much smaller than the negative electric potential energy holding them together. Thus, the total energy of the system is negative and the ions remain bound together when salt is in air.

When the salt is placed in water or blood, however, two changes occur. First, in water, there are many more collisions between molecules than there are in air. Although most of the collisions will not break an ion free from the crystal, some might. Remember, $K = (3/2)k_B T$ is the *average* kinetic energy of the molecules. A few molecules will have enough random kinetic energy to knock an ion free from the crystal during a collision.

FIGURE 18.21 The force that positive and negative ions exert on each other is reduced by the polar water molecules.

The negative ends of the dipoles effectively reduce the positive charge of the positive ion.

The positive ends of the dipoles effectively reduce the negative charge of the negative ion.

Positive ion

Negative ion

Polar water molecule

Second, any ions that do become separated from the crystal by collisions do not exert nearly as strong an attractive force on each other because of the dielectric effect of the blood. If we assume that blood is mostly water, the attractive force that the two ions exert on each other decreases by a factor of 81, the dielectric constant of water:

$$F_{\text{Na on Cl in water}} = k_C \frac{e^2}{\kappa r^2} = (9 \times 10^9 \, \text{N} \cdot \text{m}^2/\text{C}^2) \frac{(1.6 \times 10^{-19} \, \text{C})^2}{81(10^{-10} \, \text{m})^2} \approx 3 \times 10^{-10} \, \text{N}$$

The electric potential energy is also smaller by a factor of 81:

$$U_q = k_C \frac{-e^2}{\kappa r} = (9 \times 10^9 \, \text{N} \cdot \text{m}^2/\text{C}^2) \frac{-(1.6 \times 10^{-19} \, \text{C})^2}{81(10^{-10} \, \text{m})} \approx -3 \times 10^{-20} \, \text{J}$$

TIP You can think of a conductor as a medium with a dielectric constant of infinity.

This energy is almost the same as the average kinetic energy of the random motion of the water molecules. This means that the random kinetic energy of the water molecules is sufficient to keep the sodium and chlorine ions from recombining. This allows the nervous system to use the freed sodium ions to transmit information.

Dielectric breakdown

Imagine that you shuffle across a rug and then reach for a metal doorknob. Just before you touch the doorknob, a spark jumps between you and the knob. You estimate that the distance the spark jumps is about 1 cm. What happened?

Air (a dielectric under normal conditions) can turn into a conductor if free electrons in the air are accelerated to such speeds that they can knock an electron out of an air molecule on the next collision. The two electrons then move on and knock additional electrons out of other air molecules, creating a cascade called **dielectric breakdown**. When these electrons recombine with molecules that have lost an electron, light is produced—a spark. Experiments indicate that dielectric breakdown occurs in dry air when the magnitude of the $\vec{E}$ field exceeds about $3 \times 10^6 \, \text{N/C} = 3 \times 10^6 \, \text{V/m}$.

Dielectric breakdown explains why you see the spark when your hand gets relatively close to the knob, not long before. When walking on a rug, you acquire some electric charge (let's say it's negative). When you approach a metal doorknob, you polarize the metal, and an excess of positive charge appears on the side of the knob closest to you. However, the $\vec{E}$ field in the region between you and the knob is not strong enough to cause the dielectric breakdown. As you bring your hand closer to the knob, the magnitude of the field increases until it reaches the breakdown value—and the spark occurs! The value for the dielectric breakdown of air allows us to roughly estimate the potential difference between your body and the doorknob:

$$\Delta V = E \, \Delta x = (3 \times 10^6 \, \text{V/m})(0.01 \, \text{m}) = 3 \times 10^4 \, \text{V} = 30{,}000 \, \text{V}$$

This is a huge potential difference. Fortunately, the potential difference is not dangerous—it is the amount of electric charge that flows that is dangerous, and in this case it is very small.

REVIEW QUESTION 18.6 What are the differences between conducting and dielectric materials when they are placed in a region with a nonzero $\vec{E}$ field? Explain these differences.

18.7 Capacitors

We have learned about practical applications of conductors in electric fields, such as grounding and shielding. Another important application involving electric fields and conductors is storing energy in the form of electric potential energy. When positively and negatively charged objects are separated, the system's electric potential energy increases. How can this charge separation be maintained so that the electric potential energy can be stored for useful purposes? This is accomplished with a device known as a *capacitor*.

A **capacitor** consists of two conducting surfaces separated by an insulating material. Although a variety of configurations are possible, the simplest are parallel plate capacitors made of two metal plates separated by air, rubber, paper, or some other dielectric material (**Figure 18.22a**). The role of a capacitor is to store electric potential energy.

To charge a capacitor, one usually connects its plates to the terminals of a battery, a device that creates potential difference between two terminals. The positive battery terminal is connected by a conducting wire to one plate, for example, the left capacitor plate. Then an $\vec{E}$ field produced by the battery causes electrons to flow from the left capacitor plate toward the positive battery terminal (Figure 18.22b). Electrons likewise flow from the negative battery terminal to the right capacitor plate. This charge transfer continues until the capacitor plate, conducting wire, and battery terminal on each side are at the same potential (Figure 18.22c). Effectively, the battery removes electrons from one plate and deposits them on the other. The plate with a deficiency of electrons is positively charged with charge $+q$, and the plate with excess electrons is negatively charged with charge $-q$. The total charge of the capacitor is zero.

If we consider the capacitor plates to be large, flat conductors, the charges should distribute evenly on the plates. Each charged plate produces a uniform $\vec{E}$ field (**Figure 18.23a**). When the plates are close to each other, their two fields overlap between the plates and outside the plates. Outside the plates, the $\vec{E}$ field from the positively charged plate points opposite the $\vec{E}$ field from the negatively charged plate (Figure 18.23b) and they cancel each other. Between the plates, the $\vec{E}$ fields are in the same direction and add. Thus, the net $\vec{E}$ field is strong between the plates and zero outside (Figure 18.23c). The equipotential surfaces are parallel to the plates and equidistant as the $\vec{E}$ field is uniform, and thus the V field increases linearly with the distance from the $0\ V$ plate to the $+V$ plate.

According to Eq. (18.10), the magnitude of the $\vec{E}$ field between the plates relates to the potential difference ΔV from one plate to the other and the distance d separating them:

$$E = \left| -\frac{\Delta V}{\Delta x} \right| = \left| \frac{\Delta V}{d} \right|$$

Thus, if the potential difference across the plates doubles (for example, you connect the capacitor to a 12-V battery instead of a 6-V battery), the magnitude of the $\vec{E}$ field between the plates also doubles. Recall that $\vec{E}$ field lines are created by charges on the plates. Thus to double the $\vec{E}$ field, the charge on the plates has to double. We conclude that the magnitude of the charge q on the plates ($+q$ on one plate and $-q$ on the other) should be proportional to the potential difference V across the plates: $q \propto \Delta V$. Or, if we use a proportionality constant,

$$q = C|\Delta V| \qquad (18.13)$$

The proportionality constant C in this equation is called the **capacitance** of the capacitor. In the above, q is the magnitude of the charge on each plate. The unit of capacitance is 1 coulomb/volt = 1 farad (1 C/V = 1 F) in honor of Michael Faraday, whose experiments helped establish the atomic nature of electric charge.

Capacitance of a capacitor

What properties of capacitors determine their capacitance? It might seem that Eq. (18.13) answers this question. However, the capacitance of a capacitor does not change when the charge on the plates or the potential difference across them is changed. These two quantities are proportional to each other ($q \propto \Delta V$); to double the charge on the plates we need to double the potential difference across the same capacitor. But the ratio $C = q/|\Delta V|$ remains the same. This ratio is an operational definition of capacitance.

So then what does affect the capacitance of a particular capacitor? Imagine two capacitors whose plates have different surface areas and are connected to the same potential difference source. The capacitor with the larger surface area A plates should

FIGURE 18.22 Charging a capacitor.

(a) A capacitor

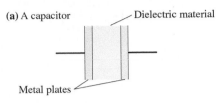

Dielectric material

Metal plates

(b) In the process of charging

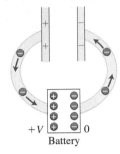

$+V$ ⬚ 0
Battery

(c) Capacitor is now fully charged.

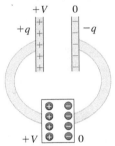

$+V$ 0
$+q$ $-q$

$+V$ ⬚ 0

FIGURE 18.23 Electric field inside a charged capacitor.

(a) The $\vec{E}$ fields produced on each plate

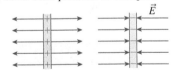

$\vec{E}$

(b) The $\vec{E}$ fields for both plates when together

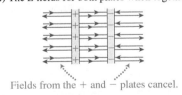

Fields from the $+$ and $-$ plates cancel.

(c) The net $\vec{E}$ field and equipotential surfaces

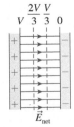

$V \quad \frac{2V}{3} \quad \frac{V}{3} \quad 0$

$\vec{E}_{net}$

FIGURE 18.24 Capacitance depends on A, d, and κ.

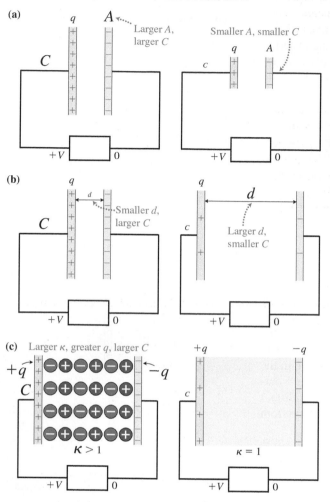

(a) Larger A, larger C; Smaller A, smaller C

(b) Smaller d, larger C; Larger d, smaller C

(c) Larger κ, greater q, larger C; $\kappa > 1$; $\kappa = 1$

be able to maintain more charge separation because there is more room for the charge to spread out (**Figure 18.24a**).

Now imagine two capacitors with the same surface area and the same potential difference across the plates but with different distances d between the plates (Figure 18.24b). Since $|\Delta V| = Ed$ (based on Equation 18.9) and ΔV is the same for both capacitors because they are connected to the same potential difference source, a larger distance d between the plates leads to a smaller magnitude $\vec{E}$ field between the plates. But since the magnitude of this $\vec{E}$ field is proportional to the amount of electric charge on the plates, a larger plate separation leads to a smaller magnitude of electric charge on the plates ($+q$ on one and $-q$ on the other). This means the capacitance of the capacitor decreases with increasing d.

In addition to surface area and distance between the plates, capacitance is affected by the dielectric constant κ of the material between the plates. The presence of a dielectric reduces the $\vec{E}$ field between the plates (Figure 18.24c). Thus the potential difference between the plates for the same amount of charge decreases. Therefore, the battery transfers additional charge to the capacitor until the potential difference across the capacitor becomes the same as across the battery again. The capacitance increases in proportion to the dielectric constant κ of the material.

Thus we conclude that the capacitance of a particular capacitor should increase if the surface area A of the plates increases, decrease if the distance d between them is increased, and increase if the dielectric constant κ of the material between them increases. A careful derivation (which we will not go through) provides the following result for a parallel plate capacitor:

$$C_{\text{Parallel plate capacitor}} = \frac{\kappa A}{4\pi k_C d} \qquad (18.14)$$

where $k_C = 9.0 \times 10^9 \, \text{N} \cdot \text{m}^2/\text{C}^2$. Let's check the units for capacitance:

$$\frac{(\text{m}^2)}{\left(\dfrac{\text{N} \cdot \text{m}^2}{\text{C}^2}\right)\text{m}} = \frac{\text{C}^2}{\text{N} \cdot \text{m}} = \frac{\text{C}^2}{\text{J}} = \frac{\text{C}}{\text{J}}\,\text{C} = \frac{\text{C}}{\text{V}} = \text{F}$$

The units check.

QUANTITATIVE EXERCISE 18.9 **Capacitance of a textbook**

Estimate the capacitance of your physics textbook, assuming that the front and back covers (area $A = 0.050 \, \text{m}^2$, separation $d = 0.040 \, \text{m}$) are made of a conducting material. The dielectric constant of paper is approximately 6.0. Second, determine what the potential difference must be across the covers in order for the textbook to have a charge separation of 10^{-6} C (one plate has charge $+10^{-6}$ C and the other has charge -10^{-6} C).

Represent mathematically The capacitance of a parallel plate capacitor is

$$C = \frac{\kappa A}{4\pi k_C d}$$

The potential difference ΔV needed to produce a charge separation of $q = 10^{-6}$ C is

$$\Delta V = \frac{q}{C}$$

Solve and evaluate

$$C = \frac{\kappa A}{4\pi k_C d} = \frac{(6.0)(0.050 \, \text{m}^2)}{4\pi(9.0 \times 10^9 \, \text{N} \cdot \text{m}^2/\text{C}^2)(0.040 \, \text{m})}$$

$$= 7.0 \times 10^{-11} \, \text{F} = 70 \, \text{pF}$$

(Here, p stands for "pico," the metric prefix for 10^{-12}.)

$$\Delta V = \frac{q}{C} = \frac{10^{-6} \, \text{C}}{7.0 \times 10^{-11} \, \text{F}} = 1.4 \times 10^4 \, \text{V} = 14 \, \text{kV}$$

This is a rather small capacitance, but reasonable given the large plate separation and the low dielectric value. The farad is a very large unit, and capacitors with picofarad or nanofarad capacitances are quite common.

Try it yourself Approximately how long must the book cover be to have a 1-F capacitance?

Answer

If we assume the same plate separation, plate width, and dielectric constant of the paper, then the covers should be about 3×10^9 m long, or about 3 million km long! One farad is a large capacitance.

The previous exercise shows that making a capacitor with a large capacitance requires a large surface area. It is possible to make capacitors of much larger capacitance by making multiple very thin layers of specially designed materials. These are called supercapacitors, and they can reach thousands of farads in capacitance.

Body cells as capacitors

Capacitors are used in all devices that require some instant action—such as camera flashes or computer keyboards. They are also present in the tuning circuits of radios, music amplifiers, etc. Biological capacitors are found inside our bodies. Cells, including nerve cells, have capacitor-like properties (see Example 18.10). The conducting "plates" are the fluids on either side of a moderately nonconducting cell membrane. In this membrane, chemical processes cause ions to be "pumped" across the membrane. As a result, the membrane's inner surface becomes slightly negatively charged while the outer surface becomes slightly positively charged.

EXAMPLE 18.10 **Capacitance of and charge on body cells**

Estimate (a) the capacitance C of a single cell and (b) the charge separation q of all of the membranes of the human body's 10^{13} cells. Assume that each cell has a surface area of $A = 1.8 \times 10^{-9} \, \text{m}^2$, a membrane thickness of $d = 8.0 \times 10^{-9}$ m, a $\Delta V = 0.070$-V potential difference across the membrane wall, and a membrane dielectric constant $\kappa = 8.0$. There is conducting fluid inside and outside the cell.

Sketch and translate A sketch of a cell is given below. Because the dielectric constant of the membrane is provided, it must be made of a dielectric material.

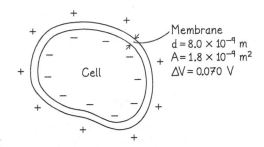

Membrane
$d = 8.0 \times 10^{-9}$ m
$A = 1.8 \times 10^{-9}$ m^2
$\Delta V = 0.070$ V

Simplify and diagram The thickness of the membrane $d = 8.0 \times 10^{-9}$ m is much smaller than the dimensions of the cell (roughly the square root of the surface area $\sqrt{A} = 4.2 \times 10^{-5}$ m). Up close, the cell membrane appears almost flat, just as Earth appears flat to people on its surface. Thus, we can use the expression for the capacitance of a parallel plate capacitor to make our estimate.

Represent mathematically The capacitance of one cell is

$$C = \frac{\kappa A}{4\pi k_C d}$$

The total charge separation q on the 10^{13} cells (biological capacitors) in the body is then

$$q_{total} = (10^{13})C|\Delta V|$$

Solve and evaluate The capacitance of one cell is

$$C = \frac{(8.0)(1.8 \times 10^{-9}\,\text{m}^2)}{4\pi(9.0 \times 10^9\,\text{N}\cdot\text{m}^2/\text{C}^2)(8.0 \times 10^{-9}\,\text{m})} = 1.6 \times 10^{-11}\,\text{F}$$

The total charge separated by the membranes of all of the cells is approximately

$$q_{total} = 10^{13}C|\Delta V| = 10^{13}(1.6 \times 10^{-11}\,\text{F})(0.070\,\text{V}) \approx 10\,\text{C}$$

Although these calculations are approximate, it is clear that the separation of electric charge (-10 C total on the inside walls of cell membranes and $+10$ C on the outside walls) is an important part of our metabolic processes. This 10-C electric charge separation is huge—about the same as the charge transferred during a lightning flash. Fortunately, our bodies' cells do not all discharge simultaneously in one surge.

Try it yourself Suppose you doubled the membrane thickness of all body cells. How would the potential difference across them have to change in order for the cells to maintain the same charge separation?

Answer Doubling the membrane thicknesses would reduce the capacitance by half. Thus, you would need to double the potential difference across the membranes to maintain the same charge separation.

Energy of a charged capacitor

A capacitor is essentially a system of positively and negatively charged objects that have been separated from each other so that the system as a whole is electrically neutral yet has electric potential energy. To determine electric potential energy in a charged capacitor, we start with an uncharged capacitor and then calculate the amount of work that must be done on the capacitor to move electrons from one plate to the other.

FIGURE 18.25 The energy stored while charging a capacitor.

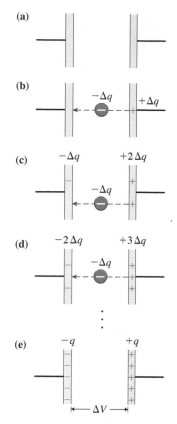

(a)

(b)

$-\Delta q$ $+\Delta q$

(c) $-\Delta q$ $+2\Delta q$

$-\Delta q$

(d) $-2\Delta q$ $+3\Delta q$

$-\Delta q$

⋮

(e) $-q$ $+q$

ΔV

FIGURE 18.26 A bar chart representing the work done to charge a capacitor.

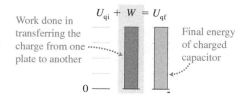

$U_{qi} + W = U_{qf}$

Work done in transferring the charge from one plate to another

Final energy of charged capacitor

0

This process is sketched in **Figure 18.25** in the form of a thought experiment. It's not possible for us to manually grab electrons and move them from one capacitor plate to the other. Usually, this work is performed by an external source such as a battery. However, in this thought experiment, we will move increments of charge $-\Delta q$ from one plate to the other. Note that the charge is negative because these charge increments are composed of electrons. After the first charge increment has been moved, the plate from which it was taken has a charge of $+\Delta q$ and the plate to which it is taken has a charge of $-\Delta q$.

Only a very small amount of work must be done to move this first $-\Delta q$ since the left plate is uncharged while moving it (Figure 18.25b). The next $-\Delta q$ is more difficult to move (Figure 18.25c) since the left plate now has a charge of $-\Delta q$ that repels the $-\Delta q$ we are trying to move there. Additionally, the $+2\Delta q$ charge on the right plate pulls back on it. The more charge increments we move from one plate to the other, the more difficult it becomes to move the next one. Eventually, with increasing effort, we will have transferred a total (negative) charge of $-q = (-\Delta q) + (-\Delta q) + \ldots$ to the left plate, leaving the right plate with a charge of $+q$. We can represent the whole process of charging the capacitor with a work-energy bar chart (**Figure 18.26**). The external object (us moving the charge increments) does work on the system (the capacitor), and the electric potential energy of the system increases.

The electric potential energy increase can be calculated using Eq. (18.8):

$$\Delta U_q = q \, \Delta V_{average}$$

where $\Delta V_{average}$ is the average potential difference from one plate to the other during the process of charging. Since the initial potential difference is zero and the final is ΔV, the average potential difference between the plates is

$$\Delta V_{average} = \frac{0 + \Delta V}{2} = \frac{\Delta V}{2}$$

Substituting this expression for $\Delta V_{average}$ into Eq. (18.8), we get

$$U_q = q \, \Delta V_{average} = q \, \frac{\Delta V}{2}$$

(We dropped the Δ, as $U_{qi} = 0$.) The above expression can be written in three different ways using Eq. (18.13), $q = C|\Delta V|$:

$$U_q = \tfrac{1}{2} q |\Delta V| = \tfrac{1}{2} C |\Delta V|^2 = \frac{q^2}{2C} \qquad (18.15)$$

Note that the process of charging a capacitor is similar to stretching a spring: at the beginning a smaller force is needed to stretch the spring by a certain amount compared to the much greater force needed when the spring is already stretched.

QUANTITATIVE EXERCISE 18.11   Energy needed to charge human body cells

In Example 18.10, we estimated that the total charge separated across all of the cell membranes in a human body was about 11 C. Recall that the potential difference across the cell membranes is 0.070 V. Estimate the work that must be done to separate the charges across the membranes of these approximately 10^{13} cells.

Represent mathematically We can use the first expression in Eq. (18.15) to answer this question:

$$U_q = \tfrac{1}{2} q \, \Delta V$$

where $q = 11$ C and $\Delta V = 0.070$ V.

Solve and evaluate

$$U_q = \tfrac{1}{2} q \, \Delta V = \tfrac{1}{2}(11 \text{ C})(0.070 \text{ V}) = 0.40 \text{ J}$$

These cells are continually being charged and discharged as part of the body's metabolic processes.

Try it yourself Estimate the energy needed to charge body cells each day assuming that each cell discharges once per second.

Answer

result is a rough estimate.
times per second, though some cells might discharge fairly infrequently. The
400 to up discharge might cells some However, much. not is This .bread
of piece a by provided energy the one-tenth about or ,kcal 8 = J 35,000

Energy density of electric field

We have found that the energy stored in a charged capacitor is

$$U_q = \tfrac{1}{2}C|\Delta V|^2$$

Where is this energy stored? The difference between a charged and an uncharged capacitor is the presence of the electric field in the region between the plates. We hypothesize that it is the electric field that possesses this energy.

What physical quantities affect the amount of this stored energy? It takes more energy to charge a larger capacitor (for example, a capacitor with larger plates) that has the same electric field between the plates. To have a measure of energy independent of the capacitor volume, we will use the physical quantity of **energy density**. Energy density quantifies the electric potential energy stored in the electric field per cubic meter of volume. The $\vec{E}$ field energy density u_E is

$$u_E = \frac{U_q}{V}$$

where U_q is the electric potential energy stored in the electric field in that region, and V is the volume of the region.

Let us rewrite the above equation for the electric field energy density in terms of the $\vec{E}$ field, using our knowledge of the electric energy of a parallel plate capacitor:

$$u_E = \frac{U_q}{V} = \frac{\tfrac{1}{2}C|\Delta V|^2}{V} = \frac{\tfrac{1}{2}\left(\dfrac{\kappa A}{4\pi k_C d}\right)(Ed)^2}{V}$$

We used Eqs. (18.14) and (18.9) to substitute for C and ΔV. A is the plate area, d is the plate separation, E is the magnitude of the $\vec{E}$ field between the plates, and κ is the dielectric constant of the material between the plates. Simplifying this, we have

$$u_E = \frac{\kappa A(Ed)^2}{8\pi k_C dV} = \frac{\kappa}{8\pi k_C} E^2 \frac{Ad}{V}$$

Since the volume of the region between the plates is $V = Ad$, this equation simplifies to

$$u_E = \frac{\kappa}{8\pi k_C} E^2 \tag{18.16}$$

The energy density depends only on the magnitude of the $\vec{E}$ field, the properties of the dielectric, and a few mathematical and physical constants. We'll use this idea in a later chapter on electromagnetic waves.

TIP Notice that here the letter V stands for volume, not for the electric potential. Sometimes similar or identical symbols are used for different physical quantities. When working with mathematical expressions, always ask yourself: "What does each of these symbols represent?"

TIP Although we derived the expression for energy density for a parallel plate capacitor, it can be applied for any electric field.

REVIEW QUESTION 18.7 A parallel plate capacitor has two oppositely charged plates, each producing its own electric field. Why is there no electric field outside the capacitor?

18.8 Electrocardiography

The human heart has four chambers. During a 1-s heartbeat, each chamber pumps in a specific sequence. For example, the lower right chamber (the right ventricle) pumps blood through the lungs, where carbon dioxide is exchanged for oxygen, and the lower left chamber (the left ventricle) pumps freshly oxygenated blood into the aorta for a new trip around the circulatory system to provide oxygen for the body cells.

An electric charge separation occurs when muscle cells in the heart contract during the pumping process. For example, when the left ventricle pumps blood into the aorta, many left ventricle muscle cells contract. As each muscle cell contracts, positive and negative charges separate (this process is represented schematically in **Figure 18.27**). The simultaneous contraction of the large number of cells in the heart muscle of the left ventricle results in a relatively large charge separation—an electric dipole. On the other

FIGURE 18.27 An electric dipole is produced by each heart muscle cell contraction.

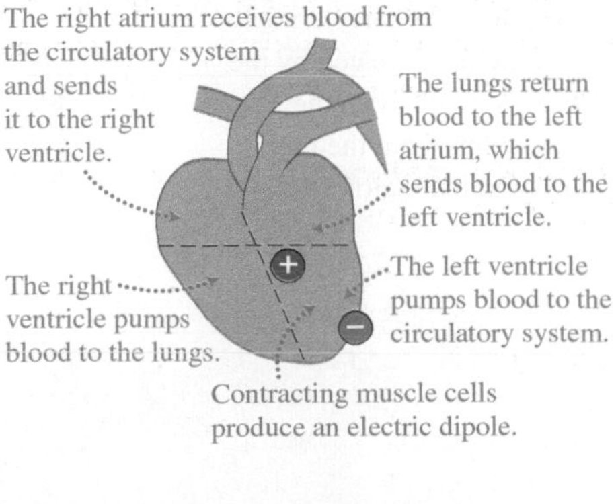

The right atrium receives blood from the circulatory system and sends it to the right ventricle.

The lungs return blood to the left atrium, which sends blood to the left ventricle.

The right ventricle pumps blood to the lungs.

The left ventricle pumps blood to the circulatory system.

Contracting muscle cells produce an electric dipole.

hand, the number of simultaneously contracting muscle cells is much smaller when the left atrium pumps blood into the left ventricle. Thus, during the 1-s cycle of a heartbeat, there is an electric dipole with continually changing magnitude and orientation that depends at each instant on the number and the orientation of the muscle cells that are contracting at that instant.

This electric dipole produces an electric field that extends outside the heart to the body tissue. A device called an electrocardiogram (ECG) detects potential difference between selected points in that field. Data from an ECG can identify abnormalities, such as enlarged left or right ventricles. How does an ECG work? Electrodes, in the form of pads, are placed on the skin. These pads transmit data to the ECG about the potential difference between the various pairs of pads. The ECG uses these data to reconstruct the changing electric dipole on the heart.

A physician looks at the varying potential difference between different pairs of electrode pads and from this can learn about the heart's changing electric dipole. This dipole in turn indicates the number of muscle cells contracting at different times. A person with left ventricular hypertrophy (an enlarged left ventricle) will have a larger than normal dipole and potential difference during contraction of the left ventricle. **Figure 18.28** shows a normal ECG.

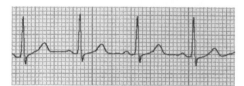

FIGURE 18.28 Electrocardiogram signals.

CONCEPTUAL EXERCISE 18.12   **Heart's electric dipole and ECG**

The figure below shows a simplified electric dipole charge distribution on a heart at one instant during a heartbeat and two ECG pads, I and II, on opposite shoulders of the person's body. (a) What do we expect to be the sign of the potential difference between the pads at that particular instant? (b) Determine the direction of the forces exerted by the electric field on a positive sodium ion (Na^+) and on a negative chlorine ion

(Cl^-) in the body tissue located exactly halfway between the ECG pads. (c) What do you expect to happen to the potential difference between the pads you found in part (a) if you assume that the ions move in the direction of these forces?

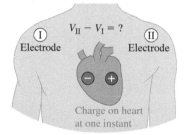

Sketch and translate The figure above gives a sketch of the situation.

Simplify and diagram We assume no other electric charges and fields are present.
(a) In the figure at right, we draw equipotential lines for the heart's dipole across the chest. Comparing the positions of the two electrodes with respect to the two outer lines of equal potential, we see that
$V_{II} > V_I \Rightarrow V_{II} - V_I > 0$.
We can also characterize the electric field created by the heart's dipole with the $\vec{E}_{heart}$ that points from the region of V_{II} to the region of V_I.

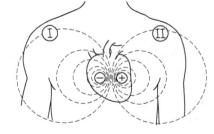

(b) In an electric field, the force exerted on a positive ion points in the direction of lower electric potential, and the force exerted on the negative ion points toward the region of higher electric potential. Adding the sodium and chlorine ions to our sketch (see figure below), and given that electrode II is at higher electric potential than electrode I, we see that the force on Na^+ points left and the force on Cl^- points right.

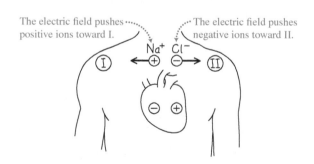

(c) The forces on the ions will cause them to move. As the ions redistribute, they create their own electric field, with the direction of $\vec{E}_{ions}$ opposite to the $\vec{E}_{heart}$; thus the potential V_I will increase and the potential V_{II} will decrease. Therefore, the measured difference $V_{II} - V_I$ will become smaller.

Try it yourself What will the potential difference between the electrodes be if the polarization of the heart is such that the dipole is vertical with the positive charge above the negative one?

Answer   The potential difference between the electrodes will be zero.

REVIEW QUESTION 18.8 Why do heart contractions produce electric dipoles?

Summary

$\vec{E}$ field The $\vec{E}$ field is a force-like quantity characterizing the electric field at a specific location. A source charge creates an electric field that exerts a force on a test charge placed in that field at that location.

The $\vec{E}$ field at the chosen location is equal to the electric force exerted by a source charge on a positively charged test object divided by the test object's charge (Section 18.1).

$$\vec{E} = \frac{\vec{F}_{Q \text{ on } q_{\text{test}}}}{q_{\text{test}}}$$ Eq. (18.2)

$$\vec{F}_{Q \text{ on } q_{\text{test}}} = q_{\text{test}}\vec{E}$$ Eq. (18.5)

Superposition principle The $\vec{E}$ field at a location is the *vector* sum of the contributions of n source charges to the $\vec{E}$ field at that location (Section 18.1).

$$\vec{E}_{\text{net}} = \vec{E}_1 + \vec{E}_2 + \vec{E}_3 + \cdots$$ Eq. (18.4)

V field (electric potential) is an energy-like quantity characterizing the electric field at a point. The V field (electric potential) at a position of interest is equal to the electric potential energy U_q of a small positively charged test charge q_{test} and the source charge divided by the test object's charge. Include the signs of charges when calculating U_q (Section 18.3).

$$V = \frac{U_{Q q_{\text{test}}}}{q_{\text{test}}} = \frac{k_C Q}{r}$$ Eq. (18.6a)

For one source charge:

$$V = \frac{k_C Q_1}{r_1}$$

$$U_{\text{Source } Q \text{ and } q} = qV_{\text{Due to source } Q}$$ Eq. (18.8)

Representing the $\vec{E}$ field and the V field $\vec{E}$ field lines and equipotential surfaces (where V is constant) represent the field. $\vec{E}$ field lines are perpendicular to equipotential surfaces (Section 18.4).

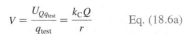

Equipotential surfaces

$\vec{E}$ field lines

$$\Delta V = -E_x \, \Delta x$$ Eq. (18.9)

or

$$E_x = -\frac{\Delta V}{\Delta x}$$ Eq. (18.10)

Dielectric constant κ A dielectric material partially cancels the $\vec{E}$ field within it produced by the environment. This reduces the magnitude of the force exerted by charged objects on each other within the material (Section 18.6).

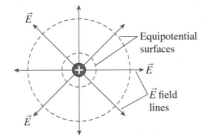

$\vec{E}_0$

$\vec{E} = \vec{E}_0 - \vec{E}_1$ $\vec{E}_1$

$$\kappa = \frac{E_0}{E}$$ Eq. (18.11)

$$F_{1 \text{ on } 2} = k_C \frac{|q_1||q_2|}{\kappa r^2}$$ Eq. (18.12)

Capacitors A capacitor is a device that (when charged) stores electric potential energy. It has two conducting surfaces separated by an insulating material. To charge a capacitor, one connects its plates to the points at different electric potential (like the terminals of a battery). The charge separation stores electric potential energy in the capacitor (Section 18.7).

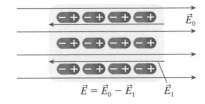
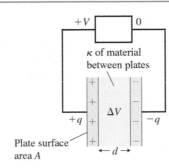

$+V$ 0

κ of material between plates

ΔV

$+q$ $-q$

Plate surface area A

$\leftarrow d \rightarrow$

$$q = C|\Delta V|$$ Eq. (18.13)

For parallel plate capacitors:

$$C = \frac{\kappa A}{4\pi k_C d}$$ Eq. (18.14)

$$U_q = \tfrac{1}{2}q|\Delta V|$$

$$= \tfrac{1}{2}C|\Delta V|^2 = \frac{q^2}{2C}$$ Eq. (18.15)

Questions

Multiple Choice Questions

1. What does the $\vec{E}$ field at point A, which is a distance d from the source charge, depend on? (More than one statement may be correct.)
   - (a) the magnitude of the test charge
   - (b) the sign of the test charge
   - (c) the magnitude and the sign of the source charge
   - (d) the magnitude and sign of the test charge

2. Why can you shield an object from an external electric field using a conductor?
   - (a) There are freely moving electric charges inside conductors.
   - (b) There are positive and negative charges inside conductors.
   - (c) Both a and b are essential.

3. If you place a block made of a conducting material in an external uniform electric field $\vec{E}_0$ so that two opposite sides of the block are perpendicular to $\vec{E}_0$, what will be the magnitude and direction of the net $\vec{E}$ field inside the block?
   - (a) magnitude: E_0; direction: same as $\vec{E}_0$
   - (b) magnitude: E_0; direction: opposite to $\vec{E}_0$
   - (c) magnitude: less than E_0 but more than zero; direction: same as $\vec{E}_0$
   - (d) magnitude: more than E_0; direction: same as $\vec{E}_0$
   - (e) magnitude: zero

4. If you place a block made of a dielectric material in an external uniform electric field $\vec{E}_0$ so that two opposite sides of the block are perpendicular to $\vec{E}_0$, what will be the magnitude and direction of the net $\vec{E}$ field inside the material?
   - (a) magnitude: E_0; direction: same as $\vec{E}_0$
   - (b) magnitude: E_0; direction: opposite to $\vec{E}_0$
   - (c) magnitude: less than E_0 but more than zero; direction: same as $\vec{E}_0$
   - (d) magnitude: less than E_0 but more than zero; direction: opposite to $\vec{E}_0$
   - (e) magnitude: zero

5. Two identical positive charges are located at a distance d from each other. Where are both the $\vec{E}$ field and the electric potential zero?
   - (a) exactly between the charges
   - (b) at a distance d from both charges (all three charges making an equilateral triangle)
   - (c) Both a and b are correct.
   - (d) None of these choices is correct.

6. An electric dipole is placed between the oppositely charged plates as shown in Figure Q18.6. At which of the marked points could the net $\vec{E}$ field be zero? Choose all possible answers.

FIGURE Q18.6

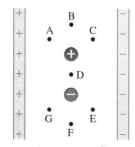

7. A positive charge is fixed at some distance d from the origin of the z-axis, as shown in Figure Q18.7. Which graph shows how E_z, the z-component of $\vec{E}$ field, depends on the position along the z-axis?

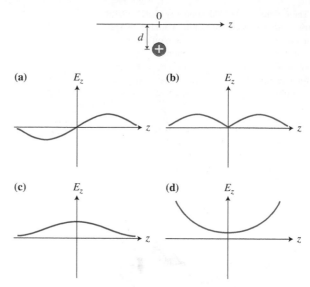

8. Figure Q18.8 shows $\vec{E}$ field lines in a region of space. Select two correct statements about the $\vec{E}$ field and V field at the points P, R, and S.
   - (a) $E_P = E_R = E_S$
   - (b) $E_P > E_S > E_R$
   - (c) $E_R > E_S > E_P$
   - (d) $V_P > V_R > V_S$
   - (e) $V_P > V_S > V_R$
   - (f) $V_R > V_S > V_P$

FIGURE Q18.8

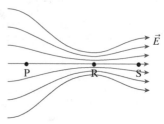

Conceptual Questions

9. How do we use the model of the electric field to explain the interaction between charges?

10. Describe a procedure to determine the $\vec{E}$ field at some point.

11. What does it mean if the $\vec{E}$ field at a certain point is 5 N/C and points north?

12. A very small positive charge is placed at one point in space. There is no electric force exerted on it. (a) What is the value of the $\vec{E}$ field at that point? (b) Does this mean there are no electric charges nearby? Explain. (c) Suggest one charge distribution (two or more charges) that would produce a zero $\vec{E}$ field at a point.

13. How do we create an $\vec{E}$ field with parallel lines and uniform density (equally spaced)?

14. Draw a sketch of the $\vec{E}$ field lines caused by (a) two positive charges of the same magnitude, (b) two negative charges of the same magnitude, and (c) two positive charges of different magnitudes.

15. Draw a sketch of the $\vec{E}$ field lines caused by (a) a positive and a negative charge of the same magnitude, (b) a positive and a negative charge of different magnitudes, and (c) a positive charge between two negative charges.

16. Jim thinks that $\vec{E}$ field lines are the paths that test charges would follow if placed in the $\vec{E}$ field. Do you agree or disagree with his statement? If you disagree, give a counter example.

17. Can $\vec{E}$ field lines cross? Explain why or why not.

18. An electron moving horizontally from left to right across the page enters a uniform $\vec{E}$ field that points toward the top of the page. Draw vectors indicating the direction of the electric force exerted by the field on the electron and its acceleration. Draw a sketch indicating the path of the electron in this field. What motion in the gravitational field is similar to the motion of the electron?

19. (a) What does it mean if the electric potential at a certain point in space is 10 V? (b) What does it mean if the potential difference between two points is 10 V?

20. Explain how grounding works.

21. Explain how shielding works.

22. Explain the difference between the microscopic structures of polar and of nonpolar dielectric materials. Give an example of each.

23. Explain why, for charged objects submerged in a dielectric material, the electric force they exert on each other needs to be divided by the dielectric constant.

24. What does it mean if the dielectric constant κ of water is 81?

25. What is the dielectric constant of a metal?

26. Describe the relation between the quantities $\vec{E}$ field and V field.

27. If the V field in a region is constant, what is the $\vec{E}$ field in this region? If the $\vec{E}$ field is constant in the region, what does the V field look like? Explain your answers.

28. Why are uncharged pieces of a dielectric material attracted to charged objects?

29. Draw equipotential surfaces and label them in order of decreasing potential for (a) one positive charge, (b) one negative charge, (c) two identical positive point charges at a distance d from each other, and (d) a negatively charged infinitely large metal plate.

30. Show a charge arrangement and a point in space where the potential produced by the charges is zero but the $\vec{E}$ field is not zero. Then repeat for the case where the potential produced by the charges is not zero but the $\vec{E}$ field is zero.

31. Explain what happens when you place a conductor in an external electric field. How can you test your explanation?

32. (a) Explain what happens when you place a dielectric material in an external electric field. How can you test your explanation?

33. Explain why the excess charge on an electrical conducting material is located on its surface.

34. Draw a microscopic representation of the charge distribution in two identical conducting plates that are touching each other and placed in an external uniform electric field as shown in Figure Q18.34. What happens if you move the plates apart (while keeping them in the external field) and then move them out of the field? How can you test your answer? Repeat the steps described above for two dielectric plates.

FIGURE Q18.34

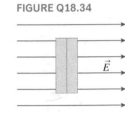

Problems

Below, **BIO** indicates a problem with a biological or medical focus. Problems labeled **EST** ask you to estimate the answer to a quantitative problem rather than derive a specific answer. Asterisks indicate the level of difficulty of the problem. Problems with no * are considered to be the least difficult. A single * marks moderately difficult problems. Two ** indicate more difficult problems.

In some of these problems you may need to know that the mass of an electron is 9.11×10^{-31} kg, the mass of a proton is 1.67×10^{-27} kg, the electric charge of the electron is -1.6×10^{-19} C, and the charge of a proton is 1.6×10^{-19} C.

18.1 A model of the mechanism for electrostatic interactions

1. * (a) Construct a graph of the magnitude of the $\vec{E}$ field-versus-position for the $\vec{E}$ field created by a point-like object with charge $+Q$, representing the $\vec{E}$ field magnitude as a function of the distance from the object. (b) Using the same set of axes, draw a graph for the field produced by an object of charge $+2Q$. (c) Using the same set of axes, draw a graph for the field produced by an object of charge $-2Q$.

2. * A uranium nucleus has 92 protons. (a) Determine the magnitude of the $\vec{E}$ field at a distance of 0.58×10^{-12} m from the nucleus (about the radius of the innermost electron orbit around the nucleus). (b) What is the magnitude of the force exerted on an electron by this $\vec{E}$ field? (c) What assumptions did you make? If the assumptions are not valid, will the magnitude in part (b) be an overestimate or underestimate of the actual value?

3. The electron and the proton in a hydrogen atom are about 10^{-10} m from each other. What quantities related to the $\vec{E}$ field can you determine using this information?

4. * Use the superposition principle to draw $\vec{E}$ field lines for the two objects whose charges are given. Consider all objects to be point-like and choose the distance you want between them: (a) $+q$ and $+q$; (b) $+q$ and $+3q$; (c) $+q$ and $-q$.

5. * Use the superposition principle to draw $\vec{E}$ field lines for the following objects whose charges are given. Consider all objects to be point-like and choose the distance you want between them: (a) $+q$ and $-3q$; (b) $+q$, $+q$, and $-3q$.

6. * $\vec{E}$ field lines for a field created by an arrangement of charged objects are shown in Figure P18.6. (a) Where are these objects located, and what are the signs of their electric charge? (b) What else can you determine using the information? Give two examples.

FIGURE P18.6

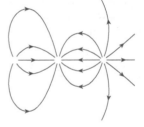

18.2 Skills for analyzing processes involving $\vec{E}$ fields

7. * Two objects with charges $+4.0 \times 10^{-9}$ C and $+3.0 \times 10^{-9}$ C are 50 cm from each other. Find a location where the $\vec{E}$ field due to these two charged objects is zero.

8. * A $+4.0 \times 10^{-9}$-C charged object is 6.0 cm along a horizontal line toward the right of a -3.0×10^{-9}-C charged object. Determine the $\vec{E}$ field at a point 4.0 cm to the left of the negative charge.

9. ** A $+4.0 \times 10^{-9}$-C charged object is 4.0 cm along a horizontal line toward the right of a -3.0×10^{-9}-C charged object. Determine the $\vec{E}$ field at a point 3.0 cm directly above the negative charge.

10. ** A distance d separates two objects, each with charge $+q$. Determine an expression for the $\vec{E}$ field at a point that is a distance d from both charges.

11. * A point-like charged object with a charge $+q$ is placed in an external horizontal uniform electric field $\vec{E}_0$ that points to the right. Determine an expression for the net $\vec{E}$ field at a distance d from the charge (a) to the right of it and along a horizontal line going through the charge $+q$ and (b) to the left of it along that same line.

12. * A 3.0-g aluminum foil ball with a charge of $+4.0 \times 10^{-9}$ C is suspended on a string in a uniform horizontal $\vec{E}$ field. The string makes an angle of 30° with the vertical. What information about the $\vec{E}$ field can you determine for this situation?

13. ** (a) If the string in the previous problem is cut, how long will it take the ball to fall to the floor 1.5 m below the level of the ball? (b) How far will the ball travel in the horizontal direction while falling? Indicate all the assumptions that you made.

14. * **EST** **Using Earth's $\vec{E}$ field for flight** Earth has an electric charge on its surface that produces a 150-N/C $\vec{E}$ field, which points down toward the center of Earth. Estimate the electric charge that a person would need so that the electric force that the field exerts on the person will support the person in the air above Earth. Indicate all of your assumptions. Is this a reasonable idea? Explain.

15. * An electron moving with a speed v_0 enters a region where an $\vec{E}$ field points in the same direction as the electron's velocity. What will happen to the electron in this field? Answer the question qualitatively and quantitatively (using symbols). Ignore the force that Earth exerts on the electron.

16. ** A 1.0-g aluminum foil ball with a charge of 1.0×10^{-9} C hangs freely from a 1.0-m-long thread. What happens to the ball when a horizontal 5000-N/C $\vec{E}$ field is turned on? Answer the question assuming the stabilized situation.

17. A 0.50-g oil droplet with charge $+5.0 \times 10^{-9}$ C is in a vertical $\vec{E}$ field. In what direction should the $\vec{E}$ field point and what magnitude should it have so that the droplet moves at constant speed? Ignore the forces that air exerts on the droplet.

18. * An electron is ejected into a horizontal uniform $\vec{E}$ field at a parallel horizontal velocity of 1000 m/s. Describe everything you can about the motion of the electron. Ignore the force that Earth exerts on the electron.

19. * **Equation Jeopardy 1** The equations below describe one or more physical processes. Solve the equations for the unknowns and write a problem statement for which the equations are a satisfactory solution.

$$-(0.001 \text{ kg})(9.8 \text{ N/kg}) + T \sin 88° = 0$$
$$(-5.0 \times 10^{-8} \text{ C})E_x + T \cos 88° = 0$$

20. * **Equation Jeopardy 2** The equations below describe one or more physical processes. Solve the equations for the unknowns and write a problem statement for which the equations are a satisfactory solution.

$$(+1.0 \times 10^{-8} \text{ C})(-4.0 \times 10^4 \text{ N/C}) = (0.20 \text{ kg})a_x$$
$$0 = (8.0 \text{ m/s}) + a_x t$$

18.3 and 18.4 The V field and Relating the $\vec{E}$ field and the V field

21. During a lightning flash, -15 C of charge moves through a potential difference of 8.0×10^7 V. Determine the change in electric potential energy of the field-charge system.

22. * (a) Construct a graph of the V field created by a point-like object with charge $+Q$, representing the V field as a function of the distance from the object. (b) Using the same set of axes, draw a graph for an object with charge $+2Q$ and for an object with charge $-2Q$.

23. * A horizontal distance d separates two objects each with charge $+q$. Determine the value of the electric potential at a point that is located at a distance d from each charge.

24. * Two objects with charges $-q$ and $+q$ are separated by a distance d. Determine an expression for the V field at a point that is located at a distance d from each charge.

25. * Four objects with the same charge $-q$ are placed at the corners of a square of side d. Determine the values of the $\vec{E}$ field and V field in the center of the square.

26. **Spark jumps to nose** An electric spark jumps from a person's finger to your nose. While passing through the air, the spark travels across a potential difference of 2.0×10^4 V and releases 3.0×10^{-7} J of electric potential energy. What is the charge in coulombs, and how many electrons flow?

27. * Two -3.0×10^{-9}-C charged point-like objects are separated by 0.20 m. Determine the potential (assuming zero volts at infinity) at a point (a) halfway between the objects and (b) 0.20 m to the side of one of the objects (and 0.40 m from the other) along a line joining them.

28. **BIO** **Electric field in body cell** The electric potential difference across the membrane of a body cell is $+0.070$ V (higher on the outside than on the inside). The cell membrane is 8.0×10^{-9} m thick. Determine the magnitude and direction of the $\vec{E}$ field through the cell membrane. Describe any assumptions you made.

29. * **BIO** **EST** **Energy used to charge nerve cells** A nerve cell is shaped like a cylinder. The membrane wall of the cylinder has a $+0.07$-V potential difference from the inside to the outside of the wall. To help maintain this potential difference, sodium ions are pumped from inside the cell to the outside. For a typical cell, 10^9 ions are pumped each second. (a) Determine the change in chemical energy each second required to produce this increase in electric potential energy. (b) If there are roughly 7×10^{11} of these cells in the body, how much chemical energy is used in pumping sodium ions each second? (c) Estimate the fraction of a person's metabolic rate used to pump these ions.

30. * **Equation Jeopardy 3** The equation below describes one or more physical processes. Solve the equation for the unknown and write a problem statement for which the equation is a satisfactory solution.

$$0 = (1/2)(1.0 \times 10^{-4} \text{ kg})(6.0 \text{ m/s})^2 + (2.0 \times 10^{-7})\Delta V$$

31. * **Equation Jeopardy 4** The equation below describes one or more physical situations. Solve the equation for the unknown and write a problem statement for which the equation is a satisfactory solution.

$$(9.0 \times 10^9 \text{ N} \cdot \text{m}^2/\text{C}^2)(+2.0 \times 10^{-8} \text{ C})/(2.0 \text{ m})$$
$$+ (9.0 \times 10^9 \text{ N} \cdot \text{m}^2/\text{C}^2)(-2.0 \times 10^{-8} \text{ C})/(1.0 \text{ m}) = \Delta V$$

32. * While a sphere with positive charge q_1 remains fixed, a second sphere with positive charge q_2 is successively placed at the points A, B, C and D (see **Figure P18.32**). Rank (a) the magnitudes of the $\vec{E}$ field at point P and (b) the V field values at P for the four positions of the second sphere.

FIGURE P18.32

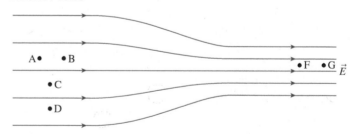

33. * **Figure P18.33** shows $\vec{E}$ field lines in a region of space. Note that the distances between A and B, between C and D, and between F and G are all equal. (a) Rank the magnitudes of the $\vec{E}$ field at points A to G from largest to smallest. (b) Rank the electric potentials at points A to G from largest to smallest. (c) A small negatively charged object is moved (I) from A to B, (II) from C to D, (III) from A to D, and (IV) from F to G. Rank the work done on the object in these four cases from largest to smallest.

FIGURE P18.33

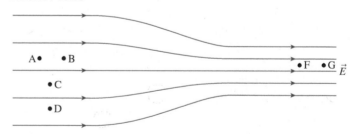

18.5 and 18.6 Conductors and dielectrics in electric fields

34. * A metal sphere has no charge on it. A positively charged object is brought near, but does not touch the sphere. Show that this object can exert a force on the sphere even though the sphere has no net charge. How can you test your answer experimentally?

35. ** **EST** A Van de Graaff generator of radius 0.10 m has a charge of about 1.0×10^{-7} C on it. The Van de Graaff generator is then turned off and grounded. How many excess electrons remain on its dome? Indicate any assumptions that you made.

36. ** A metal ball of radius R_1 has a charge Q. Later it is connected to a neutral metal ball of radius R_2. What is the fraction of the charge Q that remains on the first ball?

37. * Positively charged metal sphere A is placed near metal sphere B. B has a very small positive charge. Explain why the spheres could attract each other. Draw sketches of the spheres and their charge distribution to support your answer.

38. * Two small metal spheres A and B have different electric potentials (A has a higher potential). Describe in words and mathematically what happens if you connect them with a wire. What assumptions did you make?

39. * An electric dipole such as a water molecule is in a uniform $\vec{E}$ field. (a) Will the force exerted by the field cause the dipole to have a linear acceleration along a line in the direction of $\vec{E}$? Explain. (b) Will the field exert a torque on the dipole? Explain.

40. ** **BIO** **Electric field of a fish** An African fish called the aba has a charge $q = +1.0 \times 10^{-7}$ C at its head and an equal magnitude negative charge $-q$ at its tail (see **Figure P18.40**). Determine the magnitude and direction of the electric field at position A and the force exerted on a hydroxide ion (charge $-e$) at that point. The fish and ion are in water. Indicate any assumptions that you made.

FIGURE P18.40
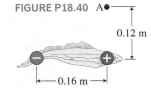

41. **BIO** **Body cell membrane electric field** (a) Determine the average magnitude of the $\vec{E}$ field across a body cell membrane. A 0.07-V potential difference exists from one side to the other and the membrane is 7.5×10^{-9} m thick. (b) Determine the magnitude of the electrical force on a sodium ion (charge $+e$) in the membrane. Assume that the dielectric constant is 1.0 (it is actually somewhat larger).

42. ** **Earth's electric field** Earth has an electric charge of approximately -5.7×10^5 C distributed relatively uniformly on its surface. Determine as many quantities as possible about the electrical properties of the space around Earth. Use any additional information that you need. Indicate any assumptions that you made.

18.7 Capacitors

43. You have a parallel plate capacitor. (a) Determine the average $\vec{E}$ field between the plates if a 120-V potential difference exists across the plates. Their separation is 0.50 cm. (b) A spark will jump if the magnitude of the $\vec{E}$ field exceeds 3.0×10^6 V/m when air separates the plates. What is the closest the plates can be placed to each other without sparking?

44. * A capacitor of capacitance C with a vacuum between the plates is connected to a source of potential difference ΔV. (a) Write an expression for the charge on each of the plates and for the total energy stored by the capacitor. (b) You then fill the capacitor with a dielectric material of dielectric constant κ while the capacitor remains connected to the same potential difference. What are the new charge on the plates and the new energy stored by the capacitor?

45. * A capacitor of capacitance C with a vacuum between the plates is connected to a source of potential difference ΔV. (a) Write an expression for the charge on each of the plates and for the total energy stored by the capacitor. (b) The capacitor is then disconnected from the potential difference source and you fill it with a dielectric that has a dielectric constant κ. What are the new charge on the plates and the new energy?

46. How does the capacitance of a parallel plate capacitor change if you double the magnitude of the charge on its plates? If you triple the potential difference across the plates? What assumptions did you make?

47. ** **BIO** **EST** **Axon capacitance** The long thin cylindrical axon of a nerve carries nerve impulses. The axon can be as long as 1 m. (a) Estimate the capacitance of a 1.0-m-long axon of radius 4.0×10^{-6} m with a membrane thickness of 8.0×10^{-9} m. The dielectric constant of the membrane material is about 6.0. (b) Determine the magnitude of the charge on the inside (negative) and outside (positive) of the membrane wall if there is a 0.070-V potential difference across the wall. (c) Determine the energy stored in this axon capacitor when charged.

48. ** **Sphere capacitance** A metal sphere of radius R has an electric charge $+q$ on it. Determine an expression for the electric potential V on the sphere's surface. Use the definition of capacitance to show that the capacitance of this isolated sphere is R/k, where k is the constant used in Coulomb's law. (Hint: Assume the other plate is infinitely far away.)

49. ** **EST** Two identical metal plates that are touching each other are placed in a uniform electric field as shown in Figure Q18.34. The electric field is created by a parallel plate capacitor with capacitance 150 pF and potential difference 3600 V between the plates. The surface area of the capacitor plates is 10 times larger than the surface area of the metal plates. You move the metal plates apart (while still in the electric field) and then move them out of the field. Estimate the final charge on each metal plate.

50. * **BIO** **EST** **Capacitance of red blood cell** Assume that a red blood cell is spherical with a radius of 4×10^{-6} m and with wall thickness of 9×10^{-8} m. The dielectric constant of the membrane is about 5. Assuming the cell is a parallel plate capacitor, estimate the capacitance of the cell and determine the positive charge on the outside and the equal-magnitude negative charge inside when the potential difference across the membrane is 0.080 V.

51. **BIO** **Defibrillator** During ventricular fibrillation the heart muscles contract randomly, preventing the coordinated pumping of blood. A defibrillator can often restore normal blood pumping by discharging the charge on a capacitor through the heart. Paddles are held against the patient's chest, and a 6×10^{-6}-F charged capacitor is discharged in several milliseconds. If the capacitor energy is 250 J, what potential difference was used to charge the capacitor?

18.8 Electrocardiography

52. * **EST** The dielectric strength of air is 3×10^6 V/m. As you walk across a synthetic rug, your body accumulates electric charge, causing a potential difference of 6000 V between your body and a doorknob. What can you can estimate using this information?

53. * **Charged cloud causes electric field on Earth** The electric charge of clouds is a complex subject. Consider the simplified model shown in **Figure P18.53**. A positive charge is near the top of the cloud and a negative charge is near the bottom. Determine the direction of the $\vec{E}$ field on Earth at point P below the cloud and explain it so that a classmate can understand why there is positive charge on the ground directly below the cloud.

FIGURE P18.53
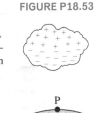

54. * **BIO** **Heart's dipole charge** The heart has a dipole charge distribution with a charge of $+1.0 \times 10^{-7}$ C that is 6.0 cm above a charge of -1.0×10^{-7} C. Determine the $\vec{E}$ field (magnitude and direction) caused by the heart's dipole at a distance of 8.0 cm directly above the heart's positive charge. All charges are located in body tissue of dielectric constant 7.0. What is the force exerted on a sodium ion (charge $+1.6 \times 10^{-19}$ C) at that point?

General Problems

55. * In a hot water heater, water warms when electric potential energy is converted into thermal energy. (a) Determine the energy needed to warm 180 kg of water by 10 °C. (b) If -10.0 C of electric charge passes through a $+120$-V potential difference in the heating coils each second, determine the time needed to warm the water by 10 °C.

56. ** **EST** **Lightning warms water** A lightning flash occurs when -40 C of charge moves from a cloud to Earth through a potential difference of 4.0×10^8 V. Estimate how much water can boil as a result of energy released during the process. Describe the assumptions you made.

57. * Four charged particles A, B, C, and D are placed in an electric field as shown in **Figure P18.57**. Vertical lines indicate equipotential surfaces of the V field. All particles are initially at rest. When they are released, they all start moving to the left. (a) Determine the signs of the charges on the particles. Rank (b) the kinetic energies of the particles and (c) the speeds of the particles as they cross the -600-V equipotential line. Relative masses and charges of the particles are indicated in the figure. Assume the particles are far from each other so they do not interact, and ignore forces exerted by Earth on the particles.

FIGURE P18.57

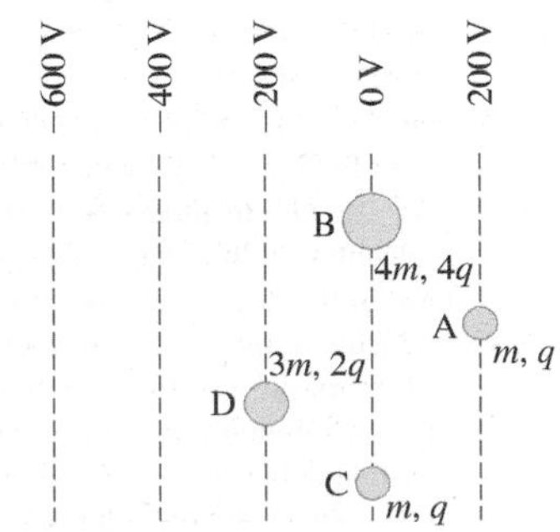

58. * EST Figure P18.58 shows a region of space with an electric field. Vertical lines indicate equipotential surfaces. A particle with charge -3.0 nC is initially at the location of the -20-V equipotential line. At time $t = 0$ the particle is released from rest. (a) Estimate the force exerted by the electric field on the particle when it passes the 0-V equipotential line and when it passes the 80-V equipotential line. (b) Draw a qualitative velocity-versus-time graph for the motion of the particle until it reaches the potential 120 V. (c) Determine the kinetic energy of the particle when it passes the 60-V equipotential line. (d) Do the data in the problem allow you to estimate the time needed for the particle to move from the -20-V line to the 120-V line? If you think the answer is no, list the additional data that are needed. Ignore the force exerted by Earth on the particle.

FIGURE P18.58

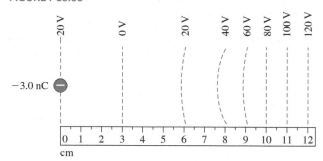

59. ** A small object of unknown mass and charge is tied to an insulating string that is connected to a sensitive force sensor. The object is placed between two large conducting plates, and the string is kept taut by charging the plates as shown in Figure P18.59. The mass of the string is 0.030 g, and the distance between the plates is 15.0 cm. As the potential difference between the plates increases, the force exerted by the force sensor on the string changes as shown in the table. Determine the mass and charge of the particle. (Note: This is a problem that requires linearization of data.)

ΔV (V)	$F_{\text{FS on S}}$ (10^{-3} N)
5000	52
5500	67
6000	82
6500	97
7000	110

FIGURE P18.59

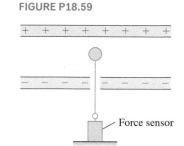

Force sensor

60. * BIO **Can a shark detect an axon $\vec{E}$ field?** A nerve signal is transmitted along the long, thin axon of a neuron in a small fish. The transmission occurs as sodium ions (Na^+) transfer like tipping dominos across the axon membrane from outside to inside. Each short section of axon gets an excess of about 6×10^8 sodium ions/mm. Determine the $\vec{E}$ field 4.0 cm from the axon produced by the excess sodium ions on the inside of the axon and an equal number of negative ions on the outside of a 1-mm length of axon. The ions are separated by the 8×10^{-9}-m-thick axon membrane. Will a shark that is able to detect fields as small as 10^{-6} N/C be able to detect that axon field? Explain.

61. ** BIO **Electrophoresis** Electrophoresis is used to separate biological molecules of different dimensions and electric charge (the molecules can have different charge depending on the pH of the solution). A particular molecule of radius R with charge q is in a viscous solution that has an $\vec{E}$ field across it. The field exerts an electric force on the molecule and the viscous solution exerts an opposing drag force $F_{\text{drag}} = DRv$, where D is a constant drag coefficient that depends on the shape and other features of the molecule and the solution, and v is the molecule's speed in the solution. When the molecule gets up to speed, the electric force exerted on it by the field is equal in magnitude and opposite in direction to the drag force. Experiments show that once the molecule reaches terminal velocity, the distance Δx traveled by the molecule during time interval Δt is determined by the expression $\Delta x = \dfrac{qE \, \Delta t}{DR}$. Describe the assumptions that were made when deriving this equation, and show the derivation.

62. ** BIO **Energy stored in axon electric field** An axon has a surface area of 5×10^{-6} m^2 and the membrane is 8×10^{-9} m thick. The dielectric constant of the membrane is 6. (a) Determine the capacitance of the axon considered as a parallel plate capacitor. (b) If the potential difference across the membrane wall is 0.080 V, determine the magnitude of the charge on each wall. (c) Determine the energy needed to charge that axon capacitor. (d) Determine the magnitude of the $\vec{E}$ field across the membrane due to the opposite sign charges across the membrane walls. (e) Calculate the energy density of that field. (f) Multiply the volume of space occupied by that field by the volume of the membrane to get the total energy stored in the field. How do the answers to (c) and (f) compare?

Reading Passage Problems

BIO **Electric discharge by eels** In several aquatic animals, such as the South American electric eel, electric organs produce 600-V potential difference pulses to ward off predators as well as to stun prey. **Figure 18.29** illustrates the key component that produces this electric shock—an **electrocyte**. The interior of an inactive electrocyte (Figure 18.29a) has an excess of negatively charged ions. The exterior has an excess of positively charged sodium ions (Na^+) on the left side and positively charged potassium ions (K^+) on the right. There is a -90-mV electric potential difference from outside the cell membrane to inside the electrocyte cell membrane, but zero potential from one exterior side to the other exterior side. How then does an eel produce the 600-V potential difference necessary to stun an intruder?

FIGURE 18.29 An electric eel's shocking system.

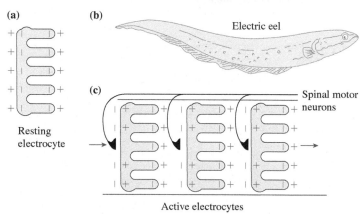

The eel's long trunk and tail contain many electrocytes placed one after the other in columns (Figures 18.29b and c). Each electrocyte contains several types of ion channels, which when activated by a nerve impulse allow sodium ions to pass through channels *on the left flat side* of each electrocyte from outside the cell to the inside. This causes the electric potential across that cell membrane to change from -90 mV to $+50$ mV. The electric potential from the left external side to the right external side of an electrocyte is now about 100 mV (Figure 18.29c). With about 6000 electrocytes placed one after the other (in series), the eel is able to produce an electric impulse of over 600 V (6000 electrocytes in series times 0.10 V per electrocyte). The discharge lasts about 2–3 ms.

63. Suppose you have a 1.0-F capacitor (very large capacitance) with a 0.10-V potential difference across the capacitor. What is the magnitude of the electric charge on each plate of the capacitor?
   (a) 0.05 C   (b) 0.10 C   (c) 0.20 C
   (d) 1.0 C   (e) 10 C

64. Suppose you place two of these 1.0-F capacitors with the charge calculated in the previous question as shown in Figure P18.64a. What is the net potential difference across the two capacitors (from one dot to the other)?
   (a) 0.05 V   (b) 0.10 V
   (c) 0.20 V   (d) none of these
   (e) not enough information

FIGURE P18.64

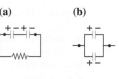

65. If both capacitors are discharged simultaneously, how much electric charge goes through the wire shown in pink in Figure P18.64a?
(a) 0.05 C
(b) 0.10 C
(c) 0.20 C
(d) 1.0 C
(e) 10 C

66. Suppose you place the two capacitors with the 0.10-V potential difference across each capacitor as shown in Figure P18.64b. What is the net potential difference across the two capacitors?
(a) 0.05 V
(b) 0.10 V
(c) 0.20 V
(d) none of these
(e) not enough information

67. Look at the electrocyte shown in Figure 18.29c. What causes the 0.10-V potential difference from the outer left to the outer right side of the cell?
(a) The membrane is thicker on the left than on the right.
(b) The ion distribution across the left membrane is different than across the right membrane.
(c) The left and right membranes have different capacitances.
(d) b and c
(e) a, b, and c

68. Suppose that one cell of the electrocyte is regarded as a small capacitor with a 0.10-V potential difference across it. How should we arrange 10 cells to get a 1.0-V potential difference across them?
(a) in series, as in Figure P18.64a
(b) in parallel, as in Figure P18.64b
(c) randomly so that they do not cancel each other
(d) not enough information

Electrostatic precipitator (ESP) Electrostatic precipitators are a common form of air-cleaning device. ESPs are used to remove particle emissions from smoke moving up smokestacks in coal and oil-fired electricity-generating plants and pollutants from the boilers in oil refineries. You can buy portable ESPs or whole-house ESPs that connect to the cold-air return on the furnace. These devices remove about 95% of dirt and 85% of microscopic particles from the air.

A basic electrostatic precipitator contains a negatively charged horizontal metal grid (made of thin wires) and a stack of large, flat, vertically oriented metal collecting plates, with the plates typically spaced about 1 cm apart (only two plates are shown in **Figure 18.30**). Air flows across the charged grid of wires and then passes between the plates. A large negative potential difference (tens of thousands of volts) is applied between the wires and the plates, creating a strong electric field that ionizes particles in the air around the thin wires. Negatively charged smoke particles flow upward between the plates. The charged particles are attracted to and stick to the oppositely charged plates and are thus removed from the moving gas.

FIGURE 18.30 An electrostatic precipitator.

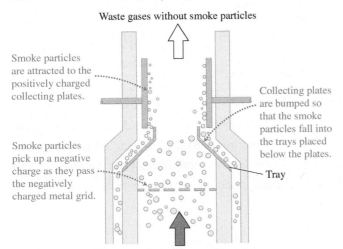

Waste gases without smoke particles

Smoke particles are attracted to the positively charged collecting plates.

Collecting plates are bumped so that the smoke particles fall into the trays placed below the plates.

Tray

Smoke particles pick up a negative charge as they pass the negatively charged metal grid.

Waste gases containing smoke particles

69. An $\vec{E}$ field of approximately 3×10^6 V/m is needed to ionize dry air molecules. If the potential difference between the thin wires and metal plates is 60,000 V, what minimum distance must the wires be from the plates in order to cause dry air to ionize?
(a) 1.0 cm
(b) 2.0 cm
(c) 4.0 cm
(d) 50 cm
(e) 50 m

70. Why are the smoke particles attracted to the closely spaced plates?
(a) The particles are negatively charged.
(b) The particles are positively charged.
(c) The plates are positively charged.
(d) The plates are negatively charged.
(e) a and c
(f) b and c

71. Suppose a 2.0×10^{-6}-kg dust particle with charge -4.0×10^{-9} C is moving vertically up a chimney at speed 6.0 m/s when it enters the $+2000$-N/C $\vec{E}$ field near the metal collecting plate, pointing perpendicularly to the plate. If the particle is 2.0 cm from the plate at that instant, which answer below is closest to the magnitude of its horizontal acceleration toward the plate?
(a) 2.0×10^{-5} m/s^2
(b) 1.2×10^{-4} m/s^2
(c) 4.0×10^{-4} m/s^2
(d) 4.0 m/s^2
(e) 24 m/s^2

72. Suppose everything is the same as in the previous problem. Which answer below is closest to the time needed for the particle to hit the plate?
(a) 0.01 s
(b) 0.10 s
(c) 0.001 s
(d) 1×10^{-6} s
(e) 1.0 s

73. What is the purpose of giving a negative charge to the particles collected by the precipitator?
(a) The particles are then attracted to the positively charged plates while moving upward.
(b) The particles are then repelled from the positively charged plates while moving upward.
(c) The extra mass slows the upward movement of the particles.
(d) The negative charge cancels the positive charge the particles have when entering the precipitator.
(e) a and c

DC Circuits

- **Why are modern buildings equipped with electrical circuit breakers?**

- **Why are LEDs replacing incandescent lightbulbs?**

- **Why is it dangerous to use a hair dryer while taking a bath?**

On a cold Sunday morning you turn on a space heater and the toaster oven. Then you start the washing machine and the dishwasher. Finally, you turn on the coffee grinder, and suddenly all the appliances turn off. What happened? A device called a circuit breaker disconnected that part of the house from the external supply of electric energy to prevent a house fire. How does a circuit breaker "know" when to turn off the energy supply, and why might keeping all of those appliances on lead to a fire? We will find out in this chapter.

BE SURE YOU KNOW HOW TO:

- Apply the concept of the electric field to explain electric interactions (Section 18.1).

- Define the V field (electric potential) and electric potential difference ΔV (Section 18.3).

- Explain the differences in internal structure between conducting materials and insulating materials (Section 17.3).

IN PREVIOUS CHAPTERS (Chapters 17 and 18) we learned to explain processes that involved electrostatic phenomena. In electric devices such as cell phones, computers, and lightbulbs, charged particles are continually moving. Similar movements occur inside the human body's nervous system. In this chapter we will learn how to explain phenomena that involve these moving, microscopic, charged particles.

19.1 Electric current

In the previous chapter we studied processes such as grounding that occur when a charged conducting object is connected with a wire to an uncharged conductor. We found that the excess electric charge on the charged object redistributes itself until the electric potential V at both conductors becomes equal. Let us look at similar experiments using a Wimshurst machine (Figure 17.21 shows one example of a Wimshurst machine). Recall that cranking the machine's handle generates opposite charges in the two metal spheres. The charge separation leads to a potential difference between the spheres (**Figure 19.1**). In the new Wimshurst machine experiments, we'll try to identify common features that might relate to transferring electric charge for useful purposes. Consider the experiments in Observational Experiment **Table 19.1**.

FIGURE 19.1 The charged Wimshurst spheres produce an $\vec{E}$ field.

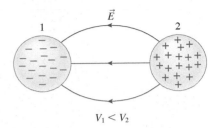

$$V_1 < V_2$$

OBSERVATIONAL
EXPERIMENT TABLE 19.1   Electric potential difference and charge transfer

VIDEO
OET 19.1

Observational experiment	Analysis
Experiment 1. Crank the handle of a Wimshurst machine and then bring the oppositely charged spheres of the machine close to each other (about 5 cm apart). You see a big spark. After the spark, when you again bring the spheres close, no further sparking occurs. 	There is charge separation and nonzero potential difference ΔV between the two spheres. The spark means that the air between the spheres becomes a conductor, leading to the rapid discharge and the production of light. The original electric potential energy is converted to light and sound energy.
Experiment 2. Crank the handle of the Wimshurst machine and hang a light aluminum foil ball from an insulating thread between the oppositely charged spheres. The ball swings back and forth from one sphere to the other for a few minutes and then stops. When you remove the foil ball and bring the spheres near each other, no spark occurs. 	The foil ball must acquire a small amount of negative charge from the negative sphere each time it touches it. The ball carries the negative charge to the positive sphere, deposits it there, and then returns to the negative sphere to repeat the process. This continues until the spheres are discharged and the potential difference between them is zero. The original electric potential energy is converted to mechanical energy and internal energy.
Experiment 3. Crank the handle of the Wimshurst machine and connect the leads of a neon bulb between the charged spheres. There is a flash of light from the bulb. When you remove the bulb and bring the spheres close, no spark occurs. 	Before the bulb touches them, the spheres are charged and at different potentials. The bulb and its leads provide a conduction path to discharge the spheres. The discharge causes a flash of light from the bulb. The original electric potential energy is converted to light energy.

Patterns

- In all three experiments, the Wimshurst machine started with negative charge on one sphere and positive charge on the other sphere. There was a potential difference ΔV between the spheres.
- This charge separation and potential difference led to a flow of charge from one sphere to the other.
- The charge flow involved different observable consequences: a spark of light, the swinging ball of foil, and the flash of the bulb.
- After the charge flow, the Wimshurst spheres were discharged and the potential difference between them was zero. No more sparking or movement could occur.

The observable events in Table 19.1 were only able to occur because of the following factors:

1. An initial charge separation and resulting potential difference ΔV between the oppositely charged spheres. After the discharges occurred, resulting in equipotential spheres, nothing further could happen.

2. The presence of a charge conduction pathway (the foil ball, the leads of the neon bulb, and the conducting neon inside). No flash or movement occurred if the charged spheres were far apart and the electric field was too weak between the spheres to ionize the air.

3. A process that converted the electric potential energy of the spheres into some other form of energy: the flash of light from the air, the mechanical energy of the foil ball, and the flash of light from the neon bulb.

Each of these three processes ended very quickly. If we want to convert electrical energy to light or mechanical energy for a substantial time interval, we need to learn how to keep the observable consequences described in Table 19.1 happening continuously.

Fluid flow and charge flow

FIGURE 19.2 A water flow analogy for electric charge flow.

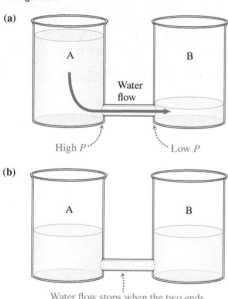

(a)

Water flow

High P ···   ··· Low P

(b)

Water flow stops when the two ends of the hose are at equal pressure.

A fluid flow analogy may help us better understand the electric potential difference and conduction pathways of these electrical processes. You have two containers with water—A and B. A is almost full, and B is almost empty. You connect a hose between the two containers (**Figure 19.2a**). Water starts flowing from A to B until the levels are the same (Figure 19.2b), at which point the water flow stops. The volume of water in container A is analogous to the excess positive charge on Wimshurst sphere 2, and the difference in water pressure in the hose is analogous to the potential difference between the spheres. The hose provides a pathway for water to flow until the pressure at both ends of the hose is the same; the electric charge flows until the electric potential at each sphere is the same.

Notice that it is the pressure difference and not the difference in the volume or mass of water in each container that makes the water flow. Imagine a large container A full of water with the water level the same as in small container B—the water will not flow (**Figure 19.3a**). Similarly, it is not the total charge difference between the spheres of the Wimshurst machine but the potential difference that makes the charge flow. Imagine a large sphere A and a small sphere B. Suppose the charge is large on A and small on B; but if the V fields (electric potentials) on the surfaces of these two spheres are the same (recall the discussion of grounding in Chapter 18),

$$V = k_C \frac{q}{R}$$

there will be no charge flow through the wire connecting them (Figure 19.3b).

FIGURE 19.3 There is no (a) water flow if $P_A = P_B$ or (b) charge flow if $V_A = V_B$.

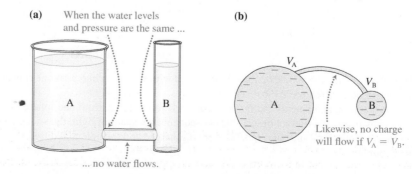

(a)   When the water levels and pressure are the same ...

A   B

... no water flows.

(b)

V_A

V_B

A   B

Likewise, no charge will flow if $V_A = V_B$.

Making the process continuous

We can make fluid flow continuously by connecting the containers with another hose attached to a pump that moves water from B back to A (**Figure 19.4**). Pumping water from B back to A maintains a pressure difference between the ends of the hose and results in a continuous flow from A to B.

To achieve a steady flow of electric charge, we need a device that can maintain a steady potential difference between, for example, the leads of the neon bulb, effectively moving the negative charges back to the negative terminal. We attach wires from the positive and negative terminals of 10 batteries connected to the leads of a neon bulb. (We use 10 because one is not enough to make a neon bulb glow) (see **Figure 19.5**). The bulb does in fact steadily glow. Thus, it appears that the battery produces a steady potential difference across its terminals, which in turn causes a steady flow of electric charge through the neon bulb. The battery is equivalent to the pump in the water analogy.

Electric current

The flow of charged particles moving through a wire between two locations that are at different electric potentials is a physical phenomenon called **electric current**. When the charged particles always move in the same direction, the current is called **direct current**. A system of devices such as a battery, wires, and a bulb that allows for the continuous flow of charge is called an **electric circuit**. An electric circuit that has a direct current in it is called a **DC circuit**. In most electric circuits, the moving charged particles are free electrons in the wires and circuit elements. However, as we will learn later in the chapter, in some circuits those moving particles can be positively or negatively charged ions.

According to a simplified model of the internal structure of a wire, the wire consists of a **crystal lattice** composed of positive metal ions and **free electrons**. Free electrons (lost by the atoms that became ions) move within the lattice structure (**Figure 19.6a**). The ions constitute the mass of the material and are relatively stationary except for minor vibrations due to their thermal motion. In this model, the electrons move chaotically inside the wire like a swarm of gnats on a day when the air is still—each gnat moves, but the swarm remains at rest above your head. When the wire is connected to the terminals of the battery, there is an electric field inside the wire, and the electrons accelerate opposite the direction of the $\vec{E}$ field. They slow down when they "collide" with the ions and then accelerate again in the same direction (Figure 19.6b). This motion, called **drift**, occurs in the same direction for all of the electrons—just as a cloud of mosquitoes collectively moves in a single direction on a windy day. The collisions of electrons with ions make the ions vibrate faster, warming the wire. You can test this phenomenon by connecting two terminals of a battery with a wire for a brief moment— both the battery and the wire get hot!

FIGURE 19.4 A pump maintains flow.

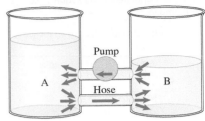

The pump returns water to container A. Water flows through the hose from A back to B.

FIGURE 19.5 Ten batteries create a constant potential difference across a neon bulb and charge flows through the bulb.

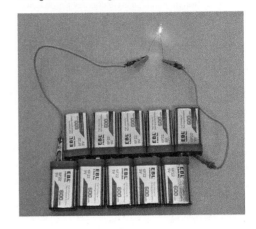

FIGURE 19.6 (a) Free electrons drift randomly, moving fast between collisions but with no preferred direction. (b) When an electric field is present, the electrons continue this random motion but now drift toward the area of higher potential, opposite the direction of the $\vec{E}$ field.

(a) Positive ions form a crystal lattice structure.

$\vec{E} = 0$ Wire

V_1 $V_2 = V_1$

In the absence of an electric field, the electrons move randomly within the wire.

(b)

$\vec{E}$

V_1 $V_2 > V_1$

In the presence of an electric field, the electrons drift toward the higher V region.

Free electrons in a wire drift toward the direction of higher V field (toward the positive terminal of a battery)—see **Figure 19.7a** on the next page. However, traditionally, the direction of electric current in a circuit is defined as opposite the direction of the electrons'

FIGURE 19.7 The direction of electric current is defined as opposite the direction that electrons move.

(a) Electrons travel around the circuit toward the positive terminal.

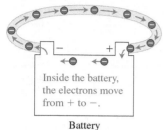

Inside the battery, the electrons move from + to −.

Battery

(b) The electric current I is defined in terms of the direction in which positive charges would move.

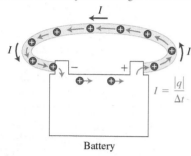

Battery

drifting motion. Imagine that positively charged particles are flowing from the higher electric potential terminal (labeled "+" on the battery) through the wires and electrical devices in the circuit to the terminal at lower electric potential (labeled "−"), then back through the battery to the positive terminal, where the cycle repeats (Figure 19.7b).

This definition of the direction of the electric current as a flow of positively charged particles is a historical quirk. When scientists first studied electric processes, they did not know about electrons and instead thought that an invisible positively charged electric fluid flowed around the circuit. For most situations involving electric circuits, it does not matter if we model the process in terms of electrons flowing one way or positive charges flowing the other way, so this historical convention is not a problem. Physicists use the term electric current to describe the drifting of electrically charged particles as well as the physical quantity that characterizes how much charge is moved through a conductor per unit time.

Electric current The magnitude of the physical quantity of electric current I in a wire equals the magnitude of the electric charge q that passes through a cross section of the wire divided by the time interval Δt needed for that charge to pass:

$$I = \frac{|q|}{\Delta t} \tag{19.1}$$

The unit of current is the **ampere** A, equivalent to 1 coulomb per second (C/s). A current of 1 A (one ampere, or amp) means that 1 C of charge passes through a cross section of the wire every second. The direction of the current is from higher potential to lower potential; therefore, it is in the direction that positive charges would move.

In SI units, the unit for electric current is a fundamental unit. The unit of electric charge (1 coulomb) is defined in terms of the ampere as $1\ C = (1\ A)(1\ s)$. In the next chapter, we will study the phenomenon that can be used to define the ampere.

QUANTITATIVE EXERCISE 19.1   **Electric current calculation**

Each second, 1.0×10^{17} electrons pass from right to left past a cross section of a wire connecting the two terminals of a battery. Determine the magnitude and direction of the electric current in the wire.

Represent mathematically The electric charge of an electron is -1.6×10^{-19} C. Since we know how many electrons (N) pass a cross section of the wire in 1 s, we can find the total electric charge passing that cross section each second and therefore the current:

$$I = \frac{|q|}{\Delta t} = \frac{|-eN|}{\Delta t}$$

Solve and evaluate

$$I = \frac{|-eN|}{\Delta t} = \frac{|-(1.6 \times 10^{-19}\,C)(1.0 \times 10^{17})|}{1\ s}$$

$$= 1.6 \times 10^{-2}\,A = 16\ mA$$

The direction of the current is from left to right, opposite the direction of electron flow.

Try it yourself If there is a 2.0-A current through a wire for 20 min, what is the total charge that moved through the wire during this time interval?

Answer   2400 C.

REVIEW QUESTION 19.1 What condition(s) is/are needed for electric charge to continuously travel from one place to another?

19.2 Batteries and emf

We have found that a battery creates a potential difference that can cause a neon bulb to glow continuously, analogous to how a water pump can cause water to flow continuously. If the current is a moving positive charge, the battery does work on the charge to move it within the battery from the lower electric potential negative terminal to the higher electric potential positive terminal. Once there, the charge can travel around the external circuit (the bulb and connecting wires) and cause the bulb to glow continuously. Batteries maintain a potential difference by means of a nonelectrostatic chemical process. This chemical process inside the battery separates neutral atoms at the negative terminal into electrons and positive ions, and moves the positive ions to the positive terminal.

The work done by a battery per unit charge moved inside the battery from one terminal to the other is the battery's **emf** $\mathcal{E}$. When first conceived, the term for this work was electromotive force, but now it is just called emf. (See the Tip in the margin.)

If you connect a neon bulb to the battery terminals, the bulb will glow and get a little warmer as work done by the battery results in the production of light and an increase in the thermal energy of the external circuit (bulb and connecting wires). If instead of a neon bulb, you were to connect a toaster to the battery, the work would be converted into thermal energy and light energy—the toaster's coils would glow red and get hot.

> **Emf $\mathcal{E}$** The emf $\mathcal{E}$ equals the work W done by a battery per coulomb of electric charge q that needs to be moved through the battery from one terminal to the other in order to maintain a potential difference at the battery terminals:
>
> $$\mathcal{E} = \frac{W}{q} \qquad (19.2)$$

The unit of emf is $J/C = V$, the same as the unit of electric potential difference. The emf of standard AAA, AA, C, and D batteries is 1.5 V. The small rectangular batteries used in smoke detectors have an emf of 9.0 V. The physical size of the battery is not related to the emf but to its storage capacity—the total charge it can move before it must be replaced or recharged. The bigger the size, the larger the product of $q = I \Delta t$, and the longer the time interval during which the battery works (assuming constant current). The storage capacity of an AA battery is about 2000 mAh (which is 2 Ah, or the charge transferred by the current of 2 amperes during 3600 s, which is about 7200 C).

TIP Electromotive force (emf), despite its name, is not a force. Emf is work done per coulomb of charge. The term *electromotive force* was coined by Alessandro Volta (1745–1827), who invented the battery. At that time the terms "force," "energy, " and "power" were used somewhat interchangeably. The linguistic distinction between force and energy had not yet been made clear. For example, kinetic energy was called "live force" and potential energy was called "dead force."

CONCEPTUAL EXERCISE 19.2   **Graphing electric potential in a circuit**

You connect a 9.0-V battery to a small motor. Describe the changes in electric potential in the circuit with a graph.

Sketch and translate A sketch of a motor connected to the battery is shown at right. The battery has two terminals, + and −. The emf of the battery is 9.0 V.

Simplify and diagram To represent the changes in electric potential in the circuit, use a V-versus-location graph for an imaginary positively charged particle moving through the circuit (see figure at right). Assume that the electric potential of the − terminal at A is zero and that of the + terminal at B is 9.0 V. The battery does work on the charged particle-circuit system, increasing its electric potential energy. This work is similar to a person lifting a ball up a hill. When the ball reaches the top of the hill and starts rolling down, the potential energy of the system begins to decrease. Similarly, after leaving the battery at B, the imaginary positively charged particle travels along a wire to the

motor at C. We don't know yet the decrease in potential in the connecting wire, but will learn later that it is very small. There is also a small decrease in potential as the charge travels along the wire from D back to the battery at A. Thus, most of the potential decrease should occur across the motor. The potential decreases back to zero at A.

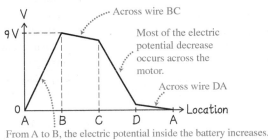

From A to B, the electric potential inside the battery increases.

Try it yourself How would the sketch and graph differ if there were two identical small motors attached one after the other to the battery?

(CONTINUED)

Answer

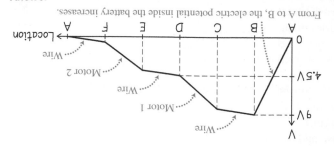

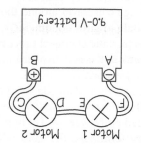

It's the same battery, so there is still a 9.0-V potential increase from its negative to its positive terminal. While passing through the two-motor circuit, about half of the decrease in potential (4.5 V) is across the first motor and the other 4.5-V decrease is across the second motor (see figure).

From A to B, the electric potential inside the battery increases.

REVIEW QUESTION 19.2 Describe the changes in electric potential through a circuit and relate them to the motion of charged particles.

19.3 Making and representing simple circuits

An electric circuit is a system of devices such as a battery, wires, and a bulb (or some other elements, such as a motor) that allows for the continuous flow of charge. However, just having those elements next to each other does not produce a circuit—the electrical connections among them are important. In this section we investigate how to make a circuit. However, first we need to learn more about lightbulbs—common indicators of current in a circuit.

Neon and incandescent bulbs

All lightbulbs have a mechanism that converts electrical energy into light and thermal energy when electric charge flows through them. Here we will consider two types of bulbs—neon and incandescent—that use different mechanisms for this energy conversion.

A neon bulb consists of a glass bulb filled with low-pressure neon gas. Two thin wires (called **electrodes**) extend from the bottom of the bulb (**Figure 19.8a**). The leads outside the bulbs are the extensions of the electrodes. If the potential difference between the electrodes is high enough, the gas inside undergoes dielectric breakdown and becomes a conductor connecting the two electrodes. As a result, ions and free electrons are now present in the space between the electrodes. As the electric field associated with the potential difference causes them to travel in opposite directions, some electrons and ions will collide with atoms and change the energies of those atoms. Atoms release this extra energy by emitting light (you will learn more about this process in Chapter 28). Even if the potential difference across the terminals of the neon bulb only lasts for a very short time, we still see a flash of light.

An incandescent bulb has a metal filament inside of it. When there is current through the filament, the interactions of free electrons with the lattice of atoms make the filament, and thus the bulb, extremely hot (Figure 19.8b). When the filament of an incandescent bulb is hot enough to glow, we see light. The bulb is filled with inert gas; this gas prevents the filament from oxidizing and burning.

Unlike incandescent bulbs, a neon bulb is relatively cool to the touch. Also unlike the neon bulb, an incandescent bulb usually has no outside leads. The filament of an incandescent bulb is connected on one side to the metal screw-like base of the bulb (one terminal) and on the other side to a separate metal contact (the second terminal) at the very bottom of the base. We will learn the importance of this feature in Observational Experiment **Table 19.2**.

FIGURE 19.8 Comparing a neon bulb and an incandescent bulb.

(a) Neon bulb

The electric field between the electrodes causes free electrons to ionize neon atoms. Light is produced when the electrons rejoin the ionized neon atoms.

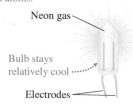

Neon gas

Bulb stays relatively cool

Electrodes

(b) Incandescent bulb

Current through the metal filament energizes the electrons in the filament, increasing thermal energy and generating light.

Filament glows when hot

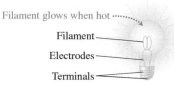

Filament

Electrodes

Terminals

OBSERVATIONAL EXPERIMENT TABLE 19.2   Making a bulb glow

VIDEO
OET 19.2

Observational experiment	Analysis
Connection 1—No light is produced. 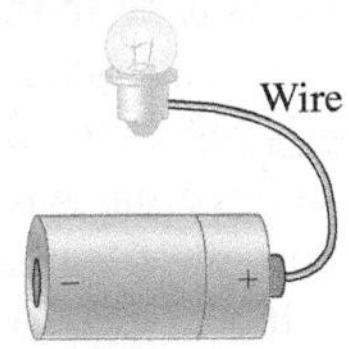	One bulb terminal is connected to one battery terminal with a wire.
Connection 2—no light 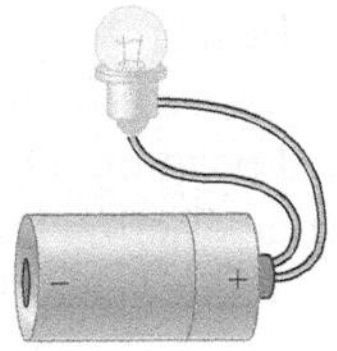	Both bulb terminals are connected to the same battery terminal with a wire.
Connection 3—no light 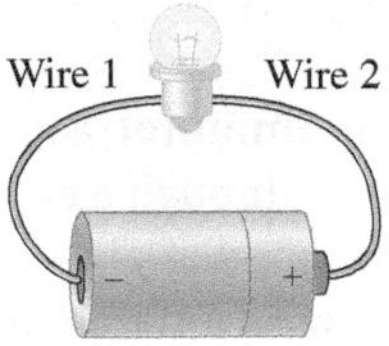	One bulb terminal is connected to both battery terminals with wires.
Connection 4—light 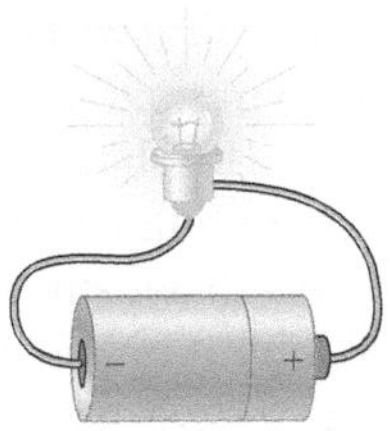	Both bulb terminals are connected to both battery terminals with wires.
Connection 5—light 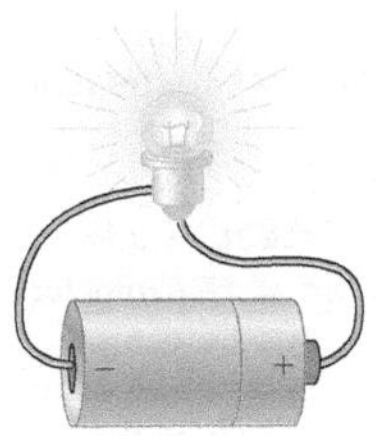	The same as connection 4, only the bulb terminals are connected to the opposite battery terminals.
Connection 6—light 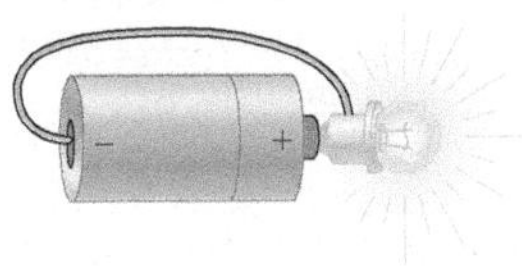	The same as connection 5, only one bulb terminal is directly touching the battery.

Pattern

The incandescent bulb glows if one battery terminal is connected to one lightbulb terminal and the other battery terminal is connected to the other bulb terminal, either with a metal wire or by direct contact, making a *complete closed* loop. This arrangement is an example of a **complete circuit**.

In Table 19.2 we identified conditions necessary to have a circuit with an electric current in it. To check whether any circuit you build is complete, trace a path of an imaginary positive charge moving from the positive terminal of the battery to the negative terminal (either in a real circuit or in the circuit you drew on paper). The path has to pass along conducting material at every location, making a complete closed loop. Check the circuits in which the bulb did not light in the table above using this rule.

FIGURE 19.9 The symbols used in circuit diagrams.

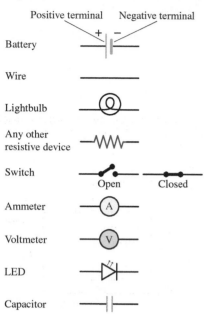

Symbols for the elements in electric circuits

It is inconvenient to represent electric circuits by drawing realistic likenesses of actual batteries and bulbs. Instead, scientists and engineers use simplified symbols for the elements in an electric circuit. Diagrams with these symbols are called **circuit diagrams**. The symbols for circuit elements used in this chapter are shown in **Figure 19.9** (some elements are already familiar to you; others we will learn about later). Notice that the positive terminal of a battery is represented with a longer line than the negative terminal. Straight lines represent connecting wires. Bulbs are represented by circles with a loop in the middle (representing the filament). A sawtooth line represents any *resistive* device, an element in a circuit that converts electric energy into some other type of energy. We will discuss resistive devices—often called **resistors**—later in the chapter. Lightbulbs and toasters are all examples of resistors. Switches—devices that allow us to insert a dielectric (such as air) into a conductive path of the circuit to stop the current—have an open and closed position. Devices that measure current through a circuit element or the potential difference across it are called ammeters and voltmeters, respectively. Their symbols are shown in Figure 19.9.

Ammeters

Recall our water analogy. An **ammeter** acts like a water flow meter. If you wish to measure how much water flows through a cross section of a pipe each second (the *flow rate*), you insert a flow meter into the pipe—just before or after the cross section that interests you. The water must pass through the flow meter. Similarly, to measure electric current through a circuit element, you place the ammeter next to that element. This means creating a break in the circuit next to that element, inserting the ammeter into the break, and reconnecting the wires to the terminals of the ammeter (**Figure 19.10a**). The circuit diagram in Figure 19.10b shows the arrangement. Ammeters have positive and negative terminals, which determine the sign of the ammeter reading. It is positive when the current is in the positive to negative terminal direction and negative when the current is in the negative to positive terminal direction.

FIGURE 19.10 Connecting an ammeter to measure electric current.

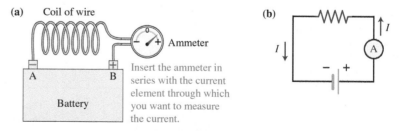

FIGURE 19.11 Connecting a voltmeter to measure potential difference.

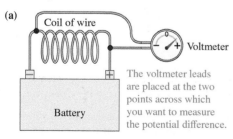

The voltmeter leads are placed at the two points across which you want to measure the potential difference.

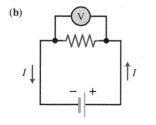

Voltmeters

A voltmeter measures the electric potential difference between two points in a circuit. Using a voltmeter to measure potential difference is analogous to using a pressure meter to measure the water pressure difference between two points in a water system. To measure the pressure difference you would submerge one terminal of the pressure meter into the water at the first point and the other terminal into the water at the second point. To measure the electric potential difference between two points in a circuit, you place a voltmeter so that its terminals are connected to those two points (**Figure 19.11a**). A circuit diagram for that connection is shown in Figure 19.11b. The voltmeter has a positive reading when the terminal labeled with a + sign is connected to the point with a higher electric potential than the terminal labeled with the − sign.

A multimeter is a device that combines the functions of an ammeter with a voltmeter. When you use a multimeter be sure you check the settings—accidentally measuring potential difference when the multimeter is on the ammeter setting may harm the device.

REVIEW QUESTION 19.3 Explain the meaning of the term "a complete circuit."

19.4 Ohm's law

Is there a relationship between the potential difference across a circuit element and the electric current through it? To determine such a relationship, we conduct an experiment (Observational Experiment **Table 19.3**). We use a wire made of constantan (an alloy of nickel and copper), the resistive device in this case, and connect the ends of the wire (wire 1) to a variable emf indicated by the symbol ↗ (**Figure 19.12**). A variable emf can change both in magnitude and in direction. We vary the potential difference across wire 1 and record the resulting current through it. We then repeat the experiment with a longer constantan wire (wire 2). We focus only on the current through and potential difference across the two constantan wires, not the connecting wires.

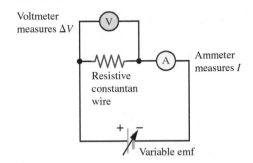

FIGURE 19.12 Measuring the current through and the potential difference across a resistive constantan wire.

OBSERVATIONAL EXPERIMENT TABLE 19.3

Relating current through and potential difference across a resistive element

Observational experiment	Analysis		
	Potential difference across the wire (volts)	Wire 1 I (amps)	Wire 2 I (amps)
The electric circuit is shown in Figure 19.12. The second constantan wire, wire 2, is twice as long as wire 1. By changing the setting of the variable emf, we vary the potential difference ΔV across the ends of the first resistive constantan wire and measure the current I through it. Then we repeat the experiment using constantan wire 2. Finally, we graph the results.	0	0.00	0.00
	1.0	0.37	0.19
	1.5	0.56	0.28
	2.0	0.74	0.37
	2.5	0.94	0.47
	−1.0	−0.37	−0.19
	−1.5	−0.56	−0.28
	−2.0	−0.74	−0.37
	−2.5	−0.94	−0.47

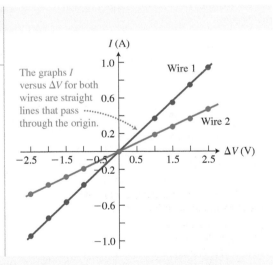

The graphs I versus ΔV for both wires are straight lines that pass through the origin.

Pattern

The graphs show that the current through both resistive wires is directly proportional to the potential difference across them. When the potential difference reverses direction, so does the current. The current through the shorter wire (wire 1) is greater than the current through the longer wire (wire 2) when the potential difference across them is the same.

The relationship we observed in Table 19.3 is reasonable given our conceptual understanding. The larger the potential difference across the wire, the larger the $\vec{E}$ field inside. This means that a larger electric force is exerted on the electrons, resulting in larger accelerations of the electrons between the collisions with the ions in the lattice and thus a larger drift velocity and a larger current through the wire. Since the pattern is the same in both directions, it seems that the electric properties of the metal are independent of the direction of current.

We notice that for the same potential difference across each wire, there is always twice as much current through wire 1 as through 2. Thus the ratio of current through and potential difference across a wire, which is called G, seems to quantify how easily the charged particles in the circuit flow (the less they interact with the ions, the less their kinetic energy is converted into the thermal energy of the lattice). The quantity G is called **conductance**. Circuit elements with higher conductance require a smaller potential difference across them to have the same current through them.

To characterize how a circuit element resists a current, we use another quantity, called **resistance R**:

$$R = \frac{1}{G}$$

We can write the relationship between current and potential difference using resistance R as

$$I = \frac{\Delta V}{R} \tag{19.3}$$

where R—resistance—is the physical quantity that characterizes the degree to which an object resists a current. The unit of R is called the **ohm** Ω (Greek omega), where 1 ohm = 1 volt/ampere (1 Ω = 1 V/A). A 1.0-ohm resistor has a one-ampere (1-A) electric current passing through it when a one-volt (1-V) potential difference is placed across it. Note that R is always positive. When we write Eq. (19.3), we usually use the absolute values of current and potential difference.

> **Ohm's law**   The current I through a circuit element (other than a battery) can be determined by dividing the potential difference ΔV across the circuit element by its resistance R:
>
> $$I = \frac{\Delta V}{R} \tag{19.3}$$

Ohmic and non-ohmic devices

Equation (19.3) is called Ohm's law in honor of 19th-century physicist Georg Ohm, who discovered it experimentally. It is a cause-effect relationship that predicts the value of the current through a resistive device when the potential difference across it and its resistance are known.

Ohm's law provides us with a method to determine the resistances of different circuit elements. We connect each circuit element into a circuit, close the circuit, and measure the current through and potential difference across the element to determine its resistance $R = \Delta V/I$. Using the above method we can determine the resistance of the two constantan wires used in the experiments in Table 19.3: $R_1 = 2.7\ \Omega$ and $R_2 = 5.3\ \Omega$. Physically, the second wire is twice as long as the first; the length of the wire must affect its resistance.

The resistance of each wire remains constant even though the potential difference across and the current through it change. Is this true for other circuit elements, such as commercial resistors that you see in many circuits inside analog and digital electronic equipment and lightbulbs? **Figure 19.13** shows graphs representing collected data for three elements: two commercial resistors and a lightbulb. Notice that the slopes of the graphs for the commercial resistors are constant. They behave in a similar fashion to constantan wires with resistances $R_1 = 25.0\ \Omega$ and $R_2 = 50.0\ \Omega$, independent of the current through them. If the resistance of a circuit element does not depend on the potential difference across it, the element is called **ohmic**. Circuit elements that cannot be modeled as ohmic devices are called **non-ohmic**.

Unlike for the commercial resistors, however, the slope of the graph for the light-bulb decreases as the potential difference increases. It takes more potential difference

FIGURE 19.13 I-vs-ΔV graphs for ohmic and non-ohmic circuit elements.

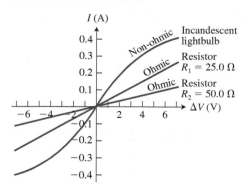

to cause the same increase in current. In other words, the resistance of the lightbulb increases with increasing potential difference and current. This relationship must be taken into account when building a circuit because the resistance of such a circuit element depends on the current through it. Note that for all of these elements, including the lightbulb, reversing the polarity of the battery does not change the shape of the curve. An incandescent lightbulb is an example of a non-ohmic device.

QUANTITATIVE EXERCISE 19.3   **Comparing an incandescent bulb and a resistor**

The graph shows how current through an incandescent lightbulb and a resistor changes with the changes in potential difference across them. The data for each were collected separately, although we present them on the same graph. (a) What are the resistances of the resistor and the lightbulb when the potential difference across each element is first 1.0 V and then 1.5 V? (b) What are the resistances of the resistor and the lightbulb when the current through each element is 50 mA?

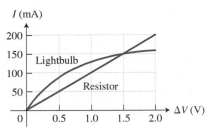

Represent mathematically We rearrange Eq. (19.3) and use data from the graphs to answer the questions:

$$I = \frac{\Delta V}{R} \implies R = \frac{\Delta V}{I}$$

Solve and evaluate (a) When $\Delta V = 1.0$ V, $I_{Res} = 100$ mA, thus $R_{Res} = 1.0$ V$/0.100$ A $= 10\ \Omega$; for the bulb, $I_{Bulb} = 0.125$ A, thus $R_{Bulb} = 1.0$ V$/0.125$ A $= 8.0\ \Omega$. When $\Delta V = 1.5$ V, $I_{Res} = 150$ mA; thus, as before, $R_{Res} = 1.5$ V$/0.150$ A $= 10\ \Omega$. For the bulb, the current and potential difference are the same as for the resistor, thus its resistance at this moment is also $R_{Bulb} = 1.5$ V$/0.150$ A $= 10\ \Omega$.

(b) When the current through each element is 50 mA, the potential difference across the resistor is 0.5 V; therefore, its resistance is $R_{Res} = 0.5$ V$/0.05$ A $= 10.0\ \Omega$. The potential difference across the bulb is 0.25 V; therefore, its resistance is $R_{Bulb} = 0.25$ V$/0.05$ A $= 5.0\ \Omega$. We see that the resistance of the resistor remains constant, while the resistance of the lightbulb increases with the current through it.

Try it yourself In which interval of potential difference is the resistance of the bulb greater than that of the resistor?

Answer Between 1.5 V and 2.0 V.

Light-emitting diodes (LEDs)

Another non-ohmic device that is present in every house is an LED–a **light-emitting diode**. LEDs are the basis of the lightbulbs that are replacing incandescent lightbulbs. How are LEDs different from incandescent bulbs? Examining an LED shows that it has two leads, one short and one long (**Figure 19.14a**). LEDs also come in different colors; we choose a green LED for our investigations in this chapter. We now connect the LED to a variable power supply, as we did with the resistor in Table 19.3, and collect data in the same way using an ammeter and voltmeter (see Figure 19.14b).

We observe several interesting patterns. First, when we connect the long lead of the LED to the *negative* terminal of the power supply, the LED does not light, no matter how much we increase the potential difference across it (as long as the potential difference across it does not exceed 10 V; above that, LED breakdown occurs, which leads to a large current and damage of the LED). Second, when we connect the long lead of the LED to the *positive* terminal of the power supply, at first the LED does not light. When the potential difference across it reaches a certain value (about +2.1 V for a green LED), it suddenly lights, and when we increase the potential difference by a small amount, its brightness increases dramatically. Can it be that the LED only allows current to pass through it in one direction, and that it glows when there is a current through it and the potential difference exceeds +2.1 V?

When we analyze the I-versus-ΔV graph from our experiment in Figure 19.14c, we find support for both aspects of our hypothesis: (1) there is no current through the LED when its long lead is connected to the negative terminal of the power supply (we will call this the "wrong" direction), and (2) the current is still zero even if the long lead of the LED is connected to the positive terminal (the "right" direction) for all potential differences below about +2.1 V.

FIGURE 19.14 A green LED. The electric circuit in (b) is used to collect the I-versus-ΔV data plotted in (c).

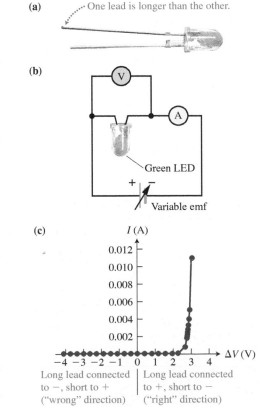

We also notice that the current "shoots up" when we increase the potential difference above +2.1 V. Note that at a certain voltage above 3 V, the LED stops glowing, even though it is connected the right way, and never lights again, even if we decrease the potential difference back to +2.1–2.3 V. The current through it becomes so large that it burns and destroys the LED.

Based on the *I*-versus-Δ*V* graph, we can say that the LED is a non-ohmic element and has asymmetric resistive properties. The resistance of the LED is infinitely large when its long lead is at a lower potential than the short lead, and it is variable (it depends on the voltage across the LED) when its long lead is at a higher potential than the short lead. For example, the resistance of a green LED is very large (infinite) below +2.1 V and becomes very small when the potential difference across it is above +2.1 V. The 2.1-V potential difference is called the **opening voltage** of the LED. Note that the opening voltage does not have a precisely defined value. When the potential difference across the LED is approaching this value, the current through the LED starts to increase, and eventually the LED starts glowing.

LEDs come in different colors. The color of an LED's light does not come from the plastic cover, but from the physical materials that comprise it. LEDs of different colors have similar properties, but different opening voltages. For example, the opening voltage for a red LED is about 1.6 V and that for a blue LED is about 2.3 V.

To represent an LED in a circuit diagram, we use the symbol shown in **Figure 19.15**. The triangle resembles an arrow and points in the direction in which the LED could conduct current. The long lead of the LED must be at higher potential than the short lead for the LED to glow (assuming that the potential difference across it is above its opening voltage).

FIGURE 19.15 The LED circuit symbol.

Resistances of connecting wires and switches

So far, we have investigated the resistances of resistors, incandescent lightbulbs, and LEDs. However, almost every electric circuit contains connecting wires and switches. What can we learn about their resistance? First, we will investigate a slightly more complicated circuit (**Figure 19.16**). A battery and switch are connected to two resistors, $R_1 = 25\ \Omega$ and $R_2 = 50\ \Omega$. There are four ammeters in the circuit and five voltmeters. When the switch is closed, the meters have the readings as shown in the figure. The current through all the elements is the same. The same current enters the 25-Ω resistor then leaves it, and that same current enters the 50-Ω resistor and then leaves it. Current is not used up!

FIGURE 19.16 Examine the ammeter and voltmeter readings.

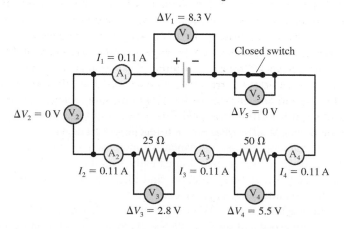

Voltmeter 2 across a connecting wire and voltmeter 5 across the closed switch both have zero readings. This is only possible if the resistance of the connecting wire and the closed switch is very small, or almost zero: $\Delta V = IR \approx I \cdot 0 = 0$. Now let's open the switch and use Ohm's law to make predictions about the meter readings.

TESTING
EXPERIMENT TABLE 19.4

Applying Ohm's law to an open circuit

VIDEO
TET 19.4

Testing experiment	Prediction	Outcome			
		Ammeter readings		**Voltmeter readings**	
Use the circuit in Figure 19.16 but with the switch open.	We predict that the current in the circuit is zero, the ammeters will measure zero current $I = 0$, and according to Ohm's law, voltmeters 2–5 will measure zero potential difference $\Delta V = IR = 0R = 0$.	Ammeter 1	0 A	Voltmeter 1	8.5 V
		Ammeter 2	0 A	Voltmeter 2	0 V
		Ammeter 3	0 A	Voltmeter 3	0 V
		Ammeter 4	0 A	Voltmeter 4	0 V
				Voltmeter 5	8.5 V

Conclusion

We predicted 0 for voltmeter 5, but it measured 8.5 V. The outcome of the experiment does not match our prediction. We need to either revise Ohm's law or examine how we apply it.

When an outcome does not match a prediction, we need to evaluate the reasoning that led to the prediction. When the switch is open, there is no current in the circuit, so on first glance Eq. (19.3) written as $\Delta V = IR$ predicts a zero potential difference. Let us look at the battery first. Even when there is no current in the circuit, the battery has potential difference across the terminals equal to its emf. Thus, the nonzero reading of voltmeter 1 makes sense. But why would there be potential difference across the switch when there is no current in the circuit? When the switch is open, it stops being a conductor; its resistance becomes infinite. We cannot apply relation $\Delta V = IR$, as we do not know the result of multiplying zero by infinity. Therefore, when we made a prediction that did not match the outcome, it was not Ohm's law that failed us, but how we applied it. How do we explain the existence of the potential difference across the switch? When the switch is open, the whole part of the circuit connected to the positive terminal of the battery is at a potential of 8.5 V. The part of the circuit connected to the negative battery terminal is at a potential of 0.0 V (**Figure 19.17**).

We can find a similarity in the behavior of a switch in a circuit and an LED. Remember that for the "wrong" direction of an LED, the current through the LED is zero. The same holds also for an LED that is connected in the "right" direction but has a potential difference across it that is less than its opening voltage. Thus we can equate an LED connected in the "wrong" direction and an LED connected in the "right" direction for voltages below opening voltage to an open switch. If this is correct and we measure the potential difference across the LED for these conditions, we should measure the same voltage as that of the battery. This is exactly what happens (see **Figure 19.18**)! However, once there is electric current through a switch, the potential difference across it is zero for any current through it, whereas the potential difference across an LED changes very little while the current might change by many factors of 10. This is due to the rapidly decreasing resistance of an LED in the range of glowing potential differences. Therefore, it is useful to remember that when an LED is glowing, the potential difference across it is always approximately the opening voltage.

TIP Any burned out lightbulb causes the current in the line with the bulb to be zero. It is like an open switch. If the light switch on the wall is on, there may be a 120-V potential difference across the contact points in the bulb canister. It is not safe to touch the contact points!

FIGURE 19.17 Electric potential in a circuit with an open switch.

Minus signs indicate that the electric potential of this part of the circuit is 8.5 V less than the potential of the positive terminal of the battery.

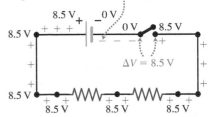

FIGURE 19.18 An LED is similar to an open switch when (a) the LED is connected in the "wrong" direction, and (b) the LED is connected in the "right" direction but the potential difference across it is less than the opening voltage.

(a)

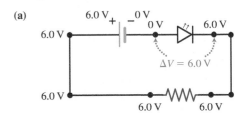

(b)

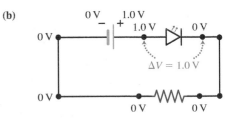

REVIEW QUESTION 19.4 Why does it make sense that all ammeters in Figure 19.16 have the same reading?

19.5 Qualitative analysis of circuits

So far, most of our analyses have applied to circuits with one element connected to a battery. Real-world circuits contain many elements. In this section we will learn about common arrangements of circuit elements and analyze the effect of these arrangements on the electric current in the circuits.

FIGURE 19.19 The bulbs are in series.

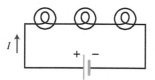

Circuits in series

In Observational Experiment **Table 19.5** we use small incandescent lightbulbs as circuit elements. We start by analyzing bulbs arranged in **series**, one after the other, as shown in **Figure 19.19**.

OBSERVATIONAL EXPERIMENT TABLE 19.5   **Multiple bulbs arranged in series**

VIDEO
OET 19.5

Observational experiment	Analysis
Experiment 1. One lightbulb connected to a battery is bright. 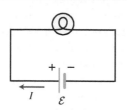	The bulb is lit when the circuit is complete.
Experiment 2. Two identical bulbs connected one after the other (in series) to the battery from Experiment 1 are each dimmer than the bulb in 1 but are equally bright. 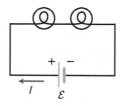	• There must be less current through the bulbs than in Experiment 1. • Each bulb must have the same lower current through it (since they are equally bright).
Experiment 3. Three identical bulbs connected in series to the same battery are each dimmer than the bulbs in Experiments 1 and 2 but are equally bright. 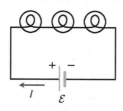	• There must be less current through the bulbs than in Experiments 1 and 2. • Each bulb must have the same even lower current through it (since they are equally bright.)

Patterns

• The brightness of all identical bulbs arranged in series is the same.
• Adding more bulbs arranged in series decreases the brightness of all bulbs.

The first pattern in Table 19.5 makes sense if we assume that the magnitude of the current through a bulb affects how bright the bulb is. Electric current is the flow of electrons. Since none of those electrons can escape the circuit or accumulate somewhere in the circuit, the current must be the same in the first bulb as in the last bulb in the series circuit—it is not used up. This reasoning is consistent with the results of the experiment in Figure 19.16 that showed us that in a similar circuit with resistors instead of the bulbs, the current was the same through all of the resistors.

The second pattern we observed in Table 19.5 indicated that adding more bulbs in series reduced the brightness of each bulb. Evidently, there is less current through series bulbs when there are more of them. Less current means that the amount of charge passing through a cross-sectional area of any element of a circuit per unit time is smaller. Adding more bulbs must increase the resistance of the circuit, reducing the electric current *everywhere* in the circuit (similar to how putting more filters in a hose slows the flow of water through the whole water system). **A battery is not a source of constant current**.

Circuits in parallel

We can also arrange bulbs in **parallel**: side by side with the terminals on each side connected together. We investigate simple parallel circuits in Observational Experiment Table 19.6.

OBSERVATIONAL EXPERIMENT TABLE 19.6 Multiple bulbs arranged in parallel

VIDEO OET 19.6

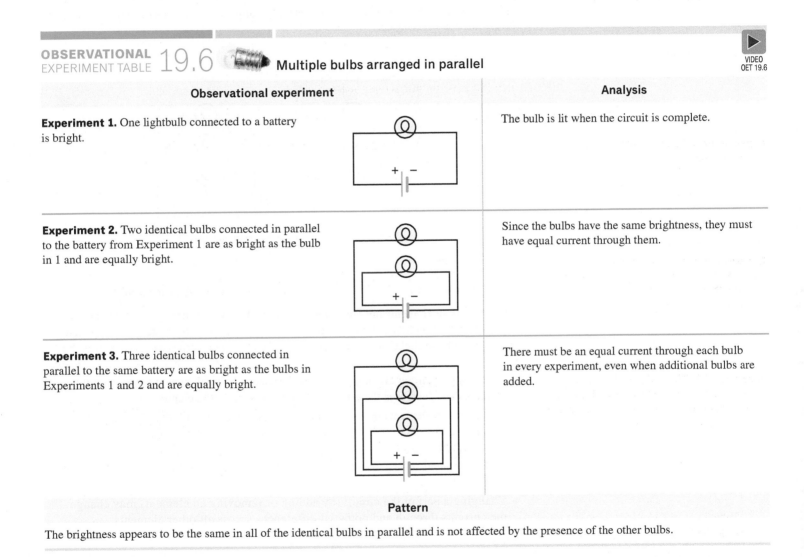

Observational experiment	Analysis
Experiment 1. One lightbulb connected to a battery is bright.	The bulb is lit when the circuit is complete.
Experiment 2. Two identical bulbs connected in parallel to the battery from Experiment 1 are as bright as the bulb in 1 and are equally bright.	Since the bulbs have the same brightness, they must have equal current through them.
Experiment 3. Three identical bulbs connected in parallel to the same battery are as bright as the bulbs in Experiments 1 and 2 and are equally bright.	There must be an equal current through each bulb in every experiment, even when additional bulbs are added.

Pattern

The brightness appears to be the same in all of the identical bulbs in parallel and is not affected by the presence of the other bulbs.

How can we explain the pattern in Table 19.6? In order for the identical bulbs to be equally bright and have the same brightness, the current must be the same through each of them. In other words, when bulbs are arranged in parallel, they behave as though they were connected to the battery all by themselves and are not affected by the presence of the other bulbs. Since the bulbs are identical, they have identical resistance (same resistance R for the same current I); the connecting wires have almost zero resistance, independent of their length. Therefore, the potential difference across all bulbs should be the same ($I = \Delta V/R$) and equal to the potential difference across the battery. The battery must drive a current through the bulbs simultaneously—the total current driven

CHAPTER 19 DC Circuits

by the battery must double for two parallel bulbs and triple for three parallel bulbs. We see again that the battery is *not* a constant current source—it sends a smaller current through bulbs arranged in series and a larger total current through bulbs in parallel. This also means that multiple bulbs connected in parallel must have less collective resistance than a single bulb because as external resistance decreases, current through the battery increases.

Conventionally, circuits such as those in Table 19.6 are represented in two different ways; see **Figure 19.20**. The two circuits in the figure are equivalent. The points that have the same potential are connected by a wire, as in the circuit on the left, or connected directly to each other, as in the circuit on the right.

FIGURE 19.20 The bulbs connected in parallel to the battery have the same potential difference across them as a single bulb.

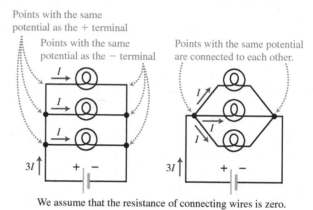

We can now summarize our qualitative investigations of electric circuits:

- When circuit elements are connected in series, the current through each element is the same; when circuit elements are connected in parallel, the battery is connected across each parallel element and the potential difference across each element is the same.

- Adding more circuit elements in series *decreases* the total current through all the elements and increases the effective total resistance of the elements.

- Adding more elements in parallel *increases* the total current through the parallel elements and reduces the effective total resistance of the elements.

- The battery is *not* a source of constant current: it drives a smaller current through circuit elements arranged in series and a larger total current through circuit elements in parallel.

- Changing a part of the circuit (by adding or removing an element) may change the currents through and potential differences across all other elements.

CONCEPTUAL EXERCISE 19.4   **Current versus charge storage**

You have two identical resistors with resistance R and a battery with a *storage capacity* Q (the total charge it can move before it becomes "dead"). Compare the current through the battery and the total time before the battery dies if (a) one resistor is connected to the battery, (b) both resistors are connected to the battery in series, and (c) both resistors are connected to the battery in parallel.

Sketch and translate The sketches of the circuits at right include the givens and the quantities to compare.

Compare I_R, I_{Ser}, I_{Par}
Compare t_R, t_{Ser}, t_{Par}

Simplify and diagram We drew the circuit diagrams in the previous step, so we use this step to reason about the process. Given that the resistors are identical, we know that the current through the resistors and the battery in series, I_{Ser}, is smaller than the current through a single resistor connected to the battery, I_R. That is, $I_{Ser} < I_R$. When the resistors are in parallel, the current through each resistor is equal to I_R. The battery needs to send these currents simultaneously; therefore, the current through the battery must be equal to $I_{Par} = 2I_R$. To summarize: $I_{Ser} < I_R < I_{Par}$.

The storage capacity of the battery is equal to the product of the current and time interval during which the charges are moving through the battery: $Q = I\,\Delta t$, or $\Delta t = Q/I$. Therefore, for a given storage capacity Q, the time interval before the battery is dead is inversely proportional to the current through the battery. Combining the relationship between the currents and the last equation, we get $t_{Par} < t_R < t_{Ser}$. This makes sense: the smaller the current through the battery, the longer the time before the battery dies.

Try it yourself You have three identical resistors with resistance R and a battery with a storage capacity Q. Compare the current through the battery and the total time before the battery dies if one resistor is connected to the battery compared to (a) when three resistors are connected to the battery in series, and (b) when three resistors are connected to the battery in parallel.

Answer (a) The current through the battery is three times smaller, and the total time before the battery dies is three times longer. (b) The current through the battery is three times larger, and the battery time is three times shorter.

How is your house wired?

Although the current in the electric circuit in our homes is different in many ways from the current we study in this chapter, there are many similarities that allow us to apply what we are learning in this book to the electrical wiring in a home.

You turn on a lightbulb, a computer, and a washing machine. Are they connected in series or in parallel? There are three possible arrangements (**Figure 19.21**): (a) all the devices are in series, (b) all are in parallel, or (c) some combination of the two. If they are all in series, when you turn one of the devices off, the circuit will open. There will be no current through the circuit, and all of the devices will be off. From your everyday experience, you know that does not happen. The devices must not be in series. If they are all in parallel, as in (b), then turning any one device on or off will not affect the others. This is what we observe, so parallel wiring is a possibility. In (c) the devices are arranged neither all in series nor all in parallel. Still, turning off one device (the computer in our case) also turns off others. This is not what happens in your house, so this sort of arrangement is not possible. The only choice that is consistent with the way devices in your house behave is (b), parallel.

REVIEW QUESTION 19.5 What experimental evidence supports the hypothesis that in your house the appliances are connected in parallel and not in series?

19.6 Joule's law

We used identical incandescent lightbulbs in our experiments in the previous section. Identical bulbs in series in a closed circuit have the same brightness. However, if we connect the two bulbs shown in **Figure 19.22a** in series to a battery, we observe that bulb 1 is brighter than 2. If we reverse the order of the bulbs, bulb 1 is still the brighter one (Figure 19.22b). We know that in a series circuit, there is the same current through each bulb. Thus, it is not just the current through a bulb that determines its brightness. In this section we will learn what other quantities affect a bulb's brightness.

Electric power

An incandescent lightbulb filament is hot and glows when there is current through it (the free electrons drift through the lattice and interact with ions, making the ions vibrate faster). Consider the bulb and connecting wires (electrons and electric field) as the system and the battery as an external object. The battery does work, increasing

FIGURE 19.21 The possible ways to wire household appliances.

(a) All in series

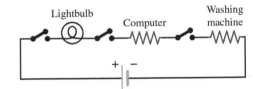

(b) All in parallel

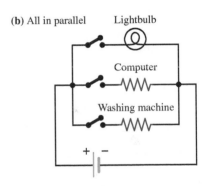

(c) A combination

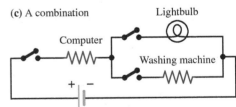

FIGURE 19.22 The bulb's brightness depends on something other than the current through the bulb.

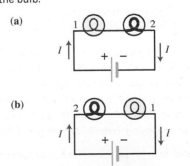

Bulb 1 is brighter than bulb 2 in both circuits.

the electric potential energy of the system. When there is current through a circuit element, the electric potential energy is continuously converted into other forms of energy (thermal energy, light energy, mechanical energy, etc.). Earlier (in Chapter 7) we learned that power is the physical quantity that describes the rate of energy conversion. In this case it is the rate at which the device converts electric energy into thermal energy (or other forms of energy):

$$P = \left| \frac{\Delta U_{th}}{\Delta t} \right|$$

This conversion leads to the decrease of the electric potential energy of the system U_q. Therefore,

$$P = \left| \frac{-\Delta U_q}{\Delta t} \right|$$

As electrons (total electric charge $-Q$) travel through a resistive device, they travel from the side that is at low electric potential to the side that is at high electric potential. This movement of electrons results in a change in the electric potential energy of the system $\Delta U_q = q\,\Delta V = -Q\,\Delta V$. Therefore,

$$P = \left| \frac{Q\,\Delta V}{\Delta t} \right| = \frac{Q}{\Delta t}|\Delta V| = I|\Delta V|$$

The above has the correct power units (watts; $1\text{ W} = 1\text{ J/s}$).

$$A \cdot V = \frac{C}{s} \cdot \frac{J}{C} = \frac{J}{s} = W$$

James Joule determined this expression in experiments using a current-carrying wire to warm water. Later, Heinrich Lenz and Alexandre-Edmond Becquerel repeated the experiments using alcohol instead of water and achieved similar results. Now the expression for the rate of electric potential energy conversion into thermal energy is called **Joule's law**.

> **Joule's law** The rate P at which electric potential energy is converted into other forms of energy ΔU_{th} in a resistive device equals the magnitude of the potential difference ΔV across the device multiplied by the current I through the device:
>
> $$P = \left| \frac{\Delta U_{th}}{\Delta t} \right| = I|\Delta V| \qquad (19.4)$$

Because potential difference and current are related through Ohm's law, $I = \Delta V/R$, Joule's law can be written in two alternate forms:

$$P = I|\Delta V| = I(IR) = I^2 R$$

$$P = I|\Delta V| = \left(\frac{\Delta V}{R} \right)|\Delta V| = \frac{(\Delta V)^2}{R} \qquad (19.5)$$

Examine the different forms of Eq. (19.5). The top version says that the power is proportional to the resistance of the circuit element; the bottom version says that it is inversely proportional. Do these contradict each other? As the current through and potential difference across the element are not independent of each other, there is no contradiction. It is usually most convenient to use the $I^2 R$ expression when comparing power in elements connected in series and $(\Delta V)^2/R$ when elements are connected in parallel.

If we assume that the brightness of a bulb is directly related to its power, we can use Joule's law to explain the experiment at the beginning of this section. The lightbulbs in Figure 19.22 had different brightnesses even though the current through them was the

TIP The word "electricity" is used in many ways. People say that electricity flows, that electricity is used to heat buildings, and that electricity comes out of a plug in the wall. When people say that electricity flows, they are referring to the flow of charged particles in wires (electric current); when they talk about heating with electricity, they mean electric potential energy being transformed into thermal energy that heats the environment; and when they talk about wall sockets, they usually refer to a 120-V effective potential difference across any appliance plugged into the socket. When you use the word electricity, make sure you understand which of these ideas you are referring to.

same. Based on Eq. (19.4), $P = I|\Delta V|$, the potential difference across bulb 1 had to be greater than across bulb 2. This could only happen if bulb 1 had a greater resistance than bulb 2, $R_1 > R_2$, as $I_1 = I_2$; thus $P_1 = I^2R_1 > P_2 = I^2R_2$. More resistance means a higher rate of conversion of electric energy to thermal energy!

If our understanding of electric power is correct, then the same two resistors also will have different relative powers when connected differently. Consider two resistors: A is a 25-Ω resistor and B is a 50-Ω resistor. They are connected to the same power supply that has potential difference of 9 V across its terminals first in series and then in parallel. When the resistors are in series, the current through them is the same; therefore, the power of resistor B is greater than that of A because

$$P = I^2R; \quad I_A = I_B; \quad R_A < R_B \implies P_A < P_B$$

When they are in parallel, the potential difference across them is the same; therefore, the power of resistor A is greater than that of B:

$$P = \frac{\Delta V^2}{R}; \quad \Delta V_A = \Delta V_B; \quad R_A < R_B \implies P_B < P_A$$

Thus the same resistor (or any electric appliance) can be more or less powerful depending on the circuit it is in; power is not a fixed characteristic. This is important to know when you buy electric appliances.

Power of incandescent lightbulbs and LEDs

In the last few years, LED lightbulbs have almost completely replaced incandescent lightbulbs in our homes. Why? A simple observation can give us a hint. When we touch a bright incandescent lightbulb, it is very hot; however, LED-based bulbs are not hot to touch. Therefore, we can hypothesize that traditional bulbs convert a significant fraction of the electric energy to thermal energy rather than to light, while LEDs convert most of the electric energy delivered to them to light energy. Both light energy and thermal energy leave the bulbs and LEDs through the process of radiation.

We can conduct an experiment to investigate this difference quantitatively using a variable power supply, a small incandescent lightbulb, and a white LED. Both light sources are enclosed in white ping-pong balls to ensure that the light they emit is spread evenly, making it easier to compare their brightness (Figure 19.23a). In order to compare the light sources more objectively, we will use a special device (a light sensor) that measures the brightness of both sources; its sensitivity to light is similar to that of our eyes. We will change the potential difference across the power supply until the light sensor has the same reading for both sources when placed at the surface of the ping-pong ball. (In Figure 19.23a, the LED appears to have a somewhat different color than the bulb, but despite this, their brightness as measured by the sensor is the same). The data collected in the experiment are shown in Figure 19.23b. A pattern is clearly visible from the data: the white LED needs about 10 times less electric power to emit light of the same brightness as an incandescent lightbulb. Thus the old-time incandescent lightbulb must be converting the rest (90%) of the energy into thermal energy or some other form of light that cannot be detected by our eyes. Figure 19.23c shows thermal images for both light sources: the LED is at room temperature and the bulb is much warmer. We will learn in Chapter 28 the mechanism behind the high efficiency of LEDs.

Paying for electric energy

Electric power companies charge their customers according to the amount of electric potential energy that is converted into other energy forms in the devices that the customer uses. Utility companies do not use the joule. Instead, they use an energy unit called the **kilowatt-hour**. A kilowatt-hour (kW·h) is the electric potential energy that a 1000-W device transforms to other energy forms in a time interval of 1 h ($1 \text{ kW·h} = 3.6 \times 10^6 \text{ J}$).

FIGURE 19.23 Comparing energy conversions for an LED versus an incandescent lightbulb.

(a) Experimental setup

LED   Incandescent lightbulb

(b) Table of measurements taken

	LED	Incandescent lightbulb
ΔV (V)	3.1	3.6
I (A)	0.020	0.190
P (W)	0.062	0.680
B (lux)*	450	450

*Brightness of light measured at the surface of the ping-pong ball

(c) Thermal images of the experiment

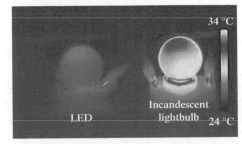

34 °C

LED   Incandescent lightbulb   24 °C

19.7 Kirchhoff's rules

Most electric circuits consist of combinations of several resistive elements, switches, one or more power supplies that may have a variable potential difference output, and a variety of other circuit elements. In this section we will develop techniques to determine the electric current through each circuit element in a complicated circuit.

Kirchhoff's loop rule

FIGURE 19.24 The changing electric potential around an electric circuit.

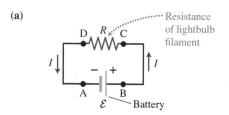

(a)

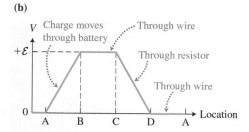

(b)

Let us first examine the changes in electric potential in a simple circuit. The current in the circuit shown in **Figure 19.24a** is counterclockwise. We trace the change in electric potential as we move with these charges around the circuit (Figure 19.24b). The potential when we return to the starting position must be the same as when we left that position (the potential cannot have multiple values at one location). We see from the graph that

Changes in electric potential

Across battery A to B		Across connecting wire B to C		Across resistor C to D		Across connecting wire D to A	
$+\mathcal{E}$	$+$	0	$+$	$(-IR)$	$+$	0	$= 0$

A pattern (that we shall call the loop pattern) accounts for the data:

> The algebraic sum of the changes in electric potential around any closed circuit loop is zero.

Let us test this pattern.

EXAMPLE 19.5 **Testing the loop pattern**

Use the loop pattern to predict the potential difference between point A and point B in the electric circuit shown in the figure.

Represent mathematically We apply the loop pattern going clockwise from A:

$$\Delta V_{10\,V} + \Delta V_{20\,\Omega} + \Delta V_{4\,V} + \Delta V_{60\,\Omega} + \Delta V_{40\,\Omega} = 0$$
$$(+10\,V) + [-I(20\,\Omega)] + (-4\,V) + [-I(60\,\Omega)] + [-I(40\,\Omega)] = 0$$

Notice that the potential difference across the 4.0-V battery was negative since we moved from the positive to the negative terminal of that battery. Also, the potential differences across the resistors were negative because we moved across these resistors in the clockwise downstream direction of the electric current.

Solve and evaluate We can now determine the current:

$$I = \frac{10.0\,V - 4.0\,V}{20\,\Omega + 60\,\Omega + 40\,\Omega} = 0.050\,A$$

To determine the potential difference from A to B, we follow the circuit along any continuous path from A to B and add the potential differences across each element in the circuit along that path. There are two possible

Sketch and translate To predict the potential difference between points A and B, we first use the loop pattern to determine the current in the circuit shown in the figure.

Simplify and diagram Assume that the wires and batteries have negligible (zero) resistance compared to that of the resistors. To determine the direction of the current, note that the 10-V battery has a larger emf than the 4-V battery. Therefore, the current in the circuit will be clockwise.

paths. We first move clockwise from A through the 10.0-V battery and 20-Ω resistor:

$$\Delta V_{AB} = \Delta V_{10\,V} + \Delta V_{20\,\Omega} = +10\text{ V} - (0.050\text{ A})(20\text{ }\Omega)$$
$$= +9.0\text{ V}$$

The V field value at B is 9.0 V higher than at A.

We can also calculate the potential change from A to B by moving counterclockwise upstream through the 40-Ω and 60-Ω resistors (opposite the current) and through the 4.0-V battery from its negative to positive terminal.

$$\Delta V_{AB} = \Delta V_{40\,\Omega} + \Delta V_{60\,\Omega} + \Delta V_{4\,V}$$
$$= +(0.050\text{ A})(40\text{ }\Omega) + (0.050\text{ A})(60\text{ }\Omega) + 4.0\text{ V}$$
$$= +9.0\text{ V}$$

The potential difference from A to B is the same no matter which path we take. When we take a voltmeter and connect its terminals to A and B (its negatively marked terminal to A and its positively marked terminal to B), we can see if the outcome and prediction are consistent. The voltmeter reads about 9 V, supporting the loop pattern.

Try it yourself Suppose that in this circuit, there was an open switch inserted in the vertical wire on the left side before position A. Apply the loop pattern to determine the potential difference across the switch.

Answer

finding in Table 19.4.
across the switch when it is open is consistent with our experimental
will be at a lower potential. The presence of potential difference
$\Delta V_{\text{switch}} = -6$ V. The top of the switch closer to the 10-V battery

---

TIP To indicate the signs of the electric potential changes across the resistors in the previous example, we chose and indicated in the circuit drawing the assumed direction of the current. The potential change across a resistor if moving in the direction of the current is $\Delta V = -IR$; if moving across the resistor opposite the current it is $\Delta V = +IR$. We *must* indicate a current direction in the circuit diagram and keep the signs of I and ΔV consistent while making a trip around the circuit.

This "loop pattern" was first developed in 1845–1846 by German physicist Gustav Robert Kirchhoff (1824–1887) and is known as **Kirchhoff's loop rule**. The rule is general: it applies to any loop in any circuit, even when a circuit has multiple loops.

Kirchhoff's loop rule The sum of the electric potential differences ΔV across the circuit elements that make up a closed path (called a loop) in a circuit is zero.

$$\sum_{\text{Loop}} \Delta V = 0 \tag{19.6}$$

EXAMPLE 19.6 **LEDs and the loop rule**

The opening voltage for a green LED is about 2.1 V, and the current that will make it glow with full brightness is about 20 mA. What resistor should we connect in series with the LED so we can safely power the LED using a 6.0-V battery?

Sketch and translate We draw a circuit diagram and label knowns and unknowns. The green LED would burn out if we connected it directly to the 6.0-V battery because the I-versus-ΔV graph for an LED is very steep once the potential difference across it is larger than the opening voltage, so the current would be too large. If we connect a suitable resistor in series with the LED, then we can limit the current to the allowed value of 20 mA.

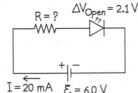

$R = ?$ $\Delta V_{\text{Open}} = 2.1$ V

$I = 20$ mA $\mathcal{E} = 6.0$ V

Simplify and diagram We assume the resistance of the connecting wires to be zero and the potential difference across the 6.0-V battery to be 6.0 V when there is current in the circuit.

Represent mathematically Based on the loop rule,

$$\Delta V_{\text{LED}} + \Delta V_R = 6.0\text{ V}$$

In addition, we know that the voltage across the normally glowing LED is about the same as the opening voltage of the LED:
$\Delta V_{\text{LED}} = \Delta V_{\text{Open}} = 2.1$ V.

We can determine the potential difference across the resistor using Ohm's law:

$$I = \frac{\Delta V_R}{R} \implies \Delta V_R = IR$$

In our case, the current is given: $I = 0.020$ A.

Solve and evaluate We combine the two equations,
$\Delta V_{\text{LED}} + \Delta V_R = 6.0$ V and $\Delta V_R = (0.020\text{ A})R$, to get

$$2.1\text{ V} + (0.020\text{ A}) \cdot R = 6.0\text{ V} \implies R = 195\text{ }\Omega$$

(CONTINUED)

We can evaluate our result using extreme-case analysis. We know that the current of a circuit with no LED should be larger than a circuit with the LED. This is exactly what we obtain: $\dfrac{6\text{ V}}{195\ \Omega} = 0.03$ A, larger than with the LED. If there were no resistor, the LED would burn out because the current would exceed 0.02 A. Resistors are therefore essential elements in LED circuits.

Try it yourself What is the voltage across the LED in this circuit if we switch the terminals of the battery?

Answer   6.0 V.

Internal resistance of a battery

FIGURE 19.25 Investigating a 9.0-V battery.

(a)

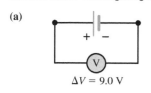

$\Delta V = 9.0$ V

(b)

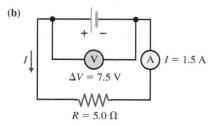

$\Delta V = 7.5$ V

$R = 5.0\ \Omega$

(c)

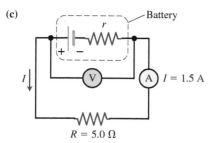

$R = 5.0\ \Omega$

In all our previous examples, the emf and potential difference across the battery terminals were assumed to be the same. Let us investigate whether this is truly the case in real life. Imagine that you take a 9.0-V battery and use a voltmeter to measure potential difference across its terminals. The voltmeter reads 9.0 V (**Figure 19.25a**). Then you build the circuit shown in Figure 19.25b. An ammeter reads a 1.5-A current through the 5.0-Ω resistor. A voltmeter reads 7.5 V across the battery and -7.5 V across the resistor. Notice that we found two unexpected results: the potential difference is smaller than the emf of the battery, and the current is less than expected $\left(I_{\text{exp}} = \dfrac{\mathcal{E}}{R} = \dfrac{9.0\text{ V}}{5.0\ \Omega} = 1.8\text{ A} \right)$. What are possible reasons for these anomalous data?

The current would be smaller than predicted if there were some other resistor connected to the battery. Perhaps the battery has its own **internal resistance** r (see Figure 19.25c). Using this hypothetical resistance, we can write the loop rule as

$$+\mathcal{E} - Ir - IR = 0$$

and use it to find the internal resistance of the battery,

$$r = \frac{\mathcal{E} - IR}{I}$$

In our case, the internal resistance of the battery is

$$r = \frac{\mathcal{E} - IR}{I} = \frac{9.0\text{ V} - (1.5\text{ A})(5.0\ \Omega)}{1.5\text{ A}} = 1.0\ \Omega$$

This explains why the measured current is less than 1.8 A:

$$I_1 = \frac{\mathcal{E}}{R + 1.0\ \Omega} = \frac{9.0\text{ V}}{5.0\ \Omega + 1.0\ \Omega} = 1.5\text{ A}$$

The idea of internal resistance is supported by the observation that batteries feel warm to the touch after they have been operating for a while. We can also explain why the voltmeter reading across the battery is smaller in a closed circuit than in an open circuit. When there is current in the circuit, the loop rule is

$$\mathcal{E} - Ir = IR \tag{19.7}$$

IR is the potential difference across the resistor, and at the same time it is the potential difference measured across the battery, because the resistor is connected directly to the battery and we assume that the resistance of the wires is negligible.

$$\Delta V_{\text{batt}} = IR = (1.5\text{ A})(5.0\ \Omega) = 7.5\text{ V}$$

We can write that the potential difference across the battery is equal to $\Delta V_{\text{batt}} = \mathcal{E} - Ir$. Notice that whenever there is a current through the battery, the potential difference across it is less than its emf; how much less depends on the current through the battery and on its internal resistance. If there is no circuit attached to the battery (no current), then the potential difference across it equals its emf. If $r = R$, then $\Delta V_{\text{batt}} \approx \mathcal{E}/2$. If $r \ll R$, then $\mathcal{E} \approx \Delta V_{\text{batt}}$.

TIP When using the loop rule for a closed circuit, follow these sign conventions:

- The potential difference across a resistor is $-IR$ when the loop traverses the resistor in the direction of the current through it and $+IR$ when the loop traverses the resistor in the direction opposite the current through it.
- The potential difference across a battery (assuming zero internal resistance) is $+\mathcal{E}$ when the loop traverses the battery from its negative terminal to its positive terminal and $-\mathcal{E}$ when the loop traverses the battery from its positive terminal to its negative terminal. The potential difference instead is $\pm(\mathcal{E} - Ir)$ if the battery's internal resistance cannot be neglected.
- Assuming the resistance of a connecting wire is zero, all points along that wire are at the same potential.

Kirchhoff's junction rule

We developed the loop rule for a circuit with only one loop. However, many electric circuits have several loops. In multiloop circuits, the charged particles moving along a wire can arrive at a junction where several wires meet. The charged particles comprising the current cannot vanish or be created out of nothing (electric charge is a conserved quantity). Thus, the sum of the currents into a junction must equal the sum of the currents out of a junction.

Consider the multiloop circuit shown in **Figure 19.26**. The battery at the top causes a counterclockwise current I in the circuit. The current entering junction A equals the sum of the currents leaving A. Expressed mathematically:

$$I = I_1 + I_2$$

When the currents I_1 (through R_1) and I_2 (through R_2) meet at junction B, the two currents recombine to form a single larger current I that leaves junction B:

$$I_1 + I_2 = I$$

For the circuit in Figure 19.26, the current I moving upward through the wire on the right side of the circuit equals the downward current I through the wire on the left side of the circuit; the current leaving the positive terminal of the battery equals the current entering the negative terminal. The current is not used up in a circuit. We can summarize this splitting and joining of currents at junctions as **Kirchhoff's junction rule**.

FIGURE 19.26 Kirchhoff's junction rule for electric currents.

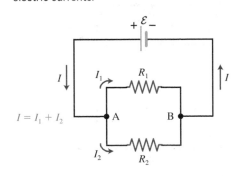

Kirchhoff's junction rule The total rate at which electric charge enters a junction equals the total rate at which electric charge leaves the junction:

Sum of currents into junction = Sum of currents out

In symbols:

$$\sum_{\text{In}} I = \sum_{\text{Out}} I \qquad (19.8)$$

Short circuit

Imagine that you accidentally connect the terminals of a battery with a connecting wire of approximately zero resistance $R = 0$ (**Figure 19.27**). The wire and the battery become very hot. Notice that when the external resistance is $R = 0$, the current through the battery is

$$I = \frac{\mathcal{E}}{R + r} = \frac{\mathcal{E}}{0 + r} = \frac{\mathcal{E}}{r}$$

FIGURE 19.27 A short circuit across a battery can cause a very large current through it.

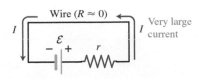

Because the internal resistance of the battery is usually small, the current becomes large; it can burn the connecting wire and destroy the battery. This situation is called a **short circuit**. Every time you build an electric circuit, before you turn it on, check that you did not connect a single wire directly to the two battery terminals without any other resistive devices in series with it.

REVIEW QUESTION 19.7 Where is the electric potential higher: where the current "enters" a resistor or where it "leaves" it? Where is the electric potential higher: where the current "enters" a conducting wire or where it "leaves" it?

19.8 Resistor and capacitor circuits

In the previous sections we learned about resistors in series and parallel. Here we will learn how to determine mathematically the equivalent resistor with which we can replace a combination of several resistors.

Resistors in series

FIGURE 19.28 A single equivalent resistor replaces the three series resistors.

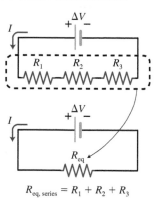

$$R_{eq, \, series} = R_1 + R_2 + R_3$$

Suppose that you arrange three resistors (R_1, R_2, and R_3) in series with a battery of emf $\mathcal{E}$. Can we replace these three resistors with a single resistor without changing the current through the battery (**Figure 19.28**)? Use Kirchhoff's loop rule and imagine traveling counterclockwise around the circuit starting at the negative terminal of the battery:

$$\Delta V - IR_1 - IR_2 - IR_3 = 0$$

where ΔV is the potential difference across the battery when there is current in the circuit.

$$\Rightarrow I = \frac{\Delta V}{R_1 + R_2 + R_3}$$

The current through each element in this circuit is the same since there are no junctions. It appears that a single resistor $R = R_1 + R_2 + R_3$ would have the same effect on the current in the circuit as the three separate resistors R_1, R_2, and R_3.

Series resistance When resistors are connected in series,

- The current through each resistor is the same: $I_1 = I_2 = I_3 = \cdots$
- The potential difference across each resistor is $\Delta V_1 = IR_1$, $\Delta V_2 = IR_2$, $\cdots$
- The potential difference across the entire series arrangement of resistors is $\Delta V = \Delta V_1 + \Delta V_2 + \Delta V_3 + \cdots$
- The equivalent resistance of the resistors arranged in series is the sum of the resistances of the individual resistors:

$$R_{eq, \, series} = R_1 + R_2 + R_3 + \cdots \qquad (19.9)$$

Resistors in parallel

When resistors are arranged in parallel, one side of each resistor is connected to one common junction and the other side of each resistor is connected to another common junction (**Figure 19.29**). We know from Kirchhoff's junction rule that the current I into the left junction equals the sum of the currents through each resistor:

$$I = I_1 + I_2 + I_3$$

The potential difference between the junctions A and B at the sides of this group of parallel resistors is the same regardless of the path taken between those two points:

$$\Delta V_1 = \Delta V_2 = \Delta V_3 = \Delta V$$

We can use Ohm's law to determine the current through each resistor:

$$I_1 = \frac{\Delta V}{R_1}, I_2 = \frac{\Delta V}{R_2}, I_3 = \frac{\Delta V}{R_3}$$

Inserting the above into Kirchhoff's junction rule, we get

$$I = I_1 + I_2 + I_3 = \frac{\Delta V}{R_1} + \frac{\Delta V}{R_2} + \frac{\Delta V}{R_3}$$

Now, suppose that we replace these three resistors by a single equivalent parallel resistor $R_{\text{eq, parallel}}$ that causes the same current I in the circuit. The current will be

$$I = \frac{\Delta V}{R_{\text{eq, parallel}}}$$

Combining these two results, we get

$$\frac{\Delta V}{R_{\text{eq, parallel}}} = \frac{\Delta V}{R_1} + \frac{\Delta V}{R_2} + \frac{\Delta V}{R_3}$$

$$\Rightarrow \frac{1}{R_{\text{eq, parallel}}} = \frac{1}{R_1} + \frac{1}{R_2} + \frac{1}{R_3}$$

$$\Rightarrow R_{\text{eq, parallel}} = \left(\frac{1}{R_1} + \frac{1}{R_2} + \frac{1}{R_3} \right)^{-1}$$

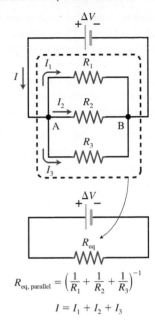

FIGURE 19.29 A single equivalent resistor replaces the three parallel resistors.

$$R_{\text{eq, parallel}} = \left(\frac{1}{R_1} + \frac{1}{R_2} + \frac{1}{R_3} \right)^{-1}$$

$$I = I_1 + I_2 + I_3$$

Parallel resistance When resistors are connected in parallel,

- The potential difference across each individual resistor is the same:

$$\Delta V_1 = \Delta V_2 = \Delta V_3 = \Delta V$$

- The sum of the currents through them equals the total current through the parallel arrangement:

$$I = I_1 + I_2 + I_3 = \frac{\Delta V}{R_1} + \frac{\Delta V}{R_2} + \frac{\Delta V}{R_3}$$

- The equivalent resistance of resistors in parallel is

$$R_{\text{eq, parallel}} = \left(\frac{1}{R_1} + \frac{1}{R_2} + \frac{1}{R_3} + \cdots \right)^{-1} \tag{19.10}$$

TIP Do not forget the (-1) exponent when you calculate the equivalent resistance of resistors in parallel.

Note that when you connect resistors in series, the equivalent resistance is larger than the largest resistance; when you connect resistors in parallel, the equivalent resistance is smaller than the smallest resistance.

Quantitative reasoning about electric circuits

In Section 19.5 we learned to reason qualitatively about currents in electric circuits. Here we analyze similar circuits quantitatively.

EXAMPLE 19.7   Power in series and parallel circuits

Assume that we have a battery and two identical resistors each of resistance R. We connect the resistors to the battery in three different configurations: (a) one resistor, (b) two resistors in series, and (c) two resistors in parallel. Determine the equivalent resistance of a single resistor that will produce the same current through the battery as produced in arrangements (b) and (c). Next, compare the total power output of the resistors in arrangements (b) and (c) with the output in (a).

Sketch and translate   We draw the three arrangements (see figures). We first determine the equivalent resistance for (b) and (c) and then use Joule's law to determine the power in each case.

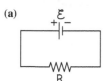
(a)

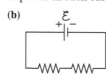
(b)

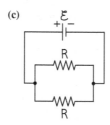
(c)

Simplify and diagram   Assume that the resistance of the connecting wires and the internal resistance of the battery do not affect the current.

Represent mathematically   The two identical resistors in series have an equivalent series resistance:

$$R_{\text{eq, series}} = R + R = 2R$$

The two identical resistors in parallel have an equivalent parallel resistance:

$$R_{\text{eq, parallel}} = \left(\frac{1}{R} + \frac{1}{R}\right)^{-1} = \left(\frac{2}{R}\right)^{-1} = \frac{R}{2}$$

The current in each simplified circuit (with the multiple resistors replaced by a single equivalent resistor) equals the battery emf divided by the circuit equivalent resistance:

$$I = \frac{\mathcal{E}}{R_{\text{eq}}}$$

The power output of the equivalent resistor (the rate at which electric potential energy is transformed into thermal energy) is

$$P = I^2 R_{\text{eq}}$$

Solve and evaluate   (a) Circuit (a) has resistance R. The current in this circuit is $I_{\text{One}} = \mathcal{E}/R$. The power output of this circuit is

$$P_{\text{One}} = I_{\text{One}}^2 R = \left(\frac{\mathcal{E}}{R}\right)^2 R = \frac{\mathcal{E}^2}{R}$$

(b) The equivalent resistance of circuit (b) is $R_{\text{Series}} = 2R$. The current in this circuit is

$$I_{\text{Series}} = \frac{\mathcal{E}}{R_{\text{eq, series}}} = \frac{\mathcal{E}}{2R} = \frac{1}{2}I_{\text{One}}$$

The power output of this circuit is

$$P_{\text{Series}} = I_{\text{Series}}^2 R_{\text{eq, series}} = \left(\frac{\mathcal{E}}{R_{\text{eq, series}}}\right)^2 R_{\text{eq, series}} = \left(\frac{\mathcal{E}}{2R}\right)^2 2R$$

$$= \frac{1}{2}\frac{\mathcal{E}^2}{R} = \frac{1}{2}P_{\text{One}}$$

(c) The equivalent resistance of circuit (c) is $R_{\text{eq, parallel}} = R/2$. The current in this circuit is

$$I_{\text{Parallel}} = \frac{\mathcal{E}}{R_{\text{eq, parallel}}} = \frac{\mathcal{E}}{R/2} = 2I_{\text{One}}$$

The power output of this circuit is

$$P_{\text{Parallel}} = I_{\text{Parallel}}^2 R_{\text{eq, parallel}} = \left(\frac{\mathcal{E}}{R_{\text{eq, parallel}}}\right)^2 R_{\text{eq, parallel}}$$

$$= \left(\frac{\mathcal{E}}{R/2}\right)^2 R/2 = 2\frac{\mathcal{E}^2}{R} = 2P_{\text{One}}$$

In general, the total power of two devices connected in parallel to the battery is four times larger than when they are connected in series.

Try it yourself   Two equally bright identical incandescent bulbs in series are shown in the first figure. What happens to the brightness of the bulbs and the power of the circuit if you place a wire (called a shorting wire) in parallel with bulb 2 (see the second figure)? Explain.

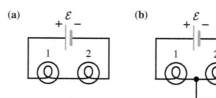
(a)   (b)

Answer   Bulb 1 becomes much brighter and bulb 2 goes out. Assuming that the shorting wire has zero resistance, all current is through the shorting wire, not through bulb 2. The resistance of the new circuit is half the original resistance. Thus, the current in the circuit doubles and the brightness of the remaining bulb quadruples; the power of the circuit doubles.

FIGURE 19.30 Investigating capacitor circuits.

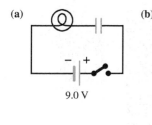
(a)

9.0 V

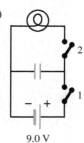
(b)

9.0 V

Capacitor circuits

In Chapter 18 we learned about capacitors, devices you can charge to store electric potential energy. What if we place a capacitor in an electric circuit? To investigate what happens, we make the circuit shown in **Figure 19.30a**, with a 9-V battery, a switch, a lightbulb, and a capacitor in series. When we close the switch, the bulb glows for a very short time interval and then goes out. Using a voltmeter, we measure the potential difference across each element after the bulb goes out. It measures 9 V across the battery, 9 V across the capacitor, and 0 V across the bulb and the switch. How can we explain this observation?

When we close the switch, the capacitor starts charging, which leads to a current through the bulb, so the bulb glows. Once the capacitor is charged, the current stops because the charged capacitor acts like a battery connected in the opposite direction to the original battery. We can test this reasoning in the following experiment: we make the new circuit shown in Figure 19.30b. When switch 1 is closed and switch 2 is open, the capacitor is connected to the battery; when 1 is open and 2 is closed, the capacitor is connected to the bulb. We first close switch 1 and wait for a few seconds. Then we open switch 1 and close switch 2. If our explanation of the first experiment is correct, then we should observe the bulb glowing for a short time once switch 2 is closed. This is exactly what happens (see the video).

VIDEO
19.1

We can investigate the phenomenon quantitatively by building the circuit shown in Figure 19.31a. We will focus on the process of discharging. Notice that we are using the same circuit as in Figure 19.30b, but we have replaced the bulb with a resistor to make sure that the resistance of the circuit remains constant. Such a circuit with a resistor and a capacitor is called an **RC circuit**. We use a voltmeter to measure potential difference across the capacitor. The potential difference-versus-time graph is shown in Figure 19.31b.

FIGURE 19.31 Discharging a capacitor in an RC circuit.

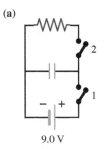

It takes about half a second for the capacitor to discharge. Changing capacitors to have different capacitance and changing resistors to have different resistance shows that the time for the potential difference to go to zero varies. Larger capacitance and larger resistance lead to a longer discharge time (when the capacitor is charged using the same battery). This makes sense: larger capacitance means that the capacitor can accumulate more charges on the plates (higher Q); larger resistance means smaller current and therefore smaller charge per unit time decrease of the original charge (smaller I).

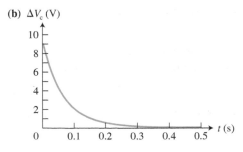

Since the time dependence of this process is repeatable and predictable with high precision, an RC circuit is often used as a timer in simple electronic circuits to periodically switch on and off the current through other devices. Examples of such timers are blinking LEDs in bicycle lights and the switching on and off of windshield wipers (depending on the intensity of the rain, the period of the wipers changes with the help of different resistor-capacitor combinations). Capacitors in DC circuits are also used when we need a large current in a short time, for example, in a camera flash. A battery cannot provide a large current due to its internal resistance, but it can slowly move a lot of charge from one capacitor plate to another. The capacitor is first slowly charged from the battery, and then it is quickly discharged through the flash lamp.

Circuit breakers and fuses

Our study of parallel circuits allows us to understand how fuses protect our houses from fires caused by electric current. Although the current through appliances connected to wall sockets (called **alternating current**, or AC) differs in some ways from the current we are studying in this chapter (**direct current**, or DC), many of the same physics ideas apply to both. As noted earlier, household appliances are connected in parallel. When we connect more appliances in parallel, the total resistance of the circuit decreases. As a result, the total current in the main line (the connection to the outside power grid) of the circuit increases. The main line provides a constant effective potential difference of 120 V. When the current in the main line becomes too large, the line can become hot enough (due to its own resistance) that the insulation around the wire melts; that hot material can set adjacent objects on fire. To prevent this from occurring, the electrical systems in buildings are installed with circuit interrupters: either **fuses** or **circuit breakers**.

A fuse is a piece of wire made of an alloy of lead and tin. This alloy melts at a relatively low temperature. Thus, when the current through the fuse gets too large to be safe, the increase in thermal energy of the fuse melts the wire. The melted fuse is a gap in the circuit, which causes the current to quickly drop to zero. In order to reconnect the circuit, the fuse must be replaced.

FIGURE 19.32 A circuit breaker panel box and a 30-A breaker switch.

Modern electrical wiring uses circuit breakers rather than fuses (Figure 19.32). There are different kinds of circuit breakers, but all of them are based on the same principle—when the current in a circuit increases to a critical value, the circuit breaker opens a switch, cutting off the current through the circuit. Circuit breakers are far more convenient than fuses because they are not self-destructive. A circuit breaker that has been tripped can quickly be reset instead of having to be replaced.

FIGURE 19.33 Review Question 19.8.

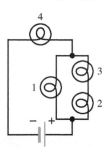

Rank the four identical bulbs shown in **Figure 19.33** from brightest to dimmest. Explain your reasoning.

19.9 Skills for solving circuit problems

The left side of the table below describes a general strategy for quantitatively solving circuit problems. The right side illustrates its application to the following problem.

PROBLEM-SOLVING
STRATEGY 19.1

Applying Kirchhoff's rules

EXAMPLE 19.8 Solving circuit problems

Determine the currents in each branch of the circuit shown in the figure. The 9.0-V battery has a 2.0-Ω internal resistance and the 3.0-V battery has negligible internal resistance.

Sketch and translate

- Draw the electric circuit described in the problem statement and label all the known quantities.
- Decide which resistors are in series with each other and which are in parallel.

- We first draw the circuit with known information.   Battery with internal resistance
- We cannot simplify the circuit using the series or parallel equivalent resistance equations because there are batteries in two of the branches and the parallel equivalent resistance rule applies for branches with resistors only.
- Since there are three unknown currents, I_1, I_2, and I_3, we need three independent equations to solve for the currents. We use the loop rule twice and the junction rule once.

Simplify and diagram

- Decide whether you can neglect the internal resistance of the battery and/or the resistance of the connecting wires.
- Draw an arrow representing the direction of the electric current in each branch of the circuit.

- The text of the problem does not provide information about the resistances of the wires, so we will assume their resistance is approximately zero.
- Add an arrow and a label for the current in each branch of the circuit. Each arrow points in the direction that we think is the direction of the current in that branch.

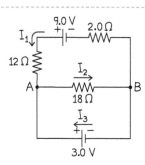

Represent mathematically

- If possible, replace combinations of resistors with equivalent resistors.
- Apply the loop rule for the potential changes as you move around one or more different loops of the circuit. Each additional loop you choose must include at least one branch of the circuit that you have not yet included.
- Once you have included branches, apply the junction rule for one or more junctions. In total, you will need the same number of independent equations as the number of unknown currents.
- If necessary, use expressions for electric power.

Apply the loop rule for two different loops, First, start at B and move clockwise around the bottom loop:

$$\Delta V_{3\,V} + \Delta V_{18\,\Omega} = 0$$
$$+3.0\ \text{V} - I_2(18\ \Omega) = 0$$

Next, start at B and move counterclockwise around the outside loop:

$$\Delta V_{2\,\Omega} + \Delta V_{9\,V} + \Delta V_{12\,\Omega} + \Delta V_{3\,V} = 0$$
$$-I_1(2\ \Omega) + 9.0\ \text{V} - I_1(12\ \Omega) - 3.0\ \text{V} = 0$$

Notice that we cross the 3.0-V battery from the positive to the negative side, a decrease in potential.

Apply the junction rule to junction A: $I_1 + I_3 = I_2$

Solve and evaluate

- Solve the equations for the unknown quantities. Check whether the directions of the current and the magnitude of the quantities make sense.

Using the first loop: $I_2 = \dfrac{3.0\ \text{V}}{18\ \Omega} = 0.17\ \text{A}$

Using the second loop: $I_1 = \dfrac{9.0\ \text{V} - 3.0\ \text{V}}{2\ \Omega + 12\ \Omega} = 0.43\ \text{A}$

Using the junction A: $I_3 = 0.17\ \text{A} - 0.43\ \text{A} = -0.26\ \text{A}$

Since I_3 turned out negative, it means our initial choice for its direction was incorrect. The magnitude is still correct.

Try it yourself Determine the potential difference from A to B in two ways: (1) through the 18-Ω resistor and (2) through the 12-Ω resistor, the 2-Ω resistor, and the 9-V battery.

Answer   Both paths give us -3.0 V.

In the next example we will use the idea of equivalent resistance to simplify a circuit and determine the current through the battery.

EXAMPLE 19.9   Using equivalent resistors to solve a circuit problem

Determine the total current I through the battery in the circuit in part (a) of the figure below.

Sketch and translate To find the current I, first simplify the circuit by combining the resistors into a single equivalent resistor.

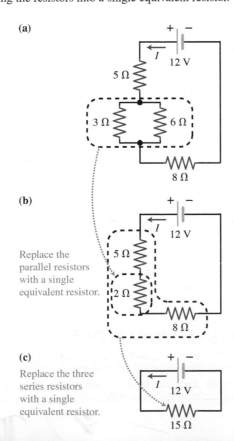

(a)

(b)

Replace the parallel resistors with a single equivalent resistor.

(c)

Replace the three series resistors with a single equivalent resistor.

Simplify and diagram Assume that the battery and the connecting wires have negligible resistance. The circuit is a combination of series and parallel parts, so simplify the circuit in steps. First, replace the parallel 3.0-Ω and 6.0-Ω resistors in (a) by the 2.0-Ω equivalent resistor in (b). Then replace the three resistors in series in (b) with the one equivalent resistor in (c). The new circuit is shown in (c). Once we have a circuit with a single resistor and a single battery, we can determine the current through the battery.

Represent mathematically First we combined the two resistors in parallel:

$$R_{\text{eq, parallel}} = \left(\frac{1}{3.0\ \Omega} + \frac{1}{6.0\ \Omega}\right)^{-1}$$

We then added this combined resistor to the other resistors in series to determine the equivalent resistance of the entire circuit:

$$R_{\text{eq}} = 5.0\ \Omega + \left(\frac{1}{3.0\ \Omega} + \frac{1}{6.0\ \Omega}\right)^{-1} + 8.0\ \Omega$$

Our simplified circuit has the single equivalent resistor attached to the 12-V battery. The current I is

$$I = \frac{\mathcal{E}}{R_{\text{eq}}}$$

Solve and evaluate Completing these calculations gives

$$R_{\text{eq, parallel}} = \left(\frac{1}{3.0\ \Omega} + \frac{1}{6.0\ \Omega}\right)^{-1} = \left(\frac{2}{6.0\ \Omega} + \frac{1}{6.0\ \Omega}\right)^{-1}$$

$$= \left(\frac{3}{6.0\ \Omega}\right)^{-1} = 2.0\ \Omega$$

$$R_{\text{eq}} = 5.0\ \Omega + 2.0\ \Omega + 8.0\ \Omega = 15.0\ \Omega$$

$$I = \frac{12.0\ \text{V}}{15.0\ \Omega} = 0.8\ \text{A}$$

This value for the current is reasonable given the resistances and battery emf.

Try it yourself Find the equivalent resistance of the combination
of resistors shown in the figure at
right.

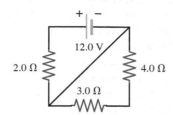

Answer   2 Ω.

QUANTITATIVE EXERCISE 19.10   Current caused by parallel appliances

A 1380-W electric heater, a 180-W computer, a 20-W LED bulb, and a
1200-W microwave are all plugged into wall sockets in the same part of
a house (see figure). When the current through the main line exceeds
20 A, the circuit breaker opens the circuit. Will the circuit breaker open
the circuit if all these appliances are in operation at the same time?

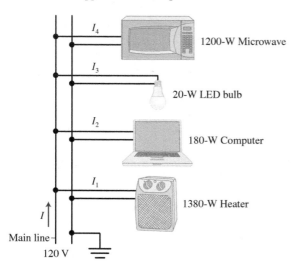

1200-W Microwave

20-W LED bulb

180-W Computer

1380-W Heater

Main line

120 V

Represent mathematically The connecting wires and the circuit
breaker have negligible resistance. Consequently, there is no electric
potential difference across them, and as a result the potential difference
across each appliance is 120 V. We can use one form of Joule's law to
determine the current in each appliance:

$$I = \frac{P}{|\Delta V|} = \frac{P}{120 \text{ V}}$$

Kirchhoff's junction rule can be used to determine the total current
through the circuit (which equals the current through the main line):

$$I = I_1 + I_2 + I_3 + I_4$$

Solve and evaluate The current through each appliance is

$$\text{Heater: } I_1 = \frac{P_1}{|\Delta V_1|} = \frac{1380 \text{ W}}{120 \text{ V}} = 11.5 \text{ A}$$

$$\text{Computer: } I_2 = \frac{P_2}{|\Delta V_2|} = \frac{180 \text{ W}}{120 \text{ V}} = 1.5 \text{ A}$$

$$\text{LED bulb: } I_3 = \frac{P_3}{|\Delta V_3|} = \frac{20 \text{ W}}{120 \text{ V}} = 0.17 \text{ A}$$

$$\text{Microwave: } I_4 = \frac{P_4}{|\Delta V_4|} = \frac{1200 \text{ W}}{120 \text{ V}} = 10.0 \text{ A}$$

The total current through the main line I is

$$I = I_1 + I_2 + I_3 + I_4$$
$$= 11.5 \text{ A} + 1.5 \text{ A} + 0.17 \text{ A} + 10.0 \text{ A} = 23.2 \text{ A}$$

Since the current I exceeds 20 A, the circuit breaker will open and cause
the current to stop.

Try it yourself Determine the resistance of each of the above
appliances.

Answer   Heater, 10 Ω; computer, 80 Ω; LED bulb, 710 Ω;
microwave, 12 Ω.

REVIEW QUESTION 19.9 What does it mean when you get a negative value for the
current in the circuit when using Kirchhoff's rules?

19.10 Properties of resistors

What properties of a resistor affect the value of its resistance? Consider again our
water flow analogy, in which a hose connects two containers of water. A wider hose
should allow more water to pass through a cross section of the hose each second, and a

longer hose should reduce the amount of water passing through per unit time (because of friction exerted by the walls of the hose on the water). By analogy, we suspect that the length and the cross-sectional area of a wire will affect its electrical resistance. Also, if a resistor is made of a dielectric material, then the current through it will be nearly zero, meaning it has a huge resistance. So it seems that resistance must also depend on the internal properties of the material of which it is made. The experimental setup in **Figure 19.34a** will help us decide how the resistance of a circuit element depends on its geometric properties (its length L and cross-sectional area $A = \pi r^2$; see Figure 19.34b.). We make measurements on wires of different lengths and radii. The measurements yield the data shown graphically in **Figure 19.35**. It appears that the resistance increases in proportion to the wire's length (Figure 19.35a) and inversely in proportion to its radius squared, that is, to the cross-sectional area of the wire (Figure 19.35b).

$$R \propto L, \text{ and } R \propto \frac{1}{r^2} \propto \frac{1}{A}$$

FIGURE 19.34 What does resistance depend on? (a) A circuit used to measure resistance. (b) Geometrical properties affect resistance.

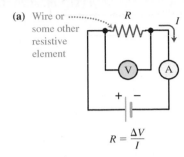

(a) Wire or .... some other resistive element

$$R = \frac{\Delta V}{I}$$

(b)

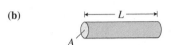

FIGURE 19.35 Wire resistance depends on (a) its length and (b) its cross-sectional area.

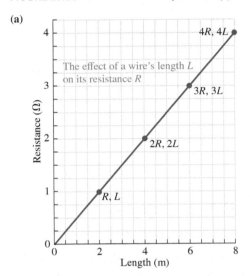

(a)

The effect of a wire's length L on its resistance R

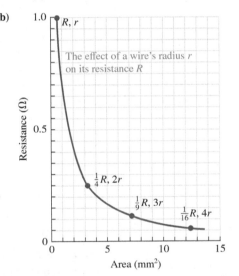

(b)

The effect of a wire's radius r on its resistance R

To investigate how the type of material affects the resistance, we measure the resistance of wires with the same length and cross-sectional area but made from different materials. We find that copper and silver wires have small resistance, and constantan and tungsten wires have relatively large resistance.

The dependence of the resistance R on the type of material of which the resistor is made is characterized in terms of a physical quantity called the **resistivity** ρ of that material. Microscopically, a material's resistivity depends on the number of free electrons per atom in the material, the degree of "difficulty" the electrons have moving through the material due to their interactions with it, and other factors. Copper has a low resistivity because of its large concentration of free electrons (two per atom) and the relative lack of difficulty they experience while moving through the copper. On the other hand, the resistivity of glass (a dielectric) is about 10^{20} times greater than that of copper because it contains almost no free electrons.

Alternatively, materials are sometimes characterized in terms of another physical quantity called **conductivity** σ. The conductivity of a material is the reciprocal of its resistivity ($\sigma = 1/\rho$).

We can assemble an equation for the resistance of a resistive circuit element by combining these three dependencies.

TIP   The Greek letter ρ is used to designate resistivity as well as the density of a substance. Do not confuse the two!

TABLE 19.7 The resistivity ρ of different materials

Material	ρ ($\Omega \cdot$ m, at 20 °C)
Metals	
Silver	1.6×10^{-8}
Copper	1.7×10^{-8}
Aluminum	2.8×10^{-8}
Tungsten	5.2×10^{-8}
Constantan	50×10^{-8}
Nichrome	130×10^{-8}
Dielectrics	
Ordinary glass	9×10^{11}
Dry wood	10^{14}–10^{16}
Hard rubber	1×10^{16}
Human tissue	
Blood	1.5
Trunk	5
Lung tissue	20
Fat	25
Skin	5000–50,000
Geological materials	
Ground water	≈ 10
Sedimentary rocks	1–10^5
Igneous rocks	10^2–10^7

Electrical resistance The electrical resistance R of a resistive circuit element depends on its geometric structure (its length L and cross-sectional area A) and on the resistivity ρ of the material of which it is made:

$$R = \rho \frac{L}{A} \qquad (19.11)$$

Eq. (19.11) helps us understand why when you connect resistors in series, the total resistance of the circuit increases, and when you connect the resistors in parallel, the total resistance decreases. Connecting the resistors in series is similar to increasing the length of the resistor ($R \propto L$), and connecting the resistors in parallel is similar to increasing the cross-sectional area of the resistor $R \propto 1/A$.

TIP **Table 19.7** lists resistivity for materials at a particular temperature. As we know, the resistance of a lightbulb is greater when it is hot than when cold; its resistivity changes with temperature. The resistivity of some other materials, such as constantan, does not change much with temperature.

In the previous sections we disregarded the resistance of connecting wires in circuits. Since wires are long and thin, according to Eq. (19.11) they could have quite high resistance. Let's investigate the resistance of such wires.

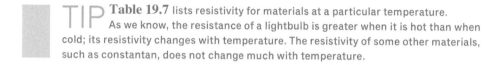

QUANTITATIVE EXERCISE 19.11   **Resistance of connecting wires**

The connecting wires that we use in electric circuit experiments are usually made of copper. What is the resistance of a 10-cm-long piece of copper connecting wire that has a diameter of 2.0 mm = 2.0×10^{-3} m?

Represent mathematically The resistance of a wire is $R = \rho \dfrac{L}{A}$. The cross-sectional area of a round wire is $A = \pi r^2$. Therefore, the resistance is

$$R = \rho \frac{L}{\pi r^2} = \rho \frac{L}{\pi (d/2)^2}$$

Solve and evaluate Insert the value for the resistivity of copper from Table 19.7 and the wire dimensions into the above:

$$R = \rho \frac{L}{\pi (d/2)^2} = (1.7 \times 10^{-8} \, \Omega \cdot \text{m}) \frac{0.1 \, \text{m}}{\pi (1.0 \times 10^{-3} \, \text{m})^2}$$

$$= 5.4 \times 10^{-4} \, \Omega$$

This is a tiny amount of resistance. If the current through this wire was 1 A, the potential difference across it would be very small:

$$\Delta V = IR = (1.0 \, \text{A})(5.4 \times 10^{-4} \, \Omega) = 5.4 \times 10^{-4} \, \text{V}$$

The filaments of old household incandescent lightbulbs have resistances of 100–300 Ω and will therefore have a potential difference of 100–300 V across them when the current through them is 1 A. Connecting wires are deliberately made of material with low resistivity so that the potential difference across them is negligible.

Try it yourself A power line from the power plant to your home has the same cross-sectional area as in this exercise but is 50 km long. What is the resistance of the power line and the potential difference across it if the current through it is 1 A?

Answer 270 Ω, 270 V. However, power lines are actually designed with much greater cross-sectional areas so that their resistance is kept low.

Microscopic model of resistivity

Recall the microscopic structure of a metal: a crystal structure of metal atoms whose outer electrons are free to roam throughout the bulk of the metal. The free electrons moving chaotically inside the metal are similar to ideal gas particles. This model suggests a microscopic explanation for resistance. As the electrons drift through the wire, they collide with the metal ions. These collisions are also the reason why the current in a wire that is connected to a battery does not increase but rather has a constant value. Although the electric field inside the metal accelerates the free electrons, they are continuously slowed down by the collisions with ions. These collisions increase the internal energy of the metal ions, causing them to vibrate more vigorously. These vibrating ions become even more of an obstacle for the electrons. That explains why a hotter metal has a higher resistivity than a cooler one and thus explains why the resistance of the lightbulb filament increases with temperature. Paul Drude first suggested this model in 1900, now known as the Drude model. The Drude model does not correctly predict the details of many properties of metals (quantum mechanical models do much better). However, the Drude model is still a concrete and convenient way to understand qualitatively the reasons for electrical resistance and the mechanism of transforming electrical energy to thermal energy.

Superconductivity

In the early 1900s Dutch physicist Kamerlingh Onnes was attempting to liquefy helium, the only gas that had not yet been liquefied at that time. In 1908 he achieved the desired result—the helium condensed from a gas to a liquid at 4.2 K. Onnes continued experimenting with helium and focused his attention on measuring the electric resistivity of materials at this low temperature. Many physicists, including Lord Kelvin himself, thought that at very low temperatures, electrons in metals should completely stop moving (as the average kinetic energy of the particles is proportional to temperature), thus making the resistivity infinite. Others, including Onnes, hypothesized that the resistivity decreased with temperature and would eventually become zero at zero K (**Figure 19.36a**).

In 1911 Onnes made an electric circuit by connecting the terminals of a battery to a sample of mercury. He measured the current through it and the potential difference across it. When Onnes submerged the mercury sample in liquid helium at 4.2 K and repeated his measurement, he found that there was still current through the sample, but the potential difference across it abruptly dropped to zero—this meant that the resistance of the sample was zero. Basically, he observed that the resistivity of mercury became zero not at 0 K, but at 4.2 K (Figure 19.36b). Onnes had discovered what became known as **superconductivity**. The temperature at which the resistivity of a material abruptly decreases to zero is called the critical temperature, T_c. Superconductivity was a surprising result that was not predicted and wasn't successfully explained until about 50 years later, using quantum mechanics. At very low temperatures, the electrons as a whole no longer transfer energy to the metal lattice ions, allowing the electrons to move through the lattice with exactly zero resistance.

In the meantime, Onnes and other physicists continued experimenting with different metals and found that many of them, including aluminum and tin, become superconductors at very low temperatures, but that others, including gold and silver, do not.

Unfortunately, the temperatures at which superconductivity was achieved were very low, never having a critical temperature above 20 K. This made the practical applications of superconductivity almost impossible. In addition, the existing theory predicted that superconductivity could not occur above about 30 K.

In 1986 a seemingly improbable event occurred. J. Georg Bednorz and K. Alex Müller observed superconductivity in a material at 35 K. The material they used was not even a metal; it was a complex copper-based ceramic compound. With this as guidance, physicists manufactured similar compounds and raised the high-temperature superconductivity record to 95 K. In 1993, a complex ceramic compound was found to have a critical temperature of 138 K. In 2008 a new family of iron-based high-temperature superconductors was discovered. While there is a long way to go, the goal of discovering a room-temperature superconductor no longer seems completely impossible.

The applications of superconductors are numerous. One is transportation by magnetic levitation, the unique property of superconductors to float suspended above magnets. Superconductors can also be used to make super-strong electromagnets. Using superconductors to transmit information will bypass limitations on data rates caused by electrical resistance. Finally, superconductors can be used as a basis for supersensitive thermometers.

FIGURE 19.36 Superconductivity.
(a) Predictions about low temperature resistivity.
(b) The measured resistivity for mercury.

(a)

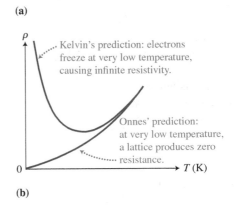

Kelvin's prediction: electrons freeze at very low temperature, causing infinite resistivity.

Onnes' prediction: at very low temperature, a lattice produces zero resistance.

(b)

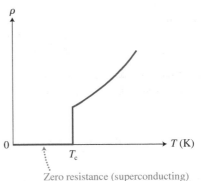

Zero resistance (superconducting) below the critical temperature (T_c)

FIGURE 19.37 Experimental setups for comparing temperature dependence of resistance of (a) an incandescent lightbulb and (b) an LED.

(a)

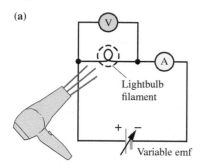

Lightbulb filament

Variable emf

(b)

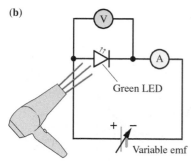

Green LED

Variable emf

Semiconductors

So far, we have learned that the resistivity of conductors made of metal (including the wires in incandescent lightbulbs) increases with temperature. Do LEDs behave in a similar way? To investigate this question, we conduct the following experiment. We make two circuits: one consisting of a battery and an incandescent lightbulb with removed glass cover, and the other made of a green LED and a battery. We measure the potential difference across and current through the bulb and the LED at room temperature. We make sure that the current through the lightbulb is so small that it does not glow and does not warm up (we know that an LED is cold even when glowing, so we don't need to worry about warming due to its own current). Then without opening the circuits, we warm up both the bulb and the LED with a hair dryer and measure the same quantities (**Figure 19.37**). Our collected data are given in **Table 19.8**.

TABLE 19.8 Data collected from the experiment in Figure 19.37

	Physical quantity	Incandescent lightbulb	Green LED
Initial conditions	T (°C)	18	18
	ΔV (V)	0.020	3.00
	I (10^{-3} A)	16.6	11.0
	R (Ω)	**1.2**	**270**
After warming up with the hair dryer	T (°C)	60	60
	ΔV (V)	0.020	3.00
	I (10^{-3} A)	14.3	12.8
	R (Ω)	**1.4** ($\approx$ 17% increase)	**230** ($\approx$ 15% decrease)

From the table, we see that the resistance of the filament of the incandescent bulb increases with temperature (as we expected). However, the resistance of the LED decreases with temperature, rather than increases, as we are used to with metals. Why? The material inside the LED is not metal. The glowing part of an LED is made of two different materials that belong to a class called **semiconductors**. Although we will not go into the details behind the mechanism through which these two semiconductors emit light in this chapter (we will do that in Chapter 27), it is important for you to learn about their basic properties. Silicon is the best example of a semiconductor; LEDs are usually made of gallium alloys.

When cooled to near 0 K, silicon has practically infinite resistivity. As its temperature increases, its resistivity decreases. Unlike in metals, the outer electrons of silicon atoms are not free—they are tightly bound to the atoms. When the temperature of a silicon crystal increases, some of the electrons do become free, leaving behind a positively charged ion in the crystal lattice. The location of this missing electron is called a **hole** (**Figure 19.38**). If charged objects outside the silicon crystal are used to produce a nonzero $\vec{E}$ field within the silicon, these newly freed conduction electrons accelerate in a direction opposite the $\vec{E}$ field, similar to the way they do in metals. In addition, the electrons of atoms next to the holes can move into the hole. The hole then moves to the atom where the electron came from. In effect, these holes behave similarly to positively charged particles, accelerating in the direction of the $\vec{E}$ field. Because the number of conduction electrons and holes increases as the temperature of the semiconductor increases, the resistivity of a semiconductor decreases with increasing temperature. Just like in metals, in semiconductors charge carriers collide

FIGURE 19.38 Migration of holes in a semiconductor in the presence of an electric field. Parts (a) to (d) show successive steps as a hole and bound electrons swap places. Part (e) shows the trajectory of the hole during the entire time interval.

with ions. The fact that resistivity decreases with temperature tells us that production of free electrons at higher temperature wins over the increased resistance due to collisions with vibrating ions.

Semiconductors are used not only in LEDs but in nearly all modern technology, including consumer electronics, medical electronic products, and solar power. We will learn later in the book how LEDs convert electric energy into light energy, and how solar cells convert light energy into electric energy.

Electrolytes

Electrolytes are substances containing free ions that make the substance electrically conductive. Several electrolytes are critical in human physiology. For example, free sodium (Na^+) and chlorine (Cl^-) ions in the body's tissues move in opposite directions, forming currents (I^+ and I^-) in response to electric fields produced by electric dipoles in the heart. Both types of ions contribute to the current ($I_{total} = I^+ + I^-$). In nerve cells, Na^+, Cl^-, and other ions move across nerve membranes, transmitting electrical signals from the brain to other parts of the body—to activate muscles, for example.

REVIEW QUESTION 19.10 Why does the resistance of a lightbulb increase as the current through it increases?

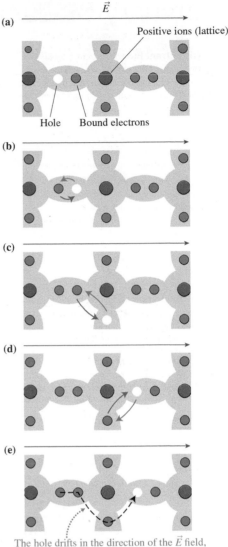

(a) $\vec{E}$

Positive ions (lattice)

Hole   Bound electrons

(b)

(c)

(d)

(e)

The hole drifts in the direction of the $\vec{E}$ field, while the electrons move in the opposite direction.

Summary

Electric current Electric current I equals the electric charge q that passes through a cross section of a circuit element divided by the time interval Δt needed for that amount of charge to pass. (Section 19.1)

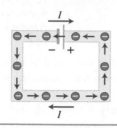

$$I = \frac{|q|}{\Delta t}$$ Eq. (19.1)

Unit: 1 A (ampere) = 1 C/s

Emf is the work done per unit charge by a power source (such as a battery) to move charge from one terminal to other. (Sections 19.2 and 19.7)

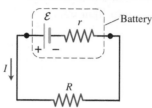

$$\mathcal{E} = \frac{W}{q}$$ Eq. (19.2)

Unit: 1 V (volt) = 1 J/C

$$\mathcal{E} - Ir = IR$$ Eq. (19.7)

Electrical resistance The electrical resistance R of a wire depends on its geometrical structure: its length L, cross-sectional area A, and the resistivity of the material ρ. (Section 19.10)

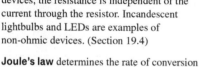

$$R = \rho \frac{L}{A}$$ Eq. (19.11)

$$I_1 = I_2 = I_3 = \cdots$$

$$R_{\text{eq, series}} = R_1 + R_2 + R_3 + \cdots$$ Eq. (19.9)

Series and parallel resistance Multiple resistors can be connected in series or in parallel in a circuit. The current through series resistors is the same; the potential difference across parallel resistors is the same. Resistor combinations can be replaced with a single resistor with equivalent resistance. (Section 19.8)

$$\Delta V_1 = \Delta V_2 = \Delta V_3 = \cdots$$

$$R_{\text{eq, parallel}} = \left(\frac{1}{R_1} + \frac{1}{R_2} + \frac{1}{R_3} + \cdots \right)^{-1}$$ Eq. (19.10)

Ohm's law The current I through a circuit element equals the potential difference ΔV across it divided by its resistance R. For ohmic devices, the resistance is independent of the current through the resistor. Incandescent lightbulbs and LEDs are examples of non-ohmic devices. (Section 19.4)

Joule's law determines the rate of conversion of electric potential energy into thermal energy. (Section 19.6)

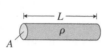

$$I = \frac{\Delta V}{R}$$ Eq. (19.3)

$$P = \left| \frac{\Delta U_{\text{th}}}{\Delta t} \right| = I |\Delta V|$$ Eq. (19.4)

$$= I^2 R = (\Delta V)^2 / R$$ Eq. (19.5)

Unit: 1 W (watt) = 1 J/s

Kirchhoff's junction rule The algebraic sum of all currents into a junction in a circuit equals the algebraic sum of the currents out of the junction. (Section 19.7)

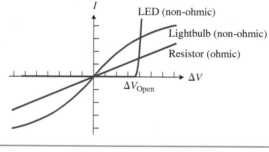

$$\sum_{\text{in}} I = \sum_{\text{out}} I$$ Eq. (19.8)

Kirchhoff's loop rule The algebraic sum of the potential differences across circuit elements around any closed circuit loop is zero. For resistors, the potential difference is $-IR$ if moving across the resistor in the direction of the current and $+IR$ if moving across the resistor opposite the direction of current. (Section 19.7)

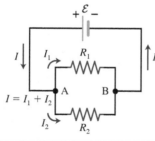

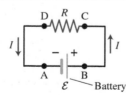

$$\sum_{\text{Loop}} \Delta V = 0$$ Eq. (19.6)

Battery: $\Delta V = \pm \mathcal{E}$, assuming no internal resistance
$\Delta V = \pm (\mathcal{E} - Ir)$ (including the internal resistance of the battery r)

Resistor: $\Delta V = \mp IR$

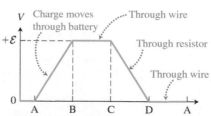

Questions

Multiple Choice Questions

1. Two identical bulbs are connected in parallel across the battery shown in **Figure Q19.1**. There is an open switch next to bulb 2. Initially, bulb 1 is bright (strong current) and bulb 2 is dark (no current). How do the electric currents change when the switch is closed? Multiple answers could be correct.

FIGURE Q19.1

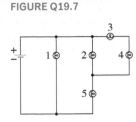

   (a) The current through the battery is the same but now splits between 1 and 2.
   (b) The current in 1 is still bigger than that in 2.
   (c) The current in 1 does not change and is the same as in 2.
   (d) The current through the battery doubles.

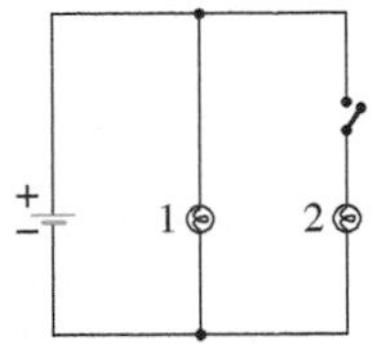

2. Compare the potential difference across bulbs 1 and 2 in Figure Q19.1 after the switch is closed. Which statement is correct?
   (a) The potential difference is greater across bulb 1.
   (b) The potential difference is greater across bulb 2.
   (c) The potential difference is the same across bulbs 1 and 2.
   (d) The potential difference across the battery doubles.
   (e) None of the above.

3. Two identical bulbs are in series as shown in **Figure Q19.3**. Which statement is correct?

FIGURE Q19.3

   (a) The current through the battery is twice as great as when one bulb was present.
   (b) The current through the battery is half as great as when one bulb was present.
   (c) The current is greater in bulb 1 than in bulb 2.
   (d) The current is greater in bulb 2 than in bulb 1.
   (e) The currents in bulbs 1 and 2 are identical.
   (f) b and e are correct.
   (g) a and c are correct.

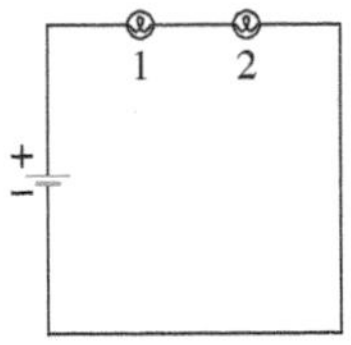

4. Which statement below about the potential difference across the battery and across the identical bulbs 1 and 2 shown in Figure Q19.3 is correct? Multiple answers could be correct.
   (a) The potential difference is greater across bulb 1 than across bulb 2.
   (b) The potential difference is greater across bulb 2 than across bulb 1.
   (c) The potential difference is the same across bulbs 1 and 2.
   (d) The potential difference across the battery is twice what it was with only one bulb present.

5. Three circuits with identical bulbs and emf sources are shown in **Figure Q19.5**. Rank the circuits in terms of the ammeter readings, with the largest ammeter reading listed first.
   (a) $A > B > C$ (b) $A > C > B$ (c) $B > A > C$
   (d) $B > C > A$ (e) $C > A > B$

FIGURE Q19.5

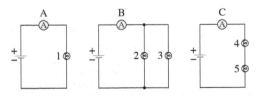

6. Rank in order the potential differences across each of the five bulbs in the three circuits in Figure Q19.5, listing the largest potential difference first. Indicate equal potential differences with equal signs. Assume that there is zero potential difference across the ammeters.
   (a) $\Delta V_1 > \Delta V_2 = \Delta V_3 > \Delta V_4 = \Delta V_5$
   (b) $\Delta V_2 > \Delta V_3 > \Delta V_1 > \Delta V_4 > \Delta V_5$
   (c) $\Delta V_2 = \Delta V_3 > \Delta V_1 > \Delta V_4 = \Delta V_5$
   (d) $\Delta V_1 = \Delta V_2 = \Delta V_3 > \Delta V_4 = \Delta V_5$
   (e) None of these

7. Rank in order the five identical bulbs in the circuit shown in **Figure Q19.7**, listing the brightest bulb first. Indicate equally bright bulbs with an equal sign.

FIGURE Q19.7

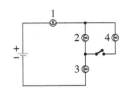

   (a) $B_1 > B_5 > B_2 > B_3 = B_4$
   (b) $B_1 = B_5 = B_2 > B_3 = B_4$
   (c) $B_1 > B_5 > B_2 = B_3 = B_4$
   (d) $B_1 = B_2 = B_3 = B_4 = B_5$
   (e) None of these is correct.

8. Four identical bulbs are shown in the circuit in **Figure Q19.8**. With the switch open, bulbs 1, 2, and 3 are equally bright and bulb 4 is dark. How does the brightness of bulbs 1, 2, and 3 change when the switch is closed?

FIGURE Q19.8

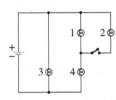

   (a) 1 and 2 stay the same and 3 gets brighter.
   (b) 1, 2, and 3 stay the same.
   (c) 1 and 3 get brighter and 2 gets dimmer.
   (d) 1, 2, and 3 all get brighter.
   (e) None of these is correct.

9. Four identical bulbs are shown in the circuit in **Figure Q19.9** with the switch open. How does the brightness of bulbs 1, 3, and 4 change when the switch is closed?

FIGURE Q19.9

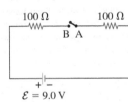

   (a) 1 gets dimmer, 4 gets brighter, and 3 stays the same.
   (b) 1 stays the same, 4 gets brighter, and 3 gets dimmer.
   (c) 1 stays the same, 4 gets brighter, and 3 stays the same.
   (d) 1, 3, and 4 all stay the same.
   (e) None of these is correct.

10. Consider the circuit in **Figure Q19.10**. The switch is open. What is the potential difference across the switch $\Delta V_{A \text{ to } B}$?
   (a) Zero   (b) 9.0 V
   (c) 4.5 V   (d) -4.5 V

FIGURE Q19.10

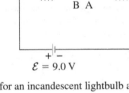

11. **Figure Q19.11** shows I-versus-ΔV graphs for an incandescent lightbulb and a resistor. Choose all correct statements.
   (a) For $\Delta V = 3$ V, the resistance of the lightbulb and the resistor are equal ($R_B = R_R$).
   (b) For $\Delta V = 3$ V, $R_B > R_R$.
   (c) For $\Delta V = 3$ V, $R_B < R_R$.
   (d) R_B increases with increasing current through it.
   (e) R_B decreases with increasing current through it.
   (f) For any potential difference $0 < \Delta V < 3$ V, $R_B > R_R$.
   (g) For any potential difference $0 < \Delta V < 3$ V, $R_B < R_R$.
   (h) For potential difference $0 < \Delta V < 3$ V, R_B can be larger than, smaller than, or equal to R_R.

FIGURE Q19.11

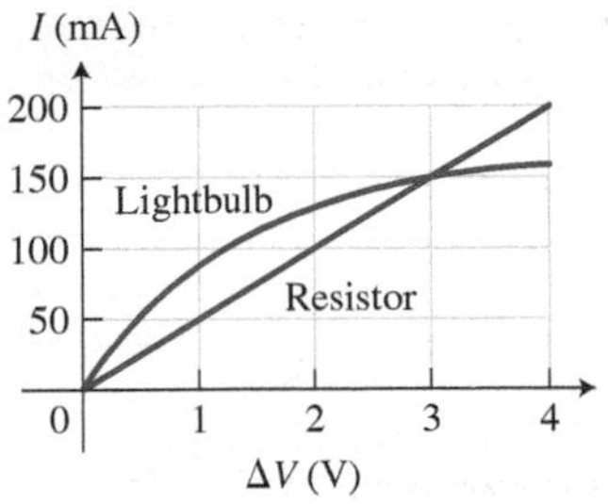

12. If an electric current were due to electrons moving only on the surface of a wire rather than through the bulk of the wire, which expression would describe the resistance of the wire? r is the wire's radius, L is its length, and C is a constant with suitable units.
   (a) $R = C_a \dfrac{L}{r^2}$ (b) $R = C_b L r$ (c) $R = C_c L$
   (d) $R = C_d \dfrac{L}{r}$ (e) None of the above

13. Three light sources (a lightbulb, a blue LED $[\Delta V_{Open\ B} = 2.3\ V]$, and a red LED $[\Delta V_{Open\ R} = 1.6\ V]$) are connected with two 50-Ω resistors and a 3.0-V battery, as shown in Figure Q19.13. Which light source will definitely not glow?
    (a) The blue LED
    (b) The red LED
    (c) The lightbulb
    (d) Both the blue LED and the red LED
    (e) All light sources will glow.

FIGURE Q19.13

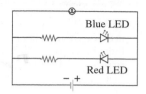

14. Figure Q19.14 shows a top view of a flat resistor that is cut from a thin metal foil of uniform thickness. Which circuit can be used as a model for this resistor? You may use a ruler to solve this problem.

FIGURE Q19.14

(a) $\quad\overset{R\quad 2R\quad 3R\quad 4R}{\bullet\text{-}\mathrm{w}\text{-}\mathrm{w}\text{-}\mathrm{w}\text{-}\mathrm{w}\text{-}\bullet}$

(b) $\quad\overset{4R\quad 3R\quad 2R\quad R}{\bullet\text{-}\mathrm{w}\text{-}\mathrm{w}\text{-}\mathrm{w}\text{-}\mathrm{w}\text{-}\bullet}$

(c) $\quad\overset{R\quad R/2\quad R/3\quad R/4}{\bullet\text{-}\mathrm{w}\text{-}\mathrm{w}\text{-}\mathrm{w}\text{-}\mathrm{w}\text{-}\bullet}$

(d) $\quad\overset{R/4\quad R/3\quad R/2\quad R}{\bullet\text{-}\mathrm{w}\text{-}\mathrm{w}\text{-}\mathrm{w}\text{-}\mathrm{w}\text{-}\bullet}$

Conceptual Questions

15. What is the role of a battery in an electric circuit? Describe a mechanical analogy.
16. Compare and contrast the physical quantities emf and potential difference.
17. **Birds on high power lines** Why can birds perch on a 100,000-V power line with no adverse effects?
18. **Preventing electric shock** When a person is repairing electrical equipment and can encounter high potential differences, he or she should keep one hand in a back pocket and work only with the other hand. Why?
19. (a) Using a voltmeter, how can you determine the electric current through a resistor? (b) Using an ammeter, how can you determine the potential difference across a resistor? (c) Using a voltmeter and an ammeter, how can you determine the resistance of a resistor?

20. (a) What does it mean if the current through a resistor is 3 A? (b) What does it mean if the potential difference across the resistor is 3 V? (c) What does it mean if the resistance of the resistor is 3 Ω?
21. Resistors become warm when there is an electric current through them. Why doesn't this phenomenon depend on the direction of the current? Give a microscopic explanation.
22. At one time aluminum rather than copper wires were used to carry electric current through homes. Which wire must have the larger radius if they are the same length and need to have the same electrical resistance? Explain.
23. How do you connect an ammeter in a circuit to measure the current? How do you connect a voltmeter in a circuit to measure the potential difference between two points in the circuit? What implications do these connections have for the resistances of these measuring instruments?
24. Why do we connect electric devices in a home in parallel rather than in series?
25. Dylan thinks that conductance is equal to the slope of the I-versus-ΔV graph at a particular point; Finn thinks resistance is equal to the ratio between the ΔV and I values at that point. Explain for which types of circuit element both Dylan and Finn are correct, and for which type only Finn is correct.
26. Construct an electric circuit that is analogous to your circulatory system. Indicate the corresponding parts of the two systems.
27. Most Christmas tree lights with incandescent bulbs are connected in parallel. (a) Describe one advantage of lights connected in parallel. (b) Suppose you connect these lights in series. Will they be brighter or darker? Explain. What modification would make the lights equally bright? Explain.
28. Two students are arguing. Student A says that Ohm's law is only valid for ohmic resistors. Student B says that Ohm's law is valid for any circuit element other than a battery. Which student do you agree with? Explain your choice.
29. Use the laws of energy and charge conservation to explain Kirchhoff's rules.
30. When you close the switch in the circuit in Figure Q19.30, the capacitor charges. When you open the switch again, the capacitor discharges. (a) Explain qualitatively what happens to the current in each process. (b) Which process takes a longer time, charging or discharging? Explain.

FIGURE Q19.30

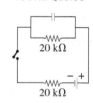

20 kΩ

20 kΩ

Problems

Below, **BIO** indicates a problem with a biological or medical focus. Problems labeled **EST** ask you to estimate the answer to a quantitative problem rather than derive a specific answer. Asterisks indicate the level of difficulty of the problem. Problems with no * are considered to be the least difficult. A single * marks moderately difficult problems. Two ** indicate more difficult problems. **If not stated otherwise, the internal resistance of batteries and the resistance of wires are negligible.**

19.1–19.3 Electric current, Batteries and emf, and Making and representing simple circuits

1. A bulb in a table lamp has a current of 0.50 A through it. Determine two physical quantities related to the electrons passing through the wires leading to the bulb.
2. A long wire is connected to the terminals of a battery. In 8.0 s, 9.6×10^{20} electrons pass a cross section along the wire. The electrons flow from left to right. What physical quantities can you determine using this information?
3. A typical flashlight battery will produce a 0.50-A current for about 3 h before losing its charge. Determine the total number of electrons that have moved past a cross section of wire connecting the battery and lightbulb.
4. * Four friends each have a battery, a bulb, and one wire. They use the materials to try to light their bulb. Two succeed and two do not. What possible circuit arrangements could they have made?

5. Draw a circuit that has a battery, a lightbulb, and connecting wires. Draw a schematic of a water flow system that allows a continual circulation of water and compare the function of each element of the circuit to each element of the water flow system.
6. Add another battery to the circuit described in Problem 19.5. In how many different ways can you do this? Draw analogous changes in the water flow system.
7. Add another lightbulb to the circuit with one battery described in Problem 19.5. In how many different ways can you connect the second bulb? What do you need to add to the water flow system to make it similar to the circuit?
8. A 9.0-V battery is connected to a resistor so that there is a 0.50-A current through the resistor. For how long should the battery be connected in order to do 200 J of work?
9. ** **Design an experiment I** When sugar is poured from a metal container, sugar particles and the container become charged with opposite charges due to rubbing. Design an experiment that will allow you to estimate the magnitude and the sign of the charge transferred by 1 kg of sugar, using only the following measuring devices: a stopwatch, a scale, and an ammeter that can measure very small current.

19.4 and 19.5 Ohm's law and Qualitative analysis of circuits

10. * A graph of the electric potential versus location in a series circuit with 1.0 A of current is shown in **Figure P19.10**. Draw a circuit in which such changes could occur.

FIGURE P19.10

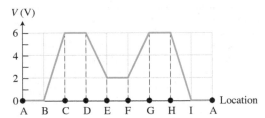

V (V)

Location
A   B   C   D   E   F   G   H   I   A

11. Sketch a potential-versus-location graph for the circuit shown in **Figure P19.11**. Start at A and move clockwise around the circuit.

FIGURE P19.11

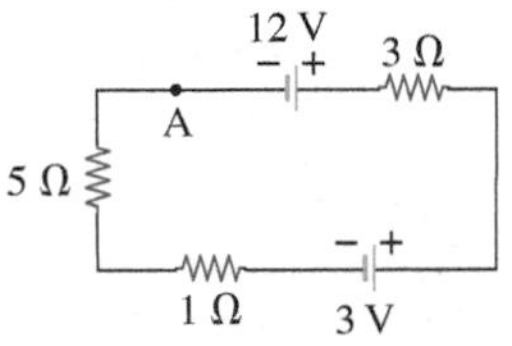

12 V
3 Ω

A

5 Ω

1 Ω   3 V

12. **BIO** **Electric currents in the body** A person accidentally touches a 120-V electric line with one hand while touching a ground wire with the other hand. Determine the current through the body when the hands are dry (100,000-Ω resistance) and when wet (5000-Ω resistance). If the current exceeds about 10 mA, muscular contractions may prevent the person from releasing the wires—a dangerous situation. Is the person in danger with dry hands? With wet hands? Explain.

13. An automobile lightbulb has a 1.0-A current when it is connected to a 12-V battery. What can you determine about the lightbulb using this information?

14. * If a long wire is connected to the terminals of a 12-V battery, 6.4×10^{19} electrons pass a cross section of the wire each second. Make a list of the physical quantities whose values you can determine using this information and determine three of them.

15. Determine the current through a 2.5-Ω resistor when connected across two 1.5-V batteries in series (a potential difference of 3.0 V).

16. * You have a circuit with a 50-Ω, a 100-Ω, and a 150-Ω resistor, and a battery connected in series. (a) Rank the current through them from highest to lowest. (b) Rank the potential difference across them from highest to lowest. Explain your rankings.

17. You have a circuit with a 50-Ω, a 100-Ω, and a 150-Ω resistor connected in parallel to the same battery. (a) Rank the current through them from highest to lowest. (b) Rank the potential difference across them from highest to lowest.

18. * A toy has two red LEDs ($\Delta V_{\text{Open R}} = 1.6$ V), two green LEDs ($\Delta V_{\text{Open G}} = 2.1$ V), and two blue LEDs ($\Delta V_{\text{Open B}} = 2.3$ V) connected in series. (a) What is the smallest number of 1.2-V rechargeable batteries needed to make all the LEDs glow? (b) How do you need to connect the batteries (in series or in parallel)? (c) For how many hours will the toy work with that number of batteries if the current through the LEDs is 20 mA and the total storage capacity of the batteries is 1200 mAh?

19. * You want to power a green LED ($\Delta V_{\text{Open G}} = 2.1$ V) and a red LED ($\Delta V_{\text{Open R}} = 1.6$ V) using a 9.0-V battery. You decide to connect the LEDs in series. In order to limit the current to the allowed value of 20 mA, you need to add a resistor in series with the LEDs. (a) Draw a circuit diagram, and (b) determine the resistance of the resistor that you need to connect to limit the current.

20. ** A circuit consists of a green LED ($\Delta V_{\text{Open G}} = 2.1$ V) and a 100-Ω resistor connected in series with a variable power supply. Draw a labeled I-versus-ΔV graph for this circuit for the interval -4.0 V $< \Delta V < +4.0$ V. (Hint: Remember that once the voltage across an LED reaches the opening voltage, an LED behaves like a closed switch, except that the voltage across it is almost constant and equal to ΔV_{Open}.)

19.6 Joule's law

21. * You connect a 50-Ω resistor to a 9-V battery whose internal resistance is 1 Ω. (a) Determine the electric power dissipated by the resistor. (b) Will a lightbulb with a 10-Ω resistance dissipate more or less power if connected to the same battery? Explain.

22. * **EST** **Making tea** You use an electric teapot to make tea. It takes about 2 min to boil 0.5 L of water. (a) Estimate the power of the heater. What are your assumptions? (b) Estimate the current through the heater. State your assumptions.

23. * If a long wire is connected to the terminals of a 12-V battery, 6.4×10^{19} electrons pass a cross section of the wire each second. What is the rate of work being done by the battery?

24. ** Three friends are arguing with each other. Adam says that a battery will send the same current to any circuit you connect to it. Alice says that the battery will produce the same potential difference at its terminals independent of the circuit connected to it. Paul says that the battery will always produce the same electric power independent of the circuit connected to it. Describe experiments that you can perform to convince all of them that their opinions are wrong in at least one situation.

25. * You have a 40-W lightbulb and a 100-W bulb. What do these readings mean? Can we say that the 100-W lightbulb is always brighter? Explain and give examples.

26. * Does a 60-W lightbulb have more or less resistance than a 100-W bulb? Explain.

19.7 and 19.8 Kirchhoff's rules and Resistor and capacitor circuits

27. * (a) Write two loop rule equations and one junction rule equation for the circuit in **Figure P19.27**. (b) Use these equations to determine the current in each branch of the circuit for the case in which $R_1 = 50$ Ω, $R_2 = 30$ Ω, $R_3 = 15$ Ω, and $\mathcal{E} = 120$ V.

FIGURE P19.27

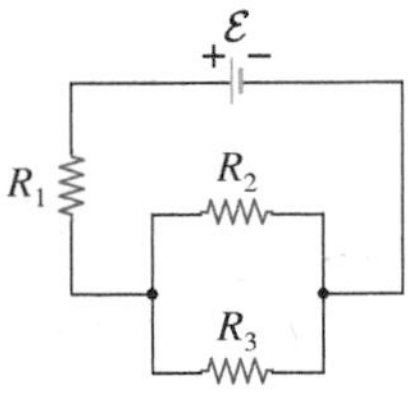

$\mathcal{E}$
+ −

R_1 R_2

R_3

28. * (a) Write Kirchhoff's loop rule for the circuit shown in **Figure P19.28** for the case in which $\mathcal{E}_1 = 20$ V, $\mathcal{E}_2 = 8$ V, $R_1 = 30$ Ω, $R_2 = 20$ Ω, and $R_3 = 10$ Ω. (b) Determine the current in the circuit. (c) Using this value of current, start at position A and move clockwise around the circuit, calculating the electric potential change across each element in the circuit (be sure to indicate the sign of each change). (d) Add these potential changes around the whole circuit. Explain your result.

FIGURE P19.28

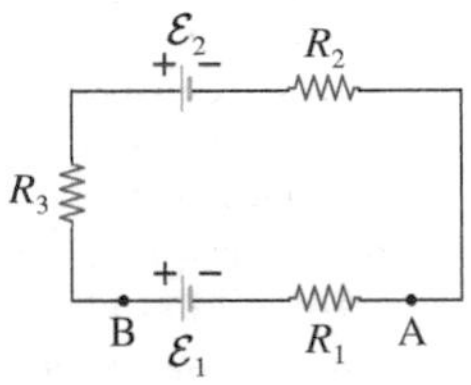

$\mathcal{E}_2$ R_2
+ −

R_3

+ −
B $\mathcal{E}_1$ R_1 A

29. * Repeat parts (a) and (b) of the previous problem for the case in which $\mathcal{E}_1 = 12$ V, $\mathcal{E}_2 = 3$ V, $R_1 = R_2 = 1$ Ω, and $R_3 = 16$ Ω. (c) Determine the potential difference from A to B. (d) Draw a potential-versus-position graph starting at any location in the circuit and returning to the same point.

30. * (a) Determine the value of $\mathcal{E}_1$ so that there is a clockwise current of 1.0 A in the circuit shown in Figure P19.28, where $\mathcal{E}_2 = 12$ V, $R_1 = 2$ Ω, $R_2 = 1$ Ω, and $R_3 = 12$ Ω. (b) Determine the potential difference from B to A. (c) Draw a graph representing the potential at different points of the circuit.

31. ** The current through resistor R_1 in Figure P19.27 is 2.0 A. Determine the currents through resistors R_2 and R_3 and the emf of the battery. $R_1 = 4$ Ω, $R_2 = 10$ Ω, and $R_3 = 40$ Ω.

32. ** Use Kirchhoff's rules to show in general that the currents I_2 and I_3 through resistors R_2 and R_3 shown in Figure P19.27 satisfy the relation $I_2/I_3 = R_3/R_2$.

33. * (a) Write the loop rule for two different loops in the circuit shown in **Figure P19.33** and the junction rule for point A. Solve the equations to find the current in each loop when $\mathcal{E}_1 = 3$ V, $\mathcal{E}_2 = 6$ V, $R_1 = 10$ Ω, $R_2 = 20$ Ω, and $R_3 = 30$ Ω. (b) Determine the potential difference from A to B. Check your answer by taking a different path from A to B. (c) Sketch a potential-versus-position graph for each of the three loops, assuming the potential is zero at B.

FIGURE P19.33

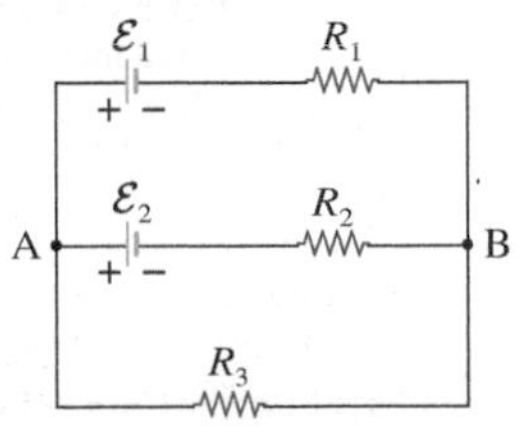

$\mathcal{E}_1$ R_1
+ −

$\mathcal{E}_2$ R_2
A   + −   B

R_3

34. ** Determine the value of R_2, shown in Figure P19.33, so that the current through R_3 equals twice that through R_2. The values of other circuit elements are $\mathcal{E}_1 = 12$ V, $\mathcal{E}_2 = 15$ V, $R_1 = 15$ Ω, and $R_3 = 30$ Ω.

35. * Determine (a) the equivalent resistance of resistors R_1, R_2, and R_3 in Figure P19.35 for $R_1 = 28$ Ω, $R_2 = 30$ Ω, and $R_3 = 20$ Ω and (b) the current through the battery if $\mathcal{E} = 10$ V.

FIGURE P19.35

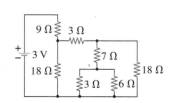

36. (a) Determine the equivalent resistance of resistors R_1, R_2, and R_3 in Figure P19.35 for $R_1 = 50$ Ω, $R_2 = 30$ Ω, and $R_3 = 15$ Ω.
(b) Determine the current through R_1 if $\mathcal{E} = 120$ V.
(c) Use Kirchhoff's loop rule and your result from part (b) to determine the current through R_2 and through R_3.

37. * Determine the equivalent resistance of the resistors shown in Figure P19.37 if $R_1 = 60$ Ω, $R_2 = 30$ Ω, $R_3 = 20$ Ω, $R_4 = 20$ Ω, $R_5 = 60$ Ω, $R_6 = 20$ Ω, and $R_7 = 10$ Ω.

FIGURE P19.37

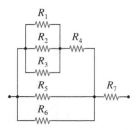

FIGURE P19.38

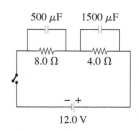

38. * Determine (a) the equivalent resistance of the resistors in the circuit in Figure P19.38 and (b) the current through the battery.

39. ** Write a problem for which the following mathematical statement can be a solution:

$$1\text{ A} = \frac{\mathcal{E} - (1\text{ A})(3\text{ Ω})}{R}$$

40. You close the switch in the circuit in Figure P19.40 and wait until the currents stop changing. (a) Determine the charge on the capacitor. (b) What happens to that charge when you open the switch again? Explain qualitatively.

FIGURE P19.40

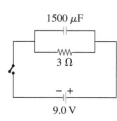

1500 μF

3 Ω

9.0 V

FIGURE P19.41

500 μF   1500 μF

8.0 Ω   4.0 Ω

12.0 V

41. * You close the switch in the circuit in Figure P19.41 and wait until the currents stop changing. Determine the charge on each capacitor.

19.9 Skills for solving circuit problems

42. * **Home wiring** A simplified electrical circuit for a home is shown in Figure P19.42. (a) Determine the currents through the circuit breaker, the lightbulb, the microwave oven, and the toaster. (b) Determine the electric power used by each appliance.

FIGURE P19.42

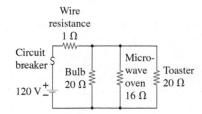

Wire resistance
1 Ω
Circuit breaker
Bulb 20 Ω
Micro-wave oven 16 Ω
Toaster 20 Ω
120 V

43. ** (a) Write Kirchhoff's rules for two loops and one junction in the circuit shown in Figure P19.43. (b) Solve the equations for the current in each branch of the circuit when $\mathcal{E}_1 = 10$ V, $\mathcal{E}_2 = 2$ V, $R_1 = 50$ Ω, $R_2 = 200$ Ω, and $R_3 = 20$ Ω. (c) Determine the potential difference across resistor R_3 from point B to point A.

FIGURE P19.43

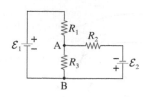

44. ** **BIO** **Electric eel** The South American eel can generate electric current that can stun and even kill nearby fish. The eel has 140 parallel rows of electric cells (0.15 V per cell). Each row has 5000 such cells for a total emf per row of $5000(0.15$ V$) = 750$ V. Each row of cells also has about 1250 Ω of internal resistance. Because each row has the same emf and the rows are connected together on each side, the eel's circuit can be represented as shown in Figure P19.44—a 750-V emf source in series with 140 1250-Ω parallel internal resistances all connected across an external resistance (the seawater from the front of the eel to its back) of about 800 Ω. Can the eel produce enough current to be dangerous to a person? Explain your answer. Identify all assumptions that you made.

FIGURE P19.44

Electric eel

750 V   1250 Ω
750 V   1250 Ω

750 V   1250 Ω

800 Ω

45. **Home wiring** A 120-V electrical line in a home is connected to a 60-W lightbulb, a 180-W television set, a 300-W desktop computer, a 1050-W toaster, and a 240-W refrigerator. How much current is flowing in the line?

46. * **Tree lights** Nine tree lights are connected in parallel across a 120-V potential difference. The cord to the wall socket carries a current of 0.36 A. (a) Determine the resistance of one of the bulbs. (b) What would the current be if the bulbs were connected in series?

47. * Two lightbulbs use 30 W and 60 W, respectively, when connected in parallel to a 120-V source. How much power does each bulb use when connected in series across the 120-V source, assuming that their resistances remain the same?

48. * Three identical resistors, when connected in series, transform electrical energy into thermal energy at a rate of 15 W (5 W per resistor). Determine the power consumed by the resistors when connected in parallel to the same potential difference.

49. ** **Impedance matching** A battery has an emf of 12 V and an internal resistance of 3 Ω. (a) Determine the power delivered to a resistor R connected to the battery terminals for values of R equal to 1, 2, 3, 4, 5, and 6 Ω. (b) Plot on a graph the calculated values of P versus the different values of R. Connect the points by a smooth curve. Confirm that the maximum power is delivered when R has the same resistance as the internal resistance of the power source (3 Ω in this example).

50. * Determine the equivalent resistance of the circuit in Figure P19.50 when (a) both switches are opened, (b) switch 1 is closed and switch 2 is opened, and (c) both switches are closed.

FIGURE P19.50

Switch 1

R_A R_B R_C

Switch 2

51. ** The resistors in the circuit shown in Figure P19.51 all have the same resistance. The current through the battery is I_0. You make the changes to the original circuit that are described in the table below, and each time record the current through the battery. Rank the currents I_0 to I_4 from largest to smallest. Assume all wires have negligible resistance and the battery has negligible internal resistance.

FIGURE P19.51

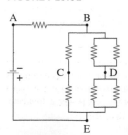

Connect the following points with a wire	Current through the battery
A and B	I_1
C and D	I_2
B and E	I_3
C and E	I_4

52. **Figure P19.52** shows a real circuit that consists of four identical lightbulbs that are mounted in yellow holders (each lightbulb holder has two terminals) and a battery. (a) Draw a diagram for this circuit. (b) Predict which lightbulb(s) will glow if you connect the free red alligator clip to the free terminal of the battery. (c) Now move the far left red alligator clip from the upper to the lower terminal of the bulb P (as explained on the figure). Draw a new circuit diagram and predict which lightbulb(s) now glows.

FIGURE P19.52

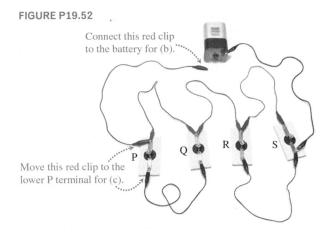

Connect this red clip to the battery for (b).

Move this red clip to the lower P terminal for (c).

P   Q   R   S

19.10 Properties of resistors

53. * A 100-m-long copper wire of radius 0.12 mm and mass 40 g is connected across a 1.5-V battery. Make a list of the physical quantities that you can determine using this information and determine the values of three of them. The resistivity of copper is $1.7 \times 10^{-8}\ \Omega \cdot m$.

54. * **BMT subway rail resistance** The BMT subway line in New York City stretches roughly 30 km from the Bronx to Brooklyn. The electrified rail on which it runs has a cross section of about 40 cm² and is made of steel with a resistivity of $10 \times 10^{-8}\ \Omega \cdot m$. Make a list of all the physical quantities describing different properties of the rail that you can determine using this information. Then calculate the values of one quantity related to the electrical properties and one quantity not related to the electrical properties.

55. * **Thermometer** A platinum resistance thermometer consists of a 0.10-mm-diameter platinum wire wrapped in a coil. Determine the length of wire needed so that the coil's resistance at 20 °C is 25 Ω. The resistivity of platinum at this temperature is $1.0 \times 10^{-7}\ \Omega \cdot m$.

56. As the potential difference in volts across a thin platinum wire increases, the current in amperes changes as follows: $(\Delta V, I) = (0, 0), (1.0, 0.112), (3.0, 0.337)$, and $(6.1, 0.675)$. Plot a graph of potential difference as a function of current and indicate whether the platinum wire is an ohmic element. Explain how you made your decision.

57. * **BIO Respiration detector** A respiration detector monitors a person's breathing. One type consists of a flexible hose filled with conductive salt water (resistivity of $5.0\ \Omega \cdot m$). Electrodes at the ends of the tube measure the resistance of the fluid in the tube. The tube is wrapped around a person's chest. When the person inhales and exhales, the tube stretches and contracts and its resistance changes. Determine the factor by which the resistance of the fluid changes when the hose is stretched so that its length increases by a factor of 1.1. The water volume remains constant.

58. * A wire whose resistance is R is stretched so that its length is tripled while its volume remains unchanged. Determine the resistance of the stretched wire.

59. * **Ratio reasoning** Determine the ratio of the resistances of two wires that are identical except that (a) wire A is twice as long as wire B, (b) wire A has twice the radius of wire B, and (c) wire A is made of copper and wire B is made of aluminum. Be sure to show clearly how you arrive at each answer.

60. ** **Electronics detective** You need to determine the mass and length of the wire inside a particular electronic device. You cannot take the wire out (to measure its length), but you can reach its free ends and connect devices to them. Devise a method to do this by using a battery, ammeter, voltmeter, and micrometer (a device that measures small distances.)

61. * A battery produces a 2.0-A current when connected to an unknown resistor of resistance R. When a 10-Ω resistor is connected in series with R, the current drops to 1.2 A. (a) Determine the emf of the battery and the resistance R. (b) What assumptions did you make? Do your assumptions make each of the values you determined greater than or less than their actual values? Explain.

62. * **BIO Resistance of human nerve cell** Some human nerve cells have a long, thin cylindrical cable (the axon) from their inputs to their outputs. Consider an axon of radius 5×10^{-6} m and length 0.6 m. The resistivity of the fluid inside the axon is $0.5\ \Omega \cdot m$. Determine the resistance of the fluid in this axon.

63. * **Conductive textiles** Metal strands can be woven into textiles to make them conductive. You have some 30-mm-wide nickel textile tape with a resistance of 3.0 Ω per meter length. You cut two rectangular strips from this tape. The first strip is 30 mm by 100 mm, and the second strip is 15 mm by 120 mm. You connect the strips in series to a 1.5-V battery as shown in **Figure P19.63**. Determine the total power output of the two strips. Assume the resistance of the wires and the internal resistance of the battery are negligible.

FIGURE P19.63

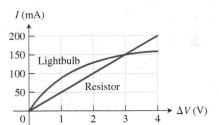

General Problems

64. * **EST** **Figure P19.64** shows an I-versus-ΔV graph for an incandescent lightbulb and a resistor (data are measured separately but presented in the same graph). In one case, the lightbulb and the resistor are connected in parallel to a battery. The current through the battery is 300 mA. Estimate the emf of the battery, ignoring the internal resistance of the battery.

FIGURE P19.64

I (mA)

200
150   Lightbulb
100
50   Resistor

0   1   2   3   4 ΔV (V)

65. * **EST** Figure P19.64 shows an I-versus-ΔV graph for an incandescent lightbulb and a resistor (data are measured separately but presented in the same graph). In one case, the lightbulb and the resistor are connected in series and the current through them is 100 mA. Estimate (a) the emf of the battery and (b) the power output of the lightbulb and of the resistor. Ignore the internal resistance of the battery.

66. * **EST** Figure P19.64 shows an I-versus-ΔV graph for an incandescent lightbulb and a resistor (data are measured separately but presented in the same graph). In one case, the lightbulb and the resistor are connected in series to a 1.5-V battery. Estimate (a) the voltage across the lightbulb and the resistor and (b) the current through the battery. Ignore the internal resistance of the battery. (Hint: The sum of the voltages across the elements is 1.5 V and the current through both elements should be the same.)

67. * **Wiring high-fidelity speakers** Your high-fidelity amplifier has one output for a speaker of resistance 8 Ω. How can you arrange two 8-Ω speakers, one 4-Ω speaker, and one 12-Ω speaker so that the amplifier powers all speakers and their equivalent resistance when connected together in this way is 8 Ω? Compare the power output of your arrangement with the power output of a single 8-Ω speaker. Express your answer as a ratio $P_{Arrange}/P_{8\Omega}$.

68. * **BIO** **EST** **Lifting forearm by electric current** You are asked to evaluate an idea for making it possible for a stroke patient to contract the biceps muscle in order to lift the forearm. The muscle will contract if a current of about 20 mA flows through the upper arm. Estimate the potential difference across electrodes, one inserted near the top of the arm and the other near the elbow, that will produce this current. Indicate any assumptions you made.

69. * **EST** **Switches** You have a power supply, a 10-W bulb, and a 15-W bulb. You also have two types of switches: a single pole single throw (SPST) switch and a single pole double throw (SPDT) switch (shown in **Figure P19.69**). Build a circuit satisfying the following conditions: when the switch is in one position, the 10-W bulb is on; when the switch is in the other position, the 10-W bulb no longer glows, but the 15-W bulb does.

FIGURE P19.69

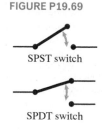

SPST switch

SPDT switch

70. ** **Wiring a staircase** Devise an electric circuit that will allow you to turn a stairway lightbulb on and off from the bottom and from the top of the stairway. You can use switches such as those shown in Figure 19.69.

71. ** **Gas or electric heating?** What information should you collect to decide whether to use electricity or gas to heat your house in the winter? Search relevant information and write a report that will convince your roommates that one way is better than the other. Make sure you include quantitative arguments.

72. ** **EST** **Electric water heater** An electric hot water heater has a 7000-W resistive heating element. (a) Determine the energy needed to warm the 120 kg of water in the heater from 20 °C to 60 °C. (b) What time interval is needed to warm the water? (c) Estimate the electric energy cost of warming the water at recent rates.

73. ** **BIO** **EST** **The hands and arms as a conductor** While doing laundry you reach to turn on the light above the washing machine. Unfortunately, you touch an exposed 120-V wire in the broken switch box. Your other hand is supporting you on the top of the grounded washing machine. Assume that the resistance is 100,000 Ω across the dry skin of each hand and that the resistivity of tissue inside your arms and body is 5 Ω·m. (a) Estimate the total resistance of the tissue inside your arms from one side to the other (ignore the resistance across your chest). (b) Estimate the total resistance across the skin of one hand through the tissue in your arms and then across the skin of your other hand. (c) Are you in danger?

74. * **Design an experiment II** You have a 1.5-V battery, a 1.0-F capacitor, a blue LED (with opening voltage 2.3 V) and connecting wires. Using only this equipment, design an experiment that will make the blue LED glow. Describe your design using words and circuit diagrams.

75. * A nickel wire of length L and a voltmeter are connected to a battery as shown in **Figure P19.75**. One terminal of the voltmeter is connected to the left end of the nickel wire and the other terminal can slide along the wire. The battery has negligible internal resistance and an emf $\mathcal{E}$. Assume the resistance of the connecting wires is negligible. Derive an expression that describes how the reading of the voltmeter ΔV depends on the position x of its right terminal, and evaluate the solution.

FIGURE P19.75

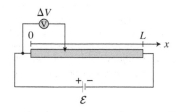

76. ** Solve the previous problem if the internal resistance of the battery is r and the resistance of the nickel wire is R. Evaluate the solution.

77. * **EST** **Figure P19.77** shows an $|I|$-versus-ΔV graph of a circuit that consists of a red LED ($\Delta V_{Open\ R} = 1.6$ V), a blue LED ($\Delta V_{Open\ B} = 2.3$ V) and a resistor. (a) Choose the circuit in the lower part that will produce the $|I|$-versus-ΔV graph of the figure and explain your choice. (b) Which terminal of the circuit (left or right) is at the higher potential and which is at the lower potential when $\Delta V = +3.0$ V? (c) Estimate the resistance of the resistor.

FIGURE P19.77

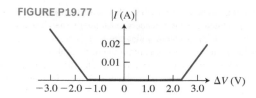

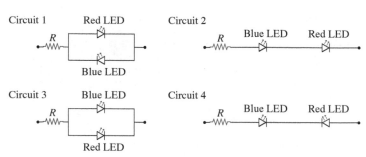

78. ** Jesse needs to determine the resistance of a resistor. Which method should he use? If you think that more than one of the described methods is useful, compare your choices and explain which would give the most accurate result. Describe any assumptions that you made. The final step in all methods is to calculate the resistance of the resistor as $R = \dfrac{\Delta V}{I}$.

(a) Connect a voltmeter to a battery's terminals. Record the voltage reading ΔV. Then (in a different experiment) connect the battery, the resistor, and an ammeter in series. Record the current reading I.

(b) Connect a battery, the resistor, and an ammeter in series and connect a voltmeter in parallel to the resistor. Record the current reading I and the voltage reading ΔV.

(c) Connect a battery, the resistor, an ammeter, and a voltmeter in series. Record the current reading I and the voltage reading ΔV.

(d) Connect a battery, the resistor, and an ammeter in series and connect a voltmeter in parallel to the ammeter. Record the current reading I and the voltage reading ΔV.

(e) Connect a battery, the resistor, and an ammeter in series. Record the current reading I. Then (in a different experiment) connect the resistor to the battery and a voltmeter in parallel to the resistor. Record the voltage reading ΔV.

79. ** **People current** Suppose that all the people on the Earth moved at the same speed around a circular track. Approximately how many times per second would each person pass the starting line if that "people current" equaled the number of electrons passing a cross section in a wire when there is a 1.0-A current through it?

80. * A 5.0-A current caused by moving electrons flows through a wire. (a) Determine the number of electrons and the number of moles of electrons that flow past a cross section each second. (b) The same number of water molecules moves along a stream. Determine the volume of water that moves each second under a bridge that passes over the stream.

81. ** **BIO** **Current across membrane wall of axon** An axon cable that connects the input and output ends of a human nerve cell is 5×10^{-6} m in radius and 0.5 m long. The thickness of the membrane surrounding the fluid in the axon is 6×10^{-9} m, its resistivity is 1.6×10^{7} Ω·m, and there is a 0.070-V potential difference across the membrane. (a) Determine the current through the membrane. (b) If the current is due to sodium ions, how many of these ions leak across the membrane wall per second? (c) How does the current change if the membrane is twice as thick?

Reading Passage Problems

BIO **Signals in nerve cells stimulate muscles** The input end of a human nerve cell is connected to an output end by a long, thin, cylindrical axon. A signal at the input end is caused by a stretch sensor, a temperature sensor, contact with another cell or nerve, or some other stimulus. At the output end, the nerve signal can stimulate a muscle cell to perform a function (to contract, provide information to the brain, etc.).

The axon of a so-called unmyelinated human nerve cell has a radius of 5×10^{-6} m and a membrane that is 6×10^{-9} m thick. The membrane has a resistivity of about $1.6 \times 10^7 \ \Omega \cdot m$. The fluid inside the axon has resistivity of about $0.5 \ \Omega \cdot m$. The membrane wall has proteins that pump three sodium ions (Na^+) out of the axon for each two potassium ions (K^+) pumped into the axon. In the resting axon, the concentration of these ions results in a net positive charge on the outside of the membrane compared to negative charge on the inside. Because of the unequal charge distribution, there is a $-70 \ mV$ potential inside compared to outside the axon.

When an external source stimulates the input end of the nerve cell so the potential inside reaches about $-50 \ mV$, gates or channels in the membrane walls near that input open and sodium ions rush into the axon. This stimulates neighboring gates to swing open and sodium ions rush into the axon farther along. This disturbance quickly travels along the axon—a nerve impulse. The potential across the inside of the membrane changes in 0.5 ms from $-70 \ mV$ to $+30 \ mV$ relative to the outside. Immediately after this depolarization, potassium ion gates open and positively charged potassium ions rush out of the axon, repolarizing the axon. Sodium and potassium ion pumps then return the axon and its membrane to their original configuration.

82. Which answer below is closest to the resistance of the fluid inside a 0.5-m-long unmyelinated axon?
    (a) $10^9 \ \Omega$ (b) $10^6 \ \Omega$ (c) $10^3 \ \Omega$ (d) $10^{-3} \ \Omega$ (e) $10^{-6} \ \Omega$
83. An electric signal transmitted in a metal wire involves electrons moving parallel to the wire. What does an electric signal in an axon involve?
    (a) Sodium ions moving across the axon membrane perpendicular to its length
    (b) Potassium ions moving across the axon membrane perpendicular to its length
    (c) Sodium and potassium ions moving inside the axon parallel to its length
    (d) Only sodium ions moving inside the axon parallel to its length
    (e) a and b
84. Which answer below is closest to the magnitude of the $\vec{E}$ field in the resting axon membrane?
    (a) $10 \ V/m$ (b) $10^3 \ V/m$ (c) $10^5 \ V/m$ (d) $10^7 \ V/m$ (e) $10^9 \ V/m$
85. The charge density on the axon membrane walls (positive outside and negative inside) is $1.0 \times 10^{-4} \ C/m^2$. Suppose the membrane of a 1.0-m-long axon discharged completely in 0.02 s. What answer below is closest to the electric current across the membrane wall of one such hypothetical axon discharge?
    (a) 0.2 A   (b) 0.02 A   (c) 2×10^{-4} A   (d) 2×10^{-5} A   (e) 2×10^{-7} A
86. The horizontal 4-Ω resistors in the two circuits in **Figure P19.86** represent the resistance of a small horizontal length of fluid inside an axon. The 15-Ω resistors represent the resistance across the axon membrane for a tiny length of axon. The 10-Ω resistor to the right side of each "ladder" is the effective resistance in front of the axon sections under consideration. Determine the resistance between B and G of the small length of axon shown in Figure P19.86a and between A and G of the slightly longer length axon shown in Figure P19.86b.
    (a) 4.2 Ω, 8.4 Ω (b) 10 Ω, 10 Ω (c) 29 Ω, 48 Ω
    (d) 10 Ω, 12 Ω (e) 4.2 Ω, 12 Ω

FIGURE P19.86

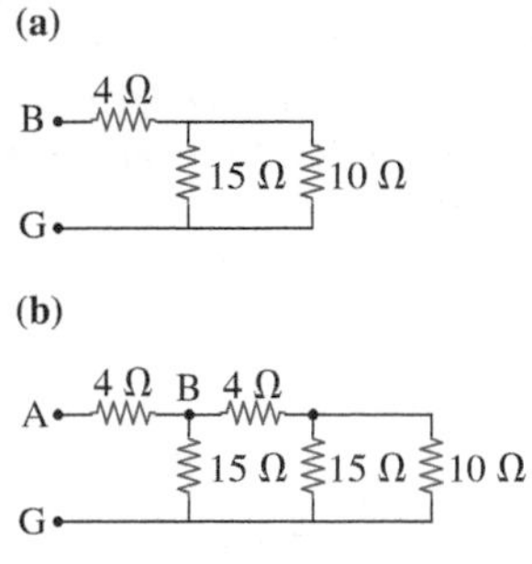

(a)

B •—WW— 4 Ω
      ⌇15 Ω ⌇10 Ω
G •—

(b)

A •—WW—•—WW— 4 Ω B 4 Ω
      ⌇15 Ω ⌇15 Ω ⌇10 Ω
G •—

87. Suppose nerve impulses travel at 100 m/s in the axons of nerve cells from your fingers to your brain and then back again to your fingers in order to stimulate muscles that lift your fingers off the hot burner of a stove. Which answer below is closest to the time interval needed for the nerve signal transmission along the axons?
    (a) 0.01 s   (b) 0.001 s   (c) 0.2 s   (d) 0.02 s   (e) 0.002 s

BIO **Effect of electric current on human body** Nerve impulses are initiated at the input end of a nerve cell, travel along a relatively long axon (cable), and then cause an effect at the output end of the cell—for example, the initiation of a muscle contraction in a muscle cell. The nerve impulse is initiated by a stimulus that lowers the potential difference from outside the cell to inside from its normal $-70 \ mV$ to about $-50 \ mV$.

A potential difference across two parts of the body (for example, the 120-V potential difference from a wall socket from one hand to the other or from the

FIGURE 19.39 The effects of electric current on the human body.

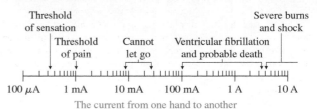

The current from one hand to another

hands to the feet) can initiate an electric current in the body that stimulates nerve endings and triggers nerve signals that cause muscular contraction. Even worse, the current in the body can upset the rhythmic electrical operation of the heart. The heart muscles might be stimulated randomly in what is called ventricular fibrillation—a random contraction of the ventricles, which can be deadly. A rough guide to the effects of electric current on the body at different current levels is provided in **Figure 19.39**. Under dry conditions, human skin has high electrical resistance. Wet skin dramatically lowers the body's resistance and makes electrocution more likely to occur.

88. The electrical resistance across dry skin is about 100,000 Ω. Suppose a person with dry skin puts one hand on a 120-V power cord from a home wall socket while the other hand is touching a metal object at 0 V (at what is called ground). Which condition described below is most likely to occur?
    (a) No sensation   (b) Threshold of pain
    (c) Cannot let go   (d) Ventricular fibrillation
    (e) Severe burns and shock
89. The electrical resistance across wet skin is about 1000 Ω. Suppose a person with wet skin puts one hand on a 120-V power cord from a home wall socket while the other hand is touching a metal object at 0 V (at what is called ground). Which condition described below is most likely to occur?
    (a) No sensation   (b) Threshold of pain
    (c) Cannot let go   (d) Ventricular fibrillation
    (e) Severe burns and shock
90. Suppose the electrical resistance across your wet skin is about 1000 Ω. Which answer below is closest to the least potential difference from one hand that will cause slight pain?
    (a) 0.1 V   (b) 1 V   (c) 10 V   (d) 100 V   (e) 1000 V
91. Suppose the electrical resistance across your wet skin is about 1000 Ω. Which answer below is closest to the least potential difference from one hand to the other that will cause ventricular fibrillation?
    (a) 0.1 V   (b) 1 V   (c) 10 V   (d) 100 V   (e) 1000 V
92. When muscular contraction caused by electrical stimulation prevents a person from releasing contact with the potential difference sources, lower current can be extremely dangerous. For example, a 100-mA current for 3 s causes about the same effect as a 900-mA current for 0.03 s. In which of these situations is the least electric charge transferred through the body?
    (a) 100 mA for 3 s
    (b) 900 mA for 0.03 s
    (c) Too little information to decide
93. Why is it dangerous to place a hair dryer, radio, or other electric appliance that is plugged into a wall socket near a bathtub?
    (a) The water provides a conductive path for current, which heats the metal cover on the appliance and can cause burns.
    (b) If the appliance is accidentally knocked into the tub while a person is bathing, large currents could pass through the person's low-resistance body because of the 120-V potential difference that powers the appliance.
    (c) There is no potential for danger, because electric appliances are grounded.
    (d) a and b
94. Occasionally, the electric circuit that produces a coordinated pumping of blood from the four chambers of the heart becomes disturbed. Ventricular fibrillation can occur—random muscle contractions that produce little or no blood pumping. To stop the fibrillation, two defibrillator pads are placed on the chest and a large current (about 14 amps) is sent through the heart, restarting its normal rhythmic pattern. The current lasts 10 ms and transfers 140 J of electric energy to the body. Which answer below is closest to the potential difference between the defibrillator pads?
    (a) 100 V   (b) 400 V   (c) 1000 V   (d) 5000 V   (e) 10,000 V

Magnetism

- What causes the northern lights?
- How does Earth protect us from solar wind and cosmic rays?
- Are we really walking northward when we follow a compass needle?

The aurora borealis—the northern lights—illuminates the autumn Arctic sky with a flickering greenish glow that is streaked with red and sometimes blue or violet. A comparable phenomenon—the aurora australis—appears in the autumn sky above southern Australia, southern South America, and the Antarctic. What causes auroras, and why do we see auroras near Earth's poles and only rarely anywhere else?

BE SURE YOU KNOW HOW TO:
- Use the electric field concept to explain how electrically charged objects exert forces on each other (18.1).
- Find the direction of the electric current in a circuit (19.1).
- Apply Newton's second law to a particle moving in a circle (5.3).

IN PREVIOUS CHAPTERS we learned that electrically charged objects do not exhibit magnetic properties. Until 1820, physicists believed that electric and magnetic phenomena were completely different. Since then, scientists have made many discoveries about the relationships between electricity and magnetism. These discoveries have made possible electromagnets, electric motors, magnetic resonance imaging (MRI), and many other devices. We begin this chapter by learning about the connections between electricity and magnetism and how scientists discovered them.

20.1 Magnetic interactions

Magnets become familiar through toys, refrigerator magnets, and countless household applications. Very early we discover that, when brought near each other, magnets can both attract and repel. The two sides of a bar magnet are called **poles**—the **north pole** and the **south pole**. If you bring the like poles of two magnets (north to north and south to south) near each other, they repel. If you bring opposite poles near each other, they attract (**Figures 20.1a** and **b**). Additionally, both sides of a magnet attract objects made from steel or iron, even if those objects aren't magnets themselves.

Magnets always have two poles. If you break a magnet into two pieces, each piece still has two poles—a north pole and a south pole (**Figure 20.2**). If you break one of those pieces again, each smaller piece has two poles, and so on. Unlike electrically charged objects that can be either negatively or positively charged, there are no magnets with a single pole (a so-called monopole).

The labels "north" and "south" originated in 11th-century China. People noticed that if they put a tiny magnet on a low-friction pivot (or attached it to a piece of wood floating in water), one end always pointed toward geographic north. This property of a magnet resulted in the names for the two ends: the north pole points toward geographical north; the south pole points toward geographical south. The device became known as a **compass**.

Since the north pole of a compass is attracted to the south pole of another magnet, and since compasses that are not near other magnets always point approximately toward Earth's geographic north pole, we can infer that Earth itself acts as a giant magnet with its magnetic south pole close to its geographic north pole and its magnetic north pole close to its geographic south pole (note the separation between the magnetic and geographical poles in **Figure 20.3**).

The magnetic interaction depends on separation

We can use a compass to investigate not only the direction of the magnetic influence on it but also the strength. **Figure 20.4** shows a sequence of four simple experiments all performed on a wooden table. In each experiment we place a compass at a specific location with respect to a magnet and observe the compass's behavior. When we put the compass at location A, it vibrates slowly and then stops in the orientation shown. Placed at location B, the compass vibrates a little faster (shorter period T) and comes to the same orientation as in A. In C, its vibrations have even a smaller period, and finally, at D, very close to the magnet, the compass vibrates very quickly until its tip points toward the magnet. From these observations, we can infer that when closer to the magnet, the interaction is stronger (in Chapter 10 we learned that the period of vibrations is proportional to the square root of the restoring force, Eq. 10.12).

TIP   In Chapter 17 we learned that although electrically charged objects attract and repel in a similar way to magnets, the interactions involved are different. Make sure you do not call magnetic poles charges, and do not confuse north and south poles with positive and negative electric charges. Review Table 17.3 before moving forward.

REVIEW QUESTION 20.1 What aspects of compass behavior near a magnet give us information about the direction of magnetic influence, and what aspects give us information about the strength?

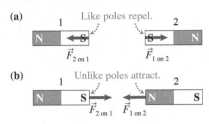

FIGURE 20.1 Like magnetic poles repel and unlike poles attract.

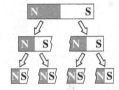

FIGURE 20.2 Breaking a magnet into pieces produces more magnets, each with a north and south pole.

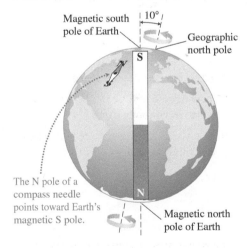

FIGURE 20.3 The magnetic and geographic poles of Earth.

FIGURE 20.4 Investigating the strength of magnetic interactions using vibrations of a compass.

$$T_A > T_B > T_C > T_D$$

FIGURE 20.5 Compasses indicate the $\vec{B}$ field directions at various locations near (a) a bar magnet and (b) a horseshoe magnet.

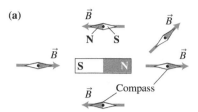
(a)

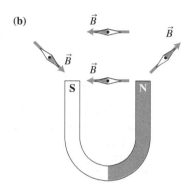
(b)

FIGURE 20.6 Compasses indicate $\vec{B}$ field lines.

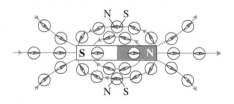

FIGURE 20.7 Iron filings indicate $\vec{B}$ field lines.

Iron filings lie on a transparent sheet lying on a bar magnet.

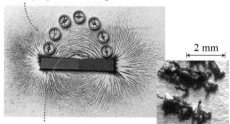

2 mm

Above the magnet, the filings align with the S-N direction of the magnet.

FIGURE 20.8 $\vec{B}$ field lines.

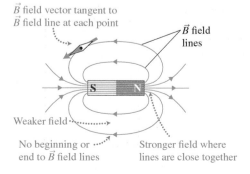

$\vec{B}$ field vector tangent to $\vec{B}$ field line at each point

$\vec{B}$ field lines

Weaker field

No beginning or end to $\vec{B}$ field lines

Stronger field where lines are close together

20.2 Magnetic field

The electric field explains how electrically charged objects interact without contact. Since magnets also interact without contact, it is reasonable to suggest the existence of some kind of field, in this case a **magnetic field**, as the mechanism behind magnetic interactions. Thus we can assume that a magnet produces a magnetic field—a magnetic disturbance with which other objects with magnetic properties (another magnet, anything made of iron, etc.) interact. In this chapter we will learn how to describe this field qualitatively and quantitatively and how to predict its effects on objects that possess their own magnetic properties.

Direction and strength of the magnetic field

To describe the magnetic field, we use a vector physical quantity called the $\vec{B}$ **field**. The $\vec{B}$ field has a magnitude and direction. We can use a compass to detect the direction of the $\vec{B}$ field at a particular location, just as we can use a test charge for the $\vec{E}$ field. The direction of the $\vec{B}$ field is defined as the direction that the north pole of the compass needle points. We can also determine the direction of the $\vec{B}$ field by drawing a vector from the south pole to the north pole of the compass inside the compass. **Figure 20.5a** shows the direction of the $\vec{B}$ field at several points near a bar magnet. Figure 20.5b shows the $\vec{B}$ field vectors near a horseshoe magnet. We can determine the relative strength of the field at a particular location by observing the vibrations of a compass needle placed there. The shorter the period of vibrations, the stronger the field.

Representing the magnetic field: field lines

To represent the magnetic field in a region, we draw $\vec{B}$ field lines (similar to $\vec{E}$ field lines). We can visualize the lines by placing multiple compasses around a bar magnet, as shown in **Figure 20.6**. Following the direction of the south-north vector for each compass as we move from compass to compass, we find that the compasses are tangent to lines surrounding the magnet—the blue lines in Figure 20.6. Instead of compasses, we can use hundreds of tiny iron sticks or needles called iron filings (which act like tiny compasses) sprinkled on a thin piece of transparent plastic placed on top of the magnet. The filings form a pattern that looks similar to the lines formed by the compasses (**Figure 20.7**; the inset shows a photo of two filings magnified under a microscope). These lines are so-called $\vec{B}$ **field lines**. They are used to represent the $\vec{B}$ field produced by the magnet in the whole region, not just at a point, as one $\vec{B}$ field vector indicates. Similar to $\vec{E}$ field lines, $\vec{B}$ field lines represent not only the direction of the $\vec{B}$ field, but also its magnitude. The lines are closer together in regions where the $\vec{B}$ field magnitude is greater (where the compass needle vibrated with shorter periods)—see the pattern shown in **Figure 20.8**.

The magnetic $\vec{B}$ field and its representation by $\vec{B}$ field lines The direction of the magnetic $\vec{B}$ field at a particular location is defined as the direction that the north pole of a compass needle points when at that location. The $\vec{B}$ field vector at a location is tangent to the direction of the $\vec{B}$ field line at that location. Both point in the same direction. The density of lines in a region represents the magnitude of the $\vec{B}$ field in that region—where the $\vec{B}$ field is stronger, the lines are closer together. Unlike $\vec{E}$ field lines, which begin on positively charged objects and end on negatively charged objects, $\vec{B}$ field lines do not have beginnings or ends. Figure 20.8 shows that they are closed loops that pass through the source of the magnetic field.

What are other ways to produce a magnetic field?

We can now explain the interaction between two magnets (let's call them magnet A and magnet B) in the following way. Magnet A creates a magnetic field around itself, which exerts a force (and possibly a torque) on magnet B. Magnet B creates its own magnetic field that exerts a force (and possibly a torque) on magnet A.

   Do any other objects besides magnets create magnetic fields? We found earlier that stationary electrically charged objects do not. Could electric currents (moving electric charges) produce magnetic fields and be affected by magnetic fields? Observational Experiment **Table 20.1** describes simplified versions of two experiments performed in 1820 by Hans Oersted that investigated this idea.

OBSERVATIONAL EXPERIMENT TABLE 20.1 Do electric currents produce magnetic fields?

VIDEO
OET 20.1

Observational experiment	Analysis
Experiment 1. Connect a battery, a switch, some wires, and a lightbulb in a circuit as shown at left below. The bulb indicates the presence of an electric current in the circuit. Note the north-south orientation of the right-hand wire. With the switch open (the bulb is off), place compasses beneath, on top of, and to the sides (not shown) of one of the wires. Notice the direction the compasses point. Close the switch (the bulb is on) as shown at right below. Notice the directions that the compasses above and beneath the wire now point. 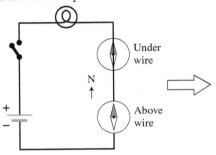 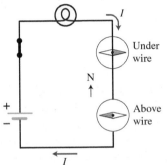	Without current in the circuit, the needles point toward geographic north (magnetic south). Thus, the wire does not produce a magnetic field. The current in the wire affects the orientation of the compasses placed nearby. Therefore, the current-carrying wire might produce a magnetic field.
Experiment 2. Reverse the direction of the current. The compasses on top of and beneath the wire point as shown. 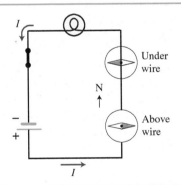	The change in the orientation of the compass indicates that the wire might produce a magnetic field and the direction of the current affects the field.

Pattern

Electric current affects the orientation of a compass. If we placed the compasses on the sides of the wire, we would have seen a complete pattern: with the current in the wire in Experiment 1, the compasses orient in the direction of the circles around the wire shown at right: toward the right when under the wire, up on the right side of the wire, toward the left on top of the wire, and down on the left side of the wire. As the $\vec{B}$ field lines are closed loops, we can use the circles to represent the effect of the current on the compasses.

In Experiment 2, the compass orientations are reversed. Circles in the opposite direction would represent the effect of this current on the compasses.

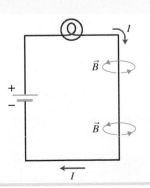

FIGURE 20.9 The right-hand rule for the direction of the $\vec{B}$ field produced by an electric current.

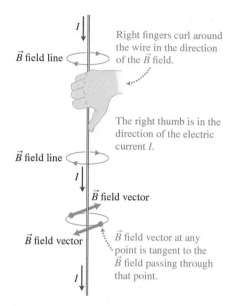

Right fingers curl around the wire in the direction of the $\vec{B}$ field.

$\vec{B}$ field line

The right thumb is in the direction of the electric current I.

$\vec{B}$ field line

$\vec{B}$ field vector

$\vec{B}$ field vector

$\vec{B}$ field vector at any point is tangent to the $\vec{B}$ field passing through that point.

We conclude that an electric current does produce a magnetic field. The direction of the $\vec{B}$ field depends on the direction of the current. The $\vec{B}$ field lines form closed circles around the current. Their direction changes when the direction of the current changes.

Since the current is made of electrically charged particles that are in collective motion with respect to the compass, this means that charged objects in motion produce a magnetic field, and stationary charged objects do not. Through the experiments in Table 20.1 we have deduced a pattern in the direction of the $\vec{B}$ field vectors and $\vec{B}$ field lines around a current-carrying wire. Imagine grasping the wire with your right hand, with your thumb pointing in the direction of the current. When you do so, your fingers wrap around the current in the direction of the $\vec{B}$ field lines (**Figure 20.9**). This method for determining the shape of the $\vec{B}$ field produced by the electric current in a wire is called the **right-hand rule** for the $\vec{B}$ field.

> **Right-hand rule for the $\vec{B}$ field** To determine the direction of the field lines produced by a current-carrying straight wire, grasp the wire with your right hand, with your thumb pointing in the direction of the current. Your four fingers wrap around the current in the direction of the $\vec{B}$ field lines. At each point on a $\vec{B}$ field line, the $\vec{B}$ field vector is tangent to the line and points in the direction of the line.

CONCEPTUAL EXERCISE 20.1   Magnetic field of a solenoid

Draw the magnetic field lines of a solenoid connected to a battery. (A **solenoid** is a wire with a large number of loops in a cylindrical shape. For simplicity in this exercise we will use a solenoid with eight loops.)

Sketch and translate To draw the $\vec{B}$ field lines produced by a current in the solenoid, we use the right-hand rule for the $\vec{B}$ field lines produced by each loop as shown in the figure. We then combine them for a complete picture of the $\vec{B}$ field produced by the solenoid.

Simplify and diagram Consider three adjacent loops near the center of the solenoid. The $\vec{B}$ field lines inside each loop point in the same direction (as shown in the figure below). In the adjacent loops the lines point in the same direction. We assume that the superposition principle applies to the $\vec{B}$ field as it does for the $\vec{E}$ field (see Chapter 18). This means that the total

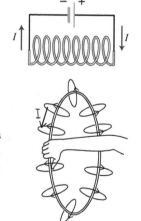

$\vec{B}$ field at any location can be found as the sum of the $\vec{B}$ field vectors due to individual loops. The final result is shown in the figure at right. To check the answer, we place a compass close to the solenoid at several locations. At each location, the compass needle points in the direction of the $\vec{B}$ field.

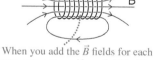

When you add the $\vec{B}$ fields for each turn, you get a uniform field inside.

Try it yourself Use the right-hand rule for the $\vec{B}$ field and the superposition principle to predict the direction of the magnetic field exactly in the middle between two straight wires. The two wires are oriented horizontally in the plane of the page, with currents as shown below.

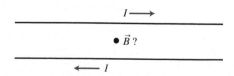

$I \longrightarrow$

$\bullet \vec{B}$?

$\longleftarrow I$

Answer

The $\vec{B}$ field contribution of each wire at the location of interest points into the page. Therefore, the $\vec{B}$ field at that location points into the page.

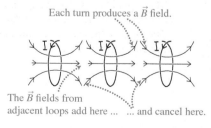

Each turn produces a $\vec{B}$ field.

The $\vec{B}$ fields from adjacent loops add here ...   ... and cancel here.

Notice that the $\vec{B}$ field produced by the current in a loop or a coil (a wire that is bent into multiple loops) (**Figure 20.10a**) and that produced by a bar magnet (Figure 20.10b) are very similar. The $\vec{B}$ field in the region to the right of the plane of the coil (or loop) looks just like the $\vec{B}$ field around the north pole region of the bar magnet. Likewise, the region to the left of the plane of the coil looks just like the $\vec{B}$ field in the south pole region of the bar magnet. Wire coils with current act magnetically in a very similar way to bar magnets and are known as **electromagnets**. We will discuss electromagnets further in Section 20.7.

FIGURE 20.10 A current loop or coil produces a $\vec{B}$ field that is similar to that of a bar magnet.

(a) $\vec{B}$ field produced by a loop

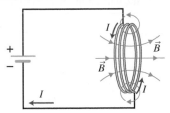

(b) $\vec{B}$ field produced by a bar magnet

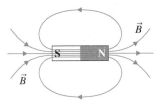

REVIEW QUESTION 20.2 What is the direction of the $\vec{B}$ field at a point halfway between two parallel wires with currents in the same direction?

20.3 Magnetic force on a current-carrying wire

If a current-carrying wire is similar in some ways to a magnet, then a magnetic field should exert a magnetic force on a current-carrying wire similar to the force it exerts on another magnet. When we place a bar magnet near a circuit with long connecting wires, we find that the magnet sometimes pulls on the wire and sometimes does not—the effect depends on the relative directions of the $\vec{B}$ field and the current in the wire.

The magnetic field exerts a force on a current-carrying wire

To investigate how the relative directions of the $\vec{B}$ field and current-carrying wire affect their interactions, we conduct the experiments described in Observational Experiment Table 20.2. We use a horseshoe magnet because between the poles of such a magnet the $\vec{B}$ field lines are almost parallel to each other, representing an example of a uniform magnetic field (see **Figure 20.11**).

FIGURE 20.11 A horseshoe magnet.

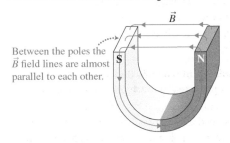

Between the poles the $\vec{B}$ field lines are almost parallel to each other.

OBSERVATIONAL EXPERIMENT TABLE 20.2 Direction of magnetic force

VIDEO OET 20.2

Observational experiment	Analysis
Experiment 1. Hang a horizontal straight wire between the poles of a horseshoe magnet using conducting support wires connected to a battery. Orient the wire parallel to the $\vec{B}$ field lines. Attach a battery and open switch in series to the wire. Turn the current on by closing the switch. The wire does not move. If you reverse the current, the wire still does not move. 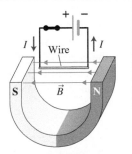	The $\vec{B}$ field lines for both experiments are parallel to the direction of the current. The magnet does not exert a force on the current-carrying wire.

(CONTINUED)

Observational experiment	Analysis
Experiment 2. Orient the wire perpendicular to the $\vec{B}$ field lines and turn the current on by closing the switch. The wire bends down. 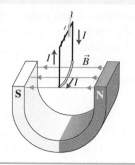	The field lines are perpendicular to the direction of the current in the wire. The force exerted on the wire by the magnetic field is downward and perpendicular to both the current and the $\vec{B}$ field.
Experiment 3. Repeat Experiment 2, but reverse the orientation of the battery, reversing the direction of the current. The wire bends up. 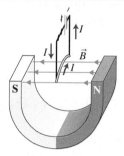	The field lines are perpendicular to the direction of the current in the wire. The force exerted on the wire by the magnetic field is upward and perpendicular to both the current and the $\vec{B}$ field.

Pattern

- The magnetic field does not exert a force on the current-carrying wire if the $\vec{B}$ field is parallel to the wire.
- When the $\vec{B}$ field is perpendicular to the direction of the current, the magnetic field exerts a force on the current-carrying wire that is perpendicular to both the direction of the current and the $\vec{B}$ field. The direction of this force depends on the direction of the current-carrying wire and the direction of the $\vec{B}$ field.

FIGURE 20.12 A method to determine the direction of the force that a magnetic field exerts on a current-carrying wire.

(a) Magnetic force

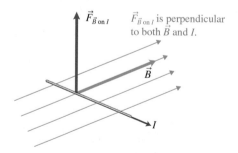

$\vec{F}_{\vec{B}\,on\,I}$ is perpendicular to both $\vec{B}$ and I.

(b) Right-hand rule for magnetic force

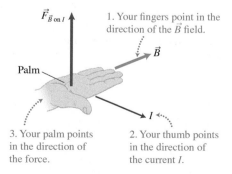

1. Your fingers point in the direction of the $\vec{B}$ field.

Palm

3. Your palm points in the direction of the force.

2. Your thumb points in the direction of the current I.

In Table 20.2 we found that the magnetic field does exert a force on a current-carrying wire when the wire is oriented perpendicular to the $\vec{B}$ field and does not exert a force when the wire is parallel. Other experiments indicate that when the current-carrying wire and the $\vec{B}$ field form any nonzero angle, the magnetic field exerts a force on the current-carrying wire that is perpendicular to both the direction of the current and the direction of the $\vec{B}$ field. The larger the angle between the current-carrying wire and the $\vec{B}$ field, the larger the magnetic force exerted on it (for angles up to 90°).

The direction of the magnetic force that a magnetic field exerts on a current-carrying wire is illustrated in **Figure 20.12a**. The right-hand rule for determining the force direction is described below and in Figure 20.12b.

Right-hand rule for the magnetic force Hold your right hand flat with your thumb extended from the four fingers. Orient your hand so that your fingers point in the direction of the $\vec{B}$ field and your right thumb points along the direction of the current. The direction of the magnetic force exerted by the magnetic field on the current is the direction your palm faces—perpendicular to both the direction of the current and the direction of the $\vec{B}$ field.

TIP   Do not confuse the two right-hand rules. The right-hand rule for the field describes the $\vec{B}$ field produced by a known current. The right-hand rule for the force describes the direction of the magnetic force exerted by a known magnetic field on a current-carrying wire.

Forces that current-carrying wires exert on each other

If a current-carrying straight wire produces a magnetic field, the field should exert a force on a second current-carrying straight wire placed nearby. Similarly, the magnetic field produced by the second wire's current should exert a force on the first wire's current. According to Newton's third law, the forces that these wires exert on each other should point in opposite directions and have the same magnitudes. Let us test this reasoning with the experiments described in Testing Experiment **Table 20.3**. In the table, the crosses X represent the $\vec{B}$ field pointing into the page, and the circled dots ⊙ represent the $\vec{B}$ field coming out of the page. This convention, which we will use throughout the book, applies to any field.

TESTING
EXPERIMENT TABLE 20.3   Do two current-carrying wires interact as magnets?
VIDEO
TET 20.3

Testing experiment	Prediction	Outcome
Two vertical strips of aluminum foil (A and D) are next to each other, each connected by wires at their ends to the terminals of their own batteries. We use aluminum strips to detect this rather weak interaction instead of wires because the strips are much lighter and more flexible. 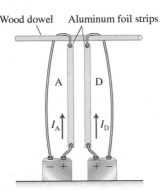 Predict what happens when the currents in the strips are in the same direction.	Choose strip D as the system. Using the right-hand rule for the $\vec{B}$ field, we find that the $\vec{B}_A$ field produced by strip A at D points into the page. Using the right-hand rule for force, we find that this field exerts a force on strip D $\vec{F}_{A \text{ on } D}$ that points to the left. 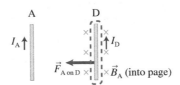 Repeat the procedure for strip A as the system. The $\vec{B}$ field $\vec{B}_D$ produced by strip D at A points out of the page. This field exerts a force $\vec{F}_{D \text{ on } A}$ that points to the right. 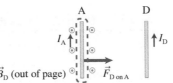 The strips should bend toward each other.	When we do the experiment, the strips bend toward each other.
Predict what happens when the currents in the strips are in the opposite directions.	We repeat the same process to make our prediction. When the current in D is in the downward direction instead of up, the same analysis shows that the strips should repel.	When the currents are in the opposite directions, the strips repel.

Conclusion

We found that two current-carrying wires exert forces on each other whose direction can be predicted using both right-hand rules. The results are consistent with Newton's third law—the forces that the strips exert on each other point in opposite directions.

TIP   In Table 20.3 the wires attract and repel each other via magnetic forces. The wires do not interact via electric forces, because the net electric charge of each wire is zero.

FIGURE 20.13 Ampere's experiment for the unit of electric current.

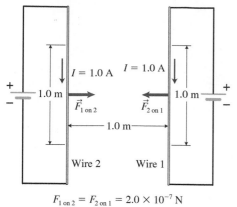

The currents are each defined as 1.0 A if the magnetic force that each meter of a long wire exerts on the other is 2.0×10^{-7} N.

$F_{1 \text{ on } 2} = F_{2 \text{ on } 1} = 2.0 \times 10^{-7}$ N

André-Marie Ampère in 1820 conducted numerous such experiments and found that the magnitude of the force that one current-carrying wire exerts on another parallel current was directly proportional to the magnitude of each current and inversely proportional to the distance between the wires. This relationship became the basis for the unit of the electric current in the SI system—the **ampere** (Figure 20.13).

> **Ampere (A)** When two long parallel wires are separated by 1.0 m (as in Figure 20.13) and there are equal-magnitude electric currents through the wires so that each meter of one wire exerts a force of 2.0×10^{-7} N on 1.0 m of the neighboring wire, the current I in each wire is defined to have a magnitude of 1.0 A.

From this definition it is apparent that the magnetic forces that current-carrying wires exert on each other are rather weak. Unless the currents are very large, the magnetic forces they exert on each other will be small and difficult to detect.

Expression for the magnetic force that a magnetic field exerts on a current-carrying wire

FIGURE 20.14 By varying the $\vec{B}$ field, we can develop an expression for the force that the magnetic field exerts on the current-carrying wire.

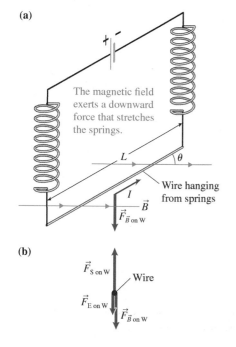

(a)

The magnetic field exerts a downward force that stretches the springs.

(b)

To determine a mathematical expression for the magnetic force that a magnetic field exerts on a current-carrying wire, we hang a horizontal wire at its ends from conducting springs. The tops of the springs are connected to a battery so that the current through the wire is in the direction shown in Figure 20.14a. We place the wire between the poles of an electromagnet (not shown), which produces an approximately uniform horizontal variable $\vec{B}$ field in the region surrounding the wire. The right-hand rule for magnetic force indicates that the magnetic field exerts a downward magnetic force $\vec{F}_{B \text{ on } W}$ on the current in the wire. Earth exerts a downward gravitational force on the wire $\vec{F}_{E \text{ on } W}$. These two forces are balanced by the net upward force of the two springs on the wire $\vec{F}_{S \text{ on } W}$, as shown in the force diagram in Figure 20.14b. Knowing the spring constant of the springs and the mass of the wire, we can use the stretch of the springs to deduce the magnitude of the magnetic force exerted on different length wires when different currents are in them. The collected data are shown in Table 20.4.

TABLE 20.4 Magnetic force data for the experiment in Figure 20.14

Current I in the wire (A)	Length L of wire (m)	Orientation angle θ between the current and the $\vec{B}$ field	Magnitude of magnetic force F exerted on the wire (N)
I	L	90°	F_1
$2I$	L	90°	$2F_1$
$3I$	L	90°	$3F_1$
I	L	90°	F_1
I	$2L$	90°	$2F_1$
I	$3L$	90°	$3F_1$
I	L	0°	0
I	L	30°	$0.5F_1$
I	L	60°	$0.87F_1$
I	L	90°	F_1

Notice that the magnitude of the magnetic force exerted on the wire is proportional to the current I through the wire (the first three rows), to the length L of the wire (the second three rows), and to the sine of the angle θ between the direction of the current and the direction of the $\vec{B}$ field (the last four rows). We can represent this mathematically:

$$F_{\vec{B} \text{ on W}} \propto IL \sin \theta \qquad (20.1)$$

When one quantity is proportional to another quantity, the ratio of the two is constant: $\dfrac{F_{\vec{B} \text{ on W}}}{IL \sin \theta} = \text{constant}$. If you place the same current-carrying wire in the magnetic field of an electromagnet in which you can increase the magnetic field by increasing the electric current in the magnet's coil windings, you find that the proportionality constant increases as the $\vec{B}$ field increases. We can hypothesize that the proportionality constant is the magnitude of the $\vec{B}$ field.

We can use the above mathematical relation to define the magnitude of the $\vec{B}$ field in a particular region as the ratio of the magnitude of the maximum force that the field exerts on a current-carrying wire to the product of the length L of the wire and the current I in the wire. This maximum force is exerted when the wire is perpendicular to the direction of the $\vec{B}$ field.

$$B = \frac{F_{\vec{B} \text{ on W max}}}{IL} \qquad (20.2)$$

Equation (20.2) allows us to define a unit for the $\vec{B}$ field, known as the **tesla, T** (named in honor of the Serbian inventor Nikola Tesla [1856–1943]). In a particular region (assuming a uniform field), the $\vec{B}$ field of magnitude 1 T exerts a force of 1 N on a 1-m-long wire with a 1-A current through it when the wire is oriented perpendicular to the $\vec{B}$ field: $1 \text{ T} = 1 \text{ N/A} \cdot \text{m}$. A 1-T $\vec{B}$ field is very strong. By comparison, the average value of the $\vec{B}$ field produced by Earth at the surface is 5×10^{-5} T. The strongest magnets, neodymium-iron-boron (NIB) magnets, can produce a $\vec{B}$ field near their poles of about 0.5 T in magnitude.

We can now use the definition of the magnitude of the $\vec{B}$ field to rewrite the expression for the magnetic force exerted by the field on a current.

> **Magnetic force exerted on a current** The magnitude of the magnetic force $F_{\vec{B} \text{ on W}}$ that a uniform magnetic field $\vec{B}$ exerts on a straight current-carrying wire of length L with current I is
>
> $$F_{\vec{B} \text{ on W}} = ILB \sin \theta \qquad (20.3)$$
>
> where θ is the angle between the directions of the $\vec{B}$ field and the current I. The direction of this magnetic force is given by the right-hand rule for the magnetic force.

EXAMPLE 20.2   **How strong is a strong magnet?**

Neodymium magnets can produce magnetic fields of about 0.5 T near their poles. They are cylindrical, about 1 cm in diameter and 1 cm high. You use a wire to short circuit a battery and achieve a 3-A current through the wire. If you want to use a neodymium magnet to lift the wire, how should you orient the magnet? Can the magnet lift a 1-g wire?

Sketch and translate The maximum force is exerted when the wire is oriented perpendicular to the $\vec{B}$ field vector, which is perpendicular to the cylinder's axis. Given that we want to lift the wire, the force exerted on it must be upward. Thus, using the right-hand rule for the magnetic force, we conclude that the poles of the magnet (i.e., the cylinder) need to be oriented along a horizontal line. The necessary orientation of the

cylinder and wire and the direction of the current are shown on the sketch below, as well as the knowns and unknowns.

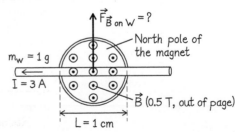

(CONTINUED)

Simplify and diagram We assume that the magnetic field at the end of the cylinder is uniform and the length of the wire in this field is 1 cm. We neglect the field on the sides of the magnet that can potentially affect the wire. From the force diagram for the wire, we can infer that to lift the wire, the magnetic force must be larger than the gravitational force.

Represent mathematically The magnitude of the force exerted by the magnet on the wire is given by Eq. (20.3). We know all the values needed to calculate the force. The angle is 90°. The magnitude of the force exerted by Earth is equal to $m_W g$.

$$\vec{F}_{\vec{B} \text{ on } W}$$

$$\vec{F}_{E \text{ on } W}$$

Solve and evaluate

$$F_{\vec{B} \text{ on } W} = ILB \approx (3\ \text{A})(10^{-2}\ \text{m})(0.5\ \text{T}) = 1.5 \times 10^{-2}\ \text{N}$$

The magnitude of the force exerted by Earth is $m_W g \approx (1 \times 10^{-3}\ \text{kg}) \times (10\ \text{N/kg}) = 10^{-2}\ \text{N}$. Therefore, the magnet can potentially lift the wire.

Try it yourself A 2.0-m-long wire has a 10-A current through it. The wire is oriented south to north and located near the equator. Earth's $\vec{B}$ field has a 5.0×10^{-5}-T magnitude in the vicinity of the wire. What is the magnetic force exerted by the field on the wire?

Answer

The wire is parallel to the $\vec{B}$ field. Thus, $\sin \theta = \sin(0°) = 0$ and the magnetic force exerted on the wire is zero.

Summary of the differences between gravitational, electric, and magnetic forces

- The gravitational and electric forces exerted on objects do not depend on the direction or the speed of motion of those objects, whereas the magnetic force does. If the direction of the electric current is parallel or antiparallel (pointing exactly in the opposite direction) to the $\vec{B}$ field, no magnetic force is exerted on it.
- While the forces exerted by the gravitational and the electric fields are always in the direction of the $\vec{g}$ or $\vec{E}$ field (or opposite that direction in the case of a negatively charged object), the force exerted by the magnetic field on a current-carrying wire is perpendicular to both the $\vec{B}$ field and the direction of the electric current (**Figure 20.15**).

FIGURE 20.15 The directions of different forces relative to their fields.

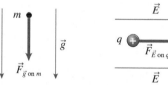

The gravitational force is parallel to the $\vec{g}$ field.

The electric force is parallel to the $\vec{E}$ field.

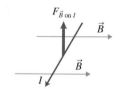

The magnetic force is perpendicular to the magnetic field and to the electric current.

The direct current electric motor

A **motor** is a device that converts electric energy into mechanical energy, specifically rotational or translational kinetic energy. In order to understand the principle behind the operation of an electric motor, we will first study the motion of a current-carrying coil in a uniform magnetic field (**Figure 20.16a**). The coil rotates around an insulating axle. How does this device convert electric energy into the rotational energy of the coil?

Let us start with the rectangular coil oriented so that the plane of the coil is parallel to the $\vec{B}$ field. The currents through sides 1 and 3 of the coil are perpendicular to the $\vec{B}$ field, so the field exerts a force on them. The currents through sides 2 and 4 are parallel to the $\vec{B}$ field: the force exerted on them is zero.

According to the right-hand rule for magnetic force, the magnetic field exerts an upward force on wire 1 of the coil and a downward force on wire 3 of the coil

(Figure 20.16b). These forces each produce a torque around the axle that causes the coil to start rotating clockwise.

As the coil turns, the orientations of the currents relative to the $\vec{B}$ field change, as do the magnetic forces exerted by the field on these sides. As the coil reaches an orientation with its surface perpendicular to the $\vec{B}$ field, the magnetic field exerts forces on each side of the coil that tend to stretch it but are not able to turn it (Figure 20.16c). Because the coil is in motion, it continues moving and turns past this orientation, reaching the one shown in Figure 20.16d. Using the right-hand rule for magnetic forces again, we find that the torques produced by the magnetic forces exerted on sides 1 and 3 cause the coil to accelerate in the counterclockwise direction, slowing down and reversing the direction of its rotation. Given that, for a motor, we need a continuous rotation of the coil, we need to reverse the direction of the current when the plane of the coil passes the perpendicular to the $\vec{B}$ field orientation (shown in Figure 20.16c) so that the net magnetic torque remains in the clockwise direction. Consequently, for the net torque produced by the magnetic force exerted on the coil to always remain clockwise, the current through the coil must change direction each time the coil passes the vertical orientation.

This reversal of the current is made possible using a device known as a **commutator** (**Figure 20.17**). A commutator consists of two semicircular rings that are attached to the rotating coil. Sliding contacts (metal brushes) connect the battery to the rotating commutator rings. Every time the rings rotate 180° (a half-turn), the brush touching the ring changes, and so does the direction of the current in the coil. Thus the direction of the torque exerted on the coil remains constant.

The motor described above, known as a **direct current electric motor**, is a much more complex version of a "motor" first invented by Michael Faraday in 1821. Faraday also invented the electric generator and was the first to describe the magnetic field. He had no formal education and did not use complex mathematics to describe his ideas.

Torque exerted on a current-carrying coil

We see that the torque produced by the magnetic forces exerted on a coil depends on the orientation of the coil relative to the $\vec{B}$ field. The magnitude of this torque depends on how far from the coil's rotation axis these magnetic forces are exerted. To derive a mathematical expression for the torque that a magnetic field exerts on a current-carrying coil, we will start with a single loop of wire. The torque that the magnetic force exerts on sides 1 and 3 of the loop in Figure 20.16b is directly proportional to the distance $D/2$ from the axis of rotation to where the force is exerted—shown in Figure 20.16a. As there are two equal-magnitude torques exerted on the loop with turning ability in the same direction, the total torque is proportional to D. The magnitude of the magnetic force exerted on wire 1 and on wire 3 depends on the length L of that side of the loop (Figure 20.16a). Therefore, the total torque should be proportional to the product of D and L, which is the area A of the loop. If we have a coil with N loops, the torque will be N times the torque exerted on each loop.

We arrive at an expression for the magnitude of the torque that magnetic forces exert on a current-carrying coil:

$$\left|\vec{\tau}_{B \text{ on Coil}}\right| = NBAI \sin \theta \qquad (20.4)$$

where N is the number of turns in the coil, B is the magnitude of the $\vec{B}$ field in the region of the coil (assumed uniform), A is the area of the coil, I is the electric current through the coil, and θ is the angle between a vector perpendicular to the coil's surface (called the **normal vector**) and the direction of the $\vec{B}$ field.

Magnetic dipole moment

The product of the current I, the area A, and the number of turns N in a coil in Eq. (20.4) is called the magnitude of the **magnetic dipole moment** $\vec{p}_m$ of the coil ($|\vec{p}_m| = NIA$). The direction of the dipole moment vector is perpendicular to the surface of the

FIGURE 20.16 A magnetic field exerts a torque on a current-carrying coil.

(a)
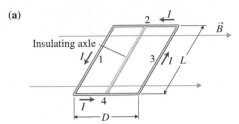

(b) The magnetic forces exerted on sides 1 and 3 cause a clockwise torque on the coil.
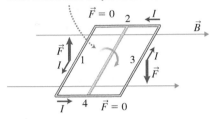

(c) The magnetic forces due to the magnetic field stretch the coil but cause zero torque.
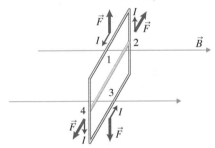

(d) The magnetic field now exerts forces on sides 1 and 3 that cause a counterclockwise torque.
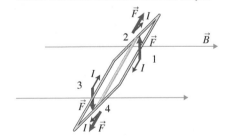

FIGURE 20.17 A commutator ring.

The commutator ring reverses the current in the coil each half turn of the coil. The torque is then always in the same direction.
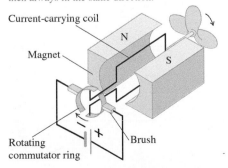

FIGURE 20.18 Magnetic dipole moment.

(a) Magnetic dipole moment vector

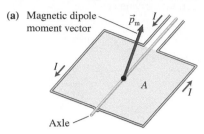

Axle

(b) The normal vector is perpendicular to the surface of the loop or coil.

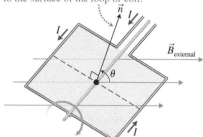

A torque rotates the loop in the clockwise direction.

FIGURE 20.20 An electron beam in a cathode ray tube is deflected by a magnetic field.

(a) Cathode ray tube

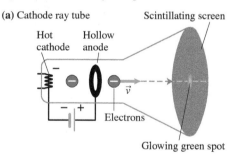

(b) $\vec{B}$ into page

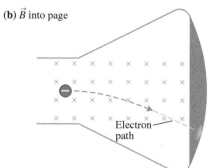

(c) $\vec{B}$ out of page

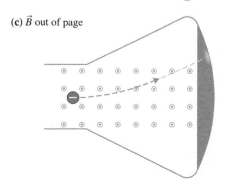

coil and in the direction of the $\vec{B}$ field produced by current at the center of the coil (**Figure 20.18a**). The bigger the dipole moment of a coil, the greater the torque that an external magnetic field exerts on it:

$$\left|\tau_{\vec{B}\text{ on Coil}}\right| = NBIA\sin\theta = Bp_{\text{m}}\sin\theta \tag{20.4}$$

where θ is the angle between the normal vector perpendicular to the surface of the current loop (same direction as the dipole moment vector) and the external $\vec{B}$ field vector (see Figure 20.18b).

The direction of the magnetic dipole moment of a bar magnet is the direction from the south to the north pole inside the magnet (**Figure 20.19**). We can use the same equation for the torque exerted by an external magnetic field on a magnet if we know its magnetic dipole moment.

FIGURE 20.19 A bar magnet's dipole moment.

REVIEW QUESTION 20.3 Equation (20.2) defines the magnitude of the $\vec{B}$ field at a point as $B = F_{\vec{B}\text{ on W max}}/IL$. Explain the meaning of every symbol in this equation. Do any quantities on the right side of the equation cause the $\vec{B}$ field to have a bigger or smaller magnitude? Explain.

20.4 Magnetic force exerted on a single moving charged particle

We have learned that the magnetic field exerts a force on a current-carrying wire. The current through a wire is the result of the collective motion of a huge number of electrically charged particles—electrons. Thus, it is reasonable to conclude that the magnetic field also exerts a force on each individual electron.

Direction of the force that the magnetic field exerts on a moving charged particle

A device called a **cathode ray tube** (**Figure 20.20a**) demonstrates how a magnetic field affects moving electrons. A cathode ray tube consists of two electrodes—a **cathode** and a hollow **anode**—in an evacuated glass enclosure. The cathode is connected to the negative terminal of a power source and the anode to the positive terminal. A current caused by a separate power source keeps the cathode hot. Due to its high temperature, the cathode emits electrons (more on this in Chapter 27). The potential difference between the cathode and the anode causes the electrons to accelerate toward the anode, through the hole in the anode, and onto a screen. The screen is treated with a material that glows green when hit by electrons. We can use the location of the green dot on the screen to infer the path of the electrons within the cathode ray tube (see the video).

VIDEO 20.1

If a magnetic field due to a hand-held bar magnet (not shown in the figure) exerts a force on individual electrons similar to the force it exerts on an electric current in a wire, then the right-hand rule for the magnetic force should predict the direction that the electrons will be deflected. The rule was formulated in terms of electric current, which by convention is the direction that positively charged particles move. The magnetic force exerted on negative electrons moving toward the screen should be in the opposite direction to the one given by the right-hand rule for the magnetic force exerted on an imaginary current-carrying wire with the current toward the screen. To conduct the testing experiment, we orient the magnet so that its $\vec{B}$ field points perpendicular to the electron path shown in Figure 20.20a (into the page, as in Figure 20.20b).

The right-hand rule for the magnetic force predicts that positively charged particles should be deflected upward, while the negatively charged electrons inside the tube should deflect downward. If we reverse the direction of the magnetic field, the electrons should be deflected upward, and the green spot does not deflect if the $\vec{B}$ field points parallel to the velocity of the electrons (Figure 20.20c). When the experiment is performed, the outcome is consistent with the prediction. This result supports the idea that the magnetic force exerted by the magnetic field on individual charged objects is similar to the one it exerts on currents. Let's construct a quantitative relation for the magnitude of the magnetic force exerted by the magnetic field on an individual charged particle.

Magnitude of the force that a magnetic field exerts on a moving charged particle

We know that a magnetic field exerts a force on a current-carrying wire of magnitude

$$F_{\vec{B} \text{ on W}} = ILB \sin\theta$$

We use this to develop an expression for the magnitude of the force that the magnetic field exerts on a single charged object with charge q moving at speed v (the current I in the wire consists of a large number of moving charged particles—**Figure 20.21**). Although we know that electrons move in the wire when current is present, for simplicity we use positively charged particles in this calculation.

Imagine that between the two dashed lines there are N moving charged particles. In a time interval Δt, all of them pass through the dashed line on the right. Thus, the electric current through this wire is

$$I = \frac{Nq}{\Delta t}$$

The speed of the charged particles is $v = L/\Delta t$ since a particle at the left dashed line takes a time interval Δt to reach the right dashed line. Rearrange this for $\Delta t = L/v$ and substitute in the above:

$$I = \frac{Nqv}{L}$$

Inserting this into $F_{\vec{B} \text{ on W}} = ILB \sin\theta$, we get

$$F_{\vec{B} \text{ on W}} = \left(\frac{Nqv}{L}\right)LB \sin\theta = N(qvB \sin\theta)$$

This is the magnitude of the force exerted by the magnetic field on all N moving charged particles. The magnitude of the force exerted by the field on a single charged particle is then given in Eq. (20.5) below.

FIGURE 20.21 Imaginary positively charged particles moving in a wire.

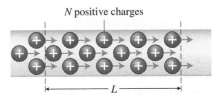

N positive charges

All charged particles between the dashed lines pass the right dashed line in a time interval Δt.

Magnetic force exerted by the magnetic field on an individual charged particle The magnitude of the magnetic force that the magnetic field of magnitude B exerts on a particle with electric charge q moving at speed v is

$$F_{\vec{B} \text{ on } q} = |q|vB \sin\theta \qquad (20.5)$$

where θ is the angle between the direction of the velocity of the particle and the direction of the $\vec{B}$ field. The direction of this force is determined by the right-hand rule for the magnetic force. If the particle is negatively charged, the force points opposite the direction given by the right-hand rule.

TIP ▸ Note that if the velocity and the $\vec{B}$ field are parallel, $\sin 0° = 0$ and the force is zero. The force is maximum when the object's velocity and the $\vec{B}$ field are perpendicular.

The direction of the magnetic force that a magnetic field exerts on a moving positively charged particle is illustrated in **Figure 20.22a**. The right-hand rule for determining the direction of the force is the same as we learned before for the current-carrying wire. It is described in Figure 20.22b.

FIGURE 20.22 The right-hand rule for the magnetic force exerted by a magnetic field on a positively charged particle.

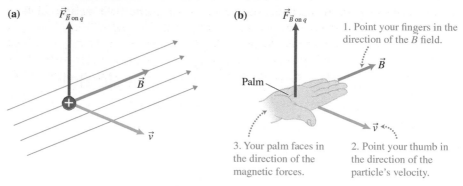

(a)

(b)

1. Point your fingers in the direction of the $\vec{B}$ field.

Palm

3. Your palm faces in the direction of the magnetic forces.

2. Point your thumb in the direction of the particle's velocity.

QUANTITATIVE EXERCISE 20.3   **Particles in a magnetic field**

Each of the lettered dots shown in the figure represents a small object with an electric charge $+2.0 \times 10^{-8}$ C moving at speed 200 m/s in the directions shown. Determine the magnetic force (magnitude and direction) that a 0.10-T $\vec{B}$ field exerts on each object. The $\vec{B}$ field points in the positive y-direction.

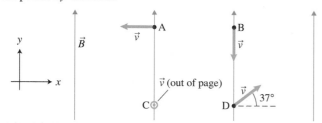

Represent mathematically First, use the right-hand rule for the magnetic force to determine the directions of the magnetic force exerted on each object. (a) For object A, point the fingers of your right hand toward the top of the page in the direction of $\vec{B}$. Then orient your hand so that your thumb points to the left in the direction of $\vec{v}$. Your palm points into the page. This is the direction of the magnetic force exerted on object A. (b) Object B moves in a direction opposite to $\vec{B}$ ($\theta = 180°$); thus, the magnetic force is zero. (c) For C, point your fingers toward the top of the page and the thumb out of the page. Your palm then faces left, so the magnetic force exerted on object C points in the negative x-direction. (d) For object D, point your fingers toward the top of the

page and your thumb parallel to the page pointing 37° above rightward. Your palm faces out of the page, the direction of the force exerted on D.

Solve and evaluate Use Eq. (20.5) to determine the magnitude of each force:

$$F_{\vec{B}\text{ on A}} = (+2.0 \times 10^{-8}\,\text{C})(200\,\text{m/s})(0.10\,\text{T})\sin(90°)$$
$$= 4.0 \times 10^{-7}\,\text{N}$$

$$F_{\vec{B}\text{ on B}} = (+2.0 \times 10^{-8}\,\text{C})(200\,\text{m/s})(0.10\,\text{T})\sin(180°)$$
$$= 0$$

$$F_{\vec{B}\text{ on C}} = (+2.0 \times 10^{-8}\,\text{C})(200\,\text{m/s})(0.10\,\text{T})\sin(90°)$$
$$= 4.0 \times 10^{-7}\,\text{N}$$

$$F_{\vec{B}\text{ on D}} = (+2.0 \times 10^{-8}\,\text{C})(200\,\text{m/s})(0.10\,\text{T})\sin(53°)$$
$$= 3.2 \times 10^{-7}\,\text{N}$$

Try it yourself The equation below represents the solution to a problem. Devise a possible problem that is consistent with the equation:

$$(1.6 \times 10^{-19}\,\text{C})v(0.50 \times 10^{-5}\,\text{T})\sin(30°) = 1.0 \times 10^{-18}\,\text{N}$$

Answer   One possible problem: A proton enters Earth's magnetic field far above the surface. The proton's velocity makes a 30° angle with the direction of the $\vec{B}$ field. The field exerts a 1.0×10^{-18} N force on the proton at the point of entry. What is the proton's speed?

Circular motion in a magnetic field

We know that the force that a magnetic field exerts on a moving charged particle is always perpendicular to the particle's velocity. This is a characteristic of circular motion (discussed in Chapter 5). Consider an example of such motion where a positively charged particle moves to the left across the top of your open textbook (position **a** in

Figure 20.23) when a uniform magnetic field pointing into the page turns on. Using the right-hand rule for the magnetic force, we find that the field exerts a magnetic force on the particle that points downward. The particle continues to move forward, but the direction of its velocity changes and now points slightly downward. The force exerted by the magnetic field always points perpendicular to the particle's velocity. This deflects the particle further. Once it has made a quarter turn and reaches position **b**, the particle is moving toward the bottom of the page, and the magnetic field exerts a force toward the right. This pattern persists—the magnetic force exerted on the particle always points toward the center of the particle's path. Thus, in a uniform $\vec{B}$ field a charged particle that initially moves perpendicular to the $\vec{B}$ field will move along a circular path in a plane perpendicular to the field. Since the force that the magnetic field exerts on a particle is perpendicular to its velocity at every instant, it is also perpendicular to its displacement during each short time interval; therefore, the work done by this force on the particle is zero. As the work done by the magnetic field on the particle is zero, its kinetic energy should not change. This is exactly what we observe: the particle continues to move in a circle at constant speed.

Cosmic rays

The deflection of charged particles in a magnetic field has an important everyday application. Charged particles zoom past and *through* us every minute, bombarding Earth and its inhabitants. These particles, called **cosmic rays**, are usually electrons, protons, and other elementary particles produced by various astrophysical processes including those occurring in the Sun and sources outside the solar system. Every minute about 20 of these fast-moving charged particles pass through a person's head. These particles can cause genetic mutations, which can lead to cancer and other unpleasant effects. Fortunately, our bodies have multiple mechanisms to repair most of the damage.

Thousands more particles would pass through our heads each minute without the protection provided by Earth itself. Earth's magnetic field serves as a shield against harmful cosmic rays, causing them to deflect from their original trajectory toward Earth. How does this shield work?

FIGURE 20.23 A charged particle moving perpendicular to a magnetic field moves in a circle.

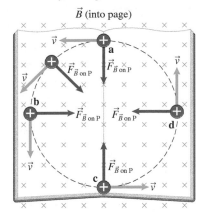

The magnetic field exerts a force perpendicular to the velocity, leading to circular motion.

EXAMPLE 20.4 **Motion of protons in Earth's magnetic field**

Determine the path of a cosmic ray proton flying into Earth's atmosphere above the equator at a speed of about 10^7 m/s and perpendicular to Earth's magnetic field. The average magnitude of Earth's $\vec{B}$ field in this region is approximately 5×10^{-5} T. The mass m of a proton is approximately 10^{-27} kg.

Sketch and translate The situation is sketched below.

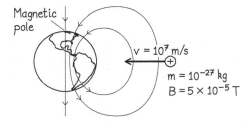

Simplify and diagram Consider a short distance that the proton travels high above Earth and assume that in this region the $\vec{B}$ field vectors are parallel to Earth's surface and have a constant magnitude of 5×10^{-5} T. Our calculations will be much easier if the magnetic force is the

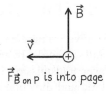

only force exerted on the proton. Can we neglect the gravitational force exerted by Earth? We do a quick estimate:

$$mg = (10^{-27}\,\text{kg})(10\,\text{N/kg}) \approx 10^{-26}\,\text{N}$$
$$qvB = (1.6 \times 10^{-19}\,\text{C})(10^7\,\text{m/s})(5 \times 10^{-5}\,\text{T}) = 8 \times 10^{-17}\,\text{N}$$

The magnetic force is more than a billion times larger than the gravitational force, so we can neglect the gravitational force.

Represent mathematically When the velocity of a charged particle is perpendicular to the $\vec{B}$ field, it moves in a circular path at constant speed. We can use the radial r component form of Newton's second law to relate the magnetic force exerted on the proton to its resulting motion. The force exerted by the magnetic field points toward the center of the proton's circular path.

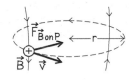

$$a_r = \frac{v^2}{r} = \frac{1}{m}\sum F_r = \frac{1}{m}(F_{\vec{B}\text{ on P}\,r}) = \frac{1}{m}(|q|vB\sin\theta)$$

$$= \frac{|q|vB\sin(90°)}{m} = \frac{|q|vB}{m}$$

(CONTINUED)

This equation can be used to determine the radius r of the proton's circular path and to determine the period T of its motion, noting that

$$v = \frac{2\pi r}{T}$$

Solve and evaluate We used Newton's second law to develop an expression describing the proton's circular motion:

$$\frac{v^2}{r} = \frac{|q|vB}{m}$$

Multiply both sides by the product of r and m and then rearrange to get an expression for r:

$$\frac{v^2 mr}{r} = \frac{|q|vBmr}{m} \Rightarrow mv^2 = |q|vBr$$

$$\Rightarrow r = \frac{mv}{|q|B}$$

$$\approx \frac{(10^{-27}\,\text{kg})(10^7\,\text{m/s})}{|1.6 \times 10^{-19}\,\text{C}|(5 \times 10^{-5}\,\text{T})} \approx 10^3\,\text{m}$$

The period T of the proton's motion is then

$$T = \frac{2\pi r}{v} \approx \frac{2\pi(10^3\,\text{m})}{10^7\,\text{m/s}} \approx 10^{-3}\,\text{s}$$

Something interesting happens if we combine the equations for r and T:

$$T = \frac{2\pi r}{v} = \frac{2\pi}{v}\left(\frac{mv}{|q|B}\right) = \frac{2\pi m}{|q|B}$$

We find that the period of the proton's motion depends only on the ratio of the mass and charge of the proton and the magnitude of the $\vec{B}$ field. It does not depend on the speed of the proton. (Do not confuse the period T with the unit for the magnetic field, the tesla T.)

Try it yourself What happens to the motion of a proton that enters a uniform magnetic field parallel to the $\vec{B}$ field?

Answer

The motion of the proton will not be affected by the magnetic field.

Earth's magnetic field acts as a shield, protecting life on the surface from cosmic ray particles. A cosmic ray proton entering the magnetic field of Earth at one-third light speed, or $1 \times 10^8\,\text{m/s}$, will move in a helical path with a radius of 1 or more km. Earth's magnetic field extends several tens of thousands of kilometers above the surface, so life on Earth is well protected. The Earth's atmosphere itself serves as an additional shield, absorbing those protons that enter it. However, astronauts traveling outside of Earth's atmosphere have neither the atmospheric shield nor the magnetic shield. They are exposed to higher levels of cosmic radiation than are present on Earth's surface. This exposure is an important consideration in planning future long missions to Mars and other planets.

The auroras

Charged particles moving in Earth's magnetic field actually follow helical paths around the $\vec{B}$ field lines and enter the Earth's atmosphere near the poles (**Figure 20.24**). These particles collide with molecules in the atmosphere, ionizing the molecules. When the electrons recombine with the ionized molecules, the excess energy is radiated as light. We see this light in the upper atmosphere in the region of the magnetic poles—the auroras mentioned at the beginning of the chapter.

Often these charged particles come from solar flares caused by the interactions of the Sun's hot ionized gas with its magnetic field. Therefore, when magnetic activity on the Sun is high, the auroras become more intense. On occasion, the auroras are visible far from the magnetic pole regions, sometimes even quite close to the equator.

FIGURE 20.24 Paths of charged particles that can cause auroras.

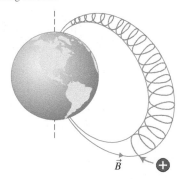

Particles entering slightly less than perpendicular to Earth's magnetic field follow helical paths toward the poles, causing auroras.

$\vec{B}$

REVIEW QUESTION 20.4 If the magnetic force is always perpendicular to the velocity of a charged particle, can it do any work on the particle? Explain your answer.

20.5 Magnetic fields produced by electric currents

So far, we have qualitatively investigated the $\vec{B}$ fields produced by current-carrying straight wires, loops, and solenoids. To analyze problems such as the danger of magnetic fields due to high-power transmission lines, we need to know how to quantitatively determine the magnitude and direction of a $\vec{B}$ field produced by a particular current configuration (the source)—the subject of this section.

The $\vec{B}$ field produced by an electric current in a long straight wire

An electric current in a long straight wire produces a magnetic field whose $\vec{B}$ field lines circle around the wire (**Figure 20.25**). What is the magnitude of that field at various locations near the wire? To answer this question we need to place some kind of a detector (test object) of magnetic field at different locations and investigate the effects of the field on this object. In Section 20.3 we learned that a magnetic field will produce a maximum torque $\left|\tau_{\vec{B}\text{ on C max}}\right| = NBAI$ on a small current-carrying coil used to probe a magnetic field at different locations. We can use such a coil as a detector to determine the relative strength of the magnetic field produced by a long straight current-carrying wire.

We place the coils at different distances from the field-producing current-carrying wire and measure the maximum torque that the magnetic field created by the wire exerts on the coil. We use this torque, the magnitude of the current through the coil, its area, and the number of turns to determine the $\vec{B}$ field surrounding a long straight wire with different currents I and at different distances r from the wire. The data are in **Table 20.5**.

From the data, we can identify a pattern: the magnitude of the $\vec{B}$ field created by a long straight current-carrying wire is directly proportional to the magnitude of the current I and inversely proportional to the distance r between the wire and the location where the field is measured.

We can express this pattern mathematically as follows:

$$B_{\text{straight wire}} \propto \frac{I_{\text{wire}}}{r}$$

Since we can experimentally measure all three of the quantities appearing in this relationship, it's possible to determine the constant of proportionality that will turn it into an equation. Traditionally, this constant is written as $\mu_0/2\pi$, where $\mu_0 = 4\pi \times 10^{-7}\ \text{T}\cdot\text{m/A}$ is called the **magnetic permeability** of a vacuum. We now have an expression for the magnitude of the $\vec{B}$ field at a perpendicular distance r from a long straight current-carrying wire:

$$B_{\text{straight wire}} = \frac{\mu_0}{2\pi}\frac{I_{\text{wire}}}{r} \tag{20.6}$$

Note that the farther you move from the current-carrying wire (larger r), the smaller the magnitude of the $\vec{B}$ field; the greater the current (larger I_{wire}), the larger the magnitude of the $\vec{B}$ field.

Figure 20.26 shows the information about magnetic fields created by different configurations of current-carrying wires. It presents the equations for a loop/coil and solenoid without the derivation. We will use the magnitude of the $\vec{B}$ field at the center of a single current-carrying loop next to analyze the magnetic field produced by the motion of an electron in a hydrogen atom.

$\vec{B}$ field due to electron motion in an atom

In an early 20th-century model of the hydrogen atom, the electron was thought to move rapidly in a tiny circular path around the nucleus of the atom (similar to the figure in Example 20.5). Although the model itself is outdated, it allows us to make reasonable predictions about magnetic properties of individual atoms and objects made from them. In this model the electron motion is like a circular electric current I that produces a $\vec{B}$ field. The $\vec{B}$ field at the center of a current-carrying loop of radius r is

$$B = \frac{\mu_0 I}{2r}$$

The electron's circular motion also produces a magnetic dipole moment of magnitude $p_m = IA$ equal to the product of the electron current I and the area A of the loop. If these dipole moments are significant in magnitude, they could potentially help

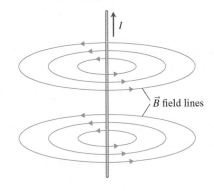
FIGURE 20.25 The magnetic field caused by a long current-carrying wire.

$\vec{B}$ field lines

TABLE 20.5 Magnitude of the $\vec{B}$ field at distance r from a straight wire carrying current I

Current I	Distance	B
I_0	r_0	B_0
$2I_0$	r_0	$2B_0$
$3I_0$	r_0	$3B_0$
I_0	$2r_0$	$B_0/2$
I_0	$3r_0$	$B_0/3$
I_0	$r_0/2$	$2B_0$

FIGURE 20.26 Expressions for the $\vec{B}$ field produced by (a) a straight wire, (b) a loop or coil, and (c) a solenoid.

(a) Straight wire

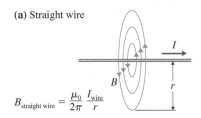

$$B_{\text{straight wire}} = \frac{\mu_0}{2\pi}\frac{I_{\text{wire}}}{r}$$

(b) Loop or coil

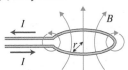

$B = \dfrac{\mu_0 I}{2r}$ at center of loop

$B = \dfrac{N\mu_0 I}{2r}$ at center of coil with N turns

(c) Solenoid

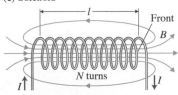

N turns

$B = \dfrac{N\mu_0 I}{l}$ inside solenoid

explain magnetic properties of materials. In the next example we will make a rough estimate of the hydrogen atom's electric current I and the $\vec{B}$ field and dipole moment produced by this current.

EXAMPLE 20.5   **Magnetic field produced by the electron in a hydrogen atom**

In the above-mentioned early 20th-century model of the hydrogen atom, the electron was thought to move in a circle of radius 0.53×10^{-10} m, orbiting once around the nucleus every 1.5×10^{-16} s. Determine the magnitude of the $\vec{B}$ field produced by the electron at the center of its circular orbit and its dipole moment.

Sketch and translate A sketch of the moving electron is shown along with the known information.

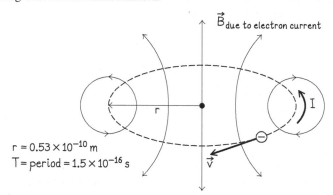

$r = 0.53 \times 10^{-10}$ m
$T = \text{period} = 1.5 \times 10^{-16}$ s

Simplify and diagram The electron's motion corresponds to a counterclockwise current (opposite the direction of travel of the negatively charged electron). This is similar to a single current loop. The direction of the $\vec{B}$ field at the center can be determined by the right-hand rule for the $\vec{B}$ field. This is also the direction of the electron's dipole moment.

Represent mathematically The magnitude of the $\vec{B}$ field at the center of a circular current I is

$$B = \frac{\mu_0 I}{2r}$$

To determine B, we have to determine the electric current due to the electron's motion. Electric current is $I = q/\Delta t$, where q is the magnitude of the total electric charge that passes a cross section of a wire in time Δt. The cross section in this case is a single point along

the electron's orbit. The single electron passes that point once every 1.5×10^{-16} s.

Solve and evaluate Using these ideas, we can calculate the magnitude of the current:

$$I = \frac{q}{\Delta t} = \frac{1.6 \times 10^{-19}\,\text{C}}{1.5 \times 10^{-16}\,\text{s}} = 1.1 \times 10^{-3}\,\text{A}$$

The magnitude of the $\vec{B}$ field at the center of the loop is then

$$B = \frac{\mu_0 I}{2r} = \frac{(4\pi \times 10^{-7}\,\text{T}\cdot\text{m/A})(1.1 \times 10^{-3}\,\text{A})}{2(0.53 \times 10^{-10}\,\text{m})} = 13\,\text{T}$$

This is a huge $\vec{B}$ field, especially if we remember that the strongest permanent magnets can produce fields only up to 1 T. The magnitude of the dipole moment of an electron is

$$p_m = IA = (1.1 \times 10^{-3}\,\text{A})[\pi(0.53 \times 10^{-10}\,\text{m})^2]$$
$$= 9.7 \times 10^{-24}\,\text{A}\cdot\text{m}^2$$

This is a tiny quantity compared to the magnetic moment of a macroscopic current-carrying loop. However, one object can easily contain a trillion trillion electrons.

Try it yourself We found in this example that a moving electron in an atom produces a very strong magnetic field. Suggest an explanation for why all materials are not strong magnets.

Answer

One possible reason is that the electron orbits of the individual atoms are oriented randomly and therefore so are their dipole moments. If this is correct, their dipole moments would cancel each other. Another reason could be that most atoms have more than one electron, and magnetic moments produced by each of these electrons add to zero so that the $\vec{B}$ field produced by each individual atom is zero.

REVIEW QUESTION 20.5 The definition of a 1-A current states that two parallel 1.0-m-long current-carrying wires separated by 1.0 m with a current of 1 A in each exert an attractive force on each other of magnitude 2.0×10^{-7} N. Where does this value of 2.0×10^{-7} N originate?

20.6 Skills for analyzing magnetic processes

Problems involving magnetic interactions are often of two main types: (1) determine the magnetic force exerted on a current or on an individual moving charged object by the magnetic field or (2) determine the $\vec{B}$ field produced by a known source such as an electric current. In this section we will develop skills needed to solve such problems.

Analyzing processes involving magnetic fields and forces

EXAMPLE 20.6 Magnetic force problem

A horizontal metal wire of mass 5.0 g and length 0.20 m is supported at its ends by two very light conducting threads. The wire hangs in a 49-mT magnetic field, which points perpendicular to the wire and out of the page. The maximum force each thread can exert on the wire before breaking is 39 mN. What is the minimum current through the wire that will cause the threads to break?

Sketch and translate

- Sketch the process described in the problem. Label the known and unknown quantities.
- Choose the system of interest.
- Show the direction of the $\vec{B}$ field and the direction of the electric current (or the velocity of a charged particle) if known.
- Decide whether the problem asks to find a $\vec{B}$ field produced by an electric current or to find a magnetic force exerted by an external field on a moving charged particle or wire with electric current.

- A sketch of the situation is shown along with the known information and the unknown quantity. We do not yet know the direction of the current.

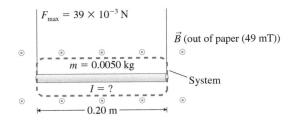

- The horizontal wire is the system. What downward force exerted by the magnetic field, when added to the gravitational force exerted by Earth, will be enough to break the threads? Then what current will produce this downward magnetic force?
- This is a problem about finding the force exerted by the external field on an object.

Simplify and diagram

- Decide whether the $\vec{B}$ field can be considered uniform in the region of interest.
- Draw a force diagram for the system (the object in the field region) if necessary. Use the right-hand rule for the magnetic force to find an unknown force, current, velocity, or field direction if needed.
- Use the right-hand rule for the $\vec{B}$ field if the problem is about the field of a known source.

- Nothing specific is mentioned about the $\vec{B}$ field, so we consider it uniform in the vicinity of the wire.
- Construct a force diagram for the wire. The wire interacts with the threads ($\vec{F}_{\text{T on W}}$), Earth ($\vec{F}_{\text{E on W}}$), and the magnetic field ($\vec{F}_{\vec{B}\text{ on W}}$). With the $\vec{B}$ field pointing out of the page, the magnetic force points down in the desired direction if the wire current is toward the right. We choose the y-axis as pointing down.

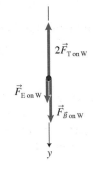

Represent mathematically

- Describe the situation mathematically using the expressions for magnetic force exerted on a current or charged particle and the expressions for the $\vec{B}$ field produced by currents.
- If necessary, use Newton's second law in component form and/or kinematics.

- The wire is in equilibrium ($\Sigma F_y = 0$). In component form,

$$F_{\text{E on W }y} + 2F_{\text{T on W }y} + F_{\vec{B}\text{ on W }y} = 0$$
$$\Rightarrow mg + 2(-F_{\text{T on W}}) + ILB = 0$$

Move the terms that do not contain I to the right side of the equation: $ILB = 2F_{\text{T on W}} - mg$. Then divide both sides by LB:

$$I = \frac{2F_{\text{T on W}} - mg}{LB}$$

(CONTINUED)

Solve and evaluate

- Use the mathematical representation of the process to determine the unknown quantity.
- Evaluate the results—units, magnitude, and limiting cases—to make sure they are reasonable.

- Inserting the appropriate values:

$$I = \frac{2(39 \times 10^{-3}\,\text{N}) - (5.0 \times 10^{-3}\,\text{kg})(9.8\,\text{N/kg})}{(0.20\,\text{m})(49 \times 10^{-3}\,\text{T})} = 3.0\,\text{A}$$

This is a large current for a thin wire but is not completely unreasonable.

- Looking at a limiting case: if the wire mass is such that $mg = 2F_{T\,\text{on w}}$, then the required current is zero—the gravitational force that Earth exerts would be enough to break the wire.

- -

Try it yourself If you want the magnetic force to lift the wire, what changes to the experiment do you need to make? Determine the minimum current.

Answer

We need to change either the direction of the current or the $\vec{B}$ field, but not both. The minimum current is 5.0 A assuming that the magnetic field remains unchanged.

EXAMPLE 20.7 **Determine the $\vec{B}$ field**

Determine the $\vec{B}$ field 5.0 cm from a long straight wire that is connected in series with a 5.0-Ω resister and a 9.0-V battery.

Sketch and translate Make a sketch of the situation including the electric circuit connected to the wire. The current in the circuit is clockwise. The problem is about finding the field produced by a known source.

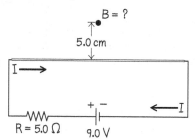

B = ?

5.0 cm

I ⟶

R = 5.0 Ω   9.0 V

Simplify and diagram Assume that the only contribution to the $\vec{B}$ field at the point of interest comes from the long wire. The other wires in the circuit contribute as well, but since they are somewhat farther away by comparison, we will neglect their contributions. Assume also that the other connecting wires and the battery have zero resistance. Using the right-hand rule for the $\vec{B}$ field, we find that below the wire the field points into the paper, and above the wire it points out of the paper.

B (out of page)

5.0 cm

I ⟶

B (into page)   5.0 cm

Represent mathematically The magnitude of the $\vec{B}$ field produced by a long straight current-carrying wire is given by Eq. (20.6):

$$B_{\text{straight wire}} = \frac{\mu_0}{2\pi}\frac{I}{r}$$

Using Ohm's law and the loop rule (traversing the loop clockwise), we can determine the current through the wire:

$$\Delta V_{\text{batt}} + \Delta V_R = 0$$
$$\Rightarrow \mathcal{E} + (-IR) = 0$$

Solve and evaluate Combining the above equations gives

$$B_{\text{straight wire}} = \frac{\mu_0}{2\pi}\frac{I}{r} = \frac{\mu_0}{2\pi}\frac{(\mathcal{E}/R)}{r} = \frac{\mu_0}{2\pi}\frac{\mathcal{E}}{rR}$$

$$= \frac{4\pi \times 10^{-7}\,\text{T}\cdot\text{m/A}}{2\pi}\frac{9.0\,\text{V}}{(0.050\,\text{m})(5.0\,\Omega)}$$

$$= 7.2 \times 10^{-6}\,\text{T}$$

We can do a limiting case analysis for the case of a discharged battery. If $\mathcal{E} = 0$, then there should be no current and the $\vec{B}$ field should be zero, consistent with the equation.

- -

Try it yourself Estimate the magnitude of the $\vec{B}$ field produced by an electron beam in a cathode ray tube at a point 1.0 m to the side of the beam. Assume that 10^{10} electrons hit the screen every second and that they move at a speed of 10^7 m/s.

Answer 3.2×10^{-16} T.

Mass spectrometer

In Section 20.4 we learned that when a charged particle's velocity is perpendicular to the direction of the $\vec{B}$ field, the particle moves in a circular path. The radius r of the circle depends on the particle's mass:

$$r = \frac{mv}{|q|B}$$

This relationship is the foundation for the **mass spectrometer**. The mass spectrometer helps determine the mass of ions, molecules, and even elementary particles such as protons and electrons. It also can determine the relative concentration of atoms of the same chemical element that have slightly different masses (known as isotopes, which we will learn about in Chapter 29).

A mass spectrometer produces ions through heating, collisions, or some other mechanism. An electric field accelerates the ions into a device called a *velocity selector*. A velocity selector consists of a region with uniform magnetic and electric fields whose vectors $\vec{B}$ and $\vec{E}$ are perpendicular to each other (**Figure 20.27a**; the $\vec{E}$ field vector points upward on the page and the $\vec{B}$ field vector points out of the page). When a positively charged particle (an ion) enters this region in the horizontal direction pointed to the right, both fields will exert forces on the particle. The electric force will accelerate the particle upward, and the magnetic force will accelerate it downward (Figure 20.27b).

If the forces have the same magnitude, the particle will pass through the region unaffected. However, the magnetic force exerted on the particle depends on its velocity; thus only particles traveling at a specific velocity will be unaffected. All others will be deflected either upward or downward on the page. To determine the speed of the unaffected particles, we will use the force diagram in Figure 20.27b. Using Newton's second law and assuming that the gravitational force exerted on the particle is negligible, we can write:

$$\vec{F}_{\vec{B} \text{ on } qy} + \vec{F}_{\vec{E} \text{ on } qy} = 0 \Rightarrow (-F_{\vec{B} \text{ on } q}) + (+F_{\vec{E} \text{ on } q}) = 0$$

$$\Rightarrow |q|vB - |q|E = 0$$

$$\Rightarrow vB - E = 0$$

The speed of unaffected particles is therefore $v = E/B$. The velocity selector will only allow particles that move with a speed equal to the ratio E/B to pass through. Note that the value of the speed does not depend on the mass of the particle or the magnitude of the charge as long as the particle is charged.

After passing through the velocity selector, the ions then enter a region with a different uniform $\vec{B}_1$ field, traveling perpendicular to the field lines. As a result, the ions move in circular paths. The radius of the circle is measured by observing the place where the ions strike a detector after moving halfway around a circle (**Figure 20.28**). The ion mass is then determined using the method in the exercise below.

Mass spectrometry has many scientific uses. For example, researchers can use the mass spectrometer to measure the concentration of two isotopes of oxygen in glacial ice: oxygen-16 and oxygen-18. The oxygen-16/18 isotope ratio changes over geologic time due to global climate conditions, and therefore can be used as a way to track climate change and to determine the age of plant and animal remains found in the ice.

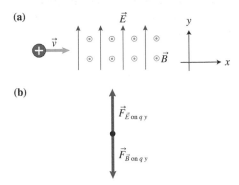

FIGURE 20.27 How a velocity selector works.

(a)

(b)

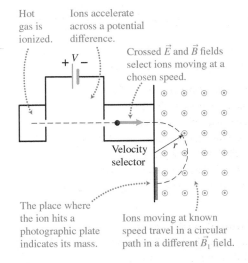

FIGURE 20.28 A mass spectrometer.

Hot gas is ionized.

Ions accelerate across a potential difference.

Crossed $\vec{E}$ and $\vec{B}$ fields select ions moving at a chosen speed.

Velocity selector

The place where the ion hits a photographic plate indicates its mass.

Ions moving at known speed travel in a circular path in a different $\vec{B}_1$ field.

QUANTITATIVE EXERCISE 20.8 **Mass spectrometer**

An atom or molecule with a single electron removed (an ion) is traveling at 1.0×10^6 m/s when it enters a mass spectrometer's 0.50-T uniform $\vec{B}$ field region. Its electric charge is $+1.6 \times 10^{-19}$ C. It moves in a circle of radius 0.20 m until it hits the detector (see figure). Determine (a) the magnitude of the magnetic force that the magnetic field exerts on the ion and (b) the mass of the ion.

Represent mathematically The magnitude of the force that the magnetic field exerts on the moving ion is determined using Eq. (20.5):

$$F_{\vec{B} \text{ on } q} = |q|vB \sin\theta$$

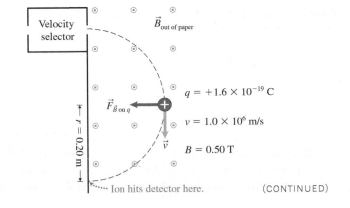

Velocity selector

$\vec{B}_{\text{out of paper}}$

$q = +1.6 \times 10^{-19}$ C

$v = 1.0 \times 10^6$ m/s

$B = 0.50$ T

$\vec{F}_{\vec{B} \text{ on } q}$

$r = 0.20$ m

Ion hits detector here.

(CONTINUED)

This force perpendicular to the ion's velocity causes its radial acceleration:

$$ma_r = |q|vB \implies m\frac{v^2}{r} = |q|vB$$

The mass of the ion is

$$m = \frac{|q|rB}{v}$$

Solve and evaluate Now insert the appropriate values to determine the two unknowns:

$$F_{\vec{B} \text{ on } q} = |q|vB \sin\theta$$
$$= (+1.6 \times 10^{-19}\,\text{C})(1.0 \times 10^6\,\text{m/s})(0.50\,\text{T}) \sin(90°)$$
$$= 8.0 \times 10^{-14}\,\text{N}$$

and

$$m = \frac{|q|Br}{v} = \frac{(+1.6 \times 10^{-19}\,\text{C})(0.50\,\text{T})(0.20\,\text{m})}{1.0 \times 10^6\,\text{m/s}}$$
$$= 1.6 \times 10^{-26}\,\text{kg}$$

The magnitude of the force exerted on the ion is small but reasonable since it is the force exerted on a single ion. The mass of the ion is also reasonable (the mass of a single proton is $1.67 \times 10^{-27}\,\text{kg}$).

Try it yourself What would be the radius of the circular path of a mass 1.7×10^{-26}-kg particle?

Answer   0.21 m.

FIGURE 20.29 An IMRT machine.

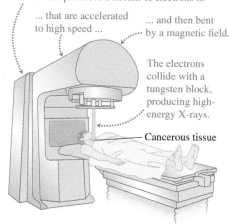

The machine produces a stream of electrons ...

... that are accelerated to high speed ...

... and then bent by a magnetic field.

The electrons collide with a tungsten block, producing high-energy X-rays.

Cancerous tissue

Intensity modulated radiation therapy

Intensity modulated radiation therapy (IMRT) is a powerful cancer-fighting technology in which a high dose of X-rays is directed at a tumor, leaving most of the surrounding tissue untouched. The IMRT machine accelerates electrons to the desired kinetic energy, then uses a magnetic field to bend them 90° into a tungsten alloy target, resulting in the production of X-rays. Movable metal leaves then shape the X-ray beam to match the shape of the tumor. The accelerator and radiation device rotate around the patient with the leaves continually changing to match the 3-D shape of the tumor (**Figure 20.29**). IMRT works well for prostate cancer as well as for tumors of the head, neck, and other delicate areas. In the next example, we consider the magnetic field that bends the electron beam inside the IMRT.

EXAMPLE 20.9   **Magnetic field that bends the electron beam in an IMRT device**

Estimate the magnitude of the $\vec{B}$ field needed for the IMRT machine. For the estimate, assume that the electrons are moving at 2×10^8 m/s, the mass of the electrons is 9×10^{-31} kg, and the radius of the turn is 5 cm = 0.05 m.

Sketch and translate The bending process is sketched in the figure below.

B (into paper) = ?

$m \approx 9 \times 10^{-31}$ kg
$q = -e = -1.6 \times 10^{-19}$ C
$v \approx 2 \times 10^8$ m/s

←r = 0.05 m→

Simplify and diagram This is an estimate, and we assume that the physics principles we will use apply without modification to these electrons traveling at a significant fraction of light speed. A force diagram for the electron part of the way around the 90° curved arc is shown in the figure at right. We neglect the gravitational force that Earth exerts on the

electron. The magnetic force points in the radial direction perpendicular to the electron's velocity at each point along its path.

Represent mathematically Apply Newton's second law to the radial direction. The magnitude of the acceleration is determined by the magnitude of the sum of the forces exerted on the electron in the radial direction and by the mass of the electron:

$$m_{\text{El}}a_r = m_{\text{El}}\frac{v^2}{r} = \Sigma F_r$$

The only force with a nonzero radial component is the magnetic force. The magnitude of that force is $F_{\vec{B} \text{ on El}} = evB$, where e is the magnitude of the electron's electric charge. We get

$$m_{\text{El}}\frac{v^2}{r} = evB$$

Solve and evaluate Divide each side of the equation by ev to determine the magnitude of the magnetic field B:

$$B = \frac{m_{\text{El}}v}{er} = \frac{(9 \times 10^{-31}\,\text{kg})(2 \times 10^8\,\text{m/s})}{(1.6 \times 10^{-19}\,\text{C})(0.05\,\text{m})} = 0.023\,\text{T} \approx 0.02\,\text{T}$$

This is an easily attained magnetic field. Thus, there should be no difficulty bending the electron beam by 90°.

REVIEW QUESTION 20.6 What is the difference between the right-hand rule for the magnetic force and the right-hand rule for the $\vec{B}$ field?

20.7 Magnetic properties of materials

Materials, even metals, have widely varying magnetic properties. For example, magnets strongly attract objects made from iron, such as paper clips, but do not exert an observable magnetic force on objects made from aluminum, such as soda cans. Iron has the ability to greatly amplify the $\vec{B}$ field surrounding it. How can we explain this? Let us examine the following experiments (be sure to watch them in the video).

Experiment 1. You place a piece of pyrolytic carbon (PC; a synthetic material similar to graphite) on a thick piece of Styrofoam (to minimize any direct interaction between the magnet and the scale), which is then placed on a high-precision scale. You hold a strong magnet right above the PC, but not touching it, and observe that the scale reading *increases* (**Figure 20.30a**). When you repeat the experiment with the magnet flipped around (reversing its poles), the outcome is the same.

To explain the increase in the reading of the scale, we consider the pyrolytic carbon piece as the system and draw a force diagram for the situation when the magnet is above it (Figure 20.30b). Given that the force that the scale exerts on the PC increases in the presence of the magnet, the magnet must exert a weak repulsive (downward) force on the PC. It seems that the PC has become a magnet with its poles oppositely aligned to the magnet above it (its north pole facing the north pole of the magnet).

Experiment 2. You place a piece of copper sulfate (CS; a blue crystal with chemical formula $CuSO_4 \cdot 5H_2O$) on the thick piece of Styrofoam, and then place the Styrofoam on the high-precision scale (**Figure 20.31a**). When you hold a strong magnet right above the CS, without touching it, the scale reading *decreases*. When the experiment is repeated with the magnet flipped around (reversing its poles), the outcome is the same.

To explain the outcome of the experiment, we consider the copper sulfate crystal as the system and draw a force diagram for the situation when the magnet is above it (Figure 20.31b). The magnet must exert a weak attractive force on the CS. It seems that the CS has become a magnet with its poles in the same orientation as the magnet above it (its south pole facing the north pole of the magnet).

Experiment 3. You place an iron nail on the thick piece of Styrofoam, which is then placed on the high-precision scale (**Figure 20.32a**). When you hold a strong magnet above the nail, the scale reading *decreases* to zero—the nail is so strongly attracted to the magnet, it flies off the scale. When the experiment is repeated with the magnet flipped around (reversing its poles), the outcome is the same.

To explain the outcome, we consider the nail as the system and draw a force diagram for the situation when the magnet is above it (Figure 20.32b). Because the nail flies off the scale, the scale does not exert a force on it anymore, but the magnet seems to exert a large attractive force on the iron nail. It seems that the iron has become a strong magnet with its poles in the same orientation as the magnet above it (its south pole facing the north pole of the magnet).

These experiments show that all of the materials have magnetic properties. Some become "magnetized" opposite to the direction of the strong external magnet and some in the same direction as the magnet. The amount of this "magnetization" varies significantly in strength.

VIDEO
20.2

FIGURE 20.30 Experiment 1. Interaction of a piece of pyrolytic carbon with a magnet.

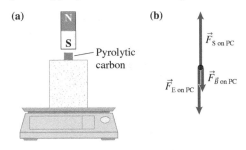

FIGURE 20.31 Experiment 2. Interaction of a copper sulfate crystal with a magnet.

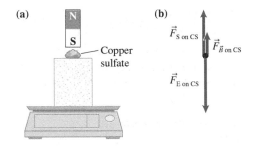

FIGURE 20.32 Experiment 3. Interaction of an iron nail with a magnet.

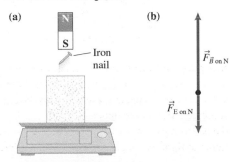

It turns out that *all* materials belong to one of the three types of materials encountered in the experiments: the materials that are repelled by magnets are called **diamagnetic** (like pyrolytic carbon or water); those that are weakly attracted are called **paramagnetic** (like copper sulfate or aluminum); and those that are strongly attracted are called **ferromagnetic** (like iron). Because copper sulfate is paramagnetic, we did not observe a significant interaction between the CS and the magnet. Ferromagnetic materials also retain their magnetization to a certain degree even after the external magnetic field that magnetized them is removed. Let's look at the mechanisms behind diamagnetism, paramagnetism, and ferromagnetism. We will learn about the quantitative analysis of magnetic materials in the next chapter.

Magnetic properties of atoms

To explain these three magnetic properties of materials, we have to understand the magnetic behavior of individual atoms. The model of an atom that we use has a pointlike positively charged nucleus at its center and point-like negatively charged electrons moving at high speeds in circular orbits around the nucleus. Due to this motion, each electron has a magnetic dipole moment (**Figure 20.33**) and acts like a tiny bar magnet. In addition to this electron orbital magnetic moment, the electron itself is like a tiny magnet, which also contributes to the total magnetic moment of the atom.

In atoms with more than one electron, the magnetic moments produced by the individual electrons often cancel each other. This happens because the electrons tend to pair up and orient themselves in opposite directions.

Diamagnetic materials

In diamagnetic materials such as water (and therefore all living organisms), graphite, bismuth, and pyrolytic carbon, the magnetic moments of individual electrons in the atoms cancel each other, making the total field produced by the atom zero. When diamagnetic materials are placed in a region with a nonzero external $\vec{B}_{ex}$ field, the motion of the electrons in the individual atoms changes slightly. This change in electron motion causes the atoms in the diamagnetic material to become magnetic dipoles, each with their own magnetic field pointing in the direction opposite to $\vec{B}_{ex}$. The change in the net magnetic field of the atoms from zero without an external field to nonzero with an external field can only be explained using quantum physics that is beyond the scope of this book. However, we can use the knowledge of such a change to qualitatively explain diamagnetism as follows.

When we place a diamagnetic sample in a uniform external magnetic field $\vec{B}_{ex}$, the field exerts a torque on each atom's magnetic moment. However, the sum of the forces exerted on each atom in the sample is zero because the external magnetic field exerts forces that are equal in magnitude but opposite in direction on the two ends of the dipole. Thus a *uniform* magnetic field has no effect on diamagnetic materials. So why did we observe the repulsion of pyrolytic carbon by a bar magnet? The reason is that the magnetic field of a bar magnet is not uniform. When we place a diamagnetic sample in a *nonuniform* $\vec{B}_{ex}$, the net force exerted by this nonuniform field on each atom in the sample points in the direction of decreasing $\vec{B}_{ex}$ magnitude, independent of the $\vec{B}_{ex}$ direction. For a bar magnet, this net force points away from the magnet, causing the observed repulsion.

Diamagnetic materials held inside an external magnetic field slightly reduce the external magnetic field in the space occupied by the material because their magnetic dipoles create their own magnetic field in the opposite direction. In other words, they try to "expel" the external magnetic field from the material. The reduction of the field is very small, except with superconductive materials. Superconductors expel all external magnetic fields—they are like an ideal diamagnetic material. This is another special property of superconductors, in addition to their zero electrical resistance, which we discussed in Chapter 19.

FIGURE 20.33 Electron motion in an atom effectively produces a tiny "bar magnet."

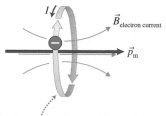

The electron motion produces a circular electric current.

Paramagnetic materials

In most atoms, the orbital magnetic moments of the electrons and the intrinsic moments of the electrons themselves usually cancel. If they don't cancel exactly, the atom will have a magnetic moment similar to that of a small bar magnet (**Figure 20.34a**). These paramagnetic materials include aluminum, sodium, and oxygen.

When a paramagnetic material is placed in a $\vec{B}_{ex}$ field, the atoms behave like tiny (but weak) compasses and tend to align with that $\vec{B}_{ex}$ field. The magnetic moments of individual atoms no longer add to zero but instead contribute a small field that adds to the $\vec{B}_{ex}$ field (Figure 20.34b). In most materials, this paramagnetic effect produces only a slight enhancement of the external $\vec{B}_{ex}$ field (about 10^{-5} times greater). The random motion of the particles causes the magnetic moments of the atoms to remain mostly randomly oriented. Therefore, as the temperature decreases, the magnetic moments become more aligned with the external field, producing a larger enhancement.

Because paramagnetic atoms in an external magnetic field possess a net magnetic moment that points in the direction of $\vec{B}_{ex}$, when we place a paramagnetic sample in a uniform external magnetic field, the field exerts a torque on the atoms. However, the sum of the forces on each atom in the sample is zero (as for diamagnetic materials). But when we place a paramagnetic sample in a nonuniform $\vec{B}_{ex}$, the net force exerted on the atoms in the sample points in the direction of *increasing* $\vec{B}_{ex}$ magnitude, independently of the $\vec{B}_{ex}$ direction. This interaction explains why both poles of a magnet attract paramagnetic materials, although very weakly.

It is important to note that diamagnetic properties are present in all materials, but in paramagnetic materials the diamagnetic effects are masked by the paramagnetic properties.

Ferromagnetic materials

Ferromagnetic materials, such as iron, nickel, and cobalt, have individual atoms with magnetic moments, just like in paramagnetic materials. However, the "magnetization" effect in an external magnetic field is thousands of times stronger in ferromagnetic materials.

Even when not in an external magnetic field, neighboring atoms in ferromagnetic materials tend to line up in small, localized regions called **domains**. Each domain may include 10^{15} to 10^{16} atoms and occupy a space less than a millimeter on a side. In a piece of iron that is unmagnetized, like a nail, the domains are oriented randomly, and the magnetic moments of the domains add to zero (**Figure 20.35a**). If an unmagnetized piece of iron is placed in a region with a nonzero $\vec{B}_{ex}$ field, the domains with magnetic moments oriented in the direction of the $\vec{B}_{ex}$ field increase in size, while those oriented in other directions decrease in size (Figure 20.35b). When the external magnetic field is removed, the magnetic moments of the domains remain aligned, and the iron now produces its own strong $\vec{B}$ field (Figure 20.35c). This is how steel nails become magnets after being placed in external magnetic fields and also how permanent magnets are created. This alignment of domains also explains why each piece of a broken magnet is still a complete magnet. Each piece has its domains aligned before splitting. Thus, each piece has its own north and south pole.

Understanding ferromagnetism helps explain how a device called an **electromagnet** works. Picture a current through a solenoid (a wire that has been shaped into a series of coils) that produces a $\vec{B}$ field that resembles the field of a bar magnet. Now, insert an iron bar into the solenoid. The $\vec{B}$ field produced by the wire causes the magnetic domains within the iron to line up. The now-magnetized iron produces its own contribution to the $\vec{B}$ field, a contribution that is up to several thousands of times stronger than the $\vec{B}$ field produced by the current. This is an electromagnet.

The permanent magnetization of ferromagnetic materials has many practical applications. Hard disk drives use them to store data. Airport metal detectors, transformers, electric motors, loudspeakers, electric generators, and permanent magnets all depend on the magnetization of ferromagnetic materials.

REVIEW QUESTION 20.7 Why is there a difference in the behavior of paramagnetic and diamagnetic materials when they are placed in a region with nonuniform $\vec{B}_{ex}$ field?

FIGURE 20.34 Paramagnetism.

(a) $\vec{B}$ field of randomly oriented atomic "magnets"

(b) In an external $\vec{B}$ field, the atoms make a slight contribution to the $\vec{B}$ field.

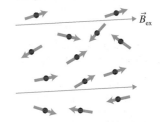

FIGURE 20.35 Ferromagnetism.

(a)

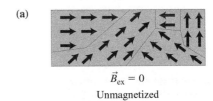

$\vec{B}_{ex} = 0$
Unmagnetized

(b)

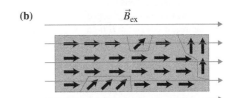

Magnetized

(c)

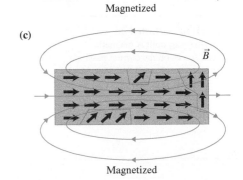

Magnetized

Summary

Magnets always have a north pole and a south pole. A magnetic field is produced by magnets or electric charges moving relative to the observer. (Section 20.1)

A magnetic field is represented by $\vec{B}$ field vectors and $\vec{B}$ field lines. The $\vec{B}$ field vector at a point is tangent to the direction of the $\vec{B}$ field line at that point. The separation of lines in a region represents the magnitude of the $\vec{B}$ field in that region—the closer the lines, the stronger the $\vec{B}$ field. (Sections 20.1–20.3, 20.5)

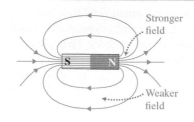

The magnitude of a magnetic field $\vec{B}$ can be found as

$$B = \frac{F_{\vec{B}\text{ on }I\text{ max}}}{IL} \qquad \text{Eq. (20.2)}$$

The fields produced by a long wire and at the center of a coil or solenoid are:

$$B_{\text{straight wire}} = \frac{\mu_0 I}{2\pi r} \qquad \text{Eq. (20.6)}$$

$$B_{\text{center of coil}} = \frac{N\mu_0 I}{2r}$$

$$B_{\text{inside solenoid}} = \frac{N\mu_0 I}{l}$$

Right-hand rule for the $\vec{B}$ field: To find the orientation of the $\vec{B}$ field produced by a current, grasp the wire with your right hand so that your thumb points in the direction of the current. Your four fingers will wrap around the wire in the direction of the $\vec{B}$ field lines. (Section 20.2)

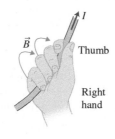

Right-hand rule for the magnetic force: Point the fingers of your open right hand in the direction of the magnetic field. Orient your hand so that your thumb points along the direction of motion of the electric current or charged particle. If the particle is positively charged (or is a current) the magnetic force exerted on it is in the direction your palm is facing. If the particle is negatively charged, it points in the opposite direction. (Section 20.3)

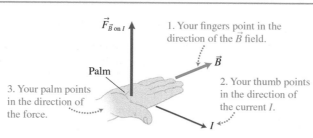

1. Your fingers point in the direction of the $\vec{B}$ field.

2. Your thumb points in the direction of the current I.

3. Your palm points in the direction of the force.

Magnitude of magnetic force depends on the object on which the magnetic field exerts a force (either a current-carrying wire or an individual moving charged particle), the magnitude of the $\vec{B}$ field vector, the relative orientation of the $\vec{B}$ field vector, and the direction of motion of charged objects. (Sections 20.3–20.5)

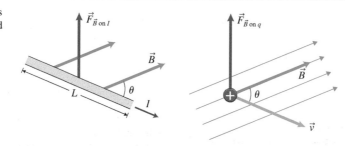

$$F_{\vec{B}\text{ on }I} = ILB\sin\theta \qquad \text{Eq. (20.3)}$$

where θ is the angle between I and $\vec{B}$.

$$F_{\vec{B}\text{ on }q} = |q|vB\sin\theta \qquad \text{Eq. (20.5)}$$

where θ is the angle between $\vec{v}$ and $\vec{B}$.

Magnetic torque exerted on current-carrying coil: A magnetic field exerts forces on the current passing through the wires of a coil, resulting in a torque on the coil. (Section 20.3)

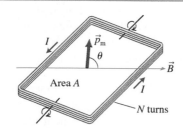

$$|\tau_{\vec{B}\text{ on C}}| = NBAI\sin\theta \qquad \text{Eq. (20.4)}$$

$$|\vec{p}_m| = NIA$$

Questions

Multiple Choice Questions

1. You place a metal bar magnet on a swivel and bring a negatively charged plastic rod near the north pole and then near the south pole. What do you observe?
    (a) The north pole turns toward the rod.
    (b) The south pole turns toward the rod.
    (c) The poles do not interact with the rod.
    (d) Both poles are attracted to the rod.

2. An electron moves at constant speed from left to right in the plane of the page when it enters a magnetic field going into the page. The acceleration of the electron is
    (a) upward.   (b) downward.
    (c) in the direction of motion.   (d) opposite to the direction of motion.
    (e) into the page.   (f) out of the page.

3. What is one tesla?
    (a) $1\,\text{N}/(1\,\text{m} \cdot 1\,\text{A})$ (b) $(1\,\text{N} \cdot \text{m})/(1\,\text{A} \cdot 1\,\text{m}^2)$
    (c) $(1\,\text{N})/(1\,\text{C} \cdot 1\,\text{m/s})$ (d) All of the above

4. Choose all that apply. Objects that produce magnetic fields include which of the following?
    (a) Current-carrying wires
    (b) Permanent magnets
    (c) A compass
    (d) A glass rod rubbed with silk sitting on the table observed by a person standing on the floor
    (e) A glass rod rubbed with silk when placed on a moving truck and observed by a person standing on the ground
    (f) Current in a solenoid

5. What is one difference between magnetic and electric field lines?
    (a) Magnetic lines start on the poles and electric field lines start on electric charges.
    (b) Magnetic lines do not start or end anywhere, whereas electric field lines do have a beginning and end.
    (c) Magnetic field lines are shorter than electric field lines.

6. Two parallel straight current-carrying wires are lying on a table, 12 cm apart. The total magnetic field produced by the currents is zero at a distance of 3 cm from the left wire. Which of the following statements are correct? Select all that apply.
    (a) The current in the left wire is larger than the current in the right wire.
    (b) The current in the left wire is smaller than the current in the right wire.
    (c) The currents are in the same direction.
    (d) The currents are in opposite directions.
    (e) The currents can be either in the same or opposite directions.

7. Choose all of the units that are fundamental, rather than derived, SI units.
    (a) Coulomb   (b) Volt
    (c) Ampere   (d) Newton
    (e) Meter   (f) Second
    (g) Gram   (h) Kilogram

8. Particles of various masses, charges, and speeds are injected into a region in which a uniform $\vec{E}$ field and a uniform $\vec{B}$ field are perpendicular to each other. All the particles are initially moving in the same direction. Which two conditions must be simultaneously fulfilled for the particles to continue moving in a straight line after entering the region?
    (a) The particles must have the same mass.
    (b) The particles must have the same charge.
    (c) The particles must have the same speed.
    (d) The particles must have the same kinetic energy.
    (e) The particles must have the same momentum.
    (f) The particles' velocities must be parallel to the $\vec{E}$ field.
    (g) The particles' velocities must be parallel to the $\vec{B}$ field.
    (h) The particles' velocities must be perpendicular to the $\vec{E}$ field and the $\vec{B}$ field.
    (i) The particles' velocities must be perpendicular to the $\vec{B}$ field.
    (j) The particles' velocities must be perpendicular to the $\vec{E}$ field.

9. When a diamagnetic material is placed in an external $\vec{B}$ field, the force exerted on the material by the magnetic field always points
    (a) in the $\vec{B}$ field direction.
    (b) in the direction opposite to the $\vec{B}$ field direction.
    (c) toward the region where the magnitude of the $\vec{B}$ field is greater.
    (d) toward the region where the magnitude of the $\vec{B}$ field is smaller.

Conceptual Questions

10. If you triple the speed of a particle entering a magnetic field, what happens to the radius of the helix that it makes? Select the correct answer, then describe the possible reasoning that could lead a student to choose each of the other answers. You do not need to agree with that reasoning.
    (a) It triples.
    (b) It becomes one-third as big.
    (c) It increases by 9 times.
    (d) It will not change if the magnetic field does not change.

11. In 1911 physicists measured a magnetic field around a beam of electrons. Draw $\vec{B}$ field lines for this field.

12. Describe two experiments that will allow you to show that electric currents create magnetic fields.

13. How can you determine if there is a magnetic field in a certain region?

14. You have a magnet on which the poles are not marked. How can you determine which pole is north and which is south if in addition to the magnet you have (a) another magnet only, (b) a current-carrying coil only, or (c) neither a magnet nor a coil, but you are familiar with the geographical data of your location? Describe what you will do so a reader can repeat the experiment and get the same results.

15. List as many ways as you can to detect a magnetic field in a particular region. Explain how they work and how you can use them to determine the magnitude and the direction of the $\vec{B}$ field.

16. List as many ways as you can to produce a magnetic field.

17. A current-carrying wire is placed in a magnetic field, as shown in Figure Q20.17. In which direction does the magnetic field exert a force on the wire?

FIGURE Q20.17

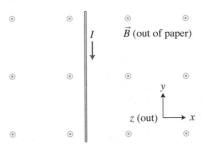

18. An electron flies through the magnetic field shown in Figure Q20.18. In which direction does the magnetic field exert a force on it? What effect does the field have on the magnitude of the velocity of the electron? Describe the path of the electron. Describe what would be different if a proton flew in the same direction through the same field.

FIGURE Q20.18   **FIGURE Q20.19**

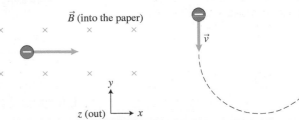

19. Find the direction of the magnetic field whose effect on a charged object is as shown in Figure Q20.19.

20. A beam of electrons is not deflected as it moves through a region of space in which a magnetic field exists. Give two explanations why the electrons might move along a straight path.

21. A beam of electrons moving toward the east is deflected upward by a magnetic field. Determine the direction in which the magnetic field $\vec{B}$ points. Repeat for the case with an $\vec{E}$ field instead of a $\vec{B}$ field. Describe the effect of each field on the speed of the electron.

22. Why are residents of northern Canada less shielded from cosmic rays than are residents of Mexico?

23. A U-shaped wire with a current in it hangs with the bottom of the U between the poles of an electromagnet (see **Figure Q20.23**). When the magnitude of the $\vec{B}$ field in the magnet is increased, does the U swing to the right or to the left, or does it remain at rest? Explain.

FIGURE Q20.23

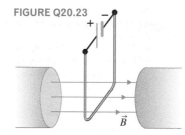

24. An electron enters a solenoid at a small angle relative to the magnetic field inside. Describe the electron's motion.

25. Two parallel wires carry electric current in the same direction. Does the moving charge in one wire cause a magnetic force to be exerted on the moving charge in the other wire? If so, in what direction is the force relative to the wires? Explain. Repeat for currents moving in opposite directions.

26. Why is the magnetic field so much greater at the end of a solenoid or long coil that has an iron core than at the end of a similar solenoid or long coil with the same current through it but with an air core?

27. Describe a situation in which an electron will be affected by an external electric field but will not be affected by an external magnetic field. Is it possible that an electron is affected by an external magnetic field but not by an external electric field? Explain.

Problems

Below, **BIO** indicates a problem with a biological or medical focus. Problems labeled **EST** ask you to estimate the answer to a quantitative problem rather than derive a specific answer. Asterisks indicate the level of difficulty of the problem. Problems with no * are considered to be the least difficult. A single * marks moderately difficult problems. Two ** indicate more difficult problems.

20.1 and 20.2 Magnetic interactions and Magnetic field

1. When a switch is closed, a compass needle deflects from the initial to final direction, as shown in **Figure P20.1**. Say everything you can about this circuit.

   FIGURE P20.1

   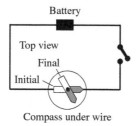

2. You have a lightbulb connected to a battery. (a) What happens if a compass is placed under the constant current cable, made of two twisty wires that connect the bulb with the battery? (b) What happens if you separate the wires and place the compass under one of the separated wires? Explain your answers for both parts.

3. The current through a circuit is shown in **Figure P20.3**. The deflection of a compass needle is shown in the figure. Is the picture correct? If not, what is wrong?

   FIGURE P20.3

   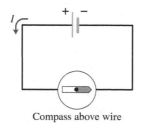

4. Draw $\vec{B}$ field lines for the magnetic field produced by the objects shown in **Figure P20.4**.

FIGURE P20.4

(a)   (b)   (c)

5. * You need to determine the direction of the $\vec{B}$ field at two points in space and compare the magnitudes of the $\vec{B}$ field at those two points. Describe the experiments that will allow you to accomplish this task.

6. * Two compass needles are fixed at the ends of a straw (the north pole of one needle is placed above the south pole of the other needle) and are hung at the end of a light string as shown in **Figure P20.6**. The needles are brought close to a long horizontal wire connected to a circuit, as shown in the figure. (a) How will the needles point after the switch is closed? Explain. (b) Your friend claims that Earth's magnetic field does not affect the two-needle device. Do you agree or disagree? Explain your reasoning.

FIGURE P20.6

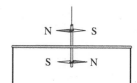

20.3 Magnetic force exerted by the magnetic field on a current-carrying wire

7. * In Houston, Earth's $\vec{B}$ field has a magnitude of 5.2×10^{-5} T and points in a direction 57° below a horizontal line pointing north. Determine the magnitude and direction of the magnetic force exerted by the magnetic field on a 10-m-long vertical wire carrying a 12-A current straight upward.

8. * A 15-g 10-cm-long wire is suspended horizontally between the poles of a horseshoe magnet. When the 0.50-A current in the wire is turned on, the wire jumps up and out of the magnet. What can you learn about the magnet using this information?

9. ** A metal rod of length l and mass m is suspended from two light wires that are connected to a battery with emf $\mathcal{E}$. When a vertical magnetic field $\vec{B}$ pointing up is turned on, the supporting wires make an angle θ with the vertical. What can you learn about the circuit using this information? Derive a mathematical expression that relates the magnitude of the magnetic field $\vec{B}$ to other relevant quantities.

10. * A metal rod is connected to a battery through two stiff metal wires that hold the rod horizontal. The rod is between the poles of a horseshoe magnet that is sitting on a mass-measuring platform scale, which reads 100 g. Draw the magnetic poles of the magnet and the battery connected to the metal rod so that when you turn the current in the circuit on (a) the reading of the scale supporting the magnet increases; (b) the reading decreases.

11. * After you turned on the current in the circuit described in the previous problem, the scale supporting the magnet read 106 g. The part of the rod that is exposed to the magnetic field is 7.0 cm long and the current through it is 1.0 A. What can you learn about the magnet using this information?

12. * **Equation Jeopardy** Describe a problem for which the following equation is a solution:

$$0.70\ A = \frac{3.0\ N}{(0.20\ m)B}$$

13. ** A square coil with 30 turns has sides that are 16 cm long. When it is placed in a 0.30-T magnetic field, a maximum torque of 0.60 N · m is exerted on the coil. What can you learn about the coil from this information?

14. A 5.0-A current runs through a 0.12 m²-area loop. The loop is in a 0.15-T $\vec{B}$ field. Determine the torque exerted on the loop (the magnitude and the direction it tends to turn the coil if initially at rest) for each orientation shown in Figure P20.14.

FIGURE P20.14

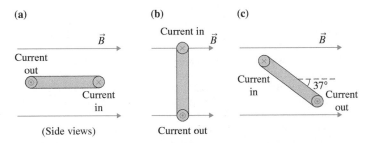

(a)

(b)

(c)

(Side views)

15. * (a) Determine the magnetic force (magnitude and direction) that a 0.15-T magnetic field exerts on segments a, b, and c of the wire shown in Figure P20.15. Segments a and c are 2.0 cm long, and segment b is 10.0 cm long. A 3.0-A current flows through the wire. (b) Determine the net torque caused by forces $\vec{F}_{\vec{B}\ on\ a}$ and $\vec{F}_{\vec{B}\ on\ c}$ and the torque caused by force $\vec{F}_{\vec{B}\ on\ b}$.

FIGURE P20.15

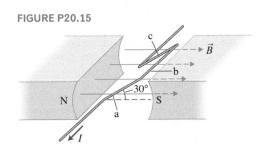

16. * A 500-turn square coil of wire is hinged to the top of a table, as shown in Figure P20.16. Each side of the coil has a length of 0.50 m. (a) In which direction should a magnetic field point to help lift the free end of the coil off the table? (b) Determine the torque caused by a 0.70-T field pointing in the direction described in part (a) when there is a 0.80-A current through the wire and when the coil is parallel to the table.

FIGURE P20.16

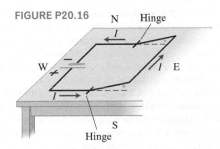

17. * **Electric motor 1** An electric motor has a square armature with 500 turns. Each side of the coil is 12 cm long and carries a current of 4.0 A. The $\vec{B}$ field inside the motor is uniform with a magnitude of 0.60 T. Choose two orientations of the coil for which you can calculate force-related and torque-related quantities. Draw pictures to show the relative orientations of the coil and the magnetic field vectors.

18. ** You make a seesaw by placing a 50-g magnet (whose poles' faces are 2-cm-by-2-cm squares) at one end of a 50-cm-long ruler and a small 50-g metal object at the other end and balancing the whole setup by placing a pencil directly under the middle of the ruler. You know that the magnet is horizontally magnetized, but you don't know which face is north and which face is south. When you place a horizontal wire close to the pole of the magnet and send a 14-A current through the wire in the direction shown in Figure P20.18, the end of the seesaw with the magnet goes up and the other end goes down. You can get the seesaw to go back to equilibrium (while keeping the current in the wire) by placing another 10-g metal object 6.0 cm from the axis of rotation, also shown in the figure. (a) Determine the polarity of the magnet and (b) the magnitude of the $\vec{B}$ field at the location of the wire. Indicate any assumptions that you made.

FIGURE P20.18

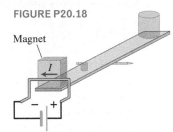

Magnet

19. * **Electric motor 2** An electric motor has a circular armature with 250 turns of radius 5.0 cm through which an 8.0-A current flows. The maximum torque exerted on the armature as it turns in a magnetic field is 1.9 N · m. Show the orientation of the armature coil in the magnetic field when this torque is exerted on it and determine the magnitude of the field.

20.4 Magnetic force exerted on a single moving charged particle

20. Each of the lettered dots a–d shown in Figure P20.20 represents a $+1.0 \times 10^{-18}$-C charged particle moving at speed 2.0×10^7 m/s. A uniform 0.50-T magnetic field points in the positive z-direction. Determine the magnitude of the magnetic force that the field exerts on each particle and indicate carefully, in a drawing, the direction of the force.

FIGURE P20.20

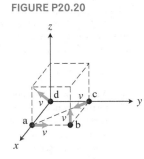

21. **Duck gets a lift** A duck accumulates a positive charge of 3.0×10^{-8} C while flying north at speed 18 m/s. Earth's magnetic field at the duck's location has a magnitude of 5.3×10^{-5} T and points in a direction 62° below a horizontal line pointing north. Determine the magnitude and direction of the force exerted by Earth's magnetic field on the duck.

22. An electron of mass 9.1×10^{-31} kg moves horizontally toward the north at 3.0×10^7 m/s. Determine the magnitude and direction of a $\vec{B}$ field that will exert a magnetic force that balances the gravitational force that Earth exerts on the electron.

23. A 1000-kg car moves west along the equator. At this location Earth's magnetic field is 3.5×10^{-5} T and points north parallel to Earth's surface. If the car carries a charge of -2.0×10^{-3} C, how fast must it move so that the magnetic force balances 0.010% of Earth's gravitational force exerted on the car?

24. * **BIO** **Magnetic force exerted by Earth on ions in the body** A hydroxide ion (OH⁻) in a glass of water has an average speed of about 600 m/s. (a) Determine the magnitude of the electrical force between the hydroxide ion (charge $-e$) and a positive ion (charge $+e$) that is 1.0×10^{-8} m away (about the separation of 30 atoms). (b) Determine the magnitude of the maximum magnetic force that Earth's 3.0×10^{-5}-T magnetic field can exert on the ion. (c) On the basis of these two calculations, does it seem likely that Earth's magnetic field has much effect on the biochemistry of the body?

25. ** **EST** **Design a magnetic shield** An alien from a planet in the galaxy M31 (Andromeda) has a ray gun that shoots protons at a speed of 2.0×10^5 m/s. Design a magnetic shield that will deflect the protons away from your body. Using rough estimates, show that your magnetic shield will, in fact, protect you. Indicate the orientation of your protective device's magnetic field relative to the direction of the proton beam and the direction in which the ions are deflected.

26. ** EST An electron beam moves toward a cathode ray tube screen, which is 30 cm away from the negative electrode. The electrons are accelerated by a potential difference of 14 kV. Estimate the maximum displacement of the electron beam caused by Earth's magnetic field.

27. * An electron and a proton, moving side by side at the same speed, enter a 0.020-T magnetic field. The electron moves in a circular path of radius 7.0 mm. Describe the motion of the proton qualitatively and quantitatively.

20.5 Magnetic fields produced by electric currents

28. An east-west electric power line carries a 500-A current toward the east. The line is 10 m above Earth's surface. Determine the magnitude and direction of the magnetic field at Earth's surface directly under the wire. What assumptions did you make?

29. * **Pigeons** A solenoid of radius 1.0 m with 750 turns and a length of 5.0 m surrounds a pigeon cage. What current must be in the solenoid so that the solenoid field just cancels Earth's 4.2×10^{-5}-T magnetic field (used occasionally by the pigeons to determine the direction they travel)?

30. * A horizontal current-carrying wire that is perpendicular to the plane of the paper is placed in a uniform magnetic field of magnitude 2 mT that is parallel to the plane of the paper and points down. At the distance 1.0 cm to the right of the wire, the total magnetic field is zero. (a) What physical quantity can you determine using this information? (b) How will this quantity change if the zero field point is 0.4 cm to the left of the wire?

31. * A coil of radius r is made of N circular loops. It is connected to a battery of known emf and internal resistance. The coil produces a magnetic field whose magnitude at the center of the coil is $\vec{B}$. Make a list of physical quantities you can determine using this information and show how to determine two of them.

32. * **Magnetic field sensor on your mobile phone** You place a long wire next to your mobile phone on a table as shown in Figure P20.32. You run an application that allows you to measure the time dependence of the component of the $\vec{B}$ field that is perpendicular to the screen of the phone (let's call it B_z). While recording the magnetic field, you repeat the following steps: connect the wire to an AA battery, disconnect the battery, flip the battery (to swap + and − terminals), again connect the wire to the battery ... and so on. Using an ammeter, you also determine the current through the wire. Average values of your measurements are summarized in the table below.

FIGURE P20.32

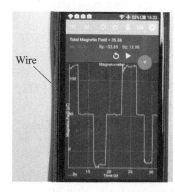

Connection	$I\,(A)$	$B_z\,(T)$
Battery $(+,-)$	4.9	122×10^{-6}
No battery	0	52×10^{-6}
Battery $(-,+)$	4.9	-19×10^{-6}

(a) Estimate the distance between the magnetic field sensor in your phone and the wire. (b) If the experiment was performed in New Jersey, determine the direction of the current in the wire (up or down) when the magnetic field reading is -19×10^{-6} T. Explain your reasoning. Indicate any assumptions that you made.

20.6 Skills for analyzing magnetic processes

33. * **Equation Jeopardy 1** The equation below describes a process involving magnetism. Solve for the unknown quantity and draw a sketch that represents a possible process described by the equation.

$$(1.6 \times 10^{-19}\,\text{C})(2.0 \times 10^7\,\text{m/s})(3.0 \times 10^{-5}\,\text{T})$$
$$= (1.7 \times 10^{-27}\,\text{kg})(2.0 \times 10^7\,\text{m/s})^2/r$$

34. * **Equation Jeopardy 2** The equation below describes a process involving magnetism. Solve for the unknown quantity and draw a sketch that represents a possible process described by the equation. Note: T is a symbol that denotes a force (for example, force exerted by a rope, tension), and T is the unit (Tesla).

$$2T - (0.020\,\text{kg})(9.8\,\text{N/kg}) + (10\,\text{A})(0.10\,\text{m})(0.10\,\text{T}) = 0$$

35. * **Equation Jeopardy 3** The equation below describes a process involving magnetism. Solve for the unknown quantity and draw a sketch that represents a possible process described by the equation.

$$100(2.0\,\text{A})(4.0 \times 10^{-2}\,\text{m}^2)(0.20\,\text{T}) - m(9.8\,\text{N/kg})(0.10\,\text{m}) = 0$$

36. ** EST The magnitude of the $\vec{B}$ field inside a long solenoid is given by the equation $B = \mu_0 I\,(N/l)$, where N is the number of turns and l is the length of the solenoid. (a) Describe an experiment that can help you test this relation. (b) Explain whether this equation is an operational definition of the magnitude of the $\vec{B}$ field or a cause-effect relationship. (c) Powerful industrial solenoids produce $\vec{B}$ field magnitudes of about 30 T. Estimate the relevant physical quantities for such solenoids.

37. * **Electron current and magnetic field in H atom** In a simplified model of the hydrogen atom, an electron moves with a speed of 1.09×10^6 m/s in a circular orbit with a radius of 2.12×10^{-10} m. (a) Determine the time interval for one trip around the circle. (b) Determine the current corresponding to the electron's motion. (c) Determine the magnetic field at the center of the circular orbit and the magnetic moment of the atom.

38. * Two long, parallel wires are separated by 2.0 m. Each wire has a 30-A current, but the currents are in opposite directions. (a) Determine the magnitude of the net magnetic field midway between the wires. (b) Determine the net magnetic field at a point 1.0 m to the side of one wire and 3.0 m from the other wire.

39. * **Minesweepers** During World War II, explosive mines were dropped by the Nazis in the harbors of England. The mines, which lay at the bottom of the harbors, were activated by the changing magnetic field that occurred when a large metal ship passed above them. Small English boats called minesweepers would tow long, current-carrying coils of wire around the harbors. The field created by the coils activated the mines, causing them to explode under the coils rather than under ships. (a) Determine the current in one long, straight wire to create a 0.0050-T magnetic field at a depth of 20 m under the water. The magnetic permeability of water is about the same as that of air. (b) How might the field be created using a smaller current?

40. An electron moves at the speed of 8.0×10^6 m/s toward the velocity selector shown in Figure 20.27a. A 0.12-T $\vec{B}$ field points into the paper. (a) Determine the magnitude and direction of the magnetic force that the magnetic field exerts on the electron. (b) What $\vec{E}$ field magnitude and direction are required so that the electric force exerted on the electron is equal in magnitude and opposite in direction to the magnetic force?

41. * **Mass spectrometer** A mass spectrometer has a velocity selector that allows ions traveling at only one speed to pass with no deflection through slits at the ends. While moving through the velocity selector, the ions pass through a 60,000-N/C $\vec{E}$ field and a 0.0500-T $\vec{B}$ field. The quantities $\vec{v}, \vec{E}$, and $\vec{B}$ are mutually perpendicular. (a) Determine the speed of the ions that are not deflected. (b) After leaving the velocity selector, the ions continue to move in the 0.0500-T magnetic field. Determine the radius of curvature of a singly charged lithium ion, whose mass is 1.16×10^{-26} kg.

42. ** **Mass spectrometer 2** One type of mass spectrometer accelerates ions of charge q, mass m, and initial speed zero through a potential difference ΔV. The ions then enter a magnetic field where they move in a circular path of radius r. How is the mass of the ions related to these other quantities?

43. * An ion with charge 1.6×10^{-19} C moves at speed 1.0×10^6 m/s into and perpendicular to the 0.30-T magnetic field of a mass spectrometer. After entering the field, the ion moves in a circular path of radius 0.31 m. What physical quantities describing the field and the ion can you determine using this information? Determine them.

General Problems

44. * A box has either an electric field or a magnetic field inside. Describe experiments that you might perform to determine which field is present and its orientation. You are allowed to make small holes in the box.

45. ** A piece of wire, shown in **Figure P20.45**, moves downward perpendicular to a magnetic field. (a) In what direction will the electrons in the wire move? (b) After the electrons are forced in one direction, leaving positive charges behind, an $\vec{E}$ field and potential difference ΔV develop from one end of the wire to the other. The electrons no longer move, since the electric and magnetic forces exerted on them balance. Show that this happens when the potential difference from one end of the wire to the other is $\Delta V = vLB$, where v is the speed of the wire, L is its length, and B is the magnitude of the magnetic field. Describe two real-world situations where this phenomenon might occur.

FIGURE P20.45

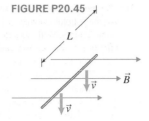

46. ** **EST** Particles in cosmic rays are mostly protons, which have energies up to about 10^{20} eV, where 1 eV is the change in the kinetic energy of an electron moving through a potential difference of 1 V. Use this information to estimate the path of these particles in Earth's atmosphere. State the assumptions that you are making in your estimate.

Reading Passage Problems

BIO **Magnetic resonance imaging** In magnetic resonance imaging (MRI), a patient lies in a strong uniform constant magnetic field $\vec{B}$ oriented parallel to the body. Typical magnitudes of this field are between 1.0 T and 3.0 T. This field is produced by a large superconducting solenoid. The MRI measurements depend on the magnetic dipole moment $\vec{\mu}$ of a proton, the nucleus of a hydrogen atom. The proton magnetic dipoles can have only two orientations: either with the field or against the field. The energy needed to reverse this orientation ("flip" the protons) from with the field to against the field is exactly $\Delta U = 2\mu B$ (like the energy needed to turn a compass needle from north to south).

The pulse of an alternating magnetic field of frequency several tens of MHz irradiates the patient's body in the region to be imaged. When this alternating field is tuned correctly so that its energy equals the $\Delta U = 2\mu B$ needed to reverse the orientation of the protons from with the external $\vec{B}$ field to against it, a reasonable number of protons will flip. When the protons return to their initial orientation and a lower energy state, they produce an alternating magnetic field with almost exactly the same frequency as that of the original radiation. This alternating magnetic field produced by the protons is detected and provides information about the spatial distribution of the concentration of protons as well as information about the structure of the material surrounding the protons in the region irradiated by the alternating magnetic field. The proton concentration as well as the protons' characteristic interactions with their surroundings differ in fat, muscle, and bone tissue, and in healthy and diseased tissue.

The MRI image of an internal body part is made by adjusting an auxiliary magnetic field, which varies the external B field over the region being examined so that the probe field energy equals the flipping energy $\Delta U = 2\mu B$ in only a thin slice of the body. A measurement is made at this slice. The external magnetic field is then adjusted to flip protons in a neighboring slice. Continual shifts in the magnetic field and detection of proton concentrations at different slices produce a map of proton concentration in the body. The MRI image of the lower back in **Figure 20.36** indicates an L45 disc that has partially collapsed— it has lost water, and because it contains fewer protons, it produces a darker MRI image.

FIGURE 20.36 An MRI image of a patient's lower back.

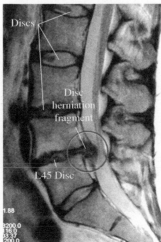

47. The uniform magnetic field in an MRI apparatus is greater than Earth's 5.0×10^{-5}-T magnetic field by about how many times?
(a) 40
(b) 400
(c) 4000
(d) 40,000
(e) 400,000

48. The energy of the probe field causes protons to flip when in a 1.50000-T magnetic field. A ± 0.001-T variation in the magnetic field causes a mismatch with the radio frequency flipping field and no flipping at a distance of 0.002 m from where the magnetic field is matched to the flipping field. Which answer below is closest to the change in the B field per unit distance?
(a) 0.08 T/m
(b) 0.8 T/m
(c) 8 T/m
(d) 0.003 T·m
(e) 0.000003 T/m

49. Using your answer to Problem 20.48, determine the quantity closest to the amount that the auxiliary magnetic field causes the magnetic field to vary over a 0.20-m region of the body being mapped by MRI.
(a) 4 T
(b) 2 T
(c) 0.002 T
(d) 0.02 T
(e) 0.2 T

50. The MRI apparatus is able to look at proton concentration (and hence hydrogen concentration) at one tiny part of the body by doing what?
(a) Aiming the probe field at the whole body
(b) Varying the probe field frequency so that only protons in one place are flipped
(c) Varying the B field over the body so $\Delta U = 2\mu B$ matches the probe field energy at one small location
(d) Placing a small hole in a body shield so the probe field reaches only one part of body
(e) All of the above

51. Which answer below is closest to the energy of the probe field needed to flip protons? The magnetic dipole moment of a proton is 1.41×10^{-26} J/T.
(a) 4 J
(b) 4×10^{-10} J
(c) 4×10^{-25} J
(d) 4×10^{-26} J
(e) 4×10^{-27} J

52. Why might a herniated disc projecting slightly out from between two vertebrae look different in an MRI image than a nonherniated disc?
(a) The vertebrae adjacent to a herniated disc are closer than vertebrae beside a nonherniated disc.
(b) There is a different concentration of hydrogen atoms in bone and in discs.
(c) Protons in the herniation produce an image that can be seen.
(d) b and c
(e) a, b, and c

BIO **Power lines—do their magnetic fields pose a risk?** Power lines produce both electric and magnetic fields. The interior of the human body is an electrical conductor, and as a result, electric fields are greatly reduced in magnitude within the body. The electric fields inside the body from power lines are much smaller than electric fields normally existing in the body.

However, magnetic fields are not reduced in the body. Earth's magnetic field, approximately 5×10^{-5} T, is very small and not regarded as a health threat. Thus, it is interesting to compare Earth's magnetic field to fields produced by high power lines. The magnetic field B produced at a distance r from a straight wire with an electric current I is

$$B = (2 \times 10^{-7} \text{ T·m/A})I/r$$

The magnetic field from a high-voltage power line located 40 m above the ground carrying a 100-A current is much smaller than Earth's $\vec{B}$ field.

Wires that provide electric power for household appliances also produce electric and magnetic fields. The current in the wire for a 500-W space heater is about 5 A. With the wire located several meters from your body, the magnetic field of such an appliance is somewhat smaller than Earth's magnetic field. By comparison, laboratory mice lived for several generations in 0.0010-T magnetic fields (20 times Earth's magnetic field) without any adverse effects.

During the last three decades, electric power use has increased the magnitudes of the $\vec{B}$ field created by power lines to which Americans are exposed by roughly a factor of 20. Yet during that same time interval, leukemia rates have slowly declined. It seems unlikely that magnetic fields produced by home appliances or high-voltage power lines are a hazard for causing leukemia.

53. Which answer below is closest to the ratio of the power line $\vec{B}$ field on the ground 50 m below a 100-A current and Earth's $\vec{B}$ field at the same location?
    - (a) 0.001
    - (b) 0.01
    - (c) 0.1
    - (d) 10
    - (e) 100

54. A 550-W toaster oven is connected to a 110-V wall outlet. Which answer below is closest to the electric current in one of the wires from the wall outlet to the oven?
    - (a) 0.2 A
    - (b) 50,000 A
    - (c) 1 A
    - (d) 5 A
    - (e) 20 A

55. Which answer below is closest to the ratio of the magnetic field produced 0.4 m from the cable for the toaster oven in the last problem and Earth's magnetic field? (This is assuming the current flows in only one direction—which it does not.)
    - (a) 0.001
    - (b) 0.003
    - (c) 0.05
    - (d) 0.5
    - (e) 5

56. Leukemia rates have declined in recent years, whereas the magnitudes of the $\vec{B}$ fields created by power lines in our environment have increased significantly. Why does this not necessarily rule out power line magnetic fields as a contributing cause of leukemia?
    - (a) Correlation studies have no cause-effect relationship.
    - (b) There is no known mechanism for power line magnetic field-induced deaths.
    - (c) The power line magnetic fields cannot penetrate clothing and skin.
    - (d) Perhaps the power line magnetic field-induced cancers have increased from 0.001 to 0.025 of the cases and other causes have decreased.
    - (e) a, b, and d
    - (f) All of the above

57. Why would scientists be more concerned about the relationship between magnetic fields and human health than about the relationship between electric fields and human health?
    - (a) The human body is a conductor.
    - (b) Many molecules are dipoles.
    - (c) There are moving electrically charged particles inside the body.
    - (d) Magnetic fields, if dangerous, are more likely to penetrate into the body.
    - (e) Electric fields are reflected from clothing but magnetic fields are not.

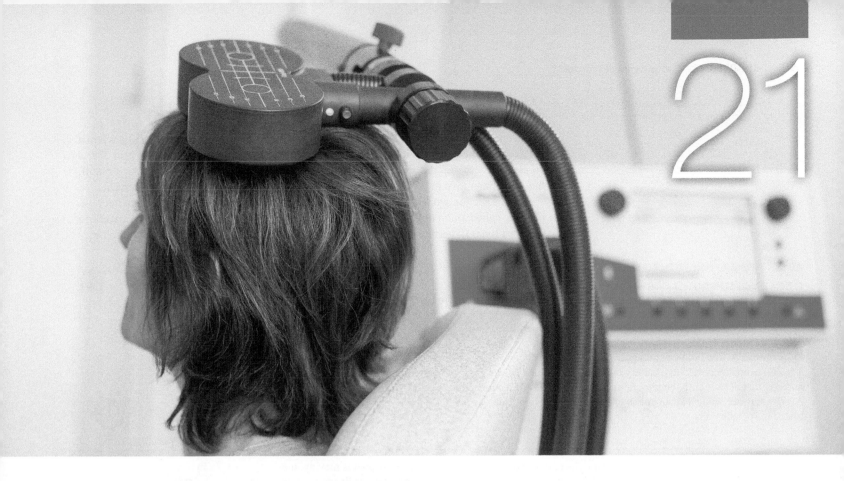

Electromagnetic Induction

Because human cells, including nerve cells, contain ions that can move, they are electrically conductive. Transcranial magnetic stimulation (TMS) is a noninvasive technology that allows doctors to electrically stimulate the brain, which helps treat certain diseases such as mood disorders, Parkinson's disease, and Huntington's disease. Given that the skull is a fairly good electrical insulator, how is it possible that changing current moving through a coil placed on the patient's scalp can produce a current in the brain without any direct electrical connection?

- How might electrical stimulation of the brain help treat certain diseases?
- How does a microphone work?
- We put metal cans and glass jars into the same recycling bin. How are they separated at the dump?

IN THE LAST CHAPTER, we learned that an electric current through a wire produces a magnetic field. Could the reverse happen? Could a magnetic field produce a current? It took scientists many years to answer this question. In this chapter, we will investigate the conditions under which this can happen.

BE SURE YOU KNOW HOW TO:
- Find the direction of the $\vec{B}$ field produced by an electric current (Section 20.2).
- Find the direction of the magnetic force exerted on moving electric charges (Sections 20.3 and 20.4).
- Explain how an electric field produces a current in a wire and how that current relates to the resistance of the wire (Sections 19.1 and 19.4).

21.1 Inducing an electric current

In the chapter on circuits (Chapter 19), we learned that an electric current results when a battery or some other device produces an electric field in a wire. The field in turn exerts an electric force on the free electrons in the wire connected to the battery. As a result, the electrons move in a coordinated manner around the circuit—an electric current.

In this section we will learn how to produce a current in a circuit without a battery—a process called *inducing* a current. We start by analyzing some simple experiments in Observational Experiment **Table 21.1**. See if you can find any patterns in the outcomes.

OBSERVATIONAL EXPERIMENT TABLE 21.1   Inducing an electric current using a magnet

VIDEO
OET 21.1

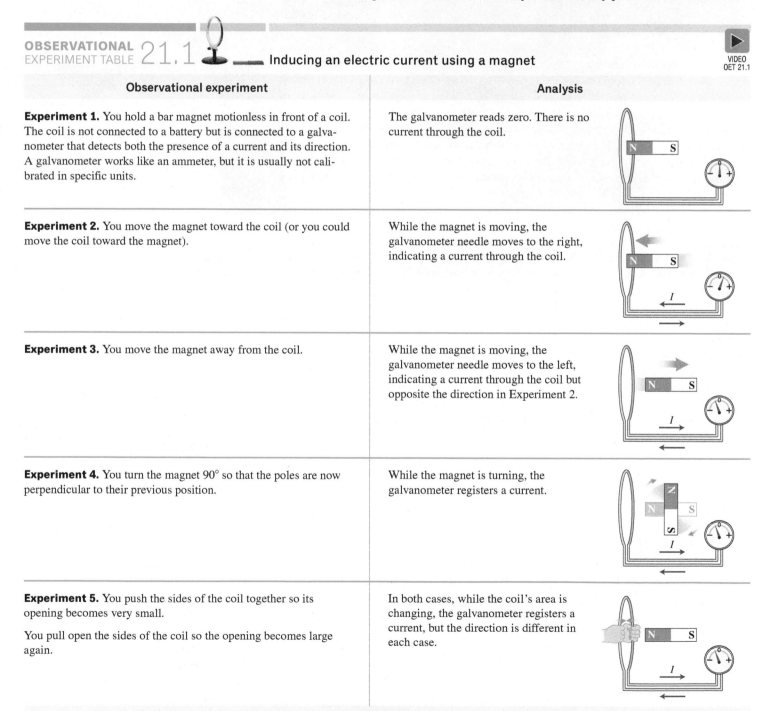

Observational experiment	Analysis
Experiment 1. You hold a bar magnet motionless in front of a coil. The coil is not connected to a battery but is connected to a galvanometer that detects both the presence of a current and its direction. A galvanometer works like an ammeter, but it is usually not calibrated in specific units.	The galvanometer reads zero. There is no current through the coil.
Experiment 2. You move the magnet toward the coil (or you could move the coil toward the magnet).	While the magnet is moving, the galvanometer needle moves to the right, indicating a current through the coil.
Experiment 3. You move the magnet away from the coil.	While the magnet is moving, the galvanometer needle moves to the left, indicating a current through the coil but opposite the direction in Experiment 2.
Experiment 4. You turn the magnet 90° so that the poles are now perpendicular to their previous position.	While the magnet is turning, the galvanometer registers a current.
Experiment 5. You push the sides of the coil together so its opening becomes very small. You pull open the sides of the coil so the opening becomes large again.	In both cases, while the coil's area is changing, the galvanometer registers a current, but the direction is different in each case.

Patterns

Although no battery was used, an electric current was induced in the coil when the magnet moved toward or away from the coil. Current was also induced when the coil's orientation relative to the magnet or the area of the coil changed.

In Table 21.1 there was no battery, yet the galvanometer registered an electric current through a coil. For the current to exist, there must be some source of emf. What produced the emf in these experiments?

Recall from our study of magnetism (in Chapter 20) that a magnetic field can exert a force on moving electrically charged particles. The force exists only if the magnetic field or a component of the field is perpendicular to the velocity of the electrically charged particles. Let's consider again Experiment 2 in Table 21.1 in which the coil moves towards the magnet; for simplicity, we use a square loop made of conducting wire (**Figure 21.1**). Inside the wire there are positively charged ions that make up the lattice of the metal (and cannot leave their locations) and negatively charged free electrons. The $\vec{B}$ field produced by the bar magnet points away from the magnet and spreads out as shown in the figure.

Notice that at the top and bottom sections of the loop, a component of the $\vec{B}$ field is perpendicular to the velocity of the loop as it moves toward the right. As a result, the magnetic field exerts a force on each electron in the wire. This force causes the electrons in the wire to accelerate clockwise around the loop as viewed from the magnet (use the right-hand rule for the magnetic force on the negatively charged electrons, as discussed in Section 20.4). Note also that electrons in the vertical section of the loop closest to us accelerate upward, while the electrons in the vertical section farthest from us accelerate downward. The overall result is that due to the relative motion between the loop and the bar magnet, the electrons start moving around the loop in a coordinated fashion—we have an induced electric current.

Thus, it seems that the currents produced in Table 21.1 can be explained using the concept of the magnetic force we have previously developed (in Chapter 20). However, the magnetic force approach does not explain how transcranial magnetic simulation (TMS) works. The magnetic force-based explanation requires motion of the loop relative to the magnetic field, but in the TMS procedure, the coil is not moving relative to the brain.

Perhaps there is another explanation. Let's examine Table 21.1 from a different perspective, one that focuses on the $\vec{B}$ field itself rather than on any sort of motion. When the magnet moved or rotated with respect to the magnetic field, the number of $\vec{B}$ field lines going through the area of the coil increased or decreased (**Figure 21.2a**). The number of $\vec{B}$ field lines through the coil also changed when the area of the coil changed (Figure 21.2b). Thus, an alternative explanation for the pattern we observed in the experiments is that when the number of $\vec{B}$ field lines through the coil's area changes, there is a corresponding emf produced in the coil, which leads to the induced current.

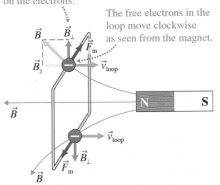

FIGURE 21.1 Magnetic field exerts a force on the electrons in the moving loop.

The component of the magnetic field that is perpendicular to the velocity exerts a magnetic force on the electrons.

The free electrons in the loop move clockwise as seen from the magnet.

FIGURE 21.2 Changing $\vec{B}$ or A causes an induced current.

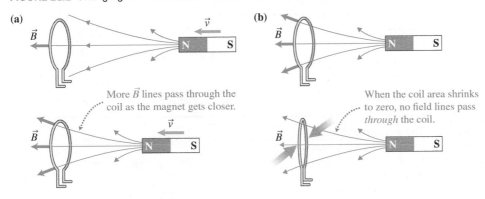

(a)

More $\vec{B}$ lines pass through the coil as the magnet gets closer.

(b)

When the coil area shrinks to zero, no field lines pass *through* the coil.

FIGURE 21.3 A current-carrying loop and a bar magnet create magnetic fields that have the same distribution of $\vec{B}$ field lines.

(a)

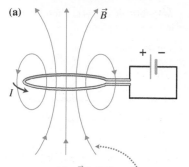

The $\vec{B}$ field lines for a current loop are about the same as those for a bar magnet.

(b)

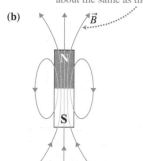

We can summarize these two explanations for the induced current as follows:

Explanation 1: The induced current is caused by the magnetic force exerted on moving electrically charged objects (for example, the free electrons in conducting wires that are moving relative to the magnet).

Explanation 2: Any process that changes the number of $\vec{B}$ field lines through a coil's area induces a current in the coil. The mechanism explaining how it happens is unclear at this point.

Let's test these explanations with an experiment involving a change in the number of $\vec{B}$ field lines through a coil's area, but with no relative motion. Recall that we can create a magnetic field using either a wire that carries current or a permanent magnet (Chapter 20). The current-carrying coil in **Figure 21.3a** has a $\vec{B}$ field that resembles the $\vec{B}$ field of the bar magnet in Figure 21.3b.

Testing Experiment **Table 21.2** uses two coils, such as those shown in **Figure 21.4**. Coil 1 on the bottom is connected to a battery and has a switch to turn the current on and off. When the switch is open, there is no current in coil 1. When the switch is closed, the current in coil 1 produces a magnetic field whose $\vec{B}$ field lines pass through coil 2's area. For each of the experiments we will use the two explanations to predict whether or not there should be an induced electric current in coil 2.

FIGURE 21.4 An arrangement of coils for the testing experiment.

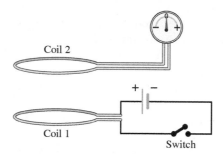

	Will a current be induced in coil 2?		
Testing experiment	**Prediction based on Explanation 1**	**Prediction based on Explanation 2**	**Outcome**
Experiment 1. The switch in the circuit for coil 1 is open, so there is no current in coil 1. Is there any current in coil 2?	There is no current in coil 1, thus there is no magnetic field at coil 2. Neither coil is moving. No current will be induced in coil 2.	There is no current in coil 1; therefore, there is no change in the number of $\vec{B}$ field lines through coil 2's area. No current will be induced in coil 2.	The galvanometer registers no current in coil 2.
Experiment 2. You close the switch in the circuit for coil 1. While the switch is being closed, the current in coil 1 increases rapidly from zero to a steady final value. Is there any current in coil 2 while the switch is being closed?	Neither coil is moving, thus no current will be induced in coil 2.	The increasing current in coil 1 produces an increasing $\vec{B}$ field. This changes the number of $\vec{B}$ field lines through coil 2's area. A brief current should be induced in coil 2.	Just as the switch closes, the galvanometer needle briefly moves to the left and then returns to vertical, indicating a brief induced current in coil 2.

TESTING
EXPERIMENT TABLE 21.2   Testing the explanations for induced current

VIDEO
TET 21.2

| Testing experiment | Will a current be induced in coil 2? | | Outcome |
	Prediction based on Explanation 1	Prediction based on Explanation 2	
Experiment 3. You keep the switch in the circuit for coil 1 closed. The current in coil 1 has a steady value. Is there current in coil 2?	Neither coil is moving, thus no current will be induced in coil 2.	There is a steady current in coil 1, which results in a steady $\vec{B}$ field. Thus, the number of $\vec{B}$ field lines through coil 2's area is not changing. Therefore, no current will be induced in coil 2.	The galvanometer registers no current in coil 2.
Experiment 4. You open the switch again. Is there any current in coil 2 while the switch is being opened?	Neither coil is moving, thus no current will be induced in coil 2.	The decreasing current in coil 1 produces a decreasing $\vec{B}$ field. This changes the number of $\vec{B}$ field lines through coil 2's area, which should induce a brief current in coil 2.	Just as the switch opens, the galvanometer needle briefly moves to the right (opposite the direction in Experiment 2) and then returns to vertical, indicating a brief induced current in coil 2.

Conclusion

The predictions based on Explanation 2 matched the outcomes in all four experiments. The predictions based on Explanation 1 did not match the outcomes in two of the four experiments.

Motion is not necessary to have an induced current. In contrast, when the number of $\vec{B}$ field lines through a coil's area changes, there is an induced current in that coil. Explanation 1 has been found not to be valid in all cases, but Explanation 2 has been found to be valid for all examined cases.

We have learned that it is possible to have a current in a closed loop of wire without using a battery. This phenomenon of inducing a current using a changing $\vec{B}$ field is called **electromagnetic induction**. The discovery of electromagnetic induction, in 1831, is credited to two scientists: a British physicist, Michael Faraday, and an American physicist, Joseph Henry. Faraday spent 10 years trying to induce electric currents using magnetic fields, and it was only the invention of wire insulation by Henry that enabled Faraday to make multi-loop coils, leading to his success.

Electromagnetic induction explains how transcranial magnetic stimulation (TMS) works. When the physician closes the switch in the circuit with the coil that rests on the outside of the skull, the increasing current in the coil produces a changing $\vec{B}$ field within the brain. A current is briefly and noninvasively induced in a small region of the brain's electrically conductive tissue.

CONCEPTUAL EXERCISE 21.1 **Moving loops in a steady magnetic field**

The figure shows four wire loops 1–4 moving at constant velocity $\vec{v}$ relative to a region with a constant $\vec{B}$ field (within the dashed lines). In which of these loops will electric currents be induced?

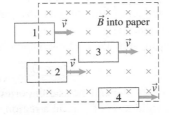

Sketch and translate An electric current is induced in a loop whenever the number of $\vec{B}$ field lines through the loop's area changes.

Simplify and diagram For each loop, the current should be as follows:

1. Loop 1 is moving into the field region, so the number of $\vec{B}$ field lines through its area is increasing. A current will be induced in the loop.

(CONTINUED)

2. A current will be induced in loop 2 for the same reasons as (1).

3. Loop 3 is completely within the field region, thus the number of $\vec{B}$ field lines is not changing. As a result, no current is induced.

4. No current is induced in loop 4 for the same reason as in (3).

Try it yourself What will happen to loops 3 and 4 as they leave the magnetic field region?

Answer The number of $\vec{B}$ field lines through each loop will change and a brief current will be induced as the loops leave the $\vec{B}$ field region.

FIGURE 21.5 A schematic of a dynamic microphone.

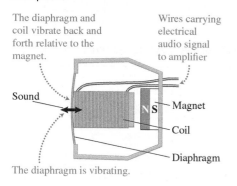

The diaphragm and coil vibrate back and forth relative to the magnet.

Wires carrying electrical audio signal to amplifier

Sound

N S — Magnet

— Coil

— Diaphragm

The diaphragm is vibrating.

FIGURE 21.6 A simple seismometer.

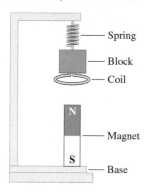

— Spring

— Block

— Coil

N

S

— Magnet

— Base

TIP Often we discuss magnetic flux through a particular area, such as the area inside a wire loop. But the loop itself is not needed for the flux to exist.

Dynamic microphones and seismometers

A dynamic microphone that converts sound vibrations into electrical oscillations employs the principle of electromagnetic induction (**Figure 21.5**). When a sound wave, such as that produced by a singer's voice, strikes the diaphragm inside the microphone, the diaphragm oscillates. This oscillation moves a coil of wire attached to the diaphragm alternately closer to and farther from a magnet in the microphone, corresponding to locations with stronger and weaker magnetic field. This changing magnetic field through the coil induces a changing current in the coil. The changes in the current mirror the sound waves that led to the production of this current; thus this current can be used to store the details of the sound electronically.

A seismometer operates by the same principle. A seismometer is a sensor that detects seismic waves during an earthquake. It has a massive base with a magnet that vibrates as seismic waves pass. At the top, a coil attached to a block hangs at the end of a spring (**Figure 21.6**). The spring acts as a shock absorber that reduces the vibrations of the hanging block and coil relative to the base. The motion of the base relative to the coil induces a current through the coil, which produces a signal that is recorded on a seismograph.

REVIEW QUESTION 21.1 Your friend thinks that relative motion of a coil and a magnet is absolutely necessary to induce current in a coil that is not connected to a battery. Support your friend's point of view with a physics argument. Then provide a counterargument and describe an experiment you could perform to disprove your friend's idea.

21.2 Magnetic flux

In the last section we found that an electric current is induced in a coil when the number of $\vec{B}$ field lines through the coil's area changes. This occurred when

- the strength of the $\vec{B}$ field in the vicinity of the coil changed, or
- the area A of the coil changed, or
- the orientation of the $\vec{B}$ field relative to the coil changed.

In this section we will construct a physical quantity describing the number of $\vec{B}$ field lines through a coil's area. Based on the analysis in the last section, changes in that quantity should cause an induced electric current in the coil. Physicists call this physical quantity **magnetic flux** Φ (Φ is the greek letter "phi").

The greater the magnitude of the $\vec{B}$ field passing through a particular two-dimensional area, the greater the number of field lines through the area. Additionally, if the area itself is larger, the number of field lines through the area is greater. If we double one or the other, the number of lines through the area should double. This suggests that the magnetic flux is proportional to the magnitude of the $\vec{B}$ field passing through the area and to the size of the area itself. Mathematically,

$$\Phi \propto BA$$

How do we include the dependence on the orientation of the loop relative to the $\vec{B}$ field lines? Imagine a rigid loop of wire in a region with a uniform $\vec{B}$ field. If the plane of the loop is perpendicular to the $\vec{B}$ field lines, then a maximum number of field lines pass through the area (**Figure 21.7a**). If the plane of the loop is parallel to the $\vec{B}$ field, zero field lines pass through the area (Figure 21.7b). In between these two extremes, the magnetic flux takes on intermediary values (Figure 21.7c).

To describe this relative orientation we use a line perpendicular to the plane of the loop—the black normal vector shown in Figure 21.7. The angle θ between this vector and the $\vec{B}$ field lines quantifies this orientation. Since the cosine of an angle is at a maximum when the angle is zero and a minimum when the angle is 90°, the magnetic flux through the area is also proportional to $\cos\theta$. This leads to a precise definition for the magnetic flux through an area.

FIGURE 21.7 Flux depends on the angle between the normal vector to the loop surface and the direction of the $\vec{B}$ field.

(a) Maximum flux: $\cos 0° = 1.0$

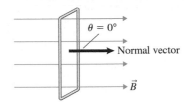

Magnetic flux Φ The magnetic flux Φ through a region of area A is

$$\Phi = BA\cos\theta \qquad (21.1)$$

where B is the magnitude of the uniform magnetic field throughout the area and θ is the angle between the direction of the $\vec{B}$ field and a normal vector perpendicular to the area. The SI unit of magnetic flux is the unit of the magnetic field (the tesla T) times the unit of area (m^2), or $\text{T}\cdot\text{m}^2$. This unit is also known as the weber (Wb).

(b) Zero flux: $\cos 90° = 0$

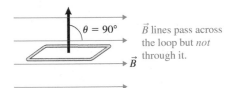

Equation (21.1) assumes that the magnetic field throughout the area is uniform and that the area is flat. For situations in which this is not the case, first split the area into small subareas within which the $\vec{B}$ field is approximately uniform; then add together the magnetic fluxes through each. This book will not address such cases.

In Section 21.1, we proposed that when the number of magnetic field lines through a wire loop changes, a current is induced in the loop. We can now refine that idea and say that *current is induced when there is a change in the total magnetic flux through the loop's area*. In other words, if the magnetic flux throughout the loop's area is steady, no current will be induced.

(c) Intermediate flux: $0 < \cos\theta < 1.0$

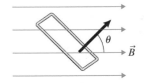

TIP In everyday life, flux often refers to something that is changing. "Our plans for the weekend are in flux." However, in physics, the word flux means the value of a particular physical quantity. When the $\vec{B}$ field, the orientation of the loop, and the area of the loop are constant, the flux through the area of the loop is constant.

QUANTITATIVE EXERCISE 21.2   **Flux through a book's cover**

A book is positioned in a uniform 0.20-T $\vec{B}$ field that points from left to right parallel to the plane of the page, shown in the figures on the right. (For simplicity, we depict the book as a rectangular loop.) Each side of the book's cover measures 0.10 m. Determine the magnetic flux through the cover when (a) the cover is in the plane of the page (figure a), (b) the cover is perpendicular to the plane of the page and the normal vector makes a 60° angle with the $\vec{B}$ field (figure b), and (c) the cover is perpendicular to the plane of the page and the normal vector points toward the top of the page (figure c).

(a)

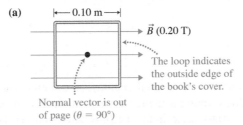

(b)

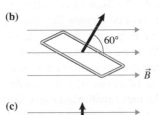

(c)

Represent mathematically The magnetic flux through an area is determined by Eq. (21.1) $\Phi = BA\cos\theta$. The angle in each of the three situations is (a) $\theta = 90°$, (b) $\theta = 60°$, and (c) $\theta = 90°$.

Solve and evaluate The magnetic flux through the cover in each case is
(a) $\Phi = (0.20\text{ T})(0.10\text{ m})^2\cos(90°) = 0$
(b) $\Phi = (0.20\text{ T})(0.10\text{ m})^2\cos(60°) = 1.0\times10^{-3}\text{ T}\cdot\text{m}^2$
(c) $\Phi = (0.20\text{ T})(0.10\text{ m})^2\cos(90°) = 0$

We can evaluate these results by comparing the calculated fluxes to the number of $\vec{B}$ field lines through the cover's area. Note that for the orientation of the book in (a) and (c), the $\vec{B}$ field lines are parallel to the book's area and therefore do not go through it. Those positions are consistent with our mathematical result. The orientation for the book in (b) is such that some $\vec{B}$ field lines do pass through the book, which is also consistent with the nonzero mathematical result.

(CONTINUED)

Try it yourself A circular ring of radius 0.60 m is placed in a 0.20-T uniform $\vec{B}$ field that points toward the top of the page. Determine the magnetic flux through the ring's area when (a) the plane of the ring is perpendicular to the surface of the page and its normal vector points to the right and (b) the plane of the ring is perpendicular to the surface of the page and its normal vector points toward the top of the page.

Answer   (a) 0; (b) 0.23 T·m².

REVIEW QUESTION 21.2 You have a bar magnet and a gold ring. How should you position the ring relative to the magnet so that the magnetic flux through the circular area inside the ring is zero?

21.3 Direction of the induced current

Recall from Section 21.1 that the galvanometer registered current in one direction for some of the experiments and in the opposite direction for others. In this section we will focus on the pattern in the direction of the induced current.

Figure 21.8 shows the results of two experiments in which the number of $\vec{B}$ field lines through a wire coil's area is changing. As the bar magnet moves closer to the coil in (a), the number of $\vec{B}$ field lines through the coil's area increases (the magnetic flux through the coil's area increases). As expected, there is a corresponding induced current. An arrow along the coil indicates the direction of this induced current as measured by a galvanometer.

FIGURE 21.8 How is the changing $\vec{B}_{ex}$ related to the direction of I_{in} and $\vec{B}_{in}$? The number of the $\vec{B}_{ex}$ field lines passing though the coil increases in (a) and decreases in (b).

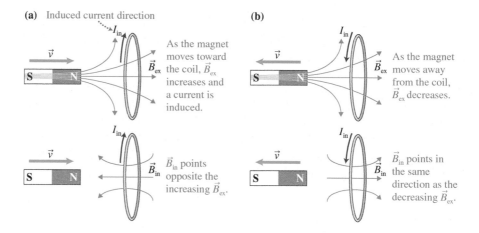

Because electric currents produce a magnetic field, the induced current in the coil must also produce a magnetic field and a corresponding magnetic flux through the coil. We call this second magnetic field $\vec{B}_{induced}$ or $\vec{B}_{in}$. The direction of $\vec{B}_{in}$ can be determined using the right-hand rule for the $\vec{B}$ field. Notice that in the case shown in Figure 21.8a, the flux through the coil due to the magnet (called $\vec{B}_{external}$ or $\vec{B}_{ex}$) is increasing and the magnetic field due to the induced current $\vec{B}_{in}$ points in the opposite direction of $\vec{B}_{ex}$.

In Figure 21.8b, the bar magnet is moving away from the coil. As a result, the number of external field lines through the coil's area (and therefore the magnetic flux through it) is decreasing. Again, there is a corresponding induced current (see Figure 21.8b). In this case, however, $\vec{B}_{in}$ (produced by the induced current) points in the same direction as $\vec{B}_{ex}$ (produced by the magnet). Can we find a pattern in these data?

Notice that in both cases $\vec{B}_{in}$ points in the direction that diminishes the change in the external flux through the coil. In the first experiment the flux through the coil was increasing. In that situation, $\vec{B}_{in}$ pointed in the opposite direction to $\vec{B}_{ex}$, as if to resist the increase. In the second experiment, the external flux through the coil was decreasing, and $\vec{B}_{in}$ pointed in the same direction as $\vec{B}_{ex}$, as if to resist the decrease. In both situations, the $\vec{B}_{in}$ due to the induced current resisted the *change* in the external flux through the coil. Put another way, $\vec{B}_{in}$ points in whatever direction is necessary to try to keep the magnetic flux through the coil's area constant.

Consider what would happen if the reverse occurred. Suppose an increasing external magnetic flux through the loop led to an induced magnetic field in the same direction as the external field. In that case, the magnetic flux due to the induced field would augment rather than reduce the total flux through the loop. This would cause an even greater induced current, which would cause yet a greater induced magnetic field and a steeper increase in magnetic flux. In other words, just by lightly pushing a bar magnet toward a loop of wire, you would cause a runaway induced current that would continually increase until the wire melted. Of course, this would violate the conservation of energy. If such a scenario were possible, we could heat water by simply moving a bar magnet over a coil in a large tank of water.

This pattern concerning the direction of the induced current was first developed in 1833 by the Russian-German physicist Heinrich Lenz.

Lenz's law The direction of the induced current in a coil is such that its $\vec{B}_{in}$ field opposes the change in the magnetic flux through the coil's area produced by other objects. If the magnetic flux through the coil is increasing, the direction of the induced current's $\vec{B}_{in}$ field leads to a decrease in the flux. If the magnetic flux through the coil is decreasing, the direction of the induced current's $\vec{B}_{in}$ field leads to an increase in the flux.

Lenz's law lets us determine the direction of the induced current's magnetic field $\vec{B}_{in}$. From there we can use the right-hand rule for the $\vec{B}$ field to determine the direction of the induced current itself. This process is summarized in Physics Tool Box 21.1, which shows how to determine the direction of the induced current in a loop of wire when the magnetic flux though the loop changes. In this Tool Box, the loop of wire is in a decreasing external field $\vec{B}_{ex}$ that is perpendicular to the plane of the loop.

Note that the magnetic flux through a loop or coil can change because of a change in the external magnetic field, a change in the area of a loop or coil, or a change in its orientation.

PHYSICS TOOL BOX 21.1 ⟩ **Determining the direction of an induced current**

1. Determine the initial external magnetic flux $\Phi_{ex\,i}$ through the coil (represented below by the number and direction of the external magnetic field lines).

2. Determine the final external magnetic flux $\Phi_{ex\,f}$ through the coil (represented below by the smaller number of external magnetic field lines).

3. The induced flux opposes the *change* in the external flux (a decrease in upward external flux in this case; $\vec{B}_{in}$ points up so that the net upward flux decreases less than it would in the absence of the induced flux).

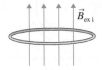

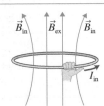

4. Use the right-hand rule for the magnetic field to determine the direction of the induced current that will produce the $\vec{B}_{in}$ and the induced flux.

A loop in the plane of the page is being pulled to the right at constant velocity out of a region of a uniform magnetic field $\vec{B}_{ex}$. The field is perpendicular to the loop and points out of the paper in the region inside the rectangular dashed line (see the first figure below). Determine the direction of the induced electric current in the loop when it is halfway out of the field, as shown in the second figure below.

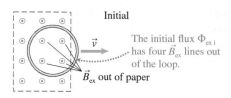

Initial

The initial flux $\Phi_{ex\,i}$ has four $\vec{B}_{ex}$ lines out of the loop.

$\vec{B}_{ex}$ out of paper

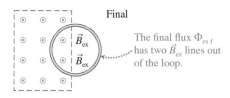

Final

The final flux $\Phi_{ex\,f}$ has two $\vec{B}_{ex}$ lines out of the loop.

Sketch and translate The number of $\vec{B}_{ex}$ lines passing through the coil's area decreases by half during the time interval from when the coil

is completely in the field to when it is halfway out. Thus, the external magnetic flux through the loop's area has decreased.

Simplify and diagram According to Lenz's law, the induced magnetic flux and induced magnetic field $\vec{B}_{in}$ should point out of the paper, as shown below, thus keeping the flux through the loop closer to the initial flux. The direction of the induced current that causes this induced magnetic field is determined using the right-hand rule for the $\vec{B}$ field; the induced electric current through the loop is counterclockwise.

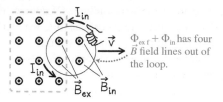

$\Phi_{ex\,f} + \Phi_{in}$ has four $\vec{B}$ field lines out of the loop.

Try it yourself Notice that the induced current in the left side of the loop is still in the external magnetic field $\vec{B}_{ex}$ (as shown in the figure above). Determine the direction of the force that the external magnetic field $\vec{B}_{ex}$ exerts on the induced current in the left side of the loop.

Answer   To the left, opposite the direction of the loop's velocity.

Eddy currents: an application of Lenz's law

Conceptual Exercise 21.3 shows an example of a phenomenon called an **eddy current**. An eddy current usually occurs when a piece of metal moves through a magnetic field. If you were to hold a sheet of aluminum or copper between the poles of a strong horseshoe magnet, you would find that neither of the poles attracts the sheet. However, when you move the sheet out from between the poles of the magnet, especially if you pull it quickly, you encounter resistance. This force is similar to the force exerted on the loop leaving the magnetic field region in the "Try it yourself" part of the exercise.

Let's examine this phenomenon. **Figure 21.9a** shows a metal sheet between the poles of an electromagnet. Pulling the sheet to the right decreases the external magnetic flux through area 1. This is similar to the situation in Conceptual Exercise 21.3, although in that exercise the external $\vec{B}$ field was out of the page instead of down.

FIGURE 21.9 Inducing an eddy current by pulling a metal sheet through a magnetic field. (a) Area 1 is leaving the field and area 2 is entering the field. (b) Changing flux through both areas induces eddy currents. The external magnetic field exerts a force on the eddy currents opposite the direction of motion of the sheet.

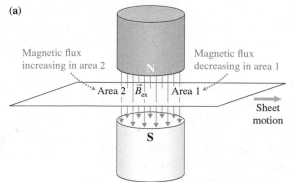

(a)

Magnetic flux increasing in area 2

Magnetic flux decreasing in area 1

Area 2 $\vec{B}_{ex}$ Area 1

Sheet motion

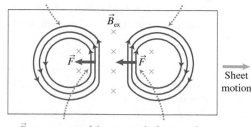

(b)   The changing flux in areas 1 and 2 causes eddy currents.

$\vec{B}_{ex}$

$\vec{F}$ $\vec{F}$

Sheet motion

$\vec{B}_{ex}$ exerts a repulsive magnetic force on the parts of the eddy currents that are in $\vec{B}_{ex}$.

This decrease in flux induces an eddy current in the metal sheet around area 1, which according to Lenz's law circles or curls clockwise in this region (see Figure 21.9b) to produce an induced field that points in the same direction as the external $\vec{B}$ field.

At the same time, area 2 of the sheet is entering the magnetic field region, so the magnetic flux through that area is increasing. This change in magnetic flux induces a counterclockwise eddy current.

How do we explain the force that points opposite the direction of the sheet's motion? The left side of the eddy current in area 1, shown in Figure 21.9b, is still in the magnetic field region. Using the right-hand rule for the magnetic force, we find that the force exerted by the magnet on the left side of the eddy current in area 1 points toward the left when the sheet is pulled to the right, in agreement with what was observed. Similarly, the magnet also exerts a force on the part of the eddy current in area 2 that is in the magnetic field region. The right-hand rule for the magnetic force determines that this force points to the left as well. Both of these forces point opposite the direction the sheet is moving, acting as a sort of "braking" force.

What would happen if you were to push the sheet to the left instead of pulling it to the right? Using Lenz's law, we find that the eddy currents reverse direction, and the magnetic forces exerted on them also reverse direction, again resulting in a magnetic braking effect.

Many technological applications rely on the phenomenon of resistance to the motion of a nonmagnetic metal material through a magnetic field. For instance, the braking system used in some cars, trains, and amusement park rides consists of strong electromagnets with poles on either side of the turning metal wheels of the vehicle (**Figure 21.10**). When the electromagnet is turned on, it exerts magnetic forces on the eddy currents in the wheels that oppose their motion. As the turning rate of the wheels decreases, the eddy currents decrease, and therefore the braking forces decrease.

Another very important application of eddy currents is the eddy current waste separator. When general waste is delivered to a dump, it is placed on a conveyer belt. First, strong magnets lift objects containing ferromagnetic materials (iron nails, for example), separating them from the rest of the waste, which continues along the belt. Under the end of the conveyer belt are new magnets (usually electromagnets that produce variable magnetic field), which induce eddy currents in all non-ferromagnetic moving metal objects (aluminum cans, for example) and therefore exert repulsive upward forces on them. These forces lift the metal objects above the belt (see **Figure 21.11**), causing them to fly off the belt horizontally while the nonmetal objects continue moving with the belt and fall straight down off its end. The metal objects land on a new conveyer belt placed below the original belt and slightly forward. Each type of waste lands in a different place, and metal objects are separated from nonmetal objects.

REVIEW QUESTION 21.3 What difficulty would occur if the $\vec{B}$ field produced by the induced current enhanced the change in the external field rather than opposed the change? Give a specific example.

21.4 Faraday's law of electromagnetic induction

In the first two sections of this chapter, we found that when the magnetic flux through a coil's area changes, there is an induced electric current in the coil. The flux depends on the magnitude of the $\vec{B}$ field, the area of the coil, and the orientation of the coil relative to the $\vec{B}$ field. Our next goal is to construct a quantitative version of this idea that will allow us to predict the magnitude of the induced current through a particular coil. We begin with Observational Experiment **Table 21.3** (on the next page), in which we will determine what factors affect the magnitude of the induced current produced by the flux change.

FIGURE 21.10 Magnetic braking is used to stop the Giant Drop ride at Dreamworld in Queensland, Australia.

FIGURE 21.11 An eddy current waste separator. Metal objects fly forward while non-metals fall off the end of the conveyer belt.

Factors affecting the magnitude of induced current

VIDEO
OET 21.3

Observational experiment	Analysis
Experiment 1. Rapidly move a bar magnet toward a coil and observe the galvanometer needle. Repeat the process, this time moving the magnet slowly.	The galvanometer registers a larger induced current when the magnet moves rapidly toward the coil compared with when it moves slowly.
Experiment 2. Rapidly rotate the magnet by 90 degrees in front of the stationary coil. Repeat the process, this time rotating the magnet slowly.	The galvanometer registers a larger induced current when the magnet rotates rapidly compared with when it rotates slowly.
Experiment 3. Use two coils, each with a different number of turns. Move the magnet toward the coil with more turns. Then move the magnet at the same speed toward the coil with fewer turns.	The galvanometer registers a larger induced current in the coil with more turns.

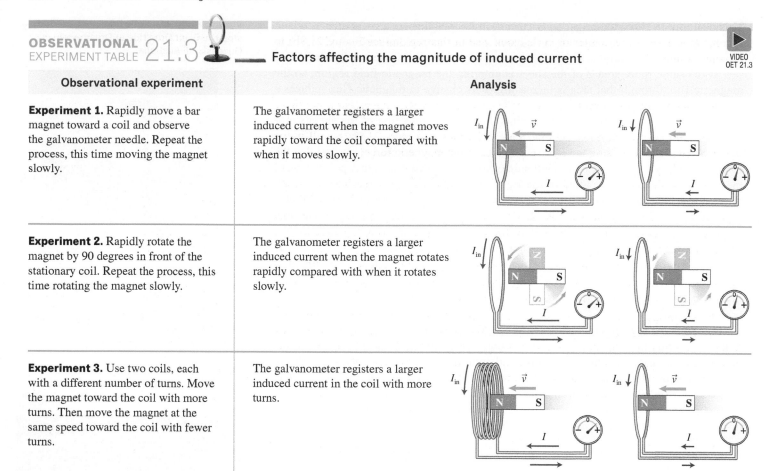

Patterns

- The speed at which the magnet moved or rotated affected the induced current. The shorter the time interval for the change of the flux through the coil, the greater the induced current.
- The induced current in a coil with a larger number of turns is greater than in a coil with a smaller number of turns.

TIP Notice that even though we were investigating the effects of two quantities on the magnitude of the induced current, we changed only one of them at a time so that we could investigate those effects separately.

In the experiments in Table 21.3, we found that the induced current was greater if the same change in magnetic flux through a coil occurred in a shorter time interval Δt. Additionally, the induced current through the coil was greater in a coil with a larger number of turns N.

Quantitatively, what is the relationship between the magnitude of the induced current in a coil and the change in magnetic flux through that coil's area? Let's connect a single circular loop with zero resistance in series with a 100-Ω resistor (**Figure 21.12a**). The loop is placed between the poles of an electromagnet with its surface perpendicular to the magnetic field. An ammeter (not shown) measures the current in the loop and resistor. The upward magnetic field (up is defined as the positive direction) decreases steadily for 2.0 s (we decrease the magnitude of the $\vec{B}$ field by decreasing the current in the coils of the electromagnet; note that this is not the induced current), after which the field and flux remain constant at a smaller positive value. The magnetic flux-versus-time and the induced current-versus-time graphs are shown in Figure 21.12b.

When the flux is *changing* at a constant rate, the current through the loop and resistor has a constant value. For example, for the first 2.0 s the *slope* of the magnetic flux-versus-time graph has a constant negative value of $-1.0 \text{ T} \cdot \text{m}^2/\text{s}$, and the induced current has a constant positive value of $+0.010$ A. If we replace the 100-Ω resistor with a 50-Ω resistor, the current-versus-time graph has the same shape, but its magnitude during the first 2.0 s doubles to $+0.020$ A. Thus, the same flux change produces different size currents in the same loop depending on the resistance of the circuit. However, the product of the current and resistance for both situations has the same value:

$$(0.010\ \text{A})(100\ \Omega) = (0.020\ \text{A})(50\ \Omega) = 1.0\ \text{A} \cdot \Omega = 1.0\ \text{V}$$

It almost seems that there must be a 1.0-V battery in series with the resistor, but there isn't.

The changing flux through the loop caused the induced current. However, the unit of the *slope* of the flux-versus-time graph is not the ampere, but the volt:

$$\frac{\Delta\Phi}{\Delta t} = \frac{-2.0\ \text{T} \cdot \text{m}^2}{2.0\ \text{s}} = -1.0\left(\frac{\text{N}}{\text{C} \cdot (\text{m/s})}\right)\frac{\text{m}^2}{\text{s}} = -1.0\ \frac{\text{N} \cdot \text{m}}{\text{C}} = -1.0\ \frac{\text{J}}{\text{C}} = -1.0\ \text{V}$$

(We used the expression for the magnetic force $F_{\text{magnetic}} = qvB$ or $B = F_{\text{magnetic}}/qv$ to convert the magnetic field in tesla (T) to other units.) Thus the induced current is a consequence of the emf induced in the coil. The changing flux acts as a "battery" that produces a 1.0-V emf that causes the electric current. You could apply Kirchhoff's loop rule to the circuit shown in Figure 21.12a. There is a $+1.0$-V potential change across the loop and a -1.0-V potential change across the resistor.

The magnitude of the emf in the loop depends only on the rate of change of flux through the loop (the slope of the flux-versus-time graph $\mathcal{E}_{\text{in}} = |\Delta\Phi/\Delta t|$). When we repeat this experiment for a coil with N loops, we find that the magnitude of the induced emf increases in proportion to the number N of loops in the coil. Thus,

$$\mathcal{E}_{\text{in}} = N\left|\frac{\Delta\Phi}{\Delta t}\right| \qquad (21.2)$$

This expression is known as **Faraday's law of electromagnetic induction**. However, the mathematical expression was actually developed by James Clerk Maxwell.

> **Faraday's law of electromagnetic induction** The average magnitude of the induced emf $\mathcal{E}_{\text{in}}$ in a coil with N loops is the magnitude of the ratio of the magnetic flux change through the loop $\Delta\Phi$ to the time interval Δt during which that flux change occurred multiplied by the number N of loops:
>
> $$\mathcal{E}_{\text{in}} = N\left|\frac{\Delta\Phi}{\Delta t}\right| = N\left|\frac{\Delta(BA\cos\theta)}{\Delta t}\right| \qquad (21.3)$$

The direction of the current induced by this emf is determined using Lenz's law. A minus sign is often placed in front of the N to indicate that the induced emf opposes the change in magnetic flux. However, we will not use this notation.

Faraday realized that the phenomenon of electromagnetic induction has the same effect on devices attached to a coil as a battery. This idea enabled him to build the first primitive electric generator that produced an induced current in a coil.

FIGURE 21.12 An experiment to quantitatively relate induced emf ($\mathcal{E}_{\text{in}}$) and induced current (I_{in}) to flux change.

(a)

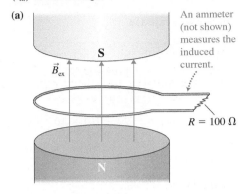

An ammeter (not shown) measures the induced current.

$\vec{B}_{\text{ex}}$ S

$R = 100\ \Omega$

N

(b) Resistance $R = 100\ \Omega$

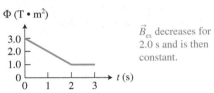

Φ (T • m^2)

3.0
2.0
1.0
0
0 1 2 3 t (s)

$\vec{B}_{\text{ex}}$ decreases for 2.0 s and is then constant.

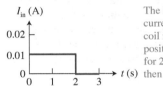

I_{in} (A)

0.02
0.01
0
0 1 2 3 t (s)

The induced current in the coil is a constant positive value for 2.0 s and is then zero.

TIP Note that the value of the emf does not depend on the properties of the material of the coil. It depends on the properties of the magnetic field and the geometry of the coil.

QUANTITATIVE EXERCISE 21.4 **Changing flux produces emf**

The figure below shows the magnetic flux-versus-time graph for the magnetic flux through a 100-turn coil that is moved with a constant speed through an external magnetic field. Draw the corresponding induced emf-versus-time graph.

Φ (10^{-7} T • m^2)

10
0
-10
-20

0.1 0.2 0.3 0.4 t (s)

I | II III

Represent mathematically The magnitude of the induced emf in the coil is

$$\mathcal{E}_{\text{in}} = N\left|\frac{\Delta\Phi}{\Delta t}\right| = 100\left|\frac{\Delta\Phi}{\Delta t}\right|$$

where $\left|\dfrac{\Delta\Phi}{\Delta t}\right|$ is equal to the absolute value of the slope of the magnetic flux-versus-time graph. The graph consists of three regions (I, II, and III); we therefore need to determine three different magnitudes. To

(CONTINUED)

determine the slope within each region, we find the change in magnetic flux for the region and divide it by the time interval during which the change occurred.

Solve and evaluate

Region I: $\Delta\Phi_I = (10 - 0) \times 10^{-7}$ T·m²; $\Delta t_I = (0.2 - 0)$ s

$$\Rightarrow \mathcal{E}_{in\,I} = 100\left|\frac{\Delta\Phi}{\Delta t}\right| = \frac{100 \times 10 \times 10^{-7}\ T\cdot m^2}{0.2\ s} = 0.5 \times 10^{-3}\ V$$

Region II: $\Delta\Phi_{II} = (-20 - 10) \times 10^{-7}$ T·m²; $\Delta t_{II} = (0.3 - 0.2)$ s

$$\Rightarrow \mathcal{E}_{in\,II} = 100\left|\frac{\Delta\Phi}{\Delta t}\right| = \frac{100 \times 30 \times 10^{-7}\ T\cdot m^2}{0.1\ s} = 3.0 \times 10^{-3}\ V$$

Region III: $\Delta\Phi_{III} = -20 - (-20) \times 10^{-7}$ T·m²; $\Delta t_{III} = (0.4 - 0.3)$ s

$$\Rightarrow \mathcal{E}_{in\,III} = 100\left|\frac{\Delta\Phi}{\Delta t}\right| = \frac{100 \times 0}{0.1\ s} = 0\ V$$

The induced emf is negative when the magnetic flux through the loop is increasing and positive when it is decreasing (by Lenz's law). Therefore, the induced emf in region I is negative and in region II is positive. The graph of $\mathcal{E}_{in}$-versus-t for the three regions is shown below.

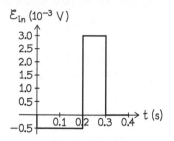

Try it yourself The previous experiment is repeated, only this time the coil is moved twice as fast. Determine the maximum induced emf in the coil.

Answer $\mathcal{E}_{in\,max} = 6.0 \times 10^{-3}$ V.

REVIEW QUESTION 21.4 Why do we write the law of electromagnetic induction in terms of emf rather than in terms of induced current?

21.5 Skills for analyzing processes involving electromagnetic induction

Faraday's law enables us to design and understand practical applications of electromagnetic induction. For example, to design an automobile ignition system that uses spark plugs, engineers must estimate how quickly the magnetic field through a coil must be reduced to zero to produce a large enough emf to ignite a spark plug. An engineer designing an electric generator will be interested in the rate at which the generator coil must turn relative to the $\vec{B}$ field to produce the desired induced emf. The general strategy for analyzing questions like these is described and illustrated in Example 21.5.

PROBLEM-SOLVING
STRATEGY 21.1   Problems involving electromagnetic induction

EXAMPLE 21.5   **Determine the $\vec{B}$ field produced by an electromagnet**

To determine the $\vec{B}$ field produced by an electromagnet, you use a 30-turn circular coil of radius 0.10 m (30-Ω resistance) that rests between the poles of the magnet and is connected to an ammeter. When the electromagnet is switched off, the $\vec{B}$ field decreases to zero in 1.5 s. During this 1.5 s, the ammeter measures a constant current of 18 mA. How can you use this information to determine the initial $\vec{B}$ field produced by the electromagnet?

Sketch and translate

- Create a labeled sketch of the process described in the problem. Show the initial and final situations to indicate the change in magnetic flux. Label knowns and unknowns.
- Determine which physical quantity is changing ($\vec{B}$, A, or θ), thus causing the magnetic flux to change.

The changing quantity (from time 0.0 s to 1.5 s) is the magnitude of the $\vec{B}$ field produced by the electromagnet. Due to this change, the flux through the coil's area changes; thus there is an induced current in the coil. We can use the law of electromagnetic induction to find the magnitude of the initial $\vec{B}$ field produced by the electromagnet.

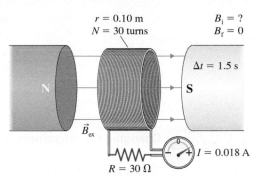

Simplify and diagram

- Decide what assumptions you are making. Does the flux change at a constant rate? Is the magnetic field uniform?
- If useful, draw a graph of the flux and the corresponding induced emf-versus-clock reading.
- If needed, use Lenz's law to determine the direction of the induced current.

Assume the following:

- The current in the electromagnet changes at a constant rate; thus the flux through the coil does also.
- The $\vec{B}$ field in the vicinity of the coil is uniform.
- The $\vec{B}$ field is perpendicular to the coil's surface.

Represent mathematically

- Apply Faraday's law and indicate the quantity ($\vec{B}$, A, or $\cos\theta$) that causes a changing magnetic flux.
- If needed, use Ohm's law and Kirchhoff's loop rule to determine the induced current.

$$\mathcal{E}_{\text{in}} = N\left|\frac{\Phi_f - \Phi_i}{t_f - t_i}\right| = N\left|\frac{0 - B_i A \cos\theta}{\Delta t}\right|$$

The number of turns N and the angle θ remain constant. The magnitude of the magnetic field changes.
Substitute $\mathcal{E}_{\text{in}} = IR$ and $A = \pi r^2$ and solve for B_i:

$$B_i = \frac{IR\Delta t}{N\pi r^2 \cos\theta}$$

Solve and evaluate

- Use the mathematical representation to solve for the unknown quantity.
- Evaluate the results—units, magnitude, and limiting cases.

The plane of the coil is perpendicular to the magnetic field lines, so $\theta = 0$ and $\cos 0° = 1$. Inserting the appropriate quantities:

$$B_i = \frac{IR\Delta t}{N\pi r^2} = \frac{0.018 \text{ A} \times 30 \text{ } \Omega \times 1.5 \text{ s}}{30 \times \pi \times (0.10 \text{ m})^2} = 0.86 \text{ T}$$

The answer is reasonable. Let's check the units:

$$\frac{A \cdot \Omega \cdot s}{m^2} = \frac{V \cdot s}{m^2} = \frac{J \cdot s}{C \cdot m^2} = \frac{N \cdot m \cdot s}{C \cdot m^2} = \frac{N}{C \cdot (m/s)} = T$$

The units match. As a limiting case, a coil with fewer turns would require a larger $\vec{B}$ field to induce the same current. Also, if the resistance of the circuit is larger, the same $\vec{B}$ field change induces a smaller current.

Try it yourself Determine the current in the loop if the plane of the loop is parallel to the magnetic field. Everything else is the same.

Answer   Zero.

In the rest of this section, we will analyze practical applications of electromagnetic induction that involve a change in (1) the magnitude of the $\vec{B}$ field, (2) the area of the loop or coil, or (3) the orientation of the coil relative to the $\vec{B}$ field. All of these processes involve the same basic idea: **a changing magnetic flux through the area of a coil or single loop is accompanied by an induced emf around the coil or loop.** In turn, this emf induces an electric current in the coil or loop.

EXAMPLE 21.6   Changing the magnitude of the $\vec{B}$ field

The magnitude of the $\vec{B}$ field from a TMS coil increases from 0 T to 0.2 T in 0.002 s. The $\vec{B}$ field lines pass through the scalp into a small region of the brain, inducing a small circular current in the conductive brain tissue in the plane perpendicular to the field lines. Assume that the radius of the circular current in the brain is 0.0030 m and that the tissue in this circular region has an equivalent resistance of 0.010 Ω. What are the direction and magnitude of the induced electric current around this circular region of brain tissue?

Sketch and translate We sketch the situation and mark knowns and unknowns. The change in magnetic flux through the circular region of brain tissue is caused by the increasing $\vec{B}$ field produced by the TMS coil (called $\vec{B}_{ex}$). This change in flux causes an induced emf, which produces an induced current in the brain tissue. The direction of the current can be determined using Lenz's law.

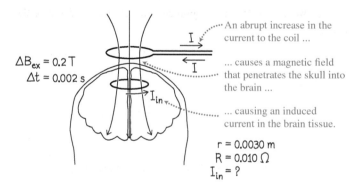

$\Delta B_{ex} = 0.2$ T
$\Delta t = 0.002$ s

An abrupt increase in the current to the coil ...

... causes a magnetic field that penetrates the skull into the brain ...

... causing an induced current in the brain tissue.

r = 0.0030 m
R = 0.010 Ω
I_{in} = ?

Simplify and diagram Assume that $\vec{B}_{ex}$ throughout the disk-like region of brain tissue is uniform and increases at a constant rate. Model the disk as a single-turn coil. Viewed from above, $\vec{B}_{ex}$ points into the page (shown as X's in the figure below). Since the number of $\vec{B}$ field lines is increasing into the page, the $\vec{B}_{in}$ field produced by the induced current points out of the page (shown as dots). Using the right-hand rule for the $\vec{B}_{in}$ field, we find that the direction of the induced current is counterclockwise.

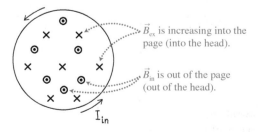

$\vec{B}_{ex}$ is increasing into the page (into the head).

$\vec{B}_{in}$ is out of the page (out of the head).

I_{in}

Represent mathematically To find the magnitude of the induced emf, use Faraday's law:

$$\mathcal{E}_{in} = N\left|\frac{\Phi_f - \Phi_i}{t_f - t_i}\right|$$

where the magnetic flux through the loop at a specific clock reading is $\Phi = B_{ex}A\cos\theta$. The area A of the loop and the orientation angle θ between the loop's normal vector and the $\vec{B}_{ex}$ field are constant, so

$$\mathcal{E}_{in} = N\left|\frac{B_{ex\,f}A\cos\theta - B_{ex\,i}A\cos\theta}{t_f - t_i}\right|$$

$$= NA\cos\theta\left|\frac{B_{ex\,f} - B_{ex\,i}}{t_f - t_i}\right|$$

Using our understanding of electric circuits, we relate this induced emf to the resulting induced current:

$$I_{in} = \frac{\mathcal{E}_{in}}{R}$$

Solve and evaluate Combine these two equations and solve for the induced current:

$$I_{in} = \frac{1}{R}\mathcal{E}_{in} = \frac{1}{R}\left(NA\cos\theta\left|\frac{B_{ex\,f} - B_{ex\,i}}{t_f - t_i}\right|\right)$$

$$= \frac{N\pi r^2\cos\theta}{R}\left|\frac{B_{ex\,f} - B_{ex\,i}}{t_f - t_i}\right|$$

$$= \frac{(1)\pi(0.0030\text{ m})^2(1)}{0.010\text{ Ω}}\left|\frac{0.2\text{ T} - 0}{0.002\text{ s} - 0}\right| = 0.28\text{ A}$$

Note that the normal vector to the loop's area is parallel to the magnetic field; therefore, $\cos\theta = \cos(0°) = 1$. This is a significant current and could affect brain function in that region of the brain.

- -

Try it yourself Derive the expression for the electric charge that has been moved around the circular region in this example.

Answer $Q = N\dfrac{\Phi_f - \Phi_i}{R}$.

EXAMPLE 21.7 **Changing the area of the loop**

A 10-Ω lightbulb is connected between the ends of two parallel conducting rails that are separated by 1.2 m, as shown in the figure below. A metal rod is pulled along the rails so that it moves to the right at a constant speed of 6.0 m/s. The two rails, the lightbulb, its connecting wires, and the rod make a complete rectangular loop circuit. A uniform 0.20-T magnitude B_{ex} field points downward, perpendicular to the loop's area. Determine the direction of the induced current in the loop, the magnitude of the induced emf, the magnitude of the current in the lightbulb, and the power output of the lightbulb.

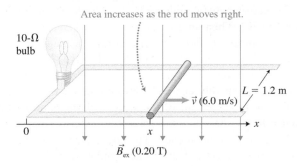

Area increases as the rod moves right.

10-Ω bulb

$L = 1.2$ m

$\vec{v}$ (6.0 m/s)

0   x x

$\vec{B}_{ex}$ (0.20 T)

Sketch and translate We sketch the situation as shown below. Choose the normal vector for the loop's area to point upward. The loop's area increases as the rod moves away from the bulb. Because of this, the magnitude of the magnetic flux through the loop's area is increasing as the rod moves to the right.

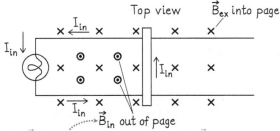

Top view $\vec{B}_{ex}$ into page

$\vec{B}_{in}$ out of page

The $\vec{B}_{ex}$ flux increases because the area with $\vec{B}_{ex}$ is increasing.

Simplify and diagram Assume that the rails, rod, and connecting wires have zero resistance. The induced field $\vec{B}_{in}$ due to the loop's induced current should point upward, resisting the change in the downward increasing magnetic flux through the loop's area. Using the right-hand rule for the $\vec{B}_{in}$ field, we find that the direction of the induced current is counterclockwise.

Represent mathematically To find the magnitude of the induced emf, use Faraday's law:

$$\mathcal{E}_{in} = N\left|\frac{\Phi_f - \Phi_i}{t_f - t_i}\right|$$

The angle between the loop's normal line and the B_{ex} field is 180°. The magnitude of the B_{ex} field is constant. Therefore,

$$\mathcal{E}_{in} = (1)\left|\frac{B_{ex}A_f\cos(180°) - B_{ex}A_i\cos(180°)}{t_f - t_i}\right|$$

$$= B_{ex}\left|\frac{A_f - A_i}{t_f - t_i}\right|$$

The area A of the loop at a particular clock reading equals the length L of the sliding rod times the x-coordinate of the sliding rod (the origin of the x-axis is placed at the bulb.) Therefore,

$$\mathcal{E}_{in} = B_{ex}\left|\frac{Lx_f - Lx_i}{t_f - t_i}\right| = B_{ex}L\left|\frac{x_f - x_i}{t_f - t_i}\right| = B_{ex}Lv$$

The quantity inside the absolute value is the x-component of the rod's velocity, the absolute value of which is the rod's speed v. Using Ohm's law, the induced current depends on the induced emf and the bulb resistance:

$$I_{in} = \frac{\mathcal{E}_{in}}{R}$$

The power output of the lightbulb is

$$P = I_{in}^2 R$$

Solve and evaluate The magnitude of the induced emf around the loop is

$$\mathcal{E}_{in} = B_{ex}Lv = (0.20\text{ T})(1.2\text{ m})(6.0\text{ m/s}) = 1.44\text{ V}$$

The current in the bulb is

$$I_{in} = \frac{1.44\text{ V}}{10\text{ Ω}} = 0.14\text{ A}$$

The lightbulb should glow: its power output is

$$P = I_{in}^2 R = (0.14\text{ A})^2(10\text{ Ω}) = 0.21\text{ W}$$

Try it yourself Suppose the rod in this example moves at the same speed but in the opposite direction, so that the loop's area decreases. The rails make a 45° angle with the direction of the $\vec{B}$ field. Determine the magnitude of the induced emf, the magnitude of current in the bulb, and the direction of the current.

Answer   1.0 V; 0.1 A; clockwise.

Limitless electric energy?

We have just found that the induced emf depended on the speed with which a metal rod is pulled along the rails. By accelerating the rod to a speed of our choosing, we could induce whatever emf we desired. Would this method of obtaining the emf by moving a conductor in the external magnetic field maintain the current indefinitely?

FIGURE 21.13 An alternative way to analyze the motion of the rod in terms of electric and magnetic forces.

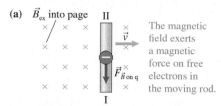

(a) $\vec{B}_{ex}$ into page

The magnetic field exerts a magnetic force on free electrons in the moving rod.

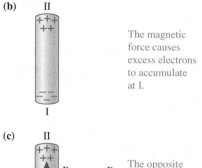

(b)

The magnetic force causes excess electrons to accumulate at I.

(c)

$F_{\vec{E} \text{ on } q} = qE$

$F_{\vec{B} \text{ on } q} = qvB$

The opposite charges at I and II produce an electric force that balances the magnetic force.

FIGURE 21.14 A simple version of an electric generator.

The steam turns the turbine, which turns the wire loop with respect to the magnet.

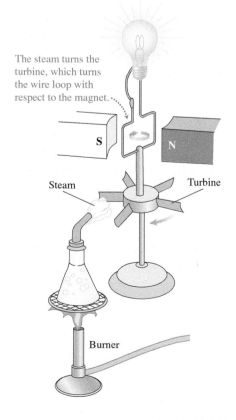

Steam

Turbine

Burner

The moving rod resulted in an emf and an induced current in the rails, bulb, and rod. However, if we use the right-hand rule for magnetic force, we find that the magnetic field exerts a force on the induced current in the rod toward the left, causing it to slow down. From an energy perspective, the kinetic energy of the rod is being transformed into light and thermal energy in the bulb. In order to keep the rod moving at constant speed, some other object must exert a force on it to the right, doing positive work on the rod.

Motional emf

The emf produced in Example 21.7 is sometimes called **motional emf**; it is caused by the motion of an object through the region of a magnetic field. We explained this emf using the idea of electromagnetic induction. Is it possible to understand it just in terms of magnetic forces? When an electrically charged object with charge q moves within a region with nonzero $\vec{B}$ field, the field exerts a magnetic force on it ($F_{\vec{B} \text{ on } q} = |q|vB \sin\theta$). Consider a metal rod, shown in **Figure 21.13a**, moving at velocity $\vec{v}$ in the horizontal direction. The external magnetic field $\vec{B}_{ex}$ points into the page. Inside the rod are fixed positively charged ions and negatively charged free electrons. When the rod slides to the right, all of its charged particles move with it. According to the right-hand rule for the magnetic force, the external magnetic field exerts a magnetic force on the electrons toward end I. The positive charges cannot move inside the rod, but the free electrons can. The electrons accumulate at end I, leaving end II with a deficiency of electrons (a net positive charge). The ends of the rod become charged, as shown in Figure 21.13b.

These separated charges create an electric field $\vec{E}$ in the rod that exerts a force of magnitude $F_{\vec{E} \text{ on } q} = qE$ on other electrons in the rod; the electric field exerts a force on negative electrons toward II (Figure 21.13c) opposite the direction of the magnetic force. When the magnitude of the electric force equals the magnitude of the magnetic force, the accumulation of opposite electric charge at the ends of the rod ceases. Then

$$qvB = qE \quad \text{or} \quad E = vB$$

Magnetic force   Electric force

An electric potential difference is produced between points I and II that depends on the magnitude of the electric field $\vec{E}$ in the rod and the distance L between ends I and II:

$$\mathcal{E}_{\text{motional emf}} = |\Delta V_{\text{I–II}}| = EL = vBL \tag{21.4}$$

This expression for motional emf is the same expression we derived in Example 21.7 using Faraday's law. Thus, for problems involving conducting objects moving in a magnetic field, we can use either Faraday's law or the motional emf expression to determine the emf produced—either method will provide the same result.

The electric generator: changing the loop's orientation

Worldwide, we convert an average of 1000 J per person of electric potential energy into other less useful energy forms every second. Electric generators make this electric potential energy available by converting mechanical energy (such as water rushing through a hydroelectric dam) into electric potential energy.

To understand how an electric generator works, consider a very simple device that consists of a loop of wire attached to a turbine (a propeller-like object that can rotate). The loop is positioned between the poles of an electromagnet that produces a steady uniform $\vec{B}$ field. A Bunsen burner next to the turbine heats a flask of water (**Figure 21.14**). The water is converted to steam, which strikes the blades of the turbine, causing the turbine to rotate. The loop of wire attached to the turbine rotates in the $\vec{B}$ field region. When the loop's surface is perpendicular to the $\vec{B}$ field, the magnetic flux through the loop's area is at a maximum. One quarter turn later, the $\vec{B}$ field lines are parallel to the loop's area, and the flux through it is zero. After another quarter turn, the flux is again at its maximum magnitude, but negative in value since the orientation of the loop's area is opposite what it was originally. This changing magnetic flux through the loop's area causes a corresponding induced emf, which produces a current that changes direction each time the loop rotates one half turn. Current that periodically changes direction in this way is known as *alternating current* (AC).

A coal-fired power plant is based on this process. Coal is burned to heat water, converting it to steam. The high-pressure steam pushes against turbine blades, causing the turbine and an attached wire coil to rotate in a strong $\vec{B}$ field. The resulting emf drives the electric power grid. Nuclear, natural gas, and geothermal power plants use the same procedure except they use different fuel to heat water.

Emf of a generator

To determine an expression for the emf produced by an electric generator, consider the changing magnetic flux through a loop's area as it rotates with constant rotational speed ω in a constant uniform $\vec{B}_{ex}$ field (**Figure 21.15**). If there is an angle θ between the loop's normal vector and the $\vec{B}_{ex}$ field, then the flux through the loop's area is

$$\Phi = B_{ex} A \cos\theta$$

Since the loop is rotating, θ is continuously changing. The loop is rotating with zero rotational acceleration ($\alpha = 0$); we can describe the motion with rotational kinematics (see Chapter 9):

$$\theta = \theta_0 + \omega t + \tfrac{1}{2}\alpha t^2 = \theta_0 + \omega t + \tfrac{1}{2}(0)t^2 = \theta_0 + \omega t$$

If we define the initial orientation θ_0 to be zero, then

$$\theta = \omega t$$

where ω is the constant rotational speed of the loop. This means that the magnetic flux Φ through the loop's area as a function of time t is

$$\Phi = B_{ex} A \cos(\omega t)$$

For a side view of the rotating loop, see **Figure 21.16a**. Figure 21.16b shows a graph of the magnetic flux through the loop's area as a function of time.

According to Faraday's law Eq. (21.2), the induced emf around a coil (a multi-turn loop) is

$$\mathcal{E}_{in} = N \left| \frac{\Delta\Phi}{\Delta t} \right| = N \left| \frac{\Phi_f - \Phi_i}{t_f - t_i} \right|$$

Since Φ is continually changing, we should use calculus to write the above equation. However, in this text, we will simply show the result:

$$\mathcal{E}_{in} = N B_{ex} A\omega \sin(\omega t) \qquad (21.5)$$

where N is the number of turns in a coil rotating between the poles of the magnet.

Figure 21.16c shows a graph of the induced emf as a function of clock reading. Comparing Figure 21.16b and c, you will see a pattern. The value of $\mathcal{E}_{in}$ at a particular clock reading equals the negative value of the slope of the Φ-versus-t graph at that same clock reading. This makes sense, since the induced emf is related to the rate of change of the magnetic flux through the loop's area. Slopes represent exactly that, rates of change.

Electric power plants in the United States produce an emf with frequency f equal to 60 Hz (Hz is a unit of frequency; 60 Hz means the emf undergoes 60 full cycles in 1 second). This corresponds to a generator coil with a rotational speed

$$\omega = 2\pi f = 2\pi(60\ \text{Hz}) = 120\pi\ \text{rad/s}$$

FIGURE 21.15 The magnetic flux through the loop changes as the angle between the normal vector to the loop and the $\vec{B}$ field lines changes.

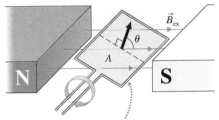

As the loop rotates in the magnetic field, the flux through the loop continually changes.

FIGURE 21.16 Three representations of the loop in Figure 21.15.

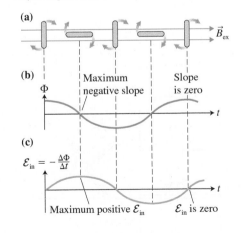

These power plants can produce a peak (maximum) emf as high as 20 kV. The peak emf produced by a generator occurs when $\sin(\omega t) = 1$ and when $\sin(\omega t) = -1$. At those times,

$$\mathcal{E}_{\text{in max}} = NB_{\text{ex}}A\omega. \qquad (21.6)$$

REVIEW QUESTION 21.5 How does the law of electromagnetic induction explain why there is an induced emf in a rotating generator coil?

21.6 AC circuits

In the previous sections we learned that electromagnetic induction allows us to produce variable emf, and thus **alternating current**, or AC (we also use the term AC when we talk about variable voltage). As we discussed above, U.S. electric power plants produce sinusoidally changing emf with a frequency of 60 Hz. Driven by the variable emf, the electric current in household wires also changes magnitude and direction sinusoidally 60 times per second. Direct current (DC) remains constant in direction.

Today, all home appliances operate using AC, but the path to this outcome was not straightforward. Why is AC better for general (massive) use than DC? The battle between AC and DC emerged at the end of the 19th century, when the understanding of electricity and magnetism and the level of related technology were sufficient to enable practical applications. There are two important figures in this story, American physicists Nikola Tesla (Serbian born) and Thomas Edison. Edison was the proponent of DC, and Tesla of AC. The battle was finally won by AC because it proved to be much more efficient in transferring electric power over large distances. We will learn how this is done in the next section, but first we need to study some basic phenomena that are characteristic of AC.

Before we begin, we will agree to call a regular battery a DC power supply and a device that produces alternating voltage an AC power supply.

Resistors in AC circuits

We start with experimental investigations comparing the properties of resistors in DC and AC circuits.

FIGURE 21.17 An incandescent lightbulb in a DC and in an AC circuit.

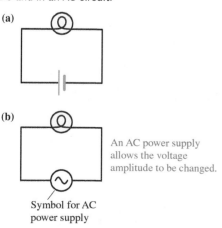

(a)

(b)

An AC power supply allows the voltage amplitude to be changed.

Symbol for AC power supply

Experiment 1. We connect an incandescent lightbulb to a DC power supply and measure the potential difference across it to be ΔV_{DC} (**Figure 21.17a**). Then we connect an identical lightbulb to an AC power supply (with a frequency of 60 Hz) and change the amplitude of the variable voltage until the bulb appears equally bright (Figure 21.17b). We use a special kind of voltmeter called an oscilloscope, which shows on the screen how voltage changes with time. We measure the amplitude of the variable voltage across the AC power supply ΔV_{AC} and find that when the bulbs appear equally bright, $\Delta V_{\text{AC}} \approx 1.4\,\Delta V_{\text{DC}}$.

Does this result make sense? Qualitatively, it does. The voltage across the DC power supply is constant all the time, while the voltage across the AC power supply is changing 60 times every second. Therefore, most of the time the voltage across the bulb is less than the amplitude. The higher amplitude AC voltage should thus produce the same warming effect as the lower DC voltage.

Can we explain the factor of 1.4? The energy transferred from the DC power supply to the lightbulb during time interval Δt is equal to the product of power and time interval:

$$P_{\text{DC}}\,\Delta t = \frac{\Delta V_{\text{DC}}^2}{R}\,\Delta t$$

We can represent this energy as the area under the plotted line on the P-versus-t graph (**Figure 21.18a**). We use the same method to determine the energy transferred to the bulb by the AC power supply:

$$P_{AC} \, \Delta t = \frac{\Delta V_{AC}^2 \sin^2(\omega t)}{R} \Delta t$$

The graph of AC power as a function of time is shown in Figure 21.18b. Note that the power is never negative because it follows the square of the sinusoidal function. The output power of the bulb is continuously changing from zero to the maximal value, 120 times a second. We do not detect these changes in brightness because our eyes can only see variations that happen about 24 or fewer times per second. For us, the bulb appears to glow with constant brightness. Therefore, it makes sense to replace this time-varying power with an equivalent average constant power (remember, we did this before in Chapter 12, when we derived the relation between the speed of gas molecules and the pressure on the container wall).

As you can see in Figure 21.18b, the area under the sinusoidal curve is equal to the area of the shaded rectangle whose height is equal to the half of the maximal power. This means that the average output power of a resistive element that is connected to an AC source is equal to half of the maximal power output of that element.

Now we can write

$$\overline{P}_{AC} \, \Delta t = \frac{\Delta V_{AC}^2}{2R} \Delta t$$

Comparing this equation with the equation

$$P_{DC} \, \Delta t = \frac{\Delta V_{DC}^2}{R} \Delta t$$

we see that when both power supplies transfer the same energy to the bulb,

$$\overline{P}_{AC} \, \Delta t = P_{DC} \, \Delta t, \, \frac{\Delta V_{AC}^2}{2R} \Delta t = \frac{\Delta V_{DC}^2}{R} \Delta t \implies \Delta V_{AC} = \sqrt{2} \, \Delta V_{DC} \approx 1.4 \, \Delta V_{DC}$$

This is in agreement with the earlier result we found experimentally. The DC power supply has to have a smaller voltage across it to produce the same average power as the AC power supply. This finding has important practical value because it reduces the analysis of AC circuits with resistive elements to the analysis of the equivalent DC circuits, which we have already learned. For this reason, we define **root-mean-square voltage** ΔV_{rms} as

$$\Delta V_{rms} = \frac{1}{\sqrt{2}} \Delta V_{AC} \qquad (21.7)$$

We could do a similar analysis for the current through the bulb using the relation $P = I^2 R$ and define **root-mean-square current** I_{rms} as

$$I_{rms} = \frac{1}{\sqrt{2}} I_{AC} \qquad (21.8)$$

Using these new quantities, we can write the average power as

$$P_{AC} = I_{rms} \, \Delta V_{rms}$$

The above equation has the same form as our DC definition of power; therefore, we can use rms values to analyze AC circuits the same way we did with DC circuits.

What is happening inside the wires and other resistive elements in AC circuits? In Chapter 19, we learned that free electrons in metals accelerate in the presence of an external electric field. When a constant potential difference is applied across a metal conductor and there is direct current through it, the free electrons, drifting in one direction (due to the DC), interact with ions of the crystal lattice, and the whole wire warms up. What happens when there is an alternating current through the wire? The changing electric field in the wire exerts a variable force on the free electrons,

FIGURE 21.18 Power-versus-time graphs for an incandescent lightbulb in a DC and in an AC circuit.

(a)

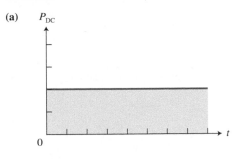

(b)

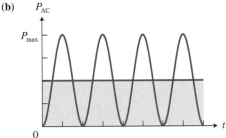

which in turn vibrate. While vibrating, the free electrons interact with the ions of the lattice and transfer some energy to them. Note that in direct current, the free electrons move along the wire from one terminal to another terminal of the power supply, while in alternating current, all free electrons just vibrate. It is due to these vibrations that the wires warm up in AC circuits.

How are current through and voltage across the resistor related? Imagine that you apply variable voltage across a metal resistor. Free electrons inside the metal immediately respond to the changes in the electric field. Therefore, current through and voltage across the resistor vibrate in phase (they have the same period T, as shown in **Figure 21.19**, where the two quantities are presented on the same time axis).

The instantaneous values of the current through and voltage across the resistor are related by Ohm's law:

$$I_R(t) = \frac{\Delta V_R(t)}{R}$$

Since the rms values are directly proportional to the peak values, we can write

$$I_{R\,rms} = \frac{\Delta V_{R\,rms}}{R}$$

FIGURE 21.19 Current through and voltage across a metal resistor in an AC circuit.

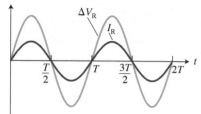

The current and the voltage are in phase.

VIDEO
21.1

Capacitors in an AC circuit

To investigate the behavior of capacitors in AC circuits, we will conduct the following experiments.

Experiment 1. Connect an incandescent lightbulb and a capacitor in series to a DC power supply. The lightbulb does not glow (**Figure 21.20a**). This makes sense—the capacitor behaves like an open switch. If we connect a wire across the capacitor, the lightbulb glows.

Experiment 2. Connect the (same) lightbulb and the capacitor to an AC power supply (see the video; the experiment in the video starts with a frequency of 9 MHz). The lightbulb glows, but not very brightly (Figure 21.20b).

Experiment 3. Increase the frequency of the AC, keeping the amplitude the same as in Experiment 2. The lightbulb becomes brighter (Figure 21.20c).

Experiment 4. Set the AC frequency to the initial value and place a dielectric between the capacitor plates (as you know, this will increase its capacitance). The bulb glows even more brightly than in Experiment 2 (Figure 21.20d).

Based on the above experiments, we can conclude that **a capacitor behaves like a resistor in an AC circuit.** The larger the value of C, the smaller the resistance of the capacitor; the larger the frequency, the smaller the resistance of the capacitor. This resistive property of a capacitor is called **capacitive reactance** and is denoted with the symbol X_C.

How can we explain why capacitors conduct AC but not DC? As we discussed earlier, the electrons do not travel but just vibrate when a wire is connected across an AC power supply. As the electrons in the conducting wires vibrate, they charge and discharge the capacitor. The fact that the electrons cannot move across the capacitor (from one plate to the other) does not prevent electrons from vibrating in other parts of the wire, including the electrons in the bulb. Hence, the bulb can glow.

Why does the bulb glow more brightly when the AC frequency increases? With all other quantities unchanged, the larger AC frequency means that the same electric charge is transferred to and from the capacitor in a shorter time; therefore, there is a larger peak current and consequently a brighter bulb.

Why does the bulb glow more brightly when we use a capacitor with a larger capacitance? With all other quantities unchanged, the larger capacitance means that

FIGURE 21.20 An incandescent lightbulb and a capacitor in series in DC and AC circuits.

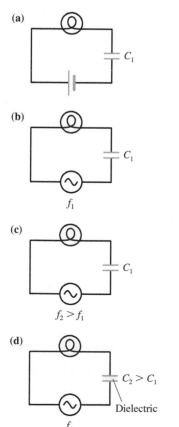

more charge is transferred to and from the capacitor plates (remember that $q = C\Delta V$), and that is why we observe a larger peak current and thus a brighter bulb.

How are current through and voltage across the capacitor related? Measurements with an oscilloscope connected to a capacitor connected to an AC power supply show time dependence of both variables: the current and voltage across the capacitor change sinusoidally with the same period T (these are both shown on the same time axis in Figure 21.21). However, the maximal values of current and voltage are achieved at different times.

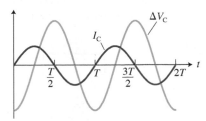

FIGURE 21.21 Current through and voltage across a capacitor in an AC circuit.

The current peaks before the voltage.

Explanation. The current to the capacitor plates is equal to the electric charge that is transferred to the plates divided by the time interval for that amount of charge to pass. In other words, the current is equal to the rate of change of charge on the capacitor plates. Because the charge on the capacitor plates is directly proportional to the voltage across the capacitor, we can say that the current through the capacitor is proportional to the rate of change of the voltage across the capacitor. The current is maximal when the rate of change of the voltage is maximal. Since the rate of change of sinusoidally changing voltage is largest every time it passes through zero (the steepest part of the sine function), the current through and the voltage across the capacitor are not in phase. They should be shifted by one quarter of a period $(T/4)$. This is exactly what is observed: Figure 21.21 shows that the peak values of the current are to the left of the peak values of the voltage.

Quantitative analysis of capacitive reactance The derivation of the cause-effect expression for capacitive reactance is beyond the scope of this book. We therefore show it without the derivation but do discuss how we can evaluate the expression. Capacitive reactance is given by

$$X_C = \frac{1}{2\pi f C} \tag{21.9}$$

where $f\ (=1/T)$ is the frequency of the AC power supply and C is the capacitance of the capacitor. To make sense of this expression, we first look at the units. Remember that the unit for frequency is Hz (1 Hz is equal to $1/s$); the unit of capacitance is F (1 F is equal to 1 C/1 V, and 1 C is equal to 1 A $\times$ 1 s). Looking at the right side of Eq. (21.9), we get the units

$$\frac{1}{\text{Hz} \cdot \text{F}} = \frac{\text{s} \cdot \text{V}}{\text{C}} = \frac{\text{V}}{\text{A}} = \Omega$$

The units are therefore correct. The expression is also consistent with our previous observations when the capacitance and frequency were increased independently: both lead to larger current in the circuit. We can next check the extreme cases. When the frequency is close to zero, the reactance should be infinite because the capacitor becomes an open switch; this is exactly what the expression predicts. When the capacitance is close to zero, the reactance should be infinite because no charge will accumulate on the plates.

We can relate the reactance X_C to $I_{C\,\text{rms}}$ through and $\Delta V_{C\,\text{rms}}$ across the capacitor with an expression that is equivalent to Ohm's law:

$$I_{C\,\text{rms}} = \frac{\Delta V_{C\,\text{rms}}}{X_C} \tag{21.10}$$

We can use Eq. (21.10) to write an operational definition for capacitive reactance:

$$X_C = \frac{\Delta V_{C\,\text{rms}}}{I_{C\,\text{rms}}} \tag{21.11}$$

Note that the *instantaneous* voltage across and current through the capacitor are not related to the reactance as described above since they are out of phase.

VIDEO
21.2

Solenoids in AC circuits

Another important element of AC circuits is a solenoid. We will investigate its behavior in the following experiments.

Experiment 1. We connect a lightbulb to a battery using a long straight wire. The lightbulb glows. We then prepare another setup using an identical lightbulb and wire, but this time we connect them to an AC voltage source. We adjust the AC voltage amplitude so the lightbulbs in both circuits are equally bright (**Figure 21.22a**). In the experiment in the video, the frequency is about 1 MHz. Now we change the shape of the long wire by winding it as a solenoid without changing its length (the solenoid is often called an *inductor*). The brightness of the bulb connected to the DC power supply does not change, while the bulb connected to the AC power supply glows more dimly (Figure 21.22b).

Experiment 2. In the same circuit, we increase the AC frequency while keeping the amplitude constant. The lightbulb connected to the AC source glows even more dimly (Figure 21.22c). Now we return the AC frequency to its initial value and insert an iron core into the solenoid. The bulb glows more dimly than with the inductor without the iron core in the last step in Experiment 1 (compare Figure 21.22d and 21.22b).

From these experiments, we can conclude that the inductor behaves like a resistor in an AC circuit. The inductor must possess resistive properties that depend on the properties of the inductor and the frequency of the AC, similar to a capacitor. This resistive property of an inductor is called **inductive reactance** and is denoted with the symbol X_L.

In order to understand how the properties of the inductor affect its inductive reactance, we need to go back to our studies of magnetic fields. In Chapter 20 we learned that the magnetic field inside a solenoid (assumed to be very long) is described by the equation

$$B_{\text{solenoid}} = \mu_0 \frac{N}{l} I$$

where N is the number of turns, l is the length of the solenoid, and μ_0 is the magnetic permeability of vacuum $(\mu_0 = 4\pi \times 10^{-7}(\text{T} \cdot \text{m})/\text{A})$. Knowing the magnetic field inside the solenoid, we can write an expression for the magnetic flux produced by the solenoid when there is current I through it:

$$\Phi = NBA = N\left(\mu_0 \frac{N}{l} I\right)A = \frac{\mu_0 N^2 A}{l} I \qquad (21.12)$$

We see that the magnetic flux depends on the current through the solenoid and a combination of physical quantities dependent only on the geometry of the solenoid and the medium inside it. We can rewrite the above equation as

$$\Phi = LI$$

where L is a quantity called **inductance** that contains only properties of the solenoid:

$$L = \frac{\mu_0 N^2 A}{l} = \mu_0 \left(\frac{N}{l}\right)^2 V \qquad (21.13)$$

Here, V is the volume of the solenoid $(V = Al)$. The symbol L for inductance is in honor of Heinrich Lenz, and the unit for inductance, the henry (H), is in honor of Joseph Henry.

As we noted above, μ_0 in Eq. (21.13) is the magnetic permeability of vacuum. If we place an iron core inside the solenoid, the magnetic flux in the solenoid increases by a factor of about 5000. This is also the constant by which we need to multiply μ_0 to find the inductance of a solenoid with an iron core. The constant is called the **relative magnetic permeability** μ_r. The relative magnetic permeability of iron is around 5000.

FIGURE 21.22 An incandescent lightbulb and wires of different shapes in series in DC and AC circuits.

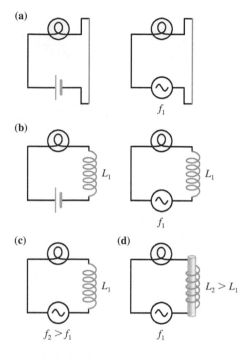

(a)

(b)

(c)

(d)

When we place a material of relative permeability μ_r inside the solenoid, Eq. (21.13) becomes

$$L = \frac{\mu_0 \mu_r N^2 A}{l} = \mu_0 \mu_r \left(\frac{N}{l}\right)^2 V$$

Equation (21.13) is the cause-effect relationship for inductance because every physical quantity in it can be varied separately. The inductive properties of a solenoid can be increased by increasing the number of turns per unit length of the solenoid, by increasing the volume of the solenoid, or by filling the inside with a material with a higher relative magnetic permeability.

Now that we have learned the new physical quantity of inductance, we can develop a microscopic explanation of the observed behavior of an inductor in an AC circuit. When an inductor is connected to an AC power supply, the electrons in the wire of the solenoid vibrate. They produce a changing magnetic field, which in turn creates an electric field (induced emf) that opposes the changes in the motion of the same electrons (by Lenz's law). Hence, the bulb glows more dimly.

To explain why the bulb glows more dimly at a larger AC frequency, we need to consider what larger frequency means for the above process. If everything else is unchanged, the larger AC frequency means faster changes of the current and a larger induced emf, which opposes the voltage produced by the AC source and therefore results in a dimmer bulb.

To explain why the bulb glows more dimly when we use an inductor with a larger L, we need to consider how the inductance affects the above process. If everything else is unchanged, the larger L of the solenoid means a larger B inside the solenoid and a larger magnetic flux produced by the solenoid for a given current through it ($\Phi = LI$). Larger magnetic flux means a larger induced emf that opposes the changes in AC source voltage, and this leads to a dimmer bulb.

How are current through and voltage across the inductor related? Using the same equipment that we used for the capacitor, we can record with an oscilloscope the time dependence of both variables (these are both presented in **Figure 21.23**, sharing the same time axis).

Explanation. We assume that the resistance of the inductor is negligible (in real life, this is often not true, which is why real inductors are more complicated to describe than real capacitors). In this case, the voltage across the inductor is equal to the induced emf due to the changing magnetic flux through the inductor. We know that the induced emf is proportional to the rate of change of the magnetic flux through the inductor (for example, a solenoid). Since the magnetic flux is directly proportional to the current through the inductor, we can say that the induced emf across the inductor is proportional to the rate of change of the current through the inductor. Since the rate of change of a sinusoidally vibrating current is largest every time it passes through zero, the voltage across and the current through the inductor are not in phase. They should be shifted by one quarter of a period ($T/4$). This is exactly what we observe: Figure 21.23 shows that the peak values of the current are to the right of the peak values of the voltage.

Quantitative analysis of inductive reactance The derivation of the cause-effect expression for inductive reactance is beyond the scope of this book. We therefore show it without the derivation, but do discuss how we can evaluate the expression. The expression for the inductive reactance is

$$X_L = 2\pi f L \tag{21.14}$$

You can check whether the units come out as ohms and whether the limiting cases predict reasonable results in the same way we did for capacitive reactance. Note that, unlike capacitive reactance, inductive reactance increases with both the frequency of current through the inductor and the inductance.

FIGURE 21.23 Current through and voltage across an inductor in an AC circuit.

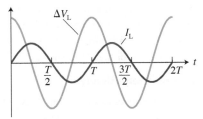

The current peaks after the voltage.

TIP   In observational experiments with capacitors and inductors, we were using high-frequency AC (frequency in the range of MHz) to make the phenomena easily observable. However, all the expressions given also hold for lower frequencies, such as 60 Hz.

The inductive reactance X_L is related to the $I_{L\,rms}$ through and $\Delta V_{L\,rms}$ across the inductor with an expression that is equivalent to Ohm's law:

$$I_{L\,rms} = \frac{\Delta V_{L\,rms}}{X_L} \tag{21.15}$$

Note that the *instantaneous* voltage across and current through the inductor are not related to the inductive reactance as described above since they are out of phase.

One important consequence of the phase shift between the current through and voltage across an ideal inductor or a capacitor is that the power output of these elements (the product of voltage and current) is changing sinusoidally. Therefore, the average power output of the capacitor or inductor connected to AC source is zero. A mechanical analogy would be a spring that is first compressed and then stretched by an external object, the average work (and thus the power) done on the spring is zero.

EXAMPLE 21.8   A capacitor in an AC circuit

You have a set of different capacitors. (a) What should be the capacitance of a capacitor that allows a 20-mA rms current to pass through when it is connected to a household electric outlet? (b) If the maximum of the current appears at $t_0 = 0$, at what later clock reading will the first maximum of voltage across the capacitor occur? (c) What is the value of the maximum voltage?

Sketch and translate We sketch the circuit at right. The knowns include $I_{C\,rms} = 20$ mA and the house AC. We need to determine the capacitance, t for the maximum voltage, and its value.

$\Delta V_{C\,rms} = 120$ V
$C = ?$
$I_{C\,rms} = 20$ mA
$f = 60$ Hz

Simplify and diagram Properties of household AC voltage depend on the country. In the United States, household power outlets provide AC voltage vibrating with a frequency of 60 Hz and an rms voltage of 120 V. In Europe, most of Africa, most of Asia, most of South America, and Australia, AC voltage vibrates at 50 Hz and rms voltage of 230 V. We will use the data for the United States.

Represent mathematically

(a) We have two equations for the capacitive reactance, an operational definition and a cause-effect relationship:

$$X_C = \frac{\Delta V_{C\,rms}}{I_{C\,rms}} \quad \text{and} \quad X_C = \frac{1}{2\pi f C}$$

We determine the reactive capacitance using the first equation and use the second to determine the capacitance.

(b) The maximum voltage across the capacitor appears one quarter of a period later than the maximum current. Remember that $T = \dfrac{1}{f}$.

(c) To find the maximum voltage, we can use the rms value $\Delta V_{C\,max} = \sqrt{2}\,\Delta V_{C\,rms}$.

Solve and evaluate

(a) $X_C = \dfrac{\Delta V_{C\,rms}}{I_{C\,rms}} = \dfrac{120\text{ V}}{0.020\text{ A}} = 6000\ \Omega$

$\Rightarrow C = (2\pi \times 60\text{ s}^{-1} \times 6000\ \Omega)^{-1} = 4.4 \times 10^{-7}$ F

(b) $T = \dfrac{1}{f} = 0.0167$ s; $t = \dfrac{1}{4}T = 0.0042$ s.

(c) The maximum voltage across the capacitor is $\Delta V_{C\,max} = \sqrt{2}\,\Delta V_{C\,rms} = 170$ V.

- -

Try it yourself Repeat the problem above but this time for an inductor (replace the words "capacitor" with "inductor" and "capacitance" with "inductance"). Assume that the resistance of the inductor is negligible.

Answer $L = 15.9$ H, $t = 0.0125$ s, and $\Delta V_{L\,max} = 170$ V.

REVIEW QUESTION 21.6 A capacitor in an electric circuit works as an open switch because its plates are separated by a dielectric material. How can there be current in a circuit with an open switch?

21.7 Transformers

Knowledge of AC helps us understand the operation of an important device, the **transformer**, which allows us to increase or decrease the maximum value of an alternating emf.

A transformer consists of two coils, each wrapped around an iron core (ferromagnetic) (**Figure 21.24**). An alternating emf across the **primary coil** is converted into a larger or smaller alternating emf across the **secondary coil**, depending on the number of loops in each coil.

Transformers are used in many electronic devices. They are also essential for transmitting electric energy from a power plant to your house. The rate of this electric energy transmission is proportional to the product of the emf across the power lines and the electric current in the lines. If the emf is low, considerable electric current is needed to transmit a considerable amount of energy. However, due to the electrical resistance of the power lines, much of the electric energy is converted into thermal energy. The rate of this conversion is $P = I^2R$. To reduce the I^2R losses, the transmission of electric energy is done at high peak emf (about 200,000 V) and low current. Transformers then reduce this peak emf to about 170 V for use in your home. How does a transformer change the peak emf?

Suppose there is an alternating current in the primary coil, the coil connected to an external power supply. The secondary coil is connected to an electrical device, but this device requires an emf that is different from what the external power supply produces. The alternating current in the primary coil produces a $\vec{B}$ field within the transformer core. Since the current is continuously changing, the magnetic flux through the primary coil's area is also continuously changing. Thus, an emf is induced in the primary coil:

$$\mathcal{E}_p = N_p \left| \frac{\Delta \Phi_p}{\Delta t} \right|$$

where the p subscript refers to the primary coil.

The primary coil is connected to an AC power supply. Assuming that the resistance of the coil is zero, the emf of the power supply and the induced emf in the coil should add to zero according to Kirchhoff's loop rule. Therefore, the emf across the primary coil is the same in magnitude as the emf of the AC power supply.

In an efficient transformer, nearly all of the $\vec{B}$ field lines passing through the primary coil's area also pass through the secondary coil's area. This is due to the presence of the ferromagnetic core around which both coils are wound. The core serves as a "guide" for the magnetic field. As a result, there is a changing magnetic flux through the secondary coil's area as well as a corresponding emf produced in it:

$$\mathcal{E}_s = N_s \left| \frac{\Delta \Phi_s}{\Delta t} \right|$$

If the transformer is perfectly efficient, then the rates of change of the magnetic flux through one turn of each coil are the same:

$$\left| \frac{\Delta \Phi_p}{\Delta t} \right| = \left| \frac{\Delta \Phi_s}{\Delta t} \right|$$

Combining the three equations above,

$$\frac{\mathcal{E}_p}{N_p} = \frac{\mathcal{E}_s}{N_s}$$

$$\Rightarrow \mathcal{E}_s = \frac{N_s}{N_p} \mathcal{E}_p \tag{21.16}$$

We see that the emf in the secondary coil can be substantially larger or smaller than the emf in the primary coil depending on the number of turns in each coil. Engineers use this result to design transformers for specific purposes. For example, a **step-down transformer** can convert the 120-V alternating emf from a wall socket to a 9-V alternating emf, which is then converted to DC to power a laptop computer.

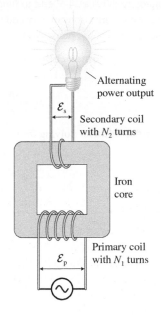

FIGURE 21.24 A transformer changes the input/output emf depending on the ratio of the turns.

Alternating power output

$\mathcal{E}_s$

Secondary coil with N_2 turns

Iron core

Primary coil with N_1 turns

$\mathcal{E}_p$

QUANTITATIVE EXERCISE 21.9   **Transformer for laptop**

Your laptop requires a 24-V emf to function. What should be the ratio of primary coil turns to secondary coil turns if this transformer is to be plugged into a standard house AC outlet (effectively a 120-V emf)?

Represent mathematically The ratio we are looking for is related to the coil emfs by Eq. (21.16):

$$\mathcal{E}_s = \frac{N_s}{N_p}\mathcal{E}_p$$

Solve and evaluate Solving for the ratio and inserting the appropriate values:

$$\frac{N_p}{N_s} = \frac{\mathcal{E}_p}{\mathcal{E}_s} = \frac{120 \text{ V}}{24 \text{ V}} = 5$$

The primary coil needs to have five times the number of turns as the secondary coil. This is a step-down transformer, since the resulting secondary coil peak emf is lower than the primary coil peak emf.

Try it yourself If the primary coil had 200 turns, how many turns should the secondary coil have to reduce the peak emf from 120 V to 6 V?

Answer   10 turns.

Some transformers are designed to increase rather than decrease emf. In a car that uses spark plugs for ignition, a transformer converts the 12-V potential difference of the car battery to the 20,000-V potential difference needed to produce a spark in the engine's cylinder (**Figure 21.25**). The battery supplies a steady current in the transformer. An electronic switching system in the circuit can open the circuit, stopping the current in the primary coil in a fraction of a millisecond. This causes an abrupt change in the magnetic flux through the primary coil of the transformer, which leads to an induced emf in the secondary coil. The secondary coil is attached to a spark plug that has a gap between two conducting electrodes. When the potential difference across the gap becomes sufficiently high, the gas between the electrodes ionizes. When the ionized atoms recombine with electrons, the energy is released in the form of light—a spark that ignites the gasoline.

FIGURE 21.25 How a transformer converts the emf from a 12-V car battery to a 20,000-V spark.

1. Current in the primary circuit produces a magnetic field in the magnetic core.

2. Opening the switch causes an abrupt decrease in the primary current and in the core magnetic field ...

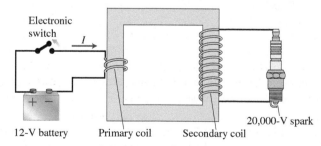

3. ... which causes a large induced potential difference in the secondary coil and dielectric breakdown (spark) across the spark plug gap.

The induced emf in the secondary coil is much greater than the 12 V in the primary coil for three reasons. First, the secondary coil has many more turns than the primary coil $(N_s \gg N_p)$. Second, the magnetic flux through the primary coil's area decreases very quickly (the Δt in the denominator of Faraday's law is very small), resulting in a large induced emf $\mathcal{E}_p$ to which $\mathcal{E}_s$ is proportional [see Eq. (21.2)]. Third, the ferromagnetic core (usually iron) passing through the two coils significantly increases the $\vec{B}$ field within it (Section 20.8). For these three reasons, it is possible for $\mathcal{E}_s \gg \mathcal{E}_p$; the 12-V car battery can produce a 20,000-V potential difference across the electrodes of a spark plug.

REVIEW QUESTION 21.7 How does a transformer achieve different induced peak emfs across its primary and secondary coils?

21.8 Mechanisms explaining electromagnetic induction

Faraday's law *describes* how a changing magnetic flux through a wire loop is related to an induced emf, but it does not *explain* how the emf comes about. In this section we will explain the origin of the induced emf.

A changing $\vec{B}_{ex}$ field has a corresponding $\vec{E}$ field

We know that a changing magnetic flux induces an electric current in a stationary loop (**Figure 21.26**). Because the loop is not moving, there is no net magnetic force exerted on the free electrons in the wire. Thus, an electric field must be present. The electric field that drives the current exists throughout the wire. This electric field is not produced by charge separation, but by a changing magnetic field.

If we were to represent it with $\vec{E}$ field lines, those lines would have no beginning or end—they would form closed loops. This electric field essentially "pushes" the free electrons along the loop. We can describe it quantitatively with the emf. But this emf is very different from the emf produced by a battery. For the battery, the emf is the result of charge separation across its terminals. For the induced emf, the electric field that drives the current does not originate on the charges but is distributed throughout the entire loop; thus the emf is also actually distributed throughout the entire loop. You might visualize it as an electric field "gear" with its teeth hooked into the electrons in the wire loop, pushing the free electrons along the wire at every point.

What do we now know about electricity and magnetism?

We have learned a great deal about electric and magnetic phenomena. We learned about electrically charged objects that interact via electrostatic (Coulomb) forces. Stationary electrically charged objects produce electric fields, and electric field lines start on positive charges and end on negative charges (**Figure 21.27a**). In our study of magnetism (Chapter 20), we learned that moving electrically charged objects and permanent magnets interact via magnetic forces and produce magnetic fields. Magnetic field lines do not have beginnings or ends (Figure 21.27b); they are continuous loops. So far, individual magnetic charges, magnetic monopoles, have not been found.

When we studied electric circuits (Chapter 19) we learned that electric fields cause electrically charged particles inside conductors to move in a coordinated way—electric currents. Later we learned that electric currents produce magnetic fields (Figure 21.27c). In this chapter, we learned about the phenomenon of electromagnetic induction and its explanation: a changing magnetic field is always accompanied by a corresponding electric field (Figure 21.27d). However, this new electric field is not produced by electric charges, and its field lines do not have beginnings or ends.

Except for the lack of magnetic charges, there is symmetry between electric and magnetic fields. This symmetry leads us to pose the following question: if in a region where the magnetic field is changing there is a corresponding electric field, is it possible that in a region where the electric field is changing there could be a corresponding magnetic field (Figure 21.27e)? This hypothesis, suggested in 1862 by James Clerk Maxwell, led to a unified theory of electricity and magnetism, a subject we will investigate in our chapter on electromagnetic waves (Chapter 25).

REVIEW QUESTION 21.8 Explain how (a) an electric current is produced when only a part of a single wire loop moves through a magnetic field and how (b) an electric current is produced when an external magnetic flux changes through a closed loop of wire.

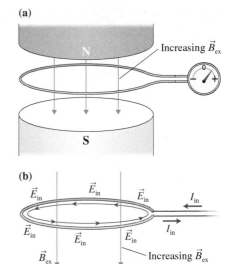

FIGURE 21.26 A changing $\vec{B}_{ex}$ creates an electric field $\vec{E}$ that induces an electric current.

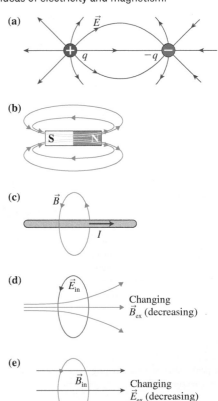

FIGURE 21.27 A summary of some of the main ideas of electricity and magnetism.

Summary

The **magnetic flux** through a loop's area depends on the size of the area, the $\vec{B}$ field magnitude, and the orientation of the loop relative to the $\vec{B}$ field. (Sections 21.2 and 21.4)

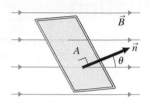

$$\Phi = BA \cos\theta \qquad \text{Eq. (21.1)}$$

when B is uniform through the loop area

Electromagnetic induction In a region with a changing magnetic field, there is a corresponding induced electric field. When a wire loop or coil is placed in this region, the magnetic flux through that loop changes, and an electric current is induced in the loop. (Section 21.4)

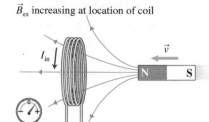

B_{ex} increasing at location of coil

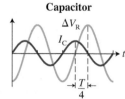

B_{ex} decreasing when switch is opened

$$\mathcal{E}_{in} = N\left|\frac{\Delta\Phi}{\Delta t}\right| = N\left|\frac{\Phi_f - \Phi_i}{t_f - t_i}\right|$$

$$\text{Eq. (21.2)}$$

Lenz's law When current is induced, its direction is such that the $\vec{B}$ field it produces opposes the change in the magnetic flux through the loop. (Section 21.3)

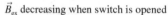
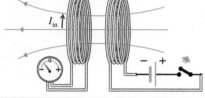
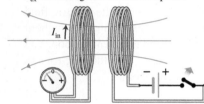

An **electric generator** produces an emf by rotating a coil within a region of strong $\vec{B}$ field, an important application of electromagnetic induction. (Section 21.5)

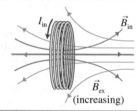

$$\mathcal{E}_{in} = NB_{ex}A\omega\sin(\omega t) \quad \text{Eq. (21.5)}$$

Alternating current (AC) circuits Alternating current is a current that varies in magnitude and direction, in contrast to direct current (DC), which remains constant in direction. AC is produced by special power supplies whose emf varies with frequency f. Resistors, capacitors, and inductors (solenoids) behave differently in DC and AC circuits. (Section 21.6)

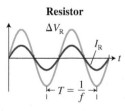
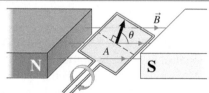
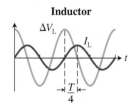

Resistor	Capacitor	Inductor

$$X_C = \frac{1}{2\pi fC} \qquad \text{Eq. (21.9)} \qquad X_L = 2\pi fL \qquad \text{Eq. (21.14)}$$

$$I_{C\,rms} = \frac{\Delta V_{C\,rms}}{X_C} \quad \text{Eq. (21.10)} \qquad I_{L\,rms} = \frac{\Delta V_{L\,rms}}{X_L} \quad \text{Eq. (21.15)}$$

$$\Delta V_{rms} = \frac{1}{\sqrt{2}}\Delta V_{AC} \qquad \text{Eq. (21.7)}$$

$$I_{rms} = \frac{1}{\sqrt{2}}I_{AC} \qquad \text{Eq. (21.8)}$$

Transformers are electrical devices used to increase or decrease the peak value of an alternating emf. (Section 21.7)

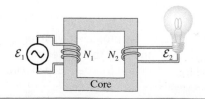

$$\mathcal{E}_s = \frac{N_s}{N_p}\mathcal{E}_p \qquad \text{Eq. (21.16)}$$

Questions

Multiple Choice Questions

1. In which of the experiments with a loop and a bar magnet shown in **Figure Q21.1** is an electric current induced in the loop at the instant shown in the figures? In experiments (d) to (f), the magnet rotates around the given axis; in (g) to (i), the loop collapses as indicated by the arrows.

FIGURE Q21.1

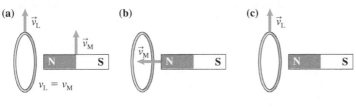

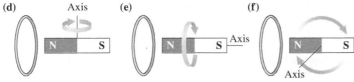

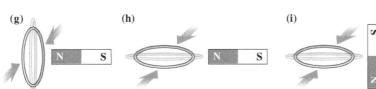

2. If you move the coil in **Figure Q21.2** toward the N pole of the large electromagnet, which produces a $\vec{B}$ field that is constant (in time) and uniform (in space), would an electric current be induced?
   (a) Yes
   (b) No
   (c) It depends on the current in the electromagnet.

FIGURE Q21.2

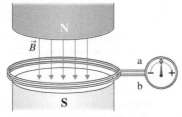

3. The magnetic flux through a 10.0-cm^2 loop is 8.0 T·cm^2, and the magnetic field at the location of the loop is uniform. Choose all correct answers.
   (a) The magnitude of the $\vec{B}$ field must satisfy the condition $B \le 0.8$ T.
   (b) The direction of the $\vec{B}$ field cannot be perpendicular to the surface of the loop.
   (c) If the direction of $\vec{B}$ field is parallel to the normal to the loop's surface, then $B = 0.8$ T.
   (d) If $B = 1.0$ T, then the angle between the $\vec{B}$ field and the normal to the loop's surface is greater than 0°.

4. Your friend says that the emf induced in a coil supports the changing flux through the coil rather than opposes it. According to your friend, what happens when the magnetic flux increases slightly?
   (a) The induced current will increase continuously.
   (b) The coil will get hot and eventually melt.
   (c) The induced emf will get larger.
   (d) None of the above.
   (e) (a), (b), and (c) will occur.

5. A metal ring lies on a table. The S pole of a bar magnet moves down toward the ring from above and perpendicular to its surface. Which answer and explanation correctly predict the direction of the induced current as seen from above?
   (a) Clockwise because the $\vec{B}$ field is down and increasing.
   (b) Clockwise because the $\vec{B}$ field is up and increasing
   (c) Counterclockwise because the $\vec{B}$ field is down and increasing
   (d) Counterclockwise because the $\vec{B}$ field is up and increasing.
   (e) There is no current; it only changes when the N pole approaches.

6. One coil is placed on top of another. The bottom coil is connected in series to a battery and a switch. With the switch closed, there is a clockwise current in the bottom coil. When the switch is opened, the current in the bottom coil decreases abruptly to zero. What is the direction of the induced current in the top coil, as seen from above while the current in the bottom coil decreases?
   (a) Clockwise   (b) Counterclockwise
   (c) Zero—the current is induced only when the coils move relative to each other
   (d) There is not enough information to answer this question.

7. Two coils are placed next to each other flat on the table. The coil on the right is connected in series to a battery and a switch. With the switch closed, there is a clockwise current in the right coil as seen from above. When the switch is opened, the current in the right coil decreases abruptly to zero. What is the direction of the induced current in the coil on the left as seen from above while the current in the right coil decreases?
   (a) Clockwise   (b) Counterclockwise
   (c) Zero because the current is only present when the coils move relative to each other
   (d) Zero because there is no magnetic field through the coil on the left

8. Two identical bar magnets are dropped vertically from the same height. One magnet passes through an open metal ring, and the other magnet passes through a closed metal ring. Which magnet will reach the ground first?
   (a) The magnet passing through the closed ring
   (b) The magnet passing through the open ring
   (c) The magnets arrive at the ground at the same time.
   (d) There is too little information to answer this question.

9. A window's metal frame is essentially a metal loop through which a magnetic field can change when the window swings shut abruptly. The metal frame is 1.0 m × 0.50 m and Earth's 5.0×10^{-5}-T magnetic field makes an angle of 53° relative to the horizontal. Which answer below is closest to the average induced emf when the window swings shut 90° in 0.20 s from initially parallel to the field? Assume that it's a vertical window that faces north when closed.
   (a) 9.9×10^{-5} V   (b) 7.5×10^{-5} V
   (c) 9.9×10^{-2} V   (d) 12.5×10^{-4} V

10. Four identical loops move at the same velocity toward the right in a uniform magnetic field inside the dashed lines, as shown in **Figure Q21.10**. Which choice below best represents the ranking of the magnitudes (largest to smallest) of the induced currents in the loops at the snapshot shown in the figure?
    (a) $I_1 = I_2 > I_3 = I_4$
    (b) $I_2 = I_1 = I_4 > I_3$
    (c) $I_3 > I_4 > I_2 > I_1$
    (d) $I_1 > I_2 > I_3 = I_4$
    (e) $I_1 = I_2 = I_3 = I_4$

FIGURE Q21.10

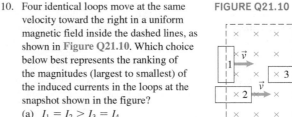

11. A 12-V automobile battery provides the thousands of volts needed to produce a spark across the gap of a spark plug. Which of the following mechanisms is involved?
    (a) The secondary loop connected to the spark plug has many more turns than the primary loop attached to the battery.
    (b) The current in the primary coil is reduced quickly.
    (c) An iron core increases the magnetic flux through the primary and secondary coils.
    (d) All three mechanisms are involved.
    (e) Only mechanisms (a) and (b) are involved.

12. A respiration detector consists of a coil placed on a person's chest and another placed on the person's back. There is a constant current in one coil. What causes an induced current to be produced in the other coil?
    (a) The person's heart beats.
    (b) The person's breathing causes the coil separation to change.
    (c) The person moves.
    (d) All three of the above occur.

13. A parallel plate capacitor and a lightbulb are connected in series to an AC power supply. Which of the following will lead to an increase in the I_{rms} through the capacitor?
    (a) Decrease the distance between the capacitor plates.
    (b) Decrease the frequency of the AC.
    (c) Insert a material with a smaller dielectric constant between the capacitor plates.
    (d) Decrease the surface area of the capacitor plates.
14. A solenoid and a lightbulb are connected in series to an AC power supply. Which of the following will lead to a decrease in the I_{rms} through the solenoid (assuming the resistance of the solenoid wires is negligible)? Multiple answers are possible.
    (a) Decrease the length of the solenoid while keeping the number of turns and the radius of the solenoid constant.
    (b) Increase the frequency of the AC.
    (c) Increase the $\Delta V_{AC\ rms}$.
    (d) Insert a paramagnetic material into the solenoid.
    (e) Increase the number of turns in the solenoid per unit length by a factor of 2 and decrease its radius by a factor of 2.

Conceptual Questions

15. A bar magnet falling with the north pole facing down passes through a coil held vertically. Sketch flux-versus-time and emf-versus-time graphs for the magnet approaching the coil and passing through it. What assumptions did you make?
16. An induction cooktop has a smooth surface. When on high, the surface does not feel warm, yet it can quickly cook soup in a metal bowl. However, it cannot cook soup in a ceramic or glass bowl. Explain how the cooktop works.

17. Describe three common applications of electromagnetic induction.
18. Two rectangular loops A and B are near each other. Loop A has a battery and a switch. Loop B has no battery. Imagine that a current starts increasing in loop A. Will there be a current in loop B? Samir argues that there will be current. Ariana argues that there will be no current. Describe experiments that support the claims of both students.
19. A simple metal detector has a coil with an alternating current in it. The current produces an alternating magnetic field. If a piece of metal is near the coil, eddy currents are induced in the metal. These induced eddy currents produce induced magnetic fields that are detected by a magnetic field detection device. Draw a series of sketches representing this process, including the appropriate directions of the magnetic fields at one instant of time, and indicate two applications for this device.
20. Construct flux-versus-time and emf-versus-time graphs that explain how an electric generator works.
21. How is it possible to get a 2000-V emf from a 12-volt battery?
22. You connect a capacitor and a lightbulb in series with an AC power supply. The lightbulb does not glow. Describe what changes you can make to the capacitor or to the AC power supply that will cause the lightbulb to glow. Explain your reasoning.
23. You connect a solenoid and a lightbulb in series with an AC power supply. The lightbulb glows. How will the brightness of the lightbulb change if you squeeze the solenoid hard between two large books so that the solenoid flattens? Explain your reasoning.

Problems

Below, BIO indicates a problem with a biological or medical focus. Problems labeled EST ask you to estimate the answer to a quantitative problem rather than derive a specific answer. Asterisks indicate the level of difficulty of the problem. Problems with no * are considered to be the least difficult. A single * marks moderately difficult problems. Two ** indicate more difficult problems.

21.1 Inducing an electric current

1. * You and your friend are performing experiments in a physics lab. Your friend claims that in general, something has to move in order to induce a current in a coil that has no battery. What experiments can you perform to support her idea? What experiments can you perform to reject it?
2. You decide to use a metal ring as an indicator of induced current. If there is a current, the ring will feel warm in your hand. You place the ring around a solenoid as shown in **Figure P21.2**, position I. Will the ring feel warm if the solenoid is connected to (a) a DC power supply or (b) an AC power supply? (c) Answer questions (a) and (b) for the case when the ring is parallel to the solenoid (position II). Explain your answers.

   FIGURE P21.2

   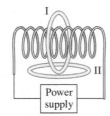

3. * To check whether a lightbulb permanently attached to a coil is still good, you place the coil next to another coil that is attached to a battery, as shown in **Figure P21.3**. Explain how or whether each of the following actions can help you determine if the lightbulb works.
    (a) Close the switch in circuit A.
    (b) Keep the switch in circuit A closed.
    (c) Open the switch in circuit A. Indicate any assumptions that you made.

   FIGURE P21.3

   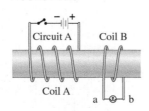

4. * **Flashlight without batteries** A flashlight that operates without batteries is lying on your desk. The light illuminates only when you repeatedly squeeze the flashlight's handle. You notice that paper clips tend to stick to the outside of the flashlight. What physical mechanism might control the operation of the flashlight?
5. You need to invent a practical application for a coil of wire that detects the vibrations or movements of a nearby magnet. Describe your invention. (The application should not repeat any described in this book.)
6. * **Detect burglars entering windows** Describe how you will design a device that uses electromagnetic induction to detect a burglar opening a window in your ground floor apartment. Include drawings and a word description.
7. * A coil connected to an ammeter can detect alternating currents in other circuits. Explain how this system might work. Could you use it to eavesdrop on a telephone conversation being transmitted through a wire?

21.2 Magnetic flux

8. * The $\vec{B}$ field in a region has a magnitude of 0.40 T and points in the positive z-direction, as shown in **Figure P21.8**. Determine the magnetic flux through (a) surface abcd, (b) surface bcef, and (c) surface adef.

   FIGURE P21.8

   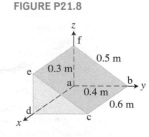

9. **EST** How do you position a bicycle tire so that the magnetic flux through it due to Earth's magnetic field is as large as possible? Estimate this maximum flux. What assumptions did you make?
10. * **EST** Estimate the magnetic flux through your head when the $\vec{B}$ field of a 1.4-T MRI machine passes through your head.
11. * **EST** Estimate the magnetic flux through the south- and west-facing windows of a house in British Columbia, where Earth's $\vec{B}$ field has a magnitude of 5.8×10^{-5} T and points roughly north with a downward inclination of 72°. Explain how you made the estimates.

21.3 Direction of the induced current

12. You perform experiments using an apparatus that has two insulated wires wrapped around a cardboard tube (Figure P21.3). Determine the direction of the current in the bulb when (a) the switch is closing and the current in coil A is increasing, (b) the switch has just closed and there is a steady current in coil A, and (c) the switch has just opened and the current in coil A is decreasing.

13. You have the apparatus shown in Figure P21.13. A circular metal plate swings past the north pole of a permanent magnet. The metal consists of a series of rings of increasing radius. Indicate the direction of the current in one ring (a) as the metal swings down from the left into the magnetic field and (b) as the metal swings up toward the right out of the magnetic field. Use Lenz's law to justify your answers.

FIGURE P21.13

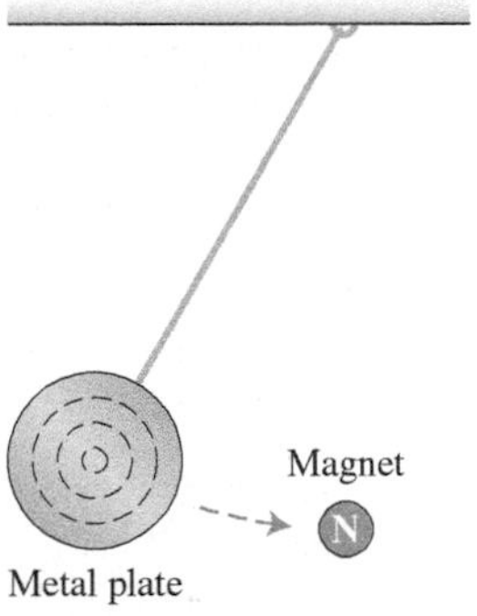

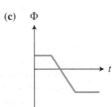

Magnet

Metal plate

14. * You suggest that eddy currents can stop the motion of a steel disk that vibrates while hanging from a spring. Explain how you can do this without touching the disk.

15. * Your friend thinks that an induced magnetic field is *always* opposite the changing external field that induces an electric current. Provide a counterexample to your friend's claim. Discuss why your friend's idea would violate conservation of energy.

21.4 and 21.5 Faraday's law of electromagnetic induction and Skills for analyzing processes involving electromagnetic induction

16. The magnetic flux through three different coils is changing as shown in Figure P21.16. For each situation, draw a corresponding graph showing qualitatively how the induced emf changes with time.

FIGURE P21.16

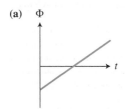

(a) Φ   (b) Φ   (c) Φ

17. The magnetic flux through three different coils is changing as shown in Figure P21.17. For each situation, draw a corresponding graph showing quantitatively how the induced emf changes with time.

FIGURE P21.17

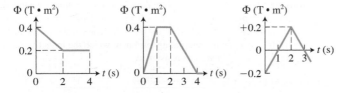

Φ (T • m²)   Φ (T • m²)   Φ (T • m²)

18. A magnetic field passing through two coils of the same diameter and length decreases from a magnitude of B_{ex} to zero in the time interval Δt. The first coil has twice the number of turns as the second. (a) Compare the emfs induced in the coils. (b) How can you change the experiment so that the emfs produced in them are the same?

19. **BIO** **Stimulating the brain** In transcranial magnetic stimulation (TMS) an abrupt decrease in the electric current in a small coil placed on the scalp produces an abrupt decrease in the magnetic field inside the brain. Suppose the magnitude of the $\vec{B}$ field changes from 0.80 T to 0 T in 0.080 s. Determine the induced emf around a small circle of brain tissue of radius 1.2×10^{-3} m. The $\vec{B}$ field is perpendicular to the surface area of the circle of brain tissue.

20. * To measure a magnetic field produced by an electromagnet, you use a circular coil of radius 0.30 m with 25 loops (resistance of 25 Ω) that rests between the poles of the magnet and is connected to an ammeter. While the current in the electromagnet is reduced to zero in 1.5 s, the ammeter in the coil shows a steady reading of 180 mA. Draw a picture of the experimental setup and determine everything you can about the electromagnet.

21. You want to use the idea of electromagnetic induction to make the bulb from your small flashlight glow; it glows when the potential difference across it is 1.5 V. You have a small bar magnet and a coil with 100 turns, each with area 3.0×10^{-4} m². The magnitude of the $\vec{B}$ field at the front of the bar magnet's north pole is 0.040 T and reaches 0 T when it is about 4 cm away from the pole. Can you make the bulb light? Explain.

22. * **BIO** **Breathing monitor** An apnea monitor for adults consists of a flexible coil that wraps around the chest (Figure P21.22). When the patient inhales, the chest expands, as does the coil. Earth's $\vec{B}$ field of 5.0×10^{-5} T passes through the coil at a 53° angle relative to a line perpendicular to the coil. Determine the average induced emf in such a coil during one inhalation if the 300-turn coil area increases by 42 cm² during 2.0 s.

FIGURE P21.22

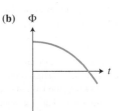

23. * A bar magnet induces a current in an N-turn coil as the magnet moves closer to it (Figure P21.23). The coil's radius is R cm, and the average induced emf across the bulb during the time interval is $\mathcal{E}$ V. (a) Make a list of the physical quantities that you can determine using this information; (b) Is the direction of the induced current from lead a to b, or from b to a? Explain.

FIGURE P21.23

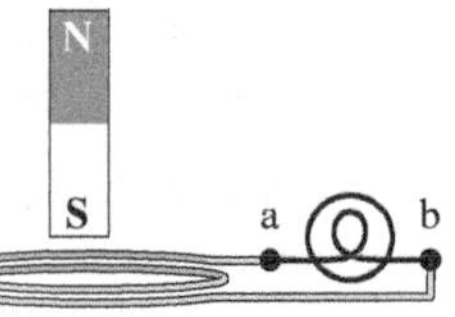

24. * You have a coil of wire with 10 turns each of 1.5-cm radius. You place the plane of the coil perpendicular to a 0.40-T $\vec{B}$ field produced by the poles of an electromagnet (Figure Q21.2). (a) Find the magnitude of the average induced emf in the coil when the magnet is turned off and the field decreases to 0 T in 2.4 s. (b) Is the direction of the induced current in the coil from lead a to b, or from b to a? Explain.

25. * An experimental apparatus has two parallel horizontal metal rails separated by 1.0 m. A 2.0-Ω resistor is connected from the left end of one rail to the left end of the other. A metal axle with metal wheels is pulled toward the right along the rails at a speed of 20 m/s. Earth's uniform 5.0×10^{-5}-T $\vec{B}$ field points down at an angle of 53° below the horizontal. Make a list of the physical quantities you can determine using this information and determine two of them.

26. A Boeing 747 with a 65-m wingspan is cruising northward at 250 m/s toward Alaska. The $\vec{B}$ field at this location is 5.0×10^{-5} T and points 60° below its direction of travel. Determine the potential difference between the tips of its wings.

27. * A circular loop of radius 9.0 cm is placed perpendicular to a uniform 0.35-T $\vec{B}$ field. You collapse the loop into a long, thin shape in 0.10 s. What is the average induced emf while the loop is being reshaped? What assumptions did you make?

28. ** **BIO** **EST** **Magnetic field and brain cells** Suppose a power line produces a 6.0×10^{-4}-T peak magnetic field 60 times each second at the location of a neuron brain cell of radius 6.0×10^{-6} m. Estimate the maximum magnitude of the induced emf around the perimeter of this cell during one-half cycle of magnetic field change.

29. * You need to test Faraday's law. You have a 12-turn rectangular coil that measures 0.20 m × 0.40 m and an electromagnet that produces a 0.25-T magnetic field in a well-defined region that is larger than the area of the coil. You also have a stopwatch, an ammeter, a voltmeter, and a motion detector. (a) Describe an experiment you will design to test Faraday's law. (b) How will you calculate the measurable outcome of this experiment using the materials available? (c) Describe how you can test Lenz's law with this equipment.

30. * You build a coil of radius r (m) and place it in a uniform $\vec{B}$ field oriented perpendicular to the coil's surface. What is the total electric charge that passes through the coil's wire loops if the $\vec{B}$ field decreases at a constant rate to zero? The resistance of the coil's wire is R (Ω).

31. * **Equation Jeopardy 1** Invent a problem for which the following equation might be a solution.

$$0.01 \text{ V} = (100)\frac{(A)\cos 0°(0.12 \text{ T} - 0)}{(1.2 \text{ s} - 0)}$$

32. * **Equation Jeopardy 2** Invent a problem for which the following equation might be a solution.

$$0.01\text{V} = 100\frac{\pi (0.10 \text{ m})^2(0.12 \text{ T})(\cos 0° - \cos 90°)}{(t - 0)}$$

33. * **Equation Jeopardy 3** Invent a problem for which the following equation might be a solution.

$$\mathcal{E} = -(35)\frac{(0.12 \text{ T})(\cos 0°)[(0)^2 - \pi (0.10 \text{ m})^2]}{(3.0 \text{ s} - 0)}$$

34. * **EST** **Generator for space station** Astronauts on a space station decide to use Earth's magnetic field to generate electric current. Earth's $\vec{B}$ field in this region has the magnitude of 3.0×10^{-7} T. They have a coil that rotates 90° in 1.2 s. The area inside the coil measures 5000 m². Estimate the number of loops needed in the coil so that during that 90° turn it produces an average induced emf of about 120 V. Indicate any assumptions you made. Is this a feasible way to produce electric energy?

35. * A toy electric generator has a 20-turn circular coil with each turn of radius 1.8 cm. The coil resides in a 1.0-T magnitude $\vec{B}$ field. It also has a lightbulb that lights if the potential difference across it is about 1 V. You start rotating the coil, which is initially perpendicular to the $\vec{B}$ field. (a) Determine the time interval needed for a 90° rotation that will produce an average induced emf of 1.0 V. (b) Use a proportion technique to show that the same emf can be produced if the time interval for one rotation is reduced to one-fourth while the radius of the coil is reduced by one-half.

36. * A generator has a 450-turn coil that is 10 cm long and 12 cm wide. The coil rotates at 8.0 rotations per second in a 0.10-T magnitude $\vec{B}$ field. Determine the generator's peak voltage.

37. * You need to make a generator for your bicycle light that will provide an alternating emf whose peak value is 4.2 V. The generator coil has 55 turns and rotates in a 0.040-T magnitude $\vec{B}$ field. If the coil rotates at 400 revolutions per second, what must the area of the coil be to develop this emf? Describe any problems with this design (if there are any).

38. * **Evaluating a claim** A British bicycle light company advertises flashing bicycle lights that require no batteries and produce no resistance to riding. A magnet attached to a spoke on the bicycle tire moves past a generator coil on the bicycle frame, inducing an emf that causes a light to flash. The magnet and coil never touch. Does this lighting system really produce no resistance to riding? Justify your answer.

39. * A generator has a 100-turn coil that rotates in a 0.30-T magnitude $\vec{B}$ field at a frequency of 80 Hz (80 rotations per second) causing a peak emf of 38 V. (a) Determine the area of each loop of the coil. (b) Write an expression for the emf as a function of time (assuming the emf is zero at time zero). (c) Determine the emf at 0.0140 s.

40. * A 10-Hz generator produces a peak emf of 40 V. (a) Write an expression for the emf as a function of time. Indicate your assumptions. (b) Determine the emf at the following times: 0.025 s, 0.050 s, 0.075 s, and 0.100 s. (c) Plot these emf-versus-time data on a graph and connect the points with a smooth curve.

41. * A rectangular wire loop is moving with constant speed through a region of uniform $\vec{B}$ field that points into the page (**Figure P21.41**). The longer side of the loop is twice the width of the magnetic field region. Draw a qualitative graph that shows how the induced current in the loop depends on time during the period in which the loop moves from the position shown in the figure to the position where the left side of the loop exits the magnetic field. Assume that a counterclockwise direction of the current is positive.

FIGURE P21.41

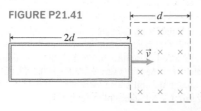

42. ** A wire loop consists of two equal parts that are connected so that they form a "rectangular figure eight." The loop is moving with constant speed through a region of uniform $\vec{B}$ field that points into the page (**Figure P21.42**). The longer side of the loop is twice the width of the magnetic field region. Draw a qualitative graph that shows how the induced current in the loop depends on time during the period in which the loop moves from the position shown in the figure to the position where the left side of the loop exits the magnetic field. Assume that a counterclockwise direction of the current in the right half of the loop is positive.

FIGURE P21.42

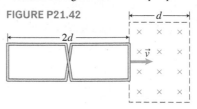

21.6 AC circuits

43. The voltage across an AC power supply is given by $\Delta V_{AC} = (100 \text{ V}) \sin (2\pi 50t)$. Determine (a) ΔV_{rms}, (b) the period with which the voltage vibrates, (c) the amplitude of the current through the 200-Ω resistor that is connected to this power supply, and (d) the average power transferred by the power supply to the resistor.

44. * The alternating current through a 2.0-μF capacitor is given by $I_C = (0.04 \text{ A}) \sin (2\pi 400t)$. Determine (a) the rms voltage across the capacitor and (b) the shortest time interval between the peak of the voltage across the capacitor and the peak of the current through the capacitor. (c) How will the answers to questions (a) and (b) change if the capacitance is doubled? Explain.

45. * The alternating current through a solenoid is given by $I_L = (3.2 \text{ A}) \sin (2\pi 180t)$, and the rms voltage across the solenoid is 190 V. Determine (a) the inductance of the solenoid and (b) the shortest time interval between the peak of the voltage across the solenoid and the peak of the current through the solenoid, assuming that the resistance of the solenoid can be ignored. (c) A nickel cylinder with a relative magnetic permeability of 300 is inserted into the solenoid. Determine the new amplitude of the current through the solenoid, assuming that the solenoid remains connected to the same power supply.

46. * The rms voltage of household AC in Europe is 230 V. Determine (a) the amplitude of the alternating voltage across and (b) the amplitude of the alternating current through a 100-W European incandescent lightbulb. Determine (c) the resistance of this lightbulb and (d) the power output of the lightbulb when you use it in the United States, where the rms voltage is 120 V. Indicate any assumptions that you made.

47. * You build a parallel plate capacitor by placing a 0.1-mm-thick sheet of paper between two 100-cm² pieces of aluminum foil. The dielectric constant of paper is 2.3. Determine (a) the capacitance of the capacitor and (b) the frequency of the alternating current at which the capacitive reactance of the capacitor is equal to 5 Ω.

48. * You have a 100-m copper wire with cross-sectional area 0.20 mm² and resistivity 1.7×10^{-8} $\Omega \cdot$ m. You make a 780-turn solenoid by winding the wire on a plastic tube (diameter 4.0 cm, length 10.0 cm). Determine (a) the resistance of the solenoid, (b) the inductance of the solenoid, and (c) the frequency of the alternating current at which the inductive reactance of the solenoid is equal to its resistance.

21.7 Transformers

49. You need to build a transformer that can step the emf up from 120 V to 12,000 V to operate a neon sign for a restaurant. What will be the ratio of the secondary to primary turns of this transformer?

50. Your home's electric doorbell operates on 10 V. Should you use a step-up or step-down transformer in order to convert the home's 120 V to 10 V? Determine the ratio of the secondary to primary turns needed for the bell's transformer.

51. A 9.0-V battery and switch are connected in series across the primary coil of a transformer. The secondary coil is connected to a lightbulb that operates on 120 V. Draw the circuit. Describe in detail how you can get the bulb to light—not necessarily continuously.

52. * You are fixing a transformer for a toy truck that uses an 8.0-V rms voltage to run it. The primary coil of the transformer is broken; the secondary coil has 30 turns. The primary coil is connected to a 120-V wall outlet. (a) How many turns should you have in the primary coil? (b) If you then connect this primary coil to a 240-V rms voltage, what will be the amplitude of the alternating voltage across the secondary coil?

21.8 Mechanisms explaining electromagnetic induction

53. * A wire loop has a radius of 10 cm. A changing external magnetic field causes an average 0.60-N/C electric field in the wire. (a) Determine the work that the electric field does in pushing 1.0 C of electric charge around the loop. (b) Determine the induced emf caused by the changing magnetic field. (c) You measure a 0.10-A electric current. What is the electrical resistance of the loop?

General Problems

54. * **Ice skater's flashing belt** You are hired to advise the coach of the Olympic ice-skating team concerning an idea for a costume for one of the skaters. They want to put a flat coil of wire on the front of the skater's torso and connect the ends of the coil to lightbulbs on the skater's belt. They hope that the bulbs will light when the skater spins in Earth's magnetic field. Do you think that the system will work? If so, could you provide specifications for the device and justification for your advice?

55. **BIO Hammerhead shark** A hammerhead shark (Figure P21.55) has a 0.90-m-wide head. The shark swims north at 1.8 m/s. Earth's $\vec{B}$ field at this location is 5.0×10^{-5} T and points 30° below the direction of the shark's travel. Determine the potential difference between the two sides of the shark's head.

FIGURE P21.55

56. * **Car braking system** You are an inventor and want to develop a braking system that not only stops the car but also converts the original kinetic energy to some other useful energy. One of your ideas is to connect a rotor coil (the rotating coil of the generator) to the turning axle of the car. When you press on the brake pedal, a switch turns on a steady electric current to a stationary coil (an electromagnet called the stator) that produces a steady magnetic field in which the rotor turns. You now have a generator that produces an alternating current and an induced emf—electric power. Make a simple drawing of the rotor and stator at one instant and determine the direction of the magnetic force exerted on the rotor. Does this force help stop the car? Explain.

57. ** Your professor asks you to help design an electromagnetic induction sparker (a device that produces sparks). Include drawings and word descriptions for how it might work, details of its construction, and a description of possible problems.

58. ** In a new lab experiment, two parallel vertical metal rods are separated by 1.0 m. A 2.0-Ω resistor is connected from the top of one rod to the top of the other. A 0.20-kg horizontal metal bar falls between the rods and makes contact at its ends with the rods. A $\vec{B}$ field of 0.50×10^{-4} T points horizontally between the rods. The bar should eventually reach a terminal falling velocity (constant speed) when the magnetic force of the magnetic field on the induced current in the bar balances the downward force due to the gravitational pull of the Earth. (a) Develop in symbols an expression for the current through the bar as it falls. (b) Determine in symbols an expression for the magnetic force exerted on the falling bar (and determine the direction of that force). Remember that an electric current passes through it, and the bar is falling in the magnetic field. (c) Determine the final constant speed of the falling bar (d) Is this process realistic? Explain.

59. ** You have a 12-V battery, some wire, a switch, and a separate coil of wire. (a) Design a circuit that will produce an emf around the coil even though it is not connected to the battery. (b) Show, using appropriate equations, why your system will work. (c) Describe one application for your circuit.

60. * You want to build a generator for a multi-day canoe trip. You have a fairly large permanent magnet, some wire, and a lightbulb. Design a generator and provide detailed specifications for it. (Ideas for the design could include cranking a handle or placing a paddle wheel that turns a coil in a nearby stream.)

61. * **EST** A sparker used to ignite lighter fluid in a barbeque grill is shown in **Figure P21.61**. You compress a knob at the end of the sparker. This compresses a spring, which when released moves a magnet at the end of the knob quickly into a 200-turn coil. The change in flux through the coil induces an emf that causes a spark across the 0.10-mm gap at the end of the sparker. (a) Estimate the time interval needed for the change in flux in order to produce this spark. Indicate any assumptions you made. (b) Is this a realistic process? Explain.

FIGURE P21.61 Note: The size of the gap is not to scale.

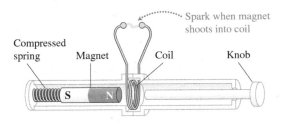

62. * **EST** **Design a magnetometer** Your friend needs to design an experiment that will allow her to measure the magnitude of the $\vec{B}$ field near a large permanent magnet. She has a small coil (50 turns, diameter 1.0 cm), a small motor that can spin at 300 revolutions per second, and an AC voltmeter that can measure voltage with 0.001-V precision. (a) Describe in detail the experimental setup (using this equipment) that will allow your friend to estimate the magnitude of the $\vec{B}$ field. (b) Estimate the smallest value of the $\vec{B}$ field magnitude that she will be able to measure. Indicate any assumptions that you made.

63. * **EST** **Can a distant strike of lightning damage your computer?** Lightning is a sudden electrical discharge that occurs during a storm. In a typical cloud-to-ground lightning flash, a discharge current changes at a rate of about 10^{11} A/s. Estimate the maximum induced emf around the metal box of a typical computer during a lightning flash, assuming the lightning strikes the ground 200 m from the computer. Describe any assumptions that you made. (Hint: Model the lightning as a long straight wire carrying current that changes in time.)

64. **BIO EST MRI** Jose needs an MRI (magnetic resonance imaging) scan. During the exam, Jose lies in a region of a very strong 1.5-T horizontal magnetic field pointing from Jose's waist toward his head. A medical assistant moves Jose from the scanner, reducing the magnetic field from 1.5 T to 0.1 T in 1.0 s. Consequently, the $\vec{B}$ field through Jose's 0.3-m by 0.4-m chest decreases. The conductive tissue inside his body around the outer part of his chest is a loop, with the chest as the area inside this loop. (a) Estimate the induced emf around this conducting loop as the $\vec{B}$ field decreases. (b) If the resistance of his body tissue around this loop is 5 Ω, what is the induced current passing around his body? (c) What is the direction of the current?

65. * **Magstripe reader** A magstripe reader used to read a credit card number or a card key for a hotel room has a tiny coil that detects a changing magnetic field as tiny bar magnets pass by the coil. Calculate the magnitude of the induced emf in a magstripe card reader coil. Assume that the magstripe magnetic field changes at a constant rate of 500 mT/ms as the region between two tiny magnets on the stripe passes the coil. The reader coil is 2.0 mm in diameter and has 5000 turns.

66. Show that when a metal rod L meters long moves at speed v perpendicular to $\vec{B}$ field lines, the magnetic force exerted by the field on the electrically charged particles in the rod produces a potential difference between the ends of the rod equal to the product BLv.

67. ** **EST** **The Tower of Terror ride** Magnetic braking brings the car of the Tower of Terror ride (similar to the one in Figure 21.10) to a stop from a speed of 161 km/h. (a) Is its 161-km/h speed what you would expect of an object after a 115-m fall? Explain. (b) Estimate the time interval for the free-fall part of its trip. (c) Estimate the average acceleration of the car while stopping due to its magnetic braking.

Reading Passage Problems

BIO **Magnetic induction tomography (MIT)** Magnetic induction tomography is an imaging method used in mineral, natural gas, oil, and groundwater exploration; as an archaeological tool; and for medical imaging. MIT has also been used to measure topsoil depth in agricultural soils. Topsoil depth is information that many farmers need: for instance, corn yield is much higher in soil that has a deep topsoil layer above the underlying, impermeable claypan. Using a trailer attached to a tractor, a farmer can map an 80,000-m² (about 20-acre) field for topsoil depth in about 1 hour.

Figure 21.28 shows how MIT works. A time-varying electric current in a source coil (Figure 21.28a) induces a changing magnetic field that passes into the region to be imaged—in this case, the soil (Figure 21.28b). This changing magnetic field induces a weak induced electric current in topsoil and a stronger induced current in the more conductive claypan soil at the same depth. (Figure 21.28c; the current direction here is drawn as though the source current and source fields are increasing). This changing induced electric current in turn produces its own induced magnetic field (Figure 21.28d). The induced magnetic field passes out of the region being mapped to a detector coil (Figure 21.28e) near the source coil. The nature of the signal at the detector (its magnitude and phase) provides information about the region being mapped. A strong signal returned to the detector coil indicates a claypan layer near the surface; a weak signal returns if the clay layer is deeper below the surface.

FIGURE 21.28 Magnetic induction tomography.

(a) A changing current in the source coil above the soil ...

(b) ... causes a changing external magnetic field in the soil ...

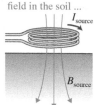

(c) ... which induces a current in the soil.

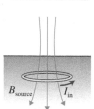

(d) The induced current causes an induced magnetic field out of the soil ...

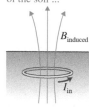

(e) ... which induces a current in a detector coil above the soil.

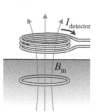

68. Why is the detected signal from an MIT apparatus greater if a moist conductive layer is near the surface?
   (a) The signal is reflected better from the top of a nearby conductive layer.
   (b) The induced current is greater if the soil is moist and conductive.
   (c) The induced magnetic field from the induced current is bigger if its source is near the detection coil.
   (d) All three of the above reasons
   (e) b and c

69. All other conditions being equal, why is the induced current greater in claypan soil than in topsoil?
   (a) Claypan soil has a higher concentration of magnetic ions compared to topsoil.
   (b) Claypan soil is partly metallic in composition.
   (c) Claypan soil has greater density than loose topsoil.
   (d) The clay is closely packed, moist, and a better electrical conductor than loose, dry topsoil.

70. Why is MIT used to search for mineral deposits (iron, copper, zinc)?
   (a) The minerals are good conductors of electricity and produce strong induced currents and strong returning magnetic fields.
   (b) The minerals absorb the incident magnetic field, indicating their presence by a lack of returning signal.
   (c) The minerals produce their own returning magnetic fields.
   (d) The minerals attract the incoming magnetic field and reflect it directly above the minerals.

71. Which of the statements below about magnetic induction tomography (MIT) and transcranial magnetic stimulation (TMS), studied in Section 21.5, are true?
   (a) Both MIT and TMS have source currents in coils, source magnetic fields, and induced currents.
   (b) MIT detects the induced magnetic field produced by the induced current, and TMS does not.
   (c) MIT provides information directly about the imaged area, whereas TMS disrupts some brain activity, and the disruption is measured in some other way.
   (d) a and c only   (e) a, b, and c

72. Describe all the changes that would occur if the source current were in the direction shown in Figure 21.28 but decreasing instead of increasing.
   (a) The induced current would be in the opposite direction.
   (b) The induced magnetic field would be in the opposite direction.
   (c) The detected current would be in the opposite direction.
   (d) a and b   (e) a, b, and c

BIO **Measuring the motion of flying insects** Studying the motion of flying animals, particularly small insects, is difficult. One method researchers use involves attaching a tiny coil with miniature electronics to the neck of an insect and another coil to its thorax (Figure 21.29). They place the insect in a strong magnetic field and observe the changing orientations and induced emfs of the two coils in the field as the insect flies. Suppose that a 50-turn coil of radius 2.0×10^{-3} m is attached to a tsetse fly that is flying in a 4.0×10^{-3}-T magnetic field. The tsetse fly makes a 90° turn in 0.020 s. Consider the average magnitude of the induced emf that occurs due to the turn of the tsetse fly and its coil.

FIGURE 21.29 The coil changes its orientation with respect to the external magnetic field as the fly makes a turn.

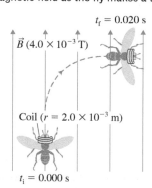

73. Which of the quantities B_{ex}, A, or θ is changing as the fly turns?
   (a) B_{ex} (b) A (c) θ
   (d) All of them   (e) None of them

74. Which answer is closest to the magnitude of the flux change?
   (a) 1×10^{-4} T·m²   (b) 2×10^{-6} T·m²
   (c) 3×10^{-7} T·m²   (d) 5×10^{-8} T·m²

75. Which answer is closest to the induced emf on the tsetse fly coil during the 90° turn?
   (a) 6×10^{-2} V   (b) 1×10^{-4} V
   (c) 4×10^{-6} V   (d) 2×10^{-7} V

76. Which of the following could double the emf produced when the fly turns 90°?
   (a) Double the number of turns in the coil.
   (b) Double the coil's area.
   (c) Double the magnitude of the external magnetic field.
   (d) Get the tsetse fly to take twice as long to turn.
   (e) a, b, and c

Reflection and Refraction

A magician breaks a glass vial with a hammer and drops the pieces of broken glass into a beaker full of oil. Then she reaches into the beaker—and takes out an intact vial! Is this really magic? No. The photo above shows the same phenomenon: the glass pipette is visible in the right-hand bottle of water, but invisible in the left-hand bottle of oil. In this chapter you will learn how to explain this trick.

- **How can you make something become invisible?**
- **What causes a mirage?**
- **What is fiber optics?**

IN THIS CHAPTER we will begin our investigation of light. Is light different from everything we have studied, or can we use the principles and tools we have already developed to understand light and its useful applications? Many of the remaining chapters in this textbook involve using and improving the model of light we will develop in this chapter as well as investigating many physical processes involving light.

BE SURE YOU KNOW HOW TO:

- Define rotational velocity (Section 9.1).
- Apply ideas of impulse-momentum to explain collisions (Section 6.4).
- Draw a wave front and apply Huygens' principle (Sections 11.1 and 11.6).

22.1 Light sources, light propagation, and shadows

Some ancient people thought that humans emitted special invisible rays from their eyes. These rays reached out toward objects and wrapped around them to collect information about the objects. The rays then returned to the person's eyes with this information. If this ancient model were accurate, humans should be able to see in total darkness. However, a simple experiment disproves this model. If you sit for a while in a room with absolutely no light sources, no matter how long you wait, you will still see nothing. There must be some other explanation for how we see things. The experiments in Observational Experiment **Table 22.1** use a laser pointer as a source of light to explore the phenomena of light and seeing.

OBSERVATIONAL EXPERIMENT TABLE 22.1   How do we see objects?

VIDEO
OET 22.1

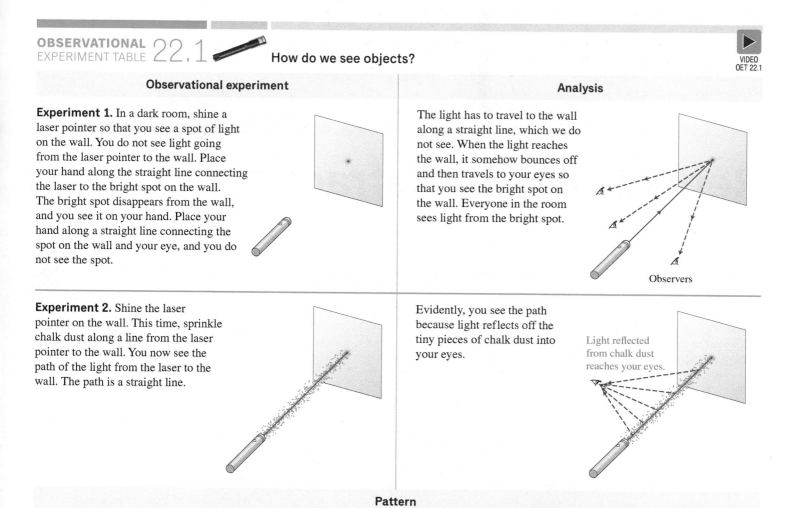

Observational experiment	Analysis
Experiment 1. In a dark room, shine a laser pointer so that you see a spot of light on the wall. You do not see light going from the laser pointer to the wall. Place your hand along the straight line connecting the laser to the bright spot on the wall. The bright spot disappears from the wall, and you see it on your hand. Place your hand along a straight line connecting the spot on the wall and your eye, and you do not see the spot.	The light has to travel to the wall along a straight line, which we do not see. When the light reaches the wall, it somehow bounces off and then travels to your eyes so that you see the bright spot on the wall. Everyone in the room sees light from the bright spot.
Experiment 2. Shine the laser pointer on the wall. This time, sprinkle chalk dust along a line from the laser pointer to the wall. You now see the path of the light from the laser to the wall. The path is a straight line.	Evidently, you see the path because light reflects off the tiny pieces of chalk dust into your eyes.

Pattern

We can see objects (even tiny ones such as dust) illuminated by light. The path of light is a straight line from the source of light to the object and then (assuming that the behavior of light does not change) another straight line of reflected light from the object to our eyes.

Table 22.1 indicates that to see something, we need a source of light and an object off which the light bounces (reflects) and then reaches the eyes of the observer. Light travels in a straight-line path between the source of the light and the object reflecting the light, then in another straight line between that object and our eyes.

How do we represent the light sent by a light source? How do objects illuminated by a light source reflect light into our eyes? We will start by investigating the first question.

Representing light emitted by different sources

A laser pointer is useful for studying light propagation because the emitted light emerges as one narrow beam. However, most light sources, such as lightbulbs and candles, do not emit light as a single beam. These **extended light sources** consist of multiple points, each of which emits light. When we turn on a lightbulb in a dark room, the walls, floor, and ceiling of the room are illuminated (**Figure 22.1**). Obviously, the bulb sends light in all directions. But does one point of the shining bulb send light in one direction, or does each point send light in multiple directions? Both of these ideas can explain why the walls, floor, and ceiling are illuminated. Let's investigate those two possible models of how extended sources emit light. To do this we will represent the travel of light from one location to another with a **light ray**, drawn as a straight line and an arrow. Diagrams that include light rays are called **ray diagrams**. A ray is not a real thing; it is a model that allows us to show the direction of light.

One-ray model: Each point of an extended light source emits light that can be represented by one outward-pointing ray (**Figure 22.2a**). Different points send rays in different directions.

Multiple-ray model: Each point on an extended light source emits light in multiple directions represented by multiple rays (Figure 22.2b).

To help us determine which model better explains real phenomena, we will use the models to make predictions about the outcome of three experiments (Testing Experiment Table 22.2). All of the experiments are conducted in an otherwise dark room.

FIGURE 22.1 A lightbulb illuminates the ceiling of a room.

FIGURE 22.2 Two models of light emission from a bulb.

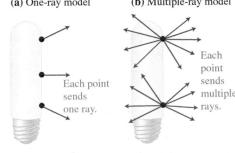

(a) One-ray model — Each point sends one ray.

(b) Multiple-ray model — Each point sends multiple rays.

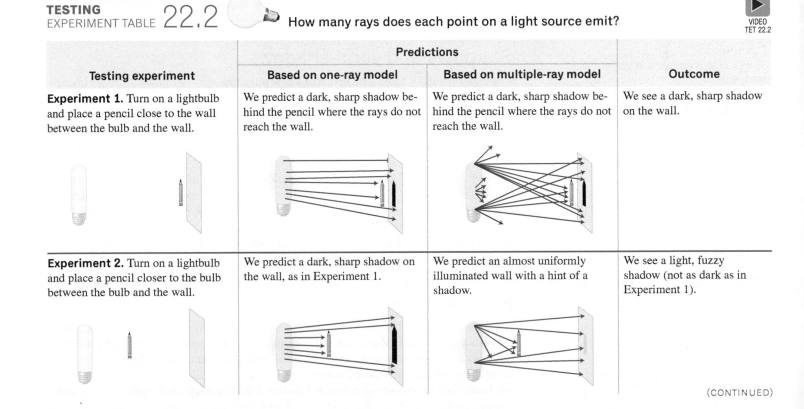

TESTING EXPERIMENT TABLE 22.2   How many rays does each point on a light source emit?

VIDEO TET 22.2

Testing experiment	Predictions		Outcome
	Based on one-ray model	**Based on multiple-ray model**	
Experiment 1. Turn on a lightbulb and place a pencil close to the wall between the bulb and the wall.	We predict a dark, sharp shadow behind the pencil where the rays do not reach the wall.	We predict a dark, sharp shadow behind the pencil where the rays do not reach the wall.	We see a dark, sharp shadow on the wall.
Experiment 2. Turn on a lightbulb and place a pencil closer to the bulb between the bulb and the wall.	We predict a dark, sharp shadow on the wall, as in Experiment 1.	We predict an almost uniformly illuminated wall with a hint of a shadow.	We see a light, fuzzy shadow (not as dark as in Experiment 1).

(CONTINUED)

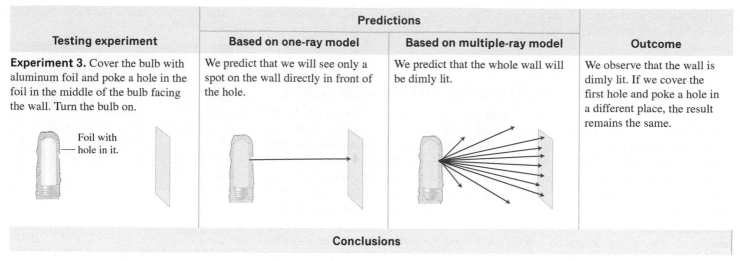

| | Predictions | | |
Testing experiment	Based on one-ray model	Based on multiple-ray model	Outcome
Experiment 3. Cover the bulb with aluminum foil and poke a hole in the foil in the middle of the bulb facing the wall. Turn the bulb on.	We predict that we will see only a spot on the wall directly in front of the hole.	We predict that the whole wall will be dimly lit.	We observe that the wall is dimly lit. If we cover the first hole and poke a hole in a different place, the result remains the same.

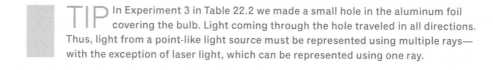

Conclusions

Experiment 1. Both models gave predictions that matched the outcome of this experiment. Neither is disproved.

Experiments 2 and 3. The predictions based on the one-ray model did not match the outcomes of the experiments; thus, that model is disproved. The predictions based on the multiple-ray model matched the outcomes of both experiments; thus, the model is supported (not proved).

The testing experiments above disprove the one-ray model and support the idea that each point on an extended light source emits light in many different directions. This light can be represented by multiple rays diverging from that point.

TIP   In Experiment 3 in Table 22.2 we made a small hole in the aluminum foil covering the bulb. Light coming through the hole traveled in all directions. Thus, light from a point-like light source must be represented using multiple rays—with the exception of laser light, which can be represented using one ray.

Shadows and semi-shadows

The experiments in Table 22.2 revealed a new phenomenon—shadows. In Experiment 1 we found a sharp **shadow**, or **umbra**, behind the pencil on the wall. A shadow is a region behind an object where no light reaches. In Experiment 2 we found that there was no extremely dark shadow anywhere on the wall. What we observed is called a **semi-shadow**, also called a **penumbra**. A semi-shadow is a region where some light reaches and some does not. It appears as a fuzzy shadow.

The shadows of our bodies on a sunny day are sharp—there are almost no semi-shadows. Why? The Sun is so *far* from Earth that the only light from the Sun that reaches a small region on Earth is the light inside a narrow cylinder that extends from the Sun to Earth (**Figure 22.3a**). We can model sunlight that reaches Earth as a collection of parallel rays (Figure 22.3b). Similarly, we can represent the laser's very narrow beam of light with just one ray (Figure 22.3c).

Pinhole camera

If you hold a candle flame about 1 m from a blank wall, you do not see a projection of the flame on the wall. Since each point on the flame is a point source of light and each point source emits light in all directions, the wall is illuminated by light coming from all of the points on the candle flame.

However, we can use a piece of cardboard with a very small hole in it to make a sharp projection of the flame on a wall in a dark room. Conceptual Exercise 22.1 explains how this projection is formed.

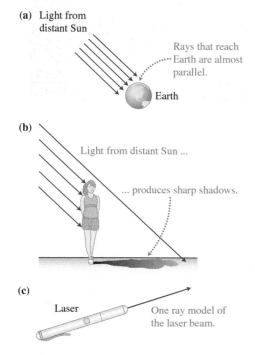

FIGURE 22.3 Ray models of sunlight and laser light.

(a) Light from distant Sun

Rays that reach Earth are almost parallel.

Earth

(b)

Light from distant Sun ...

... produces sharp shadows.

(c)

Laser   One ray model of the laser beam.

CONCEPTUAL EXERCISE 22.1 **Simulating a pinhole camera**

Use the ray model of light propagation to predict what you will see in the following experiment. You place a lit candle several meters from the wall in an otherwise dark room. Between the candle and the wall (close to the candle), you place a piece of stiff paper (or cardboard) with a small hole in it (see figure). What do you see on the wall?

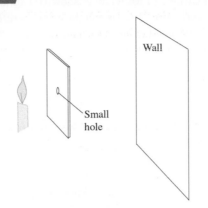

Sketch and translate
We already have a sketch of the situation, shown above. The candle flame is an extended light source. We need to predict what happens when light from the candle flame reaches the wall.

Simplify and diagram Assume that each point of the candle flame sends light in all directions. Represent this light with multiple light rays emitted from only the top and the bottom of the flame. Most of these rays are blocked by the paper and do not reach the wall, as shown in the figure at top right. However, ray 1 from the bottom of the flame passes through the hole and reaches the wall. Ray 2 from the top of the flame reaches the wall but below where ray 1 did. Because of where these

rays reach the wall, we predict that we will see an upside-down projection of the flame on the wall. Of course, many more rays reach the wall between rays 1 and 2, but these two rays help us find the top and bottom of the flame. That is what you actually see when you perform the experiment: the candle flame upside down on the wall.

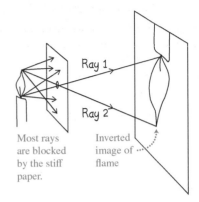

Try it yourself Use the ray diagram to predict what you will see if you (a) move the candle closer to the hole; (b) move it farther from the hole; and (c) keep the candle and the stiff paper fixed relative to each other but move them together away from the wall.

Answer

(a) The upside-down image of the flame on the wall gets bigger; (b) it gets smaller; (c) it gets bigger.

The cardboard with a small hole in it is the foundation of the pinhole camera, also called a **camera obscura.** Such a camera consists of a lightproof box with a very small hole in one wall and a photographic plate or film inside the box on the opposite wall. Before the invention of modern cameras that use lenses, pinhole cameras were used to make photographs. To photograph a person, you would shine intense light on the person for a long time. A small amount of light that reflected off the person would pass through the hole and form an inverted projection of the person on the film. In **Figure 22.4**, the outside world is projected through a pinhole onto the wall of a cardboard box. We used a GoPro camera to take a picture of the image on the wall. In the video you can see an upside-down view of a harbor.

FIGURE 22.4 A pinhole view of the outside world. Note that the woman is upside down.

REVIEW QUESTION 22.1 A light source placed in front of a screen with a small hole in it will be projected upside down on the wall behind the screen. Why?

22.2 Reflection of light

In the last section we learned that people see objects because light is reflected off of objects into our eyes. In this section we will investigate the phenomenon of reflection. We will start with a simple case of a narrow beam of light that we can represent with a single ray and a very good, smooth reflecting surface—a mirror.

TIP Light propagates along straight lines, which we represent as *rays*. We use the term "a ray of light" as a model of a very narrow beam of light that describes and helps to predict light phenomena. We use purple lines to represent light rays and red lines to represent light beams.

FIGURE 22.5 Light reflection from a mirror.

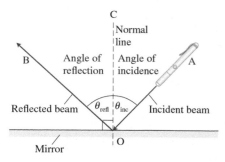

In **Figure 22.5**, light from a laser pointer at A shines on a mirror supported vertically on a horizontal surface (a top view is shown). Light striking the mirror is called **incident** light. After reflection, the laser light beam is *reflected* as shown. Notice line OC in the figure—it is perpendicular to the surface of the mirror at the point where the incident light hits the mirror. This line is called a **normal line** (recall that in mathematics and physics, "normal" means perpendicular). **Table 22.3** lists some angles related to the incident and reflected beams as measured by a protractor lying on the table.

TABLE 22.3 Angles indicating directions of incident and reflected light beams from the mirror in Figure 22.5

Angle AOB between the incident and reflected beams	Angle between the incident beam AO and the normal line CO	Angle between the reflected beam BO and the normal line CO
0°	0°	0°
40°	20°	20°
60°	30°	30°
90°	45°	45°
120°	60°	60°
160°	80°	80°

The angle between the incident beam AO and the normal line CO is the **angle of incidence**. The angle between the reflected beam BO and the normal line CO is the **angle of reflection**. The data in the table show that the angles in columns two and three (the incident angle and the reflected angle, respectively) are always equal to each other. We can formulate this pattern as a mathematical rule or law for reflection:

TIP ▸ The angle of incidence and the angle of reflection are always the angles that the light beams form with the normal line and not the angles that they form with the surface of the mirror.

Law of reflection When a narrow beam of light, represented by one ray, shines on a smooth surface such as a mirror, the angle between the incident ray and the normal line perpendicular to the surface equals the angle between the reflected ray and the normal line (the angle of reflection equals the angle of incidence). The incident beam, reflected beam, and the normal line are always in the same plane.

$$\theta_{reflection} = \theta_{incidence} \tag{22.1}$$

Specular and diffuse reflection, and absorption

When we shine a laser beam on a wall in a dark room, everyone in the room can see the bright spot. According to the law of reflection, the light beam should reflect at a particular angle determined by the angle of incidence. Does the fact that everyone sees the bright spot disprove the law of reflection?

FIGURE 22.6 Reflection from a bumpy surface.

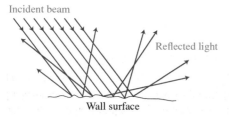

Before we reject the law of reflection, we must carefully examine the assumptions we made when developing it: the reflecting surface is very smooth (like a mirror), and the light beam shining on it can be represented by one light ray. What if the reflecting surface is bumpy, and the laser beam is wider than the bumps on the surface (**Figure 22.6a**)? In that case, we cannot represent the reflected laser light as a single ray. Various parts of the surface are oriented at many different angles, so the light is reflected in many different directions. This explains why everyone in the room can see some of the light reflected from the wall. If we repeated the experiment using a perfectly clean mirror instead of the wall, only one person would be able to see the light reflected from the spot on the mirror illuminated by the laser.

The phenomenon in which a beam of light is reflected by a smooth surface (like a mirror) is called **specular reflection**. During specular reflection, a narrow incident beam represented by parallel rays is reflected as a narrow beam of light represented by parallel rays according to the law of reflection (**Figure 22.7a**).

The phenomenon in which light is reflected by a rough surface is called **diffuse reflection**. During diffuse reflection, each of the parallel incident rays of a light beam strikes a part of the surface oriented at a different angle with respect to the incident light. Thus the reflected light spreads in multiple directions (Figure 22.7b).

FIGURE 22.7 (a) Specular light reflection from a smooth surface. (b) Diffuse light reflection from a bumpy surface.

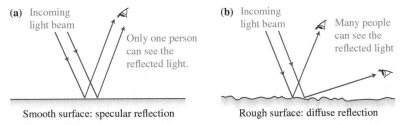

We can now explain how we saw the path of the laser beam in the experiments in Table 22.1. Each tiny speck of dust reflects incident light in multiple directions. As there are myriads of dust particles, reflected rays reach our eyes and we can see the "path" of light. Understanding diffuse reflection helps us explain the phenomenon captured in the photo in **Figure 22.8**. Sunlight coming through the church windows reflects off the dust in the air and reaches different observers below.

In addition to reflection, another process occurs when light encounters the surface of a medium. The surface warms up a little bit as some of the light energy is converted into thermal energy of the medium. This process is called **absorption**. When light interacts with a mirror's surface, very little absorption occurs, but for other media, such as black velvet, almost all of the light may be absorbed.

In the next exercise, we use diffuse and specular reflection and absorption to understand an everyday phenomenon.

FIGURE 22.8 A ray diagram represents how observers see dust particles in a church.

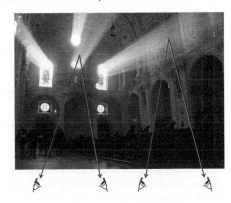

CONCEPTUAL EXERCISE 22.2 ▸ **Dark window on a sunny day**

On a sunny day, a house's uncovered window looks almost black, but its outside walls do not. How can we explain this difference?

Sketch and translate If the window looks black, it must reflect very little sunlight to your eyes. Our goal is to explain why.

Simplify and diagram When light shines on the rough walls of the house, it reflects back diffusely at different angles, and some of the light reaches your eyes—you see the walls easily. When light reaches the transparent surface of the window, most of it passes through the window and into the room and then reflects diffusely many times. During each reflection, some of the light is absorbed; because of this absorption, little comes back out. A small amount reflects off the smooth glass window as specular reflection (see the front view of the house and the top view inside the room in the figures on the right). If you are not standing in the right spot, you see very little light coming from the window—it appears dark. But if you are standing in that one location, you see a bright reflection of the Sun in the window.

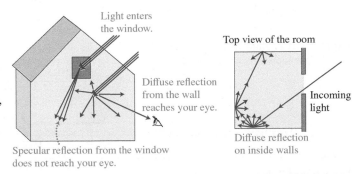

Try it yourself Why is the pupil of your eye dark?

Answer

The pupil is a hole in the eye—similar to a window in a house. Incident light enters the pupil, and very little is reflected back—the pupil looks dark.

FIGURE 22.9 The red-eye effect.

Red-eye effect

You are familiar with the so-called "red-eye" effect that occurs occasionally when using flash photography (see **Figure 22.9**). The red-eye effect is especially common at night and in dim lighting, when the pupil is wide open. When the flash illuminates the open iris, light reflects from the red blood vessels in the retina on the back of the eye. Some of this reflected light passes back out of the pupil and makes the pupil appear red.

REVIEW QUESTION 22.2 How can we test the law of reflection?

22.3 Refraction of light

At the shore of a lake, you see sunlight reflecting off the water's surface. But you also see rocks and plants under the surface. Because you see them, light must have entered the water, reflected off the rocks and plants, returned to the water surface, and then traveled from the surface to your eyes. If you have ever tried to use a stick to touch a rock under the surface of a pond or lake, you know that it is not easy. Although you carefully point the stick, you miss—you tend to extend the stick farther away from you than is needed to touch the rock. Why is that?

To answer this question, we will observe what happens when we shine a laser beam into water at different angles (see Observational Experiment **Table 22.4**).

OBSERVATIONAL EXPERIMENT TABLE 22.4 — The path of light changes when moving from air to water

VIDEO OET 22.4

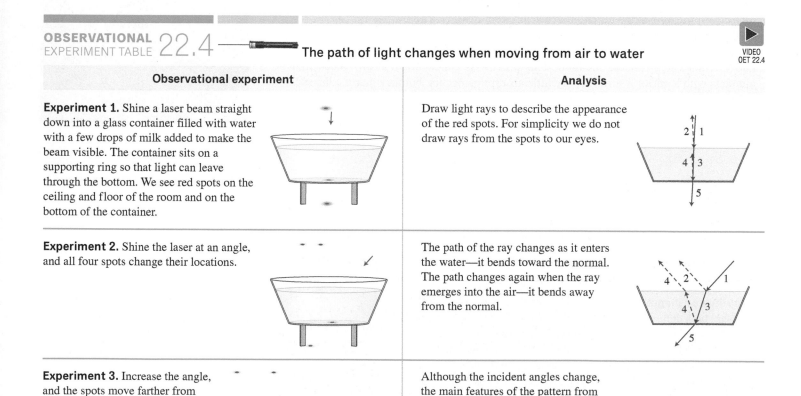

Observational experiment	Analysis
Experiment 1. Shine a laser beam straight down into a glass container filled with water with a few drops of milk added to make the beam visible. The container sits on a supporting ring so that light can leave through the bottom. We see red spots on the ceiling and floor of the room and on the bottom of the container.	Draw light rays to describe the appearance of the red spots. For simplicity we do not draw rays from the spots to our eyes.
Experiment 2. Shine the laser at an angle, and all four spots change their locations.	The path of the ray changes as it enters the water—it bends toward the normal. The path changes again when the ray emerges into the air—it bends away from the normal.
Experiment 3. Increase the angle, and the spots move farther from their original locations.	Although the incident angles change, the main features of the pattern from Experiment 2 remain.

Patterns

When light reaches the air-water boundary at the top surface, the incident light beam represented by ray 1

- partially reflects (ray 2) at the same angle as the angle of incidence and
- partially passes into the second medium (ray 3), bending at the interface *toward* the normal line.

Similar things happen to the light beam represented by ray 3. However, there are some differences. When ray 3 reaches the bottom water-air interface, it

- partially reflects (ray 4) at the same angle as the angle of incidence and
- partially passes from the water into the air below the container (ray 5), bending at the interface *away* from the normal line.

Note in Table 22.4 that when the incident light represented by rays 1 and 3 is perpendicular to the boundary of the surfaces (Experiment 1), light reflects along the same line (rays 2 and 4) and passes into the second medium without bending (ray 5). However, if the light is not perpendicular to the surface, it bends and travels in a different direction than in the previous medium. If we were to replace the water with a thick piece of transparent glass, we would observe a similar phenomenon. The change in the path of light when light travels from one medium to another is called **refraction**.

To develop a mathematical relationship between the angle of incidence and the **angle of refraction** (the angle the refracted beam makes with the normal line), we can do an experiment similar to the one we did when studying reflection. Such an experiment is shown in **Figure 22.10**. We shine a laser beam at horizontal air-water and air-glass interfaces and record the angles of incidence and refraction. The setup in Figure 22.10 allows us to do this easily because the light, after bending at the air-glass interface, always travels along the radius of the hemispheric container. Because this direction is perpendicular to the glass-air interface for the light leaving the container, light does not bend anymore. The results are shown below in **Table 22.5**.

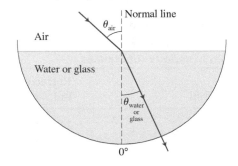

FIGURE 22.10 An experiment to measure the incident and refracted angles of light at an air-water (or air-glass) interface.

TABLE 22.5 **Angles of incidence and refraction between laser beams and the normal line**

Incident angle (ray in air) θ_{air}	Refraction angle (air into water) θ_{water}	Refraction angle (air into glass) θ_{glass}
0°	0°	0°
10°	8°	7°
20°	15°	13°
30°	22°	19°
40°	29°	25°
50°	35°	30°
60°	40°	35°

Notice that as the angle of incidence increases, the angle of refraction also increases, but not in an obvious pattern. The pattern is also not the same for all materials.

In 1621 the Dutch scientist Willebrord Snell (1580–1626) found a pattern. The ratio of the sine of the incident angle θ_1 to the sine of the refraction angle θ_2 remains constant for light traveling from air to water and for light traveling from air to glass. However, the constants are different for each pair of media (see columns four and five of **Table 22.6**, on the next page).

Notice that the ratio of the sines for air/water is about 1.3 and for air/glass is about 1.5. From these observations, Snell formulated a mathematical model for refraction phenomena:

$$\frac{\sin \theta_1}{\sin \theta_2} = n_{1 \text{ to } 2} \qquad (22.2)$$

where $n_{1 \text{ to } 2}$ is a number that depends on the two materials the light is traveling through. If we split $n_{1 \text{ to } 2}$ into a ratio of two numbers, one number that depends on the material

TABLE 22.6 **Pattern found by Snell for the ratio of the sines of the incident and refraction angles**

Air $\sin \theta_1$	Water $\sin \theta_2$	Glass $\sin \theta_2$	Air/water $(\sin \theta_1)/(\sin \theta_2)$	Air/glass $(\sin \theta_1)/(\sin \theta_2)$
0.000	0.000	0.000		
0.174	0.131	0.114	1.33	1.53
0.342	0.259	0.225	1.32	1.52
0.500	0.374	0.326	1.34	1.53
0.643	0.485	0.423	1.33	1.52
0.766	0.573	0.500	1.34	1.53
0.866	0.649	0.569	1.33	1.52

through which the incident ray travels (n_1) and the other that depends on the material through which the refracted ray travels (n_2), we get

$$\frac{\sin \theta_1}{\sin \theta_2} = \frac{n_2}{n_1}$$

We call n_1 and n_2 the **refractive indexes** (or the indexes of refraction) of medium 1 and medium 2, respectively. Rearranging the equation, we get what has become known as **Snell's law** (or the law of refraction).

TABLE 22.7 **Refractive indexes of yellow light in various substances**

Material	Refractive index n
Vacuum	1.00000
Air	1.00029
Carbon dioxide	1.00045
Water (20 °C)	1.333
Ethyl alcohol	1.362
Glucose solution (25%)	1.372
Glucose solution (50%)	1.420
Glucose solution (75%)	1.477
Glass	1.517–1.647
Diamond	2.417

Snell's law Snell's law (the law of refraction) relates the refraction angle θ_2 to the incident angle θ_1 and the indexes of refraction of the incident medium n_1 and the refracted medium n_2:

$$n_2 \sin \theta_2 = n_1 \sin \theta_1 \qquad (22.3)$$

The angles of incidence and refraction are measured with respect to the normal line where the ray hits the interface between the two media. If the refracted ray is closer to the normal than the incident ray, then medium 2 is more optically dense (with a higher refractive index) than medium 1. The incident ray, refracted ray, and normal line are in the same plane.

TIP Light going from a material with a lower to that with a higher index of refraction will bend toward the normal, but light going from a material with a higher to that with a lower index of refraction will bend away from the normal.

If we define the index of refraction of air as 1.00, then the indexes of refraction of water and of the glass used in Tables 22.5 and 22.6 are 1.33 and 1.53, respectively. Notice that glass refracted the light ray more toward the normal line than did water; the glass is more optically dense than water. The refractive indexes of several materials are given in **Table 22.7**, using yellow light. We will learn later in the chapter that light of different colors has slightly different indexes of refraction. Thus specifying the color of light in the table is important.

At the beginning of this section we discussed how it is difficult to touch an object under water with a stick. **Figure 22.11** and our knowledge of refraction help us understand why. Because the light coming from the rock bends at point A, our brain perceives that the light is coming from point B′. If we aim the stick along AB′, instead of AB, we miss the rock.

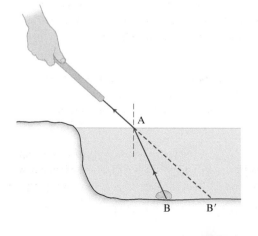

FIGURE 22.11 The bending of light makes it difficult to touch the rock.

EXAMPLE 22.3   Concentration of glucose in blood

The refractive index of blood increases as the blood's glucose concentration increases (for normal glucose content it ranges from 1.34 to 1.36; it is higher than the 1.33 for water because of all the additional organic components in the blood). Therefore, measuring the index of refraction of blood can help determine its glucose concentration.

In a hypothetical process, a hemispheric container holds a small sample of blood. A narrow laser beam enters from the bottom of the container perpendicular to the curved surface and into the sample. The light reaches the blood-air interface at a 40.0° angle relative to the normal line. The light leaves the blood and passes through the air toward a row of light detectors indicating that the light beam left the blood at a 61.7° angle. Determine the refractive index of the blood.

Sketch and translate First, we sketch the situation (see the figure). We use Snell's law to determine the index of refraction of the blood

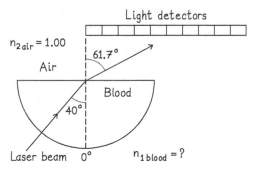

$n_{1\,blood}$. The incident angle is $\theta_{1\,blood} = 40.0°$. The refraction angle is $\theta_{2\,air} = 61.7°$. The index of refraction of the air is $n_{2\,air} = 1.00$.

Simplify and diagram The diagram in the figure can be used as a ray diagram if we assume that the laser beam is narrow.

Represent mathematically Snell's law for this situation is

$$n_{1\,blood} \sin \theta_{1\,blood} = n_{2\,air} \sin \theta_{2\,air}$$

Solve and evaluate Dividing both sides of Snell's law equation by $\sin \theta_{1\,blood}$, we have

$$n_{1\,blood} = \frac{n_{2\,air} \sin \theta_{2\,air}}{\sin \theta_{1\,blood}} = \frac{1.00 \sin 61.7°}{\sin 40.0°} = 1.37$$

This result is higher than the normal index of refraction for blood.

Try it yourself Pure blood plasma has a higher index of refraction than blood. Suppose the refractive index of the patient's plasma is 1.43 instead of 1.37. What then would the refraction angle be if the incident angle were still 40.0°?

Answer   66.8°.

QUANTITATIVE EXERCISE 22.4   Laser beams in the pool

A waterproof laser device produces three laser beams in the same vertical plane, one pointing directly upward and two at 60.0° above the horizontal. You put the device on the bottom of your swimming pool. The directions of the beams emerging from the water change as shown in the figure. At what angles relative to the horizontal do the laser beams emerge from the pool? Does the depth of the pool affect the answer?

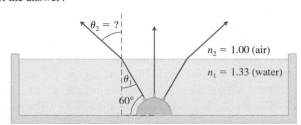

Represent mathematically The vertical beam does not change direction because it is perpendicular to the water's surface. The side beams are symmetrical, thus we can analyze one of them and apply the result to the other one.

$$n_1 \sin \theta_1 = n_2 \sin \theta_2 \implies \sin \theta_2 = \frac{n_1 \sin \theta_2}{n_2}$$

The incident angle $\theta_1 = 90° - 60° = 30°$. The angle with the horizontal is $90° - \theta_2$.

Solve and evaluate

$$\sin \theta_2 = \frac{n_1 \sin \theta_2}{n_2} = \frac{1.33 \sin 30.0°}{1.00} = 0.665$$

Since $\sin \theta_2 = 0.665$, we use a calculator to determine the angle: $\theta_2 = 41.7°$. The side beams emerge from water at angles $90° - 41.7° = 48.3°$ relative to horizontal. Because air has a smaller index of refraction than water, the rays traveling from water to air should bend away from the normal. Therefore, they should make a smaller angle with the horizontal than when in the water. Our answer therefore makes sense. The answer does not depend on the depth of water in the pool because we are only concerned with the angles, not with the trajectories of the beams. If the pool were deeper, the beams would emerge in different locations, but their directions would not change.

Try it yourself The index of refraction of water increases by 0.007 when you add 35 g of salt per 1 kg of water. How much will the angles of the beams in this exercise change as a result of such an addition of salt?

Answer   The change for the vertical beam angle will be zero. The side beams will emerge from the water at angles 48.0° relative to horizontal; therefore, their angles will decrease by 0.3°.

Restoring a broken glass

Imagine that you place a piece of glass in water. Will you be able to see it? As long as light reflects off the glass and reaches your eye, you will. Since the index of refraction of glass is different from that of water, there is an **optical boundary** between them. When light traveling in water hits the glass, it refracts at the water-glass boundary and also reflects some light back to the eyes of an observer, thus making the glass visible. Now, imagine that you have a liquid whose refractive index is exactly equal to that of the glass. A good example is vegetable oil. Light traveling through this oil will not reflect off the piece of glass submerged in it or refract at the surface of the glass because there is effectively no optical boundary between the two materials. Light will just continue traveling in straight lines as if the glass were not there. Thus a piece of glass becomes invisible in vegetable oil! This explains the magic trick described in the chapter-opening story. (Note: If you decide to try this experiment, remember to fill the unbroken vial with oil before you put it in the oil bath; otherwise, it will be full of air and the air will make it visible!)

In summary, in order for us to see things, they must radiate light (like lightbulbs or fire), reflect light (like planets, trees, and snowflakes), or refract light (like glass in water). Reflection and refraction occur off a transparent object only if the optical density of the reflecting object is different from that of the material that surrounds it.

REVIEW QUESTION 22.3 Why is the expression "light travels in a straight line" not entirely accurate?

22.4 Total internal reflection

In two examples in the last section, light traveled from a more optically dense medium to a less optically dense medium, for example, from water to air. In such cases the light bent away from the normal line when entering the less dense medium. This behavior has some very important applications in the transmission of light by optical fibers.

Consider the situation represented by the ray diagram in **Figure 22.12a**. In a series of experiments, an incident ray under water hits a water-air interface at an increasingly larger angle relative to the normal line. As the incident angle (θ_1) gets bigger, the angle in the air between the refracted ray and the normal line (θ_2) gets even bigger. At the so-called **critical angle of incidence** θ_c (Figure 22.12c), the refraction angle is 90°. The refracted ray travels along the water-air interface. Notice that when this happens, the incident angle is still less than 90°. At incident angles larger than the critical angle, $\theta_1 > \theta_c$, the light is totally reflected back into the water—none escapes into the air (Figure 22.12d).

This behavior occurs whenever light attempts to travel from a more optically dense medium 1 of refractive index n_1 to a less optically dense medium 2 of refractive index n_2 ($n_1 > n_2$). We can use Snell's law to determine the critical angle in terms of the two indexes of refraction: $n_1 \sin \theta_1 = n_2 \sin \theta_2$, where $\theta_1 = \theta_c$ and $\theta_2 = 90°$. Remember that $\sin 90° = 1$. Inserting these values and solving for $\sin \theta_c$, we have

$$n_1 \sin \theta_c = n_2 \sin 90°$$

$$\sin \theta_c = \frac{n_2 \sin 90°}{n_1} = \frac{n_2(1)}{n_1} = \frac{n_2}{n_1}$$

FIGURE 22.12 Total internal reflection.

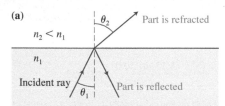

(a) $n_2 < n_1$ ... θ_2 Part is refracted. n_1. Incident ray. θ_1. Part is reflected.

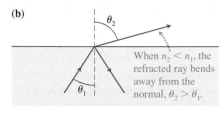

(b) θ_2 ... When $n_2 < n_1$, the refracted ray bends away from the normal, $\theta_2 > \theta_1$. θ_1

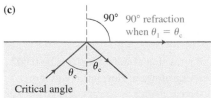

(c) 90° 90° refraction when $\theta_1 = \theta_c$. θ_c θ_c Critical angle

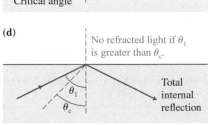

(d) No refracted light if θ_1 is greater than θ_c. θ_1 θ_c Total internal reflection

If the incident angle θ_1 is greater than θ_c, there is no solution to Snell's law, as sin θ_2 would be greater than 1.00 (the maximum value of sine is 1.00). For $\theta_1 > \theta_c$, all of the light is reflected back into the water. There is no refracted ray. Remember that this applies only for situations when $n_1 > n_2$. This phenomenon is called **total internal reflection**.

> **Total internal reflection** When light travels from a more optically dense medium 1 of refractive index n_1 into a less optically dense medium 2 of refractive index n_2 $(n_1 > n_2)$, the refracted light beam in medium 2 bends away from the normal line. If the incident angle θ_1 is greater than a critical angle θ_c, *all* of the light is reflected back into the denser medium. The critical angle is determined by
>
> $$\sin \theta_c = \frac{n_2}{n_1} \qquad (22.4)$$

Figure 22.13 shows light traveling from a less dense medium (air) into a more dense medium (Plexiglas) at an incident angle of a little less than 90°. Due to this incident angle, the light refracts into the more dense medium at an angle that is a little less than the critical angle. Given that the piece of plastic is rectangular, the incident angle at the upper edge, going from plastic to air this time, is a little *larger* than the critical angle (it is equal to 90° minus the earlier angle of refraction). We therefore observe total internal reflection at this point.

The refractive index is a fundamental physical property of a substance and is often used to identify an unknown substance, confirm a substance's purity, or measure its concentration. Refractometers, instruments that measure refractive index, have medical and industrial applications. Veterinarians use portable refractometers to measure the total serum protein in blood and urine and also to detect drug tampering in racehorses.

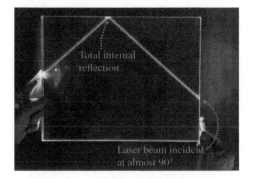

FIGURE 22.13 Total internal reflection in Plexiglas. Note the initial incident angle.

EXAMPLE 22.5   **Another method to measure glucose concentration in blood**

Let's look at another hypothetical method for determining the glucose concentration in blood by measuring its refractive index. Light travels in a narrow beam through a hemispheric block of high-refractive-index glass, as shown in the figure at right. A thin layer of blood is placed between it and another hemispheric block on top. For this example we assume that the refractive index of the blood is 1.360 and the refractive index of the glass is 1.600. What pattern of light reaching the light detectors on the top curved surface of the hemispheric block will you observe as you move the light source clockwise around the edge of the curved surface of the bottom block?

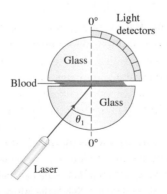

Sketch and translate We already have a sketch of the apparatus. Our goal is to predict what happens to the refracted beam if we vary the angle of incidence.

Simplify and diagram We assume that the light source is always oriented perpendicular to the curved surface on the bottom and points

toward the same point on the glass-blood interface. We also assume that the blood layer is thin. Next, we draw a ray diagram for four rays (as shown in the figure below).

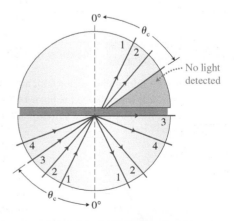

Ray 1 bends away from the normal line at the first glass-blood interface. It is partially reflected and partially refracted into the blood. On the top surface of the blood, it bends back toward the normal line as

(CONTINUED)

it moves into the hemispheric glass block above (part of it is reflected at the second interface—not shown). The net angular deflection of the ray is zero—ray 1 in the lower block is parallel to ray 1 in the top block. Ray 2 has a greater incident angle with the first glass-blood interface and also bends away from the normal line as it enters the blood. It is partially transmitted into the blood and partially reflected. On the top surface of the blood it bends back toward the normal line as it moves into the hemispheric glass block above—to the right of ray 1. Ray 3 has a greater incident angle and is refracted at 90° in the blood. Hence, the incident angle is the critical angle. Ray 4 has an even greater incident angle than the critical angle and is totally reflected back into the lower hemispheric block.

The detectors on the top surface stop detecting light when the incident angle is larger than the critical angle. Thus, you can detect the critical angle by the place where the light stops arriving at the top block.

Represent mathematically We can use Eq. (22.4) to determine the critical angle:

$$\sin \theta_c = \frac{n_2}{n_1}$$

Solve and evaluate Using the above with the glass index of refraction $n_1 = 1.600$ and the blood index of refraction $n_2 = 1.360$, we find that

$$\sin \theta_c = \frac{n_2}{n_1} = \frac{1.360}{1.600} = 0.850$$

A 58.2° angle has a sine equal to 0.850. When incident light reaches this angle, the detectors on the top hemisphere will stop detecting light. As long as the incident angle is smaller than 58.2°, the detectors on the top of the apparatus will detect light. Thus the apparatus allows us to measure the critical angle and use it to determine the index of refraction and glucose concentration of the blood.

- -

Try it yourself For a different sample of blood, you do not detect light at angles of 59.2° and larger. What is the refractive index of this new blood sample?

Answer   1.374.

REVIEW QUESTION 22.4 Why did we study total internal reflection after our investigation of refraction and not after reflection?

22.5 Skills for analyzing reflective and refractive processes

TIP When drawing ray diagrams, be sure to use a ruler.

When you solve problems involving light, use ray diagrams to help in your reasoning and quantitative work. The diagrams will also help you evaluate the final answer. In this section we will work through three exercises: a quantitative exercise, an example that requires the use of all the problem-solving steps, and, finally, an Equation Jeopardy example that shows how to interpret the mathematical description of light phenomena using ray diagrams.

QUANTITATIVE EXERCISE 22.6   **Rotating mirror**

A laser beam is incident on a mirror that rotates with rotational velocity of 1.0 rad/s around an axis that is parallel to the plane of the mirror, as shown in the figure. Determine the rotational velocity of the laser beam that reflects from the mirror.

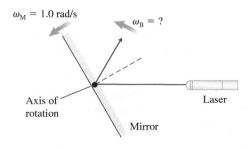

$\omega_M = 1.0$ rad/s

$\omega_B = ?$

Axis of rotation

Laser

Mirror

Represent mathematically According to the law of reflection, $\theta_{refl} = \theta_{inc}$. Suppose the initial angle of incidence of the laser beam is $\theta_{inc\ 1}$. The initial angle between the incident and reflected beam is therefore $2\theta_{inc}$. When the mirror rotates by an angle θ_M, the normal to the mirror rotates by the same angle, and because the laser does not move, the new angle of incidence becomes $\theta_{inc\ 2} = \theta_{inc\ 1} + \theta_M$. According to the law of reflection, the new reflected angle is $\theta_{refl\ 2} = \theta_{inc\ 2} = \theta_{inc\ 1} + \theta_M$. Therefore, the angle between the incident beam and reflected beam is $2\theta_{inc} + 2\theta_M$, and the change in this angle from the initial value is $\Delta\theta_B = 2\theta_M$.

Solve and evaluate Rotational velocity is given by $\omega = \frac{\Delta\theta}{\Delta t}$ (Eq. 9.2). Thus if the rotational velocity of the mirror is

$\omega_M = \dfrac{\Delta\theta_M}{\Delta t} = 1$ rad/s, the rotational velocity of the laser beam

reflected off the mirror will be $\omega_B = \dfrac{\Delta\theta_B}{\Delta t} = \dfrac{2\Delta\theta_M}{\Delta t} = 2.0$ rad/s.

Answer

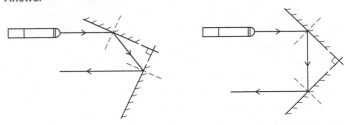

Try it yourself You send a laser beam horizontally from right to left. You have two mirrors. What are the possible arrangements of the mirrors that will allow you to send the reflected beam horizontally from left to right above the incident beam? What is the general pattern that any arrangement that achieves this goal should follow?

The figure below shows two possible arrangements of the mirrors. Any arrangement in which the mirrors are at a 90° angle to each other will work.

PROBLEM-SOLVING STRATEGY 22.1

Analyzing processes involving reflection and refraction

Sketch and translate

- Sketch the described situation or process.
- Indicate all the known and unknown quantities.
- Outline a solution for the problem as best you can.

Simplify and diagram

- Indicate any assumptions you are making.
- Draw a ray diagram directly on the sketch showing all relevant paths the light travels. Consider the light rays originating from the object and eventually reaching the observer. In the diagram you can represent the observer by an eye.

EXAMPLE 22.7 Hiding mosquito fish

A mosquito fish hides from a kingfisher bird at the bottom of a shallow lake, 0.40 m below the surface. A leaf floats above the mosquito fish. How big would the leaf need to be so that the kingfisher could not see its prey from any location above the water?

- The situation is sketched below.

- We want to know what size leaf is needed so that any light reflected from the fish and reaching the water surface undergoes total internal reflection and does not leave the water. Light incident at a smaller angle relative to the normal hits the leaf.

- Model the fish as a shining point particle and the leaf as circular.
- Draw a ray diagram. θ_c is the critical angle. For incident angles greater than θ_c, there is no refracted ray—all light is totally reflected. Note that here we show the ray diagram in a separate figure, but you only need to draw one combined sketch and ray diagram.

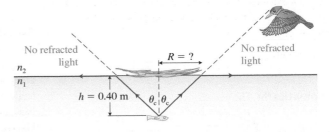

(CONTINUED)

Represent mathematically

- Use the sketch and ray diagram to help construct a mathematical description using the law of reflection, Snell's law for refraction, or the application of Snell's law for total internal reflection.

- Use Eq. (22.4), $\sin \theta_c = n_2/n_1$, where $n_1 = 1.33$ (the refractive index of water) and $n_2 = 1.00$ (the refractive index of air).

- Knowing the angle θ_c, we can determine the radius of the leaf: $R = h \tan \theta_c$.

Solve and evaluate

- Solve the equations for the unknown quantities.

- Evaluate the results to see if they are reasonable (the magnitude of the answer, its units, how the answer changes in limiting cases, and so forth).

- We find that $\sin \theta_c = 1.00/1.33 = 0.752$. The angle whose sin is 0.752 equals 48.8°.

- Thus $\tan \theta_c = \tan 48.8° = 1.14$.

- We can now determine the radius of the leaf: $R = h \tan \theta_c = (0.40 \text{ m})(1.14) = 0.46$ m.

- This is an unrealistically big leaf. The fish is not safe. Note that the ray diagram shows the large radius of the leaf compared to the depth of the fish. In addition, even if we could find such a leaf, we assumed that the fish is a point-like object with no size. For a real fish, we would need an even a bigger leaf.

Try it yourself If the leaf radius is 0.20 m, how far below the leaf would the mosquito fish have to be so that it could not be seen by the kingfisher bird?

Answer   0.18 m or less.

Equation Jeopardy

In the following example, you are asked to interpret the mathematical description of a process and construct a ray diagram that represents that process. Then you are asked to invent a word problem that the equation could be used to solve. (The problem-solving procedure is reversed in an Equation Jeopardy problem.)

EXAMPLE 22.8   Equation Jeopardy

The two equations below describe a physical process. Invent a problem for which the equations would provide a solution.

$$1.00 \sin 30° = 1.60 \sin \theta_2$$
$$1.60 \sin \theta_2 = 1.33 \sin \theta_3$$

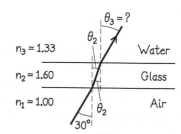

Solve and evaluate We solve for the unknown quantities:

$$\sin \theta_2 = \frac{1.00 \sin 30°}{1.60} = 0.313, \text{ or } \theta_2 = 18.2°$$

$$\sin \theta_3 = \frac{1.60 \sin \theta_2}{1.33} = 0.377, \text{ or } \theta_3 = 22.1°$$

Represent mathematically The equations are the applications of Snell's law.

Simplify and diagram Based on the indexes of refraction, the first medium in which light travels is air, then glass, then water. The ray diagram for the process is shown at top right.

Sketch and translate One possible problem is as follows. A narrow beam of light moves through air and hits the glass bottom of an aquarium at a 30° angle relative to the normal line. After passing through the glass wall, the beam hits the glass-water interface and refracts into the water, as shown in the sketch. Determine the angle relative to the normal at which the light beam propagates in water. The index of refraction of the glass is 1.60 and that of the water is 1.33.

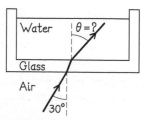

Try it yourself Show that the angle relative to the normal at which the light beam propagates in water in the previous problem does not depend on the index of refraction of glass.

Answer

$\Rightarrow n_1 \sin \theta_1 = n_3 \sin \theta_3.$

$n_1 \sin \theta_1 = n_2 \sin \theta_2 = n_3 \sin \theta_3$

REVIEW QUESTION 22.5 What is the critical angle for total internal reflection for light going from water into glass of refractive index 1.56?

22.6 Fiber optics, prisms, mirages, and the color of the sky

In this section we will consider several applications of reflection, refraction, and total internal reflection: fiber optics, prisms, mirages, and the color of the sky.

Fiber optics

Imagine light traveling inside a long, thin, flexible, glass cylinder (like a wire made of flexible glass): such a cylinder is called an **optical fiber**. Fiber optic filaments are used in telecommunications to transmit high-speed light-based data and in medicine to see inside the human body during surgery. The following example will help you understand the physics behind fiber optics.

EXAMPLE 22.9 **Trapping light inside glass**

Imagine that you have a long glass block of refractive index 1.56 surrounded by air. Light traveling inside the block hits the top horizontal surface at a 41° angle. What happens next?

Sketch and translate The situation is sketched below. As light travels from glass into air, we need to compare the incident angle to the critical angle for total internal reflection.

Air $n_2 = 1.00$

Glass $n_1 = 1.56$ $\theta_1 = 41°$

Simplify and diagram Assume that the top and bottom surfaces of the block are parallel to each other. The light is hitting the top of the glass at a 41° angle. There are three possibilities: the light leaves the block, the light refracts so it moves along the surface, or the light is totally reflected back into the block. The last possibility is shown in the sketch.

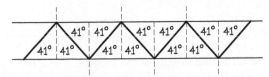

Represent mathematically The critical angle for the glass-air interface according to Eq. (22.4) is

$$\sin \theta_c = \frac{n_2}{n_1} = \frac{1.00}{1.56} = 0.64, \text{ or } \theta_c = 40°$$

If the incident angle of light in the glass is greater than this critical angle, all light is reflected at the glass-air interface. When it reaches the opposite side of the block, it hits the bottom surface at the same angle because the sides are parallel and thus reflects back again.

Solve and evaluate We found above that $\theta_c = 40°$. The given 41° angle is slightly greater than the critical angle for total internal reflection. Thus the light is totally internally reflected during the first incidence on the upper surface. From there it moves down and to the right and hits the bottom surface at a 41° angle, where it is totally internally reflected again. The process of total internal reflection continues as the light travels the length of the block. However, if the material of the block is flexible and bends, then the incident angle changes. The incident angle might become so small that total internal reflection does not occur and some light refracts out of the block.

Try it yourself What happens if the light hits the top surface at 45°? At 38°?

Answer   the intensity of light within the block to diminish.

some light refracts out of the top and bottom at each incidence, causing

incident at 40° or more). For 38° incidence (since it is less than 40°),

For 45° incidence, total internal reflection occurs (it occurs for all light

FIGURE 22.14 (a) A fiber optic view inside a knee during surgery. (b) The fiber optic endoscope used to look inside the body.

(a)

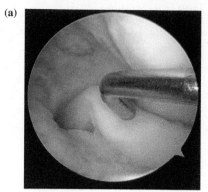

(b)

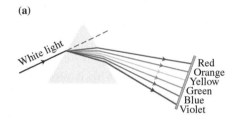

Outer fibers conduct light inside body

Inner fibers conduct image to observer

Object

Within optical fibers, light is totally internally reflected, even when the fiber is bent (unless you bend it too much), allowing rapid transmission of light and data. In surgery, fiber optics can guide the surgeon's hands (**Figure 22.14a**). A thin bundle of tiny glass fibers transmits light into the area being illuminated and then transmits the reflected light out for viewing (Figure 22.14b). A tiny tool can be inserted along with the fibers to, for example, clean cartilage or repair tendons. Only a small incision is needed to insert the fibers and the tool, reducing the amount of trauma to the joint and surrounding tissue.

Prisms

Centuries ago, Isaac Newton observed a thin beam of light coming into a room through a tiny hole in one of the shutters. By chance, this beam illuminated a prism sitting on a desk. To Newton's surprise, he saw a rainbow band of colored light on the opposite wall (**Figure 22.15a**). The band was much wider than the original beam, with violet light on the bottom and red on top. Was the prism creating these colors? To test this idea, he put an identical prism after the first one, but upside down (Figure 22.15b), and the colored band disappeared—the spot on the wall was white and small. He concluded that the prism did not create the colors; it somehow could separate the different colors out of white light and then recombine them back into white light. Newton further reasoned that the refractive index was greater for violet light and smaller for red, explaining the separation of colors (see **Table 22.8**).

FIGURE 22.15 (a) A prism separates white light into a light spectrum. (b) A second prism combines the spectrum back into white light.

(a)

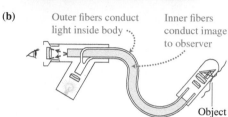

White light

Red
Orange
Yellow
Green
Blue
Violet

(b)

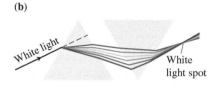

White light

White light spot

TABLE 22.8 Refractive indexes of glass for different colors

Color of light	n_{glass}
Red	1.613
Yellow	1.621
Green	1.628
Blue	1.636
Violet	1.661

FIGURE 22.16 Reflecting prisms.

(a)

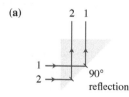

90° reflection

(b)

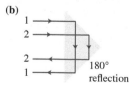

180° reflection

(c)

Since the critical angle for a glass-air interface is less than 45° (see Example 22.9), glass prisms with 45-45-90° right angles are used in many optical instruments such as telescopes and binoculars to achieve total reflection of light through 90° or 180° angles. Examples of the reflecting ability of prisms are shown in **Figure 22.16**. Prisms are better for reflection than mirrors for several reasons. First, prisms reflect almost 100% of the light incident on the prism via total internal reflection, whereas mirrors reflect somewhat less than 100%. Second, most mirrors are made of glass over a thin sheet of metal, which tarnishes and loses its reflective ability with age. Prisms retain their reflective ability. Finally, prisms reverse the order of rays, as seen in Figure 22.16.

Mirages

An interesting consequence of the refraction of light is the formation of a **mirage**, such as the distant shimmering on a roadway that looks like water (**Figure 22.17**). How can we explain this observation?

On a hot day, hot air may hover just above the pavement. The air above is cooler. The hot air is less dense and has a lower index of refraction than the cooler air above it. When light from the sky passes through air with a gradually changing index of refraction, its path gradually bends. This bending of light leads us to perceive that the source of light is at a different location than it actually is (see **Figure 22.18**).

For simplicity we will consider only one ray coming from point A in the sky (Figure 22.18a). Instead of a continuous variation in the refractive index as the light slants downward, we will assume that it passes through several layers of different refractive index (represented as parallel layers). Its path changes according to Snell's law. At points 1 and 2, the light moving into layers of lower refractive index bends away from the normal line, and its angle with the pavement decreases. At some point (point 3), the incident angle becomes so large that the ray undergoes total internal reflection and starts going up. After passing through several layers, it enters the eye of the observer, whose brain perceives the ray as traveling along the straight line that originated at point B on the ground.

FIGURE 22.17 A mirage.

FIGURE 22.18 The formation of a mirage. An observer perceives that light is coming from the pavement, when in fact the source of light is the sky.

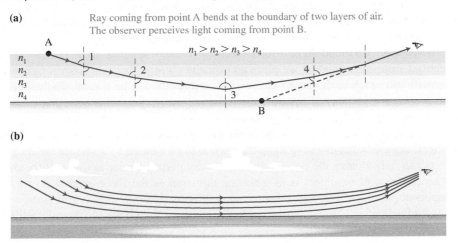

(a)

Ray coming from point A bends at the boundary of two layers of air. The observer perceives light coming from point B.

$n_1 > n_2 > n_3 > n_4$

(b)

Blue region is perceived as water, but is really formed by light from the sky.

If we now consider more rays coming from a section of the sky (Figure 22.18b), the observer would see them originating in the vicinity of B. This location will look blue and will shimmer due to convection in the air above the road surface. The result looks like water on the road (though it is actually light from the sky).

Color of the sky

We learned in this chapter that we could represent light reaching Earth from the Sun as a beam of parallel rays. Sunlight contains all visible colors. Why then does the entire sky look blue?

Molecules, dust particles, and water droplets in the atmosphere that are along the path of light from the Sun to Earth reflect sunlight in all directions, similar to the chalk dust reflecting laser light in the first experiments in this chapter. If the atmosphere reflected all colors of light similar to the way chalk dust does, the atmosphere would be the same color as the Sun.

Instead, we see a sky that is primarily blue. Due to their sizes, atmospheric particles reflect blue light more efficiently than other colors (**Figure 22.19**). Thus, all colors other than blue pass through the atmosphere without changing direction as much. Because the blue light is reflected in all directions, even when we look at the sky away from the Sun, reflected blue light from the atmosphere still reaches our eyes. This explanation is supported by probes sent to other planets, where the atmospheres have different chemical and physical compositions than ours. The skies of Venus and Mars are different colors even though illuminated by the same sunlight as Earth.

If our atmosphere reflected all colors the same way, we would see the sky as white—this is in fact what we see when the sky is covered with clouds. The water droplets in clouds reflect all colors equally.

FIGURE 22.19 Why the sky is blue. Blue light scatters off atmospheric particles and reaches our eyes from all directions. Light of other colors is not scattered as much.

The atmosphere reflects blue light better than other colors.

Blue light scatters.

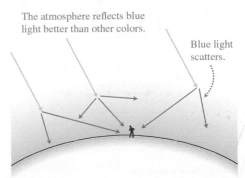

REVIEW QUESTION 22.6 Why is the sky blue? Why are most clouds white?

FIGURE 22.20 Newton's particle-bullet model of light.

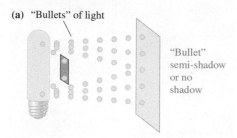

(a) "Bullets" of light

"Bullet" semi-shadow or no shadow

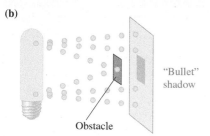

(b)

"Bullet" shadow

Obstacle

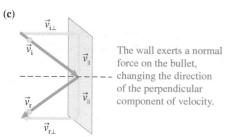

(c)

$\vec{v}_{i\perp}$

$\vec{v}_i$

$\vec{v}_{\parallel}$

$\vec{v}_r$

$\vec{v}_{\parallel}$

$\vec{v}_{r\perp}$

The wall exerts a normal force on the bullet, changing the direction of the perpendicular component of velocity.

FIGURE 22.21 Explanation of light refraction using the particle-bullet model.

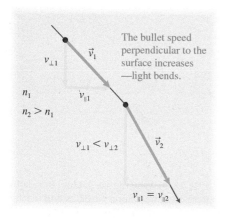

$\vec{v}_1$

$v_{\perp 1}$

The bullet speed perpendicular to the surface increases —light bends.

n_1

$v_{\parallel 1}$

$n_2 > n_1$

$v_{\perp 1} < v_{\perp 2}$

$\vec{v}_2$

$v_{\parallel 1} = v_{\parallel 2}$

FIGURE 22.22 Wavelets produce a new wave front.

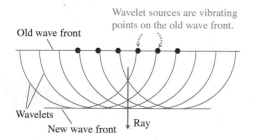

Wavelet sources are vibrating points on the old wave front.

Old wave front

Wavelets

New wave front   Ray

22.7 Explanation of light phenomena: two models of light

We have developed a ray model of light that describes the way light behaves. What other models can we use to explain the way light behaves?

Particle-bullet model

Isaac Newton modeled light as a stream of very small, low-mass particles moving at very high speeds, like bullets. According to this model, light bullets are affected by Earth's gravitational pull and move like projectiles, but since they move very quickly, their deflection from straight lines is not noticeable. Let us use this model to explain light phenomena familiar to us. We will start with shadows.

Imagine an extended light source such as a bulb sending light bullets in all directions from each point. If we place an obstacle close to the light source, bullets of light will still reach each part of the screen—the result will be a semi-shadow or no shadow (**Figure 22.20a**). If the obstacle is farther away, there will be a place on the screen where no bullets reach the screen—a shadow will form (Figure 22.20b). We can explain the reflection of light if we imagine that the bullets bounce elastically off surfaces. For the bullet model of light, the normal force exerted by a surface on the light bullets can only change the component of velocity perpendicular to the wall (because the acceleration of an object is in the same direction as the force exerted on it); the component parallel to the wall stays the same (Figure 22.20c). This is consistent with the law of reflection. Newton's particle-bullet model of light does successfully explain light propagation and reflection.

Can we explain refraction using this model of light? In order for the model to be consistent with Snell's experiments on refraction, the light bullets will have to speed up when they refract into a more optically dense medium (see **Figure 22.21**). The particle-bullet model suggested that the denser medium exerts an attractive force on the light particles, causing their speeds to increase. Accurate measurements of the speed of light turned out to be very challenging, and for many years the speed of light could only be determined in air or in a vacuum (it was found to be 299,792,458 m/s—about 3×10^8 m/s). In addition, the particle-bullet model requires an additional interaction between the surface and the light particles that causes them to speed up when entering the denser medium. Scientists prefer explanations that are as simple as possible. Another model of light that did not require this additional interaction was proposed.

Wave model of light

Simultaneously with Newton's development of the particle-bullet model of light, Christiaan Huygens was constructing a wave model of light. The motivation for the model could have come from the observations that light reflects off objects similar to the way sound reflects. Recall (from Chapter 11) that Huygens' wave propagation ideas involved disturbances of a medium caused by each point on a wave front producing a circular wavelet. Imagine that we have a wave with a wave front moving downward parallel to the page (see **Figure 22.22**). We choose six dots on this wave front, and from each dot we draw a wavelet originating from it (as shown in Figure 22.22). Each dot represents the source of a circular wavelet produced by the wave disturbance passing that point. According to Huygens' principle, each small wavelet disturbance produces its own circular disturbance that moves down the page in the direction the wave is traveling. The distance between two consecutive wavelets is equal to the wavelength. Now, note places where the wavelets add together to form bigger waves. These places are where the wave front will be once it has moved a short distance down the page. We also draw an arrow (a ray) indicating the direction the wave is traveling perpendicular to the wave fronts and consistent with the ray model of light propagation. In this model, the ray is the line perpendicular to the wave front.

When the waves emitted by the same source travel in different media, their frequency remains the source frequency, but their speed changes depending on the medium. For example, water waves travel more slowly in shallow water than in deeper water. Sound travels more slowly in cold air than in warm air. What does the wave model predict will happen when waves propagate through media with different wave speeds and with zero incidence angles? Imagine that the speed of the wave shown in **Figure 22.23** is greater in medium A than in medium B. When the wavelets from medium A reach the boundary with medium B, their radii reduce because the wave has a smaller speed in medium B. The new wave fronts are therefore closer together.

Wave model and refraction

Imagine a light wave moving in a less optically dense medium A and reaching an interface with a denser medium B (where it travels at a slower speed) at a nonzero angle of incidence (**Figure 22.24**). Using the wave model of light, we can now draw wave fronts for this wave as it travels from medium A into medium B. During a certain time interval, the wavelet that departed from the right edge of the wave front is still traveling in medium A. The wavelet that departed from the lower left edge of the wave front at the same time travels a shorter distance in medium B since it is moving slower. Once the point on a wave front reaches the boundary, the radius of the wavelet in medium B becomes smaller than in medium A, while the wavelets for the points on the same wave front that is still in medium A have larger radii. The wave front breaks at the boundary, as shown in Figure 22.24a.

What if a light wave travels from a slower to a faster medium, as shown in Figure 22.24b? Using similar reasoning, we find that the light bends away from the normal. If the wave model is going to describe refraction properly, we must conclude that light travels more slowly in water than in air.

According to the wave model of light, the speed of light should be slower in water than in the air—exactly opposite the prediction of the particle-bullet model. Which model is more accurate? An obvious way to answer this question is to measure the speed of light in water. But this experiment is difficult. Thus, we are left with two models of light that both explain the reflection and refraction of light but lead to different predictions about its speed in water. This dilemma existed in physics for a long time due to the difficulty in measuring the speed of light in different media. Ultimately, the resolution came not from measurements of the speed of light but from overwhelming experimental support for the wave model. We will discuss the wave model in more detail in a later chapter (Chapter 24).

Unanswered questions

Although we have learned a great deal about light, there are still questions that we have not answered. We do not know why different media bend light differently. We still do not know whether light propagates faster or slower in water and other media compared with air. We do not know how objects radiate light. Why do some stars look white but others look red? We also have not decided which model of light is better—the particle-bullet model or the wave model. We will investigate these questions in the coming chapters.

REVIEW QUESTION 22.7 What is the difference between the two models of light and the predictions they make for the speed of light in water?

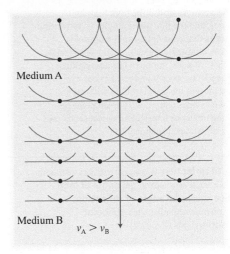

FIGURE 22.23 Wavelets crossing perpedicularly to the boundary between two regions with different wave speeds.

Medium A

Medium B $v_A > v_B$

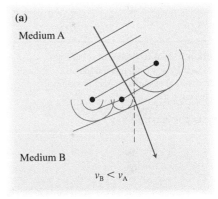

FIGURE 22.24 Huygens' principle explains why the wave changes directions when it reaches the interface between two media.

(a)

Medium A

Medium B $v_B < v_A$

The wave travels farther on the right side in the less optically dense medium A than on the left side in denser medium B.

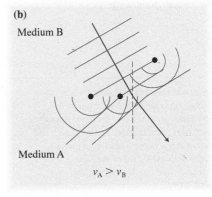

(b)

Medium B

Medium A $v_A > v_B$

Medium B is less optically dense than medium A, causing the wave to bend out rather than in.

Summary

Light sources and light rays Lightbulbs, candles, and the Sun are examples of **extended sources** that emit light. Light from such sources illuminates other objects, which reflect the light. We see an object because incident light reflects off of it and reaches our eyes. We represent the travel of light with **rays** (drawn as lines and arrows). Each point of a shining object or a reflecting object sends rays in all directions. (Section 22.1)

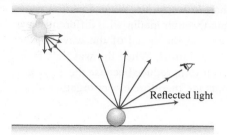

Reflected light

Law of reflection When a ray strikes a smooth surface such as a mirror, the **angle of incidence** equals the **angle of reflection**. This phenomenon is called **specular reflection**. (Section 22.2)

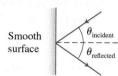

Smooth surface

$\theta_{incident}$

$\theta_{reflected}$

$$\theta_{\text{reflection}} = \theta_{\text{incidence}} \qquad \text{Eq. (22.1)}$$

Diffuse reflection If light is incident on an irregular surface, the incident light is reflected in many different directions. This phenomenon is called **diffuse reflection**. (Section 22.2)

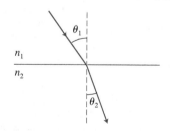

Refraction If the direction of travel of light changes as it moves from one medium to another, the light is said to refract (bend) as it moves between the media. (Section 22.3)

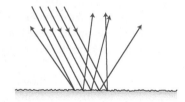

Air
$n = 1.00$

Glass
$n = 1.60$

Water
$n = 1.33$

Snell's law Light going from a lower to a higher index of refraction will bend toward the normal, but when going from a higher to a lower index of refraction it will bend away from the normal. (Section 22.3)

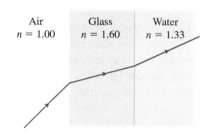

θ_1

n_1

n_2

θ_2

$$n_2 \sin \theta_2 = n_1 \sin \theta_1 \qquad \text{Eq. (22.3)}$$

Total internal reflection If light tries to move from a more optically dense medium 1 of refractive index n_1 into a less optically dense medium 2 of refractive index n_2 ($n_1 > n_2$), and the incident angle in medium 1 θ_1 is greater than a critical angle θ_c, *all* of the light is reflected back into the denser medium. (Section 22.4)

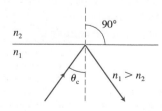

$90°$

n_2

n_1

θ_c

$n_1 > n_2$

$$\sin \theta_c = \frac{n_2}{n_1} \qquad \text{Eq. (22.4)}$$

Questions

Multiple Choice Questions

1. How can you convince your friend that a beam of light from a laser pointer travels to a wall along a straight line?
   (a) Tell him that light rays travel in straight lines.
   (b) Sprinkle chalk dust along the laser beam.
   (c) Try to move along the laser beam blocking its light and note the light path.
   (d) Both b and c will work.
   (e) All answers a–c will work.

2. Each point of a light-emitting object
   (a) sends one ray.
   (b) sends two rays.
   (c) sends an infinite number of rays.

3. To test the model mentioned in Question 22.2, what can one use? Multiple answers are possible.
   (a) A pinhole camera
   (b) A small source of light and a large obstacle near a distant screen
   (c) A large source of light and a small obstacle close to the source of light

4. What is a light ray?
   (a) A thin beam of light
   (b) A model invented by physicists to represent the direction of travel of light
   (c) A physical law

5. What is a semi-shadow?
   (a) A shadow that is half the size of the original shadow
   (b) A place where some light rays from an object reach and rays from other parts of the object do not.
   (c) A scientific term for a shadow

6. You fix a point-like light source 3.0 m away from a large screen and hold a basketball 1.0 m away from the screen so that the line connecting the center of the light source and the center of the basketball is perpendicular to the screen. You observe a shadow of the basketball on the screen. Select two correct statements.
   (a) Moving the light source away from the screen will produce a larger shadow.
   (b) Moving the basketball closer to the screen will produce a smaller shadow.
   (c) Moving the basketball and the light source away from the screen (while keeping the distance between the light source and the basketball fixed) will not change the size of the shadow.
   (d) Moving the light source up will result in moving the shadow down.
   (e) Moving the basketball up will result in moving the shadow down.

7. What is a normal line?
   (a) A line parallel to the boundary between two media
   (b) A vertical line separating two media
   (c) A line perpendicular to the boundary between two media

8. A light ray travels through air and then passes through a rectangular glass block. It exits
   (a) parallel to the original direction.
   (b) bent toward the normal line.
   (c) along the identical path that it entered the block.
   (d) bent away from the normal.

9. A right triangular prism sits on a base. A narrow light beam from a laser travels through air and then passes through the slanted side of the prism and out the vertical back side. It now travels
   (a) parallel to the original direction.
   (b) bent downward toward the base of the prism.
   (c) along the identical path that it entered the block.
   (d) bent upward away from the base.

10. A laser beam travels through oil in a horizontal direction. Which of the following objects could you put in the oil in the path of the laser beam in order to tilt the direction of the laser beam upward? Ignore partial reflections at the boundaries between media. Multiple answers are possible.
    (a) A rectangular glass plate
    (b) A mirror
    (c) A triangular plastic prism with index of refraction less than the oil's index of refraction
    (d) A triangular plastic prism with index of refraction equal to the oil's index of refraction

11. We can observe total internal reflection when light travels
    (a) from air to glass.
    (b) from water to glass.
    (c) from glass to water.
    (d) from air to water to glass.

12. You point a laser beam perpendicularly to the surface of water in a glass aquarium that is tilted at 30° (see Figure Q22.12). Where will you observe the laser beam hitting the table? Ignore any effects of the glass on the beam path. (Point P lies directly under the laser, on the extension of the incident beam in air.) (a) At point P (b) To the right of point P (c) To the left of point P (d) The laser beam will not hit the table because it will undergo total internal reflection.

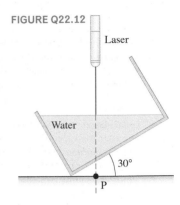

FIGURE Q22.12

Conceptual Questions

13. What effects of light radiation and reflection are responsible for the fact that we can see objects in a room?

14. Why does the inside of a well look black?

15. What do you need to do to create different size shadows of the same stick?

16. Explain how a sundial works (a sundial is just a vertical stick in the ground).

17. You are trying to see what is happening in a room through a keyhole in a door. Where should you place your eye to see the most: closer or farther away from the hole? Explain.

18. If you stand near a light pole at night, the shadow of your foot (if slightly off the ground) is rather sharp while the shadow of your head looks fuzzy. Why?

19. In what cases can you see only a semi-shadow of an object and not a shadow?

20. Why when moving away from a lightbulb, toward a screen, does your shadow on the screen become sharper and sharper?

21. Why can't you see a shadow of a pencil on the wall if you place it close to a lightbulb?

22. The visible diameters of the Moon and the Sun are almost the same. What should you see if the Moon were in the same line of sight as the Sun? Explain.

23. The shadow of the Moon on Earth is 200 km wide. Describe what you would see if you were in that area during the moment described in the previous question.

24. You stand at the side of a road. Why is light from a passing car's headlights easier to see in foggy weather than in clear weather?

25. During the day, you can see the trees in your garden through the window, but you can hardly see yourself in the window. At night, you can hardly see the trees through the same window, but you can clearly see yourself in the window. Explain.

26. You look at a fish underwater. Draw a ray diagram of the light that enters your eye from the fish.

27. Take a pencil and try to touch a penny on the bottom of a large pan of water while looking down into the pan at an angle. Explain why it is difficult to do this.

28. Will a beam of light experience total internal reflection going from glass into water or from water into glass? Explain (include drawings).

29. Explain how a prism turns a beam of white light into a rainbow.

30. On a hot sunny day, a highway sometimes looks wet. Why?

31. Why can't you see stars during the day?

32. What light phenomena can be explained using the particle-bullet model of light?

33. What phenomena can be explained using a wave model of light?

34. How is it possible that two different models can explain the same phenomenon? Give an example of another phenomenon that can be successfully explained by two different models (think of areas other than optics).

35. Oliver has finished building a wall in a house. He inspects how flat it is by placing a small lightbulb next to the wall. Explain how this method can help Oliver see how flat the wall is.

Problems

Below, **BIO** indicates a problem with a biological or medical focus. Problems labeled **EST** ask you to estimate the answer to a quantitative problem rather than derive a specific answer. Asterisks indicate the level of difficulty of the problem. Problems with no * are considered to be the least difficult. A single * marks moderately difficult problems. Two ** indicate more difficult problems.

22.1 Light sources, light propagation, and shadows

1. **Tree height** You are standing under a tree. The tree's shadow is 34 m long and your shadow is about twice your height. How tall is the tree?

2. * **Lighting for surgeon** A shadow from a surgeon's hand obstructs her view while operating. Make suggestions for an alternative light source that avoids this difficulty. Include one or more sketches for your proposed plan.

3. **Lunar eclipse** A lunar eclipse happens when the Moon, Earth, and Sun are aligned in that order (the Moon is in the shadow of Earth). Aristotle used this phenomenon to determine the shape of Earth. He proposed that Earth has a round shape. Draw a picture to describe his reasoning.

4. * **Shadows during romantic dinner** You and a friend are having a romantic candlelight dinner. You notice that the light shadows of your hands on the wall look fuzzy. However, the shadow of a glass is very sharp and crisp. Where are you and your friend sitting with respect to the candle and the wall? Where is the glass? To answer these questions, draw ray diagrams assuming that the candle is an extended light source.

5. * **Pinhole camera (camera obscura)** You want to make a pinhole camera with a blank wall as the screen and you (or a friend) as the object of interest. Draw a sketch showing the wall, the pinhole, the place you or your friend will sit, the best location for the Sun or some other light source, and the location of people viewing the wall. Will the image appear upright or inverted? Explain.

6. * **Solar eclipse** Only observers in a very narrow region on Earth (about 200 km diameter) can see a total solar eclipse. In the region of such an eclipse, there is no sunlight and a person can see stars during daytime. Draw the arrangement of the Sun, the Moon, and Earth during a total solar eclipse.

7. * **Tree height 2** Your summer ecology research job involves documenting the growth of trees at an experimental site. One day you forget your tree-height-measuring instrument. How can you determine the height of trees without it? Provide a sketch for your method.

8. An extended light source can be modeled as a group of several point-like light sources. **Figure P22.8** shows a linear light source (such as a fluorescent lamp) modeled by three point-light sources, a sheet of black paper, and a screen. Copy the sketch in your notebook. Determine the regions of shadow and semi-shadow on the screen by drawing rays from the three point-like light sources. Qualitatively compare the relative brightnesses (or darknesses) of the regions.

FIGURE P22.8

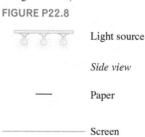

Light source

Side view

—— Paper

———————— Screen

22.2 Reflection of light

9. * You have a small mirror. While holding the mirror, you see a light spot on a wall at the same height as the mirror. At what angle are you holding the mirror if the Sun is 50° above the horizontal? Draw a ray diagram to answer the question. What assumptions did you make?

10. * **Design** Design an experiment that you can perform to test the law of reflection. Describe the instruments that you are going to use and their experimental uncertainty (half of the smallest division). How certain will you be of your results?

11. * You place one mirror on a table and a second mirror at 60° above it, as shown in **Figure P22.11**. You direct a laser beam horizontally toward the second mirror as shown in the figure.

FIGURE P22.11

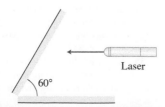

Laser

60°

(a) Draw the trajectory of the laser beam after reflection from both mirrors. (b) Determine the angle between the incoming and outgoing beams. (c) What should be the angle between the mirrors so that the outgoing beam trajectory overlaps with the incoming trajectory?

12. Design a mirror arrangement so that light from a laser pointer will travel in exactly the opposite direction after it reflects off the mirror(s), even if you change the direction the laser pointer is pointing.

13. Two mirrors are oriented at right angles. A narrow light beam strikes the horizontal mirror at an incident angle of 65°, reflects from it, and then hits the vertical mirror. Determine the angle of incidence at the vertical mirror and the direction of the light after leaving the vertical mirror. Include a sketch with your explanation.

14. * You are driving along a street on a sunny day. Only a few apartment building windows appear bright; the rest are pitch black. Explain this difference and include a ray diagram to help with your explanation.

15. A flat mirror is rotated 17° about an axis in the plane of the mirror. What is the angle change of a reflected light beam if the direction of the incident beam does not change?

22.3 Refraction of light

16. (a) A laser beam passes from air into a 25% glucose solution at an incident angle of 35°. In what direction does light travel in the glucose solution? (b) The beam travels from ethyl alcohol to air at an incident angle of 12°. Determine the angle of the refracted beam in the air. (c) Draw pictures for (a) and (b) showing the interface between the media, the normal line, the incident rays, the reflected rays, the refracted rays, and the angles of these rays relative to the normal line.

17. A beam of light passes from glass with refractive index 1.58 into water with a refractive index 1.33. The angle of the refracted ray in water is 58.0°. Draw a sketch of the situation showing the interface between the media, the normal line, the incident ray, the reflected ray, the refracted ray, and the angles of these rays relative to the normal line.

18. A beam of light passes from air into a transparent petroleum product, cyclohexane, at an incident angle of 48°. The angle of refraction is 31°. What is the index of refraction of the cyclohexane?

19. * **Moving laser beam** An aquarium open at the top has 30-cm-deep water in it. You shine a laser pointer into the top opening so it is incident on the air-water interface at a 45° angle relative to the vertical. You see a bright spot where the beam hits the bottom of the aquarium. How much water (in terms of height) should you add to the tank so the bright spot on the bottom moves 5.0 cm?

20. ** **Lifting light** You have a V-shaped transparent empty container such as shown in **Figure P22.20**. When you shine a laser pointer horizontally through the empty container, the beam goes straight through and makes a spot on the wall. (a) What happens to this spot if you fill the container with water just a little above the level at which the laser beam passes through the container? (b) What happens if you fill the container to the very top? Indicate any assumptions used and draw a ray diagram for each situation. Note: This is a multiple-possibility problem.

FIGURE P22.20

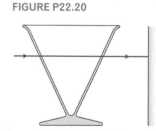

21. * A light beam hits the interface between air and an unknown material at an angle of 43° relative to the normal. The reflected ray and the refracted ray make an angle of 108° with respect to each other. What is the index of refraction of the material?

22. * A light ray passes from air through a glass plate with refractive index 1.60 into water. The angle of the refracted ray in the water is 42.0°. Determine the angle of the incident ray at the air-glass interface.

23. * **BIO** **Vitreous humor** Behind the lens of the eye is the vitreous humor, a jellylike substance that occupies most of the eyeball. The refractive index of the vitreous humor is 1.35 and that of the lens is 1.44. A narrow beam of light traveling in the lens comes to the interface with the vitreous humor at a 23° angle. What is its direction relative to the interface when in the vitreous humor?

24. * You watch a crab in an aquarium. Light traveling in air enters a sheet of glass at the side of the aquarium and then passes into the water. If the angle of incidence at the air-glass interface is 22°, what are the angles at which the light wave travels in the glass and in the water? Indicate any assumptions you made.

25. * Light moving up and toward the right in air enters the side of a cube of gelatin of refractive index 1.30 at an incident angle of 80°. Determine the angle at which the light leaves the top surface of the cube. How does the angle change if the refractive index of the gelatin is slightly greater? Explain.

26. * A laser beam is incident at 30° with respect to the normal on a stack of five transparent plates made of materials with the following indexes of refraction (from the top of the stack to the bottom): $n_1 = 1.20$, $n_2 = 1.30$, $n_3 = 1.40$, $n_4 = 1.50$, $n_5 = 1.60$. The plates touch each other. Determine the angle between the outgoing beam and the normal to the last plate. Assume the first and the last plate are in contact with air. (Hint: Before you start calculating, write the expression for the final angle and see if you can simplify it.)

22.4 Total internal reflection

27. * **Can your light be seen?** You swim under water at night and shine a laser pointer so that it hits the water-air interface at an incident angle of 52°. Will a friend see the light above the water? Explain.

28. * Light is incident on the boundary between two media at an angle of 30°. If the refracted light makes an angle of 42°, what is the critical angle for light incident on the same boundary?

29. **Diamond total reflection** Determine the critical angle for light inside a diamond incident on an interface with air.

30. Determine the refractive index of a glucose solution for which the critical angle for light traveling in the solution incident on an interface with air is 42.5°. How would the critical angle change if the glucose concentration were slightly greater? Explain.

31. * You wish to use a prism to change the direction of a beam of light 90° with respect to its original direction. Describe the shape of the prism and its orientation with respect to the original beam to achieve this goal.

32. * You aim a laser beam (in air) at 80.0° with respect to the normal of the surface of a transparent plastic cube. You observe that the refracted beam inside the plastic makes an angle of 42.8° with the normal. Determine the critical angle for light inside the plastic cube (a) incident on an interface with air and (b) incident on an interface with water.

33. * **Prism total reflection** What must be the minimum value of the refractive index of the prism shown in Figure P22.33 in order that light is totally reflected where indicated? Will some of the light make it out of the top surface? Explain.

FIGURE P22.33

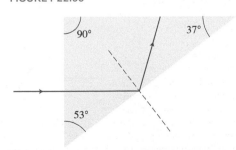

34. **Gems and critical angles** In gemology, two of the most useful pieces of information concerning an unknown gem are the refractive index of the stone and its mass density. The refractive index is often determined using a critical angle measurement. Determine the refractive index of the gemstone shown in Figure P22.34. The critical angle for the total reflection of light at the gem-air interface is 37.28°.

FIGURE P22.34

35. (a) The refractive index for the gem aquamarine is 1.57. Determine the critical angle for light traveling inside aquamarine when reaching an air interface. (b) You have a tourmaline gem and find that a laser beam in air incident on the air-tourmaline interface at a 50° angle has a refracted angle in the tourmaline of 28.2°. Determine the refractive index and critical angle of tourmaline.

36. * You have three transparent media with indexes of refraction $n_1 = 1.33$, $n_2 = 1.46$, and $n_3 = 1.66$. Rank the pairs of media according to decreasing critical angle for a light beam incident on the border between the two media.

37. (a) Rays of light are incident on a glass-air interface. Determine the critical angle for total internal reflection ($n_{glass} = 1.58$). (b) If there is a thin, horizontal layer of water ($n_{water} = 1.33$) on the glass, will a ray incident on the glass-water interface at the critical angle determined in part (a) be able to leave the water? Justify your answer.

22.5 Skills for analyzing reflective and refractive processes

38. * **Invisible in pool?** A swimming pool is 1.4 m deep and 12 m long. Is it possible for you to dive to the very bottom of the pool so people standing on the deck at the end of the pool do not see you? Explain.

39. **Equation Jeopardy 1** Tell all that you can about a process described by the following equation: $1.33 \sin 30° = 1.00 \sin \theta_2$.

40. **Equation Jeopardy 2** Tell all that you can about a process described by the following equation: $1.33 \sin \theta_c = 1.00 \sin 90°$.

41. * **Equation Jeopardy 3** Tell all that you can about a process described by the following equation: $1.00 \sin 53° = 1.60 \sin \theta_2 = 1.33 \sin \theta_3$.

42. ** When reaching a boundary between two media, an incident ray is partially reflected and partially refracted. At what angle of incidence is the reflected ray perpendicular to the refracted ray? The indexes of refraction for the two media are known.

43. * A laser beam travels from air ($n = 1.00$) into glass ($n = 1.52$) and then into gelatin. The incident ray makes a 58.0° angle with the normal as it enters the glass and a 36.4° angle with the normal in the gelatin. (a) Determine the angle of the refracted ray in the glass. (b) Determine the index of refraction of the gelatin.

44. ** The height of the Sun above the horizon is 25°. You sit on a raft and want to orient a mirror so that sunlight reflects off the mirror and travels at an angle of 45° in the water of refractive index 1.33. How should you orient the mirror?

45. ** **Rain sensor** Many cars today are equipped with a rain sensor that automatically switches on the windshield wipers when rain starts to fall. The most common modern rain sensors are based on the principle of total internal reflection. Light from an LED is beamed at a 45° angle onto the windshield glass using a semicircular piece of glass that is glued on the windshield from inside the car. When there is no rain, the light beam undergoes total internal reflection and illuminates the detector (see Figure P22.45a). If the glass is wet, the condition for total internal reflection is no longer fulfilled, and some light escapes to the outside (see Figure P22.45b). This results in a decrease in the light intensity at the detector, which is used as a signal to the car electronics to switch on the wipers. Determine the range of values of the index of refraction of the windshield glass (i.e., the minimum and the maximum values) that will allow the described operation of the sensor shown in the figure.

FIGURE P22.45

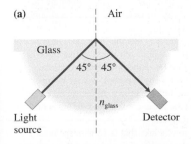
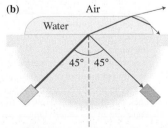

22.6 Fiber optics, prisms, mirages, and the color of the sky

46. * The prism shown in Figure P22.33 is immersed in water of refractive index 1.33. Determine the minimum value of the refractive index for the prism so that the light is totally internally reflected where shown. Will any light leave the top surface of the prism? Explain.

47. * **Light pipe** Rays of light enter the end of a light pipe from air at an angle θ_1 (Figure P22.47). The refractive index of the pipe is 1.64. Determine the greatest angle θ_1 for which the ray is totally reflected at the top surface of the glass-air interface inside the pipe.

FIGURE P22.47

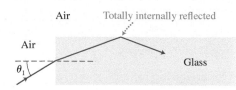

48. * A light ray is incident on a flat piece of glass. The angle of incidence is θ_1 and the thickness of the glass is d. Determine the distance that the ray travels in the glass. At what angle will the ray emerge from the glass on the other side? Draw a ray diagram to help explain your solution.

49. * **Prism** You have a triangular prism made of glass of refractive index 1.60, with angles of 30°-90°-60°. The short side is oriented vertically. A horizontal ray hits the middle of the slanted side of the prism. Draw the path of a ray as it passes into and through the prism. Determine all angles for its trip through the prism.

General Problems

50. * You have a candle and a large piece of paper with a triangular hole slightly larger than the candle flame cut in it. You place the paper between the candle and the wall. Draw ray diagrams to show what you see on the wall when the paper is placed (a) near the candle, (b) halfway between the candle and wall, and (c) near the wall.

51. ** **Design** You wish to investigate the properties of different transparent materials and decide to design a refractometer, a device used to measure the refractive index of an unknown liquid. A light beam in a liquid is refracted into a material whose refractive index is known and is less than that of the unknown liquid. The incident beam is adjusted for total internal reflection, and the equation for the angle of total internal reflection is used to determine the unknown index. Provide suggestions for a design for such a device and perform a sample calculation to show how it might work.

52. * You place a point-like source of light at the bottom of a container filled with vegetable oil of refractive index 1.60. At what height from the light source do you need to place a circular cover with a diameter of 0.30 cm so no light emerges from the liquid?

53. ** There is a light pole on one bank of a small pond. You are standing up while fishing on the other bank. After reflection from the surface of the water, part of the light from the bulb at the top of the pole reaches your eyes. Use a ray diagram to help find a point on the surface of the water from where the reflected ray reaches your eyes. Determine an expression for the distance from this point on the water to the bottom of the light pole if the height of the pole is H, your height is h, and the distance between you and the light pole is l.

54. ** **Coated optic fiber** An optic fiber of refractive index 1.72 is coated with a protective covering of glass of refractive index 1.50. (a) Determine the critical angle for the fiber-glass interface. (b) Determine the critical angle for the glass-air interface. (c) Determine the critical angle for a fiber-air interface (no glass covering). (d) Suppose a ray hits the fiber-glass interface at the angle calculated in (c). What is the angle of refraction of that ray when it reaches the glass-air interface? Will it leave the optic fiber? Explain.

55. ** You put a mirror at the bottom of a 1.4-m-deep pool. A laser beam enters the water at 30° relative to the normal, hits the mirror, reflects, and comes back out of the water. How far from the water entry point will the beam come out of the water? Draw a ray diagram.

56. ** A scuba diver stands at the bottom of a lake that is 12 m deep. What is the distance to the closest points at the bottom of the lake that the diver can see due to light from these points being totally reflected by the water surface? The height of the diver is 1.8 m and $n_{water} = 1.33$.

Reading Passage Problems

Rainbows How is a rainbow formed? Recall that the index of refraction of a medium is slightly different for different colors. When white light from the Sun enters a spherical raindrop, as shown in **Figure 22.25**, the light is refracted, or bent. After reflecting off the back surface of the drop, the light is refracted again as it leaves the front surface.

Each drop separates the colors of light. An observer on the ground with her back to the Sun sees at most one color of light coming from a particular drop (see **Figure 22.26**). If the observer sees red light from a drop (for example, the top drop in Figure 22.26), the violet light for that same drop is deflected above her head. However, if she sees violet light coming from a drop lower in the sky, the red light from that drop is deflected below her eyes onto the ground. She sees red light when her line of view makes an angle of 42° with the beam of sunlight and violet light when the angle is 40°. Other colors of light are seen at intermediate angles.

FIGURE 22.25 Different colors of light are refracted and reflected by different amounts by a spherical raindrop. The color violet is refracted more than red.

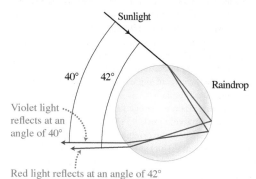

FIGURE 22.26 An observer on the ground sees only one color of light from each raindrop. The drops that send the color red are above the drops that send violet.

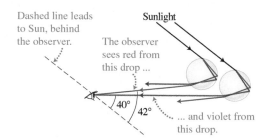

57. Why do you see rainbows only when you are between the Sun and the location of the rainbow?
    (a) The sunlight must be refracted and reflected from raindrops and needs to move back and downward from its original direction.
    (b) You need to intercept the light coming from the Sun.
    (c) If you are looking toward the Sun, rainbows would block the sunlight.
    (d) The sunlight is refracted by the raindrops and changes direction.

58. Suppose that light entered one side of a square water droplet of refractive index 1.33 at an incident angle of 20°. Some of the light reflects off the back surface and then goes back out on the same side as it entered. What is the angle of refraction of light leaving the square droplet?
    (a) 15°   (b) 18°   (c) 20°   (d) 24°   (e) 35°

59. Raindrops reflect different colors of light at different angles. Why do we see the parts of the rainbow as different colors of light rather than all colors coming from each?
    (a) The net deflection of light seen from one drop depends on its refractive index.
    (b) All of the raindrops reflecting a particular color have the same angular deflection relative to the direction of the Sun.
    (c) Your eye sees only one color coming from a particular raindrop.
    (d) The different colors are reflected and refracted at different angles.
    (e) All of the above
60. What is the minimum angle of incidence of red light against the back wall of the water drop in Figure 22.25 so that it is totally reflected at that wall?
    (a) 16°   (b) 21°   (c) 41°   (d) 45°   (e) 49°
61. Repeat the previous problem, only for the violet light.
    (a) 15°   (b) 20°   (c) 41°   (d) 45°   (e) 49°
62. Why is violet light refracted more than red light?
    (a) The violet light travels a shorter distance in the drop than the red light.
    (b) The red light travels more slowly than the violet light.
    (c) The refractive index of water for violet light is greater than that for red light.
    (d) All of the above

Earth energy balance Gases in Earth's atmosphere, such as carbon dioxide and water vapor, act like a blanket that reduces the amount of energy that Earth radiates into space. This phenomenon is called the greenhouse effect. Without the greenhouse effect, most of Earth would have a climate comparable to that of the polar or subpolar regions. What would Earth's mean surface temperature be, in the absence of the gases causing the greenhouse effect?

The Sun continually irradiates our upper atmosphere with an intensity of about 1360 W/m². About 30% of this energy is reflected back into space, leaving an average of 950 W/m² to be absorbed by the surface area of Earth that is exposed to the Sun. This exposed area is πR_{Earth}^2, the Earth's circular cross section that intercepts the sunlight. All objects, including Earth, emit radiation at a rate proportional to the fourth power of the surface temperature (T^4) in kelvins (K). Thus, the radiation is emitted from Earth's surface, which as a sphere has the area $4\pi R_{Earth}^2$.

To maintain a constant temperature, Earth's radiation rate must equal its energy absorption rate from the Sun. A fairly simple calculation indicates that the two rates are equal when the average surface temperature of Earth is 255 K, or about 0 °F. However, Earth's mean surface temperature is much warmer than that—about 288 K—because the calculation neglected the effect of greenhouse gases in the atmosphere. These gases absorb infrared radiation emitted by Earth and reflect some of it back to Earth. Thus, Earth emits less energy into space than it would without greenhouse gases. The energy absorbed from the Sun and the reduced energy emitted into space have caused the Earth to warm to its mean surface temperature of 288 K.

Over the past two centuries the concentration of carbon dioxide in our atmosphere has increased from a pre-industrial level of about 270 parts per million to 380 parts per million. This increase in carbon dioxide and other greenhouse gases has been caused by the burning of fossil fuels and the removal of forests, which absorb carbon dioxide. The carbon dioxide concentration in the atmosphere is expected to reach 600–700 parts per million by 2100. If that occurs, it will be warmer in 2100 than at any time in the last half million years.

63. Why is there essentially zero greenhouse effect on the Moon?
    (a) There is no photosynthesis on the Moon.
    (b) There is no carbon dioxide on the Moon.
    (c) There is no gaseous atmosphere on the Moon.
    (d) Answers b and c are correct.
    (e) Answers a, b, and c are correct.
64. The average Earth surface temperature without its atmosphere would be 255 K. Why?
    (a) At that temperature, the emission rate of radiation from Earth would just balance the absorption rate of radiation from the Sun.
    (b) The emission of Earth radiation is from a sphere of area $4\pi R_{Earth}^2$, whereas the absorption of radiation from the Sun is from an area πR_{Earth}^2.
    (c) Earth's cross section would be significantly reduced because of the lack of the atmosphere.
    (d) Answers a and b are correct.
    (e) Answers a, b, and c contribute about equally to this temperature calculation.
65. The Sun irradiates Earth's outer atmosphere at what rate?
    (a) 1×10^{16} J/s   (b) 4×10^{16} J/s
    (c) 2×10^{17} J/s   (d) 7×10^{17} J/s
    (e) 1400 J/s
66. Because of the greenhouse effect, Earth's average surface temperature is 288 K instead of 255 K. Because of this higher temperature, Earth's surface emits radiation at a rate that is higher by a factor of approximately which of the following?
    (a) 1.13   (b) 1.28   (c) 1.63   (d) 1.87   (e) 2.21
67. Because of the increased temperature of Earth's surface (288 K compared to 255 K) due to the greenhouse effect, its energy emission rate has increased significantly. Why hasn't Earth cooled down as a result?
    (a) It is cooling down.
    (b) The Sun absorbs much of Earth's extra radiation, and the Sun is warming and emitting more radiation.
    (c) There is a long delay between the change in temperature and the increased rate of radiation by Earth.
    (d) Earth's atmosphere absorbs some of the outgoing radiation and emits it back to Earth's surface.
    (e) All of the above are correct.
68. What will the increasing concentration of greenhouse gases do?
    (a) Increase the reflection rate of sunlight.
    (b) Increase the reflection of Earth's radiation back to the surface.
    (c) Increase the cross-sectional area of the Earth, causing more sunlight to be absorbed.
    (d) Shield Earth from dangerous ultraviolet radiation.
    (e) Provide extra nourishment for plants on Earth.

Mirrors and Lenses

- How do eyeglasses correct your vision?

- When you look in a mirror, where is the face you see?

- How is a cell phone camera similar to a human eye?

Your friend wears glasses when she is driving, while your physics professor puts her glasses on when she is using her computer. When they buy glasses, your friend gets the glasses labeled -2.0 and your professor gets the ones labeled $+1.5$. What vision problems do they have, and what do those numbers mean? In this chapter you will learn how to answer these questions.

BE SURE YOU KNOW HOW TO:

- Apply the properties of similar triangles (Appendix A).

- Draw ray diagrams and normal lines (Sections 22.1, 22.2, and 22.3).

- Use the laws of reflection and refraction (Sections 22.2 and 22.3).

IN THIS CHAPTER we will apply what we have learned about reflection and refraction to mirrors and lenses. In the process we will learn how cameras, telescopes, microscopes, human vision, and corrective lenses work.

23.1 Plane mirrors

The simplest mirror is a **plane mirror**—a flat, reflective surface, often a metal film covered in glass. When you stand in front of a plane mirror, you see a reflection of yourself. How does the second "you" appear?

Recall the model that we created to describe how extended objects emit light. Each point on a luminous object sends light rays in all directions. Some of these rays enter our eyes and we see the object at the place where those rays originate (**Figure 23.1**). In other words, to locate an object, at least two rays should come into our eyes; the object is at the intersection of those two rays. Now, suppose that a small shining object is in front of a mirror. What happens to the rays emitted by the object when they reach the mirror? Consider Observational Experiment **Table 23.1**.

FIGURE 23.1 Seeing light from an object.

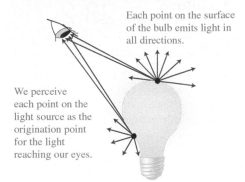

Each point on the surface of the bulb emits light in all directions.

We perceive each point on the light source as the origination point for the light reaching our eyes.

OBSERVATIONAL EXPERIMENT TABLE 23.1   Seeing a point object in a plane mirror

Observational experiment	Analysis
We place a small lightbulb on a table about 20 cm in front of a plane mirror that is held perpendicular to the table, with one side resting on the table. Observers A, B, and C place rulers on the table and point them in the direction of the bulb that they see in the mirror. 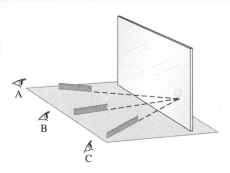 Note that the dashed extensions of the rulers all intersect at one point behind the mirror.	For observer A to see an image of the lightbulb when looking at the mirror, one or more rays of light reflected from the mirror must reach his eyes. Rays between 1 and 2 do reach observer A's eyes. 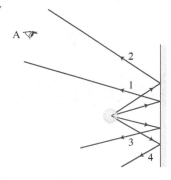

Patterns

- The rulers all point to the same spot behind the mirror. Rays between 1 and 2 reaching observer A seem to originate from that spot. This is the location of the perceived image of the bulb produced by the mirror.
- The ray diagram shows that the perceived image of the bulb is produced at the same distance behind the mirror as the bulb is in front of it.
- To locate the image, at least two rays need to reach the observer's eyes because the perceived image is at the intersection of those rays.

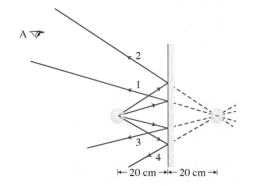

In Table 23.1 we found that the image of the real object seen in the mirror is located where light reflected from the mirror to the eye of the observer *seems* to originate. This perceived image is behind the mirror and not on the surface of the mirror. We also found, using the ray diagram, that the image is exactly the same distance behind the plane mirror as the object is in front of it. Let us test these findings in Testing Experiment **Table 23.2**, on the next page.

TESTING
EXPERIMENT TABLE 23.2

Testing the image location of a plane mirror

Testing experiment	Prediction	Outcome
We repeat the Table 23.1 experiment for observer A but cover the part of the mirror directly in front of the bulb.	If the image is due to the reflected rays, then even if we cover part of the mirror, some of the reflected rays will still reach the observer's eyes. The location of the image should not change.	The observer sees the image of the bulb exactly where it was before.

Conclusions

The position of the image formation is consistent with our previous experiment. It also disproves the common idea that the image forms on the surface of the mirror.

The image we see in a plane mirror is not real; there is no real light coming from behind the mirror. The reflected light reaching your eyes appears to originate from a point behind the mirror. What we see in the mirror is called a **virtual image**.

> **Plane mirror virtual image** A plane mirror produces a virtual image that is the same distance behind the mirror as the object is in front of it. The reflected light seems to diverge from the image behind the mirror. But no light actually leaves that image—you see light reflected from the mirror.

So far, we have learned how to find images in plane mirrors of objects that emit light. But when you stand in front of a mirror, you can see your image even though you do not seem to be a source of light. To explain this phenomenon, think back to Chapter 22. All objects reflect light diffusely. Thus, if the room is illuminated by sunlight or you have a lightbulb on, light from those sources reaches you and reflects off you, effectively making you a new (secondary) source of light. To test this hypothesis, go into an absolutely dark room and stand in front of a mirror—you will not see anything. Once you switch on a flashlight, however, you are able to see your virtual image in the mirror as well as an image of the flashlight. All objects in lit rooms become secondary sources of light due to their diffuse reflection of incident light.

CONCEPTUAL EXERCISE 23.1 ▸ **Where is the lamp?**

You place a lamp in front of a mirror and tilt it so that the top and the bottom of the lamp are at different distances from the mirror, at the position shown in the figure at right. Where do you see the image of the lamp produced by the mirror?

Mirror

Sketch and translate We use the figure in the problem statement to visualize the situation. We need to find the virtual image of the lamp produced by the mirror. The virtual image of each point of an object is

the place from which the reflected light, represented by rays, appears to diverge.

Simplify and diagram Assume that all points of the lamp (including the base) send out rays in all directions. For simplicity, we will consider the top and the bottom of the lamp and locate the images of each of these two end points. Assume that the images of all other points on the lamp are formed between the end point images. As you can see from the following ray diagram, the image of each point is behind the mirror, and each point is at the same distance behind the mirror as the point on the lamp is in front of the mirror.

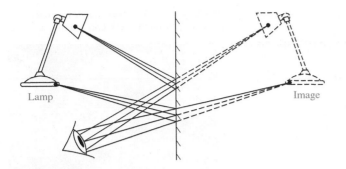

Lamp   Image

Try it yourself What happens to the distance between an observer and her image in a plane mirror if the observer moves backward, doubling her distance from the mirror?

Answer   The distance between the observer and the image doubles.

CONCEPTUAL EXERCISE 23.2   **Buying a mirror**

Your only requirement when buying a mirror is that you can see yourself from head to toe. What should be the minimum length of the mirror you buy?

Sketch and translate Draw a sketch of the situation, as shown below. Let h be your height. Express the mirror height H as a fraction or a multiple of your height. Let s be your distance from the mirror. Since the goal is to see your entire body in the mirror, rays from the top of your head and from the bottom of your feet must reflect off the mirror into your eyes.

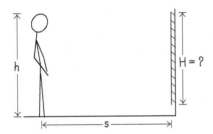

Simplify and diagram Assume that the mirror is mounted on a vertical wall and that all parts of your body are the same distance from the mirror. Also assume that you are a shining object of a particular height h. Finally, let h_1 be the vertical distance between your feet and eyes and h_2 be the vertical distance between the top of your head and your eyes. This means $h_1 + h_2 = h$.

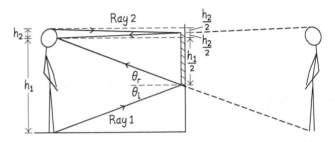

You will see your toes if ray 1 from your toes reaches your eyes after reflection. Ray 1 hits the mirror at a height $h_1/2$ above your toes. You will see the top of your head if ray 2 from the top of your head reaches your eyes after reflection. Ray 2 hits the mirror at a distance $h_2/2$ below a horizontal line from the top of your head to the mirror. From the diagram we see that you can cut off both the bottom $h_1/2$ and the top $h_2/2$ of the mirror and still see your whole body. Thus the length of the smallest mirror is $h_1/2 + h_2/2 = h/2$. Note that the size of the mirror does not depend on the distance s between you and the mirror—you can see your entire body in the mirror no matter how far you stand from it.

Try it yourself You are 1.6 m tall and stand 2.0 m from a plane mirror. How tall is your image? How far from the mirror on the other side is it? What is your image height if you double your distance from the mirror?

Answer   1.6 m; 2.0 m; the same size.

REVIEW QUESTION 23.1 A mirror is hanging on a vertical wall. Will you see more or less of your body if you step closer to the mirror?

23.2 Qualitative analysis of curved mirrors

Curved mirrors are cut from a spherically shaped piece of glass backed by a metal film. **Concave mirrors** are often used for magnification, in telescopes and cosmetic mirrors. The curve of a concave mirror bulges away from the light it reflects (see **Figure 23.2a** on the next page). In a **convex mirror**, the curve bulges toward the light it

FIGURE 23.2 Features of (a) concave and (b) convex mirrors.

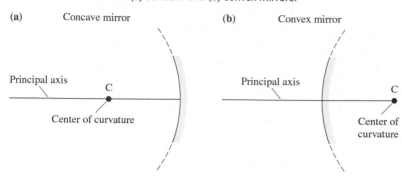

FIGURE 23.3 A convex mirror provides visibility at blind spots and around corners.

FIGURE 23.4 A concave mirror reflects a laser beam.

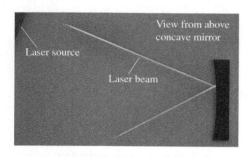

reflects (see Figure 23.2b). Convex mirrors are used as passenger-side rearview mirrors and to provide visibility at blind spots, such as hallway corners and driveway exits (**Figure 23.3**).

In both kinds of curved mirrors, the center of the sphere of radius R from which the mirror is cut is called the **center of curvature** C of the mirror. The imaginary line connecting the center of curvature with the center of the mirror's surface is called the **principal axis.**

Concave mirrors

In Observational Experiment **Table 23.3**, we cut a narrow band from a concave mirror and lay it on its edge on a piece of paper (**Figure 23.4**). The paper allows us to observe the paths of laser light incident and reflected from the mirror.

OBSERVATIONAL
EXPERIMENT TABLE 23.3   **Reflection of light from a concave mirror**

Observational experiment	Analysis
Experiment 1. We shine a laser beam parallel to the plane of the page toward a concave mirror and trace the path of the incident and reflected light on the page. 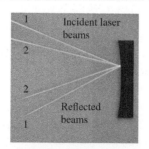	We use the law of reflection to draw a ray diagram. The dashed line from the place where the ray hits the mirror to the center of curvature is a normal line perpendicular to the surface of the mirror. The law of reflection accounts for the path of the reflected ray. 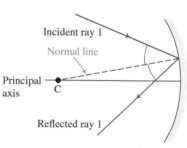
Experiment 2. We send several beams parallel to the principal axis. Reflected beams all pass near the same point on the principal axis. 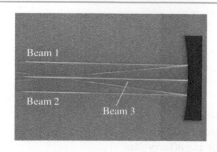	We draw a ray diagram and use the law of reflection to analyze the paths of the light. The normal lines here are radii of the mirror. We find that all reflected rays pass through the same point on the principal axis (called the *focal point* F of the mirror), matching the outcome of the experiment. 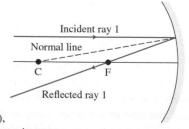

Observational experiment	Analysis
Experiment 3. This time we send several beams parallel to each other but not parallel to the principal axis. One ray passes though the center of curvature C. All reflected rays pass through the same point. 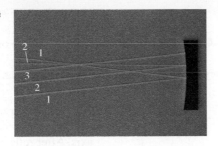	We use the law of reflection to draw a ray diagram to analyze the paths of the light. The ray passing through the center of curvature is perpendicular to the mirror and thus reflects back on itself. All other rays after reflection pass through the 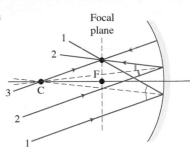 same point on a line perpendicular to the principal axis and through the focal point. This point is in the *focal plane* of the mirror. The ray diagram matches the outcome of the experiment.

Patterns

- The mirror reflects light according to the law of reflection. The normal line at any point on the mirror goes through the center of curvature. The normal lines are in the radial directions.
- After reflection, all incident rays traveling parallel to the principal axis pass through the same point on the principal axis—the focal point of the mirror.
- After reflection, all incident rays traveling parallel to each other but not parallel to the principal axis intersect at a common point. This point is on the line perpendicular to the principal axis through the focal point.

Table 23.3 shows us that a curved mirror reflects light in agreement with the law of specular reflection. But unlike a flat mirror, it causes parallel incident rays to pass through a single point after reflection. If the incident rays are also parallel to the principal axis, this point is called the **focal point** F of the mirror (see Experiment 2 in Table 23.3). If they are not parallel to the principal axis (but are still parallel to each other), they all pass through another point on the **focal plane** of the mirror (see Experiment 3 in Table 23.3). The focal plane is the plane that is perpendicular to the principal axis and passes though the focal point. The distance from the focal point to the surface of the mirror where it intersects the principal axis is called the **focal length** f (**Figure 23.5**). A concave mirror is sometimes called a *converging mirror* because incident parallel rays converge after reflection from it. Since we can consider the rays of the Sun to be almost parallel, we can easily determine the location of the focal point and the focal plane of any converging mirror if we use the Sun as the light source (be careful if you perform this experiment—you might start a fire!).

If a mirror has a large radius of curvature compared to its size and the incident rays are close to the principal axis, the focal point is approximately halfway between the center of the mirror and the center of curvature ($f = R/2$). See the geometric proof in **Figure 23.6**.

> TIP The verb *converge* does not mean that the rays reach the focal point and stop there. They continue traveling after passing through this point.

Using a ray diagram to locate the image formed by a concave mirror

We can use a ray diagram to locate the image of an object that is placed in front of a concave mirror. The method is illustrated in Physics Tool Box 23.1, on the next page. The object in this case is a lightbulb that is placed beyond the center of curvature of a concave mirror. We will represent the object with an arrow that originates on the principal axis. For the time being we only locate the image of the top of the arrow and assume that the image of the base of the object (the bottom of the arrow) is also on the axis. We will validate this assumption later.

FIGURE 23.5 The effect of a concave mirror on parallel rays.

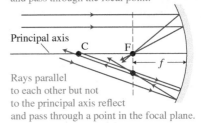

Rays parallel to the principal axis reflect and pass through the focal point.

Rays parallel to each other but not to the principal axis reflect and pass through a point in the focal plane.

FIGURE 23.6 The focal length f of a curved mirror is half the radius of curvature R.

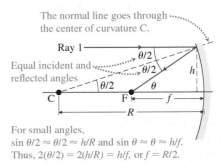

The normal line goes through the center of curvature C.

For small angles, $\sin \theta/2 \approx \theta/2 \approx h/R$ and $\sin \theta \approx \theta \approx h/f$. Thus, $2(\theta/2) = 2(h/R) = h/f$, or $f = R/2$.

PHYSICS TOOL BOX 23.1 > Constructing a ray diagram to locate the image of an object produced by a concave mirror

1. Place a vertical arrow on the principal axis to represent the object. Each point on the object emits light in all directions. We find the image location of only the tip of the object arrow.

2. Choose two or three rays from the tip of the arrow. You know the directions of these rays after reflection from the mirror.

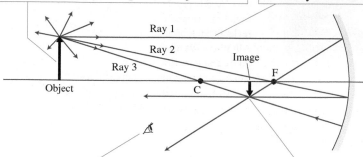

Ray 1 travels from the tip of the object toward the mirror parallel to the principal axis. After reflection, it passes through the focal point F.
Ray 2 travels from the tip of the object through the focal point F. After reflection, it travels parallel to the principal axis.
Ray 3 passes through the center of curvature C. It hits the mirror perpendicular to its surface and reflects back in the same direction.

4. The observer sees the image at the place from which the reflected rays seem to diverge.

3. A *real image* is formed where these two or three rays intersect.

TIP Remember that each point on an object emits an infinite number of rays that reflect off the mirror according to the law of reflection. Rays 1, 2, and 3 in the Physics Tool Box were chosen because we know how they reflect without needing to draw the normal line to the point of incidence and measure the angles of incidence and reflection.

In Physics Tool Box 23.1 we used three special rays to locate the image of an object produced by a concave mirror. We will use these three rays throughout our study of mirrors.

The image formed in the Tool Box is a **real inverted image**: it is **inverted**, meaning that the image points opposite the direction of the object; it is also a **real image**, meaning that the reflected rays actually pass through that image location and then reach the eyes of the observer. If you placed a small screen (a piece of paper, your hand, etc.) at that image location, you would see a sharp (focused), inverted (upside-down) image projected onto it. If you place the screen closer to the mirror or farther away than the image location, reflected light will reach it, but the image will be fuzzy because the rays pass through different points on the screen. Note that we only considered the case when the object is located behind the focal point of the mirror. We will consider the case when the object is between the mirror and the focal point below.

Not all reflected rays result in a real image. Consider the situation in **Figure 23.7a**, in which a pencil is closer to a concave mirror than the focal point F. We will use two rays to construct a ray diagram. In Figure 23.7b we see that reflected rays 1 and 2 diverge. For a person holding the pencil (the eye in the diagram), the light seems to originate from a location behind the mirror. This location is the location of the **virtual image** produced by the mirror. The image is *virtual* (similar to the images formed by plane mirrors) because the light does not actually come from that point—it just appears to. The image is magnified (bigger than the object) and upright. You see such an image when you use a concave mirror to apply makeup or shave.

FIGURE 23.7 Using a ray diagram to locate the image of an object located near a concave mirror.

(a)

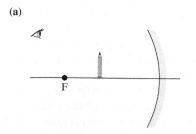

(b)

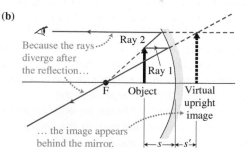

Locating the image of the base of an object produced by a concave mirror

So far, our ray diagrams have only included the image of the top of the object. We have assumed that the image of the base of the object is always on the principal axis directly below the image of the top of the object. To validate this assumption we will use a ray from the base of the object traveling along the principal axis as ray 1. It reflects back on itself. Since ray 1 stays on the principal axis, any other ray that we use to locate the image of the base will also have to intersect the principal axis. To find where on the axis the image of the base is located, first draw an arbitrary imaginary ray through the center of curvature. This is ray 3 from Physics Tool Box 23.1. That ray reflects back through C (**Figure 23.8**). Now draw another ray, which we will call ray 4, from the base of the object and oriented parallel to ray 3. Ray 4 will pass through the same point on the focal plane as ray 3. Ray 4 intersects the principal axis at the location of the image of the base of the object. This locates the image of the base. It does in fact lie directly above the image of the top of the arrow (not shown in the figure). The method that we used to locate the base of the object works for locating the image of any point of an object that resides on the principal axis.

FIGURE 23.8 Locating the base of the image produced by a curved mirror.

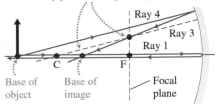

A line parallel to ray 4 that passes through C also passes through the focal plane at the same place that the reflected ray passes through that plane.

Convex mirrors

Let's investigate convex mirrors. Consider Observational Experiment **Table 23.4**.

OBSERVATIONAL EXPERIMENT TABLE 23.4   **Where is the focal point for a convex mirror?**

Observational experiment	Analysis
We send three laser beams parallel to the principal axis of a convex mirror. We observe that the reflected beams diverge. 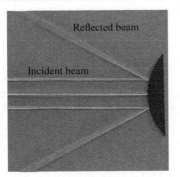	The ray diagram and the law of reflection help us analyze the situation. Using the normal lines passing though the center of curvature, we find that the reflected rays diverge. If we extend the reflected rays behind the mirror, their extensions intersect at one point, which we will call the virtual focal point F. 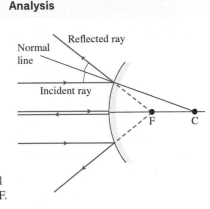

Pattern

- Light rays moving parallel to the principal axis of a convex mirror reflect and diverge away from each other.
- Lines drawn backward along the direction of the reflected light cross the axis at a focal point F behind the mirror.

Recall that for a concave mirror, rays parallel to the principal axis reflect through a focal point in *front* of the mirror. However, for a convex mirror, rays parallel to the principal axis diverge after reflection. They diverge from what is called a **virtual focal point** F behind the mirror. As with a concave mirror, the focal length f equals half the radius of curvature R of the mirror ($f = R/2$), as shown in **Figure 23.9**. Let's use a new set of ray diagrams to investigate images formed by convex mirrors.

FIGURE 23.9 A convex mirror reflects rays parallel to the principal axis.

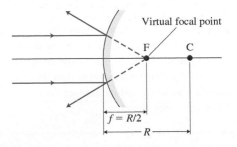

PHYSICS TOOL BOX 23.2 > Constructing a ray diagram to locate the image produced by a convex mirror

1. Place a vertical arrow on the principal axis to represent the object. Each point on the object emits light in all directions. We find the image location of only the tip of the object arrow.

2. Choose two rays from the tip of the arrow. You know the directions of these rays after reflection from the mirror. The rays diverge after reflection.

Ray 1 travels from the tip of the object toward the mirror parallel to the principal axis. After reflection, it diverges as if it were coming from the focal point F behind the mirror.
Ray 2 passes toward the center of curvature C. It hits the mirror perpendicular to its surface and reflects back in the opposite direction.

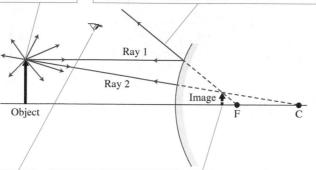

4. The observer sees the image at the place from which the reflected rays seem to diverge.

3. Extend the rays behind the mirror (the dashed lines). A *virtual image* is formed at the point behind the mirror from which these two rays seem to diverge.

In Physics Tool Box 23.2 the object was about 1.6R from the mirror. An upright virtual image was formed about 0.4R behind the mirror. What happens to the image when the object is moved closer to the mirror?

CONCEPTUAL EXERCISE 23.3 > Looking into a convex mirror

You hold a convex mirror 0.7R behind a pencil. Approximately where is the image of the pencil, and what are the properties of the image?

Sketch and translate To solve the problem, we draw a ray diagram on which we represent the pencil with an arrow. The given distance in the problem statement is the distance between the object and the mirror along the principal axis. We use the ray diagram to locate the image of the top of the pencil (the tip of the arrow).

Simplify and diagram Draw a ray diagram. We use rays 1 and 2 described in Physics Tool Box 23.2 to locate the image. The image is virtual and upright about 0.3R behind the mirror and is smaller than the object. This is what you would see if you looked at your reflection on the back of a shiny tablespoon. The images formed by convex mirrors are always upright, virtual (behind the mirror), and reduced in size compared to the objects.

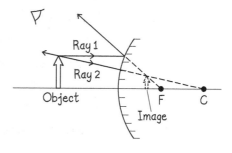

Try it yourself In order to obtain the following types of images of an object, should you use a concave or a convex mirror? Where should you place the object relative to the mirror? (a) Real, bigger than the object; (b) real, smaller than the object; (c) virtual, bigger than the object; (d) virtual, smaller than the object; (e) real and the same size as the object.

Answer

(a) Concave: the object should be between R and f from the mirror; (b) concave: the object should be farther than R from the mirror; (c) concave: the object should be closer than a focal length from the mirror; (d) convex: the object can be at any location; (e) concave: the object should be at 2f from the mirror—right at the center of curvature.

REVIEW QUESTION 23.2 You've found a concave mirror. How can you estimate its focal length?

23.3 The mirror equation

We have learned that the distance to the image from the mirror depends on where the object is located and on the mirror's type and focal length. We will use the ray diagram in **Figure 23.10** to help derive a mathematical relationship between these three quantities:

- the distance of the object from the mirror s,
- the distance of the image from the mirror s', and
- the focal length of the mirror f.

Note the following notations in Figure 23.10: AB is the object, and A_1B_1 is the image. M and N are points on the mirror where light rays strike it. We assume that the mirror isn't curved very much ($MD \approx AB$) and the rays are incident close to the principal axis. Now, using this notation, we make the following steps to complete the derivation.

- $AM \approx BD = s$
- $A_1N \approx B_1D = s'$
- $MD \approx AB$ and $ND \approx A_1B_1$.
- ABF and NDF are similar triangles. Thus,

$$\frac{ND}{DF} = \frac{A_1B_1}{DF} = \frac{AB}{BF} \quad \text{or} \quad \frac{AB}{A_1B_1} = \frac{BF}{DF} = \frac{s-f}{f}$$

- A_1B_1F and MDF are similar triangles. Thus,

$$\frac{A_1B_1}{B_1F} = \frac{MD}{DF} = \frac{AB}{DF} \quad \text{or} \quad \frac{AB}{A_1B_1} = \frac{DF}{B_1F} = \frac{f}{s'-f}$$

- Setting the previous two AB/A_1B_1 ratios equal to each other, we find that

$$\frac{s-f}{f} = \frac{f}{s'-f} \tag{23.1}$$

After some algebra (see the Tip), we get a relationship called the **mirror equation**:

$$\frac{1}{s} + \frac{1}{s'} = \frac{1}{f} \tag{23.2}$$

The mirror equation [Eq. (23.2)] allows us to predict the distance s' of the image from the mirror given the distance s of the object from the mirror and the mirror's focal length f. Before we test the equation experimentally, let us check to see if it is consistent with an extreme case. We know that a concave mirror causes rays parallel to the principal axis to pass through the focal point after reflection. If an object is extremely far away along the principal axis, we can assume that rays from the object reaching the mirror are parallel to the principal axis. *If* Eq. (23.2) is correct, and we use infinity for the object distance, *then* we should find that the image distance for that object equals the focal length distance. Using Eq. (23.2), we get

$$\frac{1}{\infty} + \frac{1}{s'} = 0 + \frac{1}{s'} = \frac{1}{f}$$

Therefore, the image is at a distance equal to the focal length away from the mirror. This result is consistent with the prediction based on the extreme case analysis.

FIGURE 23.10 A ray diagram to help develop the mirror equation.

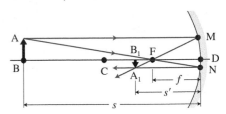

TIP ▸ To get from Eq. (23.1) to (23.2), multiply both sides by $f(s' - f)$. Then carry out the multiplication on the left side of the equation and add the quantities. You should have $ss' - sf - fs' = 0$. Now divide both sides of the equation by $ss'f$.

TIP ▸ When you divide any number (other than infinity) by infinity, the result is zero.

EXAMPLE 23.4   Where should the screen be?

You place a candle 0.80 m from a concave mirror with a radius of curvature of 0.60 m. Where should you place a paper screen to see a sharp image of the candle?

Sketch and translate Draw a sketch of the situation, as shown at right. Assemble all parts of the experiment on a meter stick so the distances can be easily measured. The known quantities are $s = 0.80$ m and $f = R/2 = 0.30$ m. To find where to place the screen, we need to determine s'.

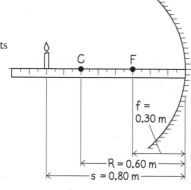

Simplify and diagram Model the candle as an arrow. We use a ray diagram to locate the image of the tip of the arrow, as shown at right. For this ray diagram, we use rays 1 and 3 as described in Physics Tool Box 23.1. You can estimate the image distance from the diagram—it looks like the image distance is a little less than two-thirds of the object distance.

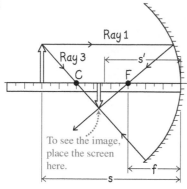

Represent mathematically We see from the ray diagram that the image is real, inverted, and closer to the mirror than the candle. Rearrange Eq. (23.2) to find s':

$$\frac{1}{s'} = \frac{1}{f} - \frac{1}{s}$$

Solve and evaluate

$$\frac{1}{s'} = \frac{1}{f} - \frac{1}{s} = \frac{1}{0.30 \text{ m}} - \frac{1}{0.80 \text{ m}}$$

$$= \frac{0.80 \text{ m} - 0.30 \text{ m}}{(0.30 \text{ m})(0.80 \text{ m})} = 2.08 \text{ m}^{-1}$$

Thus, $s' = 1/(2.08 \text{ m}^{-1}) = 0.48$ m. The image is closer to the mirror than the object, as predicted by our ray diagram. Thus, the predictions using mathematics and the ray diagram are consistent. When we place a screen where we predicted, we see an inverted real image of the flame. We will make the screen small and place it below the principal axis to avoid blocking light from the candle.

- -

Try it yourself Determine the distance of the image when the candle is 0.20 m from the mirror. Does the answer make sense to you?

Answer   the negative sign means that the image is virtual.
−0.60 m. The image is virtual and behind the mirror. Evidently,

In the "Try it yourself" part of Example 23.4, the mirror equation led to a negative value for the image distance. The image was virtual. We need to agree on some sign conventions when using Eq. (23.2).

- We put a negative sign in front of the distance s' for a virtual image that appears to be behind a mirror.
- We put a negative sign in front of the focal length f for a convex mirror.

Let us see if these sign conventions work when we use the mirror equation for a convex mirror.

EXAMPLE 23.5   Testing the mirror equation for a convex mirror

A friend's face is 0.60 m from a convex 0.50-m-radius mirror. Where does the image of her face appear to you when you look at the mirror?

Sketch and translate The situation is sketched below. The known information is $s = 0.60$ m and $f = -R/2 = -(0.50 \text{ m})/2 = -0.25$ m. Note that for a convex mirror, the focal length is a negative number.

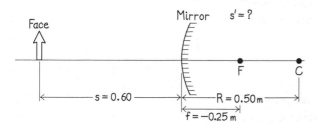

Simplify and diagram Draw a ray diagram representing your friend's face as an arrow. The image is upright, virtual, behind the mirror, and closer to the mirror than the object. Let us see if the mathematics matches this prediction.

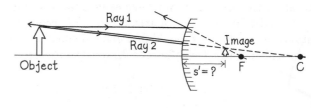

Represent mathematically Rearrange the mirror equation [Eq. (23.2)] to locate the image of your friend's face.

$$\frac{1}{s'} = \frac{1}{f} - \frac{1}{s}$$

Solve and evaluate Insert the known information:

$$\frac{1}{s'} = \frac{1}{-0.25 \text{ m}} - \frac{1}{0.60 \text{ m}} = -5.67 \text{ m}^{-1}$$

The image distance is $s' = 1/(-5.67 \text{ m}^{-1}) = -0.18$ m. We see that the image distance is negative and its magnitude is less than the magnitude of the object distance—consistent with the ray diagram.

Try it yourself The image of your friend's face in a convex mirror is upright, virtual, and 0.30 m behind the mirror when the mirror is 1.0 m from her face. Determine the radius of the sphere from which the mirror was cut.

Answer

The object distance is $s = 1.0$ m, the image distance is $s' = -0.30$ m, and the focal length is $f = -0.43$ m. Consequently, $R = -0.86$ m. The minus sign indicates a convex mirror.

It appears that the mirror equation works equally well for concave and convex mirrors provided that appropriate sign conventions are used.

> **Mirror equation** The distance s of an object from the surface of a mirror, the distance s' of the image from the surface of the mirror, and the focal length f of the mirror are related by the mirror equation:
>
> $$\frac{1}{s} + \frac{1}{s'} = \frac{1}{f} \qquad (23.2)$$
>
> The focal length is half the radius of curvature: $f = R/2$. The following sign conventions apply for mirrors:
>
> - The focal length f is positive for concave mirrors and negative for convex mirrors.
> - The image distance s' is positive for a real image and negative for a virtual image.

Magnification

You have probably noticed that the size of an image produced by a curved mirror is sometimes bigger and sometimes smaller than the size of the object. The change in the size of the image compared to the size of the object is a quantity called **linear magnification** m. Linear magnification is defined in terms of the image height h' and the object height h, where the heights are the perpendicular sizes of the image and object relative to the principal axis:

$$\text{linear magnification} = \frac{\text{image height}}{\text{object height}} = m = \frac{h'}{h} \qquad (23.3)$$

A height is positive if the image or object is upright and negative if inverted.

It is often not possible to calculate magnification as defined above since the object and image heights are sometimes unknown. However, we can determine the magnification using the mirror equation and the object distance and image distance. Consider **Figure 23.11**, which will be used to derive a relation between h' and h and s' and s. In Figure 23.11a we draw the image of the object using rays 1 and 2. Then in Figure 23.11b we draw a new ray that travels from the top of the object to the center of the mirror and then to the top of the inverted image. The reflected ray makes the same angle θ with the principal axis as does the incident ray. Thus,

$$\tan \theta = \frac{h}{s} = \left| \frac{h'}{s'} \right|$$

Therefore, the absolute value of the magnification is $|m| = |h'/h| = |s'/s|$. If the image is real, it will be inverted. Remember, the height of the inverted image is

FIGURE 23.11 Determine linear magnification.

(a)

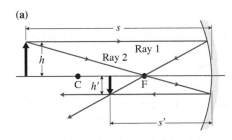

(b)

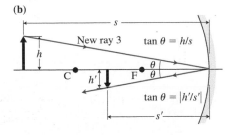

considered negative. Hence h'/h will be negative, whereas s' will be positive—the ratios have the opposite signs. To account for this, we add a negative sign. Thus, $m = (h'/h) = -(s'/s)$.

> Linear magnification m is the ratio of the image height h' and object height h and can be determined using the negative ratio of the image distance s' and object distance s:
>
> $$m = \frac{h'}{h} = -\frac{s'}{s} \qquad (23.4)$$
>
> Remember that a height is positive for an upright image or object and negative for an inverted image or object. Also note that m is a unitless physical quantity.

EXAMPLE 23.6   **Your face in a mirror**

You use a concave mirror with a radius of curvature of 0.32 m for putting on makeup or shaving. When your face is 0.08 m from the mirror, what are the image size and magnification of a 0.0030-m-diameter birthmark on your face?

Sketch and translate Sketch a ray diagram, such as the one shown below. The givens are $h = 0.30$ cm, $s = 0.08$ m, and $f = (R/2) = +0.16$ m. We need to determine the linear magnification m and then the diameter h' of the birthmark image.

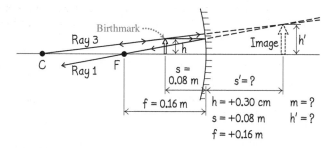

$h = +0.30$ cm	$m = ?$
$s = +0.08$ m	$h' = ?$
$f = +0.16$ m	

Simplify and diagram Represent the birthmark as an arrow. The reflected rays diverge. Extend the reflected rays back through the mirror to find where they intersect. This is the place where reflected light seems to come from and is the location of the image of the arrow's tip. The image is enlarged, upright, and virtual. We do not use ray 2 from Physics Tool Box 23.1 in this diagram because it does not hit the mirror.

Represent mathematically First use Eq. (23.2) to determine the image distance s'. Then use Eq. (23.4) to find the magnification m and the image height h'.

Solve and evaluate Rearranging Eq. (23.2), we find

$$\frac{1}{s'} = \frac{1}{f} - \frac{1}{s} = \frac{1}{(0.16 \text{ m})} - \frac{1}{(0.08 \text{ m})} = -6.25 \text{ m}^{-1}$$

or $s' = 1/(-6.25 \text{ m}^{-1}) = -0.16$ m. Use Eq. (23.4) to determine the linear magnification:

$$m = -\frac{s'}{s} = -\frac{(-0.16 \text{ m})}{(0.08 \text{ m})} = +2.0$$

The magnification is also $m = (h'/h)$, so

$$h' = mh = (+2.0)(0.30 \text{ cm}) = +0.60 \text{ cm}$$

The birthmark image is upright and two times bigger than the object.

Try it yourself You hold a 1.0-cm-tall coin 0.20 m from a concave mirror with focal length +0.60 m. Is the image of the coin upright or inverted, and what are its magnification and height?

Answer

The image is upright; $m = 1.5$ and $h' = 1.5$ cm.

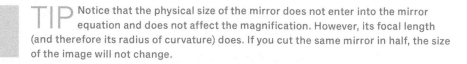

TIP   Notice that the physical size of the mirror does not enter into the mirror equation and does not affect the magnification. However, its focal length (and therefore its radius of curvature) does. If you cut the same mirror in half, the size of the image will not change.

REVIEW QUESTION 23.3 You place a concave mirror on a stand so it is vertical and slowly move a candle closer to the mirror from a relatively large distance away. What does your friend need to do to make an image appear on a paper screen as you move the candle closer to the mirror?

23.4 Qualitative analysis of lenses

Flat and curved mirrors allow us to create magnified images of objects using the reflection of light. We can also create images using a lens. A **lens** is a piece of glass or other transparent material with two curved surfaces that produces images of objects by changing the direction of light through refraction. Lenses, like curved mirrors, can be concave or convex. A convex lens (**Figure 23.12a**) is thicker in the middle than at its edges; it consists of two sections of spheres faced with the convex parts pointing out. A concave lens (Figure 23.12b) is thinner in the middle than at its edges; it consists of two sections of spheres faced with the convex parts pointing inward. In this book we will only investigate lenses in which the two surfaces are identical sections of spheres. However, there are practical uses for lenses with one curved surface and one flat, or with two differently curved surfaces. We will start with convex lenses, as they are easier to analyze.

FIGURE 23.12 Lenses are made from spherical surfaces.

(a)
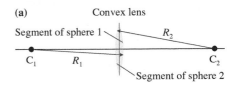

(b)
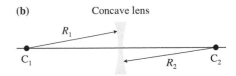

Convex lenses

In Observational Experiment Table 23.5, we show the paths of several parallel laser beams passing through a convex lens made of transparent plastic. The index of refraction of plastic is higher than that of air. So when a ray of light moves from air to plastic, it bends toward the normal line; when the ray leaves the lens, it bends away from the normal line. The overall result is that the ray bends toward the principal axis.

OBSERVATIONAL EXPERIMENT TABLE 23.5   Laser beams passing through a convex lens

Observational experiment	Analysis
Experiment 1. We shine three laser beams from the left parallel to the principal axis of a convex plastic lens. 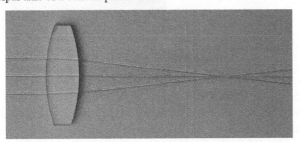	After passing through the lens, the rays pass through the same point on the principal axis. We use the law of refraction to analyze the paths of the rays. Ray 2 is perpendicular to the boundary of the two media and thus does not bend. Rays 1 and 3 bend toward the normal at the air-plastic interface and away from the normal at the plastic-air interface. 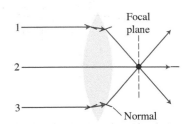
Experiment 2. We shine three laser beams from the left parallel to each other at an arbitrary angle relative to the principal axis of the lens. 	The rays pass through a point on the plane perpendicular to the principal axis that passes through the focal point found above. We use the law of refraction to analyze the paths of the rays. Although ray 2 is not perpendicular to the boundary, it passes straight through the lens—a small bending toward the normal on the air-plastic interface is cancelled by the small bending away from the normal at the plastic-air interface 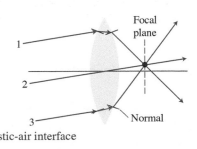

Patterns

1. In all experiments, the rays passing through the center of the lens do not bend.
2. The rays parallel to the principal axis pass through the same point after the lens. We call this point the focal point, similar to the focal point of a curved mirror.
3. The rays parallel to each other pass through a point on the plane perpendicular to the principal axis through the focal point. We call this plane the focal plane.

FIGURE 23.13 More and less curved convex lenses.

(a)

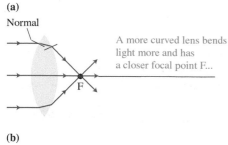

Normal

A more curved lens bends light more and has a closer focal point F...

(b)

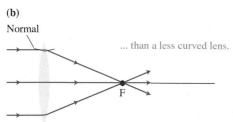

Normal

... than a less curved lens.

We have found that a convex lens made of plastic is similar to a concave mirror where incident rays parallel to the principal axis intersect at a **focal point** after passing through the lens. Note in **Figure 23.13** that the rays converge more steeply when passing through a lens with highly curved surfaces than when passing through a lens with less curvature. The focal point is closer to a lens with more curvature than for a lens with less curvature.

Therefore, a lens with two symmetrical curved surfaces has two focal points at equal distances on each side of the lens. In all of our experiments, we shine light from the left side of the lens. As a result, the focal point is on the right side of the lens. If we repeat the experiments, shining the light from the right, the rays will converge on the left side of the lens.

So far, we have used only glass and plastic lenses surrounded by air. However, other materials can act as lenses, such as a round plastic bottle filled with water. Its shape is slightly different from that of a regular lens, but if our understanding of the processes that involve lenses is correct, it should work as one. We fill the bottle with water and shine laser beams on it as shown in **Figure 23.14a**. We observe that after passing through the bottle, the beams converge at one point and then diverge again. Figure 23.14b shows why the rays converge (notice the direction of the normal at each interface). The bottle seems to work as a convex lens with a focal length of about 5 cm.

FIGURE 23.14 Experiments using a glass bottle as a lens.

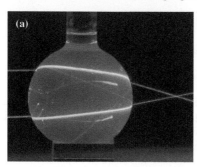

(a)

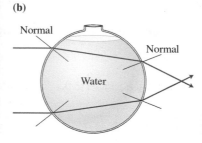

(b)

Normal

Normal

Water

(c)

What happens if we place an empty bottle in a container filled with water (Figure 23.14c)? The laser beams diverge. This experiment shows us that it is not only the shape of the lens but also the refractive index of the material from which it is made and the index of the surrounding material that determine whether the lens is converging or diverging. In addition, the focal length of a lens depends not only on the radius of curvature of the surfaces but also on the refractive indexes of the material between the surfaces and the outside matter. The closer the refractive indexes are to each other, the less the lens bends light rays, and therefore the longer its focal length.

Concave lenses

FIGURE 23.15 The focal point of a concave lens.

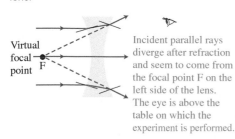

Virtual focal point F

Incident parallel rays diverge after refraction and seem to come from the focal point F on the left side of the lens. The eye is above the table on which the experiment is performed.

We can perform an experiment comparable to that in Table 23.5, sending parallel laser beams through a concave glass lens placed flat on a table. We use the rays to represent the laser beams in **Figure 23.15**. The rays parallel to the principal axis diverge after passing through the concave lens. The dashed lines extended back reveal that these rays seem to diverge from a single point on the axis. This point is called the **virtual focal point**.

Lenses are used in magnifying glasses, cameras, eyeglasses, microscopes, telescopes, and many other devices. Some of these devices produce an image that is smaller than the object (camera), and some produce an image that is larger than the object (magnifying glass). Lens ray diagrams help us understand how these devices work. The method for constructing ray diagrams for lenses is summarized in Physics Tool Box 23.3.

Convex lens

1. Draw the principal axis for the lens and a vertical line perpendicular to the axis representing the location of the lens. Label the focal points on each side of the lens.

2. Place an arrow at the object position with its base on the axis a distance s from the center of the lens.

3. *Convex lenses* Draw two or three rays.
Ray 1 moves parallel to the principal axis and after refraction through the lens passes through the focal point on the right side of the lens.
Ray 2 passes directly through the middle of the lens and its direction does not change.
Ray 3 passes through the focal point on the left and refracts through the lens so that it moves parallel to the axis on the right.

4. The place where the rays intersect on the right side at a distance s' from the lens is the location of a real image of the object.

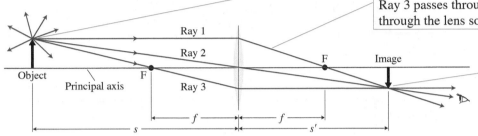

Concave lens

Steps 1 and 2 are the same as above.
3. *Concave lenses* Draw two or three rays.
Ray 1 moves parallel to the principal axis and is refracted away from the principal axis. Ray 1 appears to come from the focal point on the left side of the lens.
Ray 2 passes directly through the middle of the lens and for thin lenses its direction does not change.
Ray 3 is refracted before it reaches the focal point on the right. The ray moves parallel to the axis on the right.

4. If the rays diverge after passing through the lens, the place from which they seem to diverge is on the left side at distance s' from the lens at the location of a virtual image.

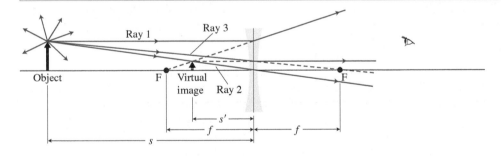

Table 23.6 (on the next page) summarizes possible combinations of the locations of objects and images produced by different types of lenses. For the convex lenses, the points that are located at twice the focal distance from the center of the lens are labeled 2F. This notation is used to emphasize the importance of these points for predicting where the image of the object will be. Specifically, when the object is behind the 2F point, the image is real, inverted, and reduced. When the object is between F and 2F, the image is real, inverted, and enlarged. When the object is between the lens and the focal point, the image is virtual, upright, and enlarged. These points are marked on both sides of the lens because the object can be on either side. Note that the 2F point is not marked for concave lenses because for an object located at any distance from the lens, the result is the same—the image is virtual, upright, and reduced.

The right-hand column of the table lists common applications for each lens type. For example, in a camera we need an image that is reduced and real. In a digital projector we need an image that is magnified and real. In a magnifying glass we want an upright enlarged virtual image.

TABLE 23.6 Ray diagrams for various lenses

Situation	Ray diagram and description of the image	Application

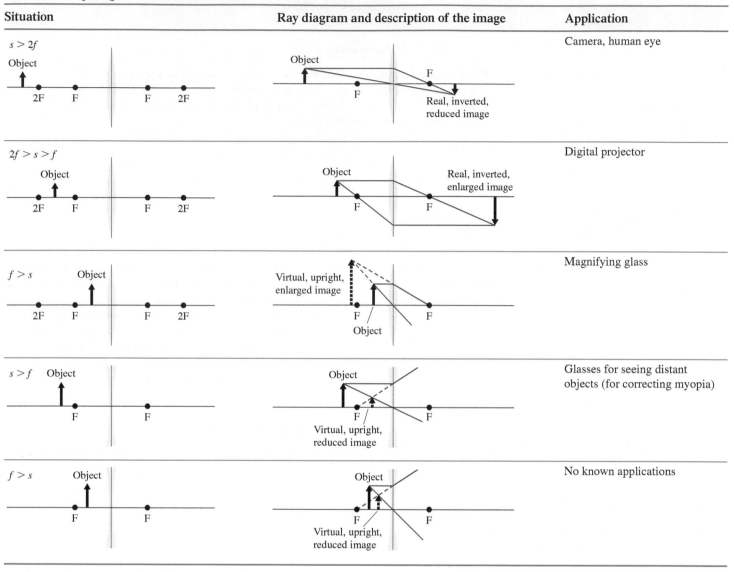

The table contains the following applications:
- $s > 2f$: Camera, human eye — Real, inverted, reduced image
- $2f > s > f$: Digital projector — Real, inverted, enlarged image
- $f > s$: Magnifying glass — Virtual, upright, enlarged image
- $s > f$: Glasses for seeing distant objects (for correcting myopia) — Virtual, upright, reduced image
- $f > s$: No known applications — Virtual, upright, reduced image

Image location of the base of the object

So far, in all of our examples we have found the image location of the top of the object and assumed that the image of the base of the object was directly beneath it on the principal axis. We will now check this assumption, using the rays that were described in Experiment 2 in Table 23.5, parallel to each other but not to the principal axis. After refraction, these rays pass through a point on the focal plane of the lens—a plane perpendicular to the principal axis and passing through the focal point (**Figure 23.16**).

Now let's draw a ray diagram for the base of an object that lies on the principal axis (**Figure 23.17a**). An infinite number of rays leave the base of the object. Ray 1, parallel to the principal axis, passes through the center of the lens and does not bend. This means that the image of the base point should be on the principal axis—but where? We can draw an imaginary ray 3 (Figure 23.17b) parallel to ray 2 passing through the center of the lens (undeflected). After leaving the lens, *parallel* rays 2 and 3 pass through the same point in the focal plane.

The image of the object appears where bent ray 2 intersects ray 1 on the principal axis (Figure 23.17b). Now we can draw the image of the arrow using the top and the bottom points (Figure 23.17c).

FIGURE 23.16 Parallel rays pass though the same point in the focal plane.

The path of each ray can be explained by refraction relative to the normal lines to each air-glass interface. The rays converge at the same distance from the lens as rays that are parallel to the principal axis.

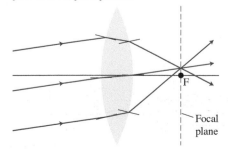

FIGURE 23.17 Locating the image of the base of an object.

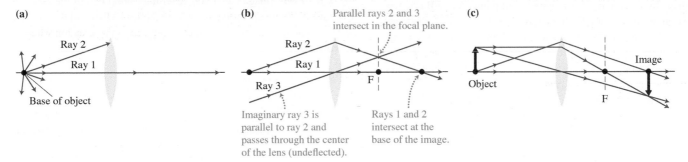

(a)

Ray 2

Ray 1

Base of object

(b)

Parallel rays 2 and 3 intersect in the focal plane.

Ray 2

Ray 1

Ray 3

F

Imaginary ray 3 is parallel to ray 2 and passes through the center of the lens (undeflected).

Rays 1 and 2 intersect at the base of the image.

(c)

Image

Object

F

CONCEPTUAL EXERCISE 23.7   **Lens Jeopardy**

The image of a shining point-like object S is produced by a lens (see figure below). The line is the principal axis of the lens. Where is the lens, what kind of lens is it, and where are its focal points?

Object

S

S′

Image

Sketch and translate Let's first decide the location of the lens that produced the image S′. We know that the shining point S sends light in all directions. Rays that reach the lens bend and pass through the image point S′. We need to find rays with paths that will help us to find that lens and its focal points.

Simplify and diagram A ray that passes through the center of the lens does not bend. This ray also passes through the image point. Thus,

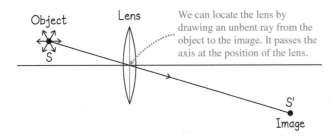

Object

Lens

S

We can locate the lens by drawing an unbent ray from the object to the image. It passes the axis at the position of the lens.

S′

Image

we draw a straight line from the object to the image. The point where the line crosses the principal axis will be the location of the center of the lens, as shown at bottom left. The lens must be convex since the image is on the opposite side of the lens from the object.

To locate a focal point, draw a ray from the object and parallel to the principal axis. This ray will refract through the lens and pass through the focal point on the right side of the lens and then through the image, as shown below. The other focal point of the lens must be the same distance to the left of the lens.

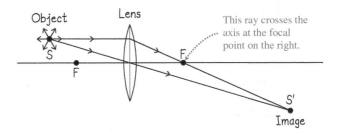

Object

Lens

S

F

This ray crosses the axis at the focal point on the right.

F

S′

Image

Try it yourself A ray from the object that passes through the focal point on the left side of the lens should bend at the plane of the lens (it actually passes below the lens) and move parallel to the axis on the right side of the lens. Does this ray also pass through the image point?

Answer   It should. Sketch it to see if it does.

CONCEPTUAL EXERCISE 23.8   **A partially covered lens**

Imagine that you have an object, a lens, and a screen. You place an object at a position $s > 2f$ from the lens and cover the top half of the lens. Half of the object is above the principal axis, half below. Describe what you will see on the screen.

Sketch and translate To answer the question, we need to draw a ray diagram and see what happens to the image with the top part of the lens covered.

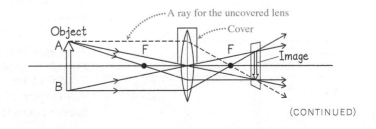

A ray for the uncovered lens

Object

A

Cover

F

F

Image

B

(CONTINUED)

Simplify and diagram We represent the object with an arrow and find the image by examining rays radiated from points A and B (the top and the bottom of the object). With the top of the lens covered, the rays moving toward the top of the lens do not pass through, but rays moving toward the bottom half do. Remember that *all* rays from a point on the object that pass through the lens pass through the corresponding image point. Covering the top half of the lens still allows some light from all points on the object to reach the image location, so the entire image will form. However, the image brightness will decrease as less light passes through the lens.

Try it yourself Imagine that you cover the central part of the lens. You have a shining object far away from the lens. How will the location of the image and its size change compared to the situation when not covered?

Answer

The size and location will not change, but the brightness will decrease.

REVIEW QUESTION 23.4   How do we know how many rays an object sends onto a lens?

23.5 Thin lens equation and quantitative analysis of lenses

We wish to derive a relation between s, s', and f that can be used to make a quantitative prediction for the location and properties of the image formed by a lens. To do this, we use the same technique that was used for curved mirrors. We use a convex lens in the derivation, but the results apply to concave lenses as well. As we did with mirrors, we will assume that the radii of curvature of the spheres from which the sides of the lens were cut are much larger than the size of the lens. Such a lens is called a **thin lens**. Both convex and concave lenses can be thin. Let's look at the geometry of an object at a known distance from a convex lens and its image. AB is the object and A_1B_1 is the image (**Figure 23.18**).

FIGURE 23.18 We can use a ray diagram to develop a relation between the location of an object, the image, and the focal distance of a lens.

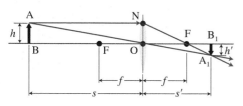

- We assume that the lens is very thin and that all of the rays are close to the principal axis. Triangles BAO and B_1A_1O are similar triangles. Thus $(AB/s) = (A_1B_1/s')$. After rearranging, we get

$$\frac{A_1B_1}{AB} = \frac{s'}{s}$$

- Triangles NOF and A_1B_1F are similar. Thus $(NO/f) = [A_1B_1/(s' - f)]$. Also, NO = AB. Thus

$$\frac{AB}{f} = \frac{A_1B_1}{s' - f}$$

- Rearranging this last equation, we get

$$\frac{A_1B_1}{AB} = \frac{s' - f}{f}$$

- Setting the above two equations with (A_1B_1/AB) on the left equal to each other, we get

$$\frac{s'}{s} = \frac{s' - f}{f} \qquad (23.5)$$

Using algebra similar to what we used to derive the mirror equation, we get a relationship that is called the **thin lens equation**:

$$\frac{1}{s} + \frac{1}{s'} = \frac{1}{f} \qquad (23.6)$$

Similar to the mirror equation, the thin lens equation will help us predict the location of the image when we know the location of the object and the focal length of the lens.

Let's see if this equation works for extreme cases. Imagine an object that is located at the focal point. Rays from this object will refract through the lens so that they are parallel. For an object at a focal point $(s = f)$:

$$\frac{1}{s} + \frac{1}{s'} = \frac{1}{f} \quad \text{or} \quad \frac{1}{f} + \frac{1}{s'} = \frac{1}{f}$$

Therefore, $(1/s') = 0$ or $s' = \infty$: the image will form at infinity. The formation of an image by a lens is summarized below, including sign conventions for lenses.

> **Thin lenses** The distance s of an object from a lens, the distance s' of the image from the lens, and the focal length f of the lens are related by the thin lens equation:
>
> $$\frac{1}{s} + \frac{1}{s'} = \frac{1}{f} \qquad (23.6)$$
>
> Several sign conventions are important when using the thin lens equation:
> 1. The focal length f is positive for convex lenses and negative for concave lenses.
> 2. The image distance s' is positive for real images and negative for virtual images.

FIGURE 23.19 The lens produces a sharp image only at s'.

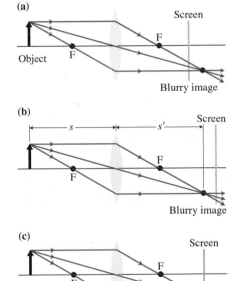

TIP ▸ Remember that all of the preceding reasoning applies to convex and concave lenses made of glass or transparent plastic when they are used in air (or made of any material of higher index of refraction than the surrounding medium).

How can we explain why a sharp, clear image of an object appears in only one place for a particular lens for a specific object-lens distance? If we place a screen closer to the lens than distance s' (**Figure 23.19a**), then the rays leaving the tip of the arrow have not yet converged to a single point. As a result, the light from the tip of the arrow is spread out on the screen and the image will be blurry. If we place the screen at a distance beyond s' (Figure 23.19b), where rays from the tip are diverging after having converged at s', we again get a blurry image. Only at distance s' from the lens do we see a sharp, clear image (Figure 23.19c). Being able to locate the position where a sharp image will appear is important for all applications involving optical instruments.

Digital projectors

A digital projector creates an enlarged real image of a small object inside the projector, an image that can be seen on an external screen. How do these devices work? Here we will focus on their optical components.

EXAMPLE 23.9 **Digital projector**

An object (a small display of the desired image inside the projector) is 0.200 m from a +0.190-m focal length lens. Where should we place the external screen in order to get a sharp image?

Sketch and translate See the labeled sketch of the situation, shown at right. The goal is to determine the correct distance s' of the screen from the projector lens.

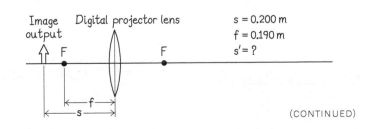

$s = 0.200 \text{ m}$
$f = 0.190 \text{ m}$
$s' = ?$

(CONTINUED)

Simplify and diagram Draw a ray diagram for the situation, such as the one shown below. Notice that the image is enlarged, real, and inverted relative to the object. We must invert what is shown on the internal display to obtain an upright image.

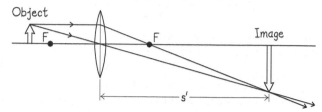

Object

F

F

Image

s'

Represent mathematically We use the thin lens equation [Eq. (23.6)] to determine the unknown image distance s':

$$\frac{1}{s} + \frac{1}{s'} = \frac{1}{f}$$

Solve and evaluate Solve for s' and substitute the known quantities to get

$$\frac{1}{s'} = \frac{1}{f} - \frac{1}{s} = \frac{1}{(0.190 \text{ m})} - \frac{1}{(0.200 \text{ m})}$$

$$= \frac{0.200 \text{ m} - 0.190 \text{ m}}{(0.190 \text{ m})(0.200 \text{ m})} = 0.263 \text{ m}^{-1}$$

Inverting, we get

$$s' = \frac{1}{(0.263 \text{ m}^{-1})} = 3.80 \text{ m}$$

The screen should be placed 3.80 m from the lens—a reasonable distance for a projector in a small conference room.

Try it yourself Suppose the screen needs to be 4.8 m from the lens. What distance should the small display inside the projector be from the same lens to produce a sharp image on the more distant screen?

Answer

$s = 0.198$ m (the lens and internal display need to be moved closer to each other).

TIP Forgetting to invert when solving for s' is a common mistake.

Linear magnification in lenses

Just as with mirrors, lenses can produce images that are bigger or smaller in size than the original objects. To characterize the relative sizes of the object and the image produced by a lens, we use the same physical quantity for magnification that we used for curved mirrors called **linear magnification**:

$$\text{linear magnification} = \frac{\text{image height}}{\text{object height}} = m = \frac{h'}{h} \tag{23.3}$$

where the heights h' and h are positive if the image or object is upright and negative if inverted. As with mirrors, we can relate the linear magnification to the image distance s' and the object distance s. Using the similar triangles in Figure 23.18, we can write

$$\text{linear magnification} = \frac{\text{image height}}{\text{object height}} = m = \frac{h'}{h} = -\frac{s'}{s} \tag{23.4}$$

EXAMPLE 23.10 Looking at an insect through a magnifying glass

You use a convex lens of focal length +10.0 cm to look at a tiny insect on a book page. The lens is 5.0 cm from the paper. Where is the image of the insect? If the insect is 1.0 cm in size, how large is the image?

Sketch and translate Draw a labeled sketch of the situation, as shown at right.

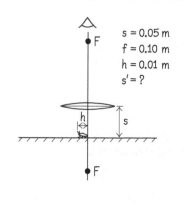

F

$s = 0.05$ m
$f = 0.10$ m
$h = 0.01$ m
$s' = ?$

h

s

F

Simplify and diagram Assume that the lens is parallel to the paper looking down at the insect. For convenience we draw the ray diagram horizontally, as shown here. From the ray diagram, we conclude that the image is virtual and is on the same side of the lens as the real insect. It is farther away, upright, and enlarged. This means that the image is *beneath* the book where the insect lies. This is where the light reaching your eye appears to originate.

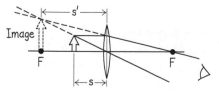

s'

Image

F

F

s

Represent mathematically Use the thin lens equation [Eq. (23.6)] to find the image distance s':

$$\frac{1}{s} + \frac{1}{s'} = \frac{1}{f}$$

Then use the linear magnification equation [Eq. (23.4)] to determine the magnification and the height of the insect's image:

$$m = \frac{h'}{h} = -\frac{s'}{s}$$

Solve and evaluate Insert $f = 0.10$ m and $s = 0.05$ m into the lens equation to get

$$\frac{1}{s'} = \frac{1}{f} - \frac{1}{s} = \frac{1}{0.10 \text{ m}} - \frac{1}{0.05 \text{ m}} = -10 \text{ m}^{-1}$$

Therefore, $s' = 1/(-10 \text{ m}^{-1}) = -0.10$ m. The minus sign indicates that the image is virtual, consistent with the ray diagram. The linear magnification is

$$m = -\frac{s'}{s} = -\frac{(-0.10 \text{ m})}{(+0.05 \text{ m})} = +2.0$$

The image is upright (the plus sign) and twice the object size, or $2.0(1.0 \text{ cm}) = 2.0$ cm, consistent with the ray diagram.

Try it yourself An image seen through a 10-cm focal length convex lens is exactly the same size as the object but inverted and on the opposite side of the lens. Where must the object and image be located?

Answer

The object must be $s = 2f = 20$ cm from the lens on one side, while the image must be $s' = 2f = 20$ cm from the lens on the other side.

EXAMPLE 23.11 ▸ **Find the image**

You place an object 20 cm to the left of a concave lens whose focal length is -10 cm (the negative sign indicates a concave lens). Where is the image located? Is it real or virtual?

Sketch and translate A labeled sketch of the situation is shown below.

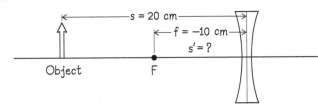

Simplify and diagram Draw a ray diagram for the situation (see below). The image is smaller than the object, virtual, and closer to the lens than the object. So, in your quantitative results, the image distance should be negative and the absolute value should be less than 10 cm.

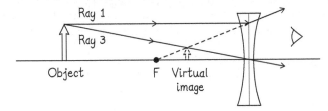

Represent mathematically We can use the thin lens equation [Eq. (23.6)] to determine the image location:

$$\frac{1}{s} + \frac{1}{s'} = \frac{1}{f}$$

Solve and evaluate Insert $f = -0.10$ m and $s = 0.20$ m into the above to get

$$\frac{1}{s'} = \frac{1}{f} - \frac{1}{s} = \frac{1}{-0.10 \text{ m}} - \frac{1}{0.20 \text{ m}} = -15 \text{ m}^{-1}$$

or $s' = 1/(-15 \text{ m}^{-1}) = -0.067$ m or -6.7 cm, consistent with the ray diagram.

Try it yourself Where will the image be if you place an object exactly at the focal point of a concave lens? How big will the height of the image be compared to the height of the object?

Answer

The image will be at a distance $f/2$, on the same side of the lens as the object. The height of the image will be half the height of the object.

The answer to the "Try it yourself" part gives us an idea of how to estimate the focal distance of a concave lens: when the image appears to be half the size of the object, the object-lens distance is equal to the focal distance.

REVIEW QUESTION 23.5 Where should you place an object with respect to a convex lens to have an image at exactly the same distance from the lens? What size will the image be?

23.6 Skills for analyzing processes involving mirrors and lenses

Example 23.12 illustrates the strategies for solving problems involving mirrors and lenses.

PROBLEM-SOLVING STRATEGY 23.1 | Processes involving mirrors and lenses

---

EXAMPLE 23.12 A camera

A camera made in the 1880s consisted of a single lens and light-sensitive film placed at the image location (**Figure 23.20**). To focus light on the image, the photographer would change the image distance—the distance from the lens to the film. Imagine that the image distance is 20 cm and that the film is a 16 cm × 16 cm square. A 1.9-m-tall person stands 8.0 m from the camera. What should be the focal length of the camera's lens for the image to be sharp? Would you be able to see the entire body of the person in the picture?

FIGURE 23.20 A single-lens camera made in the late 1800s.

Sketch and translate

- Sketch the situation in the problem statement.
- Include the known information and the desired unknown(s) in the sketch.

- The situation is sketched below.

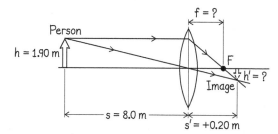

Simplify and diagram

- Assume the lens/mirror is only slightly curved and the rays are incident near the principal axis.
- Draw a ray diagram representing the situation in the problem.

- A ray diagram with the person represented as a shining arrow is included in the sketch.
- The object distance is $s > 2f$, so the image is real, inverted, and smaller than the object.

Represent mathematically

- Use the picture and ray diagram to help construct a mathematical description of the situation.

- Use the lens equation to find the focal length:

$$\frac{1}{f} = \frac{1}{s} + \frac{1}{s'}$$

- Use the linear magnification equation [Eq. (23.4)] to find the image height:

$$h' = mh = -\left(\frac{s'}{s}\right)h$$

Solve and evaluate

- Solve the equations for an unknown quantity.
- Evaluate the results to see if they are reasonable (the magnitude of the answer, its units, how the answer changes in limiting cases, and so forth).

$$\frac{1}{f} = \frac{1}{s} + \frac{1}{s'} = \frac{1}{8.0 \text{ m}} + \frac{1}{0.20 \text{ m}}$$

$$= 0.125 \text{ m}^{-1} + 5.0 \text{ m}^{-1} = 5.125 \text{ m}^{-1}$$

$$f = \frac{1}{5.125 \text{ m}^{-1}} = 0.20 \text{ m}$$

The image size is

$$h' = -\left(\frac{s'}{s}\right)h = -\left(\frac{0.20 \text{ m}}{8.0 \text{ m}}\right)(1.9 \text{ m}) = -0.048 \text{ m}$$

The inverted image size is about 5 cm and will easily fit on the film.

Try it yourself Suppose the 1.9-m-tall person stands 4.0 m from the +0.20-m focal length lens. Now where is the image formed relative to the lens, and what is the image height? Will you get a good picture? Explain.

Answer

$s' = 0.21$ m from the lens, $m = -0.053$ (the image is inverted), and $h' = -0.10$ m. The image distance is larger than 0.20 m; thus, the image will not be sharp on the screen that is 0.20 m away from the lens.

REVIEW QUESTION 23.6 If we have a mathematical equation for lenses and mirrors, what is the purpose of drawing ray diagrams when solving lens and mirror problems?

23.7 Single-lens optical systems

We have already discussed several applications for mirrors and lenses, from the magnifying glass to the digital projector. In this section we will investigate three single-lens optical systems: cameras, the human eye, and two kinds of corrective lenses.

Photography and cameras

A simple camera (**Figure 23.21**) has a lens of fixed focal length. Light from an object enters the camera through the lens, which focuses the light on a surface that has light-sensitive properties. In analog cameras, this surface is film; in digital cameras, the film has been replaced by an image sensor that consists of a two-dimensional array of light-sensitive elements—pixels. There are two types of image sensors: CCD (charge-coupled devices) and CMOS (complementary metal oxide semiconductors). To produce sharp images of objects located at different distances from the lens, the lens is moved relative to the film or the image sensor. When the picture includes multiple objects at different distances, you have to choose the object to focus on when you take the photo. But what if you could take the picture and decide what object you wanted in focus later?

A new camera provides that capability using an idea called **light field photography**. In light field photography, the image sensor records *all* the light entering the camera, not just what would produce a sharp image on the focal plane. In fact, a light field camera

FIGURE 23.21 The physics of a simple camera.

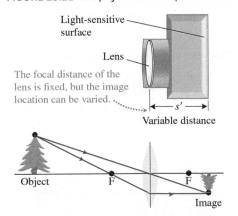

Light-sensitive surface

Lens

The focal distance of the lens is fixed, but the image location can be varied.

s'

Variable distance

Object   F   F   Image

FIGURE 23.22 A light field camera allows you to choose which object to focus on after you have taken the picture.

FIGURE 23.23 The human eye.

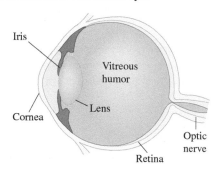

FIGURE 23.24 As the image distance is fixed in the eye, the lens changes shape to change the focal length, so that the image of any object is always on the retina.

(a)

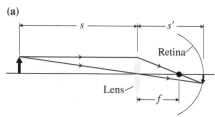

The lens changes shape to focus on objects at different distances.

(b)

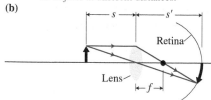

doesn't have a conventional lens at all (**Figure 23.22**). Instead, a two-dimensional array of tiny micro-lenses is placed just in front of the image sensor. Behind each micro-lens is its own personal portion of the image sensor, its own grid of pixels. Since each micro-lens has its own grid of pixels, the camera records not only how much light of each color reaches each micro-lens but also the direction of each ray reaching that micro-lens. Using this camera, a photographer can choose an object to focus on after the picture has been taken since the camera is effectively focusing on all objects at once.

Optics of the human eye

In many ways, the human eye resembles an expensive digital video camera. It is equipped with a built-in cleaning and lubricating system, an exposure meter, an automatic field finder, and about 130 million photosensitive elements that continuously send electric signals to the brain (the equivalent of a 130-megapixel digital camera CCD) in real time. Light from an object enters the **cornea** (**Figure 23.23**) and passes through a transparent lens. An **iris** in front of the lens widens or narrows, like the aperture on a camera that regulates the amount of light entering the device. The **retina** plays the role of the detector.

A normal human eye can produce sharp images of objects located anywhere from about 10–25 cm to hundreds of kilometers away (effectively at infinity). Unlike a camera, which moves a fixed focal length lens to accommodate different object distances, the eye has a fixed image distance of about 2.1 cm (the distance from the lens to the retina) and a variable focal length lens system.

The changing focal length of the eye's lens is illustrated in **Figure 23.24**. When the eye looks at distant objects, muscles around the lens of the eye relax, and the lens becomes less curved (Figure 23.24a), increasing the focal length and allowing an image to form on the retina. As the object moves closer, the corresponding image moves behind the retina, appearing blurry. To compensate, the eye muscles contract, increasing the curvature of the lens, reducing the focal length (Figure 23.24b) and moving the image back onto the retina. The contraction of these muscles causes your eyes to become tired after reading for many hours.

The **far point** is the most distant point at which the eye can form a sharp image of an object on the retina with the eye muscles relaxed. The **near point** of the eye is the nearest an object can be to the eye and still have a sharp image produced on the retina with the eye muscles tensed. For the normal human eye, the far point is effectively at infinity (we can focus on the Moon and on distant stars) and the near point is at approximately 25 cm from the eye.

When you swim with your eyes open, objects look blurry. If you wear goggles, however, you can see objects clearly. Why can't the eye produce sharp images on the retina underwater? What is the role of the goggles? The answer is that the lens is not the only optical element in the eye. The moist, curved surface of the eye (the cornea) acts as a sort of lens as well. Under normal conditions, the cornea works with the rest of the eye to form a sharp image on the retina. Without the cornea, your eye cannot bend the light passing through it to make a sharp image on your retina.

In water, the cornea stops acting as a lens because its refractive index is about the same as the refractive index of water. Therefore, the eye cannot produce sharp images on the retina, and everything looks blurry. When you wear goggles, you put air between the eye and the water, and your cornea can operate as a lens again.

Corrective lenses

Corrective lenses compensate for the inability of the eye to produce a sharp image on the retina. The two most common vision abnormalities corrected with lenses are **myopia** (nearsightedness) and **hyperopia** (farsightedness).

Myopia A person with myopia, or nearsightedness, can see close-up objects with clarity but not those that are distant. The far point of a nearsighted person may be only a few meters away rather than at infinity.

Myopia is caused by a larger-than-normal eyeball (greater than 2.5 cm in diameter) or a lens that is too curved. In such cases the image of a distant object is formed in front of the retina (**Figure 23.25a**), even when the eye muscles are relaxed. Placing a concave lens in front of the eye corrects nearsighted vision by causing rays from an object to diverge more than without the lens, so that when they pass through the eye lens, an image is formed farther back in the eye (Figure 23.25b). When an object is very distant (that is, $s = \infty$), the focal length of the concave lens is chosen so that its virtual image is formed at the far point of the eye (Figure 23.25c). Light passing through the eyeglass lens appears to come from the image at the far point, not from the more distant object, which allows the eye to produce an image on the retina. To calculate the desired focal length of the eyeglass lens, we set $s = \infty$ and $s' = -(\text{far point distance})$. The negative sign accounts for the fact that the image produced by the eyeglass is virtual. We can then calculate the value of the desired focal length f of the concave lens using the thin lens equation.

FIGURE 23.25 Correcting myopia (nearsightedness): (a) uncorrected; (b) corrected; (c) glasses help "move" the distant object to the eye's far point.

(a)
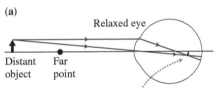
In a nearsighted eye, the image of a distant object forms in front of the retina.

(b)
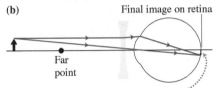
With a concave lens, the image forms farther back, on the retina.

(c)
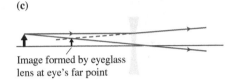
The person is actually looking at the eyeglass image at the far point of the eye.

EXAMPLE 23.13   Glasses for a nearsighted person

Alex is nearsighted, and his far point is 2.0 m from his eyes. What should be the focal length of his eyeglass lens so that he can focus on a very distant object ($s = \infty$)?

Sketch and translate A sketch of the eyeglass lens is shown in Figure 23.25b. The image of an object at infinity is to be formed at the far point 2.0 m in front of the lens.

Simplify and diagram Assume that the distance between the eyeglass lens and the eye lens is small compared to 2.0 m. The ray diagram in Figure 23.25c represents the optics of the eyeglass lens. Without glasses, the image of a distant object is formed before the light reaches Alex's retina and appears blurred. When Alex wears glasses, the object at infinity ($s = \infty$) has an image 2.0 m to the left of the eyeglass lens ($s' = -2.0$ m).

Represent mathematically We can use the given values of s and s' in the lens equation [Eq. (23.6)] to determine the focal length of the eyeglass lenses:

$$\frac{1}{s} + \frac{1}{s'} = \frac{1}{f}$$

Solve and evaluate Insert the object and image distances into the above equation:

$$\frac{1}{f} = \frac{1}{\infty} + \frac{1}{(-2.0 \text{ m})}$$

Therefore, $f = -2.0$ m. The number is negative. This means that the lens should be concave. The focal length of the lens for a nearsighted person equals the negative of the distance to the person's far point.

Try it yourself If Alex is able to see very distant objects with glasses of focal length −1.5 m, what is his far point distance?

Answer   1.5 m.

Hyperopia Unlike nearsighted people, hyperopic (farsighted) people are able to produce sharp images of distant objects on the retina but cannot do it for nearby objects, such as a book or a cell phone screen. Whereas the normal eye has a near point at about 25 cm, a farsighted person may have a near point several meters from the eye. If the

FIGURE 23.26 Correcting hyperopia (farsightedness): (a) uncorrected; (b) corrected; (c) glasses help "move" the closely positioned object to the eye's near point.

(a)

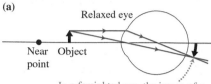

In a farsighted eye, the image of a near object forms behind the retina.

(b)

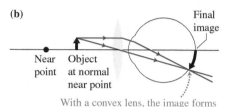

With a convex lens, the image forms farther forward, on the retina.

(c)

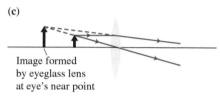

Image formed by eyeglass lens at eye's near point

The person is actually looking at an eyeglass virtual image at the near point of the eye.

object is closer than the person's near point, the image is formed behind the retina (Figure 23.26a) and appears blurred. Farsightedness may occur if the diameter of a person's eyeball is smaller than normal or if the lens is unable to curve sufficiently when the surrounding muscles contract.

Convex lenses are used to correct farsighted vision. The eyeglass lens slightly converges light from a nearby object so that the image produced by the eye lens moves forward onto the retina (Figure 23.26b). The lens produces a virtual image of the object at or beyond the person's near point (Figure 23.26c).

To determine the focal length of the eyeglass lens, we need to solve the thin lens equation for f. A farsighted person's eyeglass lens needs to produce an image at their near point, $s' = -(\text{near point distance})$. The negative sign means that the image is virtual.

> **Optical power** An optometrist prescribes glasses using a different physical quantity—the optical power P, measured in **diopters**:
>
> $$P = \frac{1}{f} \qquad (23.7)$$
>
> where f is measured in meters.

For example, the power of a concave lens of focal length $f = -0.50$ m is $P = 1/(-0.50 \text{ m}) = -2.0$ diopters. A lens of large positive power has a small positive focal length and causes rays to converge rapidly after passing through the lens. A lens of large negative power is concave. It has a small negative focal length and causes rays to diverge rapidly.

TIP Optical power P is not related in any way to energy conversion rate P, which we learned about in earlier chapters.

EXAMPLE 23.14   **Glasses for a farsighted person**

Eugenia is farsighted with a near point of 1.5 m. If you were an optometrist, what glasses would you prescribe for her so she can read a book held 25 cm from her eyes?

Sketch and translate Eugenia holds a book 0.25 m away (see Figure 23.26b) and wants her glasses to make her eyes think it is 1.5 m away at her near point. Thus $s = 0.25$ m and $s' = -1.5$ m for her glasses (on the same side of the lens as the book). We need to find their focal length f and power P.

Simplify and diagram A ray diagram for the optics of the eyeglass system is shown in Figure 23.26c.

Represent mathematically Use Eq. (23.6) to find the focal length of the lenses:

$$\frac{1}{s} + \frac{1}{s'} = \frac{1}{f}$$

Solve and evaluate Insert the known object and image distances into the previous equation to determine the focal length of the desired glasses:

$$\frac{1}{f} = \frac{1}{0.25 \text{ m}} + \frac{1}{-1.50 \text{ m}}$$
$$= 4.0 \text{ m}^{-1} - 0.67 \text{ m}^{-1} = 3.33 \text{ m}^{-1}$$

Thus,

$$f = \frac{1}{3.33 \text{ m}^{-1}} = 0.30 \text{ m}$$

and the power of the glasses is $P = (1/f) = 1/(0.30 \text{ m}) = 3.3$ diopters.

- -

Try it yourself Eugenia went to a drugstore to buy glasses, but her only choice was 4.0-diopter glasses. Where will Eugenia need to hold a book when wearing her new glasses so its image is at her 1.5-m near point?

Answer   21 cm.

REVIEW QUESTION 23.7 What is the main difference between how a camera produces sharp images of objects at different distances and how the human eye does it?

23.8 Angular magnification and magnifying glasses

The linear magnification of an optical system compares only the heights of the image and the object, but the apparent size of an object as judged by the eye depends not only on its height but also on its distance from the eye. For example, a pencil held 25 cm from your eye appears longer than one held 100 cm away (**Figure 23.27**). In fact, the pencil may appear longer than a 100-story building if the building is several miles away, even though the pencil is actually much shorter.

The impression of an object's size is quantified by its **angular size** θ, shown in Figure 23.27. The angular size depends on an object's height h and its distance r from the eye:

$$\theta = \frac{h}{r} \tag{23.8}$$

Remember that an angle θ, in radians, is the ratio of an arc length ($\approx h$ in Figure 23.27) and the radius of a circle (r in Figure 23.27: $\theta = (h/r)$). We can use h for the length of the straight line between the corners of the triangle only if h is very close to the actual arc length. This is only true for small angles. Thus Eq. (23.8) is an approximation for small angle θ.

If a person looks at an object through one or more lenses, he or she sees light that appears to come from the final image of the system of lenses. If the person's eye is close to the last lens (as in a typical telescope or microscope), then the angular size θ' of the image is

$$\theta' = \frac{h'}{s'} \tag{23.9}$$

where h' is the height of the final image and s' is the distance of the final image from the last lens in the system. The **angular magnification** M of an optical system is defined as the ratio of θ' and θ:

$$M = \frac{\theta'}{\theta} = \frac{\text{Angular size of the final image of optical system}}{\text{Angular size of object as seen by the unaided eye}} \tag{23.10}$$

A magnifying glass consists of a single convex lens. To use the lens as a magnifying glass, we position the object between its focal point and the lens, as shown in **Figure 23.28a**. A magnified virtual image is formed behind the object. The light appears to be coming from this enlarged image, which is farther away from the lens than the object. In other words, when you use a magnifying glass to look at small print on a page on a desk, the image you are viewing is located under the desk!

Note that in Figure 23.28a both the object and the image make the same angle θ' with the principal axis ($\theta' = (h'/s') = (h/s)$). The object has the same angular size when we look directly at it when it is a distance s from our eye as when we look through the magnifying glass at the image a distance s' from our eye. Using the magnifying glass lets you place the object closer than your near point (and hence have a larger angular size) and use the magnifying glass to focus it clearly on your retina (shown in Figure 23.28b).

To calculate the angular magnification of the magnifying glass, compare the angular size θ' of the image seen through the magnifying glass and the angular size θ of the object seen with the unaided eye when the object is as close as possible and still appears clear (at your near point). The angular size of the image viewed through the magnifying glass is $\theta' = (h'/s')$. By considering ray 2 in Figure 23.28a, we find that $(h'/s') = (h/s)$. Thus,

$$\theta' = \frac{h}{s}$$

where h is the actual height of the object and s is its distance from the magnifying glass.

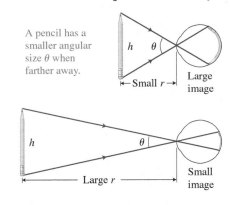

FIGURE 23.27 The angular size θ of an object.

A pencil has a smaller angular size θ when farther away.

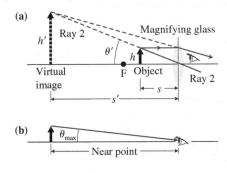

FIGURE 23.28 A magnifying glass produces an enlarged virtual image of the closely positioned object.

FIGURE 23.29 Maximum angular size θ_{max} of an object whose image is on the retina. If we bring the object closer, the image is behind the retina and what we see is blurred.

(a)

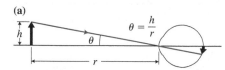

(b)

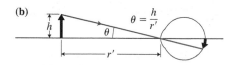

(c)

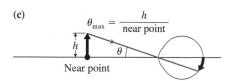

Near point

(d)

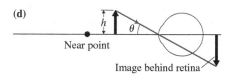

Near point

Image behind retina

FIGURE 23.30 Analyzing a two-lens optical system. The image produced by lens 1 becomes the object for lens 2.

(a) Object of first lens

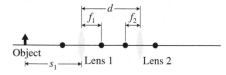

(b) Image of first lens

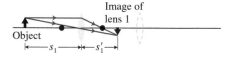

(c) Object of second lens

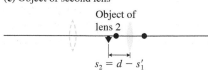

(d) Image of second lens

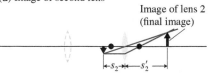

Next consider the angular size of the object as seen with the unaided eye. The angular size θ of an object of height h that we hold a distance r from the eye is $\theta = (h/r)$ (**Figure 23.29a**). As we bring the object closer to the eye (Figure 23.29b and c), its angular size increases. If we bring the object closer than the near point, the image on the retina is blurred (Figure 23.29d). The *maximum* angular size θ_{max} of an object of height h, when viewed by the unaided eye and when focused on the retina, is

$$\theta_{max} = \frac{h}{\text{Near point distance}} \tag{23.11}$$

The angular magnification of the image is the ratio of the angular size of the image θ' and the maximum angular size θ_{max} as seen with the unaided eye. Thus,

$$M = \frac{\theta'}{\theta_{max}} = \frac{h/s}{h/(\text{Near point distance})} = \frac{\text{Near point distance}}{s} \tag{23.12}$$

REVIEW QUESTION 23.8 If a person with normal vision and a farsighted person use the same magnifying glass, who will get the greater magnification? Why?

23.9 Telescopes and microscopes

Telescopes and microscopes use multiple lenses to produce images that are magnified more than is possible with a single lens. The technique for locating the final image and calculating its magnification requires careful attention to several details—provided below for a system with two lenses. This process applies to both telescopes and microscopes.

To find the final image of a two-lens system:

Step 1: Construct a sketch of the system of lenses (**Figure 23.30a**). The lenses are separated by a distance d and have focal lengths f_1 and f_2. The object is a distance s_1 from lens 1.

Step 2: Use Eq. (23.6) to determine the location s'_1 of the image formed by the first lens (Figure 23.30b). If s'_1 is positive, the image is to the right of lens 1. If s'_1 is negative, the image is to the left of lens 1. In our example, s'_1 is positive.

Step 3: Next, note that the image of lens 1 is now the object for lens 2. The object distance s_2 (Figure 23.30c) is $s_2 = d - s'_1$, where d, the separation of the lenses, is a positive number. It is possible for s_2 to be negative if the image formed by lens 1 is to the right of lens 2.

Step 4: Use the thin lens equation to determine the image distance s'_2 of the image formed by lens 2 (Figure 23.30d):

$$\frac{1}{s'_2} = \frac{1}{f_2} - \frac{1}{s_2}$$

This is the location of the final image relative to the second lens. If s'_2 is positive, the final image is real and located to the right of lens 2. If s'_2 is negative, the final image is virtual and to the left of lens 2. In our example, the final image is real and to the right of lens 2.

Step 5: The total linear magnification m of the two-lens system equals the product of the linear magnifications of each lens: $m = m_1 m_2$. If m is positive, the final image has the same orientation as the original object. If m is negative, the final image is inverted relative to the original object. Techniques for calculating the total angular magnification M are illustrated in the discussions and examples that follow.

Telescopes

Galileo is believed to have been the first to study the night sky and the Sun with a telescope. He discovered mountains and craters on the Moon, the moons of Jupiter, stars inside the Milky Way, and sunspots. His telescope consisted of two lenses—one convex and the other concave. Today, a more common version has two convex lenses separated by a distance slightly less than the sum of their focal lengths (**Figure 23.31a**). When you observe a distant object, the first lens produces a real image just beyond its own focal length (Figure 23.31b). The second lens is located so that the image from the first lens is just inside the focal point of the second lens. The image from the first lens becomes the object for the second lens. The second lens produces a magnified inverted virtual image (Figure 23.31c), which we observe by looking through the telescope.

FIGURE 23.31 A telescope.

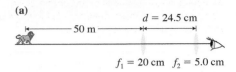

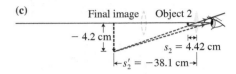

EXAMPLE 23.15 **Looking at a lion with a telescope**

A 1.2-m-tall lion stands 50 m from the first lens of a telescope (Figure 23.31a). Locate the final image of the lion. Determine the linear magnification, height, and angular magnification of the final image.

Sketch and translate Figure 23.21a shows that the givens are $f_1 = +20.0$ cm, $f_2 = +5.0$ cm, $d = 24.5$ cm, $s_1 = 50$ m $= 5000$ cm, and $h = 1.2$ m. We need to find s_2', m, and M.

Simplify and diagram The ray diagrams in Figure 23.31b and c help us locate the first and second images.

Represent mathematically Use the procedure described on the previous page to determine the locations of the first image, the second object, and the second image s_1', s_2, and s_2'. Then use the values of s_1, s_1', s_2 and s_2' to determine the linear magnification $m = m_1 m_2$ of the two-lens system, where $m_1 = -(s_1'/s_1)$ and $m_2 = -(s_2'/s_2)$. Next, determine the height of the final image of the lion $h' = mh$. To find the angular magnification, compare the angular size of the lion as seen through the optical system $\theta' = (h_2'/s_2')$ and its angular size as seen with the unaided eye $\theta = (h/r_1)$, where $r_1 = s_1 + d$. Then take the ratio of the angular sizes in order to determine the angular magnification: $M = (\theta/\theta')$.

Solve and evaluate This procedure gives us results of $s_1' = 20.08$ cm, $s_2 = 4.42$ cm, and $s_2' = -38.1$ cm. Notice that s_2' is negative, indicating a virtual image, in agreement with the ray diagram. To determine the linear magnification:

$$m = m_1 m_2 = \left(-\frac{s_1'}{s_1}\right)\left(-\frac{s_2'}{s_2}\right)$$

$$= \frac{(20.08 \text{ cm})(-38.1 \text{ cm})}{(5000 \text{ cm})(4.42 \text{ cm})} = -0.035$$

Notice that the linear magnification is less than 1; the final image is smaller than the original object. The image height is $h_2' = mh = (-0.035)(1.2 \text{ m}) = -0.042 \text{ m} = -4.2 \text{ cm}$.

Finally, determine the angular magnification M:

$$\theta' = \frac{h_2'}{s_2'} = \frac{(-4.2 \text{ cm})}{(-38.1 \text{ cm})} = 0.11 \text{ rad}$$

$$\theta = \frac{h}{s_1} = \frac{(1.2 \text{ m})}{(50 \text{ m})} = 0.024 \text{ rad}$$

$$M = \frac{\theta'}{\theta} = \frac{0.11 \text{ rad}}{0.024 \text{ rad}} = 4.6$$

Although the image of the lion is smaller than the actual lion, its angular size is almost five times bigger when viewed through the telescope than when seeing it without a telescope from 50 m away. This telescope succeeds in making the lion appear bigger than it really is.

- -

Try it yourself How can the lion appear larger when its final image is smaller?

Answer

The final image was much closer than the lion (131 times closer) and thus looked bigger through the telescope, even though it was smaller (29 times smaller).

The compound microscope

A compound microscope, like a telescope, has two convex lenses and passes light reflected from the object through two lenses to form the image. A microscope magnifies tiny nearby objects instead of large distant ones.

Both lenses have relatively short focal lengths and are separated by 10–20 cm (**Figure 23.32a**). The object is placed just outside and very close to the focal point of the first lens (called the **objective** lens). The real, enlarged, inverted image produced by the first lens is just inside the focal point of the second lens (called the **eyepiece** lens), as shown in Figure 23.32b. When an observer looks through the eyepiece, she sees an inverted, enlarged, virtual image of the object (Figure 23.32c).

FIGURE 23.32 Compound microscope.

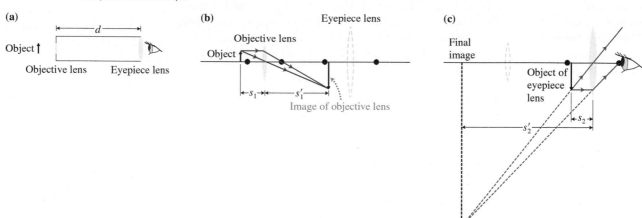

Consider the magnification of such a system. The image produced by the objective lens has linear magnification $m_1 = -(s_1'/s_1)$. The image produced by the objective lens becomes the object for the eyepiece lens. Since this object is located just inside the focal point of the eyepiece lens, the eyepiece lens acts as a magnifying glass used to view the real image produced by the objective lens (Figure 23.32c). The angular magnification of this magnifying glass is, according to Eq. (23.12), $M_2 = (\text{Near point distance}/s_2)$. The total angular magnification of the microscope is

$$M = m_1 M_2 = -\frac{s_1'}{s_1} \frac{\text{Near point distance}}{s_2}$$

FIGURE 23.33 Developing an alternative expression for angular magnification.

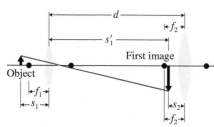

This expression for angular magnification can be rewritten in terms of the focal lengths of the lenses and their separation d. Consider **Figure 23.33**. Notice that $s_1 \approx f_1$; $s_1' \approx d - f_2$; and $s_2 \approx f_2$. By substituting these values in the previous equation, we have

$$M = \left(-\frac{d - f_2}{f_1}\right)\left(\frac{\text{Near point distance}}{f_2}\right) \qquad (23.13)$$

This expression is easy to use because the magnification depends only on the focal lengths of the lenses, their separation, and the near point distance of the eye. The very best optical microscopes provide angular magnifications of about 1000.

EXAMPLE 23.16   **Seeing a cell with a microscope**

A compound microscope has an objective lens of focal length 0.80 cm and an eyepiece of focal length 1.25 cm. The lenses are separated by 18.0 cm. If a red blood cell is located 0.84 cm in front of the objective lens, where is the final image of the cell, and what is its angular magnification? The viewer's near point is 25 cm.

Sketch and translate We have the following information: $f_1 = +0.80$ cm, $f_2 = +1.25$ cm, $d = 18.0$ cm, and $s_1 = +0.84$ cm. We need to find s_2' and the angular magnification M of the microscope. We can use Figure 23.32a as a sketch of the situation.

Simplify and diagram Assume that the red blood cell is brightly illuminated and can be modeled as a shining arrow. We use the ray diagrams in Figure 23.32b and c.

Represent mathematically First find the image produced by the first lens:

$$\frac{1}{s_1'} = \frac{1}{f_1} - \frac{1}{s_1}$$

This image becomes the object for the second lens. The object distance for the second lens is $s_2 = d - s_1'$. Use this distance to find the location of the image produced by the second lens (the final image):

$$\frac{1}{s_2'} = \frac{1}{f_2} - \frac{1}{s_2}$$

To find the angular magnification M we can use

$$M = m_1 M_2 = -\frac{s_1'}{s_1} \frac{\text{Near point distance}}{s_2}$$

Solve and evaluate

$$\frac{1}{s_1'} = \frac{1}{f_1} - \frac{1}{s_1} = \frac{1}{0.80 \text{ cm}} - \frac{1}{0.84 \text{ cm}} = 0.0595 \text{ cm}^{-1}$$

Thus, $s_1' = 1/(0.0595 \text{ cm}^{-1}) = +16.8$ cm. The object distance for the second lens is $s_2 = d - s_1' = 18.0 \text{ cm} - 16.8 \text{ cm} = 1.2$ cm. The final image distance is

$$\frac{1}{s_2'} = \frac{1}{f_2} - \frac{1}{s_2} = \frac{1}{1.25 \text{ cm}} - \frac{1}{1.20 \text{ cm}} = -0.033 \text{ cm}^{-1}$$

or $s_2' = 1/(0.033 \text{ cm}^{-1}) = -30$ cm. The total angular magnification is

$$M = m_1 M_2 = -\frac{s_1'}{s_1} \frac{\text{Near point distance}}{s_2}$$

$$= -\frac{16.8 \text{ cm}}{0.84 \text{ cm}} \frac{25 \text{ cm}}{1.2 \text{ cm}} = -(20)(21) = -420$$

The negative sign means that the final image is inverted.

Try it yourself How does the result using the magnifications for each lens compare to the approximate magnification equation [Eq. (23.13)] for a microscope?

Answer   You get the same result, −420.

REVIEW QUESTION 23.9 Why is saying that a telescope magnifies simultaneously a correct and an incorrect statement?

Summary

Plane mirror: A plane mirror produces a virtual image at the same distance behind the mirror as that of the object in front. A **virtual image** is at the position where the paths of the reflected rays seem to originate from behind the mirror. (Section 23.1)

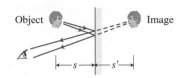

$$s = s'$$

Concave and convex mirrors: The distance s of an object from the surface of a concave or convex mirror, the distance s' of the image from the surface of the mirror, and the focal length f of the mirror are related by the mirror equation. The **focal point** F is the location through which all incident rays parallel to the principal axis pass after reflection (concave mirrors) or appear to have come from (convex mirrors). (Sections 23.2 and 23.3)

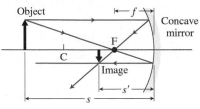

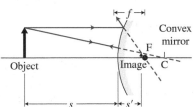

$$\frac{1}{s} + \frac{1}{s'} = \frac{1}{f} \qquad \text{Eq. (23.2)}$$

The *sign conventions* for these quantities are:

- The **focal length** f is positive for concave mirrors and negative for convex mirrors.
- The **image distance** s' is positive for real images and negative for virtual images.

Convex and concave lenses: The distance s of an object from a lens, the distance s' of the image from the lens, and the focal length f of the lens are related by the thin lens equation. (Sections 23.4 and 23.5)

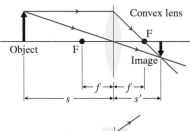

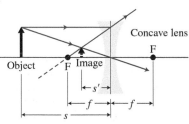

$$\frac{1}{s} + \frac{1}{s'} = \frac{1}{f} \qquad \text{Eq. (23.6)}$$

The *sign conventions* are:

- The **focal length** f is positive for convex lenses and negative for concave lenses.
- The **image distance** s' is positive for real images and negative for virtual images.

Linear magnification m is the ratio of the height h' of the image and the height h of the object. The heights are positive if upright and negative if inverted. (Sections 23.3, 23.5, 23.7, and 23.8)

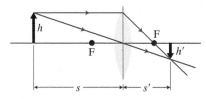

$$m = \frac{h'}{h} = -\frac{s'}{s} \qquad \text{Eq. (23.4)}$$

Angular magnification M is the ratio of the angular size θ' of an image as seen through a lens or lenses and the maximum angular size θ_{max} of the object as seen with the unaided eye. (Section 23.8)

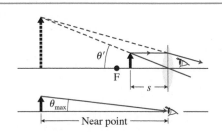

$$M = \frac{\theta'}{\theta_{max}}$$

$$= \frac{\text{Near point distance}}{s} \qquad \text{Eq. (23.12)}$$

Questions

Multiple Choice Questions

1. Where does the image of an object in a plane mirror appear?
   (a) In front of the mirror
   (b) On the surface of the mirror
   (c) Behind the mirror

2. Where does the image of an object that is s meters in front of a plane mirror appear?
   (a) s meters away from the object
   (b) $2s$ meters away from the object
   (c) There is no image formed by the plane mirror as the reflected rays do not converge.

3. A plane mirror produces an image of an object that is which of the following?
   (a) Real and enlarged
   (b) Virtual and the same size
   (c) Virtual and enlarged
   (d) Virtual and smaller

4. A concave mirror can produce an image that is which of the following?
   (a) Real, enlarged, and upright
   (b) Virtual, smaller, and inverted
   (c) Virtual, enlarged, and inverted
   (d) None of the answers is correct

5. A convex mirror can produce an image that is which of the following?
   (a) Real, enlarged, and inverted
   (b) Real, smaller, and inverted
   (c) Real, smaller, and upright
   (d) Virtual, smaller, and upright
   (e) Virtual, enlarged, and upright

6. A virtual image is the image produced
   (a) on a screen by reflected rays.
   (b) on a screen by refracted rays.
   (c) at a point where diverging rays appear to originate.

7. To see an image of an object that is enlarged, real, and inverted, you need to place the object in front of a convex lens in which region?
   (a) $s > 2f$
   (b) $2f > s > f$
   (c) $f > s$
   (d) None of these

8. To see an image of an object that is enlarged, real, and inverted, you need to place the object in front of a concave lens in which region?
   (a) $s > 2f$
   (b) $2f > s > f$
   (c) $f > s$
   (d) None of these

9. When we derived the thin lens equation [Eq. (23.6)], what did we use?
   (a) A ray diagram
   (b) Similar triangles
   (c) An assumption that the radius of curvature of the lens is much less than the distances f, s, and s'
   (d) a, b, and c
   (e) a and b only

10. When drawing images of objects produced by curved mirrors and lenses, which of the following ideas is required?
    (a) Each point of an object sends two rays.
    (b) Each point of an object sends three rays.
    (c) Each point of an object sends an infinite number of rays.

11. The focal length of a glass lens is 10 cm. When the lens is submerged in water, what is its new focal length?
    (a) 10 cm
    (b) less than 10 cm
    (c) more than 10 cm
    (d) Not enough information to answer

12. A microbiologist uses a microscope to look at body cells because the microscope will do which of the following?
    (a) Magnify the linear size of the cells
    (b) Magnify the angular size of the cells
    (c) Bring the cells closer
    (d) Make the cells brighter
    (e) Both (a) and (b) are correct.

13. The human eye works in a similar way to which of the following?
    (a) A microscope
    (b) An overhead projector
    (c) A telescope
    (d) A camera

14. Which of the following changes will result in a decrease of the focal length of a concave mirror? Select all correct answers.
    (a) Increase the radius of curvature of the mirror.
    (b) Decrease the radius of curvature of the mirror.
    (c) Immerse the mirror in water.
    (d) Reduce the diameter of the mirror without changing the curvature.

Conceptual Questions

15. When we draw a ray passing through the center of a lens, it does not bend independent of its direction. Explain why and state the implicit assumptions that we use to draw such rays.

16. You run toward a building with walls of a metallic, reflective material. Your speed relative to the ground is 1.5 m/s. How fast does your image appear to move toward you? Explain.

17. A tiny plane mirror can produce an image similar to a pinhole camera. Explain how.

18. Explain how we derived the mirror equation.

19. Explain how we derived the thin lens equation.

20. Explain the difference between a real and a virtual image.

21. You stand in front of a fun house mirror. You see that your head is enlarged and your legs and feet are shrunk. How can this be? Provide a ray diagram to support your answer.

22. A bubble of air is suspended underwater. Draw a ray diagram showing the approximate path followed by a horizontal light ray passing from the water (a) through the upper half of the bubble, (b) through its center, and (c) through its lower half. Does the bubble act like a horizontal converging or diverging lens? Explain.

23. A bubble of oil is suspended in water. Draw ray diagrams as in Question 23.22 and explain the difference between the diagrams in Questions 23.22 and 23.23.

24. A typical person underwater cannot focus clearly on another object under the water. Yet if the person wears goggles, he or she can see underwater objects clearly. Explain.

25. In a video projector, the picture that appears on a small liquid-crystal display is projected onto a screen. The lens of the video projector is adjusted so that a clear image on the display is formed on a screen. If the screen is moved farther away, must the lens be moved toward or away from the display? Explain.

26. The retina has a blind spot at the place where the optic nerve leaves the retina. Design and describe a simple experiment that allows you to perceive the presence of this blind spot. Why doesn't the blind spot affect your normal vision?

Problems

Below, **BIO** indicates a problem with a biological or medical focus. Problems labeled **EST** ask you to estimate the answer to a quantitative problem rather than derive a specific answer. Asterisks indicate the level of difficulty of the problem. Problems with no * are considered to be the least difficult. A single * marks moderately difficult problems. Two ** indicate more difficult problems.

23.1 Plane mirrors

1. You need to teach your friend how to draw rays to locate the images of objects produced by a plane mirror. Outline the steps that she needs to take.

2. Place a pencil in front of a plane mirror so that it is not parallel to the mirror. Draw an image that the mirror forms of the pencil and show using rays how the image is formed.

3. * Use geometry to prove that the virtual image of an object in a plane mirror is at exactly the same distance behind the mirror as the object is in front.

4. * You are 1.8 m tall. Where should you place the top of a mirror on the wall so you can see the top of your head? Where should you stand with respect to the wall?

5. * Two people are standing in front of a rectangular plane mirror. Each of them claims that she sees her own image but not the image of the other person. Draw a ray diagram to find out if this is possible.

6. * **Test an idea** Describe an experiment that you can conduct to test that the image produced by a plane mirror is virtual.

23.2 and 23.3 Qualitative analysis of curved mirrors and The mirror equation

7. * Describe in detail an experiment to find the image of a candle produced in a curved mirror. Draw pictures of the experimental setup and a ray diagram. Show on the ray diagram where your eye is.

8. * Explain with a ray diagram how (a) a concave mirror and (b) a convex mirror produce images of objects. Make sure that you explain the choice of rays and the location of your eye.

9. * **Test an idea** Describe an experiment to test the idea that a convex mirror never produces a real image of an object.

10. * **Test an idea** Describe an experiment to test the mirror equation. What are you going to measure? What are you going to calculate?

11. * **Tablespoon mirror** You look at yourself in the back (convex shape) of a shiny steel tablespoon. Describe and explain what you see as you bring the spoon closer to your face. Then turn the spoon around and repeat the steps.

12. * Use ray diagrams and the mirror equation to locate the position, orientation, and type of image of an object placed in front of a concave mirror of focal length 20 cm. The object distance is (a) 200 cm, (b) 40 cm, and (c) 10 cm.

13. Repeat Problem 23.12 for a convex mirror of focal length −20 cm.

14. Use ray diagrams and the mirror equation to locate the images of the following objects: (a) an object that is 10 cm from a concave mirror of focal length 7 cm and (b) an object that is 10 cm from a convex mirror of focal length −7 cm.

15. * **Sinking ships** A legend says that Archimedes once saved his native town of Syracuse by burning the enemy's fleet with mirrors. Describe quantitatively the type of mirrors that Archimedes could have used to burn ships that were 150 m away. Justify your answer.

16. * **EST** **Fortune-teller** A fortune-teller looks into a silver-surfaced crystal ball with a radius of 10 cm and focal length of −5 cm. (a) If her eye is 30 cm from the ball's surface, where is the image of her eye? (b) Estimate the size of that image.

17. * You view yourself in a large convex mirror of −1.2-m focal length from a distance of 3.0 m. (a) Locate your image. (b) If you are 1.7 m tall, what is your image height?

18. * **Seeing the Moon in a mirror** The Moon's diameter is 3.5×10^3 km, and its distance from Earth is 3.8×10^5 km. Determine the position and size of the image formed by the Hale Telescope reflecting mirror, which has a focal length of +16.9 m.

19. * You view your face in a +20-cm focal length concave mirror. Where should your face be in order to form an image that is magnified by a factor of 1.5?

20. * **Buying a dental mirror** A dentist wants to purchase a small mirror that will produce an upright image of magnification +4.0 when placed 1.6 cm from a tooth. What mirror should she order? Say everything you can about the mirror.

21. * **Using a dental mirror** A dentist examines a tooth that is 1.0 cm in front of the dental mirror. An image is formed 2.0 cm behind the mirror. Say everything you can about the mirror and the image of the tooth.

22. * **BIO** **EST** **Spooky fish** The glasshead barreleye (*Rhynchohyalus natalensis*) is a fish that lives in the oceans around the world at depths of 200 to 600 m. This fish and the brownsnout spookfish are the only two vertebrates known to use mirror-like reflective surfaces to form the images in their eyes. These fish have eyes with reflective surfaces to observe regions below them and eyes with lenses to observe regions above and beside them. **Figure P23.22** is a sketch of the cross section of an eye that uses reflection and shows how the eye focuses parallel beams of light that enter the eye at two extreme angles. Use the dimensions in the figure to estimate the range of values for the radius of curvature of the reflective surface. You will need a ruler to solve this problem.

FIGURE P23.22

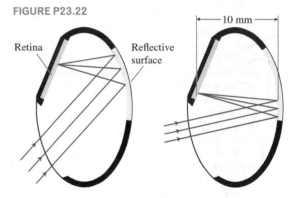

Retina   Reflective surface

23.4 Qualitative analysis of lenses

23. * If you place a point-like light source on the axis of a convex lens, you obtain on the screen the pattern shown in **Figure P23.23a**. If you repeat the experiment with a concave lens, you obtain the pattern shown in Figure P23.24b. Explain qualitatively how the patterns are formed, using ray diagrams. Note that there are no frames around the lenses.

FIGURE P23.23

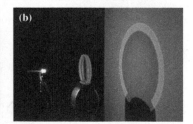

24. * You have a convex lens and a candle. Describe in detail an experiment that you will perform to find the image of the candle that this lens produces. Draw pictures of the experimental setup and a ray diagram. Show on the ray diagram where your eye is.

25. * Explain how to draw ray diagrams to locate images produced by objects in front of convex and concave lenses. Focus on the choice of rays and how you know where and what type of image is produced.

26. * Draw ray diagrams to show how a convex lens can produce (a) a real image that is smaller than the object, (b) a real image larger than the object, and (c) a virtual image.

27. * Use a ruler to draw ray diagrams to locate the images of the following objects: (a) an object that is 30 cm from a convex lens of +10-cm focal length, (b) an object that is 14 cm from the same lens, and (c) an object that is 5 cm from the same lens. (Choose a scale so that your drawing fills a significant portion of the width of a paper.) Measure the image locations on your drawings and indicate if they are real or virtual, upright or inverted.

28. * Repeat the procedure described in Problem 23.27 for the following lenses and objects: (a) an object that is 30 cm from a concave lens of −10-cm focal length, (b) an object that is 14 cm from the same lens, and (c) an object that is 5 cm from the same lens.

29. * Repeat the procedure described in Problem 23.27 for the following lenses and objects: (a) an object that is 7 cm from a convex lens of +10-cm focal length and (b) an object that is 7 cm from a concave lens of −10-cm focal length.

30. * Repeat the procedure in Problem 23.27 for the following lenses and objects: (a) an object that is 20 cm from a convex lens of +10-cm focal length, (b) an object that is 5 cm from the same lens, (c) an object that is 20 cm from a concave lens of −10-cm focal length, and (d) an object that is 5 cm from the lens in part (c).

31. * **Partially covering lens** Your friend thinks that if she covers one half of a convex lens she will only be able to see half of the object. Do you agree with her opinion? Why would she have such an opinion? Provide physics arguments. Design an experiment to test her idea.

23.5 and 23.6 Thin lens equation and quantitative analysis of lenses and Skills for analyzing processes involving mirrors and lenses

32. * Use ray diagrams to locate the images of the following objects: (a) an object that is 10 cm from a convex lens of +15-cm focal length and (b) an object that is 10 cm from a concave lens of −15-cm focal length. (c) Calculate the image locations for parts (a) and (b) using the thin lens equation. Check for consistency.

33. * Use ray diagrams to locate the images of the following objects: (a) an object that is 6.0 cm from a convex lens of +4.0-cm focal length and (b) an object that is 8.0 cm from a diverging lens of −4.0-cm focal length. (c) Calculate the image locations for parts (a) and (b) using the thin lens equation. Check for consistency.

34. Light passes through a narrow slit, and then through a lens and onto a screen. The slit is 20 cm from the lens. The screen, when adjusted for a sharp image of the slit, is 15 cm from the lens. What is the focal length of the lens?

35. * Describe two experiments that you can perform to determine the focal length of a glass convex lens. Is it a converging or a diverging lens? How do you know?

36. * **Shaving/makeup mirror** You wish to order a mirror for shaving or makeup. The mirror should produce an image that is upright and magnified by a factor of 2.0 when held 15 cm from your face. What type and focal length mirror should you order?

37. **Dentist lamps** Dentists use special lamps that consist of a concave mirror and a small, bright light source that is fixed on the principal axis of the mirror. When the light source is placed 5.0 cm from the mirror, the reflected light is focused in a bright spot at distance 70.0 cm from the mirror. Determine the radius of curvature of the mirror.

38. * A large concave mirror of focal length 3.0 m stands 20 m in front of you. Describe the changing appearance of your image as you move from 20 m to 1.0 m from the mirror. Indicate distances from the mirror where the change in appearance is dramatic.

39. * **EST** Two convex mirrors on the side of a van are shown in Figure P23.39. Estimate the ratio f_{upper}/f_{lower} of the focal lengths of the mirrors. Explain the reasoning behind your estimate, stating what assumptions you had to make.

FIGURE P23.39

40. Your friend gives you the following description of a method that apparently allows you to estimate the focal length of a concave lens: "Place a ruler on a table. Hold the lens above the ruler and parallel to the table so that you can simultaneously observe the image of the ruler (through the lens) and the part of the ruler that extends beyond the lens. Move the lens up and down until the width of the image of the ruler is half the width of the ruler that extends beyond the lens. The distance between the lens and the ruler is equal to the focal distance of the lens." Do you agree that this method gives you the focal length of the concave lens? Explain using a ray diagram.

23.7 Single-lens optical systems

41. **Camera** You are using a camera with a lens of focal length 6.0 cm to take a picture of a painting located 3.0 m from the camera lens. Where should the image sensor be positioned in relation to the lens in order to capture the image?

42. * **Camera** A camera with an 8.0-cm focal length lens is used to photograph a person who is 2.0 m tall. The height of the image on the image sensor must be no greater than 3.5 cm. (a) Calculate the closest distance the person can stand to the lens. (b) For this object distance, how far should the image sensor be located from the lens?

43. **Video projector** An LCD video projector (LCD stands for liquid-crystal display) produces real, inverted, enlarged images of a small LCD picture on a screen (the display picture is upside down in the projector). If the display is located 12.6 cm from the 12.0-cm focal length lens of the projector, what are (a) the distance between the screen and the lens and (b) the height of the image of a person on the screen who is 2.0 cm tall on the display?

44. **Photo of carpenter ant** You take a picture of a carpenter ant with an old fashioned camera with a lens 18 cm from the film. How far from the 4.0-cm focal length convex camera lens should the ant be located so you see a sharp image of the ant on the film?

45. * **Photo of secret document** A secret agent uses a camera with a 5.0-cm focal length lens to photograph a document whose height is 10 cm. At what distance from the lens should the agent hold the document so that an image 2.5 cm high is produced on the image sensor? [Note: The real image is inverted.]

46. * **Photo of landscape** To photograph a landscape 2.0 km wide from a height of 5.0 km, Joe uses an aerial camera with a lens of 0.40-m focal length. What is the width of the image on the detector surface?

47. * Make a rough graph of image distance versus object distance for a convex lens of a known focal length as the object distance varies from infinity to zero.

48. * Make a rough graph of linear magnification versus object distance for a convex lens of 20-cm focal length as the object distance varies from infinity to zero. Indicate in which regions the image is real and in which regions it is virtual.

49. * Repeat Problem 23.48 for a concave lens of −20-cm focal length.

50. **BIO** **Eye** The image distance for the lens of a person's eye is 2.10 cm. Determine the focal length of the eye's lens system for an object (a) at infinity, (b) 500 cm from the eye, and (c) 25 cm from the eye.

51. **BIO** **Lens-retina distance** Fish and amphibians accommodate their eyes to see objects at different distances by altering the distance from the lens system to the retina (to learn more about this, see the Reading Passage at the end of this chapter). If the lens system has a focal length of 2.10 cm, what is the lens-retina distance needed to view objects at (a) infinity, (b) 300 cm, and (c) 25 cm?

52. **BIO** **Nearsighted and farsighted** (a) A woman can produce sharp images on her retina only of objects that are from 150 cm to 25 cm from her eyes. Indicate the type of vision problem she has and determine the focal length of eyeglass lenses that will correct her problem. (b) Repeat part (a) for a man who can produce sharp images on his retina only of objects that are 3.0 m or more from his eyes. He would like to be able to read a book held 30 cm from his eyes.

53. * **BIO** **Prescribe glasses** A man who can produce sharp images on his retina only of objects that lie from 80 cm to 240 cm from his eyes needs bifocal lenses. (a) Determine the desired focal length of the upper half of the glasses used to see distant objects. (b) Determine the focal length of the lower half used to read a paper held 25 cm from his eyes.

54. * **BIO** **Correcting vision** A woman who produces sharp images on her retina only of objects that lie from 100 to 300 cm from her eyes needs bifocal lenses. (a) Determine the desired power of the upper half of the glasses used to see distant objects. (b) Determine the power of the lower half used to read a book held 30 cm from her eyes.

55. * **BIO** **Where are the far and near points?** (a) A woman wears glasses of 50-cm focal length while reading. What eye defect is being corrected and what approximately are the near and far points of her unaided eye? (b) Repeat part (a) for a man whose glasses have a −350-cm focal length. He wears the glasses while driving a car.

56. * BIO **Age-related vision changes** A 35-year-old patent clerk needs glasses of 50-cm focal length to read patent applications that he holds 25 cm from his eyes. Five years later, he notices that while wearing the same glasses, he has to hold the patent applications 40 cm from his eyes to see them clearly. What should be the focal length of new glasses so that he can read again at 25 cm?

23.8 Angular magnification and magnifying glasses

57. **Looking at an aphid** You examine an aphid on a plant leaf with a magnifying glass of +6.0-cm focal length. You hold the glass so that the final virtual image is 40 cm from the lens. If you assume that your near point is at 30 cm, then what is the angular magnification? How will the magnification change if you are farsighted? Nearsighted?

58. * **Reading with a magnifying glass** You examine the fine print in a legal contract with a magnifying glass of focal length 5.0 cm. (a) How far from the lens should you hold the print to see a final virtual image 30 cm from the lens (at the eye's near point)? (b) Determine the angular magnification of the magnifying glass.

59. * **Seeing an image with a magnifying glass** A person has a near point of 150 cm. (a) What is the nearest distance at which she needs to hold a magnifying glass of 5.0-cm focal length from print on a page and still have an image formed beyond her near point? (b) Determine the angular magnification for an image at her near point.

60. * **Stamp collector** A stamp collector is viewing a stamp through a magnifying glass of 5.0-cm focal length. Determine the object distance for virtual images formed at (a) negative infinity, (b) −200 cm, and (c) −25 cm. (d) Determine the angular magnification in each case.

23.9 Telescopes and microscopes

61. * You place a +20-cm focal length convex lens at a distance of 30 cm in front of another convex lens of focal length +4.0 cm. Then you place a candle 100 cm in front of the first lens. Find (a) the location of the final image of the candle, (b) its orientation, and (c) whether it is real or virtual.

62. * You place a +25-cm focal length convex lens at a distance of 50 cm in front of a concave lens with a focal length of −40 cm. Then you place a small lightbulb (2 cm tall) 30 cm in front of the convex lens. Determine (a) the location of the final image, (b) its orientation, and (c) whether it is real or virtual.

63. * EST You place a candle 10 cm in front of a convex lens of focal length +4.0 cm. Then you place a second convex lens, also of focal length +4.0 cm, at a distance of 12 cm from the first lens. (a) Use a ray diagram to locate the final image (keep the scale). (b) Using measurements on your ray diagram, estimate the linear magnification of the object. Be sure to show your rays and/or estimation technique for each step. Do not use equations!

64. * EST Repeat Problem 23.63 for an object located 6.0 cm from a convex lens of focal length 3.0 cm separated by 11 cm from another convex lens of focal length 1.0 cm.

65. ** You measure the focal length of a concave lens by first forming a real image of a light source using a convex lens. The image is formed on a screen 20 cm from the lens. You then place the concave lens halfway between the convex lens and the screen. To obtain a sharp image, you need to move the screen 15 cm farther away from the lenses. How does this experiment help you determine the focal length of the concave lens? What is the focal length?

66. ** **Telescope** A telescope consists of a +4.0-cm focal length objective lens and a +0.80-cm focal length eyepiece that are separated by 4.78 cm. Determine (a) the location and (b) the height of the final image for an object that is 1.0 m tall and is 100 m from the objective lens.

67. ** **Yerkes telescope** The world's largest telescope made only from lenses (with no mirrors) is located at the Yerkes Observatory near Chicago. Its objective lens is 1.0 m in diameter and has a focal length of +18.9 m. The eyepiece has a focal length of +7.5 cm. The objective lens and eyepiece are separated by 18.970 m. (a) What is the location of the final image of a Moon crater 3.8×10^5 km from Earth? (b) If the crater has a diameter of 2.0 km, what is the size of its final image? (c) Determine the angular magnification of the telescope by comparing the angular size of the image as seen through the telescope and the object as seen by the unaided eye.

68. * **Telescope** A telescope consisting of a +3.0-cm objective lens and a +0.60-cm eyepiece is used to view an object that is 20 m from the objective lens. (a) What must be the distance between the objective lens and eyepiece to produce a final virtual image 100 cm to the left of the eyepiece? (b) What is the total angular magnification?

69. ** **Design a telescope** You are marooned on a tropical island. Design a telescope from a cardboard map tube and the lenses of your eyeglasses. One lens has a +1.0-m focal length and the other has a +0.30-m focal length. The telescope should allow you to view an animal 100 m from the objective with the final image being formed 1.0 m from the eyepiece. Indicate the location of the lenses and the expected angular magnification.

70. * **Microscope** A microscope has a +0.50-cm objective lens and a +3.0-cm eyepiece that is 20 cm from the objective lens. (a) Where should the object be located to form a final virtual image 100 cm to the left of the eyepiece? (b) What is the total angular magnification of the microscope, assuming a near point of 25 cm?

71. ** BIO **Dissecting microscope** A dissecting microscope is designed with a larger than normal distance between the object and the objective lens. The microscope has an objective lens of +5.0-cm focal length and an eyepiece of +2.0-cm focal length. The lenses are separated by 15 cm. The final virtual image is located 100 cm to the left of the eyepiece. (a) Determine the distance of the object from the objective lens. (b) Determine the total angular magnification.

72. ** **Microscope** A microscope has an objective lens of focal length +0.80 cm and an eyepiece of focal length +2.0 cm. An object is placed 0.90 cm in front of the objective lens. The final virtual image is 100 cm from the eyepiece at the position of minimum eyestrain. (a) Determine the separation of the lenses. (b) Determine the total angular magnification.

73. ** **Microscope** Determine the lens separation and object location for a microscope made from an objective lens of focal length +1.0 cm and an eyepiece of focal length +4.0 cm. Arrange the lenses so that a final virtual image is formed 100 cm to the left of the eyepiece and so that the angular magnification is −260 for a person with a near point of 25 cm.

General Problems

74. Paul needs to determine the diameter of the wire used to make the filament of an incandescent lightbulb. He decides to project a magnified image of the filament onto a screen using a lens. (a) Describe Paul's experimental setup in words and with a sketch. (b) What additional data (other than those that can be obtained from the image on the screen) does Paul need to determine the diameter of the wire?

75. * **Figure P23.75** shows three cases of the primary axis of a lens (the lens is not shown) and the location of a shining object and its image. In each case, find the location and the type of the lens (convex or concave) that could produce the image and find the focal points of the lens.

FIGURE P23.75

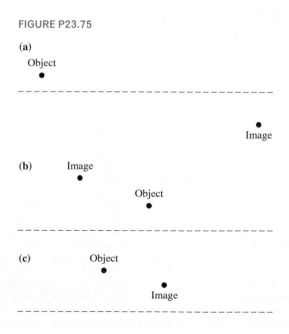

76. * **Jeopardy problem** The equations below describe a process that involves one lens. Determine the unknown quantities and write a word description of an optics situation that is consistent with the equations.

$$\frac{1}{4.0 \text{ m}} + \frac{1}{s'} = \frac{1}{0.10 \text{ m}}$$

$$h' = -\left(\frac{s'}{4.0 \text{ m}}\right)(1.6 \text{ m})$$

77. * **Jeopardy problem** The equations below describe a process involving more than one lens. Determine the unknown quantities and write a word description of an optics situation that is consistent with the equations.

$$\frac{1}{100 \text{ m}} + \frac{1}{s_1'} = \frac{1}{0.10 \text{ m}}$$

$$s_2 = (0.14 \text{ m}) - s_1'$$

$$\frac{1}{s_2} + \frac{1}{s_2'} = \frac{1}{0.042 \text{ m}}$$

78. **Image Jeopardy** Figure P23.78 shows the primary axis of a lens; the path of ray A as it approaches the lens, gets refracted, and continues to travel; and the path of ray B before it hits the lens. (a) Determine the type of lens (concave or convex) and its location, (b) determine the location of both focal points of the lens, and (c) draw the path of ray B after it passes through the lens.

FIGURE P23.78

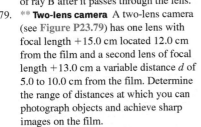

79. ** **Two-lens camera** A two-lens camera (see Figure P23.79) has one lens with focal length $+15.0$ cm located 12.0 cm from the film and a second lens of focal length $+13.0$ cm a variable distance d of 5.0 to 10.0 cm from the film. Determine the range of distances at which you can photograph objects and achieve sharp images on the film.

FIGURE P23.79

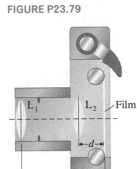

80. ** You have a small spherically shaped bottle made of plastic that is so thin that we can disregard its effects for optical processes. Show that the bottle works as a converging lens when filled with water, but when the same bottle is empty, closed, and submerged in water, it becomes a diverging lens. If you have access to the appropriate materials, test your ray diagram experimentally.

81. **BIO** **Find a farsighted person.** Design an experiment to find out what lenses will help this person read a book holding it 25 cm away. Borrow this person's glasses and invent a simple experiment to measure the approximate focal lengths of the lenses of his or her glasses. Perform the experiment. Are the real glasses close to your "prescription?"

82. **BIO** **Find a nearsighted person.** Design an experiment to find out what lenses will help this person see objects that are far away. Borrow this person's glasses and invent a simple experiment to measure the approximate focal lengths of the lenses of his or her glasses. Perform the experiment. Are the real glasses close to your "prescription?"

Reading Passage Problems

BIO **Laser surgery for the eye** LASIK (laser-assisted *in situ* keratomileusis) is a surgical procedure intended to reduce a person's dependency on glasses or contact lenses. Laser eye surgery corrects common vision problems, such as myopia (nearsightedness), hyperopia (farsightedness), astigmatism (blurred vision resulting from corneal irregularities), or some combination of these. In myopia, the cornea, the clear covering at the front of the eye, is often too highly curved, causing rays from distant objects to form sharp images in front of the retina (Figure 23.34). LASIK refractive surgery can flatten the cornea so that images of distant objects form on the retina. A knife cuts a flap in the cornea with a hinge left at one end of this flap. The flap is folded back, exposing the middle section of the cornea. Pulses from a computer-controlled laser vaporize a portion of the tissue and the flap is replaced.

FIGURE 23.34 LASIK surgery to correct nearsighted vision.

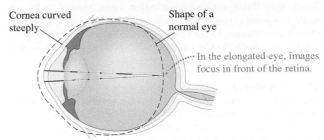

In farsighted people, an object held at the normal near point of the eye (about 25–50 cm from the eye) forms an image behind the retina (Figure 23.35). Increasing the curvature of the cornea causes rays from near objects to produce an image on the retina. Conductive keratoplasty (CK) can increase the curvature of the cornea to correct farsightedness. In CK, a tiny probe emits radio waves in a circular pattern on the outside of the cornea. Exposure to radio waves causes shrinking in the ring around the cornea, which acts like a constrictive band, increasing the overall curvature of the cornea.

FIGURE 23.35 Conductive keratoplasty (CK) surgery to correct farsighted vision.

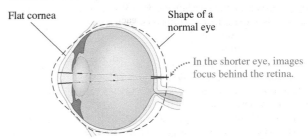

83. Your eyeball is 2.10 cm from the cornea to the retina. You look at a sign on the freeway that is effectively an infinite distance away. The image for the sign is 1.96 cm from the cornea within the eye. What is the focal length of the cornea-lens system?
    (a) $+1.96$ cm (b) $+2.10$ cm (c) $+2.26$ cm
    (d) $+2.42$ cm (e) $+27.90$ cm

84. In Problem 23.83, what should the cornea-lens focal length be in order to form a sharp image on the retina of the sign?
    (a) $+1.96$ cm (b) $+2.10$ cm (c) $+2.26$ cm
    (d) $+2.42$ cm (e) $+27.90$ cm

85. What is one surgical way to correct nearsighted vision?
    (a) Reduce the radius of curvature of the cornea, thus reducing the focal length of the eye's optics system.
    (b) Increase the radius of curvature of the cornea, thus increasing the focal length of the eye's optics system.
    (c) Increase the iris opening to allow more light to enter.
    (d) Use the ciliary muscle to increase the thickness of the lens.

86. Your eyeball is 2.10 cm from the cornea to the retina. A book held 30 cm from your eyes produces an image 2.26 cm from the entrance to the image location. What is the focal length of the cornea-lens system?
    (a) $+1.96$ cm (b) $+2.10$ cm (c) $+2.26$ cm
    (d) $+2.42$ cm (e) $+27.90$ cm

87. In Problem 23.86, what should the cornea-lens focal length be in order to form a sharp image of the book on the retina?
    (a) $+1.96$ cm (b) $+2.10$ cm (c) $+2.26$ cm
    (d) $+2.42$ cm (e) $+27.90$ cm

88. What is one surgical way to correct the farsighted vision of the person in the last two problems?
    (a) Reduce the curvature of the cornea, thus increasing the focal length of the eyes optics system.
    (b) Increase the curvature of the cornea, thus reducing the focal length of the eyes optics system.
    (c) Open the iris wider to allow more light to enter.
    (d) Cause the ciliary muscle to stretch the lens so it becomes thinner.

BIO **Bulging fish eyes** Fish eyes (see Figure 23.36) are evolutionary precursors to human eyes. Fish eyes have a dome-shaped cornea covering an iris that has a fixed opening and a spherical lens, which protrudes from the iris. Behind the lens is a retina with many rod and cone cells that detect light. Muscles attached to the lens change the lens location relative to the retina. Because water has a refractive index similar to that of the fluid inside fish eyes, there is little bending of light entering the eye at the water-cornea interface. The lens has a different refractive index than the fluid surrounding the lens and bends and focuses light so that images are formed on the retina. The fish eye lens is spherical in shape and collects light from a 180° field of view (Figure 23.37).

FIGURE 23.36 A fish eye.

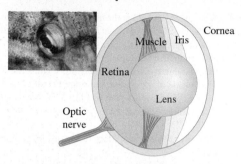

FIGURE 23.37 A fisheye lens collects light from a 180° field of view.

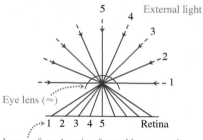

Image of ray 1 region formed here on retina.

Camera manufacturers now make fisheye lenses that produce an image from a 180° spread in front of the lens. Fisheye lens applications include IMAX projection, hemispherical photography, and peepholes that provide 180° views outside doors.

89. Why is the fish eye in Figure 23.36 black?
    (a) The lens is black.
    (b) No light reflects out from behind the lens.
    (c) The fluid behind the lens is black.
    (d) The retina is black.

90. Which of the following is the same for the fish eye and the human eye?
    (a) The muscles attached to the lenses change their shape.
    (b) The corneas help focus the light.
    (c) The lenses have a spherical shape.
    (d) The irises help focus the light.

91. Which ray in Figure P23.91 best represents the path of the incident light through the fisheye lens?
    (a) I   (b) II   (c) III   (d) IV
    (e) V

FIGURE P23.91

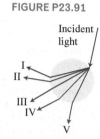

92. Which of the following statements regarding the fish eye and the human eye is not true?
    (a) They both have a cornea.
    (b) They both have an iris.
    (c) They both have a lens.
    (d) They both rely on the cornea and the lens for image formation.

93. How can the peephole lens in a hotel door have an approximately 180° view?
    (a) The light moves straight through the lens into the eye.
    (b) Light from about 90° from the straight-ahead direction is refracted into the observer's eye if very close to the lens.
    (c) Light from about 90° from the straight-ahead direction is refracted into the observer's eye if looking at the lens from inside the room.
    (d) The outside hall is illuminated so reflected light from the objects inside the 180° angle goes into the lens.
    (e) Both b and d

Wave Optics

Soap bubbles display a remarkable array of colors. Watch a bubble hanging from a bubble wand, and you will see horizontal bands appear on its surface and repeatedly change colors. The bands of color on the top of the bubble will widen until a dark band takes their place—and then the bubble pops. In this chapter you will learn why soap bubbles display such brilliant colors and why those colors disappear just before the bubble pops.

- Why do we see colors in soap bubbles?
- Which best explains the behavior of light: the particle-bullet model or the wave model?
- What limits our ability to perceive as separate two closely positioned objects?

IN OUR CHAPTER on reflection and refraction (Chapter 22) we developed two models of light: a particle-bullet model and a wave model. The particle model predicted that the speed of light in water should be greater than in air; the wave model predicted the opposite. Eventually, in 1850, Hippolyte Fizeau and Léon Foucault established that light travels more slowly in water than in air, finally disproving the particle model. However, even before Fizeau and Foucault, the wave model was gaining wide acceptance because of experiments performed by Thomas Young in the early 1800s.

BE SURE YOU KNOW HOW TO:

- Draw wave fronts for circular and plane waves (Section 11.1).
- Relate wave properties such as frequency, speed, and wavelength (Section 11.2).
- Apply Huygens' principle and the superposition principle (Section 11.6).

24.1 Young's double-slit experiment

In 1801, Englishman Thomas Young performed a testing experiment for the particle model of light. Testing Experiment **Table 24.1** describes a contemporary version of the experiment and the prediction of its outcome based on the particle model.

TESTING EXPERIMENT TABLE 24.1 ⟍ Recreating Young's experiment to test the particle model of light

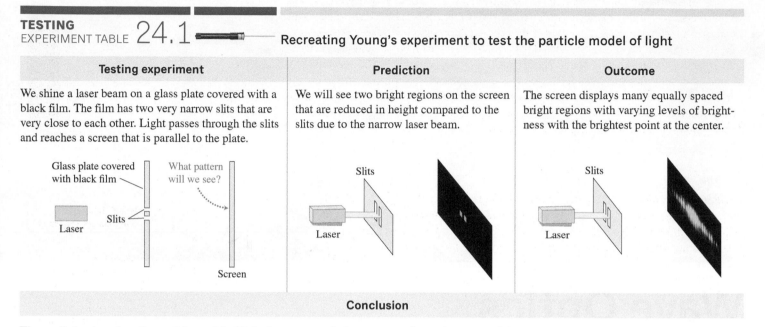

Testing experiment	Prediction	Outcome
We shine a laser beam on a glass plate covered with a black film. The film has two very narrow slits that are very close to each other. Light passes through the slits and reaches a screen that is parallel to the plate.	We will see two bright regions on the screen that are reduced in height compared to the slits due to the narrow laser beam.	The screen displays many equally spaced bright regions with varying levels of brightness with the brightest point at the center.

Conclusion

The prediction based on the particle model of light does not match the outcome, disproving the particle model.

In Table 24.1, light from the two slits produced a pattern of alternating bright regions on the screen (called **bands** or **fringes**). This outcome is inconsistent with the model of light that represents light as a stream of fast-moving bullets. In addition to alternating bright and dark regions, we also saw variation in the levels of brightness on a larger scale. We will focus on the equally spaced bright spots first, and discuss the variation in their brightness later in the chapter.

Thomas Young explained the bright fringes using a wave model and Huygens' principle (see Chapters 11 and 22). According to this principle, every point on a wave front is the source of a circular-shaped wavelet that moves in the forward direction at the same speed as the wave. A new wave front results from adding these wavelets together—the superposition (*interference*) of these wavelets.

We can use the wave model of light to explain the outcome of the testing experiment. According to this model, the laser emits some kind of wave. The brighter the light, the larger the amplitude of the wave. At the location where there is no light, we assume the amplitude of the light wave to be zero.

Because the laser beam does not spread much, we can assume that a laser produces a wave with plane wave fronts. The crests and troughs of this wave reach the two closely spaced slits and produce synchronized (in-phase) disturbances within each slit. According to Huygens' principle, the slits become sources of in-phase wavelets (**Figure 24.1a**). The new wavelets generated from each slit spread outward and eventually irradiate the screen. There, the disturbances due to both wavelets combine according to the superposition principle. At a particular time, the wavelet disturbances produced by each slit appear as shown in Figure 24.1b. (Note that the angles between the directions to the bright spots are greatly exaggerated in order to make the wave front representation easier to understand. You can infer the real angles from the photo in Figure 24.1a.) At some points, the crests from the wavelets from each slit overlap (the positions of solid dots), resulting in double-sized crests. The troughs between the crests also overlap, resulting in double-sized troughs. This *constructive interference* results in a double-amplitude wave. These alternating double crests and double troughs arrive at the screen shown in Figure 24.1b, causing a large-amplitude light wave—a bright band (or fringe). This occurs along line 0 and along the two lines labeled 1. It also occurs along several other lines not shown in the figure.

Along lines about halfway between these bright bands, you find that the crest of one wavelet overlaps with the trough of the other (the open circles in Figure 24.1c). The wavelets are out of phase along those lines and therefore cancel each other. This *destructive interference* means no light is present—a dark band (or fringe) is present.

Between the centers of the bright and dark bands, the wavelets are neither exactly in nor exactly out of phase with each other, and the wave amplitudes are between double and zero; the light has intermediate brightness. Since the pattern on the screen can be explained using the idea of interference of the wavelets from the two slits, it is called an **interference pattern**. The wave model of light explains the fringes observed by Thomas Young; the particle-bullet model does not.

> **Qualitative wave-based explanation of Young's experiment** The interference pattern of dark and bright bands produced by light passing through two closely spaced narrow slits can be explained by the addition or superposition of circular wavelets originating from the two slits. The length of the path that each wave travels to a particular spot on the screen determines the phase of the wave at that location. Locations where two waves arrive in phase are the brightest; locations where two waves arrive with different phases are darker; and locations where the waves arrive completely out of phase are dark.

What is a light wave—what is being disturbed?

Imagine that we want to draw a graphical representation of a light wave—a disturbance-versus-position graph (similar to the graphs we drew in Chapter 11 for mechanical waves). We know, for example, that a sound wave involves a moving disturbance of air pressure and density and that a wave on a Slinky involves a moving disturbance of the closeness of the Slinky coils. What is being disturbed when a light wave propagates?

In the early 1800s, physicists thought that a medium was needed for the propagation of any wave. They proposed the existence of a massless invisible medium for light called **ether**, which supposedly filled the entire universe and vibrated when light traveled through it. This vibrating ether model was accepted until the late 1800s. Later in this book (in Chapters 25 and 26), we will learn about a contemporary theory of light and the experiments that disproved the existence of ether. For now, we will assume that light is a wave-like disturbance and that the nature of that disturbance is unknown. We will also assume that the disturbances from different waves obey the superposition principle.

Quantitative analysis of the double-slit experiment

A mathematical description of the phenomenon observed by Young involves calculating the location of bright and dark bands on a screen for a particular light wavelength λ, slit separation d, and distance L of the screen from the slits. As discussed earlier, we consider the slits to be sources of synchronized circular wavelets. At all locations where the waves arrive in phase, there will be constructive interference; where they arrive completely out of phase, there will be destructive interference. For simplicity, we will represent the waves traveling to the screen from the slits as wiggles. You can think of these wiggles as cross-sections of the three-dimensional waves. The waves are in phase at a particular location if they have the same shape at this location.

FIGURE 24.1 (a) Double slits are the source of wavelets that can interfere. (b) When wave crests or troughs overlap, a large-amplitude wave occurs. (c) When a crest overlaps a trough, they cancel and there is no wave disturbance.

(a)

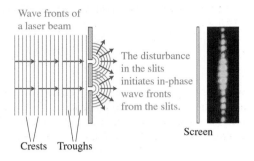

(b)

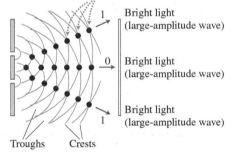

(c)

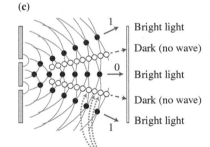

Overlap of crest and trough produces zero disturbance—destructive interference.

Now consider the brightest locations on the screen, as shown in **Figure 24.2a**. The wavelets are in phase at the location of the slits, and the waves travel the same distance from the slits to point 0. Thus the waves arrive at this point in phase no matter how far the screen is from the slits. Figure 24.2b represents the wave disturbances from the two slits at one instant of time. This bright band or fringe in the center is called the **0th order maximum** (or zeroth maximum).

FIGURE 24.2 Double-slit wave interference.

(a)

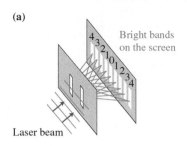

(b)

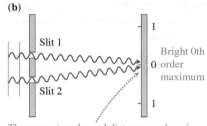

The waves travel equal distances and are in phase when they reach the screen.

(c)

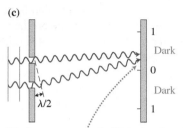

The lower wave travels $\lambda/2$ farther than the upper wave. They are out of phase when they reach the screen.

(d)

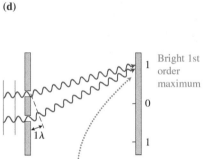

The wave from the lower slit travels one wavelength farther than the wave from the upper slit. At the screen, they are in phase and interfere constructively.

(e) **(f)**

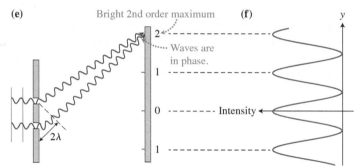

The wave from the lower slit travels two wavelengths farther.

The peaks correspond to the maximum brightness on the screen, and the troughs correspond to dark regions.

Just to either side of this location, the waves arrive slightly out of phase and the screen is dimmer. The brightness of the screen changes slowly, from a maximum at the center, where the waves arrive exactly in phase, to zero where waves arrive completely out of phase. At the two locations marked "Dark" in Figure 24.2c, the difference in the path length traveled by the two waves is one-half wavelength.

Moving away from the dark band on each side of the center, the screen becomes brighter again as the path length difference between the waves becomes larger. When the path length difference reaches one whole wavelength at location 1, there is another point of maximum brightness (Figure 24.2d). The wave from the lower slit travels exactly one wavelength farther than the distance traveled by the wave from the upper slit. This bright band is called the **1st order maximum**. A similar maximum appears at the symmetrical location 1 below location 0.

We can use similar reasoning to find the 2nd, 3rd, and 4th order maxima. The path length differences for the centers of these bright bands are 2λ, 3λ, and 4λ, respectively (see the second bright band in Figure 24.2e). We can use this thinking to predict the locations of these bands using geometry. Figure 24.2f shows the change in the intensity of light incident on the screen (measured by a photosensitive device) as a function of position for the 0th order maximum and the 1st and 2nd order maxima on both sides.

Mathematical location of the *m*th bright band

Consider the general case in **Figure 24.3**. The small shaded triangle next to the slits has a hypotenuse d equal to the slit separation. The two red lines are the distances that the two waves travel to get to the center of the *m*th bright band on the screen (position P_m in Figure 24.3). The path length difference between the waves equals an integer number of whole wavelengths—m wavelengths in this case. The side of the small shaded triangle opposite angle θ is the path difference for the two waves reaching location P_m on the screen. If there is an interference maximum at that location on the screen, then the path length difference Δ that the waves originating at the two slits travel to P_m should equal an integer number of wavelengths:

$$\Delta = m\lambda, \text{ where } m = 0, \pm 1, \pm 2, \pm 3, \text{ etc.}$$

From now on, we will use a negative sign to distinguish the maxima below the 0th band from the maxima above it. Using trigonometry: $\sin \theta = (\Delta/d) = (m\lambda/d)$. The angle θ in the small triangle equals the angle θ_m between the horizontal and the mth bright band. Thus,

$$\sin \theta_m = \frac{m\lambda}{d} \tag{24.1}$$

This derivation has led us to a quantitative relation between the slit separation d, the angle θ_m at which the mth maximum occurs, and the wavelength λ of the light. If we know the distance L between the slits and the screen, we can find the distance y_m between the 0th maximum on the screen and the mth maximum in Figure 24.3:

$$\tan \theta_m = \frac{y_m}{L} \tag{24.2}$$

If the angle θ_m is very small, the tangent of the angle equals the sine of the angle and also equals the angle itself if measured in radians: $\tan \theta_m \approx \theta_m = (y_m/L)$ and $\sin \theta_m \approx \theta_m = (m\lambda/d)$. Combining these two relationships, we find that for small angles relative to the horizontal, the bright bands on the screen are located at

$$y_m = \frac{m\lambda L}{d} \tag{24.3}$$

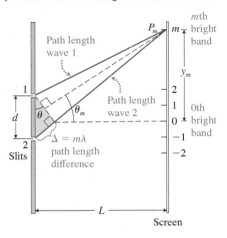

FIGURE 24.3 The geometry for finding the position of the mth bright band on a screen.

Testing expressions for bright band locations on a screen

Since it is not easy to independently measure the wavelength of light, we will instead test the inverse relationship between the location of the fringes y_m and the slit separation d (Testing Experiment Table 24.2).

TESTING EXPERIMENT TABLE 24.2 — Slit separation and locations of the bright spots

VIDEO TET 24.2

Testing experiment	Prediction	Outcome
• We shine red laser light through slits with separation $d = 0.050$ mm onto a screen at $L = 2.0$ m from the slits. The first bright fringe on the screen from the center maximum is at $y_1 = 2.75$ cm. 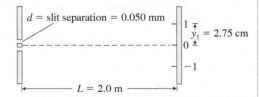 d = slit separation = 0.050 mm; y_1 = 2.75 cm; L = 2.0 m • Using the same laser, we predict the location of the first bright band on the screen if using slits with a different separation $d' = 0.030$ mm.	We solve Eq. (24.3) for λ as $L \gg y$: $$\lambda = \frac{y_m d}{mL}$$ m is 1 since we are interested in the first bright fringe. Since the same laser is used for both slit separations, we can equate the right-hand sides of this equation for each case. $$\frac{y_1 d}{(1)L} = \frac{y_1' d'}{(1)L}$$ Multiplying by L and solving for y_1', we have our prediction: $$y_1' = \frac{y_1 d}{d'} = \frac{(2.75 \text{ cm})(0.050 \text{ mm})}{0.030 \text{ mm}} = 4.6 \text{ cm}$$	We measure a distance of 4.6 ± 0.1 cm between the center $m = 0$ maximum and the $m = 1$ maxima.

Conclusion

• The outcome of the experiment matches the prediction from Eq. (24.3).
• The result supports the wave model of light.

Because the outcome of the experiment agreed with the prediction that followed from the equations we derived using the wave model of light, we gain confidence in that model. Many other experiments have also agreed with predictions based on the wave model of light.

Now that we have some trust in Eq. (24.3), we can use it to determine the wavelength of the red laser light used in Table 24.2.

$$\lambda = \frac{y_m d}{mL} = \frac{(2.75 \text{ cm})(0.050 \text{ mm})}{(1)(2.0 \text{ m})} = 6.9 \times 10^{-7} \text{ m}$$

To simplify the expression of such small numbers, physicists record wavelengths of light in nanometers; 1 nm $= 10^{-9}$ m. The wavelength for the red light described in Table 24.2 is about 690 nm. We can now summarize our findings about double-slit interference.

Double-slit interference When light of wavelength λ passes through two narrow slits separated by distance d, it forms a series of bright and dark bands on a screen beyond the slits. The angular deflection θ_m of the center of the mth bright band to the side of the central $m = 0$ bright band is determined using the equation

$$d \sin \theta_m = m\lambda \tag{24.1}$$

where $m = 0, \pm1, \pm2, \pm3$, etc. The distance y_m on a screen from the center of the central bright band to the center of the mth bright band depends on the angular deflection θ_m and on the distance L of the screen from the slits:

$$L \tan \theta_m = y_m \tag{24.2}$$

For small angles:

$$y_m = \frac{m\lambda L}{d} \tag{24.3}$$

We see from Eq. (24.1) that for light of wavelength λ, the greater the slit separation d, the smaller the angular deflection of the bright spots θ_m. Thus, to observe distinct bands, we need slits that are very close to each other. Also, the longer the wavelength, the larger the angular deflection for the same slit separation. Finally, using Eq. (24.2), we see that the greater the distance L of the screen from the slits, the greater the distance y_m of the mth bright band from the central bright band on the screen.

TIP We could have done the derivation for the minima (the locations where two waves arrive out of phase and cancel each other) instead of for the maxima and obtained a relation for the angles at which the dark bands occur. This happens when the path length difference from the two slits to the dark band equals an odd number of half wavelengths: $\Delta = (2m + 1)(\lambda/2)$, where $m = 0, \pm1, \pm2, \pm3$, etc., and consequently

$$\sin \theta = \frac{(2m + 1)\lambda/2}{d}$$

FIGURE 24.4 Pattern of white light produced by two slits on a screen.

Thomas Young used sunlight in his original experiments since lasers had not yet been invented. He found that the bands of light on the screen were colored like a rainbow, except for the central band, which was white. Each band that he observed had blue light on the inside edge (closest to the central maximum) and red light on the outside edge (**Figure 24.4**). We found that the angular deflection to the center of the mth bright band should be $\sin \theta_m = (m\lambda/d)$. If the wavelength λ is longer, the angular deflection θ_m should be greater. Evidently, red light must have a longer wavelength than blue light, with green in between. We can check this idea.

QUANTITATIVE EXERCISE 24.1 ⟩ **Wavelength of green light**

You repeat the experiment described in Table 24.2 using slits separated by $d = 0.050$ mm and a screen at $L = 2.0$ m from the slits. This time, you use a green laser instead of a red laser. The second maximum ($m = 2$) appears on the screen at $y_2 = 4.4$ cm from the center of the 0th order maximum. Determine the wavelength of this light.

Represent mathematically Since $L \gg y$, we can use Eq. (24.3) $y_m = (m\lambda L/d)$ to determine the wavelength.

Solve and evaluate To determine λ, multiply both sides by d and divide by mL to get $(y_m d/mL) = \lambda$. Now substitute the known values:

$$\lambda = \frac{y_m d}{mL} = \frac{(4.4 \times 10^{-2}\,\text{m})(5.0 \times 10^{-5}\,\text{m})}{2 \times 2.0\,\text{m}}$$

$$= 5.5 \times 10^{-7}\,\text{m} = 550\,\text{nm}$$

The order of magnitude for the wavelength is the same as that for red light, but the value is smaller, consistent with Young's observations.

- -

Try it yourself What is the angle between the direction of light going toward the central maximum and the direction to the first *dark* fringe?

Answer 5.5×10^{-3} rad.

REVIEW QUESTION 24.1 Explain why we observe multiple bright bands of light on a screen after light of a particular wavelength passes through two narrow, closely spaced slits.

24.2 Refractive index, light speed, and wave coherence

In Chapter 22, we learned that the wave model of light explains refraction of light and predicts that light should travel more slowly in a medium with a higher refractive index compared to a medium with a lower index. In this section, we will learn how to use the wave model of light to explain refraction quantitatively.

Relating the refractive index and the speed of light in a substance

In the chapter on reflection and refraction (Chapter 22), we learned that when light travels from medium 1 to medium 2, we can determine the change in the direction of light propagation using Snell's law: $n_2 \sin \theta_2 = n_1 \sin \theta_1$ (**Figure 24.5a**). Because $n_2 > n_1$, $\sin \theta_2 < \sin \theta_1$ and $\theta_2 < \theta_1$. If we are using the wave model to explain this change, we need to consider that the wave fronts that are perpendicular to the propagation of the wave also bend at the interface. For a wave to bend toward the normal, its speed must be smaller in medium 2 than in medium 1. We know that the frequency of a wave depends only on the source of the wave and does not change as it passes from medium 1 to medium 2. Therefore, the faster moving light in medium 1 has a longer wavelength than the slower moving light in medium 2.

Consider the right triangles DAB and ADC in Figure 24.5b. We know that the angle of incidence θ_1 is equal to the angle DAB because the corresponding half lines of θ_1 and of DAB are perpendicular to each other. The angle of refraction θ_2 is equal to the angle ADC for the same reason. The side opposite angle DAB (that is, DB) is one wavelength λ_1 of the light in medium 1. Similarly, the side opposite angle ADC (that is, AC) is one wavelength λ_2 of the light in medium 2. Thus, the sines of the angles are

$$\sin \text{DAB} = \sin \theta_1 = \frac{\lambda_1}{\text{DA}} \quad \text{and} \quad \sin \text{ADC} = \sin \theta_2 = \frac{\lambda_2}{\text{DA}}$$

FIGURE 24.5 (a) The wave slows as it moves from medium 1 to medium 2, causing the wavelength to decrease. (b) The angles of incidence and refraction are related to the wavelengths.

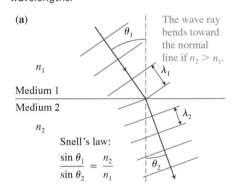

(a)

The wave ray bends toward the normal line if $n_2 > n_1$.

n_1

Medium 1
Medium 2

n_2

Snell's law:
$$\frac{\sin \theta_1}{\sin \theta_2} = \frac{n_2}{n_1}$$

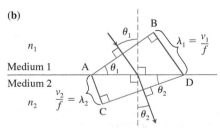

(b)

n_1

Medium 1   A
Medium 2

n_2 $\frac{v_2}{f} = \lambda_2$

$\lambda_1 = \frac{v_1}{f}$

The angle of incidence θ_1 and angle of refraction θ_2 also depend on the wave speeds.

Note that the wave frequency does not change: $f_1 = f_2 = f$.

Geometry: $\dfrac{\sin \theta_1}{\sin \theta_2} = \left(\dfrac{v_1/f}{\text{AD}}\right)\left(\dfrac{\text{AD}}{v_2/f}\right) = \dfrac{v_1}{v_2}$

Now, use the above to take the ratio of the sines of the angles θ_1 and θ_2:

$$\frac{\sin \theta_1}{\sin \theta_2} = \frac{\lambda_1/DA}{\lambda_2/DA} = \frac{\lambda_1}{\lambda_2}$$

We can use the relation between wavelength λ, frequency f, and speed v of a light wave to rewrite the above in terms of the speeds of the light in the two media:

$$\lambda_1 = \frac{v_1}{f} \quad \text{and} \quad \lambda_2 = \frac{v_2}{f}$$

After substitution, we get

$$\frac{\sin \theta_1}{\sin \theta_2} = \frac{\lambda_1}{\lambda_2} = \frac{v_1/f}{v_2/f} = \frac{v_1}{v_2}$$

From Snell's law,

$$\frac{\sin \theta_1}{\sin \theta_2} = \frac{n_2}{n_1}$$

Thus, the ratio of the speeds of light in two media 1 and 2 should be equal to the inverse ratio of their indexes of refraction:

$$\frac{v_1}{v_2} = \frac{n_2}{n_1}$$

We know that the index of refraction of water is 1.33 and that of air is 1.00. Thus the speed of light in water should equal the ratio of its speed in air c $(3.0 \times 10^8$ m/s$)$ divided by the $n = 1.33$ index of refraction of water:

$$v = \frac{c}{1.33} = \frac{(3.0 \times 10^8 \text{ m/s})}{1.33} = 2.26 \times 10^8 \text{ m/s}$$

This is exactly the speed that French physicists Hippolyte Fizeau and Leon Foucault obtained in their 1850 experiments to determine the speed of light in water.

We found that the wave model of light not only explains why light bends at the boundary of two media but also explains Snell's law by connecting the medium's index of refraction to the speed of light in that medium.

Refractive index The refractive index n of a medium equals the ratio of the speed of light c in vacuum (or in air) to the speed of light v in the medium:

$$n = \frac{c}{v} \qquad (24.4)$$

FIGURE 24.6 The refractive index n of a medium is different for different wavelengths λ of light.

(a)

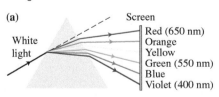

(b)

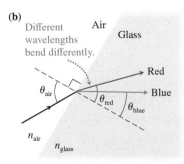

Refractive index and the color of light

We already know that when white light passes through a prism, it comes out as a set of colored bands (see **Figure 24.6a**). In other words, different colors of light are refracted differently by the prism. How does the wave model explain this? Snell's law of refraction [Eq. (22.3)] as applied to Figure 24.6b is

$$n_{\text{glass}} = \frac{n_{\text{air}} \sin \theta_{\text{air}}}{\sin \theta_{\text{glass}}}$$

Because the refracted angle in glass for blue light θ_{glass} is smaller than that for red light, the refractive index n_{glass} for blue light must be greater than that for red light. The refractive index is related to the speed of light in the medium by Eq. (24.4). The different refractive indexes for different colors mean that light of different colors or light waves of different frequencies travel at different speeds in the same medium. Only in a vacuum does light of all frequencies travel at the same speed.

Speed of light in a medium

A laser pointer emits green light that has a 550-nm wavelength when in air. (a) What is the light's frequency? (b) What are the frequency, speed, and wavelength of the laser light as it travels through glass of refractive index 1.60?

Represent mathematically The frequency of a wave can be determined using its speed and wavelength $f = (c/\lambda)$. The speed of light in glass can be found using $n = (c/v)$.

Solve and evaluate (a) We are given the wavelength of the light when in air: $\lambda = 550 \times 10^{-9}$ m. The speed of the light when in air is $c = 3.0 \times 10^8$ m/s. Thus, the frequency of the light is

$$f = \frac{c}{\lambda} = \frac{3.0 \times 10^8\,\text{m/s}}{550 \times 10^{-9}\,\text{m}} = 5.5 \times 10^{14}\,\text{Hz}$$

(b) This light has the same 5.5×10^{14}-Hz frequency when in glass because the frequency of a traveling wave is determined by its source, not by the medium. However, its speed and wavelength change:

$$v_\text{in glass} = \frac{c}{n_\text{glass}} = \frac{3.0 \times 10^8\,\text{m/s}}{1.60} = 1.9 \times 10^8\,\text{m/s}$$

$$\lambda_\text{in glass} = \frac{v_\text{in glass}}{f} = \frac{1.9 \times 10^8\,\text{m/s}}{5.5 \times 10^{14}\,\text{Hz}} = \frac{1.9 \times 10^8\,\text{m/s}}{5.5 \times 10^{14}\,\text{s}^{-1}}$$

$$= 3.5 \times 10^{-7}\,\text{m} = 350\,\text{nm}$$

Try it yourself What is the wavelength of green light traveling through water of refractive index 1.33?

Answer   410 nm.

TIP Note that when light travels into a medium with a different index of refraction, its frequency does not change, but its wavelength does.

Chromatic aberration in lenses—a practical problem in optical instruments

High-contrast photos sometimes have purple edges—an effect called *purple fringing*. Purple fringing occurs because the refractive index is wavelength dependent, resulting in a lens having a slightly smaller focal length for purple light than for red light ($f_\text{purple} < f_\text{red}$, where f stands for the focal length, not frequency). Thus, the location of an image for the purple part of an object is slightly different from the image location for the red part of that object, an effect called **chromatic aberration.** Since the image locations are slightly different, their magnifications, which depend on image location, will also be slightly different. When the magnification of the purple image is slightly greater than the other colors, purple fringing occurs (**Figure 24.7**).

Monochromatic and coherent waves

Our observations of double-slit interference (Table 24.1) involved laser light passing through two very narrow slits, each of which acted as a light source. If we replace the laser light with two small lightbulbs, we do not observe an interference pattern. Instead, we see a uniformly illuminated screen (**Figure 24.8**). Why?

FIGURE 24.7 (a) Chromatic aberration is evident in this magnification of newsprint. (b) The purple and red lights form images at different places due to differences in their refractive indexes.

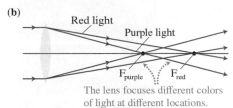

The lens focuses different colors of light at different locations.

FIGURE 24.8 Two lightbulbs do not produce an interference pattern.

(a) With the laser light, we observed bright and dark bands.

(b) With two incandescent lightbulbs, we observe a uniformly lit screen.

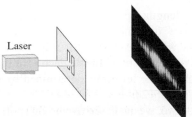

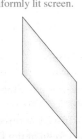

FIGURE 24.9 The vibrating beach balls are monochromatic and coherent and produce a stable interference pattern.

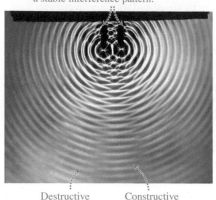

Two balls vibrating in phase produce a stable interference pattern.

Destructive interference   Constructive interference

FIGURE 24.10 A laser produces coherent and monochromatic disturbances at the slits.

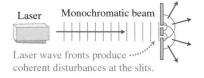

Laser   Monochromatic beam

Laser wave fronts produce coherent disturbances at the slits.

FIGURE 24.11 The bright interference bands sharpen with an increasing number of slits.

(a) Distance between the slits in all cases is 0.2 mm.

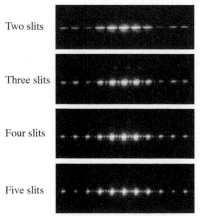

Two slits

Three slits

Four slits

Five slits

(b) Distance between the slits is 0.1 mm.

Grating

In order for two waves to interfere consistently over time, two conditions must be met. First, the waves must vibrate at the same constant frequency, producing what are called **monochromatic waves.** Second, the phases of the waves must be synchronized, meaning that as time passes they must arrive at a particular location consistently in phase, out of phase, or somewhere in between. The phase difference between the two waves must remain constant in time. When this occurs, the waves are called **coherent waves.** The synchronized vibrating beach balls shown in **Figure 24.9** produce coherent monochromatic waves and interference. If one of the balls changes phase with respect to the other but still vibrates with the same frequency, the waves produced by the balls will be monochromatic but not coherent. A laser emits plane waves that oscillate with essentially the same frequency. When such waves are incident on two narrow slits, the same wave front triggers the circular wavelets. Therefore, the circular waves behind the slits are coherent and monochromatic (**Figure 24.10**).

When two small light sources such as lightbulbs are placed at the same separation as the two slits in Figure 24.8a, the waves emitted by each source are not monochromatic (they consist of a mixture of frequencies), and they are not coherent because they produce light waves independently of each other. Thus the waves add randomly and produce no interference pattern on the wall (Figure 24.8b).

You might wonder how Young got coherent sources in his double-slit experiment when he was using sunlight, which is not monochromatic. Because the slits were very narrow and very close, they became two sources on the same wave front. Thus they produced waves that were in phase for every wavelength of sunlight. The resulting pattern on the screen had bands of different color at different locations.

Now we can summarize.

> **Coherent monochromatic waves** Only waves of constant frequency (*monochromatic*) and having constant phase difference (*coherent*) can add to produce an interference pattern. In Young's double-slit experiment, the light from the slits is coherent because the disturbance at each slit is caused by the same passing wave front.

REVIEW QUESTION 24.2 The refractive index of Crown glass is about 1.51 for a yellow color. What does this number tell us about the speed of light in glass? What does it tell us about the frequency of the light? Its wavelength?

24.3 Gratings: an application of interference

Thomas Young's double-slit experiments not only helped establish the wave model of light but also helped explain how a device invented earlier, called a **grating**, works. A grating is a plate of glass, plastic, or metal that has many very narrow parallel grooves or slits through which light can travel. Gratings allow us to analyze the wavelengths of light emitted by biological materials, particles in Earth's atmosphere, and even stars and galaxies.

How does a grating work? When light of wavelength λ passes through two narrow closely spaced slits, we observe the interference maxima at angular deflections determined by the equation $\sin \theta_m = (m\lambda/d)$, where $m = 0, \pm 1, \pm 2, \pm 3, \ldots$. In **Figure 24.11a** (two slits), notice the central five bright spots. Inside each spot there is a gradual change of brightness from the maxima at the locations determined by the equation above to zero at the locations where $\sin \theta = \left[(2m + 1)\lambda/2 \right]/d$.

If we use three slits of the same width instead of two, we again see five bright spots on the screen at the same locations. But the photo looks different (Figure 24.11a, three slits).

We see a small spot between the bright bands, and the bands are narrower and brighter. The latter is due to the fact that the total amount of light reaching the screen has increased by a factor of 1.5. If we increase to four and then to five slits, the bright bands still do not change positions (Figure 24.11a), but they become even brighter and narrower. There are more minima between the bright bands. If we increase the number of slits above 10, we will see almost no light on the screen between the bright bands. This pattern is due to the complicated interference of the coherent wavelets coming from many slits. Having such narrow bright bands allows us to distinguish more clearly between different frequencies of light passing through the slits.

A typical grating has hundreds of slits per millimeter, and the bright bands are very intense and narrow with almost complete darkness between them. Figure 24.11b shows an image of bands on a screen for a grating with 80 slits per millimeter. The locations of the bright bands are determined using Eq. (24.3). In gratings, the distance between the slits can be very small, for example, $d = 0.001$ mm. Consequently, the bright bands are much farther apart.

Thus, for a grating with spacing between adjacent slits d, the bright bands (interference maxima) occur at the angles determined from the expression

$$\sin \theta_m = \frac{m\lambda}{d}, \text{ where } m = 0, \pm 1, \pm 2, \pm 3, \ldots \quad (24.5)$$

TIP Gratings are usually labeled in terms of the number of slits per millimeter or per centimeter, for example, 100 slits/cm. In such a grating, the distance between the slits is one hundredth of a centimeter. If a grating has 1000 slits/mm, the distance between the slits is one thousandth of a millimeter. You can find the distance between the slits by dividing 1 by the number of slits per unit length. The distance comes out in the respective units (cm or mm).

QUANTITATIVE EXERCISE 24.3   **Double-slit and grating interference**

You have a slide with two narrow slits separated by 0.10 mm and a grating with 100 slits/mm. Compare the angular deflections for the $m = 1$ maximum (see figure below) using the double slit and the grating for red laser light of wavelength 670 nm. Describe the differences in the interference pattern that you observe.

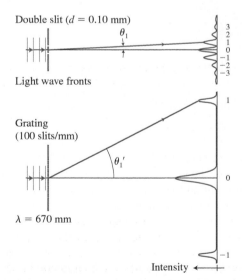

Double slit ($d = 0.10$ mm)

θ_1

Light wave fronts

Grating (100 slits/mm)

θ_1'

$\lambda = 670$ mm

Intensity

Double slit: $\sin \theta_1 = \dfrac{\lambda}{d_{slits}} = \dfrac{670 \times 10^{-9} \text{ m}}{10^{-4} \text{ m}}$

$= 6.7 \times 10^{-3}$

Grating: $\sin \theta_1' = \dfrac{\lambda}{d_{grating}} = \dfrac{670 \times 10^{-9} \text{ m}}{10^{-5} \text{ m}}$

$= 6.7 \times 10^{-2}$

The figure shows the intensity-versus-location graph. The peaks correspond to the bright regions, and the places with zero intensity correspond to completely dark regions. As expected, the bright regions are much farther apart for the grating than for the double slit. The bright regions from the grating look like sharp dots, and those from the double slit look like fuzzy bands.

- -

Try it yourself What is the angle of deflection of the 1st order maximum for a slide with three very narrow slits illuminated by the same light source as above, each separated from each other by 0.10 mm?

Represent mathematically We need to find the angle θ_1 for both cases. Use the equation $\sin \theta_m = (m\lambda/d)$ with $m = 1$.

Solve and evaluate To find the slit separation for the grating, divide 1 by the number of slits per millimeter:

$$d_{grating} = \frac{1}{100 \text{ slits/mm}} = 10^{-2} \text{ mm} = 10^{-5} \text{ m}$$

For the double slit, $d_{slits} = 10^{-4}$ m.

Answer

Adding or removing slits without changing their separation does not change the location of the maxima. With more slits, the spacing between bright bands is the same as for the two slits in this exercise; the maxima are brighter and narrower.

White light incident on grating

What happens when white light shines on a grating? After passing through the grating, light of all wavelengths produces a bright band at the center (the $m = 0$ band); thus on a screen beyond the grating, you see a central maximum of white light. However, according to $\sin \theta_m = (m\lambda/d)$, different wavelengths of the white light are deflected at different angles for each m (other than $m = 0$). The greater the wavelength, the larger the angle from the center of the $m = 0$ white band to the mth bright band. Consequently, each $m \neq 0$ band is a rainbow of colors on the screen, with the blue color always closer to the central maximum than the red, because light of smaller wavelength has a smaller deflection angle—$\sin \theta_m = (m\lambda/d)$ (Figure 24.12a). If the slit separation d is smaller, the colors are separated more. The colored band, called a **spectrum**, looks similar to the spectrum produced when white light passes through a prism, but the order of colors is reversed. The prism bends blue light the most (Figure 24.12b). However, the mechanisms by which these spectra are produced are different. A spectrum produced by a prism is a result of the different speeds with which light waves of different frequencies travel in a medium. A spectrum produced by a grating is a result of the light of different wavelengths interfering constructively at different locations.

FIGURE 24.12 A white light diffraction pattern caused by a grating.

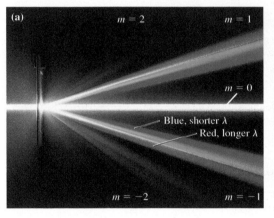

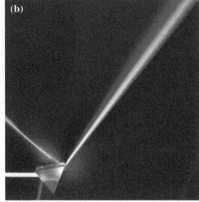

FIGURE 24.13 White light reflection from a CD.

CDs and DVDs—reflection gratings

The reflective surfaces of CDs and DVDs (Figure 24.13) consist of spirals of closely spaced grooves. This makes a CD a type of grating that reflects light instead of transmitting it. Since the distance between the grooves is constant, the grooves play the role of the slits: the reflected white light forms interference maxima for different colors at different angles. Looking at the reflected light, your eye intercepts only one color of the light reflected from each part of the CD. The grooves on DVDs are closer together (that is why they store more information than CDs), and that is why their color pattern is more spread out and their colors are less pronounced.

FIGURE 24.14 A spectrometer.

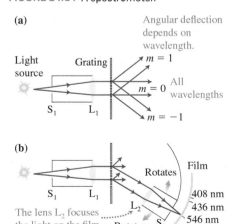

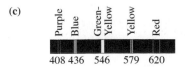

Spectrometer

Analyzing the wavelengths of light from different sources is an important tool in science. For example, astronomers determine which chemical elements are present in distant stars by identifying the wavelengths of light coming from those stars. Chemists measure the wavelengths of different chemical compounds to determine properties of the compounds. The tool they use is a **spectrometer** (Figure 24.14). Light from a source passes through a narrow slit S_1 and then through a lens L_1 (Figure 24.14a), which focuses the light into a parallel beam. This beam then passes through a grating that separates the light into its various wavelengths. The light bent at a particular angle is collected and focused by lens L_2 and slit S_2 onto a recording surface such as a film or some other light-recording device (Figure 24.14b). The detector lens and slit mechanism rotate to produce on the film the spectrum of wavelengths of light coming from the source (Figure 24.14c).

A spectrometer can be used to analyze light from any source. The Sun and incandescent lightbulbs emit a broad range of light frequencies—a continuous spectrum of light. However, hot, low-density gases do not. Instead, they emit light of specific wavelengths that indicates that specific types of atoms and molecules are present in the gas. If the light entering the spectrometer consists only of specific wavelengths, the spectrum will contain bright colored lines against a black background. For example, mercury vapor, whose spectrum is shown in Figure 24.14c, emits a pair of purple lines near 408 nm, a blue line at 436 nm, a greenish-yellow line at 546 nm, a yellowish pair of lines near 579 nm, and two dim red lines near 620 nm. We will learn more about why atoms emit specific wavelengths of light in later chapters.

Spectra also provide information about the expansion of the universe. In the spectra of light from distant galaxies, the characteristic spectral lines are shifted to longer wavelengths (lower frequencies). This so-called red shift is an example of the Doppler effect for light and indicates that distant galaxies are moving away from us—supporting the idea that the universe is expanding.

EXAMPLE 24.4 **Hydrogen spectrum**

You wish to determine the frequencies of light emitted by atoms in a hydrogen gas-discharge lamp. You use a spectrometer with a 4000 slits/cm grating. The light-sensitive detector of the spectrometer is 0.500 m from the grating, as shown. The image produced by the detector looks as shown in the upper part of the figure below. Determine the wavelengths of the hydrogen lines in the red and blue-green parts of the visible spectrum.

Sketch and translate The blue-green and red bands shown should be the $m = 1$ 1st order maxima for these colors. The y_1's in the figure are the distances between the central maximum and the 1st order $m = 1$ maxima. The slit separation is

$$d = \frac{1}{4000 \text{ lines/cm}} = 2.50 \times 10^{-4} \text{ cm} = 2.5 \times 10^{-6} \text{ m}$$

Simplify and diagram Assume that the slits of the grating are coherent sources of light. The path of the bright bands of light from the grating to the detector is depicted in the lower part of the figure below.

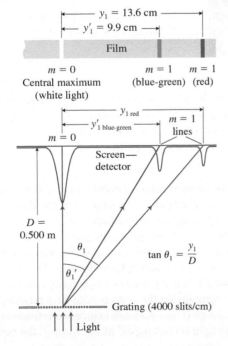

Represent mathematically Using the diagram and geometry, we get for each color:

$$\text{Red:} \qquad \tan \theta_1 = \frac{y_1}{D}$$

$$\text{Blue-green:} \qquad \tan \theta_1' = \frac{y_1'}{D}$$

Use these relations to find the corresponding angles. Then use the value for d and Eq. (24.3) $\lambda = d \sin \theta_1$ to find the wavelength of each line.

Solve and evaluate

$$\text{Red: } \tan \theta_1 = \frac{y_1}{D} = \frac{0.136 \text{ m}}{0.500 \text{ m}} = 0.272 \text{ or } \theta_1 = 15.2°$$

$$\lambda = d \sin \theta_1 = (2.50 \times 10^{-6} \text{ m})\sin 15.2°$$

$$= 6.55 \times 10^{-7} \text{ m} = 655 \text{ nm}$$

$$\text{Blue-green: } \tan \theta_1' = \frac{y_1'}{D} = \frac{0.099 \text{ m}}{0.500 \text{ m}}$$

$$= 0.198 \text{ or } \theta_1 = 11.2°$$

$$\lambda' = d \sin \theta_1' = (2.50 \times 10^{-6} \text{ m})\sin 11.2°$$

$$= 4.86 \times 10^{-7} \text{ m} = 486 \text{ nm}$$

The wavelength for red light is greater than that for blue-green, as expected.

Try it yourself Suppose a different atom emits yellow light of 580 nm. Where would it fall on the above pattern?

Answer About 12 cm from the $m = 0$ central maximum.

24.4 Thin-film interference

FIGURE 24.15 The colorful peacock tail feathers are caused by thin-film interference.

The beautiful, swirling colors that occur on the surfaces of soap bubbles also appear in the thin oil films that float on water, the shimmering patterns of some butterflies, the brilliant feathers of many hummingbirds, and the peacock's elegant tail (**Figure 24.15**). This array of color is not caused by pigment but by the interference of light waves reflected off of the boundaries of a very thin surface—a film.

Bright and dark bands due to reflected monochromatic light

Figure 24.16a shows a photo of a thin soap film with white light shining on it. Let's start by analyzing the pattern produced when we irradiate the soap film with red light only (Figure 24.16b). We see a regular pattern of bright and dark bands produced by the reflected red light, a little like a double-slit interference pattern. What causes this pattern?

FIGURE 24.16 Soap bubble patterns caused by (a) white light and (b) red light. (c) Rays that interfere to cause the colors and fringes.

(a) Pattern caused by white light

(b) Pattern caused by red light

(c)

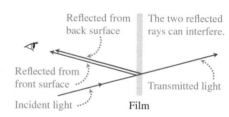

Note that we neglect refraction because the film is very thin.

We have learned that light is partially transmitted and partially reflected at interfaces between media with different refractive indexes. For a soap film and other thin films, light reflection occurs at both the front surface and the back surface (Figure 24.16c). Two factors affect the outcome of the superposition of these two reflected waves:

1. *Phase change upon reflection.* We know from our study of mechanical waves (Chapter 11) that a reflected wave changes phase by $180° = \pi$ rad if the second medium is denser than the first. For light, this would happen if the second medium had a greater refractive index than the first. However, the phase of the reflected wave remains the same if the second medium has a smaller refractive index than the first. Thus, when light traveling in air reflects off the front surface of the soap bubble film of refractive index 1.3, there is a phase change of $180° = \pi$ rad (**Figure 24.17a**). When light traveling inside the soap bubble film reflects off the back surface, there is no phase change (Figure 24.17b) since the refractive index of air is less than 1.3. So only the light reflecting off of the front of the bubble undergoes a phase change.

2. *Phase difference due to path length difference.* In Figure 24.17c we see that wave 2, which reflects from the back surface of the thin film, travels a longer distance than wave 1, which reflects from the front surface. Because of this path length difference, the two waves have different phases when they recombine after reflecting from the thin film. For example, if the film is one half wavelength thick, wave 2 travels a total of one wavelength farther than wave 1. Then the two waves should be in phase when they recombine (assuming no phase changes occur due to reflection). If the film is one-fourth of a wavelength thick, the path length difference is one-half of the wavelength. The waves are $180° = \pi$ rad out of phase when they recombine. Often those waves are said to be one half wavelength out of phase to underscore that when they add, the resulting wave has zero amplitude.

There is an additional important feature of this path length difference. The speed v of light in a medium of refractive index n is $v = (c/n)$, where c is the speed of light in air. If the refractive index n of the thin film is greater than 1.0, then the wavelength of the light in the film is

$$\lambda_{\text{medium}} = \frac{v_{\text{medium}}}{f} = \frac{c/n_{\text{medium}}}{f} = \frac{c/f}{n_{\text{medium}}} = \frac{\lambda_{\text{air}}}{n_{\text{medium}}}$$

To determine the phase difference between the waves, we have to use the light's wavelength in the film (λ_{medium}). To determine the total phase difference, we need to account for both factors (1) and (2). Let's apply this reasoning to soap bubbles.

Soap bubble The film of the soap bubble has a greater refractive index than air. Thus, light reflected from outside the air-soap bubble interface has a $180° = \pi$ rad phase change (equivalent to the one half wavelength path difference). Part of the transmitted light reflects from inside the soap bubble-air interface and has no phase change. Hence, the relative phase change from reflection between the two waves is $180° = \pi$ rad (equivalent to one half wavelength in terms of path length difference). For a *bright band* to appear on the surface of the bubble (constructive interference), the net phase change of light coming from that direction should be an integer multiple of $360° = 2\pi$ rad. Thus, the path length difference due to the light traveling back and forth through the film a distance $2t$ must also be one half wavelength, or three half wavelengths, or five half wavelengths, etc. so that when added together, those path length differences produce an integer number of whole wavelengths. Thus, $2t = (1, 3, 5, \dots)(\lambda/n)/2$, where λ/n is the wavelength of light inside the bubble film. We've assumed through all this that the incident light is almost perpendicular to the film surface.

For a *dark band* to appear on the surface of the bubble, destructive interference must occur between the two reflected waves. The net phase change needs to be an odd multiple of $180° = \pi$ rad, or an odd number of half wavelengths in terms of path lengths. Thus, the path length difference $2t$ must be an integer number of wavelengths: $0, 1, 2, \dots$ wavelengths. In short, for a dark band, $2t = (0, 1, 2, \dots)(\lambda/n)$.

Thus for a specific wavelength of light, the thickness of the bubble film determines if light will be reflected from the film. Since the film thickness varies over the bubble, some locations look shiny (constructive interference) and other locations look dark (destructive interference). We can test this reasoning experimentally. Examine the chapter-opening photo showing a soap film across a frame just before it breaks. Notice the black area at the top of the film. The black color means that no light comes from this part of the film into our eyes. Could we have predicted this phenomenon using our knowledge of thin films?

As soon as the frame is lifted out of the soap solution, the soap film across the frame starts thinning due to evaporation and dripping from the bottom. Because of gravitational effects, the film is always thinnest at the top. Just before it breaks, the thickness at the top is nearly zero (t is almost zero). When we look at the film at an almost zero degree angle with the normal line, the waves of all wavelengths (colors) reflected off the front and back of the film are out of phase because of reflection (the phase changes

FIGURE 24.17 Light incident on a film at almost zero angle with the normal. (a) and (b) Phase changes due to reflection. (c) Phase difference due to path length difference.

(a) Reflection off surface with increasing n (like air-soap interface)

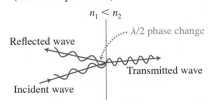

(b) Reflection off surface with decreasing n (like soap-air interface)

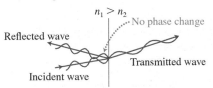

(c)

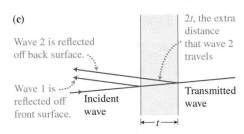

TIP When light reflected off a soap bubble interferes constructively, incident light is reflected. Soap bubbles are transparent, which means that most of the incident light passes through, even when constructive interference occurs.

by 180° during the reflection off the front of the film and does not change during the reflection off the back). When two waves exactly out of phase meet, they cancel each other out. We therefore predict that the top of the film should appear dark just before it breaks. And this is exactly what we observe in the photo!

Now that we are more confident in our reasoning about thin films, let's use it to develop a method for reducing the glare from glass surfaces, such as windows or camera lenses.

Thin film on glass surface In an optical instrument with several lenses, light reflection at each air-glass interface reduces the amount of light getting to the detector, whether that be the light detector in a digital camera or your eye looking through a microscope. To reduce this effect and increase the amount of light reaching the detector, the glass surfaces are often covered with a thin film. Waves reflecting from the film interfere destructively, minimizing reflected light. All incident light passes through.

Consider light incident almost perpendicularly on a thin film with refractive index $n_{film} = 1.4$ on the front surface of a glass lens whose refractive index is $n_{glass} = 1.6$ (**Figure 24.18**). Wave 1 undergoes a $180° = \pi$ rad phase change upon reflection, since $n_{film} > n_{air}$. Wave 2 also undergoes a $180° = \pi$ rad phase change, since $n_{glass} > n_{film}$. The net effect is that the two waves remain in phase. To minimize the amount of reflected light, we need to create destructive interference between these two waves. Thus, the path length difference must equal an odd integer multiple of $180° = \pi$ rad (an odd integer number of half wavelengths of the light in terms of path length difference). Choosing the proper thickness of the film t, so that $2t = (1, 3, 5, \dots)\left[(\lambda/n)/2\right]$, where n is the refractive index of the film, achieves this goal. If for some reason we wanted to maximize the reflected light, then the two waves must constructively interfere. This will happen when $2t = (0, 2, 4, \dots)\left[(\lambda/n)/2\right]$.

Table 24.3 summarizes the ways in which the refractive index of a thin film and its thickness combine to produce the bright and dark bands of reflected light.

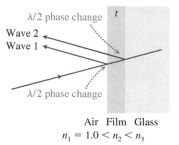

FIGURE 24.18 Light reflecting from a thin film on glass surface. Light passing through the glass is not shown.

TABLE 24.3 **Examples of thin-film interference for monochromatic incident light**

Type of thin film	Changes in path length difference	Total path length difference and outcome
Soap bubble in air or oil film on water $n_1 < n_2 > n_3$ 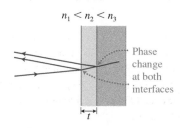	**Due to reflection** $\Delta_1 = (\lambda/2)$ for reflection on front surface of film and $\Delta_2 = 0$ for reflection on back surface of film. Net reflection path length difference is $$\Delta_{ref} = \frac{\lambda}{2}$$ **Due to thickness** $$\Delta_t = 2t$$	$\Delta = \Delta_{ref} + \Delta_t = (\lambda/2) + 2t$ • *Bright band* if twice the thickness is an odd multiple of half wavelengths: $$2t = m\frac{\lambda/n}{2} \text{ for } m = 1, 3, 5 \dots$$ • *Dark band* if twice the thickness is an even multiple of half wavelengths: $$2t = m\frac{\lambda/n}{2} \text{ for } m = 0, 2, 4 \dots$$
Thin-film coating on lens, window, windshield, or computer screen $n_1 < n_2 < n_3$	**Due to reflection** $\Delta_1 = (\lambda/2)$ for reflection on front surface of film; $\Delta_2 = (\lambda/2)$ for reflection on back surface of film. Net reflection path length difference is $\Delta_{ref} = \lambda$. **Due to thickness** $$\Delta_t = 2t$$	$\Delta = \Delta_{ref} + \Delta_t = \lambda + 2t$ • *Bright band* if twice the thickness is an even multiple of half wavelengths: $$2t = m\frac{\lambda/n}{2} \text{ for } m = 0, 2, 4, \dots$$ • *Dark band* if twice the thickness is an odd multiple of half wavelengths: $$2t = m\frac{\lambda/n}{2} \text{ for } m = 1, 3, 5 \dots$$

Reflection patterns on a soap bubble in white light

So far, we have assumed that the surface of a bubble or a camera lens film is illuminated by monochromatic light (light of one wavelength). However, usually it is white light that is incident on the thin film. Consider the soap film in **Figure 24.19**. The white light incident on the film includes light wavelengths from about 400 nm to 700 nm. When the light of different wavelengths reflects and refracts, for one wavelength the net path difference might be equal to an odd number of half wavelengths. For another wavelength it could be exactly an even number of half wavelengths. For most wavelengths, however, the interference is neither completely constructive nor completely destructive. Also, when we look at a different spot on the film, the waves arriving at our eyes are incident on the film at different angles, and the film has different thicknesses at different locations.

Because of all these factors, light of only a small wavelength range is destructively reduced in intensity at a particular location on the bubble when viewed in a particular direction. The color that is reduced differs from place to place. At other locations, other colors are destructively reduced in intensity. As a result, each observer will see a different distribution of colors reflected from each location on the film. The colors that are left when light of a small wavelength range is subtracted from the white light are called **complementary colors**. Complementary colors are different from the spectrum of rainbow colors produced by a grating or a prism. These devices separate in space the primary colors that are combined spatially inside a beam of white light. The video shows the change of colors of a soap bubble as its thickness changes with time. Watch the top of the bubble become black just before the bubble bursts.

For example, if blue light is subtracted from white light, we see the remaining light as its complement, yellow. When red is subtracted from white light, we see cyan. Long-wavelength light (red) needs a thicker bubble wall for its reflections to destructively interfere than short-wavelength light (blue). This means that as the bubble's thickness decreases, first red is canceled out, leaving blue-green; then green is canceled, leaving magenta; and finally blue is canceled, leaving yellow. Eventually, the bubble becomes so thin that all wavelengths are canceled and the bubble appears black, just as the film in the testing experiment became black at the top just before it broke.

FIGURE 24.19 A soap bubble in white light.

Colors are the result of the absence of one or more colors from white light.

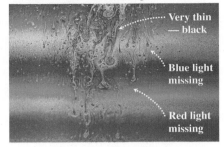

Very thin — black

Blue light missing

Red light missing

VIDEO 24.1

TIP Remember that it is the frequency of light that determines its color. The frequency does not change when light travels from one medium to another. The wavelength does change. Thus when you hear "the wavelength of green light is 550 nm," check whether this statement assumes that the light is propagating in a vacuum.

Lens coatings

As we have learned, we can reduce the reflected light of a particular wavelength if we have a film of a particular thickness. The thickness of the coating on glass lenses for cameras, microscopes, and eyeglasses is usually chosen to reduce reflection of light at a wavelength of 550 nm, the center of the visible spectrum. The film reduces the reflected light from about 4% of the intensity of the incident light (without the coating) to less than 1%. The coating is less effective at reducing the reflection at extreme visible wavelengths (red at 700 nm and violet at 400 nm). A lens with a thin-film coating has a purple hue because it reflects red and violet light more than other colors; when combined, these colors appear purple.

QUANTITATIVE EXERCISE 24.5 **How thin is the film?**

In the optical industry, a thin film of magnesium fluoride (MgF_2) with a refractive index of 1.38 is used to coat a glass lens of refractive index 1.50. What should be the thickness of the coating to eliminate the reflection of 550-nm light (the wavelength of green light in a vacuum)?

Represent mathematically The purpose of the coating (a thin film) is to produce destructive interference for the reflected light. The refractive index increases at the air-film interface and at the film-glass interface. Thus there is a net zero phase change due to reflection. For destructive interference, the path difference (twice the thickness t)

should be $2t = (\lambda_{air}/2n)$, $3(\lambda_{air}/2n)$, $5(\lambda_{air}/2n)$, and so forth, where n is the refractive index of the film.

Solve and evaluate The thinnest possible film would be when $m = 1$: $2t = (\lambda_{air}/2n)$. Thus $t = (1/4)(\lambda_{air}/n) = (1/4)(550 \text{ nm}/1.38) = 100 \text{ nm}$. This is very thin. Making the film slightly thicker might still achieve the purpose of canceling the reflected light and provide greater durability for the film. Possible thicknesses are 300 nm and 500 nm—any positive odd integer times the minimal thickness.

(CONTINUED)

Try it yourself How thick should the film in this exercise be to increase the amount of reflected light?

Answer

visors for astronauts.

We could use a 200-nm thin film of the material used in this exercise. Alternatively, we could cover a glass surface with a thin film of greater refractive index than the glass, achieving greater light reflection with the same 100-nm film. Increasing reflection is useful when building devices that allow us to look at bright light sources, such as sun

Limitations of the thin-film interference model

Window glass does not display color patterns. Why is that? Window glass is several millimeters thick, several thousand times the wavelength of light, like a *very* thick walled bubble. Natural sources of light such as the Sun do not emit light waves continuously. Instead, they emit light in short wave bursts over a very short time interval. Two consecutive bursts have a random phase difference between them. Imagine that light from one burst travels through the window glass and reflects back. While it was traveling inside the glass, a new burst reached the surface of the glass. As a result, the phase difference between the two waves is determined not only by reflections and path length differences but also by the time between bursts. The bursts are therefore not coherent and do not interfere in a way that remains constant in time. However, this limitation does not exist for laser light, which is much more coherent than natural light. You can observe interference on window glass using a laser.

Bird and butterfly colors

FIGURE 24.20 A Morpho butterfly.

Many colors in the natural world, such as those of flower petals and leaves, are caused by organic pigments that absorb certain colors and reflect others. Chlorophyll in leaves absorbs most colors but reflects green, which makes the leaves look green. However, some natural color is the result of the thin-film interference of light. Some feathers and insect bodies consist of microscopic translucent structures that act like thin films to produce destructive and constructive interference of light. We can see the colors that result from these structures in the tail feathers of peacocks, the head feathers of hummingbirds, and the wings of Morpho butterflies (**Figure 24.20**).

TIP Remember that interference patterns for light that are stable in time and thus observable appear when two coherent waves arrive at the same location at the same time. These coherent waves can be created in two ways:
1. by combining waves from two sources that produce coherent, monochromatic waves (as in double-slit interference), and
2. by dividing a monochromatic wave into two waves and combining them (as in thin-film interference).

FIGURE 24.21 Laser light reaching a screen after passing through two slits.

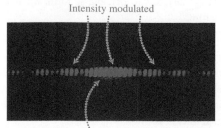

Intensity modulated

Double-slit interference bright bands

REVIEW QUESTION 24.4 If we look through a grating at a source of white light and then look at a thin film also illuminated by white light, we see different colors. What is the same and what is different about how these colors appear?

24.5 Diffraction of light

Earlier in the chapter we analyzed interference phenomena involving light that passed through two narrow slits. We modeled those slits as infinitely narrow, and when light shined on them we considered them to be point-like sources of light. This model explained the observed interference pattern on the screen (**Figure 24.21**). However,

FIGURE 24.22 The single-slit diffraction pattern widens as the slit width narrows.

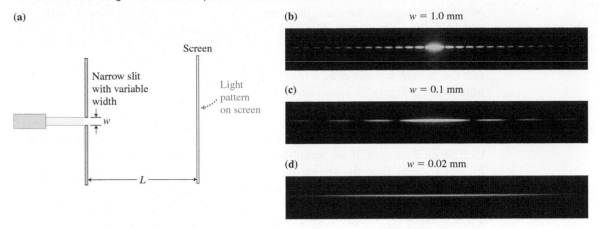

(a)

(b) $w = 1.0$ mm

(c) $w = 0.1$ mm

(d) $w = 0.02$ mm

if you look carefully at the pattern produced by light passing through two slits on the screen, you will notice that in addition to the alternating bright and dark bands, there is also an overall periodic modulation of the brightness in the pattern. The wider the slits (not their separation, but their individual widths), the more pronounced this modulation is. In addition to the interference of light coming from the two slits, something else is going on that is related to the individual slits themselves. Let us investigate this phenomenon. We start with a series of experiments in which we shine laser light through one slit. The width of the slit is adjustable so that we can change it during the experiments. **Figure 24.22a** shows the setup, and Figures 24.22b–d show the outcomes of the experiments.

We observe in all three experiments that light reaching the screen from a relatively narrow slit makes a wide interference pattern on the screen, including dark and bright bands at the sides of the central wide band. The width of the central maximum increases as the width of the slit decreases. When the slit is wider, the bright band in the center is almost equal to the width of the slit (Figure 24.22b). When the slit is narrower, the band is much wider (Figures 24.22c and d). This spreading of light combined with the additional bright and dark regions is called **diffraction**. It becomes noticeable when the slit's width is roughly 1000 times the wavelength of the light or less. It is possible that this effect might be caused by the interference of wavelets produced by different mini-slit regions within the slit—an idea we will now evaluate quantitatively.

Quantitative analysis of single-slit diffraction

When we analyzed the interference pattern in a double-slit experiment, we modeled each slit as a point-like source of secondary wavelets on the same wave front using Huygens' principle. In the single-slit situation, the slit is not infinitely narrow. It is somewhat wider and produces an interference pattern like that shown in **Figure 24.23a**. If we see an interference pattern, then several waves might be adding to produce it. Where do those waves come from if we only have one slit? As the slit has a nonzero width, we can model it as consisting of multiple imaginary tiny mini-slit regions that become sources of the secondary waves on the same wave front when light shines on the slit. Because they are on the same wave front, light waves emitted by all the mini-slits must all be in phase at the slit location. If this is the case, light emitted by those different mini-slits can interfere when reaching the screen because they travel different distances to reach the same spot on the screen.

Imagine that the screen is very far away. In this case, the waves emitted by different mini-slits are traveling nearly parallel to each other when they reach a specific location

FIGURE 24.23 Single-slit diffraction. (a) Alternating bright and dark bands observed coming from light passing through a single slit. (b) Light from different parts of the slit cancels, causing the first dark band at y_1.

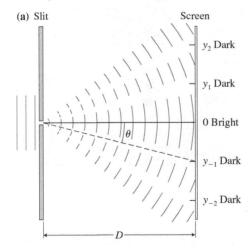

(a) Slit

(b)

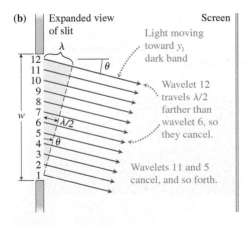

on the screen. Thus, to explain the existence of a bright or a dark spot on the screen, we only need to consider the phase difference of the parallel waves traveling from the slit to the location on the screen. Consider two wavelets coming from mini-slits to the first dark band below the central bright band (position y_1 in Figure 24.23a). One wavelet comes from the top mini-slit of the top half of the slit (mini-slit 12 in Figure 24.23b), and the second wavelet comes from the top mini-slit of the bottom half of the slit (mini-slit 6 in Figure 24.23b). The distance between these mini-slits is $(w/2)$, where w is the slit width.

If the path length difference from these two sources to the screen equals one half wavelength, then these two wavelets cancel each other, resulting in zero light at that location on the screen. If we repeat the procedure for the next pair of wavelets (11 and 5), which travel from the second mini-slit region in the top half of the slit and the second mini-slit region in the bottom half of the slit, we get the same result. The same will happen for all other pairs of waves chosen in this manner. We get total cancellation at that point on the screen—a dark band.

We see from the shaded triangle in Figure 24.23b that this first interference minimum occurs when $(w/2) \sin \theta = (\lambda/2)$, or when $w \sin \theta = \lambda$. Note that the same reasoning applies for the first dark band above the central maximum.

We designate the dark bands above the central maximum with positive signs and the dark bands below with negative signs. Thus, the first dark bands on both sides of the central bright spot occur at

$$w \sin \theta_{\pm 1} = \pm \lambda$$

We can next divide the slit into four mini-slit segments and draw lines from them to the second dark band on either side of the central maximum. We find that the angular deflection to this second dark band occurs when $(w/4) \sin \theta_2 = (\lambda/2)$, or when

$$w \sin \theta_{\pm 2} = \pm 2\lambda$$

We can repeat this process by dividing the slit into six mini-slit segments with lines drawn from them to find the third dark band on either side of the central maximum. These dark bands occur when $w \sin \theta_{\pm 3} = \pm 3\lambda$.

We can summarize this analysis as follows. The angular positions of the dark bands relative to the central maximum are determined by

$$w \sin \theta_m = m\lambda, \text{ where } m = \pm 1, \pm 2, \pm 3 \dots \qquad (24.6)$$

The first dark bands on both sides of the central maximum appear at the angles $\theta_{\pm 1}$ when $m = \pm 1$ is substituted into the above equation. All the space between these angles belongs to the central bright fringe. Therefore, we do not use $m = 0$ in the above equation.

Equation (24.6), $w \sin \theta_m = m\lambda$, was derived using a wave model of light and can be tested by performing an experiment using a shorter wavelength green laser with the same slit. The angular deflection to the first dark band on each side of the central maximum should be less, since the green laser produces light of shorter wavelength than the red laser. When you perform the experiment, you observe that the first dark bands are closer to the central bright band—the central bright band is narrower.

We can now describe the variation in the brightness of the double-slit interference pattern. Real slits have a finite width, which we did not account for in our original analysis. Thus, in addition to the two-slit interference pattern with the closely spaced alternating bright and dark bands, these bands are modulated in intensity by the additional single-slit diffraction due to the interference of light coming from different mini-slits within each slit. This single-slit pattern combines with the double-slit pattern, resulting in a variation in brightness of the double-slit maxima.

TIP ▸ Notice that $w \sin \theta_m = m\lambda$ is similar to Eq. (24.1) for double-slit interference. However, for a double slit, Eq. (24.1) describes the angles at which the maxima (bright bands) are observed, and $m = 0$ is allowed. For a single slit, Eq. (24.6), $w \sin \theta_m = m\lambda$, describes angles at which the minima (dark bands) are observed, and $m = 0$ is not allowed since there is no dark band at $\theta = 0°$.

Single-slit diffraction When monochromatic light is incident on a slit whose width is approximately 1000 wavelengths of light or less, we observe a series of bright and dark bands of light on a screen beyond the slit. The bands are caused by the interference of light from different mini-slit regions within the slit. The angle between lines drawn from the slit to the minima, the dark bands, and a line drawn from the slit to the central maximum is determined using the equation

$$w \sin \theta_m = m\lambda \qquad (24.6)$$

where w is the slit width, λ is the wavelength of the light, and $m = \pm 1, \pm 2, \pm 3, \ldots$ (not zero). The dark bands on the screen are located at positions $y_m = L \tan \theta_m$ relative to the center of the pattern. L is the slit-screen distance.

The Poisson spot—testing the wave model of light

The year 1818 was a decisive one for the acceptance of the wave model of light. Simeon-Denis Poisson, a talented mathematician and proponent of the particle model, set out to disprove the wave model of light. He suggested shining a narrow beam of light at a small round obstacle. Though Poisson did not believe in the wave model explanation, he still used it to predict the outcome of this testing experiment. His reasoning was as follows: according to the wave model, when the obstacle is just smaller than the beam, the light should illuminate the edges of it. The edges could then be considered the sources of in-phase secondary wavelets. If so, a bright spot should also appear in the middle of the obstacle's shadow because the center of the shadow is equidistant from each secondary wavelet source. This phenomenon is similar to why a bright fringe appears at the center of a double-slit interference pattern (it is equidistant from both slits). This prediction is in direct contradiction to the particle model, which does not predict such a spot. Common sense also tells us that a bright spot should not appear at the center of the dark shadow. The French Academy of Sciences conducted the experiment, and much to their (and Poisson's) surprise, they saw a bright spot right in the middle of the shadow, consistent with the wave model.

We repeated Poisson's experiment using a small ball bearing as the obstacle (**Figure 24.24a**). We expanded the laser light to make sure that the beam was wider than the bearing (Figure 24.24b). You see the resulting pattern on the screen in Figure 24.24c. Notice the tiny bright spot at the center and fringes on the outside.

Diffraction and everyday experience

Since the wavelengths of visible light are so small, we seldom observe light diffraction in everyday life. However, diffraction is a phenomenon that all waves exhibit, including sound waves. In Eq. (24.6), $w \sin \theta_m = m\lambda$, if w is nearly equal to the wavelength of the waves, the angular position of the $m = \pm 1$ minima could reach almost 90°, in which case the waves would spread into the entire region beyond the slit with no regions of destructive interference. This full-screen diffraction seldom occurs with light because slits are seldom that small. But the wavelengths of sound waves are much larger. The wavelength of concert A (frequency 440 Hz) is $\lambda = v_{sound}/f = (340 \text{ m/s})/440 \text{ Hz} = 0.77 \text{ m}$, about the width of a doorway. Thus, a 90° central maximum will fill the entire region beyond the door. Diffraction explains why we can hear a person talking around a corner in another room when the door is open.

FIGURE 24.24 Poisson's spot tested the wave theory of light.

(a)
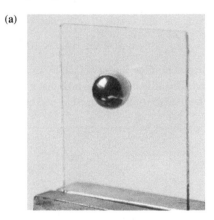

(b) Side view of Poisson's experimental design (the distance between the wave fronts is greatly exaggerated)

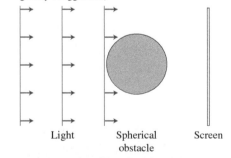

Light   Spherical   Screen
               obstacle

(c)
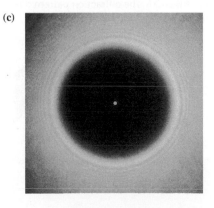

Sound of wavelength 0.85 m (approximately the wavelength of 400-Hz sound) and light of wavelength 400 nm (blue color) are incident on a 1.0-cm slit in a door. Will light and sound diffract significantly after passing through the slit?

Represent mathematically To answer the question we need to determine the angular size and the width of the central maximum for each type of wave. The angular deflection $\theta_{\pm 1}$ to the first dark band on either side of the central maximum is determined using the equation $w \sin \theta_{\pm 1} = \pm \lambda$.

Solve and evaluate Rearranging this equation and applying it to light, we get

$$\sin \theta = \frac{\lambda}{w} = \frac{400 \times 10^{-9}\,\text{m}}{1.0 \times 10^{-2}\,\text{m}} = 4.0 \times 10^{-5}$$

Taking the inverse sine of both sides,

$$\theta = \sin^{-1}(4.0 \times 10^{-5}) = 0.0023°$$

The first minima appear at 0.0023° from either side of the central bright band. The diffraction is so small that it will not be noticed. You will see a sharp image of the slit opening and no interference.

For sound, $\sin \theta = \dfrac{\lambda}{w} = \dfrac{0.85\,\text{m}}{1.0 \times 10^{-2}\,\text{m}} = 85$. This impossible result ($\sin \theta \le 1$) means that the central "bright" region for sound is spread over all angles on the other side of the opening and there will be no regions of destructive interference beyond the door. There will be almost no variation in the amplitude of sound waves behind the door due to diffraction. This phenomenon occurs whenever the wavelength is greater than the width of the opening, when $\lambda/w > 1$.

Try it yourself White light is incident on a slit that is 0.10 mm wide. A screen is 3.0 m from the slit. What is the width of the red bright band adjacent to the central white band compared to the width of the corresponding blue band? The width on the screen of the first colored bright band is the distance between the $m = 1$ and $m = 2$ dark bands. Consider the wavelength of blue light to be 480 nm and the wavelength of red light to be 680 nm.

Answer The angular positions for the first and second blue dark bands on one side of the central bright band are $\theta_1 = 0.28°$ and $\theta_2 = 0.55°$, and the distance on the screen between them is 14 mm. For red light $\theta_1 = 0.39°$ and $\theta_2 = 0.78°$, and they are separated by 20 mm.

REVIEW QUESTION 24.5 Equation (24.6), $w \sin \theta_m = m\lambda$, where $m = \pm 1, \pm 2, \pm 3 \ldots$ describes the angles at which one can see dark fringes produced by light passing through a single slit. Write a similar equation describing the angles at which one can see dark fringes produced by monochromatic light passing through two narrow slits.

24.6 Resolving power

The wave-like behavior of light limits our ability to see two distant closely spaced objects as separate objects or to discern the details of an individual distant object. Can we see the fine structure of a cell with a microscope? Can a reconnaissance aircraft take photographs of a missile site with sufficient detail? This section will help us answer such questions.

FIGURE 24.25 The diffraction pattern due to light passing through a small hole.

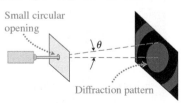

Consider light passing through the small circular hole shown in **Figure 24.25**. Light leaving the hole produces a diffraction pattern on a screen beyond the hole. The pattern resembles that formed by light passing through a narrow slit, except that alternating rings of bright light and darkness are formed rather than parallel bands of light and darkness. We can calculate the angle θ between a line drawn from the hole to the center of the pattern and a line drawn toward the first dark ring using the following equation (which we will not derive):

$$\sin \theta = \frac{1.22\lambda}{D} \tag{24.7}$$

where λ is the wavelength of the wave passing through the hole and D is the diameter of the hole. The shiny disk at the center of the pattern is called the Airy disk, named after George Airy (1801–1892), who first derived Eq. (24.7) in 1831.

The equation is similar to Eq. (24.6), used to calculate the angular position of the dark bands in a single-slit diffraction pattern. The factor 1.22 that appears in Eq. (24.7) and not in Eq. (24.6) is a result of the circular geometry of the opening. The intensities of the bright secondary rings around the central disk are much lower than that of the central bright spot.

Resolving ability of a lens

How does diffraction affect the ability of a lens to produce a sharp image? Imagine that light from a very distant object enters a lens of diameter D as plane waves and diffracts as it passes through the opening (**Figure 24.26**). The diffraction produces an angular spread given by Eq. (24.7): $\sin \theta = 1.22\lambda/D$. Thus, instead of passing through a focal *point* a distance f from the lens, parallel light rays form a central *disk* of radius $R = f \tan \theta$ at the focal point. The radius of that central diffraction pattern depends on the diameter of the lens, the focal distance of the lens, and the wavelength of the light. The wave-like properties of light make it *impossible* to form a perfectly sharp image.

This result is important for surveillance, astronomical studies, microscopy, and human vision. When looking at two closely spaced objects, we see central disks whose radii depend on the diameter of the aperture the light passes through (the pupil of an eye, the objective lens of a telescope, etc.). If the central disks of the diffraction patterns from the two objects overlap, we cannot perceive them as two distinct objects. Physicists say that we cannot **resolve** them. In **Figure 24.27a**, we see a photograph of stars as seen through a telescope with a small-diameter objective lens. How many stars are in the field of vision? It is difficult to say by looking at Figure 24.27a. Figures 24.27b and c show the same stars photographed with telescopes with larger lenses that make it easier to resolve the individual stars.

What is the smallest separation of two images that can still be perceived as distinct? Consider **Figure 24.28a**, which shows the diffraction patterns of distant point-like objects produced on a film by light of wavelength λ passing through a lens of diameter D. In (b), the objects are far enough apart (large enough angular separation) that diffraction does not prevent their images from being resolved. In (c), the images can just barely be resolved. In (d), the images cannot be resolved.

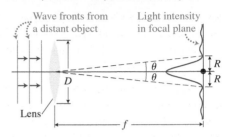

FIGURE 24.27 Telescope lenses of increasing diameter resolve two stars.

(a) Not resolved

Small lens

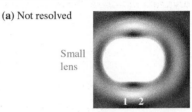

(b) Barely resolved

Medium lens

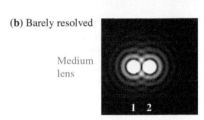

(c) Well resolved

Large lens

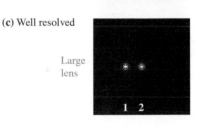

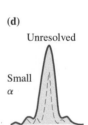

FIGURE 24.28 The angular separation of two sources affects the ability to resolve them.

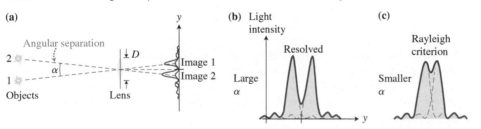

According to Lord Rayleigh (1842–1919), two objects can just barely be resolved when their angular separation α_{res} (called the **limit of resolution**) satisfies the following criterion:

Rayleigh criterion The minimal angular separation of two objects that can be resolved (perceived as separate) is the limit of resolution α_{res} of the instrument:

$$\alpha_{res} = \frac{1.22\lambda}{D} \qquad (24.8)$$

where D is the diameter of the opening through which light enters and λ is the wavelength of the light. In this equation, α_{res} is measured in radians.

TIP When the limit of resolution of the instrument increases, it means that its resolving power decreases. The larger α_{res}, the more difficult it is to resolve closely separated objects.

Equation (24.8) helps us understand the role of the opening diameter D of any instrument. Increasing the size of the opening of any instrument (such as a telescope) allows you to resolve more closely positioned objects or details on a single object.

The Rayleigh criterion is a factor in building various optical devices. Telescopes are built with objective lenses that are as large as possible. Large lenses are difficult to make; they break easily and also sag under their own weight. Thus, high-resolution telescopes use mirrors instead. Another limitation in producing sharp images of stars is the movement of the atmosphere, which results in images that continuously distort and shimmer. This is why observatories are built on mountains, where they are above a significant amount of the atmosphere.

Resolving the fine details of really small objects, such as parts of biological cells, is also challenging. We can improve the resolving power by decreasing the wavelength of the light. For example, we can use UV light instead of visible light. Much larger resolving power can be achieved with an electron microscope. Electron microscopes, discussed in Chapter 28, resolve images using the wavelengths of electrons, which are much smaller than those of visible light.

REVIEW QUESTION 24.6 Stars are so far away that they can be considered as point sources. Why then do the images of stars produced by a telescope and recorded on a film look like bright disks, not dots?

24.7 Skills for applying the wave model of light

In this section we will practice problem-solving skills while investigating some new phenomena. A problem-solving strategy is outlined on the left side of the table below and illustrated in Example 24.7.

PROBLEM-SOLVING STRATEGY 24.1   Analyzing processes using the wave model of light

EXAMPLE 24.7   **Purchase a grating**

You are to purchase a grating that will cause the deflection of the $m = 2$ 2nd order bright band for 650-nm wavelength red light to be 42°. How many lines per centimeter should your grating have? How many bright bands in total will you see on the screen?

Sketch and translate

- Visualize the situation and then sketch it.
- Identify given physical quantities and unknowns.

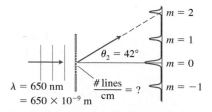

Simplify and diagram

- Decide if the sources in the problem produce coherent waves.
- Decide if the small-angle approximation is valid.
- Decide if the slit widths for multiple slits are wide enough that you have to consider single-slit diffraction as well as multiple-slit interference.
- If useful, represent the situation with a wave front diagram showing the overlapping crests and troughs of the light waves from different sources.

- Grating slits can be considered coherent light sources.
- With gratings, the angular deflection is often 10° or more, so we cannot use the small-angle approximation.
- Because the slits in gratings are very narrow compared to the distance between them, we can consider multiple-slit interference only and ignore single-slit diffraction.
- The maximum number of bright bands is determined by the maximum path length difference that the waves from two adjacent slits can have, which is equal to the distance d between the slits (see the figure at right). Thus n is equal to the integer part of d/λ.

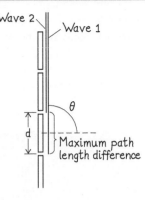

Represent mathematically

- Describe the situation mathematically.
- Use geometry if needed.

- Apply Eq. (24.5) to the second bright band to determine the slit separation: $\sin \theta_m = m(\lambda/d)$.
- Find the number of slits/cm: $\# = \dfrac{1 \text{ cm}}{d}$
- Find the number n of bright bands on each side by taking the integer part of $\dfrac{d}{\lambda}$. We then need to add one more for the central maximum; thus the total number of bands is $2n + 1$.

Solve and evaluate

- Use the mathematical description of the process to solve for the desired unknown quantity.
- Evaluate the result. Does it have the correct units? Is its magnitude reasonable? Do the limiting cases make sense?

- From Eq. (24.5) with $m = 2$, we get

$$d = m \frac{\lambda}{\sin \theta_m} = 2 \frac{650 \times 10^{-9} \text{ m}}{\sin 42°} = 1.9 \times 10^{-6} \text{ m} = 1.9 \times 10^{-4} \text{ cm}$$

- The number of slits in one centimeter is

$$\# = \frac{1 \text{ cm}}{d} = \frac{1 \text{ cm}}{1.9 \times 10^{-4} \text{ cm}} = 5100$$

The units match. This is a reasonable number of slits. A limiting case analysis is not needed, as we did not derive any new expressions.

- The number of bright bands on each side is

$$n = \frac{d}{\lambda} = \frac{1.9 \times 10^{-6} \text{ m}}{650 \times 10^{-9} \text{ m}} = 2.93$$

- We cannot round up to three because the 3rd band may not be fully visible, so we conclude that there are two bands on each side. Adding the central maximum, the total number of bands seen is five. We can check this number by using another method: the maximum angle of deflection is 90°, thus $d \sin 90° = n\lambda \Rightarrow d = n\lambda$. We arrived at the same equation. In addition, five bands is a reasonable number; if we got 50, it would be a warning that we had done something wrong in the calculation.

Try it yourself Suppose you used a spectrometer using this grating to analyze light coming from a star and found an $m = 2$ band of light at 34°. What is the wavelength of this light?

Answer $530 \times 10^{-9} \text{ m} = 530 \text{ nm}.$

EXAMPLE 24.8 **LED light passing through a slit**

Three LEDs (red, green, and blue) are mounted on a vertical plate, one above another. You observe the LEDs through a narrow vertical slit that is 2.50 m away from the LEDs. The photo below shows a magnified image of what you see (in reality, the pattern is small but clearly visible). Knowing that the red LED mostly emits light with a wavelength of about 630 nm, estimate the wavelength of light emitted by the green and the blue LEDs.

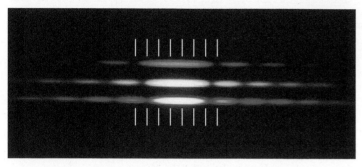

Sketch and translate We draw a sketch of the situation below and label the knowns and unknowns.

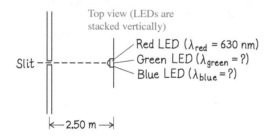

Simplify and diagram Because the LEDs emit light of clearly distinguishable colors and are far away from the slit, we can assume the light

is coherent. Since the pattern is very small, we can also assume that the angular deflections of the minima are small. In this case, the widths of the bright central bands on the photo are directly proportional to the deflection angles to the first minima for each LED.

Represent mathematically Using our last assumption, we can write

$$y_{red} = l\theta_{red} = l\frac{\lambda_{red}}{w}$$

$$y_{green} = l\theta_{green} = l\frac{\lambda_{green}}{w}$$

$$y_{blue} = l\theta_{blue} = l\frac{\lambda_{blue}}{w}$$

where l is distance between the observer and the slit. Dividing the last two equations by the first one (notice that l does not enter the final equation), we get

$$\frac{y_{green}}{y_{red}} = \frac{\lambda_{green}}{\lambda_{red}}; \quad \frac{y_{blue}}{y_{red}} = \frac{\lambda_{blue}}{\lambda_{red}}$$

Solve and evaluate Using the white grid overlaid on the photo, we estimate

$$\frac{y_{green}}{y_{red}} = \frac{6}{7} \text{ and } \frac{y_{blue}}{y_{red}} = \frac{5}{7}$$

and find $\lambda_{green} = \frac{6}{7}\lambda_{red} = \frac{6}{7}(630 \text{ nm}) = 540 \text{ nm}$ and $\lambda_{blue} = \frac{5}{7}\lambda_{red} = \frac{5}{7}(630 \text{ nm}) = 450 \text{ nm}$. These answers seem reasonable—both wavelengths are in the regions of the spectrum where these colors belong.

A final note about light

In this chapter we applied the wave model of light to explain situations in which light passes through small openings or around small objects. The particle model of light cannot explain them. We found that the effects of light's wave-like behavior are most dramatic when the sizes of the holes and obstacles are comparable to the wavelength of the light. The wave model of light also explains the interaction of light with large objects (reflection and refraction). Therefore, it is tempting to accept the wave model as the "correct" model. However, we have not yet established a mechanism for this model. What is actually vibrating when light waves propagate? That question will be the topic of the next chapter.

REVIEW QUESTION 24.7 Why is it especially important to keep track of units when solving problems in wave optics?

Summary

Speed of light and refractive index The ratio of the speed of light in air c and its speed v in a medium equals the index of refraction n of the medium. (Section 24.2)

$$\frac{c}{v} = n$$ Eq. (24.4)

Monochromatic and coherent light sources Only waves of the same constant frequency and wavelength (*monochromatic*) and of constant phase difference (*coherent*) will add to produce a constant interference pattern. (Section 24.2)

Double-slit interference In the wave model of light, double-slit interference is explained by the superposition of light waves passing through two closely spaced slits. The superposition of those waves creates a pattern of dark and bright bands on a screen beyond the slits (rather than images of the slits). (Section 24.1)

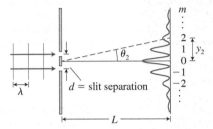

Bright bands at

$$d \sin \theta_m = m\lambda$$ Eq. (24.1)
$$m = 0, \pm 1, \pm 2, \dots$$

$$L \tan \theta_m = y_m$$ Eq. (24.2)

For small angles:
$$\tan \theta_m \approx \sin \theta_m$$

Grating A grating has many closely spaced slits, each separated by a distance d from its neighbors. The maxima produced by gratings are much narrower and brighter than those for two slits of the same slit separation. (Section 24.3)

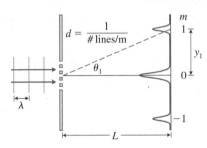

Bright bands at

$$d \sin \theta_m = m\lambda$$ Eq. (24.1)
$$m = 0, \pm 1, \pm 2, \dots$$
$$d = 1/\#\,\text{lines/m}$$

$$L \tan \theta_m = y_m$$ Eq. (24.2)

Thin-film interference Thin films such as soap bubbles and lens coatings produce interference effects by reflecting light from the front and back surfaces of the film. (Section 24.4)

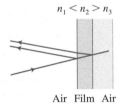

$n_1 < n_2 > n_3$

Air Film Air

See Table 24.3.

Single-slit diffraction When monochromatic light of wavelength λ is incident on a slit of width w, a series of bright and dark bands of light is formed on a screen a distance L from the slit. The bands are caused by the interference of light from different regions within the slit. (Section 24.5)

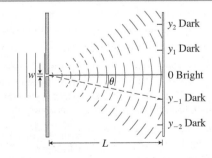

Dark bands at

$$w \sin \theta_m = m\lambda$$ Eq. (24.6)
$$m = \pm 1, \pm 2, \pm 3 \dots$$

$$y_m = L \tan \theta_m$$ Eq. (24.2)

Rayleigh criterion for resolving objects with optical instruments The minimal angular separation for resolving two objects with an instrument of aperture diameter D is α_{res}. If the objects are closer together than this, they appear as a single object (cannot be resolved). (Section 24.6)

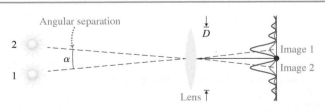

$$\alpha_{res} = \frac{1.22\lambda}{D}$$ Eq. (24.8)

Questions

Multiple Choice Questions

1. You shine a flashlight on two wide slits cut in cardboard. What do you observe on a screen beyond the slits?
   (a) A sharp image of the two slits
   (b) An interference pattern consisting of bright and dark bands
   (c) A fuzzy image of the slits
   (d) Depending on the distance from the slits to the screen, the answer could be (a) or (c).

2. When you shine a very narrow beam of white light on a grating in a dark room, what do you see on the screen behind the slits?
   (a) A sharp image of the grating's slits
   (b) A pattern of white and dark spots of approximately equal width
   (c) A colored stripe that contains rainbow colors
   (d) A set of rainbow-colored stripes separated by dark regions
   (e) A white spot in the center and a set of rainbow-colored stripes on each side of it, separated by dark regions

3. When green light travels from air to glass, what quantities change?
   (a) Speed only   (b) Frequency only
   (c) Wavelength only   (d) Speed and frequency
   (e) Speed and wavelength

4. If you add a third slit to two slits, with the same slit separation, and shine the same laser beam on the three slits, how will the three-slit pattern compare to the double-slit pattern?
   (a) It will be the same, only brighter.
   (b) It will be the same, only the spots will be farther apart.
   (c) The centers of its bright fringes will be the same distance from each other, but their width will be narrower; additional spots will appear between them.
   (d) The centers of its bright fringes will be the same distance from each other, and their width will be narrower; no additional spots will appear.

5. Why don't two flashlights about 1 m apart shining on a wall produce an interference pattern?
   (a) They are not bright enough.
   (b) The light waves are not coherent.
   (c) The fringes are too narrow to see.

6. You shine a laser beam through a grating with a known number of slits per millimeter and observe a pattern on a screen. Then you cover half of the grating with nontransparent material. What will you see?
   (a) There will be half as many bright fringes on the screen.
   (b) The bright fringes will be twice as close.
   (c) The bright fringes will be two times farther apart.
   (d) The location of the fringes will not change, but they will be a little bit wider and less bright.

7. What does the resolution limit of an optical system depend on? Choose all answers that are correct.
   (a) The wavelength of light   (b) The diameter of the aperture
   (c) The distance to the object being viewed
   (d) The distance from the aperture to the light detectors

8. You shine light of different colors on the same lens. The focal length of the lens for blue light
   (a) is shorter than that for red light.
   (b) is longer than that for red light.
   (c) is the same as that for red light because focal length is a property of the lens, not of the light shining on it.
   (d) depends on the location of the source of light.

9. You shine a green laser beam ($\lambda = 530$ nm) on a grating and observe seven bright beams behind the grating. Which of the following changes to the experiment (keeping the other parameters constant) could increase the total number of bright beams? Select all that are correct.
   (a) Using a grating with more slits per millimeter
   (b) Using a grating with fewer slits per millimeter
   (c) Using a red laser ($\lambda = 650$ nm)
   (d) Using a blue laser ($\lambda = 450$ nm)
   (e) Increasing the distance between the laser and the grating
   (f) Decreasing the distance between the laser and the grating
   (g) Immersing all of the equipment in water

Conceptual Questions

10. Describe a double-slit interference experiment for sound waves using one or more speakers, driven by an electronic oscillator, as your sources of sound waves.

11. You are investigating a pattern produced on a screen by laser light passing through two narrow slits. What would happen to the pattern if you covered one of the slits?

12. Give examples of phenomena that can be explained by a particle-bullet model of light. How can a wave model explain the same phenomena?

13. Give examples of phenomena that *cannot* be explained by a particle-bullet model of light. How can a wave model explain these phenomena?

14. How would you explain Huygens' principle to a friend who is not taking physics?

15. Draw a point-like source of light. What is the shape of wave fronts leaving that point, assuming a uniform medium? Does the distance between them change as you move away from the source? What is the direction of the rays? How are the wave fronts for laser light different from those produced by the point source of light? How are rays oriented with respect to the wave fronts?

16. Draw two coherent light sources next to each other. What will you see if you place light-sensitive detectors at several different locations around the sources? How do you know? Indicate any assumptions that you made.

17. Use the wave front representation to explain what happens to the interference pattern produced by two coherent light sources if their separation increases. Then evaluate the result using the ray representation.

18. Use the wave front representation to explain what happens to the interference pattern produced by two coherent light sources if their wavelength increases.

19. Compare the interference pattern produced by two coherent light sources in air with the case when they are both immersed in water.

20. Draw 10 coherent point-like sources of light placed along a straight line. Construct a wave front for a wave that is a superposition of the waves produced by the sources.

21. If you see green light of 520-nm wavelength when looking at the part of a soap bubble that is closest to you of uniform thickness, why will you see longer wavelength light when looking at the sides of this green region?

22. Imagine that you have a very thin uniform oil film on the surface of water. The thickness of oil is much smaller than the wavelength of blue light. White light is shining on the film. Will the film appear bright or dark if you look at it from above? Explain.

23. (a) Draw a picture of what you will see on a screen if you shine a red laser beam through a very narrow slit. (b) Redraw the picture for a wider slit. (c) Redraw the picture for a green laser for each case. Keep the scale the same.

24. Describe three situations that you can analyze using wave fronts, rays, and disturbance-versus-position graphs for a particular time. In what situation is one of the representations more helpful than the others?

25. Why can you hear a person who is around a corner talking but not see her?

26. Astronomers often called the resolution limit described by Eq. (24.8) a "theoretical resolution limit." Why do they add the adjective "theoretical" to the description? What in real life can affect the resolution limit of a telescope?

Problems

Below, **BIO** indicates a problem with a biological or medical focus. Problems labeled **EST** ask you to estimate the answer to a quantitative problem rather than derive a specific answer. Asterisks indicate the level of difficulty of the problem. Problems with no * are considered to be the least difficult. A single * marks moderately difficult problems. Two ** indicate more difficult problems.

24.1 and 24.2 Young's double-slit experiment and Index of refraction, light speed, and wave coherence

1. * **Sound interference** Two sources of sound waves are 2.0 m apart and vibrate in phase, producing sinusoidal sound waves of wavelength 1.0 m. (a) Use the wave front representation to explain what happens to the amplitude of sound along a line equidistant from each source and perpendicular to the line connecting the sources. (b) Use a graphical representation (pressure-versus-position graph) to explain what happens along that line. (c) Which representation is more helpful? Explain.

2. * Green light of wavelength 540 nm is incident on two slits that are separated by 0.50 mm. (a) Make a list of physical quantities you can determine using this information and determine three of them. (b) Describe three changes in the experiment that will each result in doubling the distance between the 0th order and the first order maximum on the screen. Explain.

3. * Blue light of wavelength 440 nm is incident on two slits separated by 0.30 mm. Determine (a) the angular deflection to the center of the 3rd order bright band and (b) its distance from the 0th order band when the light is projected on a screen located 3.0 m from the slits. (c) Draw a sketch (not to scale) that schematically represents this situation and label all known distances and angles.

4. * Red light of wavelength 630 nm passes through two slits and then onto a screen that is 1.2 m from the slits. The center of the 3rd order bright band on the screen is separated from the central maximum by 0.80 m. What can you determine using this information?

5. **Sound from speakers** Sound of frequency 680 Hz is synchronized as it leaves two speakers that are separated by 0.80 m on an open field. Draw a sketch of this arrangement and draw a line from between the speakers to a location where the sound is intense and equidistant from the two speakers (the 0th order maximum). Determine the angular deflection of a line from between the speakers to the 1st order intensity maximum to the side of this 0th order maximum. The speed of sound is 340 m/s.

6. * **EST** **Sketch a moving wave front** Draw a sketch of two narrow slits separated by 4.0 cm. On the sketch, show the crests of six waves of wavelength 1.0 cm that have left each slit. Draw lines from a point halfway between the slits in the directions of the center of the 0th order and 1st order intensity maxima. Draw a screen 7.0 cm from the slits and use your sketch to estimate the distance between the center of the central maximum on the screen and a 1st order bright spot to the side. Check your results using equations from the text. Were your sketch-based estimated results within 20% of the mathematical results? Indicate any assumptions that you made.

7. * An LED bulb emits light that looks red. After passing through a narrow single slit, the light strikes two very narrow slits separated by distance d and located a distance D from the single slit. After passing through the pair of slits, the light strikes a screen a distance L away. (a) Make a sketch of the pattern that you would expect to see on the screen if light behaves like a wave. (b) Below the interference pattern, sketch the pattern that you would expect to see if light behaves like a stream of very light particles.

8. **Characteristics of laser light when in glass** A laser light in air has a wavelength of 670 nm. What is the frequency of the light? What is the frequency of this light when it travels in glass? In water? What is the wavelength of light in these media? (Use Table 22.7 if needed.)

9. * **Prism converts white light** Use the wave model of light to explain why white light striking a side of a triangular prism emerges as a spectrum.

24.3 Gratings: an application of interference

10. Light of wavelength 520 nm passes through a grating with 4000 lines/cm and falls on a screen located 1.6 m from the grating. (a) Draw a picture (not to scale) that schematically represents the process and label all known distances and angles. (b) Determine the angular deflection of the second bright band. (c) Determine the separation of the 2nd order bright spot from the central maximum. (d) Determine the total number of constructively interfering beams that are formed after the light passes the grating.

11. **Hydrogen light grating deflection** Light of wavelength 656 nm and 410 nm emitted from a hot gas of hydrogen atoms strikes a grating with 5300 lines per centimeter. Determine the angular deflection of both wavelengths in the 1st and 2nd orders.

12. * **Purchase a grating** How many lines per centimeter should a grating have to cause a 38° deflection of the 2nd order bright band of 680-nm red light?

13. **Only half a grating** You cover the left side of a grating with black tape so that half of the slits are covered. How will it affect the location of the 2nd order bright band?

14. * **EST** **Design** Design a quick way to estimate which one of two gratings has more lines per centimeter.

15. **Laser light on grating 1** The 630-nm light from a helium-neon laser irradiates a grating. The light then falls on a screen where the first bright spot is separated from the central maximum by 0.51 m. Light of another wavelength of light produces its first bright spot 0.43 m from its central maximum. Determine the second wavelength.

16. **Laser light on grating 2** Light of wavelength 630 nm passes through a grating and then onto a screen located several meters from the grating. The 1st order bright band is located 0.28 m from the central maximum. Light from a second source produces a band 0.20 m from the central maximum. Determine the wavelength of the second source. Show your calculations. Hint: If the angular deflection is small, $\tan \theta = \sin \theta$.

17. * **EST** **Memory capacity of CD** The reflective surface of a CD consists of spirals of equally spaced grooves. If you shine a laser pointer on a CD, each groove reflects circular waves that look exactly like the circular waves transmitted by the slits in a grating. You shine a green laser pointer ($\lambda = 532$ nm) perpendicularly to the surface of a CD and observe a diffraction pattern on a screen that is 3.0 m away from the CD. You observe that the 1st order maximum ($m = 1$) appears 1.1 m away from the central maximum ($m = 0$). (a) Determine the distance between the adjacent grooves on a CD. (b) Estimate the total number of bits (units of information) on a CD, assuming that each bit occupies a square with sides that are equal to the distance between the adjacent grooves. Compare the result with the typical data storage of a CD.

18. * **EST** You shine a green laser beam ($\lambda = 530$ nm) on a grating and observe 11 bright beams (due to constructive interference) behind the grating. What can you estimate about the grating?

24.4 Thin-film interference

19. You point a green laser beam perpendicularly toward a soap film, as shown in **Figure P24.19**. Above the laser you place a curved white screen. You slowly rotate the laser through 90° while pointing the beam toward the center of the soap film, as shown in the figure. While you perform the experiment, your friend takes a long-exposure photo of the screen. Describe what you will see in the photo. Explain your answer.

FIGURE P24.19

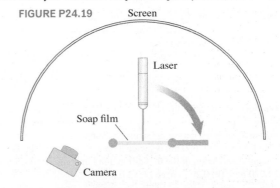

20. * **Representing thin-film interference** (a) Draw a ray diagram for a laser beam incident from air on a thin film that has air on the other side. Make sure to take into account the processes occurring on each surface of the film. (b) Discuss in words what happens to the wave phase at each boundary. (c) Under what conditions will a person observe no reflected light?

21. * **Oil film on water** A thin film of vegetable oil ($n = 1.45$) is floating on top of water ($n = 1.33$). Describe in words the processes occurring to a laser beam at the top and bottom surfaces of the oil.

22. * **Soap bubble 1** You look at a soap bubble film perpendicular to its surface. Describe the changes in colors of the film that you observe as the film thins and eventually breaks. Support your explanation with a ray diagram.

23. * **Soap bubble 2** A soap bubble of refractive index 1.40 appears blue-green when viewed perpendicular to its surface (blue-green appears when red light is missing from the continuous spectrum, where λ_{red} is about 670 nm in a vacuum). Does the light change phase when reflected from (a) the outside surface of the bubble and (b) the inside surface? (c) Determine the wavelength of red light when passing through the bubble. (d) Determine the thickness of the thinnest bubble for which the 670-nm red light reflected from the outside surface of the bubble interferes destructively with light reflected from the inside surface.

24. * **Thin-film coated lens** A lens coated with a thin layer of material having a refractive index 1.25 reflects the least amount of light at wavelength 590 nm. Determine the minimum thickness of the coating.

25. * **Thin-film coated glass plate** A film of transparent material 120 nm thick and having refractive index 1.25 is placed on a glass sheet having refractive index 1.50. Determine (a) the longest wavelength of light that interferes destructively when reflected from the film and (b) the longest wavelength that interferes constructively.

26. Two flat glass surfaces are separated by a 150-nm gap of air. (a) Explain why 600-nm-wavelength light illuminating the air gap is reflected brightly. (b) What wavelength of radiation is not reflected from the air gap?

27. * Jeff wants to solve the following problem: a light beam with $\lambda = 500$ nm in oil is incident perpendicularly on a layer of oil ($n_{oil} = 1.45$, thickness $d = \frac{\lambda}{2} = 250$ nm). The oil layer is floating on top of water ($n_{water} = 1.33$). Compare (a) the total intensity I_R of the light that *reflects* from the oil and (b) the intensity I_T of the light that goes *through* the oil into the water to the intensity of the incident light I_0. Explain your answer. Ignore any absorption of light in the oil. Jeff's solution is as follows: the total path length difference between a ray that reflects from the air-oil boundary and a ray that reflects from the oil-water boundary is equal to λ. Therefore, the reflected rays interfere constructively, giving (a) $I_R = I_0$ and (b) $I_T = 0$. Evaluate Jeff's solution, identify any errors, and, if there are any errors, provide a corrected solution.

28. ** Dawn wants to solve the following problem: a light beam with $\lambda = 500$ nm in oil is incident perpendicularly on a layer of oil ($n_{oil} = 1.45$, thickness $d = \frac{\lambda}{2} = 250$ nm). The oil layer is on top of glass ($n_{glass} = 1.55$). Compare (a) the total intensity I_R of the light that *reflects* from the oil and (b) the intensity I_T of the light that goes *through* the oil into the glass to the intensity of the incident light I_0. Explain your answer. Ignore any absorption of light in the oil. Dawn's solution is as follows: the total path length difference between a ray that reflects from the air-oil boundary and a ray that reflects from the oil-glass boundary is equal to 2λ. Therefore, the reflected rays interfere constructively, giving (a) $I_R = I_0$ and (b) $I_T = 0$. Evaluate Dawn's solution, identify any errors, and, if there are any errors, provide a corrected solution.

24.5 Diffraction of light

29. * **Explain diffraction** Draw a ray diagram and show path length differences to explain how wavelets originating in different parts of a slit produce the third dark fringe on a distant screen.

30. * **How did we derive it?** Explain how we derived the equation for the first dark fringe for single-slit diffraction. Draw a ray diagram or a disturbance-versus-position graph to help in your explanation. Show the path length difference and the phase differences.

31. * **Explain a white light diffraction pattern** White light passing through a single slit produces a white bright band at the center of the pattern on a screen and colored bands at the sides. Explain.

32. Light of wavelength 630 nm is incident on a long, narrow slit. Determine the angular deflection of the first diffraction minimum if the slit width is (a) 0.020 mm, (b) 0.20 mm, and (c) 2.0 mm.

33. * Light of wavelength of 430 nm is incident on a long, narrow slit of width 0.050 mm. Determine the angular deflection of the 5th order diffraction minimum.

34. * **Sound diffraction through doorway** Sound of frequency 440 Hz passes through a doorway opening that is 1.2 m wide. Determine the angular deflection to the first and second diffraction minima ($v_{sound} = 340$ m/s).

35. * Light of wavelength 624 nm passes through a single slit and then strikes a screen that is 1.2 m from the slit. The thin first dark band is 0.60 cm from the middle of the central bright band. Determine the slit width.

24.6 Resolving power

36. ** **Explain resolution** Explain the term "resolution limit." Illustrate your explanation with pictures and ray diagrams, and explain what characteristics of an optical device and the light passing through it affect the resolution limit.

37. **Resolution of telescope** A large telescope has a 3.00-m-radius mirror. What is the resolution limit of the telescope? What assumptions did you make?

38. * Laser light of wavelength 630 nm passes through a tiny hole. The angular deflection of the first dark band is 26°. What can you learn about the hole using this information?

39. * **Size of small bead** Infrared radiation of wavelength 1020 nm passes a dark, round glass bead and produces a circular diffraction pattern 0.80 m beyond it. You cannot see this light with the naked eye, but you can observe it with a cell phone camera. The diameter of the first dark ring is 6.4 cm. What can you learn about the bead using this information?

40. * **Resolution of telescope** How will the resolution limit of a telescope change if you take pictures of stars using a blue filter as opposed to using a red filter? Explain.

41. * **Detecting visual binary stars** Struve 2725 is a double-star system with a visual separation of 5.7 arcseconds in the constellation Delphinus. Determine the minimum diameter of the objective lens of a telescope that will allow you to resolve 400-nm violet light from the two stars.

42. **Hubble Telescope resolving power** The objective mirror of the Hubble Telescope is 2.4 m in diameter. Could it resolve the binary stars Lambda Cas in the constellation Cassiopeia? The stars have an angular separation of 0.5 arcseconds.

43. * Draw a graphical representation of Rayleigh's criterion. Explain how this criterion relates to the concept of the resolution limit.

24.7 Skills for applying the wave model of light

44. * Red light from a helium-neon gas laser has a wavelength of 630 nm and passes through two slits. (a) Draw a ray diagram to explain why you see a pattern of bright and dark bands on the screen. Show the path length difference. (b) Determine the angular deflection of the light to the first three bright bands when incident on narrow slits separated by 0.40 mm. (c) Determine the distance between the centers of the 0th and 2nd order bright bands when projected on a screen located 5.0 m from the slits. (d) List all of the assumptions that you made in your calculations.

45. * **EST** Figure P24.45 shows the diffraction pattern on a screen obtained when a narrow beam of white light is incident on a grating with 100 lines/mm. If you carefully examine the color pattern, you will notice narrow magenta bands (indicated by the arrows) that appear where the blue region from the 3rd order maximum overlaps with the red region from the 2nd order maximum (a mixture of blue and red light is perceived as magenta). Note that there are more magenta bands in higher orders, but there is no magenta between the 1st and 2nd order maxima. (a) Is there a magenta band between the 2nd and 3rd order maxima for any grating, or does the presence of the magenta color depend on the distance between the slits in the grating? Explain. (b) The 1st order maxima are about 20 cm from the central white band. Estimate how far from the screen the grating in this experiment is. Explain.

FIGURE P24.45

46. * Red light of wavelength 630 nm is incident on a pair of slits. The interference pattern is projected on a wall 6.0 m from the slits. The fourth bright band is separated from the central maximum by 2.8 cm. (a) Draw a ray diagram to represent the situation; show the path length difference. (b) What can you learn about the slit pair using this information? (c) What can you learn about the pattern on the screen using this information?

47. * Monochromatic light passes through a pair of slits separated by 0.025 mm. On a screen 2.0 m from the slits, the 3rd order bright fringe is separated from the central maximum by 15 cm. (a) Draw a ray diagram to represent the situation. Show the path length difference. (b) What can you learn about the light source using this information?

48. * **Ratio reasoning** Two different wavelengths of light shine on the same grating. The 3rd order line of wavelength A (λ_A) has the same angular deflection as the 2nd order line of wavelength B (λ_B). Determine the ratio (λ_A/λ_B). Be sure to show the reasoning leading to your solution.

49. * **Design** Design an experiment to use a grating to determine the wavelengths of light of different colors. Draw a picture of the apparatus. List the quantities that you will measure. Describe the mathematical procedure that you will use to calculate the wavelengths.

50. * **Fence acts as a grating for sound** A fence consists of alternating slats and openings, the openings being separated from each other by 0.40 m. Parallel wave fronts of a single-frequency sound wave irradiate the fence from one side. A person 20 m from the fence walks parallel to it. She hears intense sound directly in front of the fence and in another region 15 m farther along a line parallel to the fence. Determine everything you can about the sound used in this problem (the speed of sound is 340 m/s).

51. * **BIO Morpho butterfly reflection grating wings** A reflection grating reflects light from adjacent lines in the grating instead of allowing the light to pass through slits, as in a transmission grating. If we assume perpendicular incidence, then we can determine the angular deflection of bright bands the same way we did for a transmission grating. White light is incident on the wing of a Morpho butterfly (whose wings act as reflection gratings). Red light of wavelength 660 nm is deflected in the 1st order at an angle of 1.2°. (a) Determine the angular deflection in the 1st order of blue light (460 nm). (b) Determine the angular deflection in the 3rd order of yellow light (560 nm).

52. * **Ratio reasoning** Laser monochromatic light is used to illuminate two different gratings. The angular deflection of the 2nd order band of light leaving grating A equals the angular deflection of the 3rd order band from grating B. Determine the ratio of the number of lines per centimeter for grating A and for grating B.

53. * **Soap bubble interference** Light of 690-nm wavelength interferes constructively when reflected from a soap bubble having refractive index 1.33. Determine two possible thicknesses of the soap bubble.

54. * **Oil film on water** A film of oil with refractive index 1.50 is spread on water whose refractive index 1.33. Determine the smallest thickness of the film for which reflected green light of wavelength 520 nm interferes destructively.

55. * **Babinet's principle** Babinet's principle states that the diffraction pattern of complementary objects is the same. For example, a rectangular slit produces the same diffraction pattern on a screen as a rectangular obstacle of the same size as the slit, provided that the incident beam is wider than the slit or the obstacle. Determine the width of a hair that, when irradiated with laser light of wavelength 630 nm, produces a diffraction pattern on a screen with the first minimum 2.5 cm on the side of the central maximum. The screen is 2.0 m from the hair.

56. * **Diffraction from a loudspeaker** The opening of a loudspeaker can be modeled as a circular hole through which sound waves propagate into space. Determine the maximum diameter of the speaker such that the first diffraction minimum of sound is at least 45° from the direction in which the speaker points. Perform the calculations for sound waves of frequency (a) 200 Hz, (b) 1000 Hz, and (c) 10,000 Hz. The speed of sound is 340 m/s.

57. * The angular deflection of the 1st order bright band of light passing through double slits separated by 0.20 mm is 0.15°. Determine the angular deflection of the 3rd order diffraction minimum when the same light passes through a single slit of width 0.30 mm.

58. * **EST Diffraction of sound from the mouth** (a) Estimate the diameter of your mouth when open wide. (b) Determine the angular deflection of 200-Hz sound and of 15,000-Hz sound as it leaves your mouth. If, during your calculations, you find that $\sin \theta > 1$, explain the meaning of this result. The speed of sound is 340 m/s.

59. * **Determine body cell size** Light of 630 nm wavelength from a helium-neon laser passes two different-size body cells. The angular deflection of the light as it passes the cells is (a) 0.060 radians and (b) 0.085 radians. Determine the size of each cell. (See the description of the Babinet principle in Problem P24.55.)

60. * **EST** A sound of frequency 1000 Hz passes a basketball. Estimate the angular deflection from the basketball to the first ring around the central maximum in the diffraction pattern. The speed of sound is 340 m/s. (See the description of the Babinet principle in Problem P24.55.)

61. * **BIO Ability of a bat to detect small objects** Bats emit ultrasound in order to detect prey. Ultrasound has a much smaller wavelength than sound, which improves the resolution of small objects. What is the diameter of the smallest object that forms a diffraction pattern when irradiated by the 8.0×10^4-Hz ultrasound from a bat? The first diffraction minimum from the smallest object that still produces diffraction forms at 90° with respect to the direction of the incident waves.

General Problems

62. As Jess and Matt are taking an evening walk in late December, they stop by a house that is decorated with dancing red and green stars (see **Figure P24.62**). They agree that the device that projects the stars is probably using red and green lasers, but they have different explanations for how the stars are formed. Jess thinks that the stars are produced by some kind of a grating that rotates in front of the lasers. Matt thinks that

FIGURE P24.62

the device uses rotating prisms such as those sometimes used in kaleidoscope glasses. Who do you think is correct? Support your answer with data that you can estimate from the photo.

63. * Monochromatic light passes through two slits and then strikes a screen. The distance separating the central maximum and the first bright fringe at the side is 2.0 cm. Determine the fringe separation when the following quantities change simultaneously: the slit separation is doubled, the wavelength of light is increased 30%, and the screen distance is halved.

64. **Sound from speakers** Two stereo speakers separated by a distance of 0.8 m play the same musical note at frequency 1000 Hz. A listener starts from position 0 (**Figure P24.64**) and walks along a line parallel to the speakers. (a) Can the listener easily hear the sound at position 0? Explain. (b) Calculate the distance from position 0 to positions 1 and 2 where intense sound is also heard. The speed of sound is 340 m/s.

FIGURE P24.64

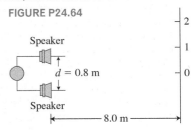

65. * **Astronomer's spectrograph** An astronomer has a grating spectrograph with 5000 lines/cm. A film 0.500 m from the grating records bands of light passing through the grating. (a) Determine the wavelength and frequency of the H_α line of hydrogen gas in a laboratory discharge tube that produces a 1st order band separated on the film by 17.4 cm from the central maximum. (b) Determine the wavelength and frequency of the H_α line coming from a galaxy in the cluster Hydra A. The 1st order band of the H_α line of this light is 18.4 cm from the central maximum. (c) Suggest a possible reason for the differences in the frequency of the H_α line from a lab source and the H_α line from the galaxy.

66. **Diffraction of water waves entering a harbor** The wavelength of water waves entering a harbor is 14 m. The angular deflection of the 1st order diffraction minimum of the waves in the water beyond the harbor is 38°. Determine the width of the opening into the harbor.

67. * **EST** You observe a green LED through a rectangular opening of width $w = 0.08$ mm and unknown height h. You see the pattern shown in Figure P24.67 (the actual pattern is very small, but you can see it clearly with the naked eye). The distance between the LED and the opening is 3.0 m. Estimate the height of the opening h. You will need a ruler to solve this problem. (Hint: A rectangular slit can be modeled as two single slits placed perpendicularly to each other.)

FIGURE P24.67

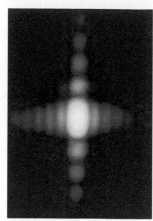

68. ** **Variable thickness wedge** A wedge of glass of refractive index 1.64 has a silver coating on the bottom, as shown in Figure P24.68. Determine the smallest distance x to a position where 500-nm light reflected from the top surface of the glass interferes constructively with light reflected from the silver coating on the bottom. The light changes phase when reflected at the silver coating.

FIGURE P24.68

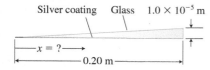

Silver coating   Glass 1.0×10^{-5} m

$x = ?$

0.20 m

69. * **EST** **Resolving car headlights** Estimate the farthest away a car can be at night so that your eyes can resolve the two headlights. Indicate any assumptions you made.

70. **Looking at Moon rocks** You have a home telescope with a 3.0-cm objective lens. Determine the closest distance between two large boulders on the Moon that you can distinguish as separate objects. Indicate any assumptions you made.

71. * **BIO** **EST** **Diffraction-limited resolving power of the eye** You look at closely spaced lines on a wall 5.0 m from your eyes. Estimate the closest the lines can be to each other and still be resolved by your eyes as separate lines. Indicate any assumptions you made in making your estimate.

72. * **Resolving sunspots** You are looking at sunspots. They usually appear in pairs on the surface of the Sun. The Sun is about 1.5×10^{11} m from Earth. How close can two sunspots be so you can distinguish them when you observe them though an amateur telescope whose aperture (objective lens) is about 20 cm? Describe all of the assumptions that you made.

73. ** **The Moon's Mare Imbrium** The outermost ring of mountains surrounding the Mare Imbrium on the Moon has a diameter of 1300 km. What diameter objective lens telescope would allow an astronomer to see the ring of mountains as a distinct feature of the Moon's landscape? What assumptions did you make? The average center-to-center distance from the Earth to the Moon is 384,403 km, which is about 30 times the diameter of the Earth. The Moon has a diameter of 3474 km.

74. * **Can you see atoms with a light-based microscope?** Explain how you can use your knowledge of the wave model of light to explain why you cannot use an optical microscope to see atoms.

75. * **Detecting insects by diffraction of sound** A biologist builds a device to detect and measure the size of insects. The device emits sound waves. If an insect passes through the beam of sound waves, it produces a diffraction pattern on an array of sound detectors behind the insect. What is the lowest-frequency sound that can be used to detect a fly that is about 3 mm in diameter? The speed of sound is 340 m/s.

Reading Passage Problems

BIO **What is 20/20 vision?** Vision is often measured using the Snellen eye chart, devised by Dutch ophthalmologist Herman Snellen in 1862 (see Figure 24.29). With normal vision (20/20 vision), you can distinguish a letter that is 8.8 mm high from other letters of similar height at a distance of 6.1 m (20 ft) (the Snellen chart in the figure is smaller than normal size). If your vision is 20/40, the letters must be twice as high to be distinguishable. Alternatively, a person with 20/40 vision could distinguish letters from 20 ft that a person with 20/20 vision can distinguish at 40 ft. Someone with 20/60 vision could distinguish letters at 20 ft that someone with 20/20 vision could distinguish at 60 ft.

Does the Rayleigh criterion limit visual resolution? Assume that the eye's pupil is 5.0 mm in diameter for 500-nm light. The Rayleigh criterion angular deflection for such light entering the eye's pupil is

$$\alpha = 1.22 \frac{500 \times 10^{-9} \text{ m}}{5.0 \times 10^{-3} \text{ m}} = 1.2 \times 10^{-4} \text{ rad}$$

If the Rayleigh criterion limited visual resolution, then from a distance of 6.1 m, you should be able to distinguish details in shapes of size

$$y = L \tan \alpha = (6.1 \text{ m}) \left[\tan(1.2 \times 10^{-4} \text{ rad}) \right]$$
$$= 0.7 \times 10^{-3} \text{ m} = 0.7 \text{ mm}$$

or about one-tenth the size of the 8.8-mm-tall letter in row 8 (20/20 vision) of the full-size Snellen eye chart. Thus, according to the Rayleigh criterion, we should easily be able to resolve different 8.8-mm-tall letters. Other factors, such as chromatic aberration, irregularities in the cornea-lens-retina shapes, the density of rods and cones in the retina, and air currents, limit visual resolution more than the Rayleigh criterion diffraction limit.

FIGURE 24.29 A Snellen eye chart used to detect visual acuity.

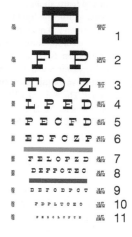

76. Which answer below is closest to the height of the letters that a person with 20/80 vision can distinguish when 20 ft from the wall chart?
    (a) 2.2 mm   (b) 4.4 mm
    (c) 8.8 mm   (d) 18 mm
    (e) 34 mm

77. Suppose that a person with 20/20 vision stands 30 ft from a Snellen eye chart. Which answer below is closest to the minimum height of the letters the person can distinguish?
    (a) 4.4 mm   (b) 6.6 mm
    (c) 8.8 mm   (d) 13 mm
    (e) 18 mm

78. What is the visual acuity of a person with 20/20 vision mainly limited by?
    (a) The Rayleigh criterion
    (b) Chromatic aberration
    (c) Focal length of the eye lens
    (d) Diameter of the eye's pupil
    (e) The density of cones and rods
    (f) A combination of these factors

79. A hawk's vision is said to be 20/5. If so, the hawk can distinguish 8.8-mm-tall letters from about what distance?
    (a) 5 ft   (b) 10 ft
    (c) 40 ft   (d) 80 ft
    (e) 120 ft

80. If the vision of a hawk is 20/5, then what is the angular resolution of the hawk?
    (a) 0.2×10^{-4} rad   (b) 0.5×10^{-4} rad
    (c) 1×10^{-4} rad   (d) 4×10^{-4} rad

Thin-film window coatings for energy conservation and comfort Thin-film coatings are applied to eyeglasses, computer screens, and automobile instrument panels to reduce glare and to binocular and camera lenses to increase the amount of light transmitted. Thin-film coatings are also used to add color to architectural glass, to reduce surface friction, and to improve the energy efficiency of windows. Consider this latter application.

FIGURE 24.30 Over 50% of solar radiation reaching Earth's surface is infrared radiation.

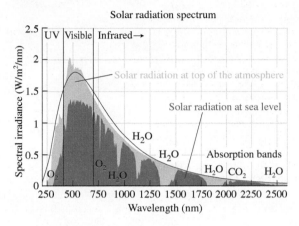

Figure 24.30 shows that over 50% of the radiation reaching Earth from the Sun is long-wavelength infrared radiation, which we perceive as warmth, not light. Thin-film coatings on windows serve two purposes. (1) On warm days, the coatings reflect almost all of the infrared radiation from outside while allowing almost all of the visible light through the window (the glass absorbs most of the ultraviolet). The rooms inside stay cooler and require less air conditioning. (2) On cold days, the same thin-film coating allows visible light to enter the room and reflects the thermal infrared radiation from the interior back into the room, keeping it warm. The reflectance of the thin-film window coating as a function of wavelength is shown in Figure 24.31.

FIGURE 24.31 A thin-film window coating causes reflectance of infrared radiation.

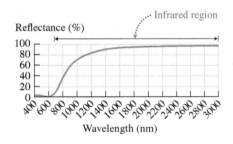

81. Which of the following are benefits of thin-film coatings on windows?
    (a) They keep ultraviolet light out of the house in summer and winter.
    (b) They keep infrared radiation out of the house in the summer.
    (c) They keep infrared radiation in the house in the winter.
    (d) b and c   (e) a, b, and c

82. A 1.25 refractive index thin film on a 1.50 refractive index glass window is made to reflect infrared radiation of wavelength 1000 nm. What is the net reflective phase change of infrared radiation reflected off the front surface of the thin film relative to the radiation reflected off the back surface of the film and returning to the front?
    (a) Zero   (b) 1/4 wavelength
    (c) 1/2 wavelength   (d) None of these

83. A 1.25 refractive index thin film on a 1.50 refractive index glass window is made to reflect infrared radiation of wavelength 1000 nm. What is the wavelength of the infrared radiation while in the thin film?
    (a) 670 nm   (b) 800 nm
    (c) 1000 nm   (d) 1250 nm
    (e) 1500 nm

84. A 1.25 refractive index thin film on a 1.50 refractive index glass window is made to reflect infrared radiation of wavelength 1000 nm. What is the desired thickness of the thin film?
    (a) 200 nm   (b) 250 nm
    (c) 400 nm   (d) 500 nm
    (e) 1000 nm

85. The actual thin film built for windows reflects about what percent of the incident infrared radiation at 1000 nm?
    (a) 10   (b) 30
    (c) 50   (d) 70
    (e) 90

25

Electromagnetic Waves

- How do polarized sunglasses work?
- How can a low-flying plane avoid radar detection?
- How is the screen of your calculator similar to the sky?

The left-hand image above is a regular photograph of the sea and the sky. Next to it is a photograph of the same scene taken with polarized sunglasses oriented in a particular direction in front of the camera lens. Comparing the two photos, you can see differences in the brightness of the sky and in the brightness of the water. But there are other important differences as well. The sailboat that is almost invisible on the left is clearly visible on the right. The same is true for the underwater rocks: in the left photo, the glare of the reflected light masks them almost completely, while on the right, there is no glare. How do the polarized sunglasses cause the right-hand image to be so different?

BE SURE YOU KNOW HOW TO:

- Explain how a capacitor and a transformer work (Sections 18.7 and 21.7).
- Explain how an electric current can be generated without a battery (Section 21.1).
- Describe wave motion using the quantities frequency, speed, wavelength, and intensity (Sections 11.2 and 11.4).

IN THE PREVIOUS THREE CHAPTERS we constructed different models of light to explain its behavior and found that a wave model explained reflection, refraction, and interference. But a mystery remains: what is vibrating in a light wave? We continue to investigate that question in this chapter. We'll resolve the question when we learn about special relativity (in Chapter 26).

25.1 Polarization of waves

In our previous study of mechanical waves, we neglected a phenomenon that is crucial for answering the question posed in the chapter opening. Consider the experiments in Observational Experiment **Table 25.1**.

OBSERVATIONAL EXPERIMENT TABLE 25.1 Rope waves and Slinky waves

VIDEO
OET 25.1

Observational experiment	Analysis
Experiment 1. A rope passes through the open ends of a narrow rectangular box. Shake the rope in a vertical plane, producing a transverse wave. The long sides of the box are parallel to the shaking direction. The rope wave is unaffected by the box. 	Vectors represent the displacement of each section of rope at one instant. They are parallel to the long sides of the box. 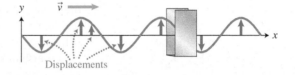
Experiment 2. Rotate the slotted box 90° with respect to the original orientation. Shake the rope the same way as in Experiment 1. The long sides of the box are perpendicular to the shaking direction. The wave does not pass through the box. 	The displacement vectors of the rope sections are perpendicular to the long sides of the box. The box exerts forces on the rope in the direction opposite to the motion of the parts of the rope inside the box, therefore decreasing the displacement of the parts of the rope. 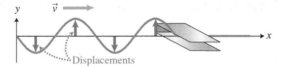
Experiment 3. Now use two slotted boxes. Rotate the first box 30° from the vertical plane. Orient box 2 perpendicular to box 1. Shake the rope as in Experiment 1. Part of the wave travels through box 1 but is eliminated by box 2. 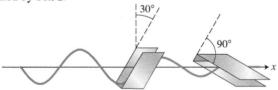	The component of the rope wave parallel to the first box travels through the first box. Once the rope wave reaches the second box, its displacement vectors are perpendicular to the box. 
Experiment 4. We achieve the same results as in Experiments 1–3 if we replace the rope with a Slinky and produce a transverse wave. However, if we compress and decompress the Slinky horizontally, producing a longitudinal wave, the wave passes through the boxes in all of the experiments. 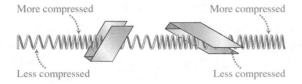	The longitudinal displacement vectors of the Slinky coil always pass through the boxes. They point along the direction of the Slinky. 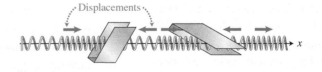

Patterns

- If the displacement vectors of a transverse wave are parallel to the slit (the box), the wave passes through undisturbed.
- If the displacement vectors are in the plane perpendicular to the slit, the wave does not pass through.
- When the displacement vectors make an angle with the slit, only the component of the vectors parallel to the slit passes through.
- The displacement vectors of a longitudinal wave always pass through the slit. Longitudinal waves are not sensitive to the orientation of the slit.

We learned in Table 25.1 that slotted boxes with the proper orientation can block transverse mechanical waves but cannot block longitudinal waves. This phenomenon is referred to as **polarization of waves**—polarization is a property of waves that describes the orientation of the oscillations of the wave. As we found in Table 25.1, only transverse waves have this property. Thus, if we observe polarization of light waves, we know that the waves are transverse.

In a **linearly polarized** mechanical wave, the individual particles of the vibrating medium vibrate along only one axis that is perpendicular to the direction the wave travels (there are other types of polarization, but we will confine our discussion to linear polarization). In our experiments, the vibrating medium was the rope. The slotted box is called a **polarizer**. A polarizer is a device that allows only a single component of transverse waves to pass through it. The component that can pass through defines the **axis** of the polarizer.

The waves in Experiments 1–3 in Table 25.1 were linearly polarized. An **unpolarized** wave is one in which the particles of the medium vibrate in all possible directions in the plane perpendicular to the direction the wave is traveling (**Figure 25.1a**), often caused by a collection of differently polarized waves. If an unpolarized wave is incident on a polarizer, the wave that emerges from the other side is linearly polarized (Figure 25.1b). A second polarizer whose axis is perpendicular to the axis of the first polarizer blocks the wave completely. Thus, a linearly polarized wave can be reduced to zero by a polarizer whose axis is perpendicular to the polarization direction of the wave. An unpolarized wave is completely blocked by two successive perpendicularly oriented polarizers.

Are light waves transverse or longitudinal? According to our observations in Table 25.1, if we can polarize light waves, they must be transverse.

We can test this hypothesis using a semiprecious crystal called tourmaline, which affects light in the same way that our slotted box affected transverse rope waves. We can compare light from a bulb (**Figure 25.2a**) to the same light after it passes through the tourmaline crystal; the light looks dimmer. Rotating the crystal does not change the intensity of the light. However, when we place a second crystal at a 45° angle to the first, we observe that the intensity of light decreases (Figure 25.2b). When the second crystal is placed perpendicular to the first, the intensity of the light reduces to zero (Figure 25.2c). To explain this experiment, we hypothesize that light is a transverse wave and that a lightbulb emits unpolarized light waves. We also hypothesize that the tourmaline crystal allows only one component of the wave to pass through the first crystal. That component is completely blocked by the second crystal.

FIGURE 25.1 The effect of polarizers on a polarized wave.

(a) The particles in an unpolarized wave vibrate in all directions in the plane perpendicular to the direction the wave travels.

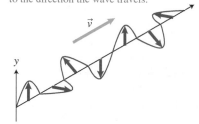

(b)

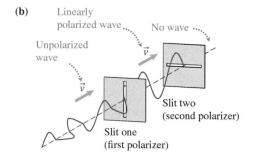

FIGURE 25.2 The polarizing effects on light of tourmaline crystal.

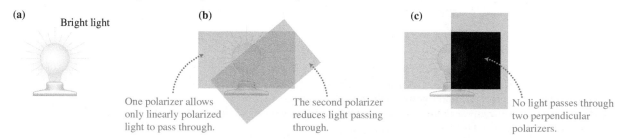

(a) Bright light

(b) One polarizer allows only linearly polarized light to pass through. The second polarizer reduces light passing through.

(c) No light passes through two perpendicular polarizers.

Most commonly used polarizers are made of a material that contains elongated molecules that serve as slits for passing light. When these molecules are oriented parallel to each other, their combined effect leads to polarization of passing light. The lenses of polarized sunglasses, for instance, are covered with a film containing such a material. Physics teaching labs often have "polarizers," sheets of plastic that are translucent and thus dark like sunglasses. Light seen through these polarizers behaves similarly to Figure 25.2. Based on these observations we can say that light's behavior is so similar to the behavior of transverse mechanical waves passing through mechanical polarizers that it supports the hypothesis that light waves are transverse waves.

Quantitative description of the effect of polarizers

Observational Experiment Table 25.2 investigates the amount of dimming caused by one or two polarizers oriented at different angles relative to each other.

OBSERVATIONAL EXPERIMENT TABLE 25.2   **Effect of polarizers on light intensity**

VIDEO
OET 25.2

Observational experiment	Analysis
We place a light meter a fixed distance from a lightbulb (a source of unpolarized light).	Light intensity $= I_0$

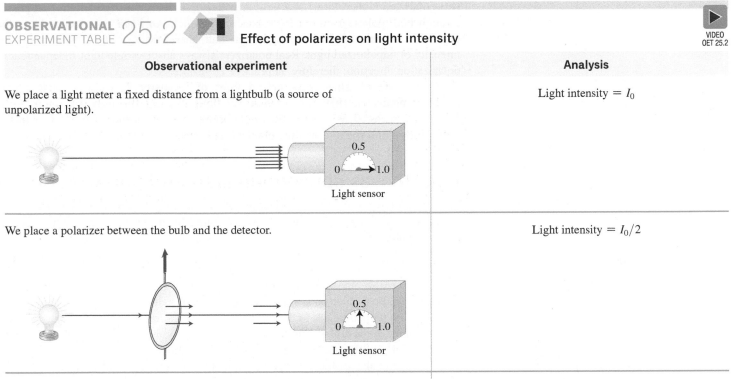

We place a polarizer between the bulb and the detector.

Light intensity $= I_0/2$

We place two polarizers between the bulb and the detector, with varying angles between their axes.

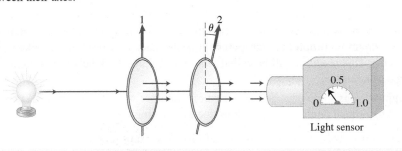

Angle	Light intensity
0°	$I_0/2$
30°	$0.75\, I_0/2$
45°	$0.50\, I_0/2$
60°	$0.25\, I_0/2$
90°	0

Patterns

- One polarizer reduces the light intensity by half: $I = (I_0/2)$.
- The second polarizer seems to further reduce the intensity depending on the angle between the polarizers. We need to determine the relation between this reduction and the angle.

The patterns in Table 25.2 indicate that the intensity of light passing through the second polarizer decreases as some function of the angle between the axes of the first and second polarizer. Looking at the experiments with waves on a rope (Table 25.1), we realize that the box polarizer allows the component of the wave amplitude that vibrates in the direction of the long side of the box to pass through the box. If the amplitude of the incident wave is A, the component that will pass through the box is $A \cos\theta$, where θ is the angle between the direction of vibration of incident waves and the direction of the long side of the box.

Can we apply this to light? Looking at the patterns in Table 25.2, we see that the dependence follows a $\cos^2 \theta$ function rather than a $\cos \theta$ function ($\cos 30° = 0.86$, but $\cos^2 30° = (0.86)^2 = 0.75$). How can we explain this? We will see later in this chapter that the intensity of light is proportional to the amplitude squared of the quantity that vibrates. What we observe and measure is the light intensity and not the amplitude. Thus we accept $\cos^2 \theta$ as a function of the angle that we are looking for, where θ is the angle between the axes of the two polarizers. The intensity of light that passes through two polarizers is therefore $I = I' \cos^2 \theta$, where I' is the intensity of light that passes through both polarizers when their axes are parallel. An *ideal* polarizer reduces the unpolarized incident light intensity by one-half; therefore, $I' = I_0/2$, where I_0 is the intensity of unpolarized light. Real polarizers always absorb some light independent of polarization direction; therefore, in practice $I' < I_0/2$.

If a beam of linearly polarized light of intensity $I_{\text{polarized}}$ is incident on an ideal polarizer, then the intensity of light passing through the polarizer is $I = I_{\text{polarized}} \cos^2 \theta$, where θ is the angle between the direction of the light's initial polarization and the axis of the polarizer. This statement is called *Malus's law*.

Hypotheses concerning light and polarization

In Table 25.1 we found that polarization effects can only be observed with transverse waves and not longitudinal waves. We just found from the light polarization experiments in Table 25.2 that light seems to behave like a transverse wave with amplitude A and intensity $I \propto A^2$. However, since light can travel through a vacuum (for example, when it travels from the Sun to Earth), what is actually vibrating in a light wave?

To answer this question, we need to look carefully at how a transverse wave travels through a medium, such as a Slinky. When one coil is displaced, it pulls the next coil in a direction perpendicular to the direction of propagation of the wave. That coil then does the same to the coil next to it, and so on. The individual vibrating coils (analogous to the particles of the medium) exert elastic forces on each other. These elastic forces point perpendicular to the propagation direction of the wave and accelerate the displaced coils back toward equilibrium.

However, light can travel through a vacuum—a medium with no particles at all. How then can it be a transverse wave? Here are two hypotheses.

1. Light is actually a mechanical vibration that travels through an elastic medium. This medium is completely transparent and has exactly zero mass (just like the vacuum). This medium will be called *ether* (not related to the chemical compound of the same name).

2. A light wave is some new type of vibration that does not involve physical particles vibrating around equilibrium positions due to restoring forces being exerted on them.

The next section will show how studies of electricity and magnetism helped test the second hypothesis.

REVIEW QUESTION 25.1 What is the difference between a linearly polarized mechanical wave and an unpolarized one?

25.2 Discovery of electromagnetic waves

Before the second half of the 19th century, the investigations of light and electromagnetic phenomena proceeded independently. However, around 1860 the work of a British physicist, James Clerk Maxwell (1831–1879), led to the unification of those phenomena and helped finally answer the question of how a transverse wave can propagate in a vacuum. We know from our study of electromagnetic induction (Chapter 21) that Michael Faraday introduced the concept of a field and the relationship between electric and magnetic fields. According to Faraday, *a changing magnetic field can produce an electric field.* Subsequently, in 1865 Maxwell suggested a new field relationship: *a changing electric field can produce a magnetic field.* This idea was motivated by a thought experiment devised by Maxwell in which he imagined what would happen in the space between the plates of a charging or discharging capacitor. He suggested that the changing electric field between the capacitor plates could be viewed as a special nonphysical current, but one that would still produce a magnetic field (**Figure 25.3**). This magnetic field was first detected in 1929 but not measured precisely until 1985 due to its extremely tiny magnitude. Maxwell summarized this new idea and other electric and magnetic field ideas mathematically in a set of four equations, now known as Maxwell's equations. The equations are written using calculus, but we can summarize them conceptually:

1. Stationary electric charges produce a constant electric field. The $\vec{E}$ field lines representing this electric field start on positive changes and end on negative charges.

2. There are no individual magnetic charges (magnetic monopoles).

3. A magnetic field is produced either by electric currents or by a changing electric field (Figures 25.3 and **25.4a**). The $\vec{B}$ field lines that represent the magnetic field form closed loops and have no beginnings or ends.

4. A changing magnetic field produces an electric field. The $\vec{E}$ field lines representing this electric field are closed loops (Figure 25.4b).

Producing an electromagnetic wave

Maxwell's equations had important consequences. First, the equations led to an understanding that a changing electric field can produce a changing magnetic field, which in turn can produce a changing electric field, and on and on in a sort of feedback loop (**Figure 25.5**). This feedback loop does not require the presence of any electric charges or currents. Maxwell investigated this idea mathematically using the four equations, and to his surprise he found they led to a wave equation similar to Eq. (11.4) in which the electric and magnetic fields themselves were vibrating. The speed of propagation of these waves in a vacuum turned out to be a combination of two familiar constants: $v = (1/\sqrt{\epsilon_0\mu_0})$, where the constants $\epsilon_0 = 8.85 \times 10^{-12}\,\mathrm{C^2/N \cdot m^2}$ and $\mu_0 = 4\pi \times 10^{-7}\,\mathrm{N/A^2}$ relate to the electric and magnetic interactions of electrically charged particles in a vacuum. The constant ϵ_0 is the **vacuum permittivity** and is related to Coulomb's constant k_C through the relationship

$$\epsilon_0 = \frac{1}{4\pi k_C}$$

(See Chapter 17 for more on vacuum permittivity.) The constant μ_0 is the **vacuum permeability**. We discussed vacuum permeability in the chapter on magnetism (Chapter 20). When Maxwell inserted the values of the constants into the expression for the speed of electromagnetic waves, he obtained

$$v = \frac{1}{\sqrt{\epsilon_0\mu_0}} = \frac{1}{\sqrt{(8.85 \times 10^{-12}\,\mathrm{C^2/N \cdot m^2})(4\pi \times 10^{-7}\,\mathrm{N/A^2})}}$$

$$= \sqrt{9.00 \times 10^{16}\,\frac{\mathrm{N \cdot m^2 \cdot A^2}}{\mathrm{C^2 \cdot N}}} = \sqrt{9.00 \times 10^{16}\,\frac{\mathrm{m^2 \cdot A^2}}{\mathrm{A^2 \cdot s^2}}} = 3.00 \times 10^8\,\mathrm{m/s}$$

At the time Maxwell did this calculation, the speed of light in air had already been measured and was consistent with this value. Could it be that light is an electromagnetic wave? This was the second testable consequence of Maxwell's model. *If* the relationship between the changing electric and magnetic fields suggested by the model is correct, *then* a change in either of these fields could generate an electromagnetic wave

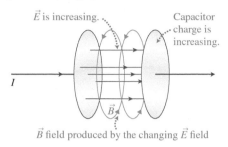

FIGURE 25.3 A changing electric field produces a magnetic field.

$\vec{E}$ is increasing.

Capacitor charge is increasing.

$\vec{B}$ field produced by the changing $\vec{E}$ field

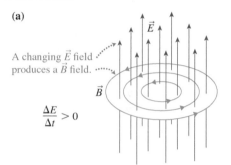

FIGURE 25.4 Changing $\vec{E}$ and $\vec{B}$ fields produce $\vec{B}$ and $\vec{E}$ fields.

(a)

$\vec{E}$

A changing $\vec{E}$ field produces a $\vec{B}$ field.

$\vec{B}$

$\dfrac{\Delta E}{\Delta t} > 0$

(b)

$\vec{B}$

A changing $\vec{B}$ field produces an $\vec{E}$ field.

$\vec{E}$

$\dfrac{\Delta B}{\Delta t} > 0$

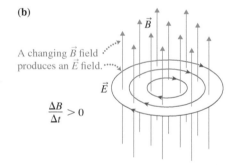

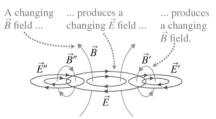

FIGURE 25.5 The changing $\vec{B}$ and $\vec{E}$ fields can spread without any charges or currents.

A changing $\vec{B}$ field ...   ... produces a changing $\vec{E}$ field ...   ... produces a changing $\vec{B}$ field.

$\vec{E}''$ $\vec{B}''$ $\vec{B}$ $\vec{B}'$ $\vec{E}'$

$\vec{E}$

The $\vec{B}$ and $\vec{E}$ fields propagate themselves.

that would propagate at the speed of light. This prediction, if consistent with the experiment, would support the hypothesis that light is an electromagnetic wave composed of vibrating electric and magnetic fields!

Testing the hypothesis that light can be modeled as an electromagnetic wave

The German physicist Heinrich Hertz (1857–1894) was the first to test the hypothesis. In 1888 Hertz built a device called a **spark gap transmitter** and used it in his experiments. His work is summarized in Testing Experiment **Table 25.3**.

TESTING
EXPERIMENT TABLE 25.3   Hertz's experiments

Testing experiment	Prediction	Outcome
Hertz used a switch to connect a transmitter with a charged capacitor to the primary coil of a transformer. He connected the secondary coil of the transformer to two metal spheres. When he discharged the capacitor, the changing current through the primary coil induced a huge emf across the secondary coil, which in turn charged the spheres, causing them to spark. A receiver that consisted of two small metal spheres separated by a small gap and connected to a loop of wire was placed some distance from the metal transmitter spheres.	A spark between the transmitter spheres would indicate a large changing electric field between the spheres of the transmitter. If Maxwell's hypothesis was correct, then this changing electric field would produce an electromagnetic wave. When the wave reached the receiver loop, it would induce a current in the loop and cause a weak spark between the spheres of the receiver.	After hours in a dark room watching for a spark, Hertz eventually saw a spark between the receiver spheres.

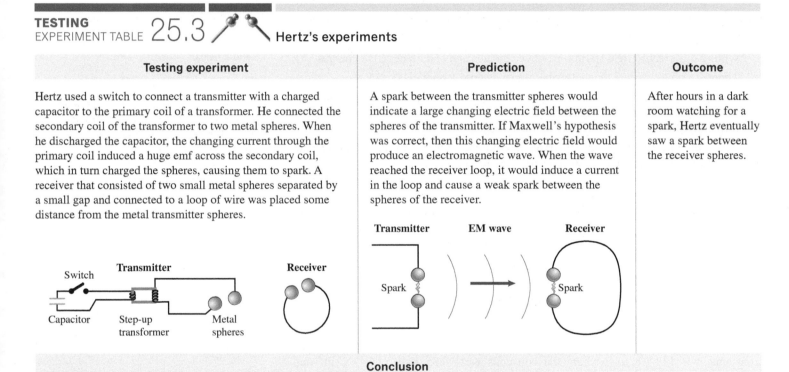

Conclusion

That Hertz could see a spark between the spheres of the receiver in response to the spark between the spheres of the transmitter meant that something had traveled from the transmitter to the receiver.

Hertz's experiment was the first supporting evidence for the idea that electromagnetic waves existed. He performed many additional experiments to determine if the electromagnetic waves generated in his experiment had the same properties as light waves. The only difference he found was in their frequency—Hertz's waves had a much smaller frequency than visible light waves. Hertz used metal sheets of different shapes to observe reflection. He let the waves pass through different media and observed refraction. He performed an analog of a double-slit experiment and observed interference. He even observed polarization of the waves using a metal fence. Finally, he performed experiments to measure the speed of the propagation of the waves, which turned out to be 3.00×10^8 m/s. These experiments supported the idea that light could be modeled as a transverse wave of vibrating electric and magnetic fields.

Maxwell's model of light as an electromagnetic wave seems to resolve the problem of what is vibrating in the light wave: the $\vec{E}$ field and the $\vec{B}$ field. Each field is vibrating perpendicular to the direction of travel of the wave. Thus we have a new model of light as a *transverse electromagnetic wave* that travels in a vacuum and in other media. In a vacuum, the speed of light is 3.00×10^8 m/s.

The waves predicted by Maxwell and experimentally discovered by Hertz soon found a practical application. In 1892 Nikola Tesla used an improved version of Hertz's transmitter to conduct the first transmission of information via radio waves, one form of electromagnetic waves. By 1899 there was successful radio wave-based communication across the English Channel.

You might be curious about what happened to the idea of ether. Although it was no longer needed to explain the nature of light, it still remained in physics until 1905. We will return to the fate of ether in Chapter 26.

Antennas are used to start electromagnetic waves

The most common device used to produce electromagnetic waves is an antenna. Cell phones, two-way radios, and over-the-air radio and television stations all use them. A simple type of antenna, called a **half-wave electric dipole antenna**, can be made from a pair of electrical conductors, one connected to each terminal of a power supply that is producing an alternating emf (**Figure 25.6**). The alternating emf leads to the continuous charging and discharging of the two ends of the antenna (somewhat similar to the action of a capacitor). How does this antenna produce an electromagnetic wave?

The source of alternating emf produces alternating current. Let the period of this oscillation be T. Assume that just after the power supply is turned on, the current is downward. This produces a $\vec{B}$ field ($t = 0$). After a very short time interval (nanoseconds for radio waves), this current causes the bottom of the antenna to become increasingly positively charged and the top increasingly negatively charged. This produces an $\vec{E}$ field ($t = T/8$). Both fields are shown in **Figure 25.7a**. Since the antenna is connected to a source of alternating emf, the current reverses direction periodically.

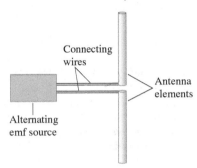

FIGURE 25.6 A half-wave dipole antenna.

FIGURE 25.7 The electric and magnetic fields produced by a dipole antenna.

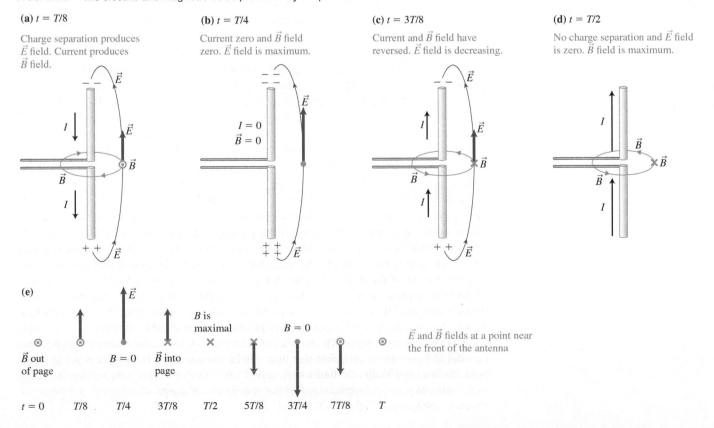

(a) $t = T/8$

Charge separation produces $\vec{E}$ field. Current produces $\vec{B}$ field.

(b) $t = T/4$

Current zero and $\vec{B}$ field zero. $\vec{E}$ field is maximum.

(c) $t = 3T/8$

Current and $\vec{B}$ field have reversed. $\vec{E}$ field is decreasing.

(d) $t = T/2$

No charge separation and $\vec{E}$ field is zero. $\vec{B}$ field is maximum.

(e)

$\vec{E}$ and $\vec{B}$ fields at a point near the front of the antenna

When this happens, the current momentarily vanishes along with its corresponding $\vec{B}$ field (Figure 25.7b, $t = T/4$). When the current starts upward, the charge separation begins to decrease along with the $\vec{E}$ field produced by it (Figure 25.7c, $t = 3T/8$). Next, the charge separation vanishes along with its $\vec{E}$ field (Figure 25.7d, $t = T/2$). The process then repeats in the opposite direction and continues to repeat billions of times each second. Figure 25.7e shows what happens at one point in space during one period T of oscillation. The $\vec{E}$ and $\vec{B}$ fields at that point are shown in sequence as time passes. Note that the second part in (e) corresponds to the fields shown in (a). The third part in (e) corresponds to the fields in (b), and so forth.

Near the antenna, the oscillating electric charge and current in the antenna produce the $\vec{E}$ and $\vec{B}$ fields as described above (near fields). Farther from the antenna, the fields co-create each other in the way Maxwell's equations describe (far fields). There is a noticeable difference between these two situations. In Maxwell's electromagnetic (EM) waves, the $\vec{E}$ field and $\vec{B}$ field oscillate in phase. That is, when the $\vec{E}$ field at a point is at a maximum, so is the $\vec{B}$ field. Close to the antenna, the $\vec{E}$ field and $\vec{B}$ field are out of phase, the $\vec{E}$ field being at a maximum while the $\vec{B}$ field is zero, and vice versa.

Both near and far $\vec{E}$ and $\vec{B}$ fields contribute to the net $\vec{E}$ and $\vec{B}$ fields at a particular distance from the antenna: (1) the fields produced by the oscillating charges and current in the antenna and (2) the electromagnetic waves produced by the oscillating $\vec{E}$ field and $\vec{B}$ field in the region surrounding the antenna. The amplitudes of the $\vec{E}$ field and $\vec{B}$ field produced by contribution 1 get weaker with distance ($1/r^2$) much faster than the amplitude of the EM waves produced by contribution 2 ($1/r$). In other words, the farther from the antenna, the less significant the out-of-phase contribution compared to the in-phase contribution. Far from the antenna, only Maxwell's in-phase EM waves are significant.

The size and shape of antennas are carefully selected so that they can efficiently send and receive EM waves with a particular range of wavelengths. In the case of the dipole antenna shown in Figure 25.6, the total length of the antenna is related to the desired wavelength of EM waves that it will produce. It is called a *half-wave* dipole antenna because its length is half the wavelength of the oscillating electric field within the antenna that drives the current. However, since this electric field exists in a different material (the metal antenna), its wavelength isn't the same as the wavelength of the EM wave produced by the antenna (those fields are in air), but they are related. Recall from our study of refraction (Chapter 22) that different materials have different indexes of refraction, which affects the wavelength of the EM waves passing through them. Knowing the index of refraction of the antenna material allows us to choose an antenna length that will very efficiently produce EM waves of the desired wavelengths. For example, most cell phones operate at frequencies between 850 and 1900 MHz (called a *frequency band*), corresponding to wavelengths of about 10–30 cm. Within the antenna the wavelength is about one-third of its value in air. As a result, cell phone antennas need to be roughly 3–10 cm in length, consistent with the size of modern cell phones.

"Reflecting" on models of light

Let us review how our understanding of light has evolved over the past four chapters. First we learned about shadows and the phenomena of reflection and refraction. To explain these, we used a particle-bullet model, which assumed that light consisted of tiny bullets that travel in straight lines through a medium. The model did not specify the nature of these bullets. Then we found new phenomena that the bullet model could not explain, such as what happens when light passes through narrow openings. So we progressed to a wave model, but it was not clear what was vibrating in a light wave. In this chapter, observations of polarization of light led us to the idea that a light wave is a transverse wave. Maxwell's and Hertz's work led to the electromagnetic wave model of light, which proposes that light can be modeled as a transverse wave in which the electric and magnetic fields themselves are vibrating. The next section will give examples of practical applications of the new model of light.

25.3 Applications of electromagnetic waves

We use electromagnetic waves every day: they are present when a police officer uses radar to catch speeding motorists, when we use GPS to find a new destination, and when we use a microwave to cook food.

Radar

The EM waves produced by Hertz in his experiments are called **radio waves**, based on their frequency (10^4–10^6 Hz). Today, these and even higher frequency waves are used to transmit speech, music, and video over long distances. Another practical application of radio waves is radar (an acronym of radio detection and ranging). **Radar** is a way of determining the distance to a faraway object by reflecting radio wave pulses off the object. Radar devices help locate airplanes, determine distances to planets and other objects in the solar system, find schools of fish in the ocean, and perform many other useful functions.

In radar, an antenna (called the **emitter**) emits a short radio wave pulse, which reflects off the surface of the object and is detected by a **receiver**. This receiver is just an antenna that is connected to a sophisticated type of ammeter that can measure the current produced when the reflected electromagnetic wave arrives. Most often, the emitting antenna and the receiving antenna are actually the same antenna.

Imagine that a radar system is being used to detect aircraft. The time interval for the round trip of the pulse indicates the distance to the aircraft. The longer it takes the pulse to return, the farther away the aircraft is. By emitting a sequence of pulses and determining how the distance to and direction of the aircraft changes from one pulse to the next, the system can determine the velocity of the aircraft.

Radar has some limitations. It is less effective in tracking objects that are very close, very far away, or flying very low to the ground. The difficulty of tracking close objects is related to the time interval τ that it takes a single pulse to be emitted (called the *pulse width*) and the time interval T that passes between the start of each successive pulse (called the *period* of the pulse; **Figure 25.8**). If the object is so close that the end of the pulse has not been emitted before the beginning of the pulse returns, then the distance to the object is difficult to determine accurately. This limitation has been alleviated somewhat by advanced electronic switching systems that produce very short pulse widths.

The source of the second limitation, measuring distant objects, is the time interval T. If the object is so far away that the reflected pulse returns after the next pulse is sent, then it becomes unclear which reflected pulse corresponds to which emitted pulse. In modern radar the period is between $T = 10^{-3}$ s and $T = 10^{-6}$ s. The overall effect of these two difficulties is that there is both a minimum and a maximum range within which an object can be accurately detected, determined by the pulse width τ and the period T.

The third limitation, tracking objects that are flying close to the ground, is a result of Earth's curved surface. Because of the curvature, the distance to the horizon is about 5 km as seen by a person standing on the ground. Thus, a target that is farther away than 5 km will not be visible to radar because the ground will block the line of sight. In order to obtain a range of 20 km, the radar system must be mounted on a tall tower in order to increase the horizon distance.

FIGURE 25.8 The radar pulse width τ and the time interval T between pulses determine the near and far distances at which a radar system can detect an airplane.

Snapshot of radar pulses at one instant

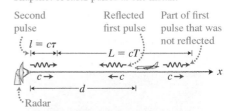

The reflected first pulse returns after the new second pulse has left.

EXAMPLE 25.1 **Radar**

Determine the range of distances to a target that can be reliably detected by a radar system with pulse width $\tau = 10.8\ \mu s$ ($1\ \mu s = 10^{-6}$ s) and a pulse emission frequency of 7.5 kHz, that is, 7500 radio wave pulses each second.

Sketch and translate The pulse width of 10.8 μs should be less than the time interval needed for the EM wave to travel to the target object and back. That way, the back of the 10.8-μs pulse will have left the transmitter before the front of the reflected pulse returns. The pulse width determines the minimum distance of the target from the radar.

The pulse repetition period is $T = 1/f = 1/(7500\ \text{Hz}) = 1.33 \times 10^{-4}$ s. This period should be more than the time interval needed for the pulse to travel to the target and back. That way, the reflected pulse returns before a new pulse leaves. The period determines the maximum distance of the target from the radar.

Simplify and diagram Assume that pulses do not reflect off other objects while traveling to or from a target. Assume also that the radio wave pulses travel at the speed of light in vacuum (they actually travel just slightly slower in air).

Represent mathematically If the target is a distance d away, the time interval for a pulse to travel to the target and back is $\Delta t = (2d/c)$. This time interval needs to be longer than the pulse width and shorter than the pulse period in order for the radar to accurately determine the distance to the target. Mathematically,

$$\tau < \frac{2d}{c} < T$$

Solve and evaluate The pulse width time interval $\tau = 10.8\ \mu s = 10.8 \times 10^{-6}$ s and the period $T = 1.33 \times 10^{-4}$ s. Solving the above inequality for d, we get

$$\frac{c\tau}{2} < d < \frac{cT}{2}$$

The range of distances is

$$\frac{(3.00 \times 10^8\ \text{m/s})(10.8 \times 10^{-6}\ \text{s})}{2}$$
$$< d < \frac{(3.00 \times 10^8\ \text{m/s})(1.33 \times 10^{-4}\ \text{s})}{2}$$

or

$$1.62 \times 10^3\ \text{m} < d < 20.0 \times 10^3\ \text{m}$$

A target outside this 1.6-km to 20-km range will not be detectable by this radar system. However, military radars have significantly shorter pulse widths and can detect objects much closer than 1.6 km.

Try it yourself What is the maximum range of a 15-kHz pulse emission rate radar system mounted on a tall tower?

Answer   10 km.

Global Positioning System

Global positioning systems enable us to determine our location using electromagnetic waves and satellites. In the United States, the Global Positioning System (GPS) is operated by the U.S. Air Force, and its network of satellites is called Navstar. GPS has three main components:

FIGURE 25.9 Twenty-four satellites circle Earth as part of the Global Positioning System. Three or four GPS satellites trilaterate to determine your location.

1. A minimum of 24 satellites that orbit 20,200 km above Earth's surface and transmit signals using **microwaves**: electromagnetic waves at frequencies 1575 MHz and 1227 MHz ($1\ \text{MHz} = 10^6\ \text{Hz}$) (**Figure 25.9**).

2. Receivers that detect signals from multiple satellites.

3. A control system that maintains accurate information about the locations of the satellites.

The receiver detects signals from at least three satellites in order to determine your position on the ground. With signals from four or more satellites, the receiver can also determine your altitude. The clocks on the satellites and a clock in your unit are synchronized to within a nanosecond (10^{-9} s). When you turn on your GPS, it starts a measurement at the same time each satellite sends a microwave signal. Your unit measures the time of arrival of the signals from the satellites. Using the known positions of the satellites, your GPS unit is able to calculate your position by a process called **trilateration**.

Trilateration is easier to understand in a two-dimensional space than in a three-dimensional space. Suppose a visitor from overseas is traveling in the United States and is unsure of his location. He comes across a sign that says Atlanta 430 miles, Minneapolis 510 miles, and Philadelphia 580 miles. The visitor has a map of the

United States and draws a circle centered on each city such that the radius of each circle is the distance to that city (**Figure 25.10**). There is only one place on the map where the three circles intersect—Indianapolis.

Your GPS system locates your current position in much the same way. The unit determines the distance between you and each of at least three satellites. The unit then constructs spheres (rather than circles) around each of them with corresponding radii equal to those distances. The spheres intersect at one unique location on Earth's surface (which is in effect a fourth sphere), your location. Most handheld GPS units can determine your position to within 3 to 10 m. More advanced receivers use a method called Differential GPS (DGPS) to obtain an accuracy of 1 m or better.

Microwave cooking

Percy Spencer accidentally discovered microwave cooking in 1945 while constructing magnetron tubes at Raytheon Corporation. While standing in front of one of the tubes, Spencer felt a candy bar in his pocket start to melt. Intrigued, he placed kernels of popcorn in front of the tube, and they popped. He placed a raw egg in front of the tube, and it exploded. How could this be explained?

Water, fat, and other substances in food absorb energy from microwaves in a process called **dielectric heating**. To understand what's happening, we have to think microscopically. Water is a polar molecule (**Figure 25.11**) and is a permanent electric dipole. When a microwave passes by a water molecule, the electric field of the microwave exerts oscillating electric forces on the positively and negatively charged regions of the molecule. This causes the water molecule to flip over billions of times each second (the frequency of microwaves). This molecular movement spreads to neighboring molecules through collisions (even to molecules that are not electric dipoles). The electric and magnetic energy in the microwaves is transformed into the internal energy of the food, which manifests as an increase in the food's temperature. Microwave heating is most efficient in liquid water and much less so in fats and sugars (which have smaller electric dipoles) and ice (where the water molecules are not as free to rotate). Substances that are composed of nonpolar molecules do not warm up (or warm up much less) in microwave ovens. Mineral oil (which you can find in a drugstore), for instance, barely warms up at all in a microwave.

REVIEW QUESTION 25.3 How are GPS and radar similar, and how are they different?

FIGURE 25.10 Surface trilateration is analogous to that used in GPS.

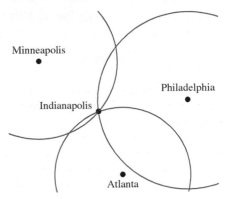

Only one place (Indianapolis) is the known distance from the other three cities.

FIGURE 25.11 A water molecule is an electric dipole with more positive charge on one end and negative charge on the other.

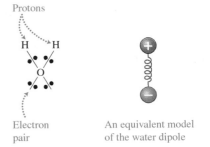

Electron pair

An equivalent model of the water dipole

25.4 Frequency, wavelength, and the electromagnetic spectrum

Recall that waves are described by frequency f, speed v, wavelength λ, and a relationship between these three quantities [Eq. (11.3)]:

$$\lambda = \frac{v}{f}$$

All electromagnetic waves travel at the speed of light c in a vacuum. However, in media other than a vacuum, the speed of electromagnetic waves changes to $v = c/n$, where n is the refractive index of the medium. In addition, light waves of different frequencies have slightly different refractive indexes and hence travel at different speeds in non-vacuum media such as glass or water. Therefore, in the equation $v = c/n$, n is not a constant for a particular medium, but has a small dependence on the frequency of the electromagnetic wave.

According to Maxwell's equations, electromagnetic waves can have essentially any frequency. AM radio stations emit relatively low-frequency EM waves (about 10^6 Hz = 1 MHz). Supernova remnants and extragalactic sources emit the highest frequency cosmic ray EM waves ever detected (up to about 10^{26} Hz).

QUANTITATIVE EXERCISE 25.2   **Wavelengths of FM radio stations**

The FM radio broadcast band is from 88 MHz to 108 MHz. What is the corresponding wavelength range?

Represent mathematically Assume that these radio waves are traveling at the speed of light in a vacuum. This is a reasonable assumption since the refractive index of air is very close to the refractive index of a vacuum (both are ≈ 1). Apply $\lambda = c/f$ to find the range of wavelengths for FM radio waves.

Solve and evaluate

$$\lambda_1 = \frac{3.0 \times 10^8 \text{ m/s}}{88 \times 10^6 \text{ s}^{-1}} = 3.4 \text{ m}$$

$$\lambda_2 = \frac{3.0 \times 10^8 \text{ m/s}}{108 \times 10^6 \text{ s}^{-1}} = 2.8 \text{ m}$$

The electromagnetic waves emitted by FM radio stations have wavelengths ranging from 2.8 m to 3.4 m.

- -

Try it yourself AM radio stations broadcast in the wavelength range from 556 m to 186 m. Determine the frequency range of AM radio station broadcasting.

Answer   540 kHz to 1610 kHz.

"Hearing" FM radio waves

Human ears are not sensitive to electromagnetic waves of any frequency. Thus, for us to hear music on the radio, the electromagnetic waves need to be converted into sound waves. This is the role of a speaker or headphones. However, even if we convert an electromagnetic signal into a sound wave, our ears can only hear sounds with frequencies ranging from about 20 Hz to 20,000 Hz, much lower than the frequency of FM or satellite radio stations. How can we hear the music from those stations? The high-frequency EM waves used by FM radio stations are known as *carrier waves*. FM stands for "frequency modulation"; the information that is converted into the sounds we hear is encoded as tiny variations in the frequency of the carrier wave. The receiver decodes the variations and converts them into an electric signal that a speaker can then convert into sound. A similar process occurs in AM (amplitude modulation) radios, only in that case the amplitude of the carrier wave is varied rather than its frequency.

The electromagnetic spectrum

The range of frequencies and wavelengths of electromagnetic waves is called the **electromagnetic spectrum** (**Figure 25.12**). As you see, visible light occupies a very narrow range of the spectrum (wavelengths of 400 nm to 700 nm). In this chapter we've already encountered radio waves and microwaves. We encountered another type of radiation briefly (in Chapter 15) when we considered heating by radiation. Historically, scientists thought that visible light and thermal radiation (now called infrared [IR] radiation) were completely different. In other words, they thought that the light emitted by a fire in a fireplace was a different phenomenon than the warmth that the fire gave off. We now understand that they are both EM waves, just with different frequency ranges.

FIGURE 25.12 The electromagnetic (EM) spectrum.

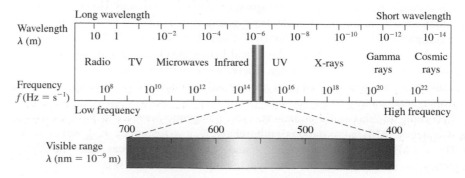

Table 25.4 lists the broad categories of EM radiation and some of the ways in which it is produced and detected. We will discuss EM radiation production in more detail in later chapters.

TABLE 25.4 **The electromagnetic spectrum**

Type of EM radiation and frequency/ wavelength range	Produced by	Detected by	Used in
Radio waves and microwaves 10^4–10^{10} Hz 10^3–10^{-3} m	Accelerated electrically charged particles in wires with oscillating current	Metal wires (antennas)	Radio and cell phone communication Microwave ovens
Infrared waves 10^{11}–10^{14} Hz 10^{-4}–10^{-5} m	Objects with a temperature between 10 K and 1000 K	Detectors whose properties (for example, electric conductivity) change with temperature	Toasters Conventional ovens
Visible light 10^{14}–10^{15} Hz 10^{-6} m	Low-density ionized gas Objects whose temperatures are between 1000 K and 10,000 K, such as typical stars	Human eyes Charge-coupled devices (CCDs) and complementary metal-oxide semi-conductor (CMOS) image sensors	Visual communication
UV light 10^{16}–10^{17} Hz 10^{-7}–10^{-8} m	Stars Black lights	Charge-coupled devices (CCDs) that can detect UV radiation	Forensics Decontamination
X-rays 10^{18}–10^{19} Hz 10^{-9}–10^{-10} m	Accelerating particles in X-ray tubes Very hot objects Energetic stellar events such as supernovas	Photographic plates Geiger counters	Medical imaging Studies of material properties Cosmology
Gamma rays 10^{20} Hz 10^{-12} m	Radioactive materials (such as uranium and its compounds) Nuclear reactions Stellar explosions	Solid-state detectors that transform gamma rays into optical or electronic signals	Detection of nuclear explosions Cosmology

Don't interpret Table 25.4 too literally. For example, most objects emit many of these types of EM waves, but in different proportions. For example, cold gas clouds in our galaxy primarily emit radio waves but also emit a small amount of infrared waves and visible light. Stars similar to our Sun emit mostly infrared, visible, and UV, but also produce some radio waves and X-rays. Even your body emits all types of EM waves: primarily infrared radiation and only very small amounts of UV, X-ray, and gamma radiation.

REVIEW QUESTION 25.4 If the frequency of one electromagnetic wave is twice that of another, how do the speeds and wavelengths of the two waves differ?

25.5 Mathematical description of EM waves and EM wave energy

When analyzing electric dipole antennas, we found that they produce electric and magnetic fields whose $\vec{E}$ and $\vec{B}$ vectors vibrate perpendicular to each other and perpendicular to the direction the wave travels (**Figure 25.13**). Maxwell's equations also predict that the amplitudes of the changing $\vec{E}$ field and $\vec{B}$ field vectors are related:

$$E_{\text{max}} = cB_{\text{max}} \qquad (25.1)$$

FIGURE 25.13 Electromagnetic wave $\vec{E}$ and $\vec{B}$ fields (a) at one time and (b) at one location.

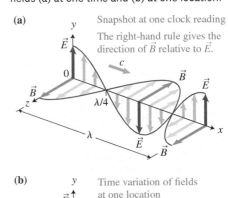

(a) Snapshot at one clock reading
The right-hand rule gives the direction of $\vec{B}$ relative to $\vec{E}$.

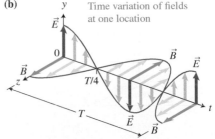

(b) Time variation of fields at one location

where $c = 3.00 \times 10^8$ m/s is the speed of electromagnetic waves in vacuum. In other words, not only do the electric and magnetic fields oscillate in phase, but also their magnitudes are related through the speed of light.

We can check the above equation with unit analysis. The units for electric field are newtons/coulomb, or, equivalently, volts/meter. The first combination comes from the definition of the $\vec{E}$ field; the second comes from the mathematical relationship between the $\vec{E}$ field and the V field (see Chapter 18). The units for magnetic field are T (tesla) = (newton · second)/(coulomb · meter). This unit can be deduced from the expression for the magnetic force exerted on a changed particle moving in a magnetic field ($F = qvB \sin \theta$ or $B = F/(qv \sin \theta)$, as discussed in Chapter 20). We can now insert the units into the right-hand side of Eq. (25.1):

$$\left(\frac{\mathrm{m}}{\mathrm{s}}\right)\left(\frac{\mathrm{N} \cdot \mathrm{s}}{\mathrm{C} \cdot \mathrm{m}}\right) = \frac{\mathrm{N}}{\mathrm{C}}$$

which are the units of the left-hand electric field side of the equation. Thus, the equation survives a unit analysis.

Mathematical description of EM waves

We now have enough background to make a quantitative description of EM waves that involves their speed c, wavelength λ, and frequency f and the relationship between the amplitudes of the vibrating $\vec{E}$ and $\vec{B}$ fields, $E_{max} = cB_{max}$. To do this, we use the mathematical descriptions of waves (developed in Chapter 11) as a guide.

Imagine the simplest EM wave—a sinusoidal wave of a single frequency. It travels along the x-axis in the positive direction and does not diminish in amplitude as it travels. Now, imagine that you place electric and magnetic field detectors at a point on the positive x-axis. As the wave passes, the detectors register the magnitude of the oscillating $\vec{E}$ field and $\vec{B}$ field as

$$E = E_{max} \cos\left(\frac{2\pi}{T}t\right) = E_{max}\cos(2\pi ft)$$

$$B = B_{max} \cos\left(\frac{2\pi}{T}t\right) = B_{max}\cos(2\pi ft)$$

We chose $t = 0$ at a moment when the fields have their maximum values E_{max} and B_{max}. The frequency f of the wave is the inverse of the period T of the wave. A convenient right-hand rule relates the direction of the $\vec{E}$ field, the direction of the $\vec{B}$ field, and the direction of propagation of the wave. Hold your right hand flat with your thumb pointing out at a 90° angle relative to your fingers. If you point your fingers in the direction of the $\vec{E}$ field, your palm will face in the direction of the $\vec{B}$ field. The direction of your thumb will determine the direction of the velocity vector of the electromagnetic wave. For example, if at a particular moment the $\vec{E}$ field points in the $+y$-direction and the $\vec{B}$ field points in the $+z$-direction, then the wave is traveling in the $+x$-direction at speed c (**Figure 25.14**).

The above two equations describe the wave vibrations at one specific point. However, the $\vec{E}$ field vectors and $\vec{B}$ field vectors are oscillating at every point along the wave's path, which means that the equation for the wave must be a function not only of time t but also of position x. The wave equation that we developed in Chapter 11 [Eq. (11.4)] takes this into account. Applied to the $\vec{E}$ field and $\vec{B}$ field, the wave equation becomes

$$E_y = E_{max} \cos\left[2\pi\left(\frac{t}{T} \pm \frac{x}{\lambda}\right)\right]$$

$$B_z = B_{max} \cos\left[2\pi\left(\frac{t}{T} \pm \frac{x}{\lambda}\right)\right]$$

(25.2)

FIGURE 25.14 (a) The $\vec{E}$, $\vec{B}$, and $\vec{v} = \vec{c}$ vectors of an electromagnetic wave. (b) The right-hand rule for the $\vec{E}$, $\vec{B}$, and $\vec{v}$ EM wave directions.

(a)

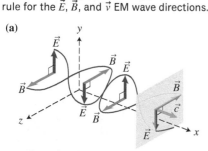

The $\vec{E}$ and $\vec{B}$ fields vibrate perpendicular to each other and in a plane perpendicular to the wave's velocity $\vec{v} = \vec{c}$.

(b)

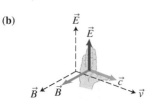

The minus signs are used to describe a wave traveling in the positive x-direction, and the plus signs are used to describe a wave traveling in the negative x-direction. We can represent these equations graphically (shown for a wave moving in the positive x-direction at $t = 0$ in (a) and at $x = 0$ in (b) in Figure 25.13).

Suppose we wish to describe light from a laser pointer that emits light of frequency 4.5×10^{14} Hz with an oscillating $\vec{B}$ field of amplitude 6.0×10^{-4} T. What else can you determine about this electromagnetic wave from this information?

(a) The period of the wave is

$$T = \frac{1}{f} = \frac{1}{4.5 \times 10^{14} \text{ Hz}} = 2.2 \times 10^{-15} \text{ s}$$

(b) The wavelength of the wave is

$$\lambda = \frac{c}{f} = \frac{3.00 \times 10^8 \text{ m/s}}{4.5 \times 10^{14} \text{ Hz}} = 6.7 \times 10^{-7} \text{ m} = 670 \text{ nm}$$

(c) A wavelength of 670 nm corresponds to red light (see Figure 25.12).

(d) The amplitude of electric field oscillation [Eq. (25.1)] is

$$E_{max} = cB_{max} = (3.00 \times 10^8 \text{ m/s})(6.0 \times 10^{-4} \text{ T}) = 1.8 \times 10^5 \text{ N/C}$$

(e) Assuming the laser light travels in the positive x-direction, we can write wave equations for it:

$$E_y = (1.8 \times 10^5 \text{ N/C}) \cos\left[2\pi \left(\frac{t}{2.2 \times 10^{-15} \text{ s}} - \frac{x}{6.7 \times 10^{-7} \text{ m}} \right) \right]$$

$$B_z = (6.0 \times 10^{-4} \text{ T}) \cos\left[2\pi \left(\frac{t}{2.2 \times 10^{-15} \text{ s}} - \frac{x}{6.7 \times 10^{-7} \text{ m}} \right) \right]$$

Energy of electromagnetic waves

In previous chapters (Chapters 17 and 18) we learned that a system of electrically charged objects possesses electric potential energy, and that same idea can be expressed through the concept of electric field energy density. This energy density u_E quantifies the energy stored in the electric field per cubic meter of volume:

$$u_E = \frac{\kappa}{8\pi k_C} E^2 \tag{18.16}$$

Rewriting $k_C = \dfrac{1}{4\pi\epsilon_0}$ and assuming that the region with electric field is a vacuum ($\kappa = 1$), the above becomes

$$u_E = \frac{1}{8\pi k_C} E^2 = \frac{1}{8\pi \left(\dfrac{1}{4\pi\epsilon_0} \right)} E^2 = \frac{1}{2} \epsilon_0 E^2$$

What about the magnetic field? Does it have energy as well? To answer this question, think of what happens to the electric current when you connect a solenoid to a DC power supply. The current through the solenoid increases from the initial value of zero to the final value I. This changing current produces a changing magnetic flux through the solenoid, and this, in turn, produces an induced emf, which (according to Lentz's law) opposes the change in the current through the circuit. Let the power supply and the circuit, including the connecting wires and the solenoid, be the system. While the current created by the DC power supply through the solenoid is increasing, the DC power supply needs to add more energy to the circuit to counteract the induced current created in the solenoid, which is in the opposite direction to the current created by the battery. This extra energy coming from the power supply results in an increase of the magnetic field energy of the system—the energy that is stored in the magnetic field created inside and around the solenoid.

We can find this energy by calculating the energy that the power supply needs to add to the circuit to increase the current from zero to the final value. Assuming that the current increases linearly with time, the total charge transferred through the solenoid can be expressed as $q = I_{average}\Delta t = \dfrac{I}{2}\Delta t$. Assuming also that the initial energy of the magnetic field is zero, the final energy of the magnetic field is

$$u_B = \mathcal{E}_{in}q = \mathcal{E}_{in}\frac{I}{2}\Delta t = \frac{\Delta\Phi}{\Delta t}\frac{I}{2}\Delta t = \frac{LI^2}{2} \tag{25.3}$$

where we used Faraday's law of electromagnetic induction and expressed the magnetic flux with the inductance of the solenoid ($\Phi = LI$, Chapter 21). Using the expression for the inductance of a long solenoid ($L = \mu_0\dfrac{N^2}{l^2}V$, where N is the number of turns, l is the length of the solenoid, and V is the volume of the solenoid) and the expression for the magnetic field in the center of the solenoid $\left(B = \mu_0\dfrac{NI}{l}\right)$, the above expression for the energy of the magnetic field of the solenoid can be expressed in the following form:

$$u_B = \frac{1}{2\mu_0}\left(\frac{\mu_0 NI}{l}\right)^2 V = \frac{B^2}{2\mu_0}V$$

Because the magnetic field of a long solenoid is approximately uniform inside the solenoid and very small outside the solenoid, we can divide this expression by the volume of the solenoid and obtain the expression for the magnetic field energy density:

$$u_B = \frac{B^2}{2\mu_0} \tag{25.4}$$

Although we derived this expression for the case of a long solenoid, it turns out that it is valid for magnetic fields in general.

We can now compare the magnetic field energy density with the electric field energy density. Because $E = cB$ and the electric field energy density is $u_E = \frac{1}{2}\epsilon_0 E^2$, $u_E = \frac{1}{2}\epsilon_0 c^2 B^2$. But we know that the speed of light is $c = \dfrac{1}{\sqrt{\epsilon_0\mu_0}}$ or $c^2 = \dfrac{1}{\epsilon_0\mu_0}$; thus,

$$u_E = \frac{1}{2\mu_0}B^2 = u_B!$$ We find that the electric and magnetic fields provide equal contributions to the total energy density of an electromagnetic wave.

The total energy density of the electric and magnetic fields in an EM wave is the sum of the energy densities of each:

$$u = u_E + u_B$$

$$\Rightarrow u = \frac{1}{2}\epsilon_0 E^2 + \frac{1}{2\mu_0}B^2$$

However, since these two terms are equal, we can write the energy density of the EM wave as twice the first term or twice the second term:

$$u = 2\left(\frac{1}{2}\epsilon_0 E^2\right) = \epsilon_0 E^2 \quad \text{or} \quad u = 2\left(\frac{1}{2\mu_0}B^2\right) = \frac{1}{\mu_0}B^2 \tag{25.5}$$

Average values of the energy densities

Remember that in the equations above, E and B are the instantaneous values of the magnitudes of the $\vec{E}$ and $\vec{B}$ fields. In an EM wave of a specific wavelength, both of those fields oscillate sinusoidally (**Figure 25.15a**). This means that the energy density at each point in the wave fluctuates as a sine-squared function since the field appears squared in the equations. During one vibration of the field, the energy density starts with a maximum value; then after one-quarter of a period the energy density is zero; and after half a period the energy is again maximal (Figure 25.15b).

FIGURE 25.15 Variation of the $\vec{E}$ and $\vec{B}$ fields at one point and (b) the energy density variation with time of an EM wave at one point.

(a)

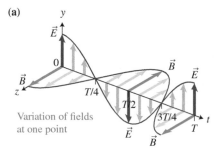

Variation of fields at one point

(b)

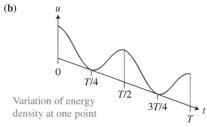

Variation of energy density at one point

Since the energy density at each point in the EM wave fluctuates with extremely high frequency, it's usually more useful to think in terms of the time-averaged energy density at each point in the EM wave. In order to do this, we need to calculate the average value of the sine-squared function shown in Figure 25.15b. We already know this value from Chapter 21: the average value of the square of a sinusoidal function (either sine or cosine) is one-half the maximum value. Therefore, the average value (indicated by the bar above u) of the energy density along the direction of the EM wave is

$$\bar{u} = \tfrac{1}{2} u_{max} = \tfrac{1}{2} (\epsilon_0 E_{max}^2) = \tfrac{1}{2} \epsilon_0 E_{max}^2 \qquad (25.6)$$

Intensity of an EM wave

Now that we understand the average energy density of the EM wave, we can understand quantitatively what is meant by its *intensity*. When you look at a source of light, it might appear dim to you, like a distant star. Or it might appear bright to you, like a car with its high beams on. The physical quantity that characterizes this brightness is called *intensity*. (We encountered this quantity in Chapter 11.)

Remember that intensity is the amount of energy (not energy density) that passes through a unit area during a unit time interval [see Eq. (11.6)].

$$\text{Intensity} = \frac{\text{Energy}}{\text{time interval} \times \text{area}} = \frac{\Delta E}{\Delta t \times A} = \frac{P}{A}$$

where P is the power of the wave. Technically, since the energy density of the field fluctuates, the intensity of the wave fluctuates as well, but this fluctuation is so rapid that it makes much more sense to think about the average intensity of the EM wave. The average intensity depends on the wave's average energy density $\bar{u} = (1/2)\epsilon_0 E_{max}^2$ and the speed c of the wave through the area (the faster the wave travels, the more energy passes through the area, though for EM waves this speed is always c). It is important to note that the intensity of all waves (sound waves, waves on rope, etc.), not just electromagnetic waves, is proportional to the square of the amplitude of vibrations.

One way to think of intensity and energy density is to imagine wind blowing into a sail. The wind speed is similar to the speed of the wave, and the kinetic energy per cubic meter of air is similar to the energy density of the wave. During a time interval Δt, the energy within the volume $V = A(c\,\Delta t)$ passes through the area A (**Figure 25.16**). The intensity equals this energy divided by the time interval and by the area:

$$I = \frac{\text{Energy}}{\text{time interval} \times \text{area}} = \frac{\bar{u}V}{\Delta t A} = \frac{\bar{u}Ac\,\Delta t}{\Delta t A} = \bar{u}c \qquad (25.7)$$

FIGURE 25.16 Intensity involves the rate of energy crossing a surface.

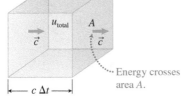

"Box" of electromagnetic energy

u_{total} A

$\vec{c}$ $\vec{c}$

Energy crosses area A.

$c\,\Delta t$

$c\,\Delta t$ = the distance the energy moves to the right during a time interval Δt.

EXAMPLE 25.3   **Solar constant**

The *solar constant* is a fundamental constant in environmental science and astronomy that describes the intensity of the Sun's EM radiation when it arrives at Earth. It is expressed in terms of the total electric and magnetic field energy incident each second on a 1.0-m^2 surface located above the atmosphere directly facing the Sun:

$$\text{Solar constant} = 1.37 \text{ kW/m}^2$$

What is the amplitude of the $\vec{E}$ field oscillation just above the atmosphere due to the Sun's illumination of Earth?

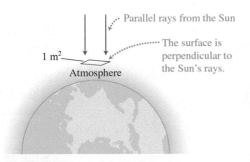

Parallel rays from the Sun

The surface is perpendicular to the Sun's rays.

1 m^2

Atmosphere

Sketch and translate   Imagine the Sun's radiation striking a 1.0-m^2 surface above Earth's atmosphere at a zero angle of incidence. Remember that we define the angle of incidence as the angle between the ray and the normal to the surface. To determine the amplitude E_{max} of the $\vec{E}$ field, we need to relate the solar constant to the average energy density of the EM radiation, which in turn depends on the $\vec{E}$ field amplitude E_{max}.

Simplify and diagram   Assume that light from the Sun travels to Earth at speed c. Assume also that all of the EM radiation is of a single wavelength. This is a poor assumption since the different wavelength EM waves arriving at Earth each have different electric field amplitudes. Therefore, what we are determining is the electric field amplitude averaged over all wavelengths.

(CONTINUED)

Represent mathematically We know from Eq. (25.5) that $I = \bar{u}c$ or $\bar{u} = (I/c)$. Using Eq. (25.4), we have $\bar{u} = (1/2)\epsilon_0 E_{max}^2$. Thus $(I/c) = (1/2)\epsilon_0 E_{max}^2$.

Solve and evaluate We can now determine E_{max} by multiplying both sides of the previous equation by 2, dividing by ϵ_0, and then taking the square root:

$$E_{max} = \sqrt{\frac{2I}{c\epsilon_0}}$$

$$= \sqrt{\frac{2(1.37 \times 10^3 \text{ W/m}^2)}{(3.00 \times 10^8 \text{ m/s})(8.85 \times 10^{-12} \text{ C}^2/(\text{N} \cdot \text{m}^2))}}$$

$$= 1010 \text{ N/C}$$

Is the number reasonable? We could estimate the $\vec{E}$ field amplitude a certain distance from a 60-W lightbulb in a room and compare it to this result. We expect the $\vec{E}$ field amplitude produced by the bulb to be significantly less since the Sun appears to be significantly brighter than a 60-W lightbulb. See the following "Try it yourself" question.

Try it yourself Estimate the amplitude of the $\vec{E}$ field produced by a 60-W electric bulb if you are 2.0 m away from it. Assume that the energy emitted by the bulb each second is completely in the form of EM waves.

Answer   produced by the Sun above Earth's atmosphere.

30 N/C. As expected, this is significantly less than the $\vec{E}$ field amplitude

---

REVIEW QUESTION 25.5 Electromagnetic waves are sinusoidal vibrations of electric and magnetic fields. The average value of a sine function is zero. How can electromagnetic waves carry any energy?

FIGURE 25.17 An antenna produces $\vec{E}$ field oscillations that far from the antenna are parallel to the antenna.

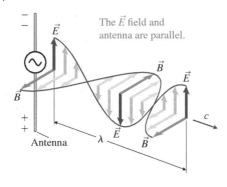

The $\vec{E}$ field and antenna are parallel.

FIGURE 25.18 Light from a bulb is unpolarized.

(a)   Side view

Waves produced by independent atomic oscillators have random polarization directions.

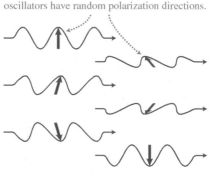

(b)   Front view

Unpolarized light

25.6 Polarization and light reflection

We now have a more powerful model of light as an electromagnetic wave. Let's use this model to understand how light is produced, how polarized sunglasses work, how to reduce glare from reflected sunlight, and the role of polarization in liquid crystal displays (LCDs) on computer and calculator screens.

Producing unpolarized light

Many sources of light produce unpolarized light. In this section we discuss the mechanism behind this phenomenon. To understand unpolarized light, we first review the production of polarized light by an antenna.

Electrons vibrating up and down in an antenna produce radio waves. Far from the antenna, these waves take the form of plane waves with the $\vec{E}$ field vector polarized in the direction parallel to the antenna (**Figure 25.17**). A polarizer can reduce the amplitude of this radio wave to zero if the polarizer is oriented in a specific direction relative to the antenna.

We can think of light coming from an incandescent lightbulb as being produced by the motion of charged particles within the filament, each acting as a tiny "antenna." The light emitted by the bulb consists of many waves originating at random times with random polarizations (**Figure 25.18a**). If we could observe the many separate EM waves as a beam of unpolarized light moving directly toward our eyes, the oscillations of both the $\vec{E}$ and $\vec{B}$ fields would look something like a porcupine (the head-on appearance of the $\vec{E}$ field vectors is shown in Figure 25.18b). These oscillations explain why light produced by an incandescent bulb is unpolarized.

Light polarizers

How do polarizers interact with light? A polarizer directly affects the electric field of the passing electromagnetic wave and does not affect the magnetic field. The reason is that the magnetic field exerts a force only on a *moving* charged object. On the other hand, the electric field exerts a force that is independent of the motion of the object. This means that the electric field has a much more significant effect on matter since, most of the time, the protons and electrons that make up matter don't have any sort of collective motion.

A polarizer absorbs one component of the $\vec{E}$ field of the EM wave passing through it, allowing the perpendicular component to pass. To understand which component is absorbed, consider unpolarized microwaves passing through a grill of narrow metal bars (**Figure 25.19**). The component of the oscillating $\vec{E}$ field of the incident microwaves that is parallel to the grill produces an oscillating electric current in the grill. The resistive heating in the metal grill rods due to this current causes this component of the $\vec{E}$ field to diminish. The $\vec{E}$ field perpendicular to the metal grill rods does not cause a current and is not diminished. So the axis of the polarizing grill is actually perpendicular to the grill rods. The microwaves leaving the metal grill polarizer are linearly polarized in the direction perpendicular to the grill rods, and that is what defines the axis of the polarizer.

The situation is similar for light, only the dimensions are much smaller. A polarizing film such as that used in polarized sunglasses is made of long polymer molecules infused with iodine. The polymer chains are stretched in one direction to form an array of aligned linear molecules. Electrons can move along the lengths of the polymer chains, but there is resistance—just like the metal wires in the microwave polarizing grill. As a result, the $\vec{E}$ field component of unpolarized light parallel to the polymer chains is diminished. The $\vec{E}$ field perpendicular to the chains is not diminished. Thus, the axis of the light polarizer is perpendicular to the direction of the stretched polymer chains.

Polarization by reflection and Brewster's angle

The left-hand chapter-opening photo shows light reflected off a body of water. If you look at this light through a polarizer, as in the right-hand photo, the intensity of reflected light varies depending on the orientation of the polarizer as we rotate it around the direction of the reflected beam (see **Figure 25.20**).

This indicates that the reflected light is partially polarized even though the incident light was completely unpolarized. When David Brewster, a Scottish physicist and mathematician who lived in the middle of the 19th century, investigated this phenomenon more deeply, he found that the intensity of reflected light that passes through the polarizer is medium bright when the axis of the polarizer is parallel to the reflective surface (water or glass, for example), but the reflected light becomes dimmer when the axis of the polarizer is rotated by 90°. We will call the orientation of the polarizing axis parallel to the reflective surface the x-direction and the direction perpendicular to the surface the y-direction (marked in Figure 25.20). Brewster found that when light was incident on the surface at a particular angle, called the **polarizing angle** θ_p, the reflected light was completely blocked by the polarizer whose axis is oriented in the y-direction. Given that the reflected light is blocked by a polarizer oriented in the y-direction, the reflected light must be completely polarized in the x-direction. Brewster also found that when the angle of incidence is θ_p, the reflected and the refracted light beam directions always make a 90° angle relative to each other.

To help you understand what happens, we set up a glass table on a balcony on a sunny day. In **Figure 25.21a** you can see metal silverware, a wooden plate filled with water, and two dark lines that are the reflections of the shadow cast on the table by the closed balcony door. The black arrow shows the orientation of the polarizer placed in front of the camera. The axis is parallel to the reflective surface of the glass table and to the water surface in the plate (it is in the x-direction marked in Figure 25.20). Notice how bright the reflected light is. The table is shiny, and you cannot see that there is something in the water. The situation changes when the polarizer is rotated 90°, that is, its axis is in the y-direction marked by the arrow in Figure 25.21b. The glare from the table surface and water is less noticeable, and you can see a slice of carrot in the water that was previously masked by the reflected light. Because the polarizer with its axis in the y-direction blocks some light, we conclude that the blocked light (the reflected light) is polarized in the x-direction.

Now examine the bottom part of the photo in Figure 25.21b. You can see that the glare from the table and the reflection of the balcony door shadows are barely noticeable.

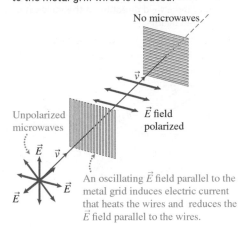

FIGURE 25.19 The $\vec{E}$ field component parallel to the metal grill wires is reduced.

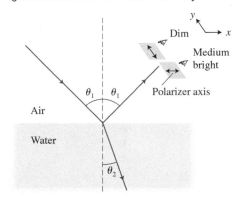

FIGURE 25.20 Investigating polarization of light reflected off an air-water boundary.

FIGURE 25.21 Photos taken on a sunny day (a) through a polarizer with its axis in the x-direction and (b) through the polarizer with its axis in the y-direction.

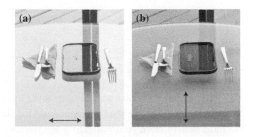

FIGURE 25.22 Reflection and refraction of light incident on a surface at the polarizing angle: the $\vec{E}$ field direction in the incident and reflected beams.

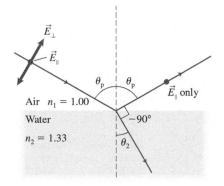

This is because light reflected into the polarizer was incident on that part of the surface at the polarizing angle θ_p. **Figure 25.22** shows the ray diagram for the situation.

There is one more important observation based on those two photos. Note that the reflections off the metal silverware did not change with the orientation of the polarizer. Light becomes partially or completely polarized only when it reflects off nonmetallic surfaces.

How can we explain these observations? We know that a polarizer blocks light in which the $\vec{E}$ field oscillates perpendicular to the axis of the polarizer. When light reflected at the angle θ_p is blocked when we hold the polarizer with its axis in the y-direction, this means that the reflected light is linearly polarized with the $\vec{E}$ field vibrating in the direction parallel to the reflective surface (Figure 25.22).

We found that when the angle of incidence is equal to the so-called polarizing angle θ_p, the reflected ray is linearly polarized in the plane parallel to the reflecting surface and the reflected and the refracted rays are perpendicular to each other.

From Figure 25.22 we see that

$$\theta_p + \theta_2 + 90° = 180°$$
$$\Rightarrow \theta_2 = 90° - \theta_p$$

We can use Snell's law to relate the angle of incidence θ_p to the angle of refraction θ_2:

$$n_1 \sin \theta_p = n_2 \sin \theta_2 = n_2 \sin(90° - \theta_p)$$

From trigonometry we know that $\sin(90° - \theta_p) = \cos \theta_p$. Substituting in the above, we get

$$n_1 \sin \theta_p = n_2 \cos \theta_p$$

or

$$\frac{\sin \theta_p}{\cos \theta_p} = \tan \theta_p = \frac{n_2}{n_1} \tag{25.8}$$

This relationship is called **Brewster's law**.

Brewster's law Light is traveling in medium 1 when it reflects off medium 2. The reflected light is totally polarized along an axis in the plane parallel to the surface when the tangent of the incident polarizing angle θ_p equals the ratio of the indexes of refraction of the two media:

$$\tan \theta_p = \frac{n_2}{n_1} \tag{25.8}$$

Brewster's law allows us to explain how polarizing sunglasses help reduce glare. When you are driving your car toward the Sun, it is often difficult to see because of the glare of reflected light from the car's hood and dashboard, from the pavement, and from other cars. This reflected light is partially polarized parallel to those surfaces. Polarizing sunglasses reduce the glare by absorbing light polarized in that direction.

EXAMPLE 25.4   **Reducing glare**

You are facing the Sun and looking at the light reflected off the ocean. At what angle above the horizon should the Sun be so that you get the most benefit from your polarizing sunglasses?

Sketch and translate The situation is illustrated on the next page. The water surface is horizontal and perpendicular to the page. The

sunlight incident on the water surface is unpolarized, represented by the dots pointing into and out of the page and the $\vec{E}$ field lines up and down perpendicular to the ray. We can answer the question by finding the polarization angle θ_p at which light reflected off the water is completely polarized parallel to the water surface—represented by the dots on the reflected ray.

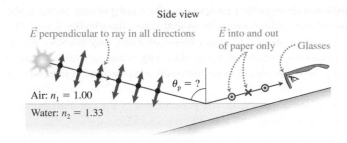

Side view

$\vec{E}$ perpendicular to ray in all directions

$\vec{E}$ into and out of paper only ·· Glasses

$\theta_p = ?$

Air: $n_1 = 1.00$

Water: $n_2 = 1.33$

Simplify and diagram The polarizing axis of your glasses and the polarization of the reflected light are shown below. Assume that the water reflects rays from the Sun as if it were a mirror. The reflected light $\vec{E}$ field oscillates parallel to the water surface—in and out of the page. The reflected light will not pass though the glasses if the polarizing axis of the glasses is perpendicular to the oscillating $\vec{E}$ field.

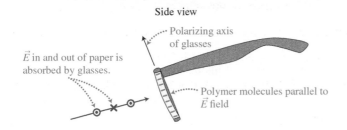

Side view

·· Polarizing axis of glasses

$\vec{E}$ in and out of paper is absorbed by glasses.

·· Polymer molecules parallel to $\vec{E}$ field

Represent mathematically We can use Brewster's law to find the angle at which parallel rays from the Sun that hit the water surface should produce reflected light that is completely polarized parallel to the air-water surface:

$$\tan \theta_p = \frac{n_2}{n_1}$$

Solve and evaluate Inserting $n_2 = 1.33$ for water and $n_1 = 1.00$ for air, we get

$$\tan \theta_p = 1.33$$

or an angle $\theta_p = 53.1°$. This is the angle between the vertical normal line and the direction of the sunlight. Thus, the direction of the sunlight above the horizontal is

$$90° - 53.1° = 36.9°$$

For all other incident angles, the reflected light will be partially polarized and the glare will not be completely eliminated by the sunglasses.

Try it yourself A light beam travels through glass ($n = 1.60$) into the air. At what angle of incidence is the light that reflects back into the glass completely polarized?

Answer   32°.

Polarization by scattering

We learned in a previous chapter (Chapter 22) that the sky appears blue from Earth's surface because the molecules in the atmosphere scatter blue light more than red into your eyes. If you look through polarized sunglasses at the clear sky in an arbitrary direction and rotate the glasses, the intensity of the light passing through the glasses changes. For example, observer A in **Figure 25.23** sees a change in the light intensity when turning polarized sunglasses while looking at the sky. This means that light scattered by the atmosphere is partially polarized. In fact, if you look in a direction that makes a 90° angle with a line toward the Sun (observer B in Figure 25.23), you can orient the glasses so the sky appears almost dark. Evidently, the sunlight scattered toward you by molecules in that direction is almost completely polarized.

To explain this observation, we need to assume that molecules in the atmosphere are like the tiny dipole antennas described at the beginning of this section. The Sun emits unpolarized light in which all $\vec{E}$ field orientations perpendicular to the direction the light travels are present in equal amounts. When this light strikes a molecule consisting of particles carrying opposite charges (electrons and nuclei), these particles vibrate as the EM wave passes. In **Figure 25.24a** (on the next page), the electric field component E_x produces molecular vibrations along the x-direction. Such a vibration is like that produced by a charge vibration in an antenna and produces an EM wave straight downward with an E_x component (it also produces an upward wave, but we are not interested in the upward wave here). The sunlight also has an electric field component E_y in the y-direction (Figure 25.24b). This component causes atoms and molecules in the atmosphere to vibrate up and down in the y-direction. However, an antenna with vibrating electric charge (the vibrating atom or molecule) cannot emit an $\vec{E}$ field along the axis that the charge vibrates since EM waves are transverse. The result is that the downward 90° wave caused by the vibrating atoms and molecules is linearly polarized in the x-direction. This explains why sunlight is almost completely polarized when you look at a point in the sky in a direction perpendicular to a line from the Sun to that point.

FIGURE 25.23 Scattered light is partially or almost completely polarized.

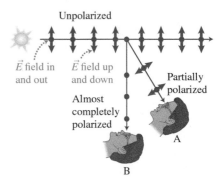

Unpolarized

$\vec{E}$ field in and out

$\vec{E}$ field up and down

Partially polarized

Almost completely polarized

A

B

FIGURE 25.24 Light scattered from a clear sky at 90° is linearly polarized.

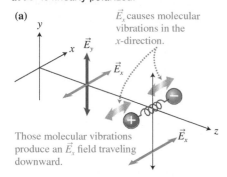

(a)

$\vec{E}_x$ causes molecular vibrations in the x-direction.

Those molecular vibrations produce an $\vec{E}_x$ field traveling downward.

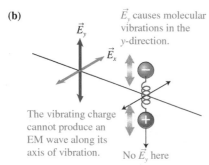

(b)

$\vec{E}_y$ causes molecular vibrations in the y-direction.

The vibrating charge cannot produce an EM wave along its axis of vibration.

No $\vec{E}_y$ here

If you look at clouds through polarized sunglasses, the light passing through the glasses is dimmer, but varies very little in intensity as you rotate the glasses. Evidently, light scattered from clouds is unpolarized. Clouds consist of many tiny water droplets. Light entering a cloud is randomly scattered many times by water droplets before it leaves. This multiple scattering keeps the light unpolarized. Photographers take advantage of this effect to take pictures of clouds using polarized "filters." The polarizers remove more of the partially polarized light from the sky and less of the unpolarized light from the clouds, making the clouds stand out (see **Figure 25.25**).

FIGURE 25.25 A polarizer affects light from the clear sky but not from the clouds.

Polarized LCDs

Nearly all computer, TV, calculator, and cell phone screens are LCDs—liquid crystal displays. Liquid crystals are made of substances that have properties of both a solid and a liquid. They can flow like a liquid, but their elongated molecules are aligned or oriented in an orderly crystal-like manner. Polarization plays an important role in the operation of these screens.

For simplicity, imagine an LCD showing an image that consists of different shades of gray. Some parts of the display screen are white, and other parts are gray or black. The display consists of a grid with many thousands of tiny pixels. Each pixel consists of five parallel layers: two polarizing filters (polarizers), two transparent electrodes, and a liquid crystal between them (**Figure 25.26a**). The axes of the polarizers are perpendicular to each other. If there is no liquid crystal between the polarizers, light from behind the display cannot pass through the polarizers. But with the liquid crystal between the polarizers, light passes through even though the two polarizers are oriented perpendicular to each other! How is this possible?

In the absence of an electric field between the electrodes, the liquid crystal (LC) molecules that are touching the electrodes tend to align with the molecules on the surface of the electrodes. Using a special procedure, these molecules are aligned so that their direction on one electrode is perpendicular to their direction on the other electrode. As a result, the LC molecules change orientation from one electrode to the other electrode, making something similar to a spiral. When polarized light passes through such a spiral, the direction of the $\vec{E}_{light}$ field of the polarized light rotates by 90° so that most of it can pass through the second polarizer even though it is perpendicular to the first (Figure 25.26b). The screen looks bright at that pixel.

FIGURE 25.26 A pixel of a liquid crystal (LC) display.

(a)

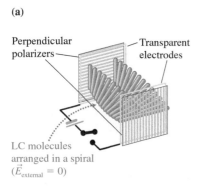

Perpendicular polarizers

Transparent electrodes

LC molecules arranged in a spiral ($\vec{E}_{external} = 0$)

(b)

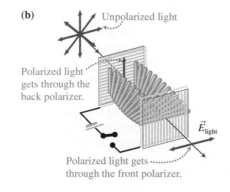

Unpolarized light

Polarized light gets through the back polarizer.

$\vec{E}_{light}$

Polarized light gets through the front polarizer.

(c)

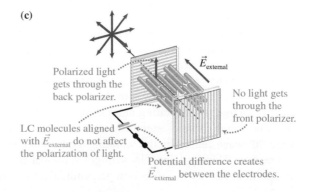

$\vec{E}_{external}$

Polarized light gets through the back polarizer.

No light gets through the front polarizer.

LC molecules aligned with $\vec{E}_{external}$ do not affect the polarization of light.

Potential difference creates $\vec{E}_{external}$ between the electrodes.

When a strong external electric field $\vec{E}_{\text{external}}$ is created between the transparent electrodes, the LC molecules align in the direction of this field, and thus the direction of the light's polarization (the direction of vibration of the $\vec{E}_{\text{light}}$ field vector) does not change as it passes through the liquid crystal (Figure 25.26c). Because the direction of $\vec{E}_{\text{light}}$ does not change, light does not pass through the second polarizer, making the screen look dark at that pixel.

Finally, if the external electric field does not have large enough magnitude E_{external}, then only some of the LC molecules align along the external $\vec{E}_{\text{external}}$ field and the rest make spirals, allowing some of the light to pass through. In this case, the pixel appears gray. When you press a key on your laptop, it turns on an electric field $\vec{E}_{\text{external}}$ that is applied to the liquid crystals of different pixels, making some parts of the screen appear black and others bright or gray. To produce colors, individual pixels are covered with red, green, or blue filters, but the intensity of light that passes through each of them is controlled in exactly the same way as described above.

Now we can answer the question at the beginning of the chapter concerning the similarities between the sky and the calculator screen: both are sources of polarized light.

3-D movies

Polarization also plays a role in one way of making a 3-D movie. Two projectors that send light of two perpendicular polarizations produce two distinct images on the screen. Each image consists of polarized light, and the polarizing axes of the two images are rotated by 90° relative to each other. The two images reflect from the metallized screen to theater viewers, who wear polarized glasses. The polarization axis of the left lens of the glasses is oriented at 90° relative to the axis of the right lens (**Figure 25.27**). Thus, each eye sees a distinct image. The two images are combined in the brain to convey the 3-D effect in much the same way that you see the real world without the help of 3-D glasses.

REVIEW QUESTION 25.6 Explain why polarizing glasses reduce the glare of sunlight reflected from the dashboard of your car or from the metal surface of another car.

FIGURE 25.27 3-D glasses to view a 3-D movie.

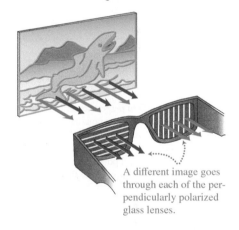

A different image goes through each of the perpendicularly polarized glass lenses.

Summary

Electromagnetic waves EM radiation consists of electric and magnetic fields whose vector $\vec{E}$ field and $\vec{B}$ field vibrate perpendicular to each other and to the direction the wave travels. Accelerating charged particles, often oscillating electric dipoles, produce the waves. Once formed, the waves are self-propagating at speed $c = 3.00 \times 10^8$ m/s in a vacuum or slightly less in air. (Sections 25.2 and 25.5)

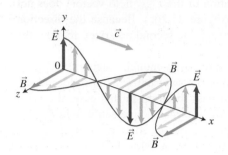

$$E_{max} = cB_{max} \qquad \text{Eq. (25.1)}$$

$$E_y = E_{max} \cos\left[2\pi\left(\frac{t}{T} \pm \frac{x}{\lambda}\right)\right]$$

$$B_z = B_{max} \cos\left[2\pi\left(\frac{t}{T} \pm \frac{x}{\lambda}\right)\right] \qquad \text{Eq. (25.2)}$$

Intensity of light The intensity I of light is the electromagnetic energy that crosses a perpendicular cross-sectional area A in a time interval Δt divided by that area and time interval. (Section 25.5)

$$u_B = \frac{B^2}{2\mu_0} \qquad \text{Eq. (25.4)}$$

$$\bar{u} = \tfrac{1}{2}u_{max} = \tfrac{1}{2}\left(\epsilon_0 E_{max}^2\right)$$

$$= \tfrac{1}{2}\epsilon_0 E_{max}^2 \qquad \text{Eq. (25.6)}$$

$$I = \frac{\text{Energy}}{\text{time interval} \times \text{area}}$$

$$= \frac{\bar{u}V}{\Delta tA} = \frac{\bar{u}Ac\,\Delta t}{\Delta tA} = \bar{u}c \qquad \text{Eq. (25.7)}$$

Polarization Randomly vibrating wave sources produce *unpolarized* light. When light travels through a polarizer, it becomes *linearly polarized*—only the component of the $\vec{E}$ field perpendicular to the chains of molecules of the polarizer material passes through. The intensity of linearly polarized light is reduced when it travels through a second polarizer oriented at an angle θ relative to the first. (Sections 25.1 and 25.6)

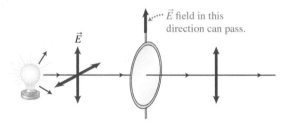

$$I' = \frac{I_0}{2}$$

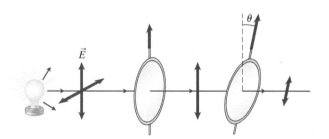

$$I = I' \cos^2\theta$$

Light polarization by reflection off a surface Light reflected off a surface is partially polarized parallel to the surface. When light is incident at the Brewster (polarization) angle θ_p, the light is totally polarized parallel to the surface. (Section 25.6)

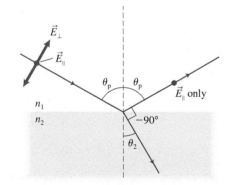

$$\frac{\sin\theta_p}{\cos\theta_p} = \tan\theta_p = \frac{n_2}{n_1} \qquad \text{Eq. (25.8)}$$

where n_1 and n_2 are the refractive indexes of the incident and reflecting media.

Questions

Multiple Choice Questions

1. The fact that light can be polarized means which of the following? (a) Light behaves like a transverse wave. (b) Light behaves like a longitudinal wave. (c) Light does not behave like a wave because it propagates in a vacuum.

2. What does a beam of unpolarized light lose after passing through one polarizer? (a) One-half of its intensity (b) One-quarter of its intensity (c) No intensity, as it was not polarized before

3. What does Faraday's law of electromagnetic induction describe? (a) A steady electric current producing a magnetic field (b) A changing magnetic field producing an electric field (c) A changing electric field producing a magnetic field

4. Maxwell's hypothesis contributed which of the following ideas? (a) A steady electric current produces a magnetic field. (b) A changing magnetic field produces an electric field. (c) A changing electric field produces a magnetic field.

5. What does a simple antenna with a changing electric current radiate? (a) An electric field whose $\vec{E}$ field vector is parallel to the antenna's orientation (b) A magnetic field whose $\vec{B}$ field vector is parallel to the antenna's orientation (c) An electric field whose $\vec{E}$ field vector is perpendicular to the antenna's orientation (d) It is impossible to say, as we do not have information about the magnitude of the current through the antenna.

6. An electrically charged particle radiates electromagnetic waves when (a) it is stationary. (b) it moves at a constant velocity. (c) it moves at changing velocity. (d) it is inside a wire with a DC current in it.

7. An electromagnetic wave propagates in a vacuum in the x-direction. Where does the $\vec{E}$ field oscillate? (a) In the x-y plane (b) In the x-z plane (c) In the y-z plane (d) Not enough information to answer

8. If the amplitude of an $\vec{E}$ field in a linearly polarized wave doubles, what will happen to the energy density of the wave? (a) It will double. (b) It will quadruple. (c) Not enough information to answer

9. You notice that unpolarized light reflected off a lake is completely blocked when observed through a polarizer. Which of the following statements are correct? Select all that apply
   (a) The $\vec{E}$ field in the reflected light is oscillating in a horizontal direction.
   (b) The $\vec{B}$ field in the reflected light is oscillating in a horizontal direction.
   (c) The $\vec{E}$ field and the $\vec{B}$ field in the reflected light are perpendicular to each other.
   (d) The axis of your polarizer is parallel to the water surface.
   (e) The axis of your polarizer is in the same plane as the plane defined by the incident and reflected beam.
   (f) The axis of your polarizer is parallel to the direction of the reflected beam.

10. You have two green lasers, A and B. A's light has twice the intensity of B's. If c is the speed of light in air and E_{max} is the amplitude of the electric field in light, then which of the following statements are correct? Select all that apply.

(a) $c_A = 2c_B$ because the intensity of light is proportional to c.
(b) $c_A = c_B$ because the speed of light in air is constant.
(c) $E_{max\,A} = 2E_{max\,B}$ because the intensity of light is proportional to E_{max}.
(d) $E_{max\,A} = \sqrt{2}E_{max\,B}$ because the intensity of light is proportional to E_{max}^2.
(e) $E_{max\,A} = 4E_{max\,B}$ because the intensity of light is proportional to E_{max}^2.

Conceptual Questions

11. Can light phenomena be better explained by a transverse wave model or by a longitudinal wave model? Explain how you know.

12. What are two models that explain how light can propagate through air? Describe their features in detail.

13. Summarize Maxwell's equations conceptually and describe a way to experimentally illustrate each of them.

14. What testable predictions followed from Maxwell's equations?

15. Describe the conditions that are necessary for the emission of electromagnetic waves.

16. Explain how radar works to determine distances to objects.

17. What determines the nearest and farthest distances of objects that can be distinguished by pulsed radar?

18. How was the hypothesis that light can be modeled as an electromagnetic wave tested experimentally?

19. What is the difference between the following two statements: Light is a wave. Light behaves like a wave. Which one do you think is more accurate and why?

20. How do polarized glasses work?

21. You bought a pair of glasses that are marketed as having polarizing filters. How can you test this claim if (a) you have another pair of identical glasses, and (b) you have only one pair of glasses and no other polarizer (and you are not allowed to break the glasses)?

22. Why, when we use polarized glasses, is the glare from reflective surfaces reduced? What parameters does the phenomenon depend on?

23. How does a polarizer for mechanical waves work, and how does a polarizer for light waves work?

24. What is an LCD and how does it work?

25. Jim does not understand the wave model of light. It does not make sense to him that light can propagate in a vacuum where there is no medium to transport vibrations from one place to another. How can you help Jim reconcile his understanding of waves with the wave model of light?

26. Make a list of phenomena that can be explained by the particle model of light. Then make a list of phenomena that can be explained using the wave model of light. Now that we have a wave model of light, does it mean that we should stop using the particle model? Explain your opinion.

27. Describe an experiment that you can perform to determine whether the light from a particular source is unpolarized or linearly polarized. If the latter, then how can you determine the polarization direction?

Problems

Below, **BIO** indicates a problem with a biological or medical focus. Problems labeled **EST** ask you to estimate the answer to a quantitative problem rather than derive a specific answer. Asterisks indicate the level of difficulty of the problem. Problems with no * are considered to be the least difficult. A single * marks moderately difficult problems. Two ** indicate more difficult problems.

25.1 and 25.2 Polarization of waves and Discovery of electromagnetic waves

1. The coils of a horizontal Slinky vibrate vertically up and down; the wave propagates in the horizontal direction. The amplitude of vibrations is 20 cm. You thread the Slinky through a board with a 50-cm slot at an angle of 60° relative to the vertical. What is the amplitude of the wave that moves past the slot? In what direction do the coils now vibrate? Indicate any assumptions that you made.

2. A 40-W lightbulb is 2.0 m from a screen. What is the intensity of light incident on the screen? What assumptions did you make?

3. * Assume that the bulb in Problem 25.2 radiates unpolarized light. You place a tourmaline crystal in front of the screen. (a) What is the intensity of light hitting the screen after passing through the crystal? (b) You place a second tourmaline crystal between the first crystal and the screen so that it is oriented at an angle of 50° relative to the axis of the first. What is the intensity of light hitting the screen? What assumptions did you make?

4. * (a) A uniform $\vec{B}$ field whose lines are oriented in the S-N direction increases steadily. Draw the $\vec{E}$ field lines of the induced electric field. (b) A uniform magnetic field whose lines are oriented E-W decreases steadily. Draw the $\vec{E}$ field lines of the induced electric field. (c) Repeat part (b) for a case when the $\vec{B}$ field changes at twice the rate.

5. * Investigate in detail how Hertz's apparatus worked and describe how it was used to produce and detect electromagnetic waves.

6. One dipole antenna points in the x-direction and the other identical antenna points in the y-direction. The currents in the antennas oscillate in phase and with the same frequency. Determine the angle between the x-axis and the direction of oscillation of the $\vec{E}$ field and $\vec{B}$ field at a point on the z-axis, far from the antennas.

7. * The current in an antenna is oscillating with a period 2.0×10^{-15} s. Using the same coordinate system, draw (a) two graphs that show how the $\vec{E}$ field and $\vec{B}$ field magnitudes at a point near the antenna change with time (label only the horizontal axis of the graph) and (b) two graphs that show how the $\vec{E}$ field and $\vec{B}$ field magnitudes at a point very far from the antenna change with time (again label only the horizontal axis of the graph).

8. **Milky Way** The Sun is a star in the Milky Way galaxy. When viewed from the side, the galaxy looks like a disk that is approximately 100,000 light-years in diameter (a light-year is the distance light travels in one year) and about 1000 light-years thick (**Figure P25.8**). What is the diameter and thickness of the Milky Way in meters? In kilometers? In miles?

FIGURE P25.8

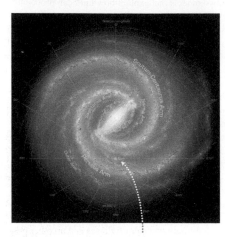

Sun and solar system's location in the Milky Way

9. We observe an increase in brightness of a star that is 5.96×10^{19} m away. When did the actual increase in brightness take place? Is the star in the Milky Way galaxy? What assumptions did you make?

10. * **Milky Way neighbors** Two neighboring galaxies to the Milky Way, the Large and Small Magellanic Clouds, are about 180,000 light-years away. How far away are they in meters? In miles?

25.3 Applications of electromagnetic waves

11. **EST** **Radar distance** Estimate the range of distances at which you can detect an object using radar with a pulse width of 12 μs and a pulse repetition of 15 kHz.

12. * **EST** **Radar antenna height** Why do you need to raise a radar emission antenna above the ground to be able to detect objects near the surface of Earth at the farthest range of the radar? Support your answer with a sketch. Estimate how high you need to raise the radar with a pulse width of 12 μs and a pulse repetition of 15 kHz.

13. * **More radar detection** Radar uses radio waves of a wavelength of 2.0 m. The time interval for one radiation pulse is 100 times larger than the time of one oscillation; the time between pulses is 10 times larger than the time of one pulse. What is the shortest distance to an object that this radar can detect?

14. * **Communicating with Mars** Imagine that you have a vehicle traveling on Mars. The shortest distance between Earth and Mars is 56×10^6 km; the longest is 400×10^6 km. What is the delay time for the signal that you send to Mars from Earth? Can you use radio signals to give commands to the vehicle?

15. * **TV tower transmission distance** The height of a TV tower is 500 m and the radius of Earth is 6371 km. What is the maximum distance along the ground at which signals from the tower can be received? Indicate any assumptions that you made.

16. ** **Radar parameters** Radar is characterized by a pulse width and pulse repetition period. (a) Explain what these words mean and how you can use them to estimate the range of distances that radar can measure. (b) If you wish to be able to detect objects at a close distance, what parameter of the radar should you change and how? If you wish to be able to detect objects at a long distance, what parameter of the radar should you change and how?

25.4 and 25.5 Frequency, wavelength, and the electromagnetic spectrum and Mathematical description of EM waves and EM wave energy

17. * (a) The amplitude of the $\vec{E}$ field oscillations in an electromagnetic wave traveling in air is 20.00 N/C. What is the amplitude of the $\vec{B}$ field oscillations? (b) The amplitude of the $\vec{B}$ field oscillations in an electromagnetic wave traveling in a vacuum is 3.00×10^{-3} T. What is the amplitude of the oscillations of the $\vec{E}$ field?

18. * **Limits of human vision** The wavelength limits of human vision are 400 nm to 700 nm. What range of frequencies of light can we see? How do the wavelength and frequency ranges change when we are underwater? The speed of light in water is 1.33 times less than in the air.

19. **AM radio** An AM radio station has a carrier frequency of 600 kHz. What is the wavelength of the broadcast?

20. * The electric field in a sinusoidal wave changes as

$$E = (25 \text{ N/C}) \cos\left[(1.2 \times 10^{11} \text{ rad/s})t + (4.2 \times 10^2 \text{ rad/m})x\right]$$

(a) In what direction does the wave propagate? (b) What is the amplitude of the electric field oscillations? (c) What is the amplitude of the magnetic field oscillations? (d) What is the frequency of the wave? (e) What is the wavelength? (f) What is the speed? (g) What other information can you infer from the equation?

21. ** A sinusoidal electromagnetic wave propagates in a vacuum in the positive x-direction. The $\vec{B}$ field oscillates in the z-direction. The wavelength of the wave is 30 nm and the amplitude of the $\vec{B}$ field oscillations is 1.0×10^{-2} T. (a) What is the amplitude of the $\vec{E}$ field oscillations? (b) Write an equation that describes the $\vec{B}$ field of the wave as a function of time and location. (c) Write an equation that describes the $\vec{E}$ field in the wave as a function of time and location. (d) What type of electromagnetic radiation is this?

22. ** For the previous problem determine (a) the frequency with which the electric energy in the wave oscillates; (b) the frequency at which magnetic field energy oscillates; (c) the maximum energy density; (d) the minimal energy density; (e) the average energy density; and (f) the intensity of the wave.

23. **BIO** **Ultraviolet A and B** UV-A rays are important for the skin's production of vitamin D; however, they tan and damage the skin. UV-B rays can cause skin cancer. The wavelength range of UV-A rays is 320 nm to 400 nm and that of UV-B rays is 280 to 320 nm. What is the range of frequencies corresponding to the two types of rays?

24. **BIO** **X-rays used in medicine** The wavelengths of X-rays used in medicine range from about 8.3×10^{-11} m for mammography to shorter than 6.2×10^{-14} m for radiation therapy. What are the frequencies of the corresponding waves? What assumption did you make in your answer?

25. ** **Power of sunlight on Earth** The Sun emits about 3.9×10^{26} J of electromagnetic radiation each second. (a) Estimate the power that each square meter of the Sun's surface radiates. (b) Estimate the power that 1 m^2 of Earth's surface receives. (c) What assumptions did you make in part (b)? The distance from Earth to the Sun is about 1.5×10^{11} m and the diameter of the Sun is about 1.4×10^9 m.

26. ** **Light from an incandescent bulb** Only about 10% of the electromagnetic energy from an incandescent lightbulb is visible light. The bulb radiates most of its energy in the infrared part of the electromagnetic spectrum. If you place a 100-W lightbulb 2.0 m away from you, (a) what is the intensity of the infrared radiation at your location? (b) What is the infrared energy density? (c) What are the approximate magnitudes of the infrared electric and magnetic fields?

27. * Explain how the information about energy radiated by a lightbulb in Problem 25.26 can be used to compare the magnitudes of $\vec{E}$ fields oscillating at different frequencies. Pose a problem that requires the use of this information.

28. * **EST** Estimate the amplitude of oscillations of the $\vec{E}$ and $\vec{B}$ fields for sunlight near the surface of the Sun. List all of the assumptions that you made.

29. ** **BIO** **New laser to treat cancer** The HERCULES pulsed laser has the potential to help in the treatment of cancer, as it focuses its power on a tiny area and essentially burns individual cancer cells. The HERCULES can be focused on a surface that is 0.8×10^{-6} m across. This pulsed laser provides 1.2 J of energy during a 27×10^{-15}-s time interval at a wavelength of approximately 800 nm. (a) Determine the power provided during the pulse. (b) Determine the magnitude of the maximum $\vec{E}$ field produced by the pulse.

30. * **Equation Jeopardy 1** Tell everything you can about the wave described by the equations below.

$$E_y = (2.4 \times 10^5 \text{ N/C}) \cos\left[2\pi\left(\frac{t}{1.5 \times 10^{-15} \text{ s}} - \frac{x}{4.5 \times 10^{-7} \text{ m}}\right)\right]$$

$$B_z = (8.0 \times 10^{-4} \text{ T}) \cos\left[2\pi\left(\frac{t}{1.5 \times 10^{-15} \text{ s}} - \frac{x}{4.5 \times 10^{-7} \text{ m}}\right)\right]$$

31. * **Equation Jeopardy 2** Tell everything you can about the wave described by the equations below.

$$E_y = (9.0 \times 10^5 \text{ N/C}) \cos\left[2\pi\left(\frac{t}{5.3 \times 10^{-15} \text{ s}} - \frac{x}{1.6 \times 10^{-6} \text{ m}}\right)\right]$$

$$B_z = B_{max} \cos\left[2\pi\left(\frac{t}{5.3 \times 10^{-15} \text{ s}} - \frac{x}{1.6 \times 10^{-6} \text{ m}}\right)\right]$$

32. * **Equation Jeopardy 3** Tell everything you can about the situation described below.

$$E_{max} = \sqrt{\frac{2 \cdot 640 \text{ kW/m}^2}{(3.0 \times 10^8 \text{ m/s})(8.85 \times 10^{-12} \text{ C}^2/\text{N} \cdot \text{m}^2)}}$$

25.6 Polarization and light reflection

33. * An unpolarized beam of light passes through two polarizing sheets that are initially aligned so that the transmitted beam is maximal. By what angle should the second polarized sheet be rotated relative to the first to reduce the transmitted intensity to (a) one-half and (b) one-tenth the intensity that was transmitted through both polarizing sheets when aligned?

34. * **BIO** **Spider polarized light navigation** The gnaphosid spider *Drassodes cupreus* has evolved a pair of lensless eyes for detecting polarized light. Each eye is sensitive to polarized light in perpendicular directions. Near sunset, the spider leaves its nest in search of prey. Light from overhead is partly polarized and indicates the direction the spider is moving. After the hunt, the spider uses the polarized light to return to its nest. Suppose that the spider orients its head so that one of these two eyes detects light of intensity 800 W/m² and the other eye detects zero light intensity. What intensities do the two eyes detect if the spider now rotates its head 20° from the previous orientation?

35. * Two polarizing sheets are oriented at an angle of 60° relative to each other. (a) Determine the factor by which the intensity of an unpolarized light beam is reduced after passing through both sheets. (b) Determine the factor by which the intensity of a linearly polarized beam with the $\vec{E}$ field oscillating at 30° relative to each polarizing sheet is reduced after passing through both sheets. Indicate any assumptions that you made.

36. * **Light reflected from a pond** At what angle of incidence (and reflection) does light reflected from a smooth pond become completely polarized parallel to the pond's surface? How do you know? In which direction does the $\vec{E}$ field vector oscillate in this reflected light wave?

37. * **Light reflected from water in a cake pan** At what angle is reflected light from a water-glass interface at the bottom of a cake pan holding water completely polarized parallel to the surface of the glass whose refractive index is 1.65?

38. * Unpolarized light passes through three polarizers. The second makes an angle of 25° relative to the first, and the third makes an angle of 45° relative to the first. The intensity of light measured after the third polarizer is 40 W/m². Determine the intensity of the unpolarized light incident on the first polarizer. Indicate any assumptions that you made.

39. ** You have two pairs of polarized glasses. Make a list of experimental questions you can answer using one of them or both. Describe the experiments you will design to answer them and discuss how you will ensure that the solutions (answers) you find make sense.

40. * A beam of unpolarized light with intensity 100 W/m² is incident on a pair of ideal crossed (perpendicular) polarizers. You insert a third ideal polarizer between the two polarizers with its polarizing axis at 45° to the others. Determine the intensity of the emerging light before and after you insert the third polarizer.

General Problems

41. ** **EST** **Density of Milky Way** What is the average density of the Milky Way assuming that it contains about 200 billion stars with masses comparable to the mass of the Sun? What additional information do you need to know to answer the question? What assumptions should you make?

42. ** **EST** **Supernova in neighboring galaxy** On February 23, 1987 astronomers noticed that a relatively faint star in the Tarantula Nebula in the Large Magellanic Cloud suddenly became so bright that it could be seen with the naked eye. Astronomers suspected that the star exploded as a supernova. Late-stage stars eject material that forms a ring before they explode as supernovas. The ring is at first invisible. However, following the explosion the light from the supernova reaches the ring. Astronomers observed the ring exactly one year after the phenomenon itself occurred (**Figure P25.42**). (a) Use geometry and the speed of light to estimate the distance to the supernova. The angular radius of the ring is 0.81 arcseconds. (b) How do you know whether your answer makes sense? (c) Use the width of the ring to estimate the uncertainty in the value of the distance.

FIGURE P25.42

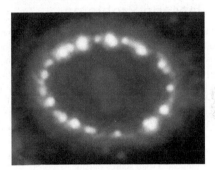

43. ** **BIO** **EST** **Human vision power sensitivity** A rod in the eye's retina can detect light of energy 4×10^{-19} J. Estimate the power of this light that the rod can detect. Indicate any assumptions you made. You will need more information than what is provided.

44. ** **Effect of weather on radio transmission** Weather affects the transmission of AM stations but does not affect FM stations. FM stations do not broadcast in remote areas. Suggest possible reasons for these phenomena. (Hint: Find information about the wavelengths of the waves and decide how they might propagate in Earth's atmosphere.)

45. ** **Measuring speed of light** In 1849 Hippolyte Fizeau conducted an experiment to determine the speed of light in a laboratory. He used an apparatus described in **Figure P25.45**. Light from the source S went through an interrupter K and after reflecting from the mirror M returned to the rotating wheel again. If light passed between the teeth of the wheel, Fizeau could see it; if light on the way back hit the tooth of the wheel, Fizeau could not see it. He could measure the speed of light by relating the time interval between teeth crossing the beam and the distance light traveled during those time intervals. The information that Fizeau had about the system was L = the distance between the wheel and the mirror; T = the period of rotation of the wheel; and N = the number of teeth in the wheel (the width of one tooth was equal to the width of the gap between the teeth). Using these parameters Fizeau derived a formula for calculating the speed of light: $c = (4LN/T)$. Explain how he arrived at this equation.

FIGURE P25.45

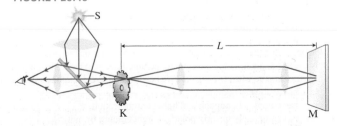

46. ** **Speed of light** In Fizeau's experiment (described in Problem 25.45) the distance between the wheel and the mirror was 3.733 km, the wheel had 720 teeth, and the wheel made 29 rotations per second. What was the speed of light determined by Fizeau? What were the uncertainties in this value?

47. * A sinusoidal electromagnetic wave in air has a 20-N/C $\vec{E}$ field amplitude. Determine everything you can about this wave.

48. * **EST** A microwave oven produces electromagnetic waves with a frequency of 2.45 GHz and emits energy at a rate of 700 W. The microwaves enter the cooking compartment through a rectangular waveguide of dimensions 5 cm × 10 cm. (a) Determine the wavelength of the microwaves, (b) estimate the amplitude of the $\vec{E}$ field at the waveguide opening, and (c) estimate the amplitude of the $\vec{E}$ field when the waves spread over the 30 cm × 20 cm cross section of the oven.

49. ** A beam of unpolarized light of intensity I_0 is incident on an ideal pair of crossed (perpendicular) polarizers. You insert a third ideal polarizer between the two polarizers, with its polarizing axis at an angle θ with respect to the axis of the first polarizer. Derive the expression for the intensity of the emerging light as function of θ.

Reading Passage Problems

BIO **Amazing honeybees** The survival of a bee colony depends on the ability of bee scouts to locate food and to convey that information to the hive. After finding a promising food source, a honeybee scout returns to the hive and uses a waggle dance to tell its worker sisters the direction and distance to the food.

Recall that light coming directly from the Sun is unpolarized. Bees use the direction of the Sun as a reference for their travel. In the hive, the scout bee's waggle dance is in the shape of a flat figure eight (Figure 25.28). The upward direction in the vertical hive represents the direction toward the Sun. The middle line of the scout's figure eight points in the direction of the food relative to the direction of the Sun. Thus a 50° middle waggle dance to the right of the vertical in the hive would indicate a food source outside that is 50° to the right of the direction toward the Sun. The distance to the food depends on the length of time the scout takes while wiggling her tail and wings through this middle of the figure eight. The other bees know the direction to the food even if the Sun is behind a cloud—they can detect the degree of polarization in other open parts of the sky and deduce the angle to the Sun from that position, and they learned from the scout the angle from the Sun's direction to the food.

FIGURE 25.28 A scout bee's waggle dance.

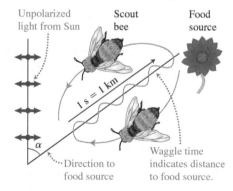

50. Bees know the direction to the Sun because
   (a) direct sunlight is linearly polarized.
   (b) direct sunlight is unpolarized.
   (c) they detect infrared radiation from the sun.
   (d) a and c.   (e) b and c.

51. Suppose that the scout bee in a waggle dance indicates that a good food source is at a 90° angle from the direction to the Sun. Bees who have never been to the food source when leaving the hive head toward a region of the sky that
   (a) has a flowery odor.
   (b) has completely unpolarized light coming from it.
   (c) has completely linearly polarized light coming from it.
   (d) has partially unpolarized light coming from it.
   (e) None of these answers

52. If the direction of the middle line of the scout's waggle dance figure eight points 53° to the left side of the upward direction in the hive, in what direction to the left should bees leaving the nest for the food source head relative to light coming from the Sun?
   (a) Where the light is 36% linearly polarized to the left of the Sun's direction
   (b) Where the light is 60% linearly polarized to the left of the Sun's direction
   (c) Where the light is 64% linearly polarized to the left of the Sun's direction
   (d) Where the light is 80% linearly polarized to the left of the Sun's direction
   (e) Where the light is 100% linearly polarized to the left of the Sun's direction

53. Polar molecules are caused to vibrate in all directions perpendicular to the direction of travel of sunlight. When the Sun is rising in the east and you look at the molecules directly overhead, the light from these molecules is
   (a) unpolarized.
   (b) linearly polarized along an axis between you and the molecules overhead.
   (c) linearly polarized parallel to a plane with you, the Sun, and the molecules.
   (d) linearly polarized perpendicular to a plane with you, the Sun, and the molecules.
   (e) All of these are correct.

54. In what direction(s) is the light from the sky almost completely unpolarized?
   (a) Looking directly at the Sun
   (b) Looking directly away from the Sun
   (c) Looking at 90° relative to the Sun
   (d) a and b   (e) b and c

Incandescent lightbulbs—soon to disappear Australia, Canada, New Zealand, and the European Union started phasing out incandescent lightbulbs in 2009. The United States phased them out in 2014. These bulbs have provided light for the world for more than 90 years. What's the problem?

Incandescent lightbulbs produce light when electrons (electric current) move through the filament and collide with tungsten atoms and ions in the filament, causing them to vibrate violently. Accelerating charged particles emit electromagnetic radiation. This is true for vibrating and accelerating particles in the filament, in your skin, and on the Sun. The 35 °C skin of a person emits most of its radiation in the low-frequency, long-wavelength infrared—sometimes called thermal radiation. At 2500 °C, the filament of an incandescent lightbulb emits most of its radiation as higher frequency infrared radiation and just 10% as visible light (very inefficient if the goal is to produce light). At 6000 °C, the Sun emits about 45% in the infrared range and 40% in the visible light range.

Banning incandescent bulbs will reduce energy usage. According to the Department of Energy, about 10^{18} J of electric energy is used each year in the United States for household lighting. Compact fluorescent lightbulbs (CFLs) and light-emitting diode (LED) bulbs use about one-fourth the energy of incandescent bulbs. Switching to CFLs and LEDs would also reduce greenhouse gas emissions by reducing the need for power produced by coal-burning electric power plants.

55. Why are the United States and other countries banning the use of incandescent lightbulbs?
   (a) The bulbs get too hot.
   (b) The bulbs have a tungsten filaments.
   (c) Ninety percent of the electric energy used is converted to thermal energy.
   (d) The bulbs are only 10% efficient in converting electric energy to light.
   (e) c and d

56. What is the rate of visible light emission from a 100-W incandescent lightbulb?
   (a) 10 W   (b) 20 W   (c) 50 W
   (d) 100 W   (e) Not enough information to make a determination

57. What does the surface of the body at about 35 °C primarily emit?
   (a) No electromagnetic radiation
   (b) Long-wavelength infrared radiation
   (c) Short-wavelength infrared radiation
   (d) Light   (e) Invisible ultraviolet light

58. Suppose you changed all incandescent lightbulbs to more energy-efficient bulbs that used one-fourth the amount of energy to produce the same light. About how many 3.3×10^9-kW·h/year electric power plants could be removed from the power grid?
   (a) 3   (b) 10   (c) 20   (d) 50   (e) 100

59. How much money will you save on your electric bill each year if you replace five 100-W incandescent bulbs with five CFL or LED 25-W bulbs that produce the same amount of light? Assume that the bulbs are on 3.0 h/day and that electric energy costs $0.12/kW·h.
   (a) $5   (b) $10   (c) $20   (d) $50   (e) $100

Special Relativity

At the end of the 19th century, Albert Michelson measured the speed of light quite accurately, but thought it needed a medium (called ether) through which to travel, just as sound needs air to propagate. Michelson and Edward Morley built a device that would demonstrate the existence of this invisible substance. Despite their efforts, they failed to show that ether exists. This negative result became the experimental foundation for a new theory—the subject of this chapter, and one that most of us use on a daily basis with our GPS devices.

- Does ether, the proposed medium for light waves, exist?

- If you were in a spaceship moving at near light speed away from a light source, at what speed would light pass you?

- How can atomic clocks on satellites help determine your location on Earth?

SO FAR IN THIS BOOK, we have assumed that time passes in the same way for all observers. When a second passes for me, the same second passes for you. Similarly, we have assumed that length measurements of objects are the same for all observers. These assumptions sound so reasonable that it is tempting to think of them as "absolute truths." In this chapter we will find that we need to reconsider our ideas—not just of time and space, but also of momentum and energy.

BE SURE YOU KNOW HOW TO:

- Distinguish between inertial and noninertial reference frames (Section 3.4).

- Explain why observers in noninertial reference frames cannot use Newton's laws to explain mechanics phenomena (Section 3.4).

- Calculate the change in kinetic energy of a charged particle that travels across a potential difference ΔV (Section 18.3).

FIGURE 26.1 Two identical length boat trips on a river.

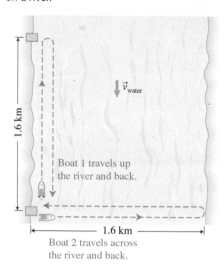

Boat 1 travels up the river and back.

Boat 2 travels across the river and back.

FIGURE 26.2 The velocity of boat 1.

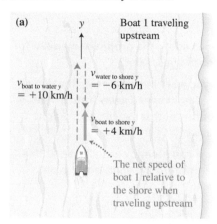

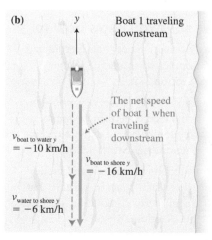

26.1 Ether or no ether?

Recall (from Chapter 24) that in 1801, Thomas Young performed the double-slit experiment that strongly supported the wave model of light. All waves known at that time required a medium to travel through: ocean waves through water, sound waves through air or other materials, and so on. The role of the medium is to transfer the disturbance produced by a source at one location to other locations. The particles of the medium interact with each other, allowing this to occur. Light, then, it was reasoned, must also require a medium in which to travel. In the 19th century, physicists thought that this medium was *ether*.

Ether was considered to be invisible but able to undergo shear deformations so that transverse light waves could travel through it. We know (from Chapter 25) that the work of Maxwell and Hertz led to the conclusion that light propagation could be explained by changing $\vec{E}$ and $\vec{B}$ fields that do not require any medium to travel. However, that work was done in the late 1880s. Before that, physicists were searching for ether. This search, like many investigations in physics, produced an unexpected outcome that eventually changed the way we think about space and time.

An analogy: a boat race

To understand how one of those ether-testing experiments was done, we will use a mechanics analogy. Consider a process involving two identical boats in a race on a wide river. Each boat is to travel a distance of 1.6 km relative to the shore, then 1.6 km back (Figure 26.1). Boat 1 travels from a starting dock to another dock 1.6 km upstream from the first, then back again to where it started. Boat 2 starts at the same dock but instead travels 1.6 km across the river to a dock on the opposite side, then travels back to where it started. Each boat travels at speed 10 km/h relative to the water, and the water travels at speed 6 km/h relative to the shore. Which boat returns to the starting dock first?

You may think that the boats will arrive back at the dock at the same time. Since all the speeds are the same and are constant, won't everything just balance out? As you will see shortly, the reality is something different.

Travel time for boat 1 going upstream and back The displacement of boat 1 with respect to the shore as it moves upstream $\vec{d}_{\text{boat 1 to shore}}$ is the result of two displacements—the displacement of the boat with respect to the water $\vec{d}_{\text{boat 1 to water}}$ and the displacement of the water with respect to the shore $\vec{d}_{\text{water to shore}}$. Mathematically:

$$\vec{d}_{\text{boat 1 to shore}} = \vec{d}_{\text{boat 1 to water}} + \vec{d}_{\text{water to shore}}$$

The time intervals during which these three displacements occur are all the same:

$$\Delta t_{\text{boat to shore}} = \Delta t_{\text{boat to water}} = \Delta t_{\text{water to shore}} = \Delta t$$

If we divide all three parts of the displacement equation by the time interval Δt during which the displacements occurred, we obtain

$$\frac{\vec{d}_{\text{boat 1 to shore}}}{\Delta t} = \frac{\vec{d}_{\text{boat 1 to water}}}{\Delta t} + \frac{\vec{d}_{\text{water to shore}}}{\Delta t}$$

or

$$\vec{v}_{\text{boat 1 to shore}} = \vec{v}_{\text{boat 1 to water}} + \vec{v}_{\text{water to shore}}$$

Note that these are velocities, not speeds, which is why they are labeled with vector symbols. In **Figure 26.2** we summarize the situation using vector components. The upstream velocity of the boat relative to the shore is then (with the upstream direction positive)

$$v_{\text{boat 1 to shore } y} = +10 \text{ km/h} + (-6 \text{ km/h}) = +4 \text{ km/h}$$

You can see that the water moving against the boat's direction of travel slows the boat down relative to the shore. The time interval it takes the boat to reach the upstream dock is

$$\Delta t_{\text{upstream}} = 1.6 \text{ km}/(4 \text{ km/h}) = 0.4 \text{ h}$$

The downstream velocity relative to the shore is

$$v_{\text{boat 1 to shore } y} = -10 \text{ km/h} + (-6 \text{ km/h}) = -16 \text{ km/h}$$

The time interval it takes the boat to return to the downstream dock is

$$\Delta t_{\text{downstream}} = -1.6 \text{ km}/(-16 \text{ km/h}) = 0.1 \text{ h}$$

As you would expect, the return trip takes less time because the velocity of the water relative to the shore is in the same direction as that of the boat. The total time interval for boat 1's round trip is therefore

$$\Delta t = \Delta t_{\text{upstream}} + \Delta t_{\text{downstream}} = 0.4 \text{ h} + 0.1 \text{ h} = 0.5 \text{ h}$$

Travel time for boat 2 going across the stream and back The velocity of boat 2 relative to the shore is perpendicular to the velocity of the water with respect to the shore (Figure 26.3). In order for the boat to travel directly across the river, the 10-km/h boat velocity relative to the water has to point at an angle so that it has a 6-km/h upstream component to cancel the downstream velocity of the water relative to the shore. The magnitudes of the three velocities shown are related through the Pythagorean theorem:

$$v_{\text{boat 2 to water}}^2 = v_{\text{boat 2 to shore}}^2 + v_{\text{water to shore}}^2$$

or

$$v_{\text{boat 2 to shore}}^2 = v_{\text{boat 2 to water}}^2 - v_{\text{water to shore}}^2 = (10 \text{ km/h})^2 - (6 \text{ km/h})^2$$
$$= 64 \text{ (km/h)}^2$$
$$\Rightarrow v_{\text{boat 2 to shore}} = 8 \text{ km/h}$$

The crossing time interval is then

$$\Delta t_{\text{crossing}} = 1.6 \text{ km}/(8 \text{ km/h}) = 0.2 \text{ h}$$

This crossing velocity is the same in both directions. The time interval for the complete round trip, therefore, is just twice this:

$$\Delta t = 2\Delta t_{\text{crossing}} = 2(0.2 \text{ h}) = 0.4 \text{ h}$$

We have shown that boat 1 takes longer to make its round trip than boat 2, even though both boats travel a distance of 3.2 km relative to the shore. The moving water affects the time intervals differently depending on the direction of the boat's travel.

Testing the existence of ether

An experiment analogous to the boat experiment can be used to test for the existence of ether. Imagine that ether fills the solar system and is stationary with respect to the Sun. Because Earth moves around the Sun at a speed of about 3.0×10^4 m/s, ether should be moving past Earth at this speed. Shining light waves parallel and perpendicular to the ether's motion relative to Earth is similar to sending boats parallel and perpendicular to a flowing river. Therefore, the travel time for light waves (1) up and back along the direction of the ether's motion and (2) across and back perpendicular to the ether's motion should differ if they travel the same distance. If one could measure this time interval difference experimentally, it would serve as strong support for the existence of ether as a medium for light wave propagation.

In 1887, Albert Michelson and Edward Morley set up such an experiment (although they themselves firmly believed in ether and were trying to measure Earth's speed relative to it). They used a device called an **interferometer** to detect

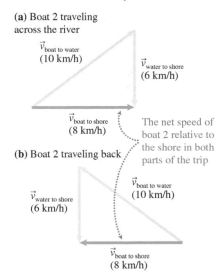

FIGURE 26.3 The velocity of boat 2.

(a) Boat 2 traveling across the river

$\vec{v}_{\text{boat to water}}$ (10 km/h)

$\vec{v}_{\text{water to shore}}$ (6 km/h)

$\vec{v}_{\text{boat to shore}}$ (8 km/h)

The net speed of boat 2 relative to the shore in both parts of the trip

(b) Boat 2 traveling back

$\vec{v}_{\text{boat to water}}$ (10 km/h)

$\vec{v}_{\text{water to shore}}$ (6 km/h)

$\vec{v}_{\text{boat to shore}}$ (8 km/h)

ether (see Testing Experiment **Table 26.1**). A light beam is sent to a beam splitter (a half-silvered mirror) that causes the light beam to split into two beams. One beam continues forward and the other beam reflects perpendicularly, each beam moving along a separate arm of the device. Mirrors at the end of the arms reflect the two beams back to the half-silvered mirror, where they recombine and are detected.

Because the two beams are formed by splitting a single light beam, the two beams are coherent (have a constant phase difference with respect to each other). When the beams recombine after reflection, an interference pattern results. The details of this interference pattern depend on the difference in the travel time intervals between the two beams along each arm of the interferometer. *If* the ether hypothesis is correct and the interferometer is carefully rotated (causing the direction of each beam of light to change relative to the ether), *then* the time interval it takes each beam to travel along each arm and back should change. As a result, the interference pattern produced by the recombined light should change as the device rotates.

TESTING
EXPERIMENT TABLE 26.1   **Testing the existence of ether**

Testing experiment	Prediction	Outcome
Assume that the ether is moving to the left past the interferometer. We shine a beam of light from the left onto the beam splitter. One beam moves parallel to the ether's velocity and the other beam moves perpendicular to its velocity.	It should take the light longer to travel against and with the ether than to travel sideways across the moving ether. Rotating the interferometer in the plane of the page should lead to a change in the interference pattern formed by the two light beams at the detector.	No matter how we change the interferometer orientation, the interference pattern does not change.
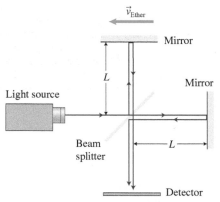	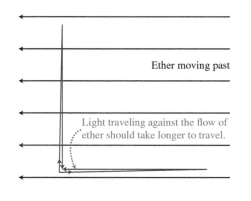	

Possible conclusions

1. There is no ether through which light travels.
2. There is ether, but it is "stuck" to Earth's surface and does not move relative to the interferometer.

Michelson and Morley repeated the experiment several times, but the phase difference between the two beams never changed. It seemed that light traveled at the same speed independent of its direction. This appeared to clearly indicate that ether as a medium for light to propagate did not exist.

Physicists, including Michelson, were reluctant to accept this no-ether result. They tried to revise their model to fit the experimental data. Perhaps the ether was dragged along through space by Earth. This would mean that the ether was at rest relative to the interferometer, explaining the null result of their experiment. However, if the ether is attached to Earth in some way, then as Earth rotates around its axis and orbits the Sun, it should cause the ether in the solar system to become twisted. This would cause light coming from stars to be slightly deflected on its way to Earth (like a water wave would be deflected by a whirlpool in the water). However, no one observed such an effect. Another explanation was that the length of the side of the interferometer parallel to

Earth's motion changed due to the motion of Earth relative to ether, thus causing the change in the interference pattern. No such change was measured. It seemed that light did not require a medium to travel through.

REVIEW QUESTION 26.1 Why is the historical role of the Michelson-Morley experiment that of a testing experiment, not an observational experiment?

26.2 Postulates of special relativity

By the end of the 19th century, the laws of Newtonian physics had been well developed and accepted. Physicists knew that using Newton's laws to analyze a process would yield consistent results, regardless of the inertial (nonaccelerating) reference frames used. This feature is known as **invariance**.

Invariance is a result of acceleration (rather than velocity) being proportional to the net force exerted on an object. For example, when you are on an airplane moving smoothly at constant velocity, you can not detect the airplane's motion; all the events around you occur as if the airplane is at rest on the ground. If you drop a book, it lands at your feet; if you push a suitcase, it accelerates in the same way as if the airplane were parked on the runway. However, for observers in noninertial reference frames (accelerating frames), such as airplane passengers when the plane is taking off, Newton's laws do not explain observed phenomena, such as a loose food cart starting to roll without being pushed.

Invariance did not apply to electromagnetism and in particular to the observation of light in different reference frames. The speed of light in different reference frames was addressed by a teenager named Albert Einstein, who became a master at designing *thought experiments*. In 1895, at age 16, he tried to visualize himself traveling beside a beam of light. If he traveled through air at speed 3×10^8 m/s, it would be possible to stay beside one of the crests of the light wave, which would appear as a static electromagnetic field oscillating in space but not in time. Such a field would contradict the mechanism that explains EM waves. Changing fields create each other; these are not changing. We can visualize an even more disturbing difficulty. If Einstein then turned on a lamp held in his hand, the light from this lamp should move away from him at speed 3×10^8 m/s. Thus, to another observer at rest with respect to the source of the original light wave, there would be two beams of light: the one beside Einstein traveling at 3×10^8 m/s and the one from Einstein's lamp moving at Einstein's speed plus the speed of the light leaving his lamp, that is, at 3×10^8 m/s $+ 3 \times 10^8$ m/s $= 6 \times 10^8$ m/s (**Figure 26.4**). These thought experiments meant that Maxwell's equations (which we discussed at the end of Chapter 21) had to be written differently for these different observers, with a different speed of light in each case.

FIGURE 26.4 A thought experiment. Imagine a person traveling at light speed beside the crest of a light wave. His lamp is shining forward in the direction of travel. Would an observer detect one light wave traveling at c and light from the moving light source traveling at 2c?

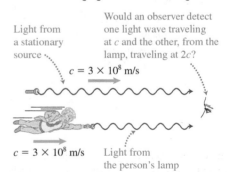

Einstein's two postulates

Einstein believed that the invariance principle was fundamental to any physics idea. He also believed that Maxwell's equations were correct. To resolve this contradiction, in 1905 Einstein proposed a new theory, called **the special theory of relativity**. He could have based it on the results of the Michelson-Morley experiment that suggested that the speed of light in air or a vacuum was independent of the motion of the observer, but he did not. In fact, it is not certain whether Einstein was aware of Michelson and Morley's results at all. Although there is some historical controversy about this, the general consensus is that Einstein based his ideas on his thought experiments.

Einstein's theory started with the following two *postulates*. A postulate is a statement that is assumed to be true (it is not derived from anything). It is usually used as the starting point for a logical argument.

1. *The laws of physics are the same in all inertial reference frames.* This was not a new idea in mechanics. Newton's second law,

$$\vec{a}_O = \frac{\sum \vec{F}_{on\ O}}{m_O}$$

remains the same regardless of the inertial reference frame in which you choose to apply it. But it was a new idea for Maxwell's equations of electromagnetism.

2. *The speed of light in a vacuum is measured to be the same in all inertial reference frames.* The speed of light in a vacuum measured by observers in different inertial reference frames is the same regardless of the relative motion of those observers.

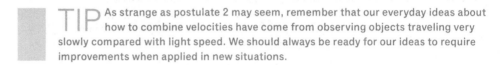

TIP   As strange as postulate 2 may seem, remember that our everyday ideas about how to combine velocities have come from observing objects traveling very slowly compared with light speed. We should always be ready for our ideas to require improvements when applied in new situations.

Although postulate 2 is supported by the Michelson-Morley experiment and is consistent with Maxwell's equations, it is still counterintuitive. Think back to the boats in Section 26.1. The speed with respect to the shore of the boat traveling parallel to the shore depended on its velocity with respect to the water and the velocity of the water with respect to the shore. Now, suppose that an intense pulse of laser light is shot from Earth toward a spaceship moving away. A stationary observer on Earth measures the light's speed c as 3.0×10^8 m/s (**Figure 26.5a**). If the spaceship is moving away from Earth at a speed v that is just a little less than light speed, then our intuition tells us that the laser light will have a difficult time reaching the ship. After all, isn't the laser light's speed just $c - v$ relative to the ship (similar to the boat going upstream)? However, according to Einstein's second postulate, an observer on the spaceship should also measure the laser light speed to be 3.0×10^8 m/s (Figure 26.5b). Although it sounds impossible, this is exactly the outcome of the Michelson-Morley experiment!

FIGURE 26.5 All inertial reference frame observers measure the same speed for light c.

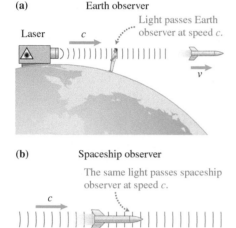

FIGURE 26.6 Bob and Alice sending pulses.

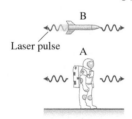

REVIEW QUESTION 26.2 Alice is standing on the ground; Bob is above her in a high-speed rocket. They each shoot laser pulses to the right and left when Bob is directly above Alice (see **Figure 26.6**). Draw a sketch that shows these pulses a moment later (draw the relative locations of Alice, Bob, and their four laser pulses). Explain your reasoning.

26.3 Simultaneity

In our boat analogy, we assumed that the time interval during which each boat traveled during its trip was the same independent of the reference frame (shore or water) $(\Delta t_{boat\ to\ shore} = \Delta t_{boat\ to\ water} = \Delta t)$. However, this seemingly reasonable way of thinking about time turns out to contradict the second postulate proposed by Einstein (and therefore it contradicts the results of the Michelson-Morley experiment).

The second postulate of the special theory of relativity made physicists completely rethink their ideas about time. Consider the following thought experiment.

Simultaneity of events in two different inertial reference frames

Suppose Alice is standing on the ground, and Bob is in a rocket (**Figure 26.7a**). They each have two clocks that they want to synchronize by sending light pulses to them. The clocks start when a pulse reaches them. If they start at the same time, the clocks are considered synchronized. All the clocks are *attached to the ground* and are equidistant from Alice (Bob's clocks are raised on poles so that his pulses reach them as they travel in straight lines to the right and left). When Bob is directly above Alice, they both send their signals. Who ends up with a synchronized pair of clocks, Alice or Bob? Let's consider the two perspectives.

Because the clocks are equidistant from Alice and the clocks are at rest with respect to her (that is, both Alice and the clocks are in the same inertial reference frame), she notes that the light reaches her front clock (on the right) and her back clock (on the left) at the same time (Figure 26.7b). However, Bob sees his front clock moving toward him while the light pulse he sent is moving toward it. He also sees his back clock moving away from him while the light pulse is moving toward it (Figure 26.7c). Therefore, the light pulse headed to Bob's (approaching) front clock travels a shorter distance before reaching the clock than the light pulse headed to Bob's (receding) back clock. Because the speed of light is finite and is the same in all directions for all observers, according to Bob, the light pulse reaches his front clock before it reaches his back clock. We conclude that the pulses arrive at the clocks at the same time for Alice, but do not arrive at the clocks at the same time for Bob. Alice's clocks are synchronized; Bob's are not.

Implications of the difference in observations

Think about what this means. We are focusing on two events for each person: (1) their light pulse reaching their front clock and (2) their light pulse reaching their back clock. In Alice's reference frame (attached to the ground), these two events are simultaneous. In Bob's reference frame (the flying rocket), these two events are not simultaneous.

This really occurs! Events that happen at the same time in one reference frame do not necessarily occur at the same time in another. At regular rocket speeds, however, the times of the events won't vary by more than 10^{-12} to 10^{-13} s between observers. This difference gets larger the faster the rocket travels relative to the ground, but the effect only becomes significant if the rocket travels at a substantial fraction of light speed.

Phenomena that only become significant in high-speed circumstances like this are called **relativistic effects**. This so-called **breakdown of simultaneity** is just the first of many relativistic effects that we will investigate in this chapter.

> **Breakdown of simultaneity** Events that occur simultaneously for an observer in one inertial reference frame do not necessarily occur simultaneously for observers in other inertial reference frames.

REVIEW QUESTION 26.3 You hear in your physics class the following statement: "Time is relative." How would you explain the meaning of this statement to a friend who is not taking physics?

26.4 Time dilation

In this section we will examine how the time interval between events differs in different inertial reference frames. An event is anything that happens: a person's departure for a trip, the person's return from that trip, light pulses reaching clocks, as in the example from the previous section, and so on. To specify an event quantitatively, we need to indicate where it occurs (a position) and when it occurs (a clock reading).

Our everyday experience supports the idea that the time interval between two events is something that all observers will agree on. For example, if your physics class lasts 60 min according to your clock, won't astronauts on the International Space Station watching the webcast find that 60 min passes on their clocks as well? It shouldn't matter that the astronauts are moving at 7700 m/s (17,000 mi/h) relative to you, right? The thought experiment in the previous section, however, suggests that it does matter.

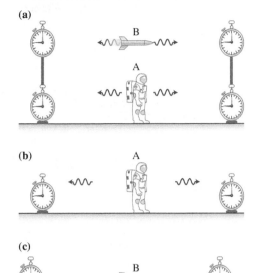

FIGURE 26.7 Laser pulses used to synchronize clocks. (a) Alice, Bob, and the clocks. (b) Pulses arrive at the clocks at the same time for Alice. (c) Pulses do not arrive at the clocks at the same time for Bob.

FIGURE 26.8 An experiment for two observers (a) to detect the time interval between two events. The time interval between events as seen (b) in the proper reference frame, where they occur at the same place and (c) in another reference frame, where the events occur at different places.

(a)

Laser light travels upward, reflects off of a mirror, and is detected below.

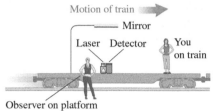

Observer on platform

(b)   You on train

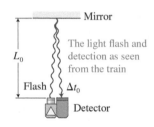

(c)   Observer on platform

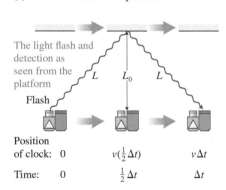

We can use another thought experiment to further investigate this. Suppose you are standing on a moving train. Your friend stands on the platform at the station. On the train is a laser pointing upward, a mirror located directly above the laser, and a light detector next to the laser (**Figure 26.8a**). The train is moving to the right at speed v, a significant fraction of light speed relative to the platform. A flash of light is emitted from the laser (event 1), reflects off the mirror, and is then detected (event 2).

In your reference frame on the train, you see the laser light travel vertically upward, then downward (Figure 26.8b). In your friend's reference frame on the platform, however, the laser light must move upward at an angle in order to reflect off the mirror, then downward at an angle to hit the detector—because the train is moving (Figure 26.8c).

What is the time interval between the flash and its detection (the two events), as measured by each of the two observers? For you, the light travels at speed c a distance $2L_0$, where L_0 is the distance from the light source to the mirror. The time interval between the flash and its detection is then

$$\Delta t_0 = \frac{2L_0}{c}$$

According to your friend on the platform, the light travels a longer distance, $2L$. Using the Pythagorean theorem, the distance L is the square root of the sum of the lengths squared of the two sides of the triangle with L as its hypotenuse:

$$L = \sqrt{L_0^2 + \left(\tfrac{1}{2}v\,\Delta t\right)^2}$$

The base of the triangle $(1/2)v\,\Delta t$ is one half the distance the train travels at speed v during the time interval Δt during which the light travels from its source up to the mirror and back down to the receiver. The time interval between the flash and its detection is

$$\Delta t = \frac{2L}{c} = \frac{2\sqrt{L_0^2 + \left(\tfrac{1}{2}v\,\Delta t\right)^2}}{c}$$

Rearrange $\Delta t_0 = \dfrac{2L_0}{c}$ to $L_0 = \dfrac{c\Delta t_0}{2}$ and substitute this into the above equation to get

$$\Delta t = \frac{2\sqrt{\dfrac{c^2\Delta t_0^2}{4} + \dfrac{v^2\Delta t^2}{4}}}{c}$$

Solving for Δt, we get

$$\Delta t = \frac{\Delta t_0}{\sqrt{1 - \dfrac{v^2}{c^2}}}$$

The term

$$\sqrt{1 - \frac{v^2}{c^2}}$$

is always less than or equal to 1. (It is equal to 1 only when the train is at rest with respect to the platform.) Therefore, the time interval Δt between the flash and its detection measured by your friend on the platform is greater than the time interval Δt_0 measured by you on the train.

Notice that we used $c = 3.00 \times 10^8$ m/s for the speed of light in both reference frames (according to Einstein's second postulate). This analysis results in a striking conclusion—the time interval during which the laser light was in flight differs depending on the reference frame in which it is measured. This relativistic effect is called **time dilation**.

Notice that your reference frame on the train is the one reference frame where the two events—the flash and the detection—occur at the same position. The time interval Δt_0 between the events measured in this reference frame is called the **proper time interval**, and this reference frame is known as the **proper reference frame**. The time interval Δt between these same two events measured in any other reference frame will always be longer. That is, $\Delta t > \Delta t_0$. The term "proper" comes from a mistranslation of "proprietary," which implies that only the person holding the clock would experience that much time.

We certainly don't observe this occurring in everyday life—if a particular boat trip takes 2 h according to the captain on the boat, won't it also take 2 h according to people on shore? Not if the idea of time dilation is correct. Actually, time dilation is happening all the time, though the effect only becomes substantial when the two reference frames are moving relative to each other at significant fractions of light speed. For example, for Δt and Δt_0 to be one-millionth of a percent different, the two reference frames would have to be moving relative to each other at over 4×10^4 m/s (over 90,000 mi/h).

Time dilation can be tested using what is known as a muon decay experiment, described in Testing Experiment **Table 26.2**. A muon is an elementary particle that occurs naturally in the streams of particles coming to us from the Sun and distant stars (called cosmic rays); it also can be produced during reactions in high-energy particle accelerators. Once produced in an accelerator, muons "live" for a very short time. If 1000 muons are produced, then 1.5×10^{-6} s later, approximately 500 muons remain. This time is called the **half-life** because half of the particles remain after this time.

Consider an experiment in which 1000 muons are produced and move across the lab at $0.95c$. Event 1 will be the production of the muons. Event 2 will be the decay of the 500th muon.

TIP ▸ Relativistic effects are important when objects are observed to move at speeds that are significant fractions of light speed and are not important at "everyday" speeds. A good rule of thumb is that if an object is observed to move at a speed $0.1c$ or greater, then relativistic effects should be included.

TESTING EXPERIMENT TABLE 26.2 Testing the idea of time dilation

Testing experiment	Prediction	Outcome
A beam of 1000 muons is produced by a source (event 1) and emitted at a speed of $0.95c$. How far from the source will the muons travel before their number is reduced to 500 (event 2)? 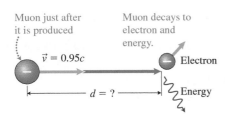	*If time passes the same way for all observers (no time dilation),* then in the laboratory reference frame, the muons should travel a distance equal to the product of their speed with respect to the lab reference frame and their half-life: $$d = (0.95c)(1.5 \times 10^{-6} \text{ s})$$ $$= (0.95)(3.0 \times 10^8 \text{ m/s})(1.5 \times 10^{-6} \text{ s}) = 430 \text{ m}$$ *If time passes differently for observers in different reference frames and obeys the time dilation equation,* then in the laboratory reference frame, one half of the muons should travel a distance equal to the product of their speed with respect to the lab reference frame and their half-life *in the lab reference frame,* which will be different from their half-life in the proper reference frame. Their half-life in the lab reference frame is $$\Delta t = \frac{1.5 \times 10^{-6} \text{ s}}{\sqrt{1 - \frac{(0.95c)^2}{c^2}}} = 4.8 \times 10^{-6} \text{ s}$$ Because they move at speed $0.95c$ in the lab frame, one half of the muons travel $$d = (0.95c)(4.8 \times 10^{-6} \text{ s})$$ $$= (0.95)(3.0 \times 10^8 \text{ m/s})(4.8 \times 10^{-6} \text{ s}) = 1400 \text{ m}$$ Thus if time dilation occurs, one half of the muons should travel a significantly greater distance through the lab than they would if there were no time dilation.	The muons actually traveled about 1400 m before their number was reduced to one-half.

Muon just after it is produced — $\vec{v} = 0.95c$. Muon decays to electron and energy. → Electron, Energy. $d = ?$

Conclusion

The outcome matches the prediction based on the time dilation equation. Many experiments confirm that short-lived particles travel much farther than would be possible given their proper half-lives. The idea of time dilation is not rejected. We have gained more confidence in it.

Muons are produced in nature as well. They are created when cosmic rays (high-energy protons) hit molecules in the atmosphere approximately 60,000 m above Earth's surface. If there were no time dilation, these muons would have a 1 in 10^{40} chance of reaching Earth from that altitude. However, because of time dilation and the fact that they are moving at nearly the speed of light, about 1 out of every 10 muons reaches Earth's surface.

Time dilation The reference frame in which two events occur at the same position is called the *proper reference frame* for these events. The time interval between these events measured by an observer in the proper reference frame is the *proper time interval*, Δt_0. The time interval Δt between the same two events as measured by an observer moving at constant velocity $\vec{v}$ relative to the proper reference frame is

$$\Delta t = \frac{\Delta t_0}{\sqrt{1 - \dfrac{v^2}{c^2}}} \qquad (26.1)$$

where v is the speed of the observer relative to the proper reference frame.

The words "moving at constant velocity $\vec{v}$ relative to the proper reference frame" are especially important. Consider as an example a spaceship on which a light flashes at regular time intervals. The time interval between flashes as measured by a passenger on the ship is the proper time interval because the flashes occur at the same place relative to the passenger. For an observer on Earth, the flashes occur at different places as the spaceship moves past Earth. Therefore, Earth is not the proper reference frame. We say that Earth moves at speed v relative to the proper reference frame of the spaceship.

REVIEW QUESTION 26.4 You are on a train eating an apple. When you start eating, the train is passing a station, and a clock there reads 5 pm. When you finish, the train is passing the next station, where the clock reads 5:05 pm. Assuming that these two clocks are synchronized, did it take you 5 min to eat the apple? More than 5 min? Less than 5 min? Explain.

FIGURE 26.9 A moving arrow (a) starts and (b) stops a clock.

(a)

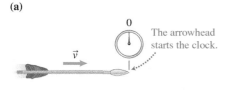

The arrowhead starts the clock.

(b)

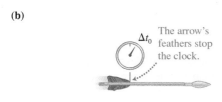

The arrow's feathers stop the clock.

26.5 Length contraction

The measured time interval between two events depends on the reference frame from which the measurement is made. In addition, other physical quantities that are normally thought to be independent of reference frame, such as the length of an object, in fact do depend on the frame of reference.

Consider an arrow flying across a lab that moves past a clock at rest with respect to the lab (**Figure 26.9**). As the arrow's head passes the clock (event 1) it triggers the clock to start (Figure 26.9a). As the arrow's feathers pass the clock (event 2) the clock stops (Figure 26.9b). The two events occurred at the same place in the lab reference frame, so that is the proper reference frame, and the time interval Δt_0 measured by the clock is the proper time interval between the two events. In contrast, if the arrow had a tiny clock attached to it, the time interval Δt between the events would be

$$\Delta t = \frac{\Delta t_0}{\sqrt{1 - \dfrac{v^2}{c^2}}}$$

where v is the speed at which the lab moves past the arrow.

Could the length of the arrow depend on the reference frame from which it is measured? We'll call the arrow's **proper length**, L_0, its length in a reference frame in which the arrow is stationary. In this case, it's the reference frame defined by the arrow itself. Imagine that a fly has landed on the arrow and is traveling with it. Because the arrow is at rest relative to the fly, the fly would measure the length of the arrow to be the arrow's proper length L_0 (**Figure 26.10a**).

To an observer in the clock's (and lab's) reference frame, however, the length L of the moving arrow may be different (Figure 26.10b). We can relate L to L_0 by noting that the arrow's speed relative to the lab is the same as the lab's speed relative to the arrow (although their velocities are in opposite directions). We can calculate this speed in two different ways. In the arrow's reference frame, its speed is the proper length L_0 divided by the time interval Δt:

$$v = \frac{L_0}{\Delta t}$$

In the lab reference frame, the speed is the length L divided by the proper time interval Δt_0:

$$v = \frac{L}{\Delta t_0}$$

Because these two speeds are the same, we can set them equal to each other:

$$\frac{L}{\Delta t_0} = \frac{L_0}{\Delta t}$$

Using the time dilation equation, we can substitute for Δt:

$$\frac{L}{\Delta t_0} = \frac{L_0 \sqrt{1 - \dfrac{v^2}{c^2}}}{\Delta t_0}$$

Multiplying both sides of the above by Δt_0, we get a relationship between the proper length of the arrow, L_0, and its length L when measured in another reference frame:

$$L = L_0 \sqrt{1 - \frac{v^2}{c^2}} \qquad (26.2)$$

The square root that appears in the above equation will always have a value less than 1. As a result, the length of the arrow measured in a reference frame where the arrow is moving will always be less than the arrow's proper length. In other words, the arrow will appear shorter along the direction that it is moving. This is known as **length contraction**.

Length contraction An object has a length L_0 when measured in a reference frame at rest relative to the object. This length is called the object's *proper length*. In a reference frame moving at speed v relative to the proper reference frame, the object's length L is

$$L = L_0 \sqrt{1 - \frac{v^2}{c^2}} \qquad (26.2)$$

FIGURE 26.10 Measuring the arrow length (a) in the proper reference frame, where the arrow is at rest, and (b) in the clock's reference frame, where the arrow is moving.

(a)

A fly on the arrow measures the proper length, L_0.

(b)

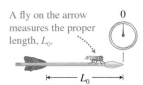

An observer in the clock's reference frame measures the length as L.

TIP Notice that length contraction only occurs parallel to the direction the object is moving. In the perpendicular direction, this effect does not occur.

QUANTITATIVE EXERCISE 26.1   **Muons travel far!**

In Table 26.2 we found that muons survive longer as a result of time dilation. However, that explanation does not work in the frame of the muon. How can a muon created in the upper atmosphere make it all the way to Earth if it should travel only about 450 m, given that it lives only 1.5 μs? The upper atmosphere is approximately 60 km deep. Cosmic rays are very energetic and create muons that move at $0.995c$.

Represent mathematically In the frame of the muon, its average lifetime is 1.5 μs and Earth is moving toward it at $0.995c$. But some variable has to change for it to reach Earth, with the long distance involved. Specifically, according to the muon, the distance to Earth is shorter according to length contraction. Using Eq. (26.2), we can write

$$L = L_0\sqrt{1 - \frac{v^2}{c^2}}$$

Solve and evaluate

$$L = L_0\sqrt{1 - \frac{v^2}{c^2}} = (60\text{ km})\sqrt{1 - \frac{(0.995c)^2}{c^2}} = 6.0\text{ km}$$

This distance is much closer to 450 m than 60 km and, given the statistics, an average muon now has a high probability of making it to Earth.

Try it yourself Suppose you live in a strange universe in which the speed of light is constant, but is only 20 m/s. Your car is 3.0 m long when parked by the curb in front of your house. How long will it look to your friend who is driving by your car, traveling at 10 m/s relative to your car?

Answer   2.6 m.

REVIEW QUESTION 26.5 You hold a 1-m-long stick. Describe those conditions under which an observer will measure the stick to be shorter than 1 m and those conditions under which an observer will measure the stick to be longer than 1 m.

26.6 Spacetime diagrams

FIGURE 26.11 World lines for two objects and two light beams drawn on a spacetime diagram.

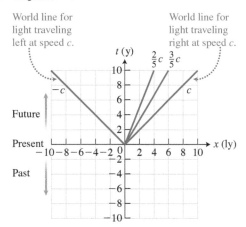

In relativity, as in other areas, a graph is useful for keeping track of information. This is especially true when there are multiple objects. There is a tool from Chapter 2 that, with only slight modification, comes in handy here. Recall that a position-versus-time graph shows where an object is at any given time. To represent the motion of a stationary object on a position-versus-time graph, we draw a horizontal line. If a position-versus-time graph is a straight line with a positive slope, it means that the object is moving in a positive direction with constant speed. On these graphs, time is represented on the horizontal axis and position on the vertical axis. When physicists are accounting for relativity, the axes are traditionally switched, so that time is represented on the vertical axis and position on the horizontal axis. This new representation is often referred to as a **spacetime diagram** to emphasize that space and time are intertwined. The line showing the time-versus-position graph of an object is called a **world line** for that object (**Figure 26.11**). In Figure 26.11 you see world lines for two objects and two light beams sent in opposite directions. Notice that the two objects move at speeds less than the speed of light and that their world lines are more steeply angled. In general, the slower the object, the steeper its world line. Thus the line for the object moving at $\frac{2}{5}c$ is steeper than that for the object moving at $\frac{3}{5}c$.

Because relativistic effects do not typically appear until close to the speed of light, we will be choosing units that highlight this aspect. And because the ultimate speed limit is the speed of light c, the units are also selected with this in mind. The motion of light is represented by a line with a slope of $\pm 45°$ depending on which direction the light is traveling. There are several systems of units that would give these results. For example, if the unit on the time axis is years (y), then the unit on the horizontal axis will be a **light-year** (ly), defined as the distance light travels in a year. Alternatively, we could have the unit of time as a second and the unit of distance the distance that light travels in 1 s—one *light-second*.

One result of switching the axes is that the slope of the line is no longer velocity; rather, it is the *inverse* of velocity. However, because we chose years (or seconds) for our time units and light-years (or light-seconds) for distance units, a line with the slope of ± 1 disregarding the units ($\pm 45°$) corresponds to the motion of an object traveling at the speed of light. Thus, an object that is moving at $\frac{1}{2}c$ would have a world line with slope 2. This leads to an important tip for building spacetime diagrams: **the slope of the world line for an object moving at speed v is c/v.**

CONCEPTUAL EXERCISE 26.2   **Drawing spacetime diagrams**

Draw a spacetime diagram for the situation in Review Question 26.2, with Bob traveling at 86.6% the speed of light. Include all objects (including light pulses) and draw the diagram with Alice as the observer.

Sketch and translate We have a sketch in Figure 26.6. It is important to remember that we are considering Alice's reference frame.

Simplify and diagram The problem statement does not say where Alice is standing, so we assume for simplicity that she is standing at the origin. Alice's world line is therefore vertical (her position does not change as time elapses). Bob is traveling to the right at $0.866c$. The slope of his world line is therefore $\frac{c}{(0.866c)} = 1.15$. In other words, his world line makes an angle to the horizontal of slightly more than 45°.

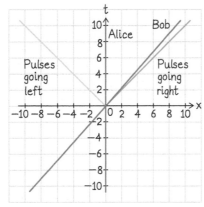

The four light pulses are created when Bob is directly above Alice. The pulses keep up with one another because of postulate 2. Because the graph only represents one dimension, each line represents two pulses. They each have a slope of ± 1. (Notice we have not specified the units yet. This diagram would work no matter which units we chose.) The spacetime diagram from Alice's perspective is shown in the left column.

Try it yourself Draw the spacetime diagram with Bob as the observer. Explain your reasoning.

Answer

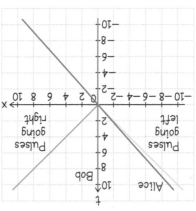

To get Bob's perspective, we first have to realize that Bob does not consider himself to be moving, so he would draw his world line as a vertical line, say, at the origin. Bob would say Alice is moving to the left. And Bob would still say that the light pulses travel away from him at the speed of light. This gives us the diagram at right.

Spacetime diagrams deserve one more note. When switching frames of reference, there are two effects that we have not yet incorporated: time dilation and length contraction. In principle, you could calculate either of these to adjust the amount of time between two events or distance between two objects. It turns out that by adjusting the distance between objects with length contraction, the amount of time that elapses between events always works itself out. This will also account for any mismatches in simultaneity experienced by the two observers. The following example shows how the process works.

EXAMPLE 26.3   **Using spacetime diagrams to analyze simultaneity**

Explain what you need to do to add world lines for the two clocks equidistant from Alice in Figure 26.7 to the spacetime diagrams in the previous exercise (including that in the "Try it yourself" section) and do the required calculations.

Sketch and translate The situation in the problem is sketched in Figures 26.7a and b. We need to represent the same process with spacetime diagrams.

Simplify and diagram The problem does not specify the distance between Alice and the clocks, so we will choose a distance consistent with our previous diagrams; let's say 6 units. It is helpful to pick a unit at this point, so let's agree that 1 unit of distance is 300 m. It takes light $1\ \mu s = 10^{-6}$ s to go that far, so our choice of 1 unit of time is 10^{-6} s . This puts each set of clocks at (6 units) $\times$ (300 m)/unit $= 1800$ m away from Alice. The clocks are not moving, so they should be

(CONTINUED)

represented by vertical lines on the diagram corresponding to Alice's perspective, as shown below.

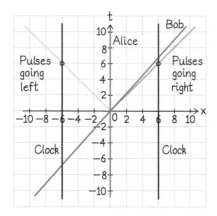

On the above diagram, the clocks are represented as black lines, and the times of the intersection of these lines with the lines of the light pulses indicate when those clocks register the reading (that is, when the pulses arrive at the clocks). You can see that the clocks register the arrival of the pulses simultaneously at a time of 6.0 units.

The spacetime diagram for Bob will be different. Once again, we need to add the lines for the clocks to the diagram, but this time the clocks are moving at the same speed as Alice and are equidistant from Alice. This would make the clocks' lines parallel to Alice's world line, which is tilted to the left. However, we do not know the distance between Alice and the clocks as seen by Bob; thus we cannot draw the diagram before we do the calculation.

Represent mathematically and solve Given that the clocks and Alice are moving with respect to Bob, he registers that there are not

6 units of length between them, but a smaller length because of length contraction. This length is calculated using the expression for length contraction:

$$L = L_0\sqrt{1 - \frac{v^2}{c^2}} = (6 \text{ units})\sqrt{1 - \frac{(0.866c)^2}{c^2}} = \pm(6 \text{ units}) \times 0.5$$

$$= \pm 3 \text{ units}$$

Evaluate The distance to the clocks is smaller for Bob than for Alice. We can now draw the spacetime diagrams for the clocks as seen by Bob. Notice that the clock that is approaching Bob (on the right) starts where the dot is—this is where the line for the light pulse crosses the line for the clock (just under 2 time units)—while the other light pulse is still chasing the receding clock. It doesn't catch it until sometime after the graph ends (where the yellow line for light pulses going left crosses the left-hand black line).

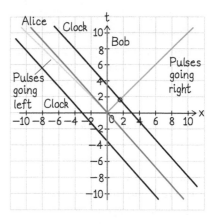

Try it yourself At what time (in units of time) will the pair of right clocks fly by Bob? Show on the spacetime diagram and confirm with the calculation.

Answer   3.5 units of time.

A few final comments about spacetime diagrams are in order. First, each observer has a simple explanation for what happened. Alice says the clocks are synchronized because they were equidistant and not moving, whereas Bob declares that they are not synchronized because they were moving. Einstein would say that both Alice and Bob are correct! Each person just needs to explain what happened in their own reference frame. When we switch between reference frames, we encounter length contraction (we also experience time dilation).

Below is a summary of our knowledge of spacetime diagrams:

1. The slope of the world line for an object moving at speed v is c/v.

2. The world line for light is tilted at $\pm 45°$ (this means that the units on the axes need to be chosen for this to be true).

3. When you switch reference frames, length contraction needs to be accounted for. (Time dilation needs to be considered as well, but getting length contraction correct takes care of that.) If an object is not moving in one frame, it will be moving in the new frame, and its length will be contracted. Similarly, if two objects are stationary, the distance between them will be contracted.

REVIEW QUESTION 26.6 We encounter vertical and tilted world lines on spacetime diagrams, but never horizontal lines. Why?

26.7 Velocity transformations

In this section we will learn how to reconcile the idea that light speed is the same for all observers even if they are moving relative to each other at nearly the speed of light. Imagine you are on a skateboard moving to the left at velocity $\vec{v}_{\text{skateboarder to ground}}$ relative to the ground. You throw a ball you are carrying in the same direction at velocity $\vec{v}_{\text{ball to skateboarder}}$. Your friend (the ball catcher) standing on the ground detects the ball moving at a different velocity: $\vec{v}_{\text{ball to catcher}} = \vec{v}_{\text{ball to skateboarder}} + \vec{v}_{\text{skateboarder to catcher}}$ (**Figure 26.12**). This result (similar to the result for the boat traveling upstream or downstream) is based on the assumption that the time interval of the ball in flight is the same for the skateboarder and for the ball catcher on the ground (the same assumption that we made in the boat example in Section 26.1). We can generalize this equation as

$$\vec{v}_{\text{OS}} = \vec{v}_{\text{OS}'} + \vec{v}_{\text{S}'\text{S}} \qquad (26.3)$$

Here $\vec{v}_{\text{OS}}$ is the velocity of the object (the ball) with respect to reference frame S (the ball catcher); $\vec{v}_{\text{OS}'}$ is the velocity of the object with respect to a second reference frame S′ (the skateboarder); and $\vec{v}_{\text{S}'\text{S}}$ is the velocity of the reference frame S′ relative to the reference frame S. Equation (26.3) is called the **classical** (or **Galilean**) **velocity transformation equation**. It is consistent with our everyday experience. The derivation of the equation is based on the assumption that the time interval during which the motion occurs is the same for all observers.

Difficulties with the classical velocity transformation equation

According to the ideas of special relativity and the results of the Michelson-Morley experiment, the speed of light is the same in all inertial reference frames. This idea contradicts the classical velocity transformation equation. Suppose you are in reference frame S′ moving toward the right at speed $v_{\text{S}'\text{S}} = 0.99c$ relative to another reference frame S. You shine a laser beam straight ahead. The light moves at speed $v_{\text{LS}'} = c$ forward in your S′ reference frame. According to the classical velocity transformation equation, the speed v_{LS} of the light in reference frame S should be

$$v_{\text{LS}} = v_{\text{LS}'} + v_{\text{S}'\text{S}} = c + 0.99c = 1.99c$$

This result contradicts Einstein's second postulate. We need an improved velocity transformation equation that respects the postulates of relativity.

Relativistic velocity transformation equation

Deriving this equation requires techniques beyond the scope of this book, so we'll just provide the result for one-dimensional processes.

Relativistic velocity transformation Suppose that inertial reference frame S′ is moving relative to inertial reference frame S at velocity $\vec{v}_{\text{S}'\text{S}}$ (positive to the right and negative to the left). An object O moves at velocity $\vec{v}_{\text{OS}'}$ relative to reference frame S′ (also positive to the right and negative to the left). The velocity $\vec{v}_{\text{OS}}$ of O in reference frame S is then

$$v_{\text{OS}} = \frac{v_{\text{OS}'} + v_{\text{S}'\text{S}}}{1 + \dfrac{v_{\text{OS}'}v_{\text{S}'\text{S}}}{c^2}} \qquad (26.4)$$

Notice that in Eq. (26.4) the numerator is the same as in the classical equation, but the denominator is different. Does the above equation satisfy postulate 2 of special relativity and agree with the classical equation for low speeds? Let's check some limiting cases.

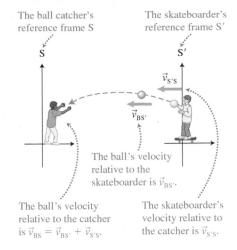

FIGURE 26.12 The classical velocity transformation. Velocities are simply additive in everyday situations.

The ball catcher's reference frame S

The skateboarder's reference frame S′

The ball's velocity relative to the skateboarder is $\vec{v}_{\text{BS}'}$.

The ball's velocity relative to the catcher is $\vec{v}_{\text{BS}} = \vec{v}_{\text{BS}'} + \vec{v}_{\text{S}'\text{S}}$.

The skateboarder's velocity relative to the catcher is $\vec{v}_{\text{S}'\text{S}}$.

TIP Because Eq. (26.3) is a vector equation, we must break it into components before using it. We have to choose a coordinate system that determines the signs of the individual vector components.

Limiting case analysis of Eq. (26.4)

Case 1: Suppose that an object O moves slowly in reference frame S' ($v_{OS'} \ll c$), or that reference frame S' moves slowly relative to reference frame S ($v_{S'S} \ll c$). In either case, the denominator of Eq. (26.4) is approximately 1 and the equation becomes the classical velocity transformation Eq. (26.3).

Case 2: Suppose that a rocket (reference frame S') passing Earth emits a laser pulse in the forward direction. The light travels at velocity $v_{LS'} = c$ relative to the rocket. Suppose the rocket (reference frame S') travels at velocity $v_{S'S} = 0.99c$ relative to the Earth (reference frame S). What is the velocity v_{LS} of the laser pulse relative to Earth?

$$v_{LS} = \frac{v_{LS'} + v_{S'S}}{1 + \dfrac{v_{LS'}v_{S'S}}{c^2}} = \frac{c + 0.99c}{1 + \dfrac{(c)(0.99c)}{c^2}} = \frac{(1 + 0.99)c}{(1 + 0.99)} = c$$

The laser pulse is observed to travel at speed c in both inertial reference frames, consistent with the postulates of special relativity.

REVIEW QUESTION 26.7 What is the meaning of the term "velocity transformation," and why is the classical velocity transformation different from the relativistic transformation?

26.8 Relativistic momentum

We know (from Chapter 6) that if the net impulse exerted on a system is zero, the momentum of the system is constant. This principle is very useful for analyzing collisions of various kinds. Our past use of the principle involved objects moving at nonrelativistic speeds ($v \ll c$). Can we use the classical expression for the momentum of an object

$$\vec{p} = m\vec{v} = m\frac{\Delta\vec{x}}{\Delta t}$$

and the impulse-momentum principle to analyze situations where objects are moving at relativistic velocities?

When we use the above classical definition of momentum to analyze collisions of elementary particles moving at high speed, we find that even for an isolated system, the momentum of the system is constant in some reference frames but not in others. Thus, we have to either give up momentum as a conserved quantity or find a new expression that restores consistency across all reference frames. We'll take the second approach and find a new relativistic expression for momentum.

To redefine momentum, consider the classical expression ($\vec{p} = m\vec{v} = m(\Delta\vec{x}/\Delta t)$). In that expression, m is the mass of the object and $\Delta\vec{x}$ is the displacement of the object during the time interval Δt, both measured in some inertial reference frame. To get an improved relativistic expression for momentum, we try replacing Δt with the proper time interval Δt_0, where the relevant events are the object at the beginning and end of its displacement:

$$\vec{p} = m\vec{v} = m\frac{\Delta\vec{x}}{\Delta t_0}$$

If we substitute the expression for Δt_0 shown in Eq. (26.1) into the above definition, we get

$$\vec{p} = m\frac{\Delta\vec{x}}{\Delta t\sqrt{1 - (v/c)^2}} = \frac{m}{\sqrt{1 - (v/c)^2}}\frac{\Delta\vec{x}}{\Delta t} = \frac{m\vec{v}}{\sqrt{1 - (v/c)^2}}$$

Note that the $\vec{v}$ in the numerator is the velocity of the object, while the v in the denominator is its speed. This new expression reduces to the classical expression if the speed of the object is much less than the speed of light. Using this new expression for momentum, it is possible to show (though we will not do it here) that when the net impulse exerted on a system is zero, the system's momentum is constant regardless of inertial reference frame.

Relativistic momentum The relativistic momentum of an object of mass m in a reference frame where the object is moving at velocity $\vec{v}$ is

$$\vec{p} = \frac{m\vec{v}}{\sqrt{1 - \left(\dfrac{v}{c}\right)^2}}$$

(26.5)

TIP Equation (26.5) is an improved expression for the momentum of objects traveling at all speeds rather than just speeds that are small compared with light speed. However, if the speed of the object is not a significant fraction of light speed, the simpler nonrelativistic expression can be used.

In Eq. (26.5) the denominator gets smaller and smaller as the object's speed gets closer and closer to the speed of light. This means that as speed increases toward light speed, the momentum of the object approaches infinity. This result is consistent with the idea that no object can travel faster than light speed in a vacuum. No such restriction was present in the nonrelativistic expression for the momentum of an object ($\vec{p} = m\vec{v}$).

EXAMPLE 26.4 ▸ **Cosmic ray hits nitrogen**

Some cosmic rays are high-energy protons produced during the explosive collapse of stars near the ends of their lives. Suppose a cosmic ray proton traveling at $0.90c$ enters our atmosphere and collides with a nitrogen atom. After the collision, the proton recoils at speed $0.70c$ in the opposite direction. The mass of the nitrogen atom is 14 times greater than the proton's mass. How fast is the nitrogen atom moving after the collision?

Sketch and translate We first sketch the process (see below). The system of interest is the proton and the nitrogen atom. Choose an inertial reference frame that is at rest relative to the nitrogen atom before it is hit by the proton and choose the positive x-axis to be the direction the proton is initially traveling. In that case, the initial and final velocity components of the proton and nitrogen atom along that axis are $v_{pix} = +0.90c$, $v_{pfx} = -0.70c$, and $v_{Nix} = 0$, and v_{Nfx} is the unknown final velocity component of the nitrogen atom. Let $m_p = m$; then $m_N = 14m$.

Initial situation

Final situation

Simplify and diagram Assume that there are no interactions between the system and the environment so that the momentum of the system is

constant. Because of the high speeds of the proton before and after the collision, we need to use the relativistic expression for the proton momentum. But we'll have to wait and see whether we need to treat the nitrogen atom relativistically; it depends on its speed relative to c. A momentum bar chart represents the process.

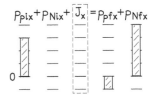

Represent mathematically The bar chart helps us apply the conservation of momentum principle along the x-axis:

$$p_{pix} + p_{Nix} + J_x = p_{pfx} + p_{Nfx}$$

or

$$\frac{m_p v_{pix}}{\sqrt{1 - (v_{pi}/c)^2}} + 0 + 0 = \frac{m_p v_{pfx}}{\sqrt{1 - (v_{pf}/c)^2}} + \frac{m_N v_{Nfx}}{\sqrt{1 - (v_{Nf}/c)^2}}$$

Solve and evaluate Substitute the known information and solve for v_{Nf}:

$$\frac{m(0.90c)}{\sqrt{1 - (0.90c/c)^2}} = \frac{m(-0.70c)}{\sqrt{1 - (0.70c/c)^2}} + \frac{14 m v_{Nf}}{\sqrt{1 - (v_{Nf}/c)^2}}$$

Cancel factors of m and c and simplify to get

$$\frac{0.90}{\sqrt{1 - (0.90)^2}}c = \frac{-0.70}{\sqrt{1 - (0.70)^2}}c + \frac{14 v_{Nf}}{\sqrt{1 - (v_{Nf}/c)^2}}$$

$$\Rightarrow \frac{14 v_{Nf}}{\sqrt{1 - (v_{Nf}/c)^2}} = \left(\frac{0.90}{\sqrt{1 - (0.90)^2}} + \frac{0.70}{\sqrt{1 - (0.70)^2}}\right)c$$

$$= (2.06 + 0.98)c$$

(CONTINUED)

Finally, we get

$$\frac{v_{Nf}}{\sqrt{1 - (v_{Nf}/c)^2}} = 0.22c$$

To solve for v_{Nf}, multiply both sides by $\sqrt{1 - (v_{Nf}/c)^2}$ and square both sides:

$$v_{Nf}^2 = (0.22c)^2(1 - (v_{Nf}/c)^2)$$

Now collect all terms with v_{Nf} on the left side:

$$v_{Nf}^2 + (0.22v_{Nf})^2 = (0.22c)^2$$

Now factor out v_{Nf}^2 and solve for it:

$$v_{Nf}^2 = \frac{(0.22c)^2}{1 + 0.22^2}$$

v_{Nf} is then

$$v_{Nf} = \frac{0.22}{\sqrt{1 + 0.22^2}}c = 0.21c = 6.4 \times 10^7 \text{ m/s}$$

This is a significant fraction of light speed, but not at all unreasonable for a cosmic ray collision.

Try it yourself Calculate the speed of the nitrogen atom after the collision using the nonrelativistic expression for the momentum of the proton and nitrogen. How does it compare to the relativistic value?

Answer   Relativistic effects are clearly relevant in this situation. 3.4×10^7 m/s, compared to the relativistic value of 6.4×10^7 m/s.

REVIEW QUESTION 26.8 Why must the classical expression of momentum be modified in special relativity?

26.9 Relativistic energy

Our ideas about space, time, velocity, and now momentum have been modified to include relativistic effects. A simple thought experiment shows that ideas concerning energy must also be modified. Imagine that we accelerate an electron through a potential difference of 300,000 V (large, but still achievable). The kinetic energy of this electron will then be $(1/2)mv^2 = e\,\Delta V$. This means the speed of the electron once it has finished accelerating will be

$$v = \sqrt{\frac{2e\,\Delta V}{m}} = \sqrt{\frac{2(1.6 \times 10^{-19}\text{ C})(300,000\text{ V})}{9.11 \times 10^{-31}\text{ kg}}} = 3.2 \times 10^8 \text{ m/s}$$

This speed is faster than the speed of light. Something is wrong. Before we begin resolving this, let's introduce a little notation that will save time in our future analysis.

Special notations used in relativity

You've probably noticed that certain expressions commonly show up in relativistic equations. For example, the speed v of an object relative to the speed of light c often occurs as the ratio v/c. It is common to define this ratio as beta β:

$$\beta = \frac{v}{c} \tag{26.6}$$

Relativistic effects start becoming significant when $\beta > 0.1$, that is, at speeds $v > 0.1c$.

The quantity $\sqrt{1 - (v/c)^2}$ appears frequently in relativistic expressions as well, most often in the denominator. To make relativistic equations more compact and easier to work with, physicists define the symbol gamma γ as

TIP Notice that γ is always >1 and β is always <1.

$$\gamma = \frac{1}{\sqrt{1 - (v/c)^2}} = \frac{1}{\sqrt{1 - \beta^2}} \tag{26.7}$$

Using these abbreviations, the time dilation, length contraction, and momentum equations become $\Delta t = \gamma \Delta t_0$, $L = L_0/\gamma$, and $p = \gamma mv$, respectively.

Relativistic kinetic energy

Deriving the relativistically correct expression for kinetic energy from first principles is rather complicated. Instead, we will provide the accepted expression and evaluate its low-speed limiting case to check for consistency with $(1/2)mv^2$.

The system of interest is a point-like object of mass m moving at constant speed v and not interacting with any other objects. The system therefore has no potential energy and no internal energy but does have kinetic energy. The correct relativistic expression for the energy of the system is

$$\gamma mc^2 = \frac{mc^2}{\sqrt{1-\beta^2}} = \frac{mc^2}{\sqrt{1-(v/c)^2}}$$

Note that mc^2 (mass times velocity squared) has energy units. Gamma $\gamma = 1/\sqrt{1-(v/c)^2}$ and beta $\beta = v/c$ are dimensionless numbers.

To see if this expression is consistent with $K = (1/2)mv^2$ at low speeds ($\beta \ll 1$), we use the binomial expansion to rewrite the part of the expression that involves the square root. For situations where the quantity x is small,

$$\frac{1}{\sqrt{1-x}} = (1-x)^{-1/2} \approx 1 + \tfrac{1}{2}x$$

In our case $x = \beta^2$. Therefore, for small $\beta = v/c$, the expression becomes

$$\gamma mc^2 = \frac{mc^2}{\sqrt{1-\beta^2}} = mc^2(1-\beta^2)^{-1/2} \approx mc^2(1 + \tfrac{1}{2}\beta^2)$$

$$= mc^2\left(1 + \frac{v^2}{2c^2}\right) = mc^2 + \tfrac{1}{2}mv^2$$

Notice that the last term on the right is just the classical kinetic energy of an object of mass m moving at speed v. Based on the above, it appears that

$$K = \gamma mc^2 - mc^2$$

This is the relativistically correct expression for kinetic energy K that a precise derivation produces.

What is the meaning of the mc^2 term? Apparently, it is a kind of energy that depends only on the mass of the object. This is a new and very profound idea—an object has energy just because of its mass. The term mc^2 is called the **rest energy** of the object. The expression γmc^2 represents the total energy (rest plus kinetic) of a point-like object. The kinetic energy of the object equals its total energy minus its rest energy. These three ideas are summarized below.

Relativistic energy A point-like object of mass m has so-called rest energy because it has mass:

$$\text{Rest energy } E_0 = mc^2 \tag{26.8}$$

The total energy of the object is

$$E = \gamma mc^2 = \frac{mc^2}{\sqrt{1-(v/c)^2}} \tag{26.9}$$

The object's kinetic energy is the object's total energy minus its rest energy:

$$\text{Kinetic energy } K = \gamma mc^2 - mc^2 = (\gamma - 1)mc^2 \tag{26.10}$$

You can see from Eq. (26.9) that when the speed of an object approaches the speed of light in a vacuum, its kinetic energy approaches infinity. Thus the relativistic expression for kinetic energy explains why no object can move faster than the speed of light in a vacuum—the work that needs to be done to accelerate the object to this speed is infinite.

FIGURE 26.13 One electron volt (1 eV) is the kinetic energy gained by an electron that accelerates across a 1-V potential difference.

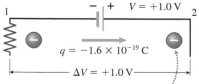

The kinetic energy of the electron is 1 eV.

Relativistic energy effects are most commonly relevant in the context of atomic and nuclear processes. The joule is an inconvenient energy unit to use when analyzing these processes because the energies involved are so small. Instead, we use another energy unit—the electron volt. One electron volt is the kinetic energy of an electron that accelerates from rest across a potential difference of 1 V (from lower potential at position 1 to higher potential at position 2 in **Figure 26.13**). Using the work-energy principle:

$$K_1 + U_{q1} = K_2 + U_{q2}$$

$$\Rightarrow K_2 = U_{q1} - U_{q2} = q(V_1 - V_2) = (-1.6 \times 10^{-19}\,\text{C})(0.0\,\text{V} - 1.0\,\text{V})$$

$$= 1.6 \times 10^{-19}\,\text{J}$$

> **Electron volt** An electron volt (1 eV) is the increase in kinetic energy of an electron when it moves across a 1.0-V potential difference:
>
> $$1\,\text{eV} = 1.6 \times 10^{-19}\,\text{J}$$

In nuclear and elementary particle physics, energies much higher than this are common (but still small compared with everyday-life energies). In those cases the mega-electron volt $(1\,\text{MeV} = 1.6 \times 10^{-13}\,\text{J})$ and giga-electron volt $(1\,\text{GeV} = 1.6 \times 10^{-10}\,\text{J})$ are commonly used.

Rest energy of particles

Any object with mass has rest energy, from elementary particles such as electrons to massive objects such as the Sun. Rest energy can be converted into other forms of energy. In fact, the rest energy of the Sun is being slowly converted via nuclear fusion reactions into internal energy, which accounts for the Sun's high temperature and brightness.

QUANTITATIVE EXERCISE 26.5 **Mass equivalent of energy to warm and cool house**

On average, each year about 2×10^{10} J of electric and chemical potential energy are converted to cool and warm your home. If rest energy could be converted for this purpose, how much mass equivalent of rest energy would be needed?

Represent mathematically According to Eq. (26.8):

$$m = \frac{E_0}{c^2}$$

Solve and evaluate Inserting the appropriate values:

$$m = \frac{2 \times 10^{10}\,\text{J}}{(3.00 \times 10^8\,\text{m/s})^2} = 2 \times 10^{-7}\,\text{kg}$$

This is approximately one-tenth the mass of one of the hairs on your head.

Try it yourself Estimate the mass equivalent of the chemical potential energy that is transformed into other energy types when you drive your car for 1 year. Note that 3.8 L (1 gallon) of gasoline produces about 1.2×10^8 J.

Answer

Assume that you drive 13,000 km (8000 mi) in one year and get 15 km/L (35 mi/gallon). You would use 870 L (230 gallons) of gasoline, or about 3×10^{10} J; this is about 0.3 μg of rest energy (far less than the mass of the gasoline itself).

Relationship between mass and energy

Earlier (in Chapter 6) we learned that mass is a conserved quantity. Now we understand that that statement is not exactly correct. Rest energy (mass) can be converted into other energy forms, so only the *total* energy of the system, $E_{\text{total}} = \Sigma E_{\text{particles}} = \Sigma \gamma mc^2$, is a conserved quantity. However, when objects are not moving at relativistic speeds, mass and energy are conserved separately almost exactly.

EXAMPLE 26.6   **Electron particle accelerator**

An electron in a particle accelerator accelerates through a potential difference of 10^6 V. What is its final speed?

Sketch and translate We sketch the process to help visualize the motion of the electron, marking knowns and unknowns. We know the electric charge and the mass of the electron and need to find its final speed.

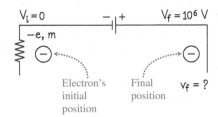

Simplify and diagram Assume that the electron starts at rest. The system is the electron and the electric field. The initial energy of the system is the system's rest energy; the final energy is the sum of the electron's positive kinetic energy, the system's negative electric potential energy, and the system's rest energy. Because no work is done on the system by the environment, the total energy of the system is constant. The energy bar chart below represents the process.

$$mc^2 + K_i + U_{qi} + \boxed{W} = mc^2 + K_f + U_{qf}$$

Represent mathematically We can now convert the energy bar chart into an equation:

$$E_{0i} + 0 + 0 + 0 = E_{0f} + K_f + U_{qf}$$

Because the rest energy of the system doesn't change during this process, we can subtract it from both sides of the equation. Inserting the appropriate expressions for each type of energy (including Eq. (26.10) for relativistic kinetic energy), we get

$$0 = (\gamma m_e c^2 - m_e c^2) + (-eV_f)$$

where m_e is the mass of the electron. We can simplify this equation by dividing both sides by $m_e c^2$ and solving for γ:

$$\gamma = \frac{eV_f}{m_e c^2} + 1$$

Remember:

$$\gamma = \frac{1}{\sqrt{1 - (v_f/c)^2}}$$

Solve and evaluate Now we need to solve for the final speed of the electron. We usually derive the expression for the needed quantity first and then substitute in the numbers. In this more complicated case, however, it is easier to first determine a numerical value for γ and then use it to solve for v_f. Insert the appropriate values in the first expression for γ above:

$$\gamma = \frac{eV_f}{m_e c^2} + 1$$

$$= \frac{(1.6 \times 10^{-19}\ \text{C})(10^6\ \text{V})}{(9.11 \times 10^{-31}\ \text{kg})(3.00 \times 10^8\ \text{m/s})^2} + 1 = 3.0$$

Now we can find the electron's final speed:

$$\frac{1}{\sqrt{1 - (v_f/c)^2}} = 3.0$$

Invert both sides of the equation:

$$\sqrt{1 - (v_f/c)^2} = \frac{1}{3.0}$$

Solve for v_f/c:

$$1 - (v_f/c)^2 = \frac{1}{9.0} \quad \Rightarrow \quad (v_f/c)^2 = 1 - \frac{1}{9.0}$$

$$\Rightarrow v_f/c = \sqrt{1 - \frac{1}{9.0}} = 0.94$$

or

$$v_f = 0.94c$$

Our result is reasonable—although the speed is very high, it is less than the speed of light.

Try it yourself Protons are accelerated from rest to a speed of $0.993c$ in the "small" 150-m-diameter booster accelerator that is part of the Fermi National Accelerator Laboratory in Batavia, Illinois. (a) Determine the proton's total energy when it leaves the booster accelerator. (b) Determine its kinetic energy as it leaves the accelerator.

Answer   (a) $E_f = 1.27 \times 10^{-9}$ J; (b) $K_f = 1.12 \times 10^{-9}$ J.

REVIEW QUESTION 26.9 If we did not use the relativistic expression for kinetic energy to calculate the speeds of elementary particles in accelerators, would the calculated nonrelativistic speeds be greater or less than the calculated relativistic speeds? (Hint: Estimate the values of any physical quantities that you need.)

26.10 Doppler effect for EM waves

The observed frequency of a mechanical wave (for example, a sound wave) is different from the frequency of the wave source if the observer and/or the source are moving relative to each other. A similar phenomenon exists for light (EM) waves; however, due to relativistic effects, the Doppler effect for light is somewhat different.

In the Doppler effect for sound [Eq. (11.17)], if the source or observer moves through the medium at a speed equal to or faster than the speed of sound, or if their motion relative to each other is equal to or faster than this speed, then the sound waves might not reach the observer at all. But the speed of EM waves, $c = 3.00 \times 10^8$ m/s, is the same for all observers no matter how they are moving relative to the source. As a result, the equation that describes the Doppler effect for EM waves differs from the equation for the Doppler effect for sound.

We won't derive the equation for the Doppler effect for EM waves. Instead, we just provide the equation and use it to analyze several processes.

Doppler effect for EM waves

$$f_O = f_S \sqrt{\frac{1 + v_{\text{rel}}/c}{1 - v_{\text{rel}}/c}} \qquad (26.11)$$

where f_O is the EM wave frequency detected by an observer, f_S is the frequency emitted by the source, and v_{rel} is the component of the relative velocity between the observer and the source along the line connecting them (the component is positive when the observer and source are approaching each other and negative when moving apart).

The major difference between this and the Doppler effect equation for sound is the square root. The need for this alteration comes about due to time dilation: the source and observer are in different inertial reference frames. The square root can be simplified for the Doppler effect for EM waves if v_{rel} is small compared to the speed of light. Using the binomial expansion, the equation becomes the following.

Doppler effect for EM waves (slow observer-source relative motion):

$$f_O = f_S \left(1 + \frac{v_{\text{rel}}}{c} \right) \qquad (26.12)$$

The sign conventions for Eq. (26.11) apply here.

When the observer and source are approaching, the observed frequency is higher than the source frequency. When they are moving apart, the observed frequency is lower.

Suppose, for example, that the source emits light in the blue part of the spectrum and that the source and observer are moving apart. The light detected by the observer could be in the yellow or red part of the spectrum. In astronomy this is known as a **red shift**, and it was important in discovering that the universe is expanding. Conversely, if the source and observer are moving toward each other, the light detected by the observer would be shifted toward the blue end of the spectrum.

Modern police radar and sports radar measure speeds by knowledge of the source frequency and the ability to measure the frequency of a reflected wave. Let's look at such practical examples of the Doppler effect for EM waves.

Doppler radar and speeding tickets

Police radar uses microwaves, one type of electromagnetic wave. Consider the next example.

EXAMPLE 26.7 **The speeding ticket**

Physicist Dr. R. Wood ran a red light while driving his car and was pulled over by a police officer. Dr. Wood explained that because he was driving toward the red light, he observed it as green due to the Doppler effect. Dr. Wood was then given a very expensive speeding ticket. Should Dr. Wood go to court to fight the ticket?

Sketch and translate The traffic light is the source, and Dr. Wood is the observer. The situation is sketched below. We are interested in determining the speed of Dr. Wood's car relative to the traffic light. Dr. Wood was traveling fast enough so that the red light emitted by the traffic signal was observed by him to be green. Because he is moving toward the source, the relative velocity v_{rel} will be positive and equal to Dr. Wood's car's speed relative to the traffic light.

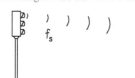

Frequency f_s of red light traveling toward Dr. Wood's car.

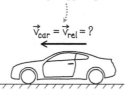

What relative velocity would make the light appear green?

$\vec{v}_{car} = \vec{v}_{rel} = ?$

Simplify and diagram Assume that the traffic light emits red monochromatic light at a frequency of about $f_S = 4.5 \times 10^{14}$ Hz and that the frequency of green light is about $f_O = 5.4 \times 10^{14}$ Hz. Assume also that the car moves with constant velocity.

Represent mathematically Because we are talking about a low-speed car, we can use the low-speed Doppler effect $f_O = f_S(1 + (v_{rel}/c))$ to determine the speed of the car. Multiply both sides of the equation by c and then solve for v_{rel}:

$$cf_O = cf_S\left(1 + \frac{v_{rel}}{c}\right) = f_S(c + v_{rel}) = f_S c + f_S v_{rel}$$

$$\Rightarrow f_S v_{rel} = cf_O - f_S c$$

$$\Rightarrow v_{rel} = \frac{c(f_O - f_S)}{f_S} = c\left(\frac{f_O}{f_S} - 1\right)$$

Solve and evaluate Inserting the appropriate values:

$$v_{rel} = c\left(\frac{f_O}{f_S} - 1\right) = (3.0 \times 10^8 \text{ m/s})\left(\frac{5.4 \times 10^{14} \text{ Hz}}{4.5 \times 10^{14} \text{ Hz}} - 1\right)$$

$$= 6.0 \times 10^7 \text{ m/s}$$

This speed is approaching the speed of light. Dr. Wood was clearly just trying to get out of a ticket. His explanation that the red light appeared green to him isn't reasonable. Note also that the speed we found contradicts our assumption for using the low-speed Doppler effect. However, even if we used the equation appropriate for such high speeds, we would still obtain a speed close to the speed of light.

Try it yourself Can you detect the change in frequency of a red light if you are driving at a speed of 34 m/s (76 mi/h)?

Answer The change is 5×10^7 Hz, which is

$$\left(\frac{5 \times 10^7 \text{ Hz}}{4.5 \times 10^{14} \text{ Hz}}\right) \times 100\% = 1 \times 10^{-5}\%$$

which is not detectable by the human eye (although detectors do exist that can register this small change).

Measuring car and ball speeds using the Doppler effect

A modern radar gun used by police emits microwaves (EM waves with frequencies much smaller than visible light) in the frequency range 33.4–36.0 GHz. Sports radar guns use a frequency of 10.525 GHz. The speed detection process involves three distinct steps outlined below for a baseball in flight.

1. The radar gun emits source microwaves at frequency $f_S = 10.525$ GHz. These microwaves are "detected" at frequency f_D by the ball (**Figure 26.14a**).

2. The waves reflect from the ball, making the ball act as a source of microwaves at frequency f_D. Some of these waves are reflected back toward the radar gun, which observes or detects them at frequency f_O (Figure 26.14b).

3. The observed frequency f_O and source frequency f_S are often combined, resulting in a beat frequency $f_{beat} = |f_O - f_S|$, which is then used to determine the speed of the baseball.

FIGURE 26.14 How a radar detector determines ball speed.

(a)

Microwaves at source frequency f_S

Ball "detection" frequency f_D

(b)

Ball reflects waves at frequency f_D

Observer at source detects reflected waves at frequency f_O

FIGURE 26.15 The expanding universe is analogous to an expanding loaf of raisin bread.

Bread loaf expands while baking in an oven.

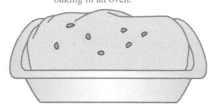

All raisins move farther apart as the dough expands.

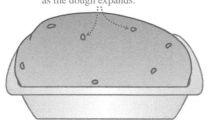

Hubble's discovery of the expansion of the universe

At the beginning of the 20th century it was believed that the universe was static. Planets moved around the Sun, but the distances between galaxies remained constant. This static universe model seemed consistent with observations. However, the gravitational interaction is always attractive, so why didn't the universe collapse? Even Einstein's ideas about gravity (which you will learn about in the next section) had trouble accommodating a static universe.

During this same time, astronomers Edwin Hubble and Milton Humason worked for years studying the light emitted by stars in other galaxies. They found that these stars emitted light wavelengths that were nearly the same as the light emitted from similar nearby stars—but with one major difference. With few exceptions, the light emitted by the distant galaxies was shifted toward longer wavelengths (smaller frequencies). Hubble concluded that the shift was not because the stars in other galaxies were different, but because significant relative motion existed between those stars and Earth, and nearly all those stars and the galaxies they were in were moving away from us. The universe was expanding.

The idea of an expanding universe can be visualized as a loaf of raisin bread while it's rising. The galaxies are the raisins and the space between the galaxies is the dough (Figure 26.15). As the bread expands, the raisins embedded in the dough all move farther and farther apart from each other.

Hubble's law

Hubble used the EM wave Doppler effect equations to determine the speeds of the other galaxies. They turned out to range from 0 km/s to 1500 km/s, and all but the closest galaxies were moving away from us. Additionally, Hubble found that the farther away the galaxy, the faster it was moving away. Notice in the expanding bread loaf in Figure 26.15 that the nearest two raisins on the top right changed positions from about 1 cm apart to 2 cm apart (a change in separation of 1 cm). However, the raisin farthest away from the top right raisin changed from about 2 cm apart to 4 cm apart (a change in separation of 2 cm). Hubble found a similar pattern for galaxies and described it mathematically as

$$v = Hd \qquad (26.13)$$

where v is the galaxy's recessional speed away from Earth, d is the distance between the galaxy and Earth, and H is a constant, now known as the **Hubble constant**. It has units of kilometers per second per unit of distance between the galaxies. In astronomy, distances to other galaxies are usually measured in megaparsecs (Mpc), with 1 parsec equal to 3.09×10^{16} m, or about 3.3 light-years. The current accepted value for the Hubble constant, coming from multiple experiments, is about 70 ± 6 (km/s)/Mpc.

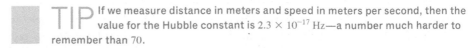

TIP If we measure distance in meters and speed in meters per second, then the value for the Hubble constant is 2.3×10^{-17} Hz—a number much harder to remember than 70.

It might seem that Hubble's result implies that our galaxy is the center of the universe and all other galaxies are moving away from that center. We can see it another way, however. Going back to the analogy of the raisin bread, if Earth were one of the raisins, we would observe all other raisins moving away from us. But that's exactly what each of the other raisins would observe as well! There is no center of the universe.

Figure 26.16 helps us visualize the expansion of the universe. It shows spacetime diagrams (one-dimensional representations) of several galaxies as observed by two different observers for a time interval between 0 and 1 time units, assuming that $H = 1$ with the units of 1/unit of time. The galaxies of the two observers are marked in red and blue. Notice that the observer in the red galaxy (Figure 26.16a) sees other galaxies moving away, as does the observer in the blue galaxy (Figure 26.16b). Although a particular galaxy (any black dot) moves at a different speed for these two observers, both observers describe the relation between the galaxy's speed and distance to them using the same Hubble constant, $H = 1$. Two galaxies that are separated by a certain distance will move away from each other at the same relative speed no matter where in the universe they are or where the observer is (follow three pairs of dots separated by half a unit of distance at $t = 0$ in Figures 26.16a and b).

FIGURE 26.16 Expansion of the universe described by different observers.

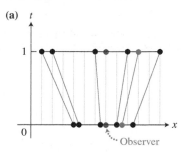

(a)

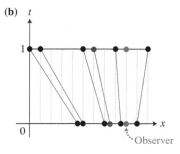

(b)

Hubble's law The average recessional speed v of two distant galaxies a distance d apart is determined by

$$v = Hd \qquad (26.13)$$

where $H = 70 \pm 6 \ (\text{km/s})/\text{Mpc}$ is Hubble's constant. One megaparsec (Mpc) equals 3.3×10^6 light-years (ly). Nearby galaxies do not necessarily obey this rule.

Age of the universe

If galaxies are moving away from each other at measurable speeds, and we can measure the distances to them, then we can estimate how long it took the galaxies to move apart to their present positions. This would give an estimate for the age of the universe. Choose two distant galaxies a distance d apart. Assuming they have moved away from each other at a constant speed $v = Hd$, then the time interval Δt since they were together is

$$\Delta t = \frac{d}{v} = \frac{d}{Hd} = \frac{1}{H}$$

The approximate age of the universe is the inverse of the Hubble constant. Substituting $70 \pm 6 \ (\text{km/s})/\text{Mpc}$ for H and converting to years from seconds, we obtain the following:

$$\Delta t = \frac{1}{H} = \frac{1}{\left(70 \ \dfrac{\text{km/s}}{\text{Mpc}}\right)\left(\dfrac{1000 \text{ m}}{1 \text{ km}}\right)\left(\dfrac{1 \text{ Mpc}}{10^6 \text{ pc}}\right)\left(\dfrac{1 \text{ pc}}{3.09 \times 10^{16} \text{ m}}\right)\left(\dfrac{3.16 \times 10^7 \text{ s}}{1 \text{ year}}\right)}$$

$$= 14.0 \text{ billion years}$$

This result is uncertain by $(\pm 6/70) \times 100\% = \pm 8.6\%$. This result is compatible with geological studies that estimate the age of Earth at approximately 4.5×10^9 years old (4.5 billion years); that is, geological studies do not conclude that Earth is older than the universe. Rather, the findings imply that Earth formed approximately 9.5 billion years after the universe came into being.

Returning briefly to our galaxies in Figure 26.16, the age of the universe in this model is $\frac{1}{H} = 1$ because $H = 1$. Therefore, our universe was born at $t = -1$. You can verify this by extending backward the world lines in Figure 26.16a that show expansion—they all meet at the same point, at time $t = -1$ (**Figure 26.17**). If you repeat the same procedure for Figure 26.16b you will again see that both observers determine the same age of the universe.

The assumption about the universe expanding at a constant rate is questionable. In fact we have learned in the last 15 years that the universe is accelerating slightly in its expansion. The 2011 Nobel Prize in physics was awarded to Saul Perlmutter, Brian P. Schmidt, and Adam G. Riess for this stunning and unexpected discovery.

FIGURE 26.17 How to determine the age of the universe.

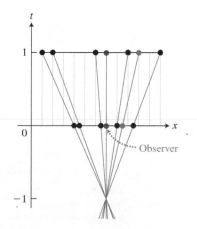

REVIEW QUESTION 26.10 Why can't we use the equations for the Doppler effect for sound to describe the Doppler effect for electromagnetic waves?

26.11 General relativity

The special theory of relativity allows us to compare measurements made by observers in two reference frames who move at constant velocity relative to each other. In these inertial reference frames, the laws of physics have the same form (that is, they are said to be **invariant**). Einstein was able to generalize this invariance to all reference frames, including noninertial (accelerated) reference frames. The result is the **general theory of relativity**, which is also an improved theory of the gravitational interaction. The cornerstone of general relativity is the **principle of equivalence**.

> **Principle of equivalence** No experiment can be performed that can distinguish between an accelerating reference frame and the presence of a uniform gravitational field.[1]

Consider a spaceship at rest or drifting at constant velocity far from any stars or planets. A passenger inside the ship would float freely (**Figure 26.18a**). Suppose now that the spaceship's rockets fired, causing the ship to accelerate at 9.8 m/s^2. Initially, the floating passenger "falls" to the spaceship's floor (Figure 26.18b) with acceleration 9.8 m/s^2. From there, the passenger would be able to stand, jump "up and down," and throw a ball "upward" (away from the floor) only to have it come back "down." The passenger would have exactly the same sensations as if he or she were standing on Earth's surface.

If the spaceship had no windows, and the engines provided a perfectly smooth ride, the passenger would have no way of knowing whether the ship was accelerating at 9.8 m/s^2 or was resting on Earth's surface in the presence of a nearly uniform gravitational field. The principle of equivalence says that there is no experiment that the passenger can perform to distinguish between these two situations.

From the principle of equivalence, Einstein reasoned that gravitation should be understood by a different mechanism entirely. He suggested that objects with mass cause the space and time around them to become curved and that objects moving in this curved space will not travel along straight paths at constant speed. These objects do not actually have a gravitational force exerted on them; rather, they are simply moving along natural paths through a curved spacetime. The Moon orbits Earth not because Earth exerts a force on it, but because Earth curves space and the Moon then naturally moves in a curved path. This path in the new spacetime is the shortest line connecting two points (called a *geodesic line*, similar to a straight line in Euclidean geometry). This also very neatly explains why all objects, regardless of size, mass, or composition, move identically in a vacuum. It also explains why astronauts do not notice any forces being exerted on them as they orbit Earth: because there aren't any!

An early testing experiment

An object of large mass such as the Sun causes space to curve more than an object of smaller mass such as Earth does (see **Figure 26.19**). In 1915 Einstein used general relativity to predict that light from distant stars that passes close to the Sun's surface would be deflected by an angle of $\theta = 4.86 \times 10^{-4}$ degrees. During a total solar eclipse, the observed position of those stars will be different because light bends as it passes close to the Sun (in **Figure 26.20**, the dashed line is the star's light path when the Sun is not present).

In 1919, astronomers under the leadership of Sir Arthur Eddington tried to measure such a deflection. They announced that the observations were in total agreement with

FIGURE 26.18 Acceleration produces the same effect as gravity.

(a)

When moving at constant velocity very far from massive objects, the person and the ball float in the spaceship.

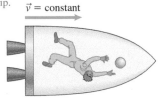

(b)

When the ship is accelerating toward the right at $\vec{a} = \vec{g}$, the left floor of the ship presses against the person, causing the same sensation as if standing on Earth's surface.

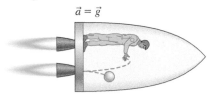

FIGURE 26.19 The Sun's mass causes curvature of space.

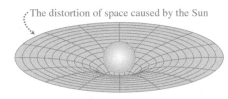

The distortion of space caused by the Sun

FIGURE 26.20 Starlight deflection caused by Sun's space curvature.

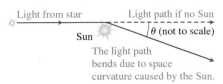

Light from star   Light path if no Sun

Sun θ (not to scale)

The light path bends due to space curvature caused by the Sun.

---

[1]Einstein's statement of the principle of equivalence is more general than this, but this restricted version is sufficient for our discussion.

the prediction of general relativity. Einstein and his theory of general relativity became famous almost overnight. However, despite the announcement, Eddington's results were not as conclusive as he had wanted them to be. It took many years and multiple experiments with solar eclipses until finally, by the 1960s, the measurements were convincingly in agreement with the predictions of general relativity.

Another testing experiment

General relativity solved another problem that had plagued astronomers since the early 1800s. The elliptical path of Mercury around the Sun was known to slowly rotate (**Figure 26.21**), a phenomenon called **precession**. This effect is so small that it takes about 3 million years for the elliptical orbit to complete one full precession. Newton's theory of gravitation predicts a precession from slight gravitational tugs due to the other planets (mostly Jupiter). However, Mercury's precession rate had been observed to be slightly larger than that predicted by Newton's theory. But if general relativity is used to predict the precession rate, the result is consistent with the observations. The additional precession results from the motion of Mercury through the curved space produced by the Sun.

FIGURE 26.21 Mercury's orbit about the Sun.

Precession of Mercury's elliptical orbit about the Sun

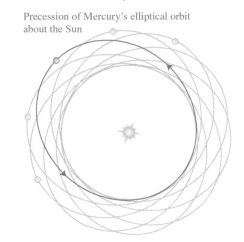

Gravitational time dilation and red shift

Another prediction of general relativity is **gravitational time dilation**. Objects do not just curve space around them, but also alter the rates at which time passes around them as well. Time passes at different rates at different points near a massive object. The closer a point is to the massive object, the slower time passes there. For example, a person working in an office on the first floor of a tall building would age more slowly relative to an observer on the top floor. However, this effect would be only about 10^{-5} s over an entire human lifetime.

One consequence of gravitational time dilation is an effect known as **gravitational redshift**. If EM waves are emitted from a region closer to a massive object and observed in a region farther away, the observed frequency will be lower than the emitted frequency. In 1960, R. V. Pound and G. A. Rebka Jr. at Harvard University tested this prediction. They observed a small reduction in the vibration frequency of gamma rays emitted from radioactive nuclei on the bottom floor of a laboratory building at Harvard compared to the frequency of gamma rays emitted from the top floor—yet another testing experiment supporting general relativity.

Gravitational waves and black holes

General relativity predicts that both space and time are curved by the presence of objects with mass. Experiments support this idea. These two ideas are unified by general relativity into the idea that space and time form a single entity known as **spacetime**. Newton imagined that space and time were a rigid background for the processes of the universe. Einstein showed that spacetime is curved by the presence of mass and changes shape when that mass moves. These changes in the curvature of spacetime can propagate as gravitational waves that ripple through the vacuum at the speed of light. In 1915, Einstein predicted the existence of such waves and what the properties of these waves should be, and after 100 years of searching, scientists found the first direct evidence of their existence in 2015.

General relativity also predicts the existence of black holes (objects we first encountered in Chapter 7). In Newtonian mechanics, black holes are hypothetical objects whose escape speed is equal to or greater than light speed. In general relativity, black holes are regions of spacetime that are so extremely curved that all matter and even light within that region cannot escape (see **Figure 26.22**). The most common black holes form when large stars run out of nuclear fuel and collapse under their own gravity to extreme densities. Much larger black holes have been detected at the centers

FIGURE 26.22 Extreme space curvature near a black hole does not allow passing objects (including light) to escape.

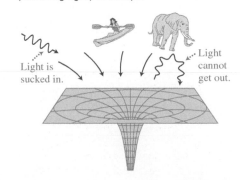

Light is sucked in.

Light cannot get out.

of most galaxies. Astronomers cannot detect black holes directly but can find them by the extreme influence they have on their immediate environment.

It turns out that black holes are the objects that helped scientists detect gravitational waves. Imagine that two massive black holes orbit each other. According to general relativity, they will continuously emit gravitational waves until they collide and join together, making one rotating super-massive black hole. When they collide, even more of their energy is converted into gravitational waves. The gravitational waves from such a collision were detected on September 14, 2015 by the two interferometers of the Laser Interferometer Gravitational-Wave Observatory (LIGO) in Louisiana and Washington. Based on the detected signals, scientists estimated that the two black holes were about 29 and 36 times more massive than the Sun and that the collision occurred around 1.3 billion years ago. The energy of the gravitational waves emitted during the event was equal to the rest energy of an object with the mass of three Suns, and it was emitted during a time interval of milliseconds. This was the first direct observation of the disturbances in space and time that gravitational waves passing through Earth make.

The discovery of gravitational waves by the LIGO experiment was the result of the joint effort of thousands of scientists from almost a hundred universities all over the world. Such significant collaboration was needed because to find gravitational waves, scientists needed to be able to design an experiment that would detect an incredibly tiny change, 10^{-19} m, in the length of the arm of an interferometer (similar to that of Michelson and Morley, but much larger, at 2.5 miles long) while eliminating all other factors that might contribute to much bigger changes in the length. To understand this difficulty in design, imagine that you need to measure a change in the distance from our solar system to the nearest star of the width of one hair—this is the relative distance change caused by gravitational waves that scientists in the LIGO experiment detected.

REVIEW QUESTION 26.11 What is the general relativity explanation for why Earth orbits the Sun?

26.12 Global Positioning System (GPS)

You will likely never travel at speeds close to the speed of light, never travel to a black hole, and never worry about your biological clock becoming out of sync by a thousandth of a second over the course of your life because you live on the top floor of a building. In later chapters, however, we will explore many examples of the impact of relativistic effects on our daily lives. Meanwhile, here's an everyday example of both special and general relativity that affects your life: determining your location using the Global Positioning System (GPS).

Each GPS satellite is about 20,000 km from the surface of Earth and has an orbital speed of about 3900 m/s. An atomic clock on each satellite "ticks" every nanosecond $(1 \text{ ns} = 10^{-9} \text{ s})$, which allows it to transmit precisely timed signals. A GPS receiver in a car or on an airplane determines its position on Earth to within 5 to 10 m using these signals.

To achieve this level of precision, the clock ticks from the GPS satellites must be measured to an accuracy of 20 to 30 ns. Because the satellites move with respect to us, their clocks tick slightly slower than ours (special relativity's time dilation). Using the time dilation equation [Eq. (26.1)], we can estimate by how much the clocks on the satellites should fall behind during 1 day.

The reference frame of the satellite is the proper reference frame, and a 1-ns tick is the proper time interval. The satellite, as stated, moves at a speed of 3900 m/s. This time interval as measured from Earth is

$$\Delta t = \frac{\Delta t_0}{\sqrt{1 - \dfrac{v^2}{c^2}}} = \frac{1 \text{ ns}}{\sqrt{1 - \dfrac{(3.90 \times 10^3 \text{ m/s})^2}{(3.00 \times 10^8 \text{ m/s})^2}}} = 1.0000000001 \text{ ns}$$

The difference between the proper time interval and the measured time interval is

$$\Delta t - \Delta t_0 = 1.0000000001 \text{ ns} - 1.0000000000 \text{ ns} = 1.0 \times 10^{-10} \text{ ns}$$

This means that with each 1-ns tick of the satellite's atomic clock, that clock falls behind clocks on Earth by 1.0×10^{-10} ns. The number N of nanosecond ticks in each day is

$$N = \left(\frac{3600 \text{ s}}{\text{h}}\right)\left(\frac{24 \text{ h}}{\text{day}}\right)\left(\frac{10^9 \text{ ticks}}{\text{s}}\right) = 8.64 \times 10^{13} \text{ ticks}$$

Therefore, in 1 day, time dilation will cause the satellite clock to fall behind Earth clocks by a total of

$$(\Delta t - \Delta t_0)N = (1.0 \times 10^{-10} \text{ ns})(8.64 \times 10^{13} \text{ ticks}) \approx 8.6 \times 10^3 \text{ ns} = 8.6 \text{ }\mu\text{s}$$

This result is just an estimate because the GPS satellites are not in exactly inertial reference frames. But because their accelerations are low, they are very close to inertial.

In addition, the satellites are in orbits high above Earth. There, the curvature of spacetime is smaller than on Earth's surface, and thus the clocks will tick faster than clocks on Earth. General relativity predicts that the space clocks get ahead by about 45 μs per day.

Together, the two effects result in the GPS satellite clocks running at about $(+45 - 8.6)$ μs per day, or about 36 to 37 μs per day faster than Earth clocks, or about 0.014 s/year. Thus, in 1 year, the satellite positions (moving at about 3900 m/s) would be off by about (0.014 s)(3900 m/s) = 50 m. The position the receiver on Earth would report would be off by about the same amount.

The precise locations of the satellites are determined by tracking stations around the world. Those data are collected, processed, and regularly transmitted up to the satellites, which then update their internal orbit models. Along with the orbit updates, transmissions are also sent to synchronize the atomic clocks on the satellites. This is where relativistic effects are taken into account. GPS receivers use the time and position data from multiple satellites to trilaterate their location. An understanding of both special and general relativity is essential for the GPS system to work.

REVIEW QUESTION 26.12 What are two relativistic effects that must be accounted for so that the GPS system can function accurately?

Summary

Postulates of special relativity

1. The laws of physics are the same in all inertial reference frames.
2. The speed of light in a vacuum is the same in all inertial reference frames. (Section 26.2)

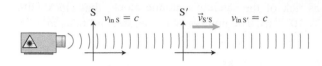

Time dilation In the **proper reference frame**, events 1 and 2 occur at the same place. The time interval between events 1 and 2 measured by an observer in the proper reference frame is the **proper time interval** Δt_0. The time interval between events 1 and 2 as measured by an observer moving at constant speed v relative to the proper reference frame is Δt. (Section 26.4)

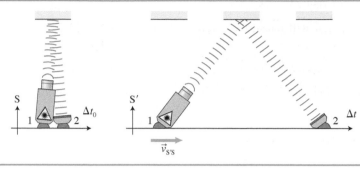

$$\Delta t = \frac{\Delta t_0}{\sqrt{1 - \dfrac{v^2}{c^2}}} \qquad \text{Eq. (26.1)}$$

Length contraction An object's **proper length** L_0 is its length measured in an inertial reference frame in which the object is at rest. The length of the object when measured in a reference frame moving at speed v relative to the proper reference frame is L. (Section 26.5)

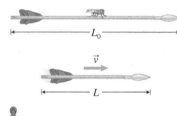

$$L = L_0\sqrt{1 - \frac{v^2}{c^2}} \qquad \text{Eq. (26.2)}$$

A **spacetime diagram** is a representation of relativistic motion on a time-versus-position graph. The motion of an object is shown by its **world line**. The world line for a light beam is tilted at 45° in the first and second quadrants, representing two light beams traveling in opposite directions. All other objects have steeper world lines.

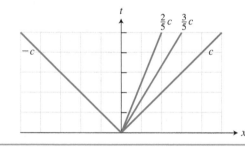

Relativistic momentum The relativistic momentum $\vec{p}$ of an object of mass m must be used when an object is moving at speed $v \geq 0.1c$. (Section 26.8)

$$\vec{p} = \frac{m\vec{v}}{\sqrt{1 - (v/c)^2}} \qquad \text{Eq. (26.5)}$$

Relativistic energy
- **Rest energy** An object of mass m has rest energy E_0 that depends only on the mass m of the object and the speed of light c.
- **Kinetic energy** The relativistic kinetic energy K of an object is its total energy minus its rest energy.
- **Total energy** is the total relativistic energy E of an object moving at speed v. (Section 26.9)

$$E_0 = mc^2 \qquad \text{Eq. (26.8)}$$

$$K = \frac{mc^2}{\sqrt{1 - (v/c)^2}} - mc^2 \qquad \text{Eq. (26.10)}$$

$$E = \frac{mc^2}{\sqrt{1 - (v/c)^2}} \qquad \text{Eq. (26.9)}$$

Doppler effect for electromagnetic radiation If a source emits EM waves at frequency f_S and the source and observer are moving at speed v_{rel} relative to each other (+ if toward each other and − if away from each other), then the observer detects EM waves at frequency f_O. (Section 26.10)

$$f_O = f_S\sqrt{\frac{1 + v_{rel}/c}{1 - v_{rel}/c}} \quad \text{for } v_{rel} \geq 0.1c \quad \text{Eq. (26.11)}$$

$$f_O = f_S\left(1 + \frac{v_{rel}}{c}\right) \quad \text{for } v_{rel} < 0.1c \quad \text{Eq. (26.12)}$$

Questions

Multiple Choice Questions

1. The Michelson-Morley experiment had as its goal to
   (a) measure the speed of light in different reference frames.
   (b) find out whether the Earth moves with respect to ether.
   (c) find whether ether exists.
   (d) All of the above
   (e) a and b only

2. On what did Michelson and Morley base their conclusions?
   (a) The measurement of the time intervals light takes to travel the same distance in different directions with respect to the moving ether
   (b) The analysis of an interference pattern for two beams of light
   (c) Comparison of the intensities of two perpendicular light beams

3. Physicists explained the results of the Michelson-Morley experiment by saying that
   (a) the motion of Earth through ether is undetectable.
   (b) there is no ether.
   (c) the length of experimental apparatus parallel to Earth's motion changes.
   (d) All of the above
   (e) None of the above

4. What is a proper time interval?
   (a) A correctly measured time interval
   (b) The shorter time interval of the two measured times
   (c) The time interval measured by observers with respect to whom the events occurred in the same location

5. You stand on a train platform and watch a person on a passing train through a train window. The person is eating a sandwich. The time interval it takes her to eat the sandwich as measured by you
   (a) is longer than measured by her.
   (b) is shorter than measured by her.
   (c) is the same.
   (d) depends on the orientation of the sandwich.

6. In Question 26.5, the length of the sandwich as measured by you
   (a) is longer than measured by the passenger.
   (b) is shorter than measured by the passenger.
   (c) is the same.
   (d) depends on the orientation of the sandwich.

7. A supernova explodes in a distant galaxy. The galaxy is moving away from us. The speed of light from the explosion with respect to Earth is
   (a) smaller than the speed of light with respect to the galaxy.
   (b) the same as light speed with respect to the galaxy.
   (c) larger than light speed with respect to the galaxy.

8. We used the muon experiment to test which of the following hypotheses? Multiple answers may be correct.
   (a) The existence of ether
   (b) Time dilation
   (c) Length contraction

9. The measurement of the coordinates of a star during a solar eclipse can be used as a testing experiment for
   (a) the existence of ether.
   (b) time dilation.
   (c) length contraction.
   (d) space-time curvature.

10. Suppose an early GPS system did not take the laws of special relativity into account. A 1-ns time interval measured by clocks on the GPS satellite would be
    (a) smaller than intervals measured on Earth.
    (b) larger than intervals measured on Earth.
    (c) exactly the same for both.

11. What are cosmic rays?
    (a) Light rays sent by cosmic objects
    (b) Electromagnetic radiation produced during supernova explosions
    (c) Signals sent by extraterrestrial civilizations
    (d) Subatomic particles such as protons and neutrons

12. Which of the blue world lines in the spacetime diagrams in Figure Q26.12 represent motion of an object that is impossible? The straight black line on each diagram represents a light pulse traveling in the +x-direction.

FIGURE Q26.12

(a) 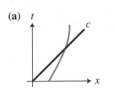   (b) 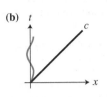   (c)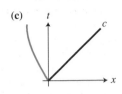

(d) 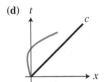   (e)

Conceptual Questions

13. What is an inertial reference frame? How can you convince someone that Newton's laws work the same way for observers in two different inertial reference frames?

14. Give an example of a phenomenon that an observer in a noninertial reference frame cannot explain using Newton's laws.

15. Explain the difference between a proper reference frame and other inertial reference frames in which the time interval between two events is measured.

16. What are the two postulates of the special theory of relativity? How would you explain them to a friend using your own words?

17. What does it mean to say that the speed of something is the same in different reference frames? Give an example.

18. You move toward a star at a speed of $0.99c$. At what speed does light from the star pass you? What if you are moving away from the star?

19. You pass Earth in a spaceship that moves at $0.99c$ relative to Earth. Do you notice a change in your heartbeat rate? Does an observer on Earth think your heart is beating at the normal rate? Do you think the heartbeat of Earth's observer is normal? Explain all three of your answers.

20. It takes light approximately 10^{10} years to reach Earth from the edge of the observable universe. Would it be possible for a person to travel this distance during a lifetime? Explain.

21. A person holds a meter stick in a spaceship traveling at $0.95c$ past Earth. The person rotates the meter stick so that it is first parallel and then perpendicular to the ship's velocity. Describe its changing length to
    (a) an observer at rest on the Earth and
    (b) the person in the spaceship.

22. Name several ways in which your life would be different if the speed of light were 20 m/s rather than its actual value.

23. If the speed of light were infinite, how would time dilation, length contraction, and the ways we calculate momentum and energy be affected? Justify your answer carefully.

24. The classical equation for calculating kinetic energy [Eq. (7.5)] and the relativistic equation for calculating kinetic energy [Eq. (26.10)] appear quite different. Under what conditions is each equation appropriate? Invent a simple example to show that they produce the same result at speeds less than $0.001c$.

25. How did the Doppler effect for light help scientists estimate the age of the universe?

26. What is the principle of equivalence? How would you explain this principle to a friend?

27. Can any of the blue world lines in Figure Q26.12 represent the motion of an object that is slowing down at constant acceleration? Explain.

Problems

Below, **BIO** indicates a problem with a biological or medical focus. Problems labeled **EST** ask you to estimate the answer to a quantitative problem rather than derive a specific answer. Asterisks indicate the level of difficulty of the problem. Problems with no * are considered to be the least difficult. A single * marks moderately difficult problems. Two ** indicate more difficult problems.

You may need to know the following masses to work some of these problems: $m_{electron} = 9.11 \times 10^{-31}$ kg and $m_{proton} = 1.67 \times 10^{-27}$ kg.

26.1 and 26.2 Ether or no ether? and Postulates of special relativity

1. **Relative motion on an airport walkway** A person is walking on a moving walkway in the airport. Her speed with respect to the walkway is 2 m/s. The speed of the walkway is 1 m/s with respect to the floor. What is her speed with respect to a person walking on the floor in the opposite direction at the speed of 2 m/s?

2. **Running on a treadmill** Explain how you can run on a treadmill at 3 m/s and remain at the same location.

3. * Describe the important parts of the Michelson-Morley experimental setup and explain how this setup could help them determine whether Earth moves with respect to ether. Explain which reference frames are involved and how the setup relates to the previous problem.

4. * Describe what Michelson and Morley would have observed when they rotated their interferometer if Earth were moving through ether compared to what they would have observed if Earth were not moving through ether.

5. **Person on a bus** A person is sitting on a bus that stops suddenly, causing her head to tilt forward. (a) Explain the acceleration of her head from the point of view of an observer on the ground. (b) Explain the acceleration of her head from the point of view of another person on the bus. (c) Which observer is in an inertial reference frame?

6. * **Turning on a rotating turntable** A matchbox is placed on a rotating turntable. The turntable starts turning faster and faster. At some instant the matchbox flies off the turning table. (a) Draw a force diagram for the box when still on the rotating turntable. (b) Draw a force diagram for the box just before it flies off. (c) Explain why the box flies off only when the turntable reaches a certain speed. In what reference frame are you when providing this explanation? (d) How would a bug sitting on the turntable explain the same situation?

26.3–26.6 Simultaneity, Time dilation, Length contraction, and Spacetime diagrams

7. * A particle called Σ^+ lives for 0.80×10^{-10} s in its proper reference frame before transforming into two other particles. How long does the Σ^+ seem to live according to a laboratory observer when the particle moves past the observer at a speed of 2.4×10^8 m/s?

8. A Σ^+ particle discussed in the previous problem appears to a laboratory observer to live for 1.0×10^{-10} s. How fast is it moving relative to the observer?

9. A person on Earth observes 10 flashes of the light on a passing spaceship in 22 s, whereas the same 10 flashes seem to take 12 s to an observer on the ship. What can you determine using this information?

10. A spaceship moves away from Earth at a speed of 0.990c. The pilot looks back with a telescope and measures the time interval for one rotation of Earth on its axis. What time interval does the pilot measure? What assumptions did you make?

11. **Extending life?** Free neutrons have an average lifetime of about 1000 s before transforming into an electron, a proton, and an anti-neutrino. If a neutron leaves the Sun at a speed of 0.999c, (a) how long does it live according to an Earth observer? (b) Will such a neutron reach Pluto (5.9×10^{12} m from the Sun) before transforming? Explain your answers.

12. * A Σ^+ particle lives 0.80×10^{-10} s in its proper reference frame. If it is traveling at 0.90c through a bubble chamber, how far will it move before it disintegrates?

13. * **Extending the life of a muon** A muon that lives 2.2×10^{-6} s in its proper reference frame is created 10,000 m above Earth's surface. At what speed must it move to reach Earth's surface at the instant it disintegrates?

14. * **Effect of light speed on the time interval for a track race** Suppose the speed of light were 15 m/s. You run a 100-m dash in 10 s according to the timer's clock. How long did the race last according to your watch?

15. * Explain why an object moving past you would seem shorter in the direction of motion than when at rest with respect to you. Draw a sketch to illustrate your reasoning.

16. ** Explain why the length of an object that is oriented perpendicular to the direction of motion would be the same for all observers whose velocity is parallel to the object's velocity. Draw a sketch to illustrate your reasoning.

17. * You sit in a spaceship moving past the Earth at 0.97c. Your arm, held straight out in front of you, measures 50 cm. How long is it when measured by an observer on Earth?

18. **Length of a javelin** A javelin hurled by Wonder Woman moves past an Earth observer at 0.90c. Its proper length is 2.7 m. What is its length according to the Earth observer?

19. At what speed must a meter stick move past an observer so that it appears to be 0.25 m long?

20. **Changing the shape of a billboard** A billboard is 10 m high and 15 m long according to a person standing in front of it. At what speed must a person in a fast car drive by parallel to the billboard's surface so that the billboard appears to be square?

21. * A classmate says that time dilation and length contraction can be remembered in a simple way if you think of a person eating a foot-long "sub" sandwich on a train (the sandwich is oriented parallel to the train's motion). The person on the train finishes the sandwich in 20 min. You, standing on the platform, observe the person eating a shorter sandwich but for a longer time interval. Do you agree with this example? Explain your answer.

22. * **A pole and barn "paradox"** A barn of width 4.0 m is at rest in reference frame S. Let us take a pole of length 8.0 m and accelerate it to a velocity such that its length in frame S becomes equal to 4.0 m. Then at a certain moment the pole, flying through the barn (let's say from left to right), fits entirely within it. However, for an observer in reference frame S′ attached to the pole, the barn is moving to the left, and so it is the barn width that becomes reduced by half, and consequently the pole (8.0 m) does not fit in the barn (2.0 m). (a) Determine the speed of the pole relative to the barn. (b) Determine the time interval between the following two events, as observed by the observer in reference frame S and the observer in reference frame S′: *event 1*: the right end of the pole meets the right side of the barn; *event 2*: the left end of the pole meets the left side of the barn. (c) Why is there no paradox in this situation? Explain. (d) Represent the situation with spacetime diagrams for the observer in the reference frame S and the observer in the reference frame S′.

23. * **Space travel** An explorer travels at speed 2.90×10^8 m/s from Earth to a planet of Alpha Centauri, a distance of 4.3 light-years as measured by an Earth observer. (a) How long does the trip last according to the Earth observer? (b) How long does the trip last for the explorer on the spaceship? (c) Represent the situation with spacetime diagrams for the person on Earth and the explorer in the spaceship.

26.7 Velocity transformations

24. * Give examples of cases in which two observers record the motion of the same object to have different speeds. Sketch each example and explain how each observer arrives at the value of the measured speed.

25. * Now repeat Problem 26.24, only this time instead of a moving object, use a light flash. Describe what speeds of light different observers should measure according to the second postulate of special relativity.

26. * **Life in a slow-light-speed universe** Imagine that you live in a universe where the speed of light is 50 m/s. You sit on a train moving west at speed 20 m/s relative to the track. Your friend moves on a train in the opposite direction at speed 15 m/s. What is the speed of his train with respect to yours?

27. * **More slow-light-speed universe** In the scenario described in Problem 26.26, you and your friend listen to music on the same radio station. What is the speed of the radio waves that your antenna is registering compared to the speed of the waves that your friend's antenna registers if the station is 100 miles to the west of you?

28. * You are on a spaceship traveling at $0.80c$ with respect to a nearby star sending a laser beam to a spaceship following you, which is moving at $0.50c$ in the same direction. (a) What is the speed of the laser beam registered by the second ship's personnel according to the classical addition of the velocities? (b) What is the speed of the laser beam registered by the second ship's personnel according to the relativistic addition of the velocities? (c) What is the speed of the second ship with respect to yours according to the classical addition of the velocities? (d) What is the speed of the second ship with respect to yours according to the relativistic addition of the velocities?

29. ** Your friend says that it is easy to travel faster than the speed of light; you just need to find the right observer. Give physics-based reasons for why your friend would have such an idea. Then explain whether you agree or disagree with him.

26.8 Relativistic momentum

30. An electron is moving at a speed of $0.90c$. Compare its momentum as calculated using a nonrelativistic equation and using a relativistic equation.

31. * Explain why a relativistic expression is needed for fast-moving particles. Why can't we use a classical expression?

32. * If a proton has a momentum of 3.00×10^{-19} kg · m/s, what is its speed?

26.9 Relativistic energy

33. Determine the ratio of an electron's total energy to rest energy when moving at the following speeds: (a) 300 m/s, (b) 3.0×10^6 m/s, (c) 3.0×10^7 m/s, (d) 1.0×10^8 m/s, (e) 2.0×10^8 m/s, and (f) 2.9×10^8 m/s.

34. **Solar wind** To escape the gravitational pull of the Sun, a proton in the solar wind must have a speed of at least 6.2×10^5 m/s. Determine the rest energy, the kinetic energy, and the total energy of the proton.

35. * At what speed must an object move so that its total energy is 1.0% greater than its rest energy? 10% greater? Twice its rest energy?

36. * A person's total energy is twice his rest energy when he moves at a certain speed. By what factor must his speed now increase to cause another doubling of his total energy?

37. * A proton's energy after passing through the accelerator at Fermilab is 500 times its rest energy. Determine the proton's speed.

38. * A rocket of mass m starts at rest and accelerates to a speed of $0.90c$. Determine the change in energy needed for this change in speed.

39. * Determine the total energy, the rest energy, and the kinetic energy of a person with 60-kg mass moving at speed $0.95c$.

40. * An electron is accelerated from rest across 50,000 V in a machine used to produce X-rays. Determine the electron's speed after crossing that potential difference.

41. * A particle originally moving at a speed $0.90c$ experiences a 5.0% increase in speed. By what percent does its kinetic energy increase?

42. * An electron is accelerated from rest across a potential difference of 9.0×10^9 V. Determine the electron's speed (a) using the nonrelativistic kinetic energy equation and (b) using the relativistic kinetic energy equation. Which is the correct answer?

43. ** A particle of mass m initially moves at speed $0.40c$. (a) If the particle's speed is doubled, determine the ratio of its final kinetic energy to its initial kinetic energy. (b) If the particle's kinetic energy increases by a factor of 100, by what factor does its speed increase?

44. * Determine the mass of an object whose rest energy equals the total yearly energy consumption of the world (5×10^{20} J).

45. * **Mass equivalent of energy to separate a molecule** Separating a carbon monoxide molecule CO into a carbon and an oxygen atom requires 1.76×10^{-18} J of energy. (a) Determine the mass equivalent of this energy. (b) Determine the fraction of the original mass of a CO molecule 4.67×10^{-26} kg that was converted to energy.

46. * **Hydrogen fuel cell** A hydrogen-oxygen fuel cell combines 2 kg of hydrogen with 16 kg of oxygen to form 18 kg of water, thus releasing 2.5×10^8 J of energy. What fraction of the mass has been converted to energy?

47. ** **Mass to provide human energy needs** Determine the mass that must be converted to energy during a 70-year lifetime to continually provide electric power for a person at a rate of 1000 W. The production of the electric power from mass is only about 33% efficient.

48. * **EST** An electric utility company charges a customer about 6–7 cents for 10^6 J of electrical energy. At this rate, estimate the cost of 1 g of mass if converted entirely to energy.

49. * **Mass to produce electric energy in a nuclear power plant** A nuclear power plant produces 10^9 W of electric power and 2×10^9 W of waste heating. (a) At what rate must mass be converted to energy in the reactor? (b) What is the total mass converted to energy each year?

50. * **BIO** **EST** **Metabolic energy** Estimate the total metabolic energy you use during a day. (You can find more on metabolic rate in the first Reading Passage in Chapter 7.) Determine the mass equivalent of this energy.

51. ** **Energy from the Sun** (a) Determine the energy radiated by the Sun each second by its conversion of 4×10^9 kg of mass to energy. (b) Determine the fraction of this energy intercepted by Earth, which is 1.50×10^{11} m from the Sun and has a radius of 6.38×10^6 m.

26.10 Doppler effect for EM waves

52. **Why no color change?** Why don't the colors of buildings and tree leaves change when we look at them from a flying plane? Shouldn't the trees ahead look more bluish when you are approaching and reddish when you are receding?

53. * **Change red light to green** In a parallel universe the speed of light in a vacuum is 70,000 m/s. How fast should a driver's car move so that a red light looks green?

54. * **Effect of the Hubble constant on age and radius of the universe** How would the estimated age of the universe change if the new accepted value of the Hubble constant became 100 (km/s)/Mpc? How would the visible radius of the universe change?

55. * **Expanding faster** New observations suggest that our universe does not expand at a constant rate but instead is expanding at an increasing rate. How does this finding affect the estimation of the age of the universe using Hubble's law?

56. **Baseball Doppler shift** In September of 2010 Aroldis Chapman threw what may be the fastest baseball pitch ever recorded at 105 mi/h (47 m/s). What would the observed frequency of microwaves reflected from the ball be if the source frequency were 10.525 GHz? What would be the beat frequency between the source frequency and the observed frequency?

57. ** **Were you speeding?** A police officer stops you in a 29 m/s (65 mi/h) speed zone and says you were speeding. The officer's radar has source frequency 33.4 GHz and observed a 3900-Hz beat frequency between the source frequency and waves reflected back to the radar from your car. Were you speeding? Explain.

General Problems

58. * **Boat trip** A boat's speed is 10 m/s. It makes a round trip between stations A and B and then another between stations A and C. Stations A and B are on the same side of the river 0.5 km apart. Stations A and C are on the opposite sides of the river across from each other and also 0.5 km apart. The river flows at 1.5 m/s. What time interval is the round trip between stations A and B and then between A and C?

59. * **Space travel** An explorer travels at speed 2.90×10^8 m/s from Earth to a planet of Alpha Centauri, a distance of 4.3 light-years as measured by an Earth observer. (a) How long does the trip last according to an Earth observer? (b) How long does the trip last for the person on the ship?

60. ** **EST** **Extending life** Suppose that the speed of light is 8.0 m/s. You walk slowly to all of your classes during one semester while a classmate runs at a speed of 7.5 m/s during the time you are walking. Estimate your classmate's change in age, as judged by you, and your change in age according to you during that walking time. Indicate how you chose any numbers used in your estimate.

61. ** **Racecar when c is 100 m/s** Suppose that the speed of light is 100 m/s and that you are driving a racecar at speed 90 m/s. What time interval is required for you to travel 900 m along a track's straightaway (a) according to a timer on the track and (b) according to your own clock? (c) How long does the straightaway appear to you? (d) Notice that the speed at which the track moves past is your answer to part (c) divided by your answer to part (b). Does this speed agree with the speed as measured by the stationary timer?

62. ** **EST** Cherenkov radiation is electromagnetic radiation emitted when a fast-moving particle such as a proton passes through an insulator at a speed faster than the speed of light in that insulator. The Cherenkov radiation looks like a blue glow in the shape of a cone behind the particle. The radiation is named after Soviet physicist Pavel Cherenkov, who received a Nobel Prize in 1958 for describing the radiation. Estimate the smallest speed of a proton moving in oil that will produce Cherenkov radiation behind it.

63. ** A pilot and his spaceship of rest mass 1000 kg wish to travel from Earth to planet Scot ML, 30 light-years from Earth. However, the pilot wishes to be only 10 physiological years older when he reaches the planet. (a) At what constant speed must he travel? (b) What is the total energy of his spaceship and the rest energy, according to an Earth observer, while making the trip?

64. * Alice's friends Bob and Charlie are having a race to a distant star 10 light-years away. Alice is the race official who stays on Earth, and her friend Darien is stationed on the star where the race ends. Bob is in a rocket that can travel at $0.7c$, whereas Charlie's rocket can reach a speed of $0.866c$. Bob and Charlie start to move at the same time. Draw a spacetime diagram from each person's perspective (four diagrams altogether).

65. * Determine how long it takes each person in the previous problem to finish the race according to each observer. Compare all answers with the spacetime diagrams. Assume that both rockets can reach their speed instantly. (Hint: Don't forget that you can always use the definition of velocity, $v = \Delta x / \Delta t$.)

66. ** **Space travel** A pilot and her spaceship have a mass of 400 kg. The pilot expects to live 50 more Earth years and wishes to travel to a star that requires 100 years to reach even if she were to travel at the speed of light. (a) Determine the average speed she must travel to reach the star during the next 50 Earth years. (b) To attain this speed, a certain mass m of matter is consumed and converted to the spaceship's kinetic energy. How much mass is needed? (Ignore the energy needed to accelerate the fuel that has not yet been consumed.) Indicate any assumptions that you made.

67. ** (a) A container holding 4 kg of water is heated from 0 °C to 60 °C. Determine the increase in its energy and compare this to the rest energy when at 0 °C. (b) If the water, initially at 0 °C, is converted to ice at 0 °C, determine the ratio of its energy change to its original rest energy.

68. ** Your friend argues that Einstein's special theory of relativity says that nothing can move faster than the speed of light. (a) Give physics-based reasons for why your friend would have such an idea. (b) What examples of physical phenomena do you know of that contradict this statement? (c) Restate his idea so it is accurate in terms of physics. (Hint: See Problem 26.62 on Cherenkov radiation.)

Reading Passage Problems

Venus Williams's record tennis serve Venus Williams broke the women's tennis ball serving speed record at the European Indoor Championships at Zurich, Switzerland, on October 16, 1998 with a 57 m/s (127 mi/h) serve. She has since broken her own record more than once. The Doppler radar gun used in 1998 to measure the speed had a source microwave frequency of 10.525 GHz. Answer the following questions about the radar gun used on that serve.

69. Which principle can we use to determine the frequency f_D "detected" by the ball as it moved toward the source waves from the radar?
    (a) The beat frequency equation
    (b) The high-speed Doppler effect equation
    (c) The low-speed Doppler effect equation
    (d) The time dilation equation
    (e) The relationship between wave speed, frequency, and wavelength

70. Which frequency is closest to the frequency f_D detected by the ball as it moved toward the radar source waves?
    (a) 10.525 GHz (b) 10.525 GHz + 2.0 × 10⁻⁶ GHz
    (c) 10.525 GHz − 2.0 × 10⁻⁶ GHz (d) 3 × 10⁻⁷ Hz

71. Which principle can we use to determine the frequency f_O detected by the radar from waves reflected from the ball?
    (a) The beat frequency equation
    (b) The high-speed Doppler effect equation
    (c) The low-speed Doppler effect equation
    (d) The time dilation equation
    (e) The relationship between wave speed, frequency, and wavelength

72. Which answer is closest to the frequency f_O detected by the radar from the waves reflected from the ball?
    (a) Exactly 10.525 GHz (b) 10.525 GHz + 4.0 × 10⁻⁶ GHz
    (c) 10.525 GHz − 4.0 × 10⁻⁶ GHz (d) 3 × 10⁻⁷ Hz

73. Which principle is used to determine the frequency that the radar measures of the combined source and observed waves?
    (a) The beat frequency equation
    (b) The high-speed Doppler effect equation
    (c) The low-speed Doppler effect equation
    (d) The time dilation equation
    (e) The relationship between wave speed, frequency, and wavelength

74. Which answer is closest to the frequency that the radar measures of the combined source and observed waves?
    (a) 2.0 × 10³ Hz (b) 4.0 × 10³ Hz
    (c) 10.525 GHz (d) 3 × 10⁻⁷ Hz
    (e) 6 × 10⁻⁷ Hz

Quasars In 1963, Maarten Schmidt of the California Institute of Technology found the most distant object that had ever been seen in the universe so far. Called 3C 273, it emitted electromagnetic radiation with a power of 2×10^{13} Suns, or 100 times that of the entire Milky Way galaxy! Schmidt called 3C 273 a "quasi-stellar radio source," a name that was soon shortened to "quasar." Since then, astronomers have found many quasars, some substantially more powerful and distant than 3C 273.

Quasar 3C 273 is moving away from Earth at about speed $0.16c$. The Sun is 1.5×10^{11} m from Earth and emits energy at a rate of 3.8×10^{26} J/s caused by the conversion of 4.3×10^9 kg/s of mass into light and other forms of electromagnetic energy. The Hubble constant is

$$H = 70.8 \, \frac{\text{km/s}}{\text{Mpc}} = 70.8 \, \frac{10^3 \, \text{m/s}}{3.09 \times 10^{22} \, \text{m}} = 22.9 \times 10^{-19} \, \text{s}^{-1}$$

75. What principle would you use to estimate the distance of 3C 273 from Earth?
    (a) The high-speed Doppler effect equation
    (b) The low-speed Doppler effect equation
    (c) Hubble's law
    (d) The time dilation equation
    (e) The relationship between wave speed, frequency, and wavelength

76. Which answer below is closest to the distance of 3C 273 from the Earth in terms of the distance of the Sun from Earth?
    (a) ≈10³ Sun distances (b) ≈10⁶ Sun distances
    (c) ≈10⁹ Sun distances (d) ≈10¹² Sun distances
    (e) ≈10¹⁴ Sun distances

77. Which answer below is closest to the power of light and other forms of radiation emitted by 3C 273?
    (a) ≈10⁸ J/s (b) ≈10¹⁸ J/s
    (c) ≈10²⁵ J/s (d) ≈10³² J/s
    (e) ≈10⁴⁰ J/s

78. Which answer below is closest to the mass of 3C 273 that is converted to light and other forms of radiation each second? By comparison, the mass of Earth is 6 × 10²⁴ kg.
    (a) ≈10¹¹ kg/s (b) ≈10¹⁵ kg/s
    (c) ≈10¹⁹ kg/s (d) ≈10²³ kg/s
    (e) ≈10²⁹ kg/s

79. What is the speed of light emitted by 3C 273 as detected by an observer on 3C 273?
    (a) $1.15c$ (b) c
    (c) $0.85c$ (d) None of these is correct.

80. If 3C 273 is moving away from Earth at $0.16c$, what speed below is closest to the light speed we on Earth detect coming from 3C 273?
    (a) $1.15c$ (b) c
    (c) $0.85c$ (d) None of these is correct.

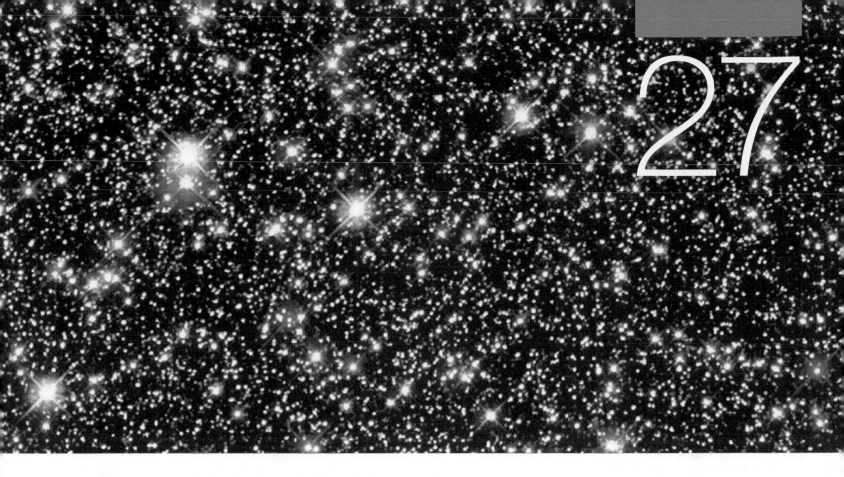

Quantum Optics

Max Planck was a gifted young musician and composer. However, he chose physics as a field of study when he entered the University of Munich in 1874 at age 16. One of his professors discouraged him from studying physics because "almost everything is already discovered, and all that remains is to fill a few holes." Planck disregarded the advice; by 1879 he had defended his Ph.D. thesis, and in 1900 he made a discovery that totally changed how physicists understood the world.

- **What investigations started the transition from classical physics to quantum physics?**
- **Why do some stars look red and some look blue?**
- **How does a solar cell work?**

WE HAVE STUDIED the behavior of light and investigated various models of light, ultimately arriving at an electromagnetic wave model. That model explained the propagation of light in different media and its interactions with other objects, but it did not explain the mechanisms of light emission and absorption. To explain those, a new model of light was needed, and Planck was among those who helped develop it. This new model initiated the revolutionary *quantum physics* that is the subject of this and subsequent chapters.

BE SURE YOU KNOW HOW TO:

- Connect an ammeter and a voltmeter in a circuit to measure current and potential difference (Section 19.3).
- Relate the change of kinetic energy of an electrically charged particle to the potential difference across which it travels (Section 18.3).
- Relate the wavelength of a wave to its frequency and speed (Section 11.2).

27.1 Black body radiation

In the last quarter of the 19th century, physicists became interested in how hot objects such as stars and the coals of a fire emit thermal radiation (what we now call **infrared radiation**). Scientists studied the absorption and emission of infrared radiation and visible light from a hot object by modeling that object as a so-called **black body**. A black body absorbs all incident (incoming) light and converts the light's energy into thermal energy (no light is reflected). The black body then radiates electromagnetic (EM) waves based solely on its temperature.

What characterizes a black body?

To understand how the black body model works, imagine a small window in a house whose lights are off, or a small opening in a box, or the pupil of the human eye. Small openings like these look black to an outside observer because light entering the openings is not reflected—it is trapped inside. A small window, for example, admits sunlight. That light is absorbed inside the room, making the room warmer. As a result, more and more infrared radiation also leaves the room through the window. Eventually, the rate at which sunlight energy enters the room through the window equals the rate at which infrared radiation energy leaves the room. At this point the window becomes "a black body." In **Figure 27.1** an infrared camera has detected the infrared radiation being emitted through the windows of an unlit house.

We will model a black body as the surface of a small hole in one side of a closed container (**Figure 27.2a**). Imagine that this container also has a thermometer inside. The hole looks black, and the box is at temperature $T_1 = 310$ K. If you measure the power output per unit radiating surface area of the EM radiation coming from the hole at different wavelengths, you find that the hole emits a continuous EM spectrum (the lower curve in Figure 27.2b).

The graph in Figure 27.2b is called a **spectral curve**. The quantity on the vertical axis is the power output per unit radiating surface area of the black body per small wavelength interval in units of $W/m^2/m$. On the horizontal axis is the radiation's wavelength. The *total* power output per unit radiating surface area is the area under the black body spectral curve; this quantity is also known as the **intensity** (I) of the EM radiation. (Notice that at a particular wavelength we see the greatest intensity of EM radiation coming from the hole.)

Now imagine that you place the box on a hot stove—the box's temperature rises and the thermometer in the box indicates a higher temperature $T_2 = 373$ K. If we again measure the spectrum of the EM radiation being emitted from the hole, we find that

- the total power output (intensity multiplied by the area of the hole) from the hole is now greater,
- the spectral curve rises (the upper curve in Figure 27.2b) at all wavelengths, and
- the peak of the spectral curve shifts to a shorter wavelength.

As the temperature of the box increases further, these patterns persist. The higher the temperature, the shorter the wavelength at which the maximum of the black body spectral curve occurs, and the greater the power radiated at all wavelengths.

A black body is not a real object; it is a model. It turns out, however, that many real objects emit EM radiation similar to that of a black body. For example, the metal filament of an incandescent lightbulb gets very warm when electric current is passed through it. The bulb radiates some visible light, a lot of invisible infrared radiation, and a little bit of UV. The spectral curve of a lightbulb is very close to the spectral curve of a black body of the same temperature. Thus, studying black body radiation might allow us to understand how objects emit and absorb light and how the radiation they emit might relate to the objects' temperatures.

FIGURE 27.1 An infrared photo taken at night with the house lights off.

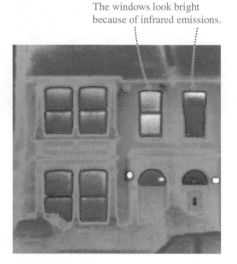

The windows look bright because of infrared emissions.

FIGURE 27.2 A black body spectrum.

(a)

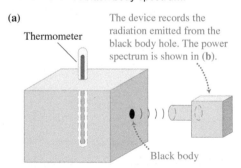

Thermometer

The device records the radiation emitted from the black body hole. The power spectrum is shown in (b).

Black body

(b)

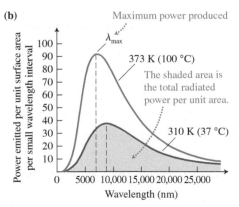

Maximum power produced

λ_{max}

373 K (100 °C)

The shaded area is the total radiated power per unit area.

310 K (37 °C)

Power emitted per unit surface area per small wavelength interval

100
90
80
70
60
50
40
30
20
10

0   5000 10,000 15,000 20,000 25,000

Wavelength (nm)

Studies of black body radiation

Several physicists in the late 1800s used the black body model to determine characteristics of EM radiation. Among their findings were the total power output of a black body and the wavelength at which the peak of the emitted radiation power occurs.

Total power output In 1879 a Slovenian physicist, Joseph Stefan (1835–1893), using his own experiments and data collected by John Tyndall, investigated the power of light emission from a black body—the total power output per unit of surface area—which is equal to the area under the black body spectral curve (see **Figure 27.3**). As the graph lines for different temperatures show, the total radiation output (power per unit area of the emitting object) increases dramatically as the temperature of the black body increases. Stefan found that the total power output per unit of surface area in Figure 27.3 is proportional to the fourth power of the temperature. For example, if you double the temperature of the object (in kelvins), the total radiation power increases by 16 times. The area under the 6000-K curve in Figure 27.3 is 16 times the area under the 3000-K curve.

The total energy emitted by a 1-m² black body every second can be written as $I \propto T^4$ or $I = \sigma T^4$, where σ (Greek letter sigma) is the coefficient of proportionality. To find the power P radiated by the whole object, we multiply the intensity of its emitted radiation at its surface by the radiating surface area A of the object. In 1884 Ludwig Boltzmann (1844–1906) derived theoretically the empirical relationship found by Stefan by modeling a black body as a thermodynamic engine with light instead of gas.

The work of Stefan and Boltzmann resulted in the following relationship between the total emitted power P in watts (W), the temperature of the black body T in kelvins, and its surface area A in square meters (**Stefan-Boltzmann's law**):

$$P = IA = \sigma T^4 A \qquad (27.1)$$

where sigma (σ) stands for the coefficient of proportionality (Stefan-Boltzmann's constant): $\sigma = 5.67 \times 10^{-8}\ \text{W/m}^2 \cdot \text{K}^4$.

Wavelength at which maximum intensity occurs In 1893, the German physicist Wilhelm Wien (1864–1928) quantified a second aspect of black body radiation using data similar to those in the graphs in Figures 27.2 and 27.3. The wavelength λ_{max} at which a black body emits radiation of maximum power per wavelength depends on the temperature of the black body:

$$\lambda_{max} = \frac{2.90 \times 10^{-3}\ \text{m} \cdot \text{K}}{T} \qquad (27.2)$$

This became known as **Wien's law.** Note that the wavelength λ_{max} at which the black body emits maximum power per wavelength interval becomes shorter as the black body temperature T becomes greater.

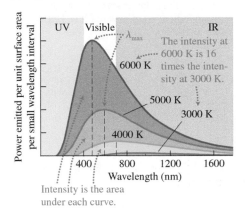

FIGURE 27.3 Black body radiation at different temperatures. The total power is proportional to the fourth power of the temperature in kelvins.

TIP ► A black body (the system) emits electromagnetic waves with power given by Stefan-Boltzmann's law even if it is at the same temperature as the environment—but it also absorbs energy at the same rate. If the body is hotter than the environment, it radiates more energy than it absorbs. If it is cooler, it absorbs more than it radiates. The black body continues to emit and radiate different amounts of energy until it reaches thermal equilibrium with the environment. After that, emission and absorption continue, but their rates become the same.

QUANTITATIVE EXERCISE 27.1 ▷ **The surface temperature of the Sun**

The maximum power per wavelength interval of light from the Sun is at a wavelength of about 510 nm, which corresponds to yellow light. What is the surface temperature of the Sun?

Represent mathematically Rearrange Wien's law solving for T in terms of λ_{max}:

$$T = \frac{2.90 \times 10^{-3}\ \text{m} \cdot \text{K}}{\lambda_{max}}$$

Solve and evaluate The temperature of a black body with its maximum power per wavelength interval at 510 nm = 510×10^{-9} m is

$$T = \frac{2.90 \times 10^{-3}\ \text{m} \cdot \text{K}}{510 \times 10^{-9}\ \text{m}} = 5.69 \times 10^3\ \text{K}$$

This is about the temperature of the object in Figure 27.3 with the highest intensity (the uppermost curve).

At this temperature, matter is in a gaseous state. But the huge mass (2.0×10^{30} kg) of the Sun produces a strong gravitational field that prevents the gaseous atoms and molecules from easily leaving.

- -

Try it yourself The surface temperature of Earth is about 288 K. What is the wavelength at which Earth radiates EM waves with maximum power per wavelength interval?

Answer   Infrared radiation with a wavelength of 10,000 nm.

As we saw in the "Try it yourself" question of Quantitative Exercise 27.1, the Earth radiates EM waves primarily in the infrared part of the electromagnetic spectrum, invisible to our eyes. Thus, when we see Earth and most objects on its surface, we are seeing sunlight reflected from them. The same mechanism (reflection of sunlight) makes the Moon visible to us. The Sun and other stars, however, do not need an external source of EM waves to illuminate them.

A star that has a lower surface temperature than the Sun's radiates with a maximum intensity wavelength that is longer than 510 nm, and the star looks red instead of yellow. A star that is hotter than the Sun radiates with maximum intensity at shorter wavelengths, a combination of blue and other visible wavelengths that cause it to look whiter. Very hot stars (around 10,000 K) look blue. Stars even hotter than these are generally invisible to our eyes, but they can be detected with instruments sensitive to ultraviolet radiation.

Figure 27.4 shows a pattern characteristic of the majority of the stars. You can see that hotter stars are bigger and cooler stars are smaller. This relationship is true for stars of ages similar to that of the Sun. Older stars can have lower temperatures and large sizes (red giants) or very high temperatures and smaller sizes (white dwarfs). The fascinating topic of stellar evolution is unfortunately beyond the scope of this book.

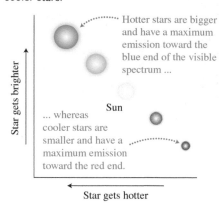

FIGURE 27.4 The color of stars. The color of a star is an indication of how hot it is. Hotter stars are bigger and appear bluer than cooler stars.

EXAMPLE 27.2 | **What is the power of surface radiation from the Sun?**

The radius of the Sun is about 7×10^8 m. Estimate the total amount of electromagnetic energy emitted by the Sun every second if its average surface temperature is about 6×10^3 K.

Sketch and translate We can visualize the Sun as a shining sphere of a given radius. If we know the total surface area of the Sun, we can use Stefan-Boltzmann's law to calculate the total energy radiated per unit time (power).

Simplify and diagram To apply Stefan-Boltzmann's law, assume that the Sun radiates as a black body. To find the surface area, assume that the Sun is a perfect sphere whose every square meter emits electromagnetic waves equally.

Represent mathematically According to Stefan-Boltzmann's law, the power radiated by a black body with a surface area A is

$$P = \sigma A T^4$$

The temperature is given, and the surface area of a sphere is $A = 4\pi R^2$. Thus the total power radiation by the Sun's surface is

$$P = \sigma(4\pi R^2)T^4$$

Solve and evaluate Now insert the appropriate values. Because we are doing an estimation, we will only calculate to one significant digit:

$$P \approx (6 \times 10^{-8}\,\text{W/m}^2 \cdot \text{K}^4)4\pi(7 \times 10^8\,\text{m})^2(6 \times 10^3\,\text{K})^4$$
$$= 5 \times 10^{26}\,\text{W}$$

The actual number is close to 4×10^{26} W. Either way, this is a huge power output. Power plants on Earth typically have an output of 10^{12}–10^{13} W. The Sun has maintained (approximately) this power output for billions of years. You'll learn more about the mechanism of the Sun's power output in a later chapter (Chapter 29).

- -

Try it yourself Estimate how much solar energy reaches Earth every second. The effective surface area of Earth facing the Sun is a disk of radius 6.4×10^6 m (the radius of Earth). The average Sun-Earth distance is about 1.5×10^{11} m.

Answer $2 \times 10^{17}\,\text{W} = 2 \times 10^{17}\,\text{J/s}.$

The ultraviolet catastrophe

By the end of the 19th century, physicists had devised a mechanism to explain how objects emitted electromagnetic radiation, namely, that the atoms and molecules inside objects are made of charged particles, and when these charged particles vibrate, they radiate electromagnetic waves. The higher the vibration frequency, the higher the frequency of the electromagnetic waves that are emitted. The atoms and molecules in a black body at temperature T vibrate at many different frequencies centered on the frequency of the peak radiation on the black body spectrum curve.

Maxwell's electromagnetic wave theory (Chapter 25), together with some assumptions about the structure of matter, predicts that the intensity of the radiation emitted by a black body should increase with the frequency of the radiation—greater intensity for ultraviolet light than for visible light; greater intensity for X-rays than for ultraviolet light. However, actual spectral curves show that at high frequencies (short wavelengths) the emitted intensity actually drops off (**Figure 27.5**). No models built on the physics of the late 19th century made predictions that were consistent with this drop. This problem became known as the **ultraviolet catastrophe**.

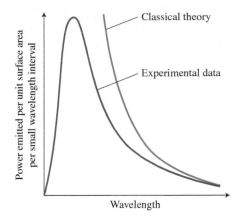

FIGURE 27.5 Classical black body radiation theory failed to match the experimental data at short wavelengths.

Planck's hypothesis

After working on the problem for several years, Max Planck (1858–1947) devised an ingenious model that correctly predicted the spectral curve for a black body. In Planck's model, the charged particles inside radiating objects were still responsible for producing electromagnetic radiation. However, they could radiate energy only in discrete portions called **quanta** (one portion is one quantum). Each quantum of emitted energy was equal to a multiple of some fundamental portion of energy: E_0, $2E_0$, $3E_0$, $4E_0$, etc. This idea is known as **Planck's hypothesis**.

Although the idea of this so-called **energy quantization** was completely revolutionary, the idea of quantization itself was not new to physics. Think of mechanical waves. When waves travel back and forth on a string of finite length with fixed ends, the string can vibrate only at discrete standing wave frequencies:

$$f_n = n\,\frac{v}{2L}$$

where $n = 1, 2, 3, \ldots$. The frequencies depend on the speed v of the waves on the string and the string length L. The standing wave frequencies are therefore said to be **quantized**. The energy of these standing waves can take on any positive value, however, because the energy of the wave depends on the *amplitude* of the wave and not on its frequency.

Planck's hypothesis is different. He suggested that it was *the energy of the oscillator* that was quantized, rather than its frequency.

Using his energy quantization model, Planck was able to derive a mathematical function describing the black body spectral curve. The derivation is complicated, so we won't show it. When he matched the calculated values from the function to experimental results, he was able to determine the constant of proportionality between the oscillating frequency of a charged particle f and its smallest fundamental energy E_0. The relationship is

$$E_0 = hf \tag{27.3}$$

The proportionality constant h became known as **Planck's constant**. The units of h in the SI system are $\text{J}/\text{Hz} = \text{J}/(1/\text{s}) = \text{J} \cdot \text{s}$. Its value was determined experimentally to be $h = 6.63 \times 10^{-34}\ \text{J} \cdot \text{s}$.

Planck viewed his energy quantization model as a mathematical trick of some sort rather than an explanation of actual physical phenomena. How could something oscillate smoothly at a particular frequency yet emit waves of that frequency only in discrete energy portions, as though it were emitting some sort of energy "particle"? He was almost embarrassed to have suggested such a model, and for much of his life he harbored doubts that real molecules or atoms radiated energy in this way. However, other scientists, including Albert Einstein, successfully applied this idea to explain other phenomena.

QUANTITATIVE EXERCISE 27.3 **Energy of light quanta**

Estimate the energy of a quantum of radio waves (frequency of 10^6 Hz), infrared radiation (3×10^{13} Hz), visible light (5×10^{14} Hz), and UV radiation (10^{15} Hz).

Represent mathematically According to Planck's hypothesis, the energy of a quantum of EM radiation is $E_0 = hf$.

Solve and evaluate The estimated energies are

$$E_{0\,\text{radio}} = hf_{\text{radio}} = (6.63 \times 10^{-34}\ \text{J} \cdot \text{s})(1 \times 10^6\ \text{Hz})$$
$$= 7 \times 10^{-28}\ \text{J}$$

$$E_{0\,\text{infrared}} = hf_{\text{infrared}} = (6.63 \times 10^{-34}\ \text{J} \cdot \text{s})(3 \times 10^{13}\ \text{Hz})$$
$$= 2 \times 10^{-20}\ \text{J}$$

$$E_{0\,\text{visible}} = hf_{\text{visible}} = (6.63 \times 10^{-34}\ \text{J} \cdot \text{s})(5 \times 10^{14}\ \text{Hz})$$
$$= 3 \times 10^{-19}\ \text{J}$$

$$E_{0\,\text{ultraviolet}} = hf_{\text{ultraviolet}} = (6.63 \times 10^{-34}\ \text{J} \cdot \text{s})(1 \times 10^{15}\ \text{Hz})$$
$$= 7 \times 10^{-19}\ \text{J}$$

These are very small energies. For perspective, we can estimate how many quanta a 60-watt incandescent lightbulb emits every second. Its filament (at about $T = 2500$ K) produces a spectrum very close to a black body spectrum. Using Wien's law we can estimate the wavelength of the maximum intensity radiation emitted by the filament and the corresponding frequency of that radiation:

$$\lambda_{\text{max}} = \frac{c}{f} = \frac{2.90 \times 10^{-3}\ \text{m} \cdot \text{K}}{T}$$

or

$$f = \frac{c}{\lambda_{\text{max}}} = \frac{cT}{2.90 \times 10^{-3}\ \text{m} \cdot \text{K}}$$
$$= \frac{(3.0 \times 10^8\ \text{m/s})(2500\ \text{K})}{2.90 \times 10^{-3}\ \text{m} \cdot \text{K}} \approx 2.6 \times 10^{14}\ \text{Hz}$$

This frequency corresponds to infrared radiation. The energy of one quantum of infrared radiation at this frequency is approximately 2×10^{-19} J. If we assume that all of the filament's radiated power is radiated at this frequency, we can estimate the number n of quanta emitted every second as the total energy per second divided by the energy of one quantum:

$$n = \frac{60\ \text{J/s}}{2 \times 10^{-19}\ \text{J/quantum}} \approx 3 \times 10^{20}\ \text{quanta/s}$$

Try it yourself Estimate how many quanta of electromagnetic radiation from the Sun reach Earth each second (use the data from the "Try it yourself" question in Example 27.2) and assume that all of the Sun's energy is radiated as visible light.

Answer

About 10^{36} quanta each second—equivalent to the output of about 3×10^{15} 60-watt bulbs.

Planck's quantum idea added a new twist to the study of light, and this divergence made him uncomfortable even though he knew the mathematics was correct. **Table 27.1** presents a summary of that history. Note that as of the early 20th century, there was no explanation for the absorption of light.

TABLE 27.1 **Evolution of ideas concerning the nature of light**

Approximate time period	Model	Experimental evidence that the model explains
1600s	*Particle (bullet) model* Light can be modeled as a stream of bullet-like particles.	Reflection, shadows, and possibly refraction
Early 1800s	*Wave model* Particle model insufficient. Light modeled as wave. Wave mechanism unknown.	Double-slit interference, diffraction, light colors, reflection, and refraction
Middle 1800s	*Electromagnetic wave (ether model)* Light modeled as transverse vibrations of ether medium.	Polarization and electromagnetic wave transmission and all of the above evidence
Late 1800s	*Electromagnetic wave (no-ether model)* Light modeled as transverse vibrations of electric and magnetic fields. No medium required.	Michelson-Morley experiment
Early 1900s	*Quantum model* Electromagnetic wave model insufficient. Light modeled as stream of discrete energy quanta.	Emission of light by black bodies

Today, Planck's quantum model is understood as the birth of quantum physics. To understand how this transformation from classical to quantum physics occurred and what model of light is currently accepted, we now turn the discussion to a new phenomenon.

REVIEW QUESTION 27.1 Planck's hypothesis of light emission, like the classical explanation, stated that objects emit light through the accelerated motion of electrically charged particles. What in Planck's hypothesis brought the theory in line with observed values of light emission?

27.2 Photoelectric effect

Planck's quantum hypothesis contradicted the wave model of light and remained untested for several years. However, in the interim the hypothesis proved instrumental in understanding an interaction between metals and electromagnetic radiation. In order to understand this interaction, we will first review what we know about the structure of metals.

The structure of metals

In an earlier chapter (Chapter 19) you learned that metals are made of atoms whose outermost one or two electrons do not remain bound to the nucleus but are loosely bound to the whole crystal (**Figure 27.6a**). The positively charged ions (the metal atoms minus their outermost electrons) form a crystal lattice. The negatively charged electrons that have left the atoms move freely through the lattice (Figure 27.6b). These free electrons have positive kinetic energy $K = +(1/2)mv^2$. Why don't they leave the metal?

An electron and a positive ion have opposite electric charges ($-e$ for the electron and $+e$ for the positive ion). If we simplify the situation and consider a single electron and a single ion as the system of interest, the electric potential energy U_q of this system is negative (Section 17.5):

$$U_q = k\frac{q_e q_{\text{ion}}}{r} = k\frac{(-e)(+e)}{r} = -k\frac{e^2}{r} < 0$$

In addition to the electric potential energy of the electron-lattice system, the electrons possess kinetic energy. This energy must be smaller in magnitude than the electric energy because the electrons do not spontaneously leave the metal; thus the total energy of the system must be negative.

Because the energy of the system is negative, energy must be added to the system to remove the free electron from the metal (Figure 27.6c). The minimum energy needed to remove a free electron from a metal is called the **work function** ϕ (Greek letter phi) of the metal (Figure 27.6d).

The work function ϕ has a specific, known value for many metals. Although the work function has units of energy, it is not measured in joules because typical work function values are very small. Instead, we use the **electron volt** (eV), where $1\text{ eV} = 1.6 \times 10^{-19}$ J. As you may recall (from Chapter 26), one electron volt is the magnitude of the change in kinetic energy of an electron of charge $e = -1.6 \times 10^{-19}$ C when it moves across a potential difference of 1 V.

For most metals (silver, tin, aluminum, etc.), the value of ϕ ranges from about 4.30 eV to 4.60 eV. However, the ϕ values of some metals, such as cesium and lithium, are in the range of 2.10 eV to 2.90 eV. Thus it is significantly easier to remove electrons from these metals. The greater the work function of a particular metal, the more strongly the free electrons are bound to the metal as a whole, and the more energy that must be added to the electron-metal system to separate them.

Detection of the photoelectric effect

Now that we have an understanding of how metals hold on to their free electrons, let's discuss some experiments that eventually helped support Planck's hypothesis. We learned from our investigation of black body radiation that EM radiation is emitted as quanta. Is EM radiation absorbed as quanta as well? We investigate this idea in

FIGURE 27.6 The structure of metals. (a and b) Positive ions and free electrons. (c) Energy bar chart for a lattice-free electron system. (d) Adding the energy equal to the work function makes the total energy of the system zero.

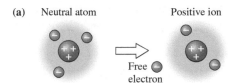

(a) Neutral atom   Positive ion

In a metal, a neutral atom becomes a positive ion and a free electron.

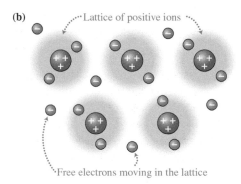

(b) ⋯⋯Lattice of positive ions⋯⋯

⋯Free electrons moving in the lattice

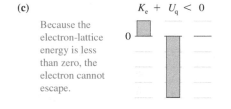

(c) $K_e + U_q < 0$

Because the electron-lattice energy is less than zero, the electron cannot escape.

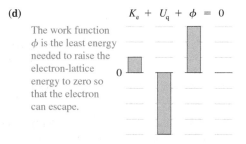

(d) $K_e + U_q + \phi = 0$

The work function ϕ is the least energy needed to raise the electron-lattice energy to zero so that the electron can escape.

TIP Even though ϕ is called the work "function," it's not a function in the mathematical sense. It is a constant that differs from material to material. It is also not work because it is a property of a particular metal. Finally, although some electrons in metals are called "free" electrons, they are not free to leave the metal; they are free to move around within the metal, however. From this analysis of the vernacular, you can see how important it is to question the terms we use in order not to be confused.

Observational Experiment **Table 27.2**. These experiments were performed originally by Wilhelm Hallwachs (1859–1922) in Germany in 1888. He used a gold leaf electroscope, in which two thin "leaves" of gold foil repel each other if they carry a like charge and hang straight down if they are uncharged. The first experiment is similar to one we performed earlier (in Chapter 17).

OBSERVATIONAL EXPERIMENT TABLE 27.2   **Discharging an electroscope**
VIDEO
OET 27.2

Observational experiment	Analysis
Experiment 1. Take a plastic rod, rub it with felt, and touch it to the metal (zinc) sphere of an electroscope. Repeat the experiment using a glass rod rubbed with silk and another electroscope. Observe that the leaves of both electroscopes are deflected and remain deflected. 	The negatively charged plastic rod rubbed with felt transfers electrons to the electroscope, making it negatively charged.  The positively charged glass rod rubbed with silk takes free electrons away from the electroscope, leaving it positively charged. 
Experiment 2. Shine visible light from an incandescent bulb on both the positively charged and the negatively charged electroscope. Nothing happens to them. 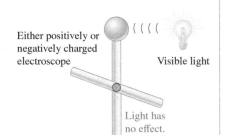	Visible light does not affect the charge of the electroscope.
Experiment 3. Shine UV radiation on the positively charged electroscope. The leaves of the electroscope stay deflected. Shine UV radiation on the negatively charged electroscope. The leaves quickly return to their original uncharged position. 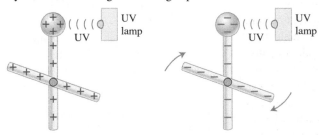	The UV light discharges a negatively charged electroscope but does not discharge a positively charged electroscope.

Patterns

- UV radiation makes the negatively charged electroscope discharge but has no effect on the positively charged electroscope.
- Lower frequency visible light does not discharge either electroscope.

The sphere on the top of the electroscope in Table 27.2 was made of zinc. It turns out that if the same sphere had been covered with a thin layer of cesium-antimony, green light would have had an effect on it similar to the effect the UV light had in the experiments above. It is the combination of the frequency of light and the material on which it shines that allows the negatively charged electroscope to discharge. A positively charged electroscope does not discharge in the presence of visible or UV light.

The interaction described in Table 27.2 is an example of the **photoelectric effect**. Let's try to invent an explanation for the patterns that we've observed.

Explanation 1: The light's vibrating electric field interacts with the electric charges in the molecules of air surrounding the electroscope. It shakes them back and forth until the molecules ionize into electrons and positively charged ions. The positive ions are attracted to the negative electroscope and attach to it, making it neutral overall.

This sounds reasonable. However, if it is correct, then both the positively and negatively charged electroscopes should discharge, because the newly freed electrons in the air should neutralize the positively charged electroscope. But this was not observed. We can rule out this explanation for the photoelectric effect.

Explanation 2: The light's vibrating electric field interacts with the electrons in the electroscope and can cause them to be ejected from it, but it cannot cause positive ions to be ejected. This hypothesis makes sense when we recall that free electrons in metals are negatively charged particles, and when we charge an object by touching, we either add or remove these electrons. The positive ions in the metal make up its crystal lattice and probably cannot be removed easily.

Explanation 2 seems reasonable, but it does not explain why the frequency of light is important: in Table 27.2 light from the incandescent lightbulb did not discharge the electroscope, whereas UV radiation did. We need more observational experiments.

More experiments

Additional observational experiments were conducted by a German physicist, Philipp von Lenard (1862–1947), in 1902. He used an apparatus similar to that shown in **Figure 27.7**. The apparatus, called a *phototube*, consists of an evacuated glass container with a window that allows both visible light and UV radiation to shine on a metal plate. This plate is normally connected to the negative terminal of a battery and is at an electric potential of zero; it is known as a **cathode**. A corresponding plate, termed an **anode**, is usually connected to the positive terminal of the battery at potential V. However, the connection can be reversed by connecting the plates to the opposite terminals of the battery. When reversed, the potential of the former anode potential is negative relative to the zero cathode potential. The ammeter shown in Figure 27.7 can detect changes in the current in the circuit.

We can change the frequency of light passing through the window by inserting color filters in front of the window. We can also change the material of the cathode and the anode. In all further experiments, we will assume that the anode is made of the same material as the cathode. When they are made of different materials, the difference in the work functions of the two metals makes the analysis more complicated and beyond the scope of this book.

When there is no light shining on the cathode, the ammeter detects no current. When light is shining and we vary the filters, allowing light of different frequencies to illuminate the cathode, the ammeter detects current in the circuit for some filters but not for others. For example, when the cathode is made of zinc, only UV light leads to current in the circuit (visible light does not), whereas for a cesium-antimony cathode, green light leads to current (but red light does not).

Observational Experiment **Table 27.3** allows us to analyze the data that von Lenard collected.

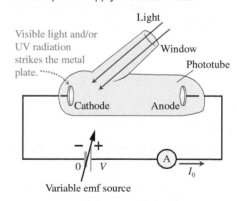

FIGURE 27.7 A phototube connected to a variable power supply and an ammeter.

Observational experiment	Analysis
Experiment 1. Connect the cathode on which light from a light source with a variable aperture shines to a negative battery terminal. Find the frequency of light (using filters) for which the ammeter detects current. Keep the potential of the anode at a constant positive value relative to the zero potential at the cathode ($V_{anode} > V_{cathode} = 0$). Without changing the frequency, steadily increase the intensity of radiation shining on the cathode by opening the aperture. The ammeter measures current.	Electric current from light exposure (the **photocurrent**) registered by the ammeter is directly proportional to the intensity of the light radiation. When no light is shining on the cathode, no current is measured by the ammeter. 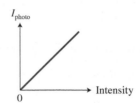
Experiment 2. Use the frequency of EM radiation for which the photoelectric effect occurs. Using the same cathode and the same light filter, detect current in the circuit. Keeping the intensity and the frequency of the light or UV radiation constant, vary $V_{anode} - V_{cathode}$. Notice that a negative value means that the electric potential at the anode is lower than at the cathode. The ammeter measures current.	Assume that the potential at the cathode is 0. The photocurrent increases as the positive potential at the anode increases. After the anode potential reaches a particular positive value, the current stops increasing. At a particular negative anode potential $-V_s$ (the **stopping potential**), the current stops. Note that V_s is the magnitude of the stopping potential. 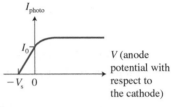
Experiment 3. Use the frequency of EM radiation for which the photoelectric effect occurs. Repeat Experiment 2 for an increased intensity of light or UV radiation by opening the aperture on the light source.	When the intensity of UV radiation increases, the maximum value of the photocurrent increases, but the negative stopping potential $-V_s$ does not change. 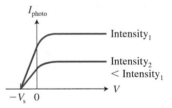
Experiment 4. Repeat Experiment 2 with radiation in both the visible light range and UV range, using different frequencies, and each time recording the stopping potential.	The stopping potential (now shown on the y-axis) is plotted against frequency (the x-axis). When the frequency of electromagnetic radiation increases, the negative stopping potential $-V_s$ becomes more negative. For higher frequency light, it takes a greater opposing potential difference to stop the photocurrent. Also, there is no photocurrent when the frequencies are less than a value termed the **cutoff frequency**. 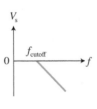

Patterns

The photocurrent in the circuit depends on three factors: the potential difference across the electrodes (we assume that the potential of a cathode is zero), the intensity of the light, and the frequency of the light. If for a particular frequency of light, photocurrent is present in the circuit, then

1. For constant positive anode potential V, the photocurrent is directly proportional to the intensity of the radiation.
2. When the V_{anode} is zero, a nonzero photocurrent I_0 is still present in the circuit.
3. For the same intensity and frequency of radiation, the photocurrent increases with increasing positive anode potential V until it reaches some maximum value.
4. A well-defined negative stopping potential $-V_s$ stops the photocurrent.
5. Increasing the intensity of radiation increases the photocurrent I_0 and the maximum value of the photocurrent, but does not affect the negative anode stopping potential $-V_s$.
6. $-V_s$ is affected only by the frequency of the EM radiation incident on the cathode. The relationship is linear. When the frequency is below a certain cutoff value f_{cutoff}, the photocurrent is zero for any potential difference between the cathode and anode.

Von Lenard noted another interesting outcome. He saw that the ammeter recorded a current in the circuit immediately after the light was turned on.

Attempting to explain these patterns with the EM wave model

Let's try to use the EM wave model of light to explain the patterns reported for the frequencies of light for which the photoelectric effect occurred in Table 27.3. Remember that in all experiments, the cathode and the anode are made of the same material.

Pattern 1 For constant $V_{anode} - V_{cathode} > 0$, the photocurrent is directly proportional to the intensity of the radiation.

EM wave model explanation: Evidently, electrons in the cathode absorb enough energy from EM radiation that they are ejected from the cathode. The greater the intensity of the light, the more energy is transferred to the cathode. More electrons are freed from the cathode, causing a greater photocurrent. The wave model explains pattern 1.

Pattern 2 When $V_{anode} - V_{cathode} = 0$, there is still a nonzero photocurrent I_0 in the circuit.

EM wave model explanation: Evidently, the electrons in the metal gain enough energy from the EM radiation that they can escape the metal and have energy left over to reach the anode. The wave model explains pattern 2.

Pattern 3 For the same intensity of EM radiation, the photocurrent increases with increasing positive $V_{anode} - V_{cathode}$ until the current reaches a maximum value.

EM wave model explanation: It seems that with constant intensity EM radiation, the same number of electrons is ejected from the metal independent of $V_{anode} - V_{cathode}$. The electric force pushing the electrons to the anode increases as $V_{anode} - V_{cathode}$ increases. However, if the wave model is correct, it seems that a minimum $\vec{E}$ field is needed to draw all of the electrons to the anode. Increasing the $\vec{E}$ field beyond that would have no effect. The wave model explains pattern 3.

Pattern 4 A well-defined negative stopping potential $-V_s$ stops the photocurrent.

EM wave model explanation: With a negative $-V_s$, the anode is at a lower electric potential than the cathode. The electric field now opposes the motion of electrons across the tube to the anode (Figure 27.8a). Suppose that $V_{anode} - V_{cathode} = (-V_s) - 0$ is the smallest negative value it can be while still preventing electrons from reaching the anode. This process can be quantified using the generalized work-energy principle—see the bar chart in Figure 27.8b. Mathematically:

$$K_i = U_{qf}$$

Inserting the expressions for kinetic energy and electric potential energy gives

$$\tfrac{1}{2} m_e v_{e\,i\,max}^2 = (-e)(-V_s)$$

where $-e$ and m_e are the electron charge and mass, respectively, and $v_{e\,i\,max}$ is the maximum speed of the electron emitted from the metal. The magnitude of the stopping potential is then

$$V_s = \frac{m_e v_{e\,i\,max}^2}{2e}$$

The electrons have *one* particular $v_{e\,i\,max}$ and one stopping potential difference for low and high EM radiation intensity. It seems that with high-intensity EM radiation, electrons absorb more energy and escape at a higher speed. Thus the wave model explains pattern 4 qualitatively but does not explain why there is no intensity dependence in the expression for the stopping potential.

Pattern 5 Increasing the intensity of EM radiation increases the photocurrent I_0 and the maximum value of the photocurrent but does not affect the value of $-V_s$.

FIGURE 27.8 (a) The negative potential stops the electron. (b) A work-energy bar chart represents the stopping process.

(a)

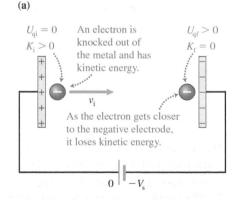

$U_{qi} = 0$ An electron is $U_{qf} > 0$
$K_i > 0$ knocked out of $K_f = 0$
the metal and has kinetic energy.

v_i

As the electron gets closer to the negative electrode, it loses kinetic energy.

$0 \quad -V_s$

(b)

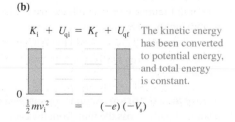

$K_i + U_{qi} = K_f + U_{qf}$ The kinetic energy has been converted to potential energy, and total energy is constant.

$\tfrac{1}{2} m v_i^2 = (-e)(-V_s)$

EM wave model explanation: The greater light intensity results in more electrons knocked out of the cathode and therefore a greater photocurrent. According to the EM wave model of light, a higher intensity electromagnetic wave should also transfer more energy to electrons, thus making it more difficult to stop them from reaching the anode. This is not what happens. The light intensity has no effect on the stopping potential. Thus the wave model does not explain pattern 5.

Pattern 6 The value of $-V_s$ is only affected by the frequency f of the EM radiation incident on the anode. The relationship is linear. When the frequency is below a certain cutoff value f_{cutoff}, the photocurrent does not appear at all.

EM wave model explanation: According to the wave model, the energy of an EM wave increases with the intensity of the wave, so it should be possible to use very intense but low-frequency light to knock electrons out of the metal. This does not happen. The EM wave model cannot explain pattern 6 at all and predicts something that contradicts the experiments.

In addition, for frequencies above the cutoff frequency, no delay occurs between the electromagnetic waves hitting the cathode and the release of the electrons, even for a very low intensity of EM radiation (for the frequencies above the cutoff frequency for a particular cathode). This is somewhat puzzling. Shouldn't it take a while for sufficient energy to be transferred to electrons in the cathode before they start being released, especially when the EM radiation intensity is low?

To summarize, the EM wave model of light can explain some aspects of the photoelectric effect, but it can't explain the following: (a) the lack of dependence of the stopping potential $-V_s$ on the intensity of the light, (b) the presence of the cutoff frequency f_{cutoff} below which there is no photocurrent (despite high intensity of light), and (c) the appearance of a photocurrent immediately after the EM radiation first strikes the cathode, even for low EM radiation intensity.

The photoelectric effect became the second phenomenon that made scientists question whether the EM wave model of light was sufficient to explain all EM wave behavior. In the next section you'll learn how to explain the photoelectric effect in its entirety. First, work through the following example to become comfortable with the notion of stopping potential and an electron's kinetic energy in the photoelectric effect.

EXAMPLE 27.4   Kinetic energy of the electrons and the stopping potential

In one of von Lenard's experiments, the value of the stopping potential was -0.20 V. What is the maximum kinetic energy of one of the electrons leaving the cathode?

Sketch and translate Figure 27.7 shows a schematic of von Lenard's experiment with electrons being released from the cathode by incident UV. The situation with negative potential across the electrodes is shown in Figure 27.8a. The left electrode (the cathode) is at zero potential and the right electrode (the anode) is at $-V_s$. The electric force exerted on the electrons is toward the left, causing them to slow down and stop. The electron has zero kinetic energy when it reaches the anode.

Simplify and diagram Assume for simplicity that the electrons travel in a straight path across the tube from the cathode to the anode. Assume also that the cathode from which an electron is ejected is at zero electric potential, and the anode that the electron flies toward is at -0.20 V. The work-energy bar chart in Figure 27.8b represents the process.

The electron and electric field produced by the electrodes are the system of interest. The initial state is when the electron has maximum kinetic energy just after being knocked out of the cathode and the system has zero electric potential energy; the final state is when the electron reaches the anode with zero kinetic energy. Thus, the final electric potential energy of the system is equal to the initial kinetic energy (Figure 27.8b).

Represent mathematically Using the generalized work-energy principle and the bar chart, we have

$$K_i = (-e)(-V_s)$$

Solve and evaluate The above equation can be used to determine the kinetic energy of the electron when it leaves the cathode; the energy of the electron is given in coulombs times volts (C · V):

$$K_i = (-e)(-V_s) = (-1.6 \times 10^{-19} \, \text{C})(-0.20 \, \text{V})$$
$$= 3.2 \times 10^{-20} \, \text{C} \cdot \text{V}$$

Note that $1\ V = 1\ J/C$. Therefore, the electron's initial kinetic energy is

$$K_i = 3.2 \times 10^{-20}\ C \cdot \left(\frac{J}{C}\right) = 3.2 \times 10^{-20}\ J$$

This is a very tiny amount of energy, but is reasonable for a single electron. Converting this value to electron volts, we have

$$K_i = (3.2 \times 10^{-20}\ J)\left(\frac{1\ eV}{1.6 \times 10^{-19}\ J}\right) = 0.20\ eV$$

Try it yourself What is the stopping potential for the electrons starting with the same kinetic energy if the distance between the cathode and anode is halved?

Answer The stopping potential does not depend on the distance between the electrodes; it depends only on the change in the kinetic energy of the individual electrons. The potential will be the same: $-0.20\ V$ (with respect to the zero potential of the cathode).

REVIEW QUESTION 27.2 What features of the photoelectric effect could not be explained by the electromagnetic wave model of light?

27.3 Quantum model explanation of the photoelectric effect

Philipp von Lenard thought that the absorption of UV radiation (and in some cases visible light) by the electron-metal system (the cathode) increased the energy of the system in accordance with the generalized work-energy principle. If the energy is large enough to raise the negative total energy of the system to zero $(-\phi + E_{light} = 0)$, then an electron could just barely be ejected from the metal (see the bar chart representing the process in **Figure 27.9a**). If the system gained more energy from the EM radiation than it needed to eject the electron, then the electron would have a nonzero kinetic energy after leaving the cathode, and it might reach the anode even if the electric field produced by the electrodes pushed it in the opposite direction (Figure 27.9b). Mathematically:

$$-\phi + E_{light} = K_{ef} \qquad (27.4)$$

where K_{ef} is the kinetic energy of the electron immediately after leaving the metal. However, von Lenard could not explain why the stopping potential did not depend on the intensity of the incident light. In these experiments, light could mean either visible light or UV—whichever is causing the photoelectric effect. According to his understanding, increased light intensity should lead to greater energy of the light and cause the electron to leave with more kinetic energy, thus requiring a greater stopping potential to prevent it from reaching the anode.

In 1905, Albert Einstein (1879–1955) made a breakthrough similar to Planck's hypothesis. At that time working as a clerk in a Swiss patent office, Einstein suggested that light can be viewed as a stream of quanta of energy not only when it is *emitted* by objects (as proposed by Planck), but also when it is *absorbed* by them. According to Einstein's hypothesis, an electron-lattice system (a metal) can only absorb light energy one whole quantum at a time. If the value of the light energy of a particular quantum is more than the magnitude of the metal's work function, an electron could leave the lattice. If the quantum of light energy is less than that, the electron would stay even if the number of incident quanta is very high (corresponding to high-intensity light.)

According to Planck, the energy of a quantum of EM radiation equals $E_0 = hf$. Using this idea, Einstein rewrote Eq. (27.4) as

$$-\phi + hf = K_{ef} \qquad (27.5)$$

where K_{ef} is the kinetic energy of a single electron just after leaving the metal. This equation explains why for most metals UV radiation causes the photoelectric effect, whereas visible light does not—each quantum of higher frequency UV radiation has

FIGURE 27.9 The electron-lattice system gains energy from light and UV. (a) The energy gained from light offsets the negative electron-lattice energy, barely allowing the electron to escape. (b) If the energy gained from the light (or UV) is more than the amount needed to escape, then the electron has kinetic energy when it escapes.

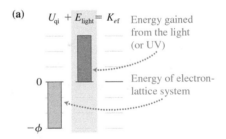

(a) $U_{qi} + E_{light} = K_{ef}$ Energy gained from the light (or UV)

Energy of electron-lattice system

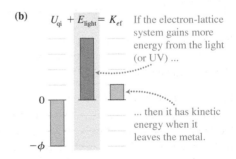

(b) $U_{qi} + E_{light} = K_{ef}$ If the electron-lattice system gains more energy from the light (or UV) ...

... then it has kinetic energy when it leaves the metal.

sufficient energy to remove an electron from the metal, whereas a lower frequency visible light quantum does not.

To check the feasibility of Einstein's hypothesis, let's calculate the energy of a quantum of UV radiation and compare it to typical work functions of metals. As mentioned in Section 27.2, the work function for most metals (silver, tin, aluminum, etc.) falls within the range $\phi = 4.30$ eV–4.60 eV. The energy of one UV radiation quantum is about

$$E_{0\,\text{UV}} = hf_{\text{UV}} = (6.63 \times 10^{-34}\,\text{J} \cdot \text{s})(1 \times 10^{15}\,\text{Hz}) \approx 7 \times 10^{-19}\,\text{J} \approx 4.4\,\text{eV}$$

This is encouraging! The frequency lies within the right range. Similarly, the energy of one green light quantum is about

$$E_{0\,\text{visible}} = hf_{\text{visible}} = (6.63 \times 10^{-34}\,\text{J} \cdot \text{s})(6 \times 10^{14}\,\text{Hz}) = 4 \times 10^{-19}\,\text{J} \approx 2.5\,\text{eV}$$

which is close to the magnitude of the work function for a few metals (cesium, lithium, sodium). This supports Einstein's hypothesis. We will test it more carefully in Testing Experiment **Table 27.4**.

TESTING EXPERIMENT TABLE 27.4   Testing Einstein's hypothesis using the photoelectric effect

Testing experiment	Prediction	Outcome
We repeat von Lenard's experiments, but this time we use a sodium cathode and anode with a work function of 2.3 eV and use red and blue visible light.	If Einstein's hypothesis is correct, then • 1.9 eV red quanta of light (650-nm wavelength) will not have enough energy to eject electrons, and thus no photocurrent will be detected in the circuit. • 2.7 eV blue photons (450-nm wavelength) will be able to eject electrons from the sodium and thus will produce a photocurrent in the circuit.	• When red light shines on the cathode, the ammeter registers no photocurrent. • When blue light shines on the cathode, the ammeter registers a photocurrent.

Conclusion

The outcome of the experiment matches the prediction. Thus we have not disproved Einstein's hypothesis. Instead, we have more confidence in it.

The outcome of the testing experiment supports Einstein's hypothesis. Let's now consider other features of the photoelectric effect that the wave model could not explain, as described on pages 857 and 858: (a) the lack of dependence of the stopping potential $-V_s$ on the intensity of the light, (b) the presence of the cutoff frequency f_{cutoff} of light, and (c) the immediate appearance of the photocurrent once light is incident on the cathode.

(a) *Stopping potential is independent of intensity:* The negative stopping potential $(-V_s)$ of the anode with respect to the zero cathode potential depends on the kinetic energy of the electrons leaving the metal:

$$(-e)\left[(-V_s) - 0\right] = \tfrac{1}{2} m_e v_{ei}^2$$

According to Eq. (27.5), the kinetic energy of an electron knocked out of the metal is

$$\tfrac{1}{2} m_e v_{ei}^2 = -\phi + hf$$

Thus, the kinetic energy of the electron depends only on the energy hf of a single quantum and not on the intensity of EM radiation incident on the cathode. Since the stopping potential depends on the initial kinetic energy of a freed electron, the stopping potential also depends only on the frequency of the radiation and not on the intensity of the radiation. Einstein's hypothesis explains this pattern.

(b) *No photocurrent below cutoff frequency:* The cutoff frequency occurs if the energy hf of a light quantum equals the work function ϕ, that is, if $hf = \phi$. We can then express the cutoff frequency in terms of the work function of the metal and Planck's constant:

$$f_{\text{cutoff}} = \frac{\phi}{h} \qquad (27.6)$$

Under these circumstances the electron leaves the metal with zero kinetic energy. If the frequency of light is such that $hf < \phi$, then the energy of one quantum is less than the work function of the metal, and no photocurrent is produced. Einstein's hypothesis explains this pattern, too.

(c) *Photoelectric current produced immediately:* Absorbing one quantum of light of sufficient energy could instantly eject an electron out of the metal. Because the photoelectric effect is the result of an interaction between a single quantum of EM radiation and a single electron, energy does not need to accumulate in the metal before we see an effect. Here again, Einstein's hypothesis accounts for the experimental evidence.

Einstein's hypothesis explains the features of the photoelectric effect that the EM wave model could not. Let's see if it accounts for features that the wave model did explain.

CONCEPTUAL EXERCISE 27.5   **Applying Einstein's hypothesis**

Use Einstein's hypothesis to explain why the photocurrent in von Lenard's experiments is directly proportional to the intensity of the EM radiation incident on the cathode.

Sketch and translate According to Einstein's hypothesis we can visualize EM radiation as a stream of quanta, each with a specific amount of energy.

Simplify and diagram
We represent the quanta as little "bubbles" in the figure on the right. The greater the intensity of the light, the greater the density (number per unit volume) of bubbles. Each bubble can potentially liberate one electron

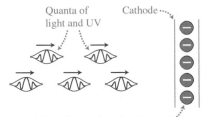

Quanta of light and UV   Cathode

Each free electron can absorb one quantum and leave the cathode.

from the metal; thus more bubbles lead to more liberated electrons and therefore a larger photocurrent. Einstein's hypothesis does indeed predict that the photocurrent is directly proportional to the intensity of EM radiation.

Try it yourself Explain why there is a photocurrent even when the potential difference between the electrodes in the tube is zero.

Answer The kinetic energy of a liberated electron is $K_{e'} = hf - \phi$. If the energy of one quantum is more than the magnitude of the work function, then the electron will have nonzero kinetic energy after being separated from the metal. There is a chance that some of these electrons will arrive at the anode and produce the photocurrent even in the absence of a potential difference between the electrodes.

Einstein's hypothesis was successful in explaining the features of the photoelectric effect that the EM wave model for light could not explain. Einstein's hypothesis also leads to another testable prediction. Equation (27.6) $f_{\text{cutoff}} = \phi/h$ implies that for every metal, the ratio of its work function to its cutoff frequency should equal Planck's constant:

$$h = \frac{\phi}{f_{\text{cutoff}}}$$

Robert Millikan (1868–1953) used this equation to determine the value of Planck's constant experimentally. Millikan's goal was to disprove Einstein's hypothesis and show that the absorption of light was not explained by a quantum hypothesis.

Millikan's reason for opposing Einstein's hypothesis was very similar to Planck's reason for not believing in his own quantum idea. The model of light as an electromagnetic wave could explain so many different phenomena that returning to a particle-like model seemed inconceivable. As Millikan wrote himself in a paper published in 1916: "Einstein's photoelectric equation . . . cannot in my judgment be looked upon at present as resting upon any sort of a satisfactory theoretical foundation," even though "it actually represents very accurately the behavior" of photoelectricity.[1]

Later, when accepting the Nobel Prize in 1923 for this work and for his measurement of the charge of the electron, Millikan said: "this work resulted, contrary to my own expectation, in the first direct experimental proof . . . of the Einstein equation, and the first direct photoelectric determination of Planck's h."

The quanta of energy first invented by Planck to explain the emission of light and later applied by Einstein to explain the absorption of light later became known as **photons**.

> **Photon** A photon is a discrete portion of electromagnetic radiation that has energy
>
> $$E_{\text{photon}} = hf \qquad (27.7)$$
>
> where f is the frequency of the electromagnetic radiation and $h = 6.63 \times 10^{-34} \, \text{J} \cdot \text{s}$ is Planck's constant.

Using the language of photons, we can now summarize Einstein's photoelectric effect equation.

> **Einstein's equation for the photoelectric effect** During the photoelectric effect, the kinetic energy K_e of the emitted electron equals the difference in the energy of the photon $E_{\text{photon}} = hf$ absorbed by the metal and the metal's work function ϕ:
>
> $$-\phi + hf = K_e \qquad (27.5)$$

It is worth noting that Einstein received the Nobel Prize for the explanation of the photoelectric effect in 1921, two years before Millikan did. In the next example, we use Einstein's explanation of the photoelectric effect to understand some additional experimental results.

---

[1]Millikan, R. A. (1916). A direct photoelectric determination of Planck's "h." *Phys. Rev.* 7, 355. http://focus.aps.org/story/v3/st23

EXAMPLE 27.6   **Ejecting electrons from different metals**

Light and UV shine on three different metals, each metal being a cathode in a photoelectric tube. The anode in each experiment is made of the same material as the cathode. Graph lines representing the stopping potential $-V_s$ versus the incident light or UV frequency are shown in the figure at right. The work functions of the metals are sodium (Na), 2.3 eV; iron (Fe), 4.7 eV; and platinum (Pt), 6.4 eV. Use Einstein's hypothesis of light absorption to explain the results.

Sketch and translate The graph lines all have the same slope but have different intercepts on the horizontal frequency axis. Does Einstein's explanation of the photoelectric effect produce a mathematical equation that relates the observed stopping potential $-V_s$ and the frequency f of the incident light and UV? The experimental setup is shown below.

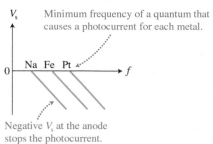

Minimum frequency of a quantum that causes a photocurrent for each metal.

Negative V_s at the anode stops the photocurrent.

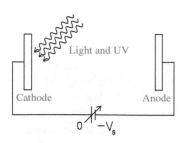

Note that the cathode electric potential is at zero and the anode potential V is negative (necessary to stop the photoelectrons). The system of interest is the electron and the metal lattice—the cathode. The energy of each absorbed photon of incident light may cause an electron to be ejected from the metal. If the energy of the photon is greater than the work function of the metal, then the ejected electron has some kinetic energy after leaving the metal.

Simplify and diagram We represent the process with a bar chart, as shown in the first figure below. After ejection, the electron travels to the anode. We adjust the potential difference between the electrodes so that the electron stops just as it reaches the anode. The anode is at potential $-V_s$, less than the cathode (which is at zero). The travel of the ejected electron to the anode is represented by the second bar chart below.

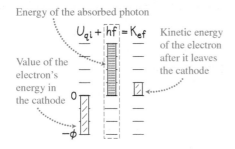

Energy of the absorbed photon

$U_{qi} + hf = K_{ef}$ Kinetic energy of the electron after it leaves the cathode

Value of the electron's energy in the cathode

$-\phi$

Kinetic energy leaving the cathode $K_{ei} = U_{qf}$ Electrical potential energy reaching the anode

Represent mathematically Use the first bar chart above to represent the electron ejection process mathematically as

$$-\phi + hf = K_{ef}$$

The kinetic energy K_{ef} of the electrons after ejection equals their kinetic energy K_{ei} at the beginning of the next process (the movement of the electron from the cathode to the anode). Use the bar chart to represent that process mathematically:

$$K_{ei} = U_{qf} = qV_f = (-e)(-V_s)$$

where $-V_s$ is the negative anode potential. We combine the two equations to get

$$-\phi + hf = (-e)(-V_s)$$

Here $-V_s$ is the specific value of the stopping potential.

Solve and evaluate If we want to put the previous equation in a slope-intercept form so it can be compared with the graph, we need to replace the specific value of the stopping potential $(-V_s)$ with the variable V_s that we are plotting on the y-axis:

$$V_s = \left(-\frac{h}{e}\right)f + \left(\frac{\phi}{e}\right) \tag{27.8}$$

According to this equation, regardless of the metal used to make the cathode and anode, the slope will be $-h/e$. This explains why all the graph lines have the same slope, as seen in the first figure. Millikan measured these slopes to determine Planck's constant h. Because the photon energy hf is greater in magnitude than the work function ϕ, the stopping potential, $-V_s$, must be negative.

- -

Try it yourself If blue light of frequency 6.7×10^{14} Hz is incident on a sodium target, what is the value of the stopping potential?

Answer -0.5 V.

In the above example, the x-intercept of each graph line is different. According to Eq. (27.8), the intercept when $V_s = 0$ depends on the work function of each metal. Put $V_s = 0$ into Eq. (27.8) to get

$$0 = \left(-\frac{h}{e}\right)f + \left(\frac{\phi}{e}\right)$$

Solving for f, we get

$$f_{cutoff} = \frac{\phi}{h}$$

where f_{cutoff} is the cutoff frequency for a particular metal, the lowest frequency photon needed to eject an electron with no extra kinetic energy. The electron in that case is stopped with zero stopping potential $V_s = 0$. Measuring this cutoff frequency allows us to determine the work function of the metal. Or, if we know the work function, we can determine the cutoff frequency. We do this in Table 27.5 (on the next page) for the three metals shown in Example 27.6.

TABLE 27.5 **Cutoff frequencies for selected metals**

Metal (ϕ)	Cutoff frequency	Part of EM spectrum
Sodium (2.3 eV)	$f_{Na} = \dfrac{\phi_{Na}}{h} = \dfrac{(2.3\text{ eV})\left(\dfrac{1.6 \times 10^{-19}\text{ J}}{1\text{ eV}}\right)}{6.63 \times 10^{-34}\text{ J} \cdot \text{s}} = 5.6 \times 10^{14}\text{ Hz}$	Visible
Iron (4.7 eV)	$f_{Fe} = \dfrac{\phi_{Fe}}{h} = \dfrac{(4.7\text{ eV})\left(\dfrac{1.6 \times 10^{-19}\text{ J}}{1\text{ eV}}\right)}{6.63 \times 10^{-34}\text{ J} \cdot \text{s}} = 1.1 \times 10^{15}\text{ Hz}$	Ultraviolet
Platinum (6.4 eV)	$f_{Pt} = \dfrac{\phi_{Pt}}{h} = \dfrac{(6.4\text{ eV})\left(\dfrac{1.6 \times 10^{-19}\text{ J}}{1\text{ eV}}\right)}{6.63 \times 10^{-34}\text{ J} \cdot \text{s}} = 1.5 \times 10^{15}\text{ Hz}$	Ultraviolet

The results in Table 27.5 make sense because the x-intercept has a greater cutoff frequency for a metal with a larger work function.

REVIEW QUESTION 27.3 How does Einstein's hypothesis explain the cutoff frequency observed for a particular metal cathode in a photoelectric experiment?

27.4 Photons

In the first three decades of the 20th century, scientists faced a serious dilemma. The EM wave model of light explained a wide variety of phenomena, but not black body radiation and not the photoelectric effect. Explaining those required the construction of a new quantum model of light in which light was emitted and absorbed in discrete portions. Physicists started to think of light as composed of particle-like photons (the quanta of light). However, to explain interference and diffraction phenomena, photons had to have wave-like properties as well. The dual particle-wave properties of photons are summarized in Table 27.6.

TABLE 27.6 **Wave-like and particle-like properties of photons**

Experimental evidence	Wave-like properties f, λ	Particle-like properties $E = hf$
Double-slit interference	Superposition (including destructive interference)	
Diffraction	Single-slit interference and bending around obstacles	
Doppler effect	Change in frequency and wavelength	
Black body radiation		Photon (quanta) emission
Photoelectric effect		Photon (quanta) absorption

If we accept this model, photons must have both wave-like and particle-like behaviors. Experiments performed in 1932 tested the dual nature of photons in a dramatic way.

Low-intensity experiments support the hypothesis of the dual nature of photons

Experiments by Soviet physicists Sergei Vavilov and Eugeny Brumberg supported the hypothesis that photons have both particle and wave properties. Vavilov and Brumberg used a double-slit setup with light of extremely low intensity. Their goal was to reduce the number of photons that pass through the slits each second to make them easier to detect as individual photons. For detection, they used the extreme sensitivity of human eyes to low intensities of light. A person sitting in a dark room for extended periods of time develops increasing visual sensitivity to the extent that he or she can eventually see the individual photons hitting a screen that has been covered with a special material.

In Testing Experiment **Table 27.7** we describe the experimental setup, including the predictions for the outcome made by the particle model of light, by the wave model, and by the dual wave-particle model.

TESTING EXPERIMENT TABLE 27.7 Do photons have both wave-like and particle-like properties?

Testing experiment	Prediction	Outcome
Use a light source with variable intensity. Shine the light on two narrow slits. Place a fluorescent screen beyond the slits that indicates where light hits the screen. **Top view** 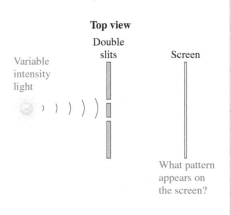	**Particle model:** Only two bright bands should appear—images of the slits themselves. As intensity decreases, we expect to see individual flashes caused by single photons at the slit image locations. High intensity / Low intensity 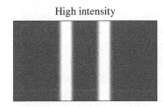 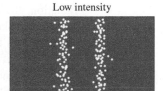 **Wave model:** Many alternating bright and dark bands should appear. As the intensity decreases, we expect all of the bright bands to become uniformly dimmer until they disappear. High intensity / Low intensity 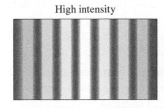 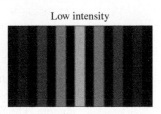 **Dual wave-particle model:** Many alternating bright and dark bands should appear. As the intensity decreases, we expect to see individual flashes due to single photons at bright band locations only. At the locations of the dark bands no flashes should be seen. High intensity / Low intensity 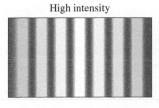 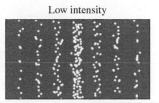	The experimental outcome is identical to the dual wave-particle model's prediction.

Conclusion

- Photons hit the screen, indicated by the flashes at low intensity.
- The photons only reach the screen at places where constructive wave interference occurs.
- Photons are simultaneously exhibiting both wave-like and particle-like behaviors.

Vavilov and Brumberg's experimental results were astounding—individual photons somehow "know" that they should only reach screen locations where coherent waves from the two slits would interfere constructively. It is as if each individual photon passes through both slits, interfering with itself and thereby producing a flash only at a location on the screen where constructive interference occurs.

Earlier in this chapter we constructed the idea of a photon as a single quantum of EM radiation. If it is a single quantum, then it seems impossible that it could pass through both slits simultaneously. However, Vavilov and Brumberg's experiment supports this seemingly strange idea.

Photon momentum

We have seen that photons participate in collisions with electrons inside metals (the photoelectric effect). This suggests that photons must have momentum. Otherwise, how could they knock an electron out of the cathode of a phototube during a collision? Because photons travel at light speed, they must be treated relativistically. The relativistic expression for the magnitude of the momentum of a particle of mass m moving at speed v is

$$p = \frac{mv}{\sqrt{1 - (v/c)^2}}$$

This expression has difficulties when applied to a photon. For a photon $v = c$, and the denominator becomes zero. In addition, if photons are quanta of electromagnetic radiation, then they would be composed of nothing but electric and magnetic fields. Their mass should be zero, and the numerator is also zero. So we get $p = 0/0$. That clearly doesn't work. Is there anything we can do to get a reasonable expression for the momentum of a photon?

We need an expression that does not have that square root factor in the denominator. Einstein resolved this by looking at the relativistic equations for the total energy and momentum of an object with nonzero mass:

$$E = \frac{mc^2}{\sqrt{1 - (v/c)^2}}$$

$$p = \frac{mv}{\sqrt{1 - (v/c)^2}}$$

Einstein eliminated v from these equations by solving for $(v/c)^2$ in the second equation and then substituting the result into the first equation. This takes a fair amount of algebra. The final result is

$$E = \sqrt{(pc)^2 + (mc^2)^2} \qquad (27.9)$$

What we have here is a new equation for the total energy of an object. It is equivalent to

$$E = \frac{mc^2}{\sqrt{1 - (v/c)^2}}$$

but has an important new feature: the mass of the object can be set to zero without the equation misbehaving:

$$E = \sqrt{(pc)^2 + (0 \cdot c^2)^2} = \sqrt{(pc)^2} = pc$$

In other words, according to Eq. (27.9), the total energy of an object with zero mass (like a photon) is $E = pc$. But we already know from Einstein's explanation of the photoelectric effect that the energy of a photon is $E = hf = (hc/\lambda)$. Therefore,

$$\frac{hc}{\lambda} = pc$$

TIP Notice what happens when Eq. (27.9) is used to determine the total energy of an object at rest. Then $p = 0$ and $E = mc^2$. The total energy of an object at rest just equals its rest energy, $E_0 = mc^2$.

Solving for p, we get an expression for the momentum of a photon:

$$p = \frac{h}{\lambda}$$

(27.10)

Following is an application of this expression.

QUANTITATIVE EXERCISE 27.7 ⟩ Revisiting $E_0 = mc^2$ with the Sun's changing mass

Each second the Sun emits 3.8×10^{26} J of energy, mostly in the form of visible, infrared, and UV photons. Estimate the change in mass of the Sun as a result of the energy these photons carry away each second.

Represent mathematically Use the idea of energy conservation for the system that includes the Sun and the emitted photons:

$$E_{0i} = E_{0f} + E_{photons}$$

The rest energy of an object is $E_0 = mc^2$. Therefore,

$$m_i c^2 = m_f c^2 + E_{photons}$$

Solve and evaluate Solving for $\Delta m = m_f - m_i$:

$$m_i c^2 - m_f c^2 = E_{photons}$$

$$\Rightarrow m_f c^2 - m_i c^2 = -E_{photons}$$

$$\Rightarrow (m_f - m_i)c^2 = -E_{photons}$$

$$\Rightarrow \Delta m = -\frac{E_{photons}}{c^2}$$

Inserting the appropriate values gives

$$\Delta m = -\frac{3.8 \times 10^{26} \text{ J}}{(3.00 \times 10^8 \text{ m/s})^2} = -4.2 \times 10^9 \text{ kg}$$

The minus sign indicates that the Sun loses more than 4 billion kg each second! Won't the Sun "disappear" quickly while losing billions of kilograms each second?

The mass of the Sun is $m_{Sun} = 2.0 \times 10^{30}$ kg. How long will it take for the Sun to "disappear"?

$$\frac{2.0 \times 10^{30} \text{ kg}}{4.2 \times 10^9 \text{ kg/s}} = 4.8 \times 10^{20} \text{ s}$$

There are about 3.2×10^7 s in 1 year, which means the Sun will last

$$\frac{4.8 \times 10^{20} \text{ s}}{3.2 \times 10^7 \text{ s/year}} = 1.5 \times 10^{13} \text{ years}$$

The age of the universe (as we know from Chapter 26) is approximately 13.8×10^9 years, which means the Sun will last more than 1000 times the current age of the universe if it continues to radiate energy at this rate (assuming that the mechanism behind it can continue for that long).

Try it yourself Estimate the order of magnitude of the number of photons the Sun emits every second.

Answer   Over 10^{45} photons/s are emitted from the Sun. Assumption: All photons have a frequency of 10^{14} Hz. If we make a different assumption, the estimated number will be different.

Our understanding of the momentum of photons can explain an important aspect of the photoelectric effect that we have so far overlooked. This aspect is described by the **quantum efficiency** of the photoelectric effect. Quantum efficiency is the number of electrons ejected per single incident photon. The efficiency depends on many factors. The most important is the work function of the metal—if the energy of the photon is less than the magnitude of the work function, the efficiency is zero. However, even if the energy of the photon is larger than the work function, it might still fail to eject an electron if the photon is reflected by the metal or absorbed by the lattice. Finally, due to the initial momentum of the incident photons being directed into the lattice, electrons that absorb the photons mostly move within the lattice, and only a few leave the metal.

REVIEW QUESTION 27.4 Explain how the outcome of the Vavilov-Brumberg experiment supports the idea that a photon has both wave-like and particle-like behaviors.

27.5 X-rays

The experiments and theory described in the previous section supported the idea that photons have momenta of magnitude $p = h/\lambda$. We will test this expression using X-rays in the next section. First, we need to learn about X-rays.

Cathode ray tubes

The story of X-rays is another example of how persistence and attention to detail often lead to groundbreaking discoveries. At the beginning of Section 27.2 we assumed that free electrons reside inside metals, and we used this idea to help explain the photo-electric effect. In this section we'll find how the electron was discovered and how its discovery led to the accidental discovery of X-rays.

In the middle of the 19th century, physicists did not know that metals consisted of a crystal lattice and free electrons; furthermore, they did not know that electric current in metals was caused by the movement of those free electrons.

At that time, some physicists started experimenting with what are called cathode ray (Crookes) tubes—evacuated glass tubes with two electrodes embedded inside (**Figure 27.10**). The electrodes were connected to a battery that produced a potential difference across the electrodes. The tubes were similar in some ways to the tube that von Lenard used to study the photoelectric effect. The difference was that the cathode (connected to the negative terminal of the battery) could be heated to high temperatures, rather than being illuminated by light.

Physicists discovered that although a physical gap was present between the cathode and the anode, when the cathode was heated, a current appeared in the circuit and the tube would glow. Because the tube's interior was a vacuum, they thought that the cathode must emit some kind of rays that traveled from the cathode to the anode. These rays were called *cathode rays*.

J. J. Thomson (1856–1940), at the Cavendish Laboratory at Cambridge University, was one of the scientists studying cathode rays. He observed that when the rays hit a metal target, the target became negatively charged. Through numerous experiments, Thomson found that the rays could be deflected by electric and magnetic fields as though they consisted of negatively charged particles. He determined their charge-to-mass ratio (q/m) and found it to be over 1000 times greater than that of a hydrogen ion (the modern value being $q/m = -1.76 \times 10^{11}$ C/kg), suggesting that the cathode rays either were very highly charged or had a very small mass. Later experiments by other scientists supported the latter hypothesis. The cathode rays' charge-to-mass ratio was also independent of the choice of cathode material. This led Thomson to conclude that there was only one type of cathode ray rather than many different types.

Let's model a cathode ray as a stream of charged particles and see if we can explain the behavior of the cathode ray tubes. Remember, the cathode in the tube is hot. This means that the particles inside the cathode have a large amount of kinetic energy of random motion (thermal energy) and could knock each other out of the cathode. Once outside the cathode, the electric field between the cathode and anode would cause them to accelerate toward the anode.

At about the same time, chemists were trying to find out what comprised electric currents. What was moving inside metals? They concluded that the charge carriers were tiny particles, and in 1894 George Johnstone Stoney called them "electrons." Experiments in electrochemistry helped determine the value of the smallest charge carried by these particles to be -1.6×10^{-19} C.

We can use the value of the charge of the electron found by Stoney and the charge-to-mass ratio found by Thomson to determine the mass of the electron:

$$m_e = \frac{1}{q_e/m_e}q_e = \frac{1}{-1.76 \times 10^{11}\,\text{C/kg}}(-1.6 \times 10^{-19}\,\text{C}) = 9.1 \times 10^{-31}\,\text{kg}$$

The mass of the electron is about 1/2000th the mass of a hydrogen atom.

The accidental discovery of X-rays

Experiments with cathode ray tubes led to the accidental discovery of X-rays. In 1895 Wilhelm Roentgen (1845–1923), like von Lenard, was experimenting with the rays outside the vacuum tubes by letting them pass through special windows

FIGURE 27.10 A cathode ray tube. (a) An actual cathode ray tube. (b) The electrical schematic of a cathode ray tube.

(a)

(b)

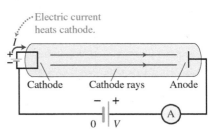

Electric current heats cathode.

Cathode   Cathode rays   Anode

$0 \quad V$

and observing them when they hit a fluorescent screen. The screen glowed when the cathode rays hit it. One night when Roentgen was ready to leave the lab, he covered the cathode ray tube with cardboard and turned off the lights. He noticed a faint glow on the fluorescent screen that was lying nearby on the lab table (Figure 27.11a). This was strange; the room was dark, and the cathode ray tube was covered with cardboard. Roentgen turned the lights back on and checked all of the equipment. He turned the tube off and turned off the lights again. The fluorescent screen was dark.

At this point most people would have breathed a sigh of relief and gone home. But Roentgen went back to the tube with the lights still off and turned it on. The glow on the screen appeared again even though the tube was completely covered with cardboard! When Roentgen placed his hand near the screen, he could see a strange shadow made by his hand. The shadow looked as though the flesh of his hand was completely transparent, and only his bones blocked whatever was making the screen glow.

Roentgen stayed in the lab all night trying to figure out what rays could possibly come out of the completely covered tube and make the fluorescent screen glow. He called these mysterious rays X-rays since at the time he couldn't explain what they were. Roentgen took the first X-ray image of a human on December 22, 1895; he photographed his wife's hand (Figure 27.11b). Although we now understand what X-rays are, they are still called X-rays.

Explanation of X-rays

Roentgen and others continued their experiments and found the following about X-rays:

1. X-rays are not deflected by either electric or magnetic fields.
2. X-rays do not cause the viewing screen to become electrically charged.
3. X-rays cause photographic paper to darken.
4. X-rays produce a diffraction pattern of dark and bright bands on the screen after passing through a single narrow slit.
5. X-rays can be polarized.
6. X-rays can go through many materials that are opaque to visible light.
7. X-rays can ionize gases.

Analysis of the single-slit diffraction patterns of the rays showed that the wavelengths of X-rays were about 10^{-10} to 10^{-11} m, a much smaller wavelength than any other EM radiation known at the time—much smaller than visible light. Given all of these characteristics, it is reasonable to hypothesize that X-rays are streams of very high energy photons. But how are these photons produced?

Producing X-ray photons

Consider Figure 27.12. The hot metal cathode is a source of negatively charged electrons. Due to their high random motion (the metal is very hot), the kinetic energy of some of the free electrons in the metal exceeds the metal's work function. These electrons escape the cathode and accelerate toward the anode because of the electric field within the evacuated cathode ray tube. (The electric field is produced by the negatively charged cathode and positively charged anode.) When the electrons reach the anode, they collide with it and stop abruptly. Recall that when a charged object accelerates (such as an electron stopping when it collides with the anode), it emits electromagnetic radiation (Chapter 25). The radiation emitted in this case might be in the form of X-ray photons. Let's estimate the energy of the photons produced by this mechanism and compare it to the energy of photons of visible light. We want to determine whether the energy of the electrons stopping in the cathode ray tube is enough to produce an X-ray photon.

FIGURE 27.11 Roentgen discovers X-rays. (a) Although the cathode ray tube was covered, the fluorescent screen still glowed in the dark. (b) Roentgen took an X-ray photo of his wife's hand.

(a) High voltage on

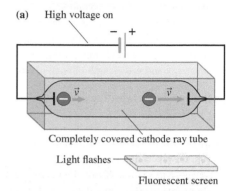

Completely covered cathode ray tube

(b)

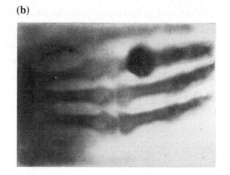

FIGURE 27.12 Diagram of an X-ray tube. The collision of high-energy electrons with the anode produces X-ray photons.

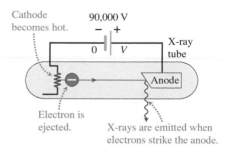

EXAMPLE 27.8   **X-ray photons**

In a cathode ray tube used to produce X-rays for imaging in a hospital, the potential difference between the cathode and anode is 90.0 kV (90,000 V). If the X-rays are generated by electrons in the tube that stop abruptly when they collide with the anode and emit photons, what are the energy and wavelength of each photon?

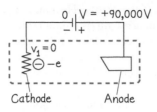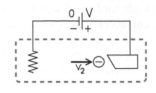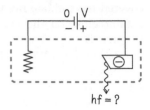

Sketch and translate The process is sketched in three steps (states) at top right.

Simplify and diagram We consider the system to consist of a single electron, the electric field produced by the electrodes, the electrodes themselves, and a single X-ray photon. We represent each part of the process using a work-energy bar chart. The initial state shown below is just as the electron leaves the cathode. We assume that the electron has zero kinetic energy and the system has zero electric potential energy—that is, the electric potential at the cathode is zero.

The intermediate state is just before the fast-moving electron smashes into the anode. The final state is when the electron has stopped inside the anode and just after the X-ray photon has been emitted. We assume that none of the kinetic energy of the electron is converted into the internal energy of the cathode.

$$K_1 + U_{q1} + hf \ = \ K_2 + U_{q2} + hf \ = \ K_3 + U_{q3} + hf$$

Represent mathematically Use the first and second state bar charts to apply the generalized work-energy principle:

$$0 + 0 + 0 = \tfrac{1}{2}mv_2^2 + (-e)V$$

Use the second and third state bar charts to write

$$\tfrac{1}{2}mv_2^2 + (-e)V = (-e)V + hf$$

where $h = 6.63 \times 10^{-34}$ J·s is Planck's constant. Combining these two results, we get

$$0 = (-e)V + hf$$

Solve and evaluate The photon's energy is

$$hf = eV = (1.6 \times 10^{-19}\,\text{C})(90.0 \times 10^3\,\text{V})$$
$$= 90.0 \times 10^3\,\text{eV}$$

This is much greater than the energy characteristic of visible light photons. The photon's frequency is

$$f = \frac{eV}{h} = \frac{(1.6 \times 10^{-19}\,\text{C})(90.0 \times 10^3\,\text{V})}{6.63 \times 10^{-34}\,\text{J·s}}$$
$$= 2.2 \times 10^{19}\,\text{Hz}$$

which is much higher than visible light frequencies. The photon's wavelength is

$$\lambda = \frac{c}{f} = \frac{3.00 \times 10^8\,\text{m/s}}{2.2 \times 10^{19}\,\text{Hz}} = 1.4 \times 10^{-11}\,\text{m}$$

- - - - - - - - - - - -

Try it yourself What potential difference between the electrodes in a cathode ray tube would produce 10^{-10}-m wavelength X-rays?

Answer   12,000 V.

Based on the result of our calculation we can say that in a cathode ray tube, an electron can acquire enough kinetic energy to produce an X-ray photon when it stops.

Are X-rays dangerous?

Short-wavelength UV radiation, X-rays, and shorter wavelength (higher frequency) gamma ray photons are called ionizing radiation because each photon has enough energy to knock electrons out of atoms or to break bonds that hold atoms together in molecules. Ionizing radiation can damage genetic material (DNA), increasing the risk of mutations that lead to cancer. Fortunately, the body has potent DNA repair mechanisms, reducing the chance of serious harm from UV and X-ray exposure. However, people who work around ionizing radiation (dental hygienists and radiology technicians, for example) must take precautions to avoid overexposure.

QUANTITATIVE EXERCISE 27.9   Number of photons absorbed in a medical X-ray

Determine the number of photons absorbed during a single chest X-ray if the body absorbs a total of 10^{-3} J of 0.025-nm wavelength X-ray photons.

Represent mathematically The energy of a single X-ray photon is

$$E_1 = hf = \frac{hc}{\lambda}$$

The number of photons is then

$$N = \frac{E_{total}}{E_1}$$

Solve and evaluate Each photon has energy

$$E_1 = \frac{(6.63 \times 10^{-34}\,\text{J} \cdot \text{s})(3.00 \times 10^8\,\text{m/s})}{0.025 \times 10^{-9}\,\text{m}} = 8 \times 10^{-15}\,\text{J}$$

Thus, the number of photons absorbed in one chest X-ray is approximately

$$N = \frac{10^{-3}\,\text{J}}{8 \times 10^{-15}\,\text{J}} \approx 10^{11}\,\text{photons}$$

Try it yourself Suppose the wavelength of the X-rays was three times as long. Then approximately how many photons would pass through your body in a 10^{-3}-J X-ray exam?

Answer   About 3×10^{11} photons.

QUANTITATIVE EXERCISE 27.10   High-energy photons absorbed from environment

Your body is continuously exposed to natural radiation from sources such as radioactive materials in the environment and in the food you eat, and from cosmic rays coming from supernovae and other objects in the universe. Each year on average you absorb about 10^{-3} J of this radiation per kilogram of body mass. Estimate the number of these ionizing photons absorbed each second.

Represent mathematically Assume that your body mass $m = 70.0$ kg. The rate at which energy from this radiation is absorbed is $R = 10^{-3}$ (J/kg)/year. Assume for example that the radiation consists of $\lambda = 6 \times 10^{-13}$ m gamma ray photons each with energy $E_1 = 3 \times 10^{-13}$ J. The number N of photons absorbed in 1 year is then

$$N = \frac{Rm}{E_1}$$

We can then determine the number absorbed each second by dividing by the number of seconds in 1 year:

$$1\,\text{year} \left(\frac{365\,\text{days}}{1\,\text{year}}\right)\left(\frac{24\,\text{h}}{1\,\text{day}}\right)\left(\frac{3600\,\text{s}}{1\,\text{h}}\right) = 3.2 \times 10^7\,\text{s}$$

Solve and evaluate Inserting the appropriate values gives

$$N = \frac{(10^{-3}\,(\text{J/kg})/\text{year})(70.0\,\text{kg})}{3 \times 10^{-13}\,\text{J}}$$

$$= 2.3 \times 10^{11}\,\text{photons/year}$$

The number of photons absorbed each second is then

$$\frac{2.3 \times 10^{11}\,\text{photons/year}}{3.2 \times 10^7\,\text{s/year}} \approx 7 \times 10^3\,\text{photons/s}$$

As mentioned earlier, the body has ample repair mechanisms to take care of such radiation exposure.

Try it yourself Exposure to a 1.0-s burst of 0.010-nm gamma rays depositing 1.0 J of energy per kilogram into a human body has a 10% chance of being fatal within 30 days. This level of radiation is 10 billion times greater than the ambient natural radiation our bodies are continuously exposed to. How many gamma ray photons are absorbed by the body during this exposure? (Some workers present at the 1986 Chernobyl nuclear power plant disaster experienced as much as 16 times this exposure.)

Answer $3-4 \times 10^{15}$ photons.

X-ray interference

We have mostly focused on the particle-like behavior of X-ray photons. However, X-rays can exhibit wave-like behavior if they can be made to interact with objects whose size is comparable to the wavelength of the X-rays. The spacing between atoms and ions in a crystal is the same order of magnitude as the wavelength of X-ray photons. When X-rays shine on a crystal, the X-rays reflected off the crystal lattice form a pattern very similar to that of visible light passing through an interference grating.

FIGURE 27.13 A photocell.

Photocell

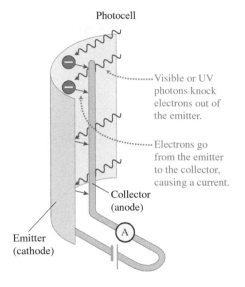

Visible or UV photons knock electrons out of the emitter.

Electrons go from the emitter to the collector, causing a current.

Collector (anode)

Emitter (cathode)

FIGURE 27.13 A photocell.

FIGURE 27.14 A photoelectric smoke detector.

No smoke

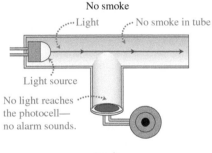

Light ······ No smoke in tube

Light source

No light reaches the photocell— no alarm sounds.

Smoke

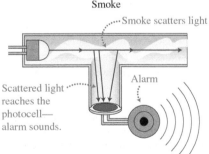

······ Smoke scatters light

Scattered light reaches the photocell— alarm sounds.

Alarm

FIGURE 27.15 Light shining light on silicon produces free electrons and "holes."

Effect of light on silicon

Photon knocks valence electron out of a silicon atom, leaving a hole ...

No light

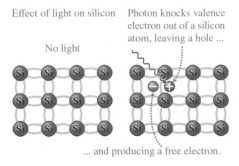

... and producing a free electron.

The details of the interference pattern allow scientists to determine the internal structure of the crystal. X-ray scattering can also be used to determine the structure of proteins and DNA. X-ray diffraction images produced by British biophysicist Rosalind Franklin played a key role in determining the helical structure of DNA in the early 1950s.

REVIEW QUESTION 27.5 What physical mechanism produces X-rays in a cathode ray tube?

27.6 Photocells, solar cells, and LEDs

The photoelectric effect has numerous applications. One of them is the **photocell**, which functions similarly to the tube that von Lenard used for his experiments. The cathode of the photocell (also called the *emitter*) is connected to the negative terminal of a battery and therefore is at a lower electric potential than the anode (called the *collector*). The cathode emitter is shaped like a cylindrical concave mirror to focus emitted electrons toward the anode, which has the shape of a thin wire that collects the electrons (**Figure 27.13**). When light of a frequency greater than the cutoff frequency shines on the emitter, electrons are ejected and absorbed by the collector, causing an electric current in the circuit. The magnitude of the current is a measure of the visible light or UV intensity being directed at the photocell. Products that use photocells include light meters (used in photography to adjust the size of a camera lens aperture), motion detectors, and photoelectric smoke detectors (**Figure 27.14**). In such a smoke detector, light from the source shoots across the top of the T-shaped detector. A cathode sits at the bottom of the T. If the air is clear, no light is scattered to the cathode and hence no photocurrent will be produced. However, smoke particles scatter light out of the beam, causing it to reach the cathode of the photocell and producing a photocurrent. When this photocurrent reaches a certain level, the alarm is triggered.

Solar cells

Solar cells, such as the rooftop panels used to generate electric current, are another application of the photoelectric effect. Solar cells are based on the semiconductor technology we discussed in an earlier chapter (Chapter 19). Semiconductors are materials that act as insulators under some conditions and conductors under others.

Silicon is a commonly used semiconductor. A silicon atom has four valence electrons that form covalent chemical bonds with neighboring atoms (each electron is shared between them). As a result, a silicon atom has no free electrons; therefore, silicon is an electric insulator. However, when silicon is heated or when light of sufficient frequency shines on it, some of the electrons gain enough energy to become free to roam around the crystal. The spaces vacated by these electrons lack negative charge and become positively charged "holes." These holes can be filled by other bound electrons that are not energetic enough to become free but can nevertheless move among the still-bound valence electrons. Both free electrons and holes can potentially contribute to a current if the silicon is placed in an external electric field. **Figure 27.15** shows this process— light photons deliver energy to the silicon, increasing its energy and allowing some of its valence electrons to become free. Whether electrons will be freed depends on the frequency of the light.

Adding impurities to silicon—a process called doping—is another way to turn it into a conductor. There are two types of impurities: electron donors and electron acceptors. Phosphorus, an example of an electron donor, contains five valence electrons, one more than silicon's four. The result is one "extra" electron that does not form a strong bond with a neighboring silicon atom and is essentially free to move about the bulk of

FIGURE 27.16 Doping of silicon. (a) Phosphorus is an n-type donor of free electrons and (b) boron is a p-type acceptor of bound electrons—a creator of "holes." (c) The contact of p- and n-type semiconductors causes charge transfer across the junction. (d) Charge transfer and recombination across the junction produces an internal electric field $\vec{E}_{int}$.

(a) n-type silicon when doped with phosphorus

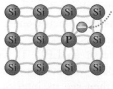

Phosphorus has five valence electrons, four bonding to silicon atoms, leaving one free electron.

(b) p-type silicon when doped with boron

Boron has three valence electrons, three bonding to silicon atoms, leaving a hole at the unfilled bond.

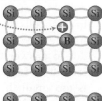

After the free electron drifts away, the phosphorus atom becomes a positive ion.

After the hole drifts away, the boron atom becomes a negative ion.

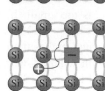

We use square symbols to distinguish ions (that come from doped atoms) from Si atoms.

(c) p-n junction right after p and n are brought together

Boron ions   Phosphorus ions

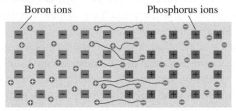

Free electrons and holes move randomly. When they meet, they recombine. Those near the junction recombine first.

(d) p-n junction after some time, when equilibrium is reached

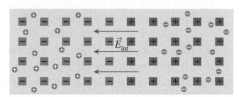

In the region with no electrons and holes, the remaining ions create the internal electric field that eventually prevents electrons and holes from entering the region.

the crystal (**Figure 27.16a**). The opposite is true for an electron acceptor. Boron, for example, has only three valence electrons. It can capture nearby electrons to form a strong fourth bond with the adjacent silicon atoms, creating extra holes (Figure 27.16b). These holes act as free positively charged particles (they lack the presence of a negatively charged electron). Silicon doped with phosphorus or any other electron donor is called an **n-type semiconductor**; the boron-doped version is called a **p-type semiconductor**. Note that these crystals are still electrically neutral, since they are still completely composed of neutral atoms.

When a p-type semiconductor and an n-type semiconductor are brought into contact, they form a **p-n junction** (Figure 27.16c). Free electrons in the n-type silicon migrate toward the p-type silicon, while the holes in the p-type silicon migrate toward the n-type silicon. When the free electrons and holes meet, they recombine. Therefore, in the n-type we have positively charged ions (the result of electrons leaving), and in the p-type we have negatively charged ions (the result of the disappearance of positively charged holes). In the region near the boundary, the n-type silicon becomes positively charged and the p-type silicon becomes negatively charged. This charge separation produces an electric field. As a result, an internal electric field appears in the boundary region (Figure 27.16d).

If we shine light of sufficient frequency on the p-n junction, both halves of the junction will absorb photons whose energy is transferred to valence electrons that then become free. These electrons leave behind holes—virtual positively charged particles (**Figure 27.17**). The electric field created by the p-n junction then separates the freed electrons and holes by accelerating them in opposite directions. The p-n junction in the presence of light results in an electric current—it behaves like a battery! If we now connect a lightbulb to the two ends of the p-n junction, we will observe a current in the circuit, and the bulb will light. Thus the energy of light photons is converted into electric energy.

FIGURE 27.17 The free electrons and the holes produced by absorbing the Sun's photons are separated by the electric field of the p-n junction. The p-n junction in the presence of light acts like a battery.

A silicon atom absorbs a photon, producing a hole and a free electron.

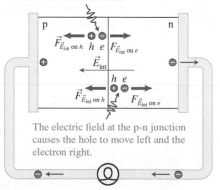

The electric field at the p-n junction causes the hole to move left and the electron right.

FIGURE 27.18 What is inside
an LED?

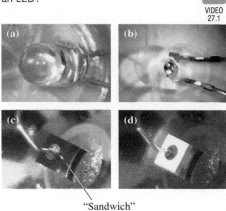

"Sandwich"

Light-emitting diodes

The process described above is reversed in light-emitting diodes (LEDs), which you studied in Chapter 19. They also consist of two doped semiconductor materials that are joined together and form a p-n junction. However, the semiconductor material that is doped is not silicon; it is usually gallium arsenide (a semiconductor alloy). In contrast to solar cells, which transform light into electric energy, LEDs transform electric energy into light. Which part of the LED emits light, and how does the process work?

To answer the first question, we need to examine the inside of an LED. If we look at an LED with the naked eye, we do not see anything inside—whatever is there is too small. If we place an LED under a microscope, will we then see its interior? **Figure 27.18a** shows that a microscope does not help—the plastic dome refracts light and makes the interior hard to see. To get rid of refraction, we can submerge the LED in a liquid with the same index of refraction as the plastic (in this case, glycerin). Figure 27.18b shows the result. We can see an object inside a dome. If we increase the magnification, we see the details more clearly. Figure 27.18c shows the interior of the LED when it is off. We see two different materials in contact with each other that look like a sandwich: a thicker layer of one material covered with a thinner layer of another material. Figure 27.18d shows the same LED when the potential difference between its leads is just enough to make the LED glow. We see that light comes from the side of the "sandwich," where the materials join together. The top layer is the p-doped material and the bottom layer is n-doped. The video shows this process step by step.

As you know from the solar cells, when a p-n junction is formed, the free electrons and holes near the junction recombine, and the region near the junction becomes free of charge carriers (this region is called the *depletion region*). As a result, positive and negative ions (that came from the doped atoms) are the only charged particles in the depletion region—positive on the n side and negative on the p side. These ions create an $\vec{E}$ field that points from the n-type to the p-type. The more electrons and holes that recombine, the thicker the depletion region and the larger the magnitude of this $\vec{E}$ field. Because the forces due to this electric field push electrons back toward the n-type and holes toward the p-type of the semiconductor, the growth of the region eventually stops and the $\vec{E}$ field reaches its final value $\vec{E}_{int}$.

When an LED is connected to a battery as shown in **Figure 27.19a**, the external $\vec{E}_{ex}$ field points in the same direction as the $\vec{E}_{int}$ field of the junction and prevents electrons and holes from crossing the junction. There is no current through the p-n junction (note that in this figure the positive and negative ions are not shown, although the internal field is produced by the ions). However, when you connect the LED as shown in Figure 27.19b, the external $\vec{E}_{ex}$ field points in the opposite direction, and if its magnitude is large enough, it cancels the $\vec{E}_{int}$ field and accelerates free electrons toward the p-type semiconductor and holes toward the n-type semiconductor. When the free electrons and holes meet, they recombine, reducing the energy as they become bound electrons, not free. This excess energy is emitted as photons of light. For this process to happen, the magnitude of $\vec{E}_{ex}$ must be larger than the magnitude of $\vec{E}_{int}$; thus the potential difference across the LED created by the power supply must be larger than a certain value (the opening voltage) to make the LED glow. This explains the opening voltage of LEDs. Interestingly, if you do not connect an LED to a battery but instead shine light on it, it behaves like a solar cell and produces some voltage across its terminals.

Our knowledge of the internal structure of LEDs explains not only their electric properties but also the optical properties that we used in Chapters 2 and 3 when we attached a blinking LED to a moving object to keep track of the object's motion. The rapid switching on and off of an LED is possible because the processes inside LEDs (the movement and recombination of electric charges) happen almost instantly, unlike the warming up and cooling down of an incandescent lightbulb filament, which takes time. This is why incandescent bulbs are not used in processes where changes in brightness need to occur quickly.

FIGURE 27.19 A microscopic explanation of how an LED emits light.

(a)

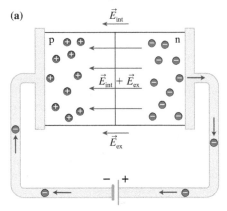

(b)

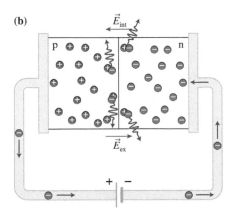

REVIEW QUESTION 27.6 What characteristics would a material need in order to be used to make cathodes in photocells?

Summary

Stefan-Boltzmann's law for total radiating power The total black body radiation power P (all wavelengths) from an object with surface area A and kelvin temperature T is proportional to the fourth power of the temperature. (Section 27.1)

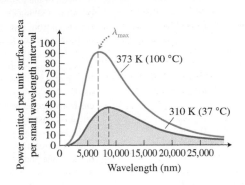

$$P = \sigma T^4 A \qquad \text{Eq. (27.1)}$$
where $\sigma = 5.67 \times 10^{-8}\,\text{W/m}^2 \cdot \text{K}^4$

Wien's law The wavelength λ_{max} at which a black body emits the maximum intensity radiation in a small wavelength interval depends on the temperature T in kelvins of the black body. (Section 27.1)

$$\lambda_{max} = \frac{2.90 \times 10^{-3}\,\text{m} \cdot \text{K}}{T} \qquad \text{Eq. (27.2)}$$

A **photon** is a discrete quantum of electromagnetic radiation that has both wave-like properties and particle-like properties. The energy of a photon depends on its frequency f and Planck's constant h. (Sections 27.3 and 27.4)

c

c

$E = hf$

$$E_{photon} = hf \qquad \text{Eq. (27.7)}$$
where $h = 6.63 \times 10^{-34}\,\text{J} \cdot \text{s}$

Photoelectric effect A photon of energy hf hits a metal cathode of work function ϕ (the minimum energy needed to remove a free electron from the metal). The electron-lattice system absorbs the photon to free the electron from the metal. If the photon energy is greater than ϕ, the electron has nonzero kinetic energy when it leaves the metal. (Sections 27.2 and 27.3)

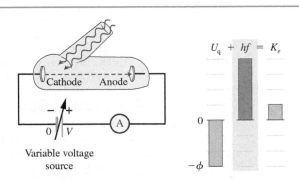

Cathode   Anode

$0 \mid\mid V$ A

Variable voltage source

$U_q + hf = K_e$

0

$-\phi$

$$-\phi + hf = K_e \qquad \text{Eq. (27.5)}$$

Stopping potential $-V_s$ is the minimum negative electric potential that stops the maximum speed electron from reaching the anode. (This assumes that the potential at the cathode is zero.) (Section 27.2)

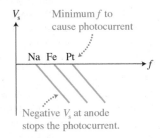

V_s

Minimum f to cause photocurrent

Na Fe Pt

f

Negative V_s at anode stops the photocurrent.

$$\tfrac{1}{2} m_e v_{e\,i\,max}^2 = (-e)(-V_s)$$

where V_s is the positive magnitude of the stopping potential

Cutoff frequency f_{cutoff} is the minimum frequency that a photon must have to remove an electron from the metal. (Sections 27.2 and 27.3)

$$f_{cutoff} = \frac{\phi}{h} \qquad \text{Eq. (27.6)}$$

Photon momentum The magnitude of a photon's momentum p_{photon} depends on its frequency f (or wavelength λ), Planck's constant h, and the speed of light c. (Section 27.4)

p_{photon}

$$p_{photon} = \frac{h}{\lambda} = \frac{hf}{c} \qquad \text{Eq. (27.10)}$$

Questions

Multiple Choice Questions

1. What is a black body? (a) An object painted with black, non-reflective paint (b) An object that does not radiate any electromagnetic waves (c) A model of an object that emits electromagnetic waves with a characteristic frequency distribution

2. If you triple the temperature of a black body (in kelvins), the total amount of energy that it radiates increases how many times? (a) 3 times (b) 9 times (c) 27 times (d) 81 times

3. If you triple the temperature of a black body, the wavelength at which it emits the maximum amount of energy (a) increases 3 times. (b) increases 9 times. (c) increases 27 times. (d) increases 81 times. (e) decreases 3 times.

4. Among the items listed below, choose all of the photoelectric effect observations that could not be explained by the wave model of light. (a) Photocurrent appears instantly after the light is turned on. (b) The magnitude of the photocurrent is directly proportional to the intensity of light. (c) For each intensity of light there is a maximum value of the photocurrent. (d) The stopping potential difference does not depend on the intensity of light. (e) The cutoff frequency does not depend on the intensity of light.

5. What is the work function equal to? (a) The positive work one needs to do to remove an electron from a metal (b) The positive potential energy of interaction of an electron and a metal (c) The negative potential energy of interaction of an electron and a metal (d) a and b (e) a and c

6. What is a photon? (a) A physical quantity (b) A phenomenon (c) A model (d) A law of physics

7. What did the Vavilov-Brumberg experiments show? (a) Photons behave like waves. (b) Photons behave like particles. (c) Photons simultaneously behave like waves and like particles.

8. Which statement describes most accurately our present understanding of light? (a) Light is like a stream of particles. (b) Light is like an electromagnetic wave. (c) Light is a complex phenomenon that cannot be modeled by only one of the above models.

9. Which graph in **Figure Q27.9** correctly shows the dependence of photocurrent on the frequency of EM radiation that is incident on the cathode of a photocell, assuming that the intensity of the EM radiation and the anode potential are kept constant? Assume that quantum efficiency is independent of the frequency of light.

FIGURE Q27.9

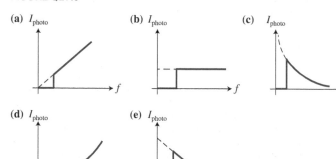

10. In which of the following cases does an electron emit EM radiation? (a) A fast-moving electron hits metal and slows down. (b) An electron speeds up from rest in a uniform electric field. (c) An electron enters a uniform magnetic field; the initial velocity of the electron is perpendicular to the field. (d) An electron enters a uniform magnetic field; the initial velocity of the electron is parallel to the field.

11. Which of the following options correctly describes how an increase in the frequency of light above the cutoff frequency affects the number of electrons ejected per unit time from a cathode when the intensity of light remains constant? Assume that each photon ejects one electron from the cathode independent of photon frequency as long as its frequency is larger than the cutoff frequency. (a) The number of electrons remains the same because one photon can only eject one electron. (b) The number of electrons increases because each photon can be absorbed by two neighboring electrons. (c) The number of electrons increases because a photon can collide with an electron, transfer a part of its energy to the electron as it scatters, and then collide with other surrounding electrons. As long as the photon has enough energy, it will eject an electron at each collision. (d) The number of electrons decreases because we need fewer photons with larger frequency to maintain the same intensity of the light. (e) The number of electrons decreases because photons with a higher frequency have a lower probability of colliding with electrons than low-frequency photons.

12. A light source is shining light on a cathode in a phototube. The potential of the cathode is 0 V, and the potential of the anode is 10 V. The current through the phototube is zero. Which of the following changes could produce a current through the phototube due to photoelectrons? (a) Use more light sources aimed at the same position on the cathode. (b) Increase the anode potential. (c) Decrease the anode potential. (d) Replace the metal of the cathode and the anode with a metal with a smaller work function. (e) Replace the metal of the cathode and the anode with a metal with a larger work function. (f) Use light with a smaller wavelength. (g) Use light with a larger wavelength. (h) Decrease the distance between the cathode and the anode. (i) Increase the distance between the cathode and the anode.

Conceptual Questions

13. How do we know that photons possess wave-like properties?
14. How do we know that photons possess particle-like properties?
15. What is the difference between the photon model of light and the wave model of light?
16. What is the difference between the photon model of light and the particle model of light?
17. The Sun and other celestial objects emit X-rays. Why are we, on the Earth's surface, not concerned about them?
18. When photographic film is being developed, technicians work in a darkroom in which a special red light is used for illumination, not a regular white light. Explain why.
19. Why is the photon model of light supported by the fact that light below a certain frequency, when striking a metal plate, does not cause electrons to be freed from the metal?
20. Explain how we know that cathode rays are low-mass negatively charged particles. Draw pictures and field representations to illustrate your explanation.

Problems

Below, **BIO** indicates a problem with a biological or medical focus. Problems labeled **EST** ask you to estimate the answer to a quantitative problem rather than derive a specific answer. Asterisks indicate the level of difficulty of the problem. Problems with no * are considered to be the least difficult. A single * marks moderately difficult problems. Two ** indicate more difficult problems.

27.1 Black body radiation

1. **Wavelength of radiation from a person** If a person could be modeled as a black body, at what wavelength would his or her surface emit the maximum energy?

2. * (a) A surface at 27 °C emits radiation at a rate of 100 W. At what rate does an identical surface at 54 °C emit radiation? (b) Determine the wavelength of the maximum amount of radiation emitted by each surface.

3. **Maximum radiation wavelength from star, Sun, and Earth** Determine the wavelengths for the following black body radiation sources where they emit the most energy: (a) A blue-white star at 40,000 K; (b) the Sun at 6000 K; and (c) Earth at about 300 K.

4. * **EST** **Star colors and radiation frequency** The colors of the stars in the sky range from red to blue. Assuming that the color indicates the frequency at which the star radiates the maximum amount of electromagnetic energy, estimate the surface temperature of red, yellow, white, and blue stars. What assumptions do you need to make about white stars to estimate the surface temperature?

5. * **EST** Estimate the surface area of a 60-watt lightbulb filament. Assume that the surface temperature of the filament when it is plugged into an outlet of 120 V is about 3000 K and the power rating of the bulb is the electric energy it consumes (not what it radiates). Incandescent lightbulbs usually radiate in visible light about 10% of the electric energy that they consume.

6. * **EST** **Photon emission rate from human skin** Estimate the number of photons emitted per second from 1.0 cm^2 of a person's skin if a typical emitted photon has a wavelength of 10,000 nm.

7. * **Balancing Earth radiation absorption and emission** Compare the average power that the surface of Earth facing the Sun receives from it to the energy that Earth emits over its entire surface due to it being a warm object. Assume that the average temperature of Earth's surface is about 15 °C. The distance between Earth and the Sun is about 1.5×10^{11} m.

27.2 and 27.3 Photoelectric effect and Quantum model explanation of the photoelectric effect

8. (a) Explain how you convert energy in joules into energy in electron volts. (b) The kinetic energy of an electron is 2.30 eV. What is its kinetic energy in joules?

9. Determine the average kinetic energy of an atom in an ideal gas at room temperature (20 °C) and convert the result into electron volts. Compare the result with the energy of a photon of visible light. Indicate any assumptions that you made.

10. * The stopping potential for an ejected photoelectron is −0.50 V. What is the maximum kinetic energy of the electron ejected by the light?

11. Light shines on a cathode and ejects electrons. Draw an energy bar chart describing this process. Explain why the frequency of incident light determines whether the electrons will be ejected or not.

12. * What is the cutoff frequency of light if the cathode in a photoelectric tube is made of iron?

13. The work function of cesium is 2.1 eV. (a) Determine the lowest frequency photon that can eject an electron from cesium. (b) Determine the maximum possible kinetic energy in electron volts of a photoelectron ejected from the cesium that absorbs a 400-nm photon.

14. * Visible light shines on the metal surface of a phototube having a work function of 1.30 eV. The maximum kinetic energy of the electrons leaving the surface is 0.92 eV. Determine the light's wavelength.

15. * **Equation Jeopardy 1** Solve for the unknown quantity in the equation below and write a problem for which the equation could be a solution.

$$-3.9 \text{ eV} + f(6.63 \times 10^{-34} \text{ J} \cdot \text{s}) \left(\frac{1 \text{ eV}}{1.6 \times 10^{-19} \text{ J}} \right) = (-e)(-1.0 \text{ V})$$

16. **Camera film exposure** In an old-fashioned camera, the film becomes exposed when light striking it initiates a complex chemical reaction. A particular type of film does not become exposed if struck by light of wavelength longer than 670 nm. Determine the minimum energy in electron volts needed to initiate the chemical reaction.

27.4 Photons

17. **Breaking a molecular bond** Suppose the bond in a molecule is broken by photons of energy 5.0 eV. Determine the frequency and wavelength of these photons and the region of the electromagnetic spectrum in which they are located.

18. * A l.0-eV photon's wavelength is 1240 nm. Use a ratio technique to determine the wavelength of a 5.0-eV photon.

19. **BIO** **Tanning bed** In a tanning bed, exposure to photons of wavelength 300 nm or less can do considerable damage. Determine the lowest energy in electron volts of such photons.

20. * Determine the number of 650-nm photons that together have energy equal to the rest energy of an electron. (Hint: See Section 26.9.)

21. * **EST** You create a light pulse by switching on a green laser ($\lambda = 532$ nm) for 1 μs. The pulse contains 10^{10} photons. (a) Determine the power output of the laser. (b) Estimate the distance between two neighboring photons, assuming the photons travel as localized particles and the light pulse has uniform intensity.

22. ** **BIO** **Laser surgery** Scientists studying the use of lasers in various surgeries have found that very short 10^{-12}-s laser pulses of power 10^{12} W with 65 pulses every 200×10^{-6} s produced much cleaner welds and ablations (removals of body tissues) than longer laser pulses. Determine the number of 10.6-μm photons in one pulse and the average power during the 65 pulses delivered in 200×10^{-6} s.

23. * A laser beam of power P in watts consists of photons of wavelength λ in nanometers. Determine in terms of these quantities the number of photons passing a cross section along the beam's path each second.

24. What is the total momentum of the photons passing the cross section every second in the previous problem?

25. * **BIO** **Pulsed laser replaces dental drills** A laser used for many applications of hard surface dental work emits 2780-nm wavelength pulses of variable energy (0–300 mJ) about 20 times per second. Determine the number of photons in one 100-mJ pulse and the average power of these photons during 1 s.

26. ** **Light hitting Earth** The intensity of light reaching Earth is about 1400 W/m^2. Determine the number of photons reaching a 1.0-m^2 area each second. What assumptions did you make?

27. * **EST** **Lightbulb** Roughly 10% of the power of a 100-watt incandescent lightbulb is emitted as visible light, the rest being emitted as infrared radiation. Estimate the number of photons of light coming from a bulb each second. What assumptions did you make? How will the answer change if the assumptions are not valid?

28. * **BIO** **Human vision sensitivity** To see an object with the unaided eye, the light intensity coming to the eye must be about 5×10^{-12} J/m$^2 \cdot$ s or greater. Determine the minimum number of photons that must enter the eye's pupil each second in order for an object to be seen. Assume that the pupil's radius is 0.20 cm and the wavelength of the light is 550 nm.

29. * A laser produces a short pulse of light whose energy equals 0.20 J. The wavelength of the light is 694 nm. (a) How many photons are produced? (b) Determine the total momentum of the emitted light pulse.

30. * **Levitation with light** Light from a relatively powerful laser can lift and support glass spheres that are 20.0×10^{-6} m in diameter (about the size of a body cell). Explain how that is possible.

27.5 X-rays

31. * Explain how a cathode ray tube works. Draw a picture and an electric circuit. Label the important elements and explain how they work together to produce cathode rays.

32. * An X-ray tube emits photons of frequency 1.33×10^{19} Hz or less. (a) Explain how the tube creates the X-ray photons. (b) Determine the potential difference across the X-ray tube.

33. * Electrons are accelerated across a 40,000-V potential difference. (a) Explain why X-rays are created when the electrons crash into the anode of the X-ray tube. (b) Determine the frequency and wavelength of the maximum-energy photons created.

34. * A small 1.0×10^{-5}-g piece of dust falls in Earth's gravitational field. Determine the distance it must fall so that the change in gravitational potential energy of the dust-Earth system equals the energy of a 0.10-nm X-ray photon.

35. **BIO** **X-ray exam** While being X-rayed, a person absorbs 3.2×10^{-3} J of energy. Determine the number of 40,000-eV X-ray photons absorbed during the exam.

36. * **BIO** **Body cell X-ray** (a) A body cell of 1.0×10^{-5}-m radius absorbs 4.2×10^{-14} J of X-ray radiation. If the energy needed to produce one positively charged ion is 100 eV, how many positive ions are produced in the cell? (b) How many ions are formed in the 3.0×10^{-6}-m-radius nucleus of that cell (the place where the genetic information is stored)? Indicate any assumptions that you made.

27.6 Photocells, solar cells, and LEDs

37. * BIO EST **Light detection by human eye** The dark-adapted eye can supposedly detect one photon of light of wavelength 500 nm. Suppose that 100 such photons enter the eye each second. Estimate the intensity of light in W/m^2. Assume that the diameter of the eye's pupil is 0.50 cm.

38. * Light of wavelength 430 nm strikes a metal surface, releasing electrons with kinetic energy equal to 0.58 eV or less. Determine the metal's work function.

39. * **Solar cell** A 0.20-m × 0.20-m photovoltaic solar cell is irradiated with 800 W/m^2 sunlight of wavelength 500 nm. (a) Determine the number of photons hitting the cell each second. (b) Determine the maximum possible electric current that could be produced. (c) Explain how a solar cell converts the energy of sunlight into electric energy.

40. * Mark and Kat are discussing doped semiconductors. Mark says that p-doped semiconductors are positively charged and n-doped semiconductors are negatively charged. Kat disagrees; she says that both types are neutral. Discuss the reasoning that could lead each of them to their opinions. Who do you agree with? Explain.

41. * Figure Q27.41 shows the sketch of the processes in a solar cell. Evaluate the sketch and make suggestions for improvements.

FIGURE P27.41

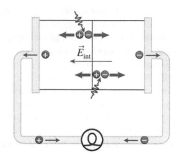

General Problems

42. * EST Estimate the temperature of the Sun's surface knowing the solar constant (1.37 kW/m^2). Decide what other data you need to make the estimate.

43. * If you shine UV radiation on the positively charged zinc sphere on top of an electroscope, the leaves of the electroscope stay deflected and do not move. Lisa and Shumaila are trying to explain the outcome of this experiment.
   Lisa: "The positive charges increase the work function of the zinc. As a result, the photons of UV light do not have enough energy to knock electrons from the zinc sphere."
   Shumaila: "UV light knocks electrons from the sphere, but as soon as they leave the positively charged sphere they slow down and return to the sphere." Who do you think is correct? Propose a testing experiment whose outcome might reject one of the explanations. Predict the outcome of your testing experiment based on each student's hypothesis.

44. A light source consists of a green (530 nm) and a blue (430 nm) LED. Both LEDs emit light of the same intensity. Of 10^3 photons emitted by the light source, (a) how many are photons of green light and (b) how many are photons of blue light?

45. * A light source consists of a green (530 nm) and a blue (430 nm) LED. The intensity of light emitted by the green LED is two times larger than the intensity of the light emitted by the blue LED. The light from this source is incident on a cathode with work function 2.2 eV. (a) Determine the stopping potential between the anode and the cathode that will stop the current through the phototube. (b) When the potential of the anode is set to +10.0 V with respect to the cathode and the cathode is illuminated with the light source, you measure a photocurrent of 5.0 μA. Determine the photocurrent if the green LED in the light source is turned off. Assume that the number of photons required to eject an electron from the cathode does not depend on the frequency of light as long as the frequency is larger than the cutoff frequency, and that the cathode and the anode are made of the same metal.

46. * BIO EST **Fireflies** Fireflies emit light of wavelengths from 510 nm to 670 nm. They are about 90% efficient at converting chemical energy into light (compared to about 10% for an incandescent lightbulb). Most living organisms, including fireflies, use adenosine triphosphate (ATP) as an energy molecule. Estimate the number of ATP molecules a firefly would use at 0.5 eV per molecule to produce one photon of 590-nm wavelength if all the energy came from ATP.

47. * BIO EST **Owl night vision** Owls can detect light of intensity 5×10^{-13} W/m^2. Estimate the minimum number of photons an owl can detect. Indicate any assumptions you used in making the estimate.

48. ** BIO **Photosynthesis efficiency** During photosynthesis in a certain plant, eight photons of 670-nm wavelength can cause the following reaction: $6CO_2 + 6H_2O \rightarrow C_6H_{12}O_6 + 6O_2$. During respiration, when the plant metabolizes sugar, the reverse reaction releases 4.9 eV of energy per CO_2 molecule. Determine the ratio of the energy released (respiration) to the energy absorbed (photosynthesis), a measure of photosynthetic efficiency.

49. * Suppose that light of intensity 1.0×10^{-2} W/m^2 is made of waves rather than photons and that the waves strike a sodium surface with a work function of 2.2 eV. (a) Determine the power in watts incident on the area of a single sodium atom at the metal's surface (the radius of a sodium atom is approximately 1.7×10^{-10} m). (b) How long will it take for an electron in the sodium to accumulate enough energy to escape the surface, assuming it collects all light incident on the atom?

50. * **Force of light on mirror** A beam of light of wavelength 560 nm is reflected perpendicularly from a mirror. Determine the average force that the light exerts on the mirror when 10^{20} photons hit the mirror each second. (Hint: Refer to the impulse-momentum equation (Chapter 6). You may assume that the magnitude of the photons' momenta is unchanged by the collision, but their directions are reversed.)

51. * EST **Sirius radiation power** Sirius, a star in the constellation of Canis Major, is the second brightest star of the northern sky (the brightest is the Sun). Its surface temperature is 9880 K and its radius is 1.75 times greater than the radius of the Sun. Estimate the energy that Sirius emits every second from its surface and compare this energy to the energy that the Sun emits. The radius of the Sun is about 7.0×10^8 m and the energy emitted per second is about 3.9×10^{26} W.

52. ** **Force of sunlight on Earth** We wish to determine the net force on Earth caused by the absorption of light from the Sun. (a) Determine the net area of the surface of Earth exposed to sunlight (Earth's radius is 6.38×10^6 m). (b) The solar radiation intensity is 1400 $J/s \cdot m^2$. Determine the momentum of photons hitting Earth's surface each second. (c) Use the impulse-momentum equation to determine the average force of this radiation on Earth.

53. ** EST **Levitating a person** Suppose that we wish to support a 70-kg person by levitating the person on a beam of light. (a) If all of the photons striking the person's bottom surface are absorbed, what must be the power of the light beam, which is made of 500-nm-wavelength photons? (b) Estimate the person's temperature change in 1 s.

54. ** You use an incandescent lightbulb to shine light onto a cesium cathode in a phototube. How will increasing the light intensity affect the kinetic energy of photoelectrons? Explain. Assume the quantum efficiency is independent of the frequency of light.

Reading Passage Problems

BIO **Capturing energy from sunlight—photosynthesis** Green plants capture and store the energy of photons from the Sun to help build complex molecules such as glucose. The process starts with the photoelectric effect. Chloroplasts in plant cells contain many photosynthetic units, each of which has about 300 pigment "antenna" molecules that absorb sunlight (Figure 27.20).

Suppose that one of the antenna molecules absorbs a photon. The electrons in the molecule are now in an excited state, which means that they are temporarily storing the extra energy gained from the photon. Typically, such a molecule reemits a photon and returns to its original energy state in about 10^{-8} s. If the energy of the photon is simply released, the plant has lost the Sun's energy.

In the photosynthetic units, the antenna molecules are linked closely to each other. When a photon is absorbed by one antenna molecule, its excitation energy is transferred to a neighbor antenna molecule. This excited neighbor in turn excites

FIGURE 27.20 Photosynthetic unit inside plant chloroplasts. "Antenna" molecules absorb photons and become excited. The energy of this excitation is passed from one antenna to another until it arrives at an "acceptor" molecule. The acceptor passes the energy to an electron transport chain, where it is captured and stored in biomolecules like ATP.

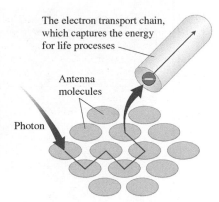

another of its neighbors, in a more or less random fashion. Eventually, an electron in an excited antenna molecule is transferred to a primary acceptor, carrying the extra energy of the photon along with it. This energetic electron now passes through a complex electron transport chain. At several places along this pathway, the energy is given up to help form stable chemical bonds (Figure 27.21).

FIGURE 27.21 The electron transport chain. Electrons are passed between intermediate molecules, and energy is captured at certain steps for the ultimate production of ATP, the cellular energy molecule.

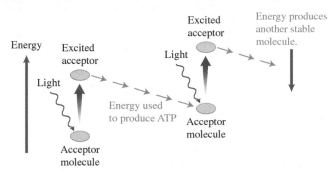

Life on Earth depends on the absorption of the photon, the "random walk" of the excitation energy from antenna molecule to antenna molecule, and the capture of the excited electron energy, all in a fraction of a second.

55. What is the number of antenna molecules that can absorb light in a photosynthetic unit?
 (a) 1   (b) About 10   (c) Over 100
 (d) Over 10,000   (e) About 10^8

56. Suppose an antenna molecule absorbs a 430-nm photon and that this energy is transferred directly to the acceptor molecule. Which answer below is closest to the energy that the photoelectron brings to the electron transport chain?
 (a) 1 eV   (b) 2 eV   (c) 3 eV   (d) 4 eV   (e) 5 eV

57. Suppose that the excited energy of one antenna molecule is transferred 100 times between neighboring antenna molecules. Which answer below is closest to the maximum time interval for the transfer between neighboring antenna molecules?
 (a) 10^{-3} s   (b) 10^{-6} s   (c) 10^{-8} s   (d) 10^{-10} s   (e) 10^{-14} s

58. Suppose that neighboring antenna molecules are separated by about 10^{-10} m and that the excitation energy is transferred 100 times between neighboring antenna molecules before the photoelectric transfer of an electron to the electron transport chain. Which answer below is closest to the minimum speed of the excitation energy through the photosynthetic unit?
 (a) 1 m/s   (b) 10^2 m/s   (c) 10^4 m/s
 (d) 10^{-2} m/s   (e) 10^{-4} m/s

59. The high-energy electron that transfers into an electron transport chain from a photosynthetic unit
 (a) comes from the antenna molecule that absorbed the photon.
 (b) comes from the acceptor molecule, which absorbed the photon.
 (c) comes from the acceptor molecule that is excited by a nearby antenna molecule.
 (d) is produced by the oxidation of a water molecule.
 (e) is produced in the electron transport chain as other molecules react.

BIO **Radiation from our bodies** A person's body can be modeled as a black body that radiates electromagnetic radiation. Wien's law gives the peak wavelength of this radiation: $\lambda_{max} = (2.90 \times 10^{-3}\ \text{m} \cdot \text{K}/T)$, which is mostly infrared radiation.

 The net radiated power is the difference between the power emitted and the power absorbed ($P_{net} = P_{emit} - P_{absorb}$) and is given by Stefan-Boltzmann's law: $P_{net} = A\sigma\epsilon(T^4 - T_0^4)$, where $\sigma = 5.67 \times 10^{-8}\ \text{W/m}^2 \cdot \text{K}^4$, T is the absolute temperature of the skin, and T_0 is the absolute temperature of the surroundings. The surface area A of a human body is about 2 m^2, and the emissivity ϵ of the skin is about 1 (it is an almost perfect emitter and perfect absorber of infrared radiation). The skin temperature for a nude person is about 306 K and the temperature in a room is about 20 °C (293 K). Thus, for a nude person, $P_{net} \approx 160$ W. For a clothed person, the effective skin temperature is cooler, about 28 °C (301 K). The net radiated power is then $P_{net} \approx 100$ W.

60. The wavelength of maximum light emission from the body is closest to
 (a) 500 nm.   (b) 700 nm.   (c) 1200 nm.
 (d) 4500 nm.   (e) 9500 nm.

61. During one day, the total radiative energy loss by a clothed person having a 2 m^2 surface area in a 20 °C room in kcal (1 kcal = 4180 J) is closest to
 (a) 0.5 kcal.   (b) 100 kcal.   (c) 400 kcal.
 (d) 2000 kcal.   (e) 3000 kcal.

62. Photographs taken with a regular camera and with an infrared camera are shown in Figure Q27.62. The man's arm is covered with a black plastic bag. Why is his arm visible in the infrared picture?
 (a) Light does not pass through black plastic, but infrared radiation does.
 (b) His arm is very warm under the black plastic bag and emits much more infrared radiation.
 (c) The bag temperature is similar to his arm temperature.
 (d) The black bag absorbs light and becomes warm and is a good thermal emitter.
 (e) None of the above

FIGURE P27.62

Visible light photo   Infrared photo

63. The man's glasses appear clear with the regular camera photo and black with the infrared camera photo in Figure P27.62. Why?
 (a) Light does not pass through glass, but infrared radiation does.
 (b) Light passes through glass, but infrared radiation does not.
 (c) The lenses of the glasses are cool compared to the man's face, and thus they emit little infrared radiation.
 (d) a and c   (e) b and c

64. What is the ratio of the emitted radiative power from a 310 K surface and the same surface at 300 K?
 (a) 0.86 (b) 0.97 (c) 1.03 (d) 1.07 (e) 1.14

28

Atomic Physics

- How do scientists determine the chemical composition of stars?

- How do lasers work?

- Why is it impossible to know exactly where an electron is?

The photo above shows a spectrum of the Sun obtained using a spectrometer. It looks like a continuous spectrum with lots of dark lines. The dark lines mean that little light of those wavelengths reaches Earth. What causes these dark lines in the spectrum, and how do scientists use them to identify chemical elements present in the Sun's atmosphere as well as to study the Sun's magnetic field?

BE SURE YOU KNOW HOW TO:

- Write an expression for the rotational momentum of a point-like object of mass m moving at constant speed v in a circular orbit of radius r (Section 9.4).

- Apply Coulomb's law for the interaction between charged point-like objects (Section 17.4).

- Explain how a spectrometer works (Section 24.3).

PREVIOUSLY, you learned about cathode ray experiments that led to the discovery of the electron, a very low mass, negatively charged particle. However, most forms of matter we encounter on Earth are electrically neutral, which implies that there must be a positively charged component as well. The relationship between these positively charged components and electrons is the subject of this chapter.

28.1 Early atomic models

The ancient Greek philosopher Democritus proposed that "atoms" were the smallest indivisible components of matter. Indeed, much of what we now know about matter supports some of Democritus's claims. From experiments involving evaporation, we have learned that substances are made of individual point-like objects that move randomly and are surrounded by empty space. We also have learned that both dielectrics and conductors contain electrically charged particles. These particles move freely inside conductors and only a little inside dielectrics. How did these ideas motivate early models of the atomic nature of matter? By the end of the 19th century scientists had several hypotheses.

Dalton model: atom as billiard ball

In 1803 John Dalton proposed a billiard ball model of the atom as a small solid sphere. According to Dalton, a sample of a pure element was composed of a large number of atoms of a single kind. Compounds were substances that contained more than one kind of atom in specific integer ratios. In this model, each atom, regardless of type, contains an equal amount of positively and negatively charged subcomponents so that each atom is electrically neutral.

Thomson model: atom as plum pudding

In 1897 J. J. Thomson's experiments supported the model that cathode rays were electrons. Because cathodes made of different materials emitted the same electrons, Thomson hypothesized that these electrons were the negatively charged components of atoms.

According to his model, an atom is similar to a spherically shaped plum pudding. The plums represent electrons embedded in a massive sphere of diffuse positive charge represented by the pudding (**Figure 28.1**). This model explained the electrical neutrality of an atom and provided a qualitative explanation of how atoms could emit light by assuming that electrons could absorb and emit light through oscillations. At the time, it looked like a perfect model.

FIGURE 28.1 The "plum pudding" model of the atom.

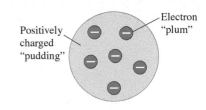

Positively charged "pudding"

Electron "plum"

Rutherford model: atom as planetary system

Alpha particles are emitted by uranium and other radioactive elements. These particles have a mass four times greater than a hydrogen atom and a charge twice that of an electron, but positive, not negative. According to the Thomson model, alpha particles beamed at a thin layer of metal foil should pass nearly straight through the foil because the atoms are soft and their mass and positive charge are evenly distributed. In 1909, Ernest Rutherford, a New Zealand-born British physicist, noticed that such a beam came out of the foil much wider than it had been when it entered. Thomson's model could not explain this phenomenon.

The atomic nucleus Rutherford and his postdoctoral colleague Hans Geiger hypothesized that the atom's mass and positive charge were not spread uniformly throughout the atom but were instead concentrated in a small region. If this were true, then the beam should spread as observed. This new model also predicted that the alpha particles that come very close to the small dense region inside the atom should recoil backward at very large angles, possibly even in the reverse direction. However, Rutherford and Geiger did not believe that they would find this recoil. After all, how could an atom cause a very fast-moving alpha particle to almost instantaneously reverse direction? The idea seemed absurd. Geiger and Ernest Marsden set out to perform a testing experiment to disprove this hypothesis. In Testing Experiment **Table 28.1** (on the next page) we look at their experiment.

TESTING
EXPERIMENT TABLE 28.1

Structure of the atom

Testing experiment	Prediction	Outcome
The experimental apparatus is shown below. 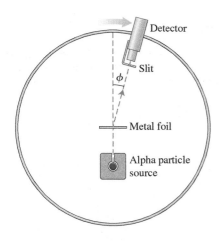	**Thomson model:** If the atom's mass and positive charge are dispersed evenly, the alpha particles will only undergo small deflections. 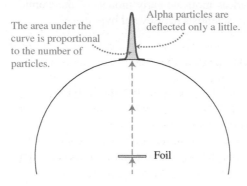 **Rutherford model:** If the atom's mass and positive charge are concentrated in a small region, then most of the alpha particles will pass straight through. But those that pass near the region of concentrated mass will recoil at larger angles, resulting in a wider beam. Some will recoil at very large angles. 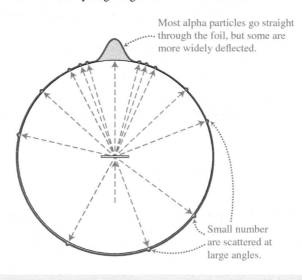	Geiger and Marsden recorded a significant widening of the beam and a small number of alpha particles recoiling at large angles.

Conclusion

The outcome is not consistent with the prediction based on the Thomson model. The outcome is consistent with the prediction based on the Rutherford model. It seems that the mass and positive charge of the atom are concentrated in a small region, with the rest of the atom mostly empty space.

Rutherford and his colleagues were surprised to find that some alpha particles recoiled backward at such large angles. Reflecting on these experiments, Rutherford once said, "It was quite the most incredible event that happened to me in my life. It was almost as incredible as if you fired a 15-inch shell at a piece of tissue paper and it came back and hit you."[1] Following these experiments, Rutherford developed a new model

---

[1]Rutherford, E. *Background to Modern Science*. New York: Macmillan, 1938.

of the atom in which a tiny **nucleus** contained nearly all of the mass of the atom and all of its positive charge. To be consistent with experiments, the diameter of the nucleus would have to be $1/100{,}000$ times the diameter of the atom, which was known at that time to be about 10^{-10} m. This made the nucleus extremely tiny—about 10^{-15} m.

Rutherford used this quantitative model to make predictions about the number of alpha particles that would be scattered at a particular angle for a specific thickness of foil, charge of the nucleus, and velocity of alpha particles. Geiger and Marsden performed experiments to test Rutherford's model and found the outcomes consistent with the model's predictions.

The location of the electrons Now that the nuclear model of the atom had emerged, questions remained about the electrons. Where in the atom were the electrons? Rutherford understood that electrons could not just sit stationary at the edge of the atoms, far from the nucleus. The positive nucleus would attract the negative electrons, causing the atom to collapse. Rutherford avoided this problem by proposing what became known as the **planetary model** of the atom. In this model, the electrons orbited the nucleus just as the planets orbit the Sun (**Figure 28.2**). The speed of the electrons could then be calculated using Newton's second law and the mathematical description of circular motion.

In the hydrogen atom, the nucleus exerts an electric force on the atom's lone electron. We can determine the magnitude of this force using Coulomb's law:

$$F_{\text{N on }e} = k_{\text{C}} \frac{q_{\text{N}}q_e}{r^2}$$

where r is the distance between the electron and the nucleus, which in this case is equal to the radius of the atom. The notations q_{N} and q_e are the magnitudes of the charges of the nucleus and the electron, respectively. In uniform circular motion, the radial component of Newton's second law is

$$a_r = \frac{v^2}{r} = \frac{1}{m_e} \sum F_r$$

where m_e is the mass of the electron, v is the magnitude of the tangential component of the electron's velocity (in this case, it equals the electron's speed, v) and a_r is its radial acceleration. As the only force exerted on the electron in the radial direction is the electric force, we can write

$$\frac{v^2}{r} = \frac{1}{m_e}\left(k_{\text{C}}\frac{q_{\text{N}}q_e}{r^2}\right)$$

Here q_{N} is the magnitude of the electric charge of the nucleus, q_e is the magnitude of the electron's charge, r is the radius of the electron's orbit, and k_{C} is the Coulomb's law constant. Multiplying both sides of the equation by r and taking the square root, we solve for v:

$$v = \sqrt{\frac{k_{\text{C}}q_{\text{N}}q_e}{m_e r}}$$

The above expression gives the constant speed of an electron moving in a circle at a distance r from the nucleus. Thus if the electron is moving fast enough, it can avoid falling onto the nucleus.

A difficulty with the planetary model The total energy of the nucleus-electron system, E, is the sum of its electric potential energy (U_{q}) and the kinetic energy of the electron (K_e):

$$E = U_{\text{q}} + K_e$$

Assuming the electron is not moving relativistically:

$$E = k_{\text{C}} \frac{q_{\text{N}}q_e}{r} + \tfrac{1}{2}m_e v^2 = -k_{\text{C}}\frac{e^2}{r} + \tfrac{1}{2}m_e v^2$$

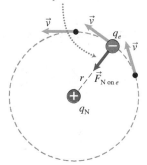

FIGURE 28.2 Rutherford's orbital model of the atom. The sizes of the electron and the nucleus are not to scale.

The inward force that the electric charge of the nucleus exerts on the electron keeps it in a circular orbit.

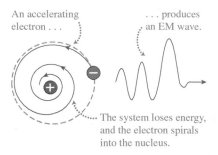

An accelerating
electron . . .

. . . produces
an EM wave.

The system loses energy,
and the electron spirals
into the nucleus.

FIGURE 28.4 A continuous spectrum and a line spectrum.

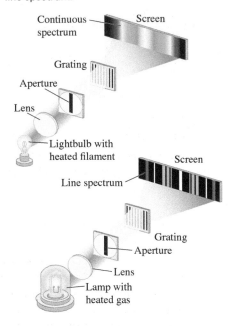

Continuous
spectrum

Screen

Grating

Aperture

Lens

Lightbulb with
heated filament

Screen

Line spectrum

Grating

Aperture

Lens

Lamp with
heated gas

FIGURE 28.5 A hydrogen atom line spectrum.

For a hydrogen atom $q_N = +e$ and $q_e = -e$. Because the atom is a bound system, its total energy should be negative (as we learned in Chapter 7, the zero level of potential energy is at infinity; if the total energy were zero or positive, the system would fly apart).

As discussed earlier (in Chapter 25), an accelerating electron emits electromagnetic (EM) radiation, and this radiation carries energy. Because the electron orbiting the nucleus is continually accelerating, the electron-nucleus system should continuously lose energy. As the electron emits electromagnetic radiation, the electron should get closer to the nucleus (**Figure 28.3**). EM theory predicts that the electron would spiral into the nucleus in about 10^{-12} s. This definitely contradicts our experience. The Rutherford model could not explain the stability of atoms. It also could not explain another phenomenon that was well known to physicists at that time—the light emitted by low-density gases.

Spectra of low-density gases and the need for a new model

By the end of the 19th century, physicists knew that many objects emit a continuous spectrum of light whose properties depend on the temperature of the object—that is, the random motion of their particles. Room-temperature objects primarily emit infrared photons. Objects at several thousand degrees kelvin (such as an incandescent lightbulb filament) emit a larger number of visible light photons. We learned about this type of spectrum in Chapter 27 when studying black body radiation.

If atoms are in a gaseous form and the gas's temperature is increased so that it emits visible light, we observe a **line spectrum** (**Figure 28.4**), in which only specific wavelengths of light ("lines") are present. For example, if you throw a dash of table salt into a flame and observe the emitted light with a spectrometer, you see a line spectrum—two closely positioned yellow lines.

Another way to observe a gas spectrum is to put a strong electric field across a tube filled with gas. Neon signs use this method; the spectrum of neon gas is linear.

Observations show that different gases produce different sets of spectral lines. These line combinations do not depend on how the gases are made to glow—whether heated or placed in a strong electric field. When gases are mixed together, the spectral lines characteristic of each individual gas are present.

Scientists could not explain where the lines came from, but they could determine which chemical elements emitted which lines. The most carefully studied element was hydrogen because of the simplicity of its line spectra: just four distinct lines in the visible light region of the spectrum (**Figure 28.5**).

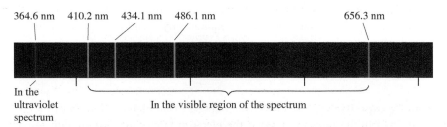

364.6 nm   410.2 nm   434.1 nm   486.1 nm   656.3 nm

In the
ultraviolet
spectrum

In the visible region of the spectrum

In 1885 Johann Balmer found a pattern in the wavelengths of the visible spectral lines produced by hydrogen and represented it with the following equation:

$$\frac{1}{\lambda} = R\left(\frac{1}{2^2} - \frac{1}{n^2}\right) \tag{28.1}$$

where λ is the wavelength of the spectral line, n has integer values of 3, 4, 5, and 6, and R is a constant experimentally determined to equal $R = 1.097 \times 10^7 \text{ m}^{-1}$ when the wavelength is measured in meters. The constant R is called the **Rydberg constant**. The Balmer equation produces the four wavelengths shown in Figure 28.5. However, Balmer could not explain why this pattern existed.

The Thomson model could explain the emission of light by assuming that electrons can vibrate inside the atom, but it did not provide a quantitative account for the specific frequencies. The Rutherford model predicted that individual atoms should radiate a continuous spectrum until the electron spiraled into the nucleus. Physicists needed a new atomic model that was stable and that could explain the line spectra; we will discuss this model in Section 28.2.

QUANTITATIVE EXERCISE 28.1 **Calculate photon energies**

Use Balmer's formula [Eq. (28.1)] to determine the energies in joules and in electron volts of the possible visible photons emitted by hydrogen atoms.

Represent mathematically Balmer's formula [Eq. (28.1)] provides an expression for the inverse of the emitted photon wavelength:

$$\frac{1}{\lambda} = R\left(\frac{1}{2^2} - \frac{1}{n^2}\right)$$

The four visible spectral lines correspond to $n = 3, 4, 5,$ and 6. The energy of a photon is

$$E = hf = \frac{hc}{\lambda} = hc\,\frac{1}{\lambda}$$

Solve and evaluate Combine these two equations to get

$$E = hcR\left(\frac{1}{2^2} - \frac{1}{n^2}\right)$$

Inserting the appropriate values for Planck's constant h, the speed of light c, and Rydberg's constant R, we get for the combined constant hcR

$$hcR = (6.63 \times 10^{-34}\ \text{J} \cdot \text{s})(3.00 \times 10^{8}\ \text{m/s})(1.097 \times 10^{7}\ \text{m}^{-1})$$
$$= 2.18 \times 10^{-18}\ \text{J}$$

This combined constant shows up frequently when we deal with line spectra, and it is worth remembering for convenience.

Inserting the four different n values and converting to electron volts ($1\ \text{eV} = 1.6 \times 10^{-19}\ \text{J}$), we get the following:

$$n = 3:\ \ E_3 = (2.18 \times 10^{-18}\ \text{J})\left(\frac{1}{2^2} - \frac{1}{3^2}\right) = 3.03 \times 10^{-19}\ \text{J}$$
$$= (3.03 \times 10^{-19}\ \text{J})\left(\frac{1\ \text{eV}}{1.6 \times 10^{-19}\ \text{J}}\right) = 1.89\ \text{eV}$$

$$n = 4:\ \ E_4 = (2.18 \times 10^{-18}\ \text{J})\left(\frac{1}{2^2} - \frac{1}{4^2}\right) = 4.09 \times 10^{-19}\ \text{J}$$
$$= (4.09 \times 10^{-19}\ \text{J})\left(\frac{1\ \text{eV}}{1.6 \times 10^{-19}\ \text{J}}\right) = 2.55\ \text{eV}$$

$$n = 5:\ \ E_5 = (2.18 \times 10^{-18}\ \text{J})\left(\frac{1}{2^2} - \frac{1}{5^2}\right) = 4.58 \times 10^{-19}\ \text{J}$$
$$= (4.58 \times 10^{-19}\ \text{J})\left(\frac{1\ \text{eV}}{1.6 \times 10^{-19}\ \text{J}}\right) = 2.86\ \text{eV}$$

$$n = 6:\ \ E_6 = (2.18 \times 10^{-18}\ \text{J})\left(\frac{1}{2^2} - \frac{1}{6^2}\right) = 4.84 \times 10^{-19}\ \text{J}$$
$$= (4.84 \times 10^{-19}\ \text{J})\left(\frac{1\ \text{eV}}{1.6 \times 10^{-19}\ \text{J}}\right) = 3.03\ \text{eV}$$

To determine whether these results are reasonable, complete the "Try it yourself" question below and see if the wavelengths correspond to the observed visible light wavelengths for hydrogen.

- -

Try it yourself Determine the wavelengths in nanometers (nm) of the above spectral lines, and identify the colors.

Answer

Using the Balmer equation, we get 656 nm for $n = 3$ (orange), 486 nm for $n = 4$ (greenish blue), 434 nm for $n = 5$ (blue), and 410 nm for $n = 6$ (violet). All of these wavelengths fall within the visible part of the EM spectrum. Compare them to the values in Figure 28.5.

REVIEW QUESTION 28.1 Rutherford made the following statement: "I was perfectly aware when I put forward the theory of the nuclear atom that . . . the electron ought to fall onto the nucleus." Why must the electrons in Rutherford's model fall into the nucleus?

28.2 Bohr's model of the atom: quantized orbits

In 1913, Danish physicist Niels Bohr succeeded in creating a new model that explained the line spectra and the stability of the atom. He kept the structure of Rutherford's model in terms of the small nucleus and electrons moving around it, but imposed a restriction on the electron orbits.

TIP Before you read the description of Bohr's ideas, review the concept of rotational (angular) momentum (Chapter 9).

Bohr's model applies only to hydrogen and to one-electron ions. Here are the fundamental ideas of Bohr's model, known as **Bohr's postulates**.

1. The atom is made up of a small nucleus and an orbiting electron. The electron can occupy only certain orbits, called **stable orbits**, which are labeled by the positive integer n. When in these orbits, the electron moves around the nucleus but *does not* radiate electromagnetic waves. All other orbits are prohibited. Each of these stable orbits results in a specific value of the total energy (kinetic plus electric potential) of the atom, designated E_n.

2. When an electron transitions from one stable orbit to another, the atom's energy changes. When the energy of the atom decreases, the atom emits a photon whose energy equals the decrease in the atom's energy ($hf = E_i - E_f$, where "i" is the energy state number for the initial state and "f" is the energy state number for the final state). For the atom's energy to increase, the atom must absorb some energy, often by absorbing a photon whose energy equals the increase in the atom's energy. Because the stable orbits are discrete, the atom can radiate or absorb only certain specific amounts of energy.

3. The stable electron orbits are the orbits where the magnitude of the electron's rotational (angular) momentum L is given by

$$L = mvr = n\frac{h}{2\pi}, \quad n = 1, 2, 3, \ldots \qquad (28.2)$$

In this equation m is the mass of the electron, v is its speed, r is the radius of its orbit, h is Planck's constant, and n is any positive integer. In this equation, rotational momentum L is **quantized**, meaning that it can have only specific discrete values (multiples of $h/2\pi$ in this case).

While Bohr's model explained the line spectra and stability of the hydrogen atom, it does not explain why electrons do not emit EM waves when accelerating in a circular orbit. It also does not explain why the only stable orbits are those with the above rotational momenta [Eq. (28.2)]. Let's see if Bohr's model correctly predicts the size of the hydrogen atom and the specific wavelengths present in its line spectrum.

Size of the hydrogen atom

To calculate the size of the atom using Bohr's postulates, assume that the electron and nucleus obey Newton's laws and interact only via the electrostatic (Coulomb) force, that gravitational forces exerted on the atom are negligible, and that the nucleus has charge $q_N = +e = 1.6 \times 10^{-19}$ C. A sketch of the atom and a force diagram for the electron are shown in **Figure 28.6**. The nucleus exerts an electric force on the electron, $F_{N \text{ on } e}$, of this magnitude:

$$F_{N \text{ on } e} = k_C \frac{|q_N q_e|}{r_n^2} = k_C \frac{|(+e)(-e)|}{r_n^2}$$

$$\Rightarrow F_{N \text{ on } e} = k_C \frac{e^2}{r_n^2} \qquad (28.3)$$

This force points toward the nucleus, resulting in the electron's uniform circular motion. Recall that an object's radial acceleration a_r for uniform circular motion is given by the expression $a_r = (v^2/r)$. The radial r component of Newton's second law is

$$a_r = \frac{F_{N \text{ on } e}}{m_e} = \frac{1}{m_e}\left(k_C \frac{e^2}{r_n^2}\right) \Rightarrow \frac{v_n^2}{r_n} = \frac{k_C e^2}{m_e r_n^2}$$

Here m_e is the mass of the electron and a_r is the radial acceleration. We use the symbol e for the magnitudes of the charge of both the electron and the nucleus. The radius of

TIP   Note that an italicized f represents the physical quantity of frequency, and roman f is an abbreviation for "final."

FIGURE 28.6 Force-acceleration analysis of the Bohr atomic model.

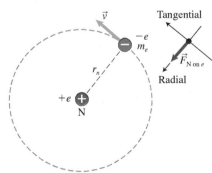

The electrical force that the nucleus exerts on the electron is much greater than any gravitational force that the nucleus exerts on the electron.

the electron's orbit is r_n, and v_n is the corresponding speed; k_C is the Coulomb's law constant. Multiplying both sides by r_n, we get

$$v_n^2 = \frac{k_C e^2}{m_e r_n} \tag{28.4}$$

Equation (28.4) relates two unknown quantities, v_n and r_n. Bohr's third postulate involving the rotational momentum L_n of the electron in the nth orbit provides a second relationship between these same two quantities:

$$L_n = m_e v_n r_n = n\frac{h}{2\pi}, \quad \text{with } n = 1, 2, 3, \ldots$$

Solving for v_n, we find

$$v_n = \frac{nh}{2\pi m_e r_n}, \quad n = 1, 2, 3, \ldots$$

Insert this expression for v_n into Eq. (28.4) to get

$$\left(\frac{nh}{2\pi m_e r_n}\right)^2 = \frac{k_C e^2}{m_e r_n}$$

Solving the above for r_n, we get

$$r_n = \frac{h^2}{4\pi^2 k_C e^2 m_e} n^2, \quad \text{with } n = 1, 2, 3, \ldots \tag{28.5}$$

Let's check the units of this result. The units for h are J·s and for k_C are $(\text{N}\cdot\text{m}^2/\text{C}^2)$. Thus, the unit of the right side of the equation is

$$\frac{(\text{J}^2\,\text{s}^2)}{\left(\dfrac{\text{N}\cdot\text{m}^2}{\text{C}^2}\right)\text{C}^2\,\text{kg}} = \frac{(\text{N}\cdot\text{m})^2\,\text{s}^2}{\text{N}\cdot\text{m}^2\,\text{kg}} = \frac{\text{N}\cdot\text{s}^2}{\text{kg}} = \frac{\left(\dfrac{\text{kg}\cdot\text{m}}{\text{s}^2}\right)\cdot\text{s}^2}{\text{kg}} = \text{m}$$

which is the correct unit for a radius. Substituting the values of the known constants into Eq. (28.5), we get

$$r_n = \frac{(6.63 \times 10^{-34}\,\text{J}\cdot\text{s})^2}{4\pi^2\left(9.00 \times 10^9\,\dfrac{\text{N}\cdot\text{m}^2}{\text{C}^2}\right)(1.6 \times 10^{-19}\,\text{C})^2(9.11 \times 10^{-31}\,\text{kg})} n^2$$

or

$$r_n = (0.53 \times 10^{-10}\,\text{m})n^2, \quad \text{for } n = 1, 2, 3, \ldots \tag{28.6}$$

The smallest possible electron orbit radius corresponds to the $n = 1$ energy state:

$$r_1 = 0.53 \times 10^{-10}\,\text{m}$$

This result is in agreement with the measurements of atomic sizes, about 10^{-10} m.

This smallest radius orbit is called the **Bohr radius** r_1. The other stable orbit radii can be calculated using Eq. (28.6). Notice that the radii depend on the square of n:

$$r_2 = 4r_1; \ r_3 = 9r_1; \ r_4 = 16r_1; \ \text{therefore, } r_n = n^2 r_1$$

As a result, the radius increases dramatically as n increases. The term n is known as a **quantum number** and must be a positive integer. Because of this, only certain radii represent stable electron orbits (for example, $4r_1$ and $9r_1$, which correspond to $n = 2$ and $n = 3$, are stable orbits, but $5r_1$ is not); a stable radius is said to be **quantized**. The radii of allowed electron orbits in the Bohr model of the hydrogen atom are represented in **Figure 28.7**. (Note that the way the nucleus is drawn in the figure dramatically exaggerates its size in comparison to the electron orbits.)

FIGURE 28.7 Radii of stable Bohr electron orbits.

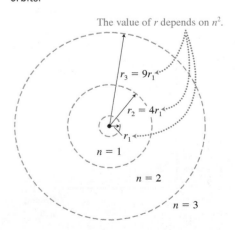

The value of r depends on n^2.

$r_3 = 9r_1$

$r_2 = 4r_1$

r_1

$n = 1$

$n = 2$

$n = 3$

Energy states of the Bohr model

Now that we have gained some confidence in Bohr's model, we can use it to determine the energy states of the hydrogen atom (or, put another way, the total energy of the atom for each energy state n). We'll make the reasonable assumption that the nucleus remains at rest as the electron orbits because of its much larger mass. We'll also assume that relativistic effects are negligible. The total energy of the electron-nucleus system is the sum of its electric potential energy plus the electron's kinetic energy (we neglect the motion of the nucleus due to the interactions with the electron). The electric potential energy U_{qn} is

$$U_{qn} = k_C \frac{q_N q_e}{r_n} = k_C \frac{(+e)(-e)}{r_n} = -k_C \frac{e^2}{r_n}$$

The total energy of the system in the nth energy state is then

$$E_n = U_{qn} + K_{en} = -k_C \frac{e^2}{r_n} + \frac{1}{2} m_e v_n^2 \tag{28.7}$$

From Eq. (28.4), note that $m_e v_n^2 = k_C \left(e^2/r_n\right)$. Therefore, Eq. (28.7) becomes

$$E_n = -k_C \frac{e^2}{r_n} + \frac{1}{2}\left(k_C \frac{e^2}{r_n}\right) = -k_C \frac{e^2}{2r_n}$$

The Bohr model has predicted that the total energy of the atom is negative. This is good news because the total energy of a bound system should be negative, as explained earlier. Now use Eq. (28.5) for the radius of the atom to write an equation for its total energy:

$$E_n = -k_C \frac{e^2}{2r_n} = -k_C \frac{e^2}{2\left(\dfrac{h^2}{4\pi^2 k_C e^2 m_e} n^2\right)} = -\left(\frac{2\pi^2 e^4 k_C^2 m_e}{h^2}\right)\frac{1}{n^2} \tag{28.8}$$

Evidently, the energy of the atom is inversely proportional to the square of the quantum number n. Substituting the values of the constants into the above and converting from joules to electron volts, we get

$$E_n = \frac{-13.6 \text{ eV}}{n^2} \quad \text{for } n = 1, 2, 3, \ldots \tag{28.9}$$

To help visualize the energy states corresponding to these possible total energy values, we use a new representation—an energy state diagram (**Figure 28.8**). On this diagram the vertical axis represents the total energy of the hydrogen atom. Each line represents a specific energy state of the atom labeled by the quantum number n. The -13.6 eV horizontal line at the very bottom corresponds to the $n = 1$ lowest energy state, also called the **ground state**. The next horizontal line $(-13.6 \text{ eV})/2^2 = -3.4$ eV corresponds to the $n = 2$ energy state, also called the first **excited state**. This pattern repeats for the different integral values of n. This representation allows us to see that the higher the quantum number, the less negative the energy of the system. Equation 28.7 shows that the less negative the energy of the system, the larger the radius of the electron's orbit. The $n = \infty$ state has an infinite radius and zero energy, indicating the electron is unbound, an ionized hydrogen atom $(\text{H}^+ + \text{e}^-)$.

The energy state diagram in Figure 28.8 also allows us to represent the process of emission of light by the atom. If the atom is currently in the $n = 2$ energy state and makes a transition to the $n = 1$ energy state, the energy of the atom decreases. According to Bohr's second postulate, the energy is carried away by a newly created photon. This **emission process** is represented in **Figure 28.9** by a downward arrow pointing from the initial (higher energy) state of the atom to the final (lower energy) state. This energy, equal to the energy decrease of the atom, is indicated with the symbol E_γ. The subscript γ (Greek letter gamma) is often used to indicate a photon.

The atom's energy can increase if it absorbs a photon. This **absorption process** is represented on the energy diagram with an upward arrow pointing from the initial

FIGURE 28.8 Energy state diagram. The atom's energy is inversely proportional to the square of n.

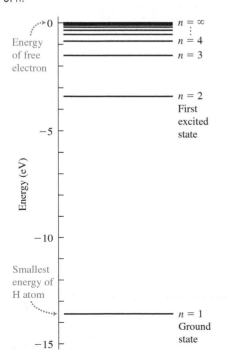

FIGURE 28.9 An example of photon emission. An atom transitions from the first excited state to the ground state, and a photon is emitted.

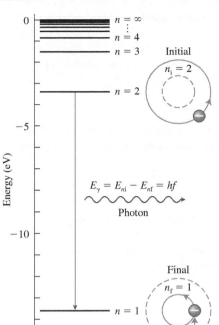

FIGURE 28.10 An example of photon absorption. An atom transitions from the ground state to the first excited state by absorbing the energy of a photon.

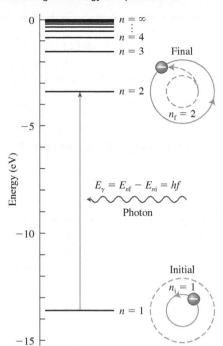

state of the atom to its final state (**Figure 28.10**). An atom can also move from one energy state to another (increasing or decreasing its energy) when it collides with another atom.

We have not yet tested whether Bohr's model correctly predicts the visible light photons emitted by hydrogen; this test is the goal of Example 28.2.

TIP Remember that according to Bohr's first postulate, the hydrogen atom can exist in states whose energies can be calculated by Eq. (28.8). States with energies between those values are not allowed.

EXAMPLE 28.2 ▸ **Photons emitted by hydrogen atoms**

Use the Bohr model to predict the energies and wavelengths of photons that a group of hydrogen atoms would emit if the atoms were all initially in the $n = 3$ state. In which parts of the electromagnetic spectrum are each of these photons?

Sketch and translate According to Bohr's model, an atom emits photons when the energy of the atom decreases. This emission occurs when the electron in the atom transitions from an orbit with a higher quantum number (n_i) to an orbit with a lower number (n_f). Three possible transitions could occur: $3 \rightarrow 1$, $3 \rightarrow 2$, and $2 \rightarrow 1$ (after the atom has already made the $3 \rightarrow 2$ transition).

Simplify and diagram The three processes are represented in the energy state diagram at right. Downward vertical arrows represent the energy transitions.

Represent mathematically Using the principle of energy conservation, the initial energy E_{n_i} of the atom must equal the final energy E_{n_f} plus the photon's energy E_γ:

$$E_{n_i} = E_{n_f} + E_\gamma$$

or

$$E_\gamma = E_{n_i} - E_{n_f}$$

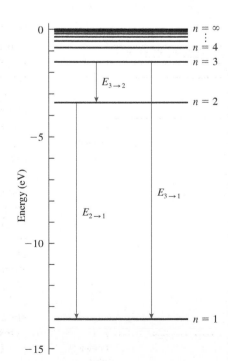

(CONTINUED)

Inserting the expressions for the total energy of the atom, we have

$$E_\gamma = \frac{-13.6 \text{ eV}}{n_i^2} - \frac{-13.6 \text{ eV}}{n_f^2} = -13.6 \text{ eV}\left(\frac{1}{n_i^2} - \frac{1}{n_f^2}\right)$$

Once we determine the energies E_γ of the emitted photons, we can determine their wavelengths λ using

$$E_\gamma = hf = \frac{hc}{\lambda} \text{ or } \lambda = \frac{hc}{E_\gamma}$$

Solve and evaluate We can now determine the energies in electron volts and joules of the photons emitted in each of the three processes. Remember that $1 \text{ eV} = 1.6 \times 10^{-19}$ J.

$$n_i = 3 \text{ to } n_f = 1: \quad E_\gamma = -13.6 \text{ eV}\left(\frac{1}{3^2} - \frac{1}{1^2}\right)$$

$$= 12.1 \text{ eV} = 1.94 \times 10^{-18} \text{ J}$$

$$n_i = 3 \text{ to } n_f = 2: \quad E_\gamma = -13.6 \text{ eV}\left(\frac{1}{3^2} - \frac{1}{2^2}\right)$$

$$= 1.89 \text{ eV} = 3.02 \times 10^{-19} \text{ J}$$

$$n_i = 2 \text{ to } n_f = 1: \quad E_\gamma = -13.6 \text{ eV}\left(\frac{1}{2^2} - \frac{1}{1^2}\right)$$

$$= 10.2 \text{ eV} = 1.63 \times 10^{-18} \text{ J}$$

We can determine the photon wavelengths from their energies:

$n_i = 3 \text{ to } n_f = 1$:

$$\lambda = \frac{hc}{E_\gamma} = \frac{(6.63 \times 10^{-34} \text{ J} \cdot \text{s})(3.00 \times 10^8 \text{ m/s})}{1.94 \times 10^{-18} \text{ J}}$$

$$= 1.03 \times 10^{-7} \text{ m} = 103 \text{ nm}$$

$n_i = 3 \text{ to } n_f = 2$:

$$\lambda = \frac{hc}{E_\gamma} = \frac{(6.63 \times 10^{-34} \text{ J} \cdot \text{s})(3.00 \times 10^8 \text{ m/s})}{3.02 \times 10^{-19} \text{ J}}$$

$$= 6.59 \times 10^{-7} \text{ m} = 659 \text{ nm}$$

$n_i = 2 \text{ to } n_f = 1$:

$$\lambda = \frac{hc}{E_\gamma} = \frac{(6.63 \times 10^{-34} \text{ J} \cdot \text{s})(3.00 \times 10^8 \text{ m/s})}{1.63 \times 10^{-18} \text{ J}}$$

$$= 1.22 \times 10^{-7} \text{ m} = 122 \text{ nm}$$

The 3 to 2 transition is in the visible (red) part of the electromagnetic spectrum, and the 2 to 1 and 3 to 1 transitions are in the ultraviolet part of the spectrum.

- -

Try it yourself Show that the wavelengths of emission lines calculated using Balmer's empirical formula [Eq. (28.1)] are consistent with Bohr's model. The transitions described by Balmer's equation are indicated in **Figure 28.11**. Series of emission lines corresponding to other possible transitions are also shown.

Answer

Balmer's equation is

$$\frac{1}{\lambda} = R\left(\frac{1}{2^2} - \frac{1}{n^2}\right)$$

which produces the same wavelength transitions as calculated above.

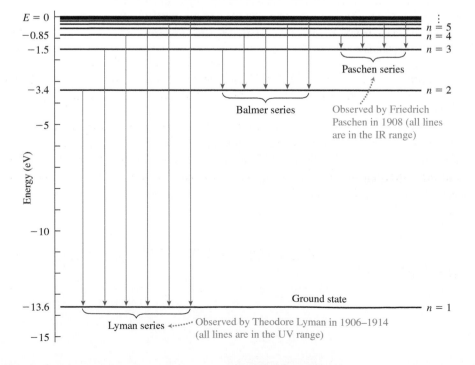

FIGURE 28.11 Three series of emission transitions for H atoms.

Consistency of Balmer's equation with Bohr's model

Let's find whether Balmer's result is consistent with the Bohr model prediction for emission transitions.

$$E_\gamma = -13.6 \text{ eV}\left(\frac{1}{n_i^2} - \frac{1}{n_f^2}\right)$$

If we multiply both sides of the Balmer equation by hc, we get

$$\frac{hc}{\lambda} = hcR\left(\frac{1}{2^2} - \frac{1}{n^2}\right)$$

The left-hand side is the energy E_γ of the emitted photon. The constant hcR must then have units of energy. This constant in electron volts is

$$hcR = (6.63 \times 10^{-34} \text{ J} \cdot \text{s})(3.00 \times 10^8 \text{ m/s})(1.097 \times 10^7 \text{ m}^{-1})\left(\frac{1 \text{ eV}}{1.6 \times 10^{-19} \text{ J}}\right)$$

$$= 13.6 \text{ eV}$$

Balmer's equation becomes

$$E_\gamma = 13.6 \text{ eV}\left(\frac{1}{2^2} - \frac{1}{n^2}\right) = -13.6 \text{ eV}\left(\frac{1}{n^2} - \frac{1}{2^2}\right)$$

This is exactly the result from the Bohr model with $n_i = n$ and $n_f = 2$.

Because there are infinitely many energy states (n can be any positive integer), there are infinitely many possible transitions. However, only four of these transitions emit a visible wavelength photon. The 3 to 2 transition is one. The others are 4 to 2, 5 to 2, and 6 to 2. All transitions from states 7 and higher to 2 result in ultraviolet photons. Any transition to the 1 state also results in an ultraviolet photon.

Notice that the photon emitted by the 3 to 2 transition closely matches the 656-nm line shown in the hydrogen line spectrum in Figure 28.5. The 4 to 2, 5 to 2, and 6 to 2 transitions produce the other visible light spectral lines in Figure 28.5.

Limitations of the Bohr model

We saw in Example 28.2 that the wavelengths of the hydrogen spectral lines predicted by Bohr's model are in good agreement with the observational evidence. Unfortunately, the model does not provide predictions that account for the spectral lines emitted by other atoms. It does, however, work well if the atoms are ionized and have only one electron (such as singly ionized helium, He^+, or doubly ionized lithium, Li^{2+}).

Also, note that Bohr's model arbitrarily imposes restrictions on the motion of the electron (its quantized rotational angular momentum), which results in a restriction on the allowed energy states of the atom. However, Bohr's model provides no explanations for these restrictions. Therefore, physicists continued searching for a new model of the atom that could provide them.

REVIEW QUESTION 28.2 Why is an atom's total energy negative? Why does this make sense?

28.3 Spectral analysis

Bohr's model, as you have seen, helps us understand the **emission spectra** of gases (see the examples of spectra in **Figure 28.12**). These spectra allow scientists to analyze the chemical composition of different materials according to the light they emit. So far we have discussed only the emission spectrum of hydrogen. We can understand the spectra of other atoms and molecules in a qualitatively similar way.

FIGURE 28.12 Emission spectra of elements in a gaseous state. Each type of atom has a unique spectrum, allowing its identification.

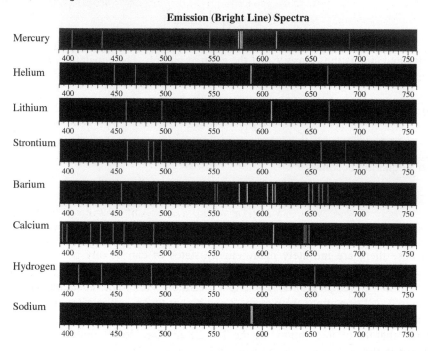

When an atom transitions from a higher energy state to a lower energy state, a photon of a specific wavelength is emitted. Thus, to observe an emission spectrum, some mechanism is needed to put the atom in an excited state. We observed the spectra of heated gases and gases placed in electric fields. What are the mechanisms that explain their glow in these conditions?

Thermal excitation

Heating can significantly increase the random kinetic energy of atoms (**Figure 28.13a**). If the kinetic energies of the atoms are high enough, a collision can cause one or both of the atoms to transition to an excited state (Figure 28.13b). When the atom returns to a lower energy state, it emits a photon of a characteristic wavelength (Figure 28.13c). The bright yellow, red, and blue colors of fireworks come from photon emission by excited sodium, strontium, and copper compounds, respectively.

FIGURE 28.13 Thermal excitation and production of a photon.

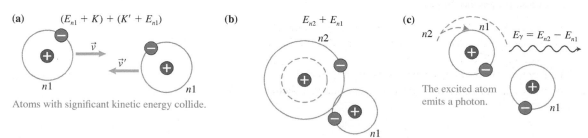

Discharge tube excitation

Another process that causes atoms to transition to higher energy states involves filling a discharge tube similar to a cathode ray tube with gas. There are always some free electrons (or other charged particles) present in our atmosphere due to cosmic rays. These free electrons are accelerated across an electric potential difference from the cathode to the anode (**Figure 28.14a**). The electrons then collide with atoms of the gas in the tube, causing the atoms to reach excited states (Figure 28.14b). When the excited atoms return to the ground state, they emit photons whose energy equals the difference in the energy of the atom between its initial and final states (Figure 28.14c). Looking at a discharge tube, you see a glow that appears to be of a single color. But if you look at the light through a grating, you see a line spectrum.

FIGURE 28.14 A discharge tube produces light.

(a)

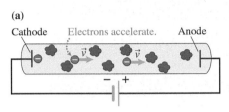

Cathode   Electrons accelerate.   Anode

(b)

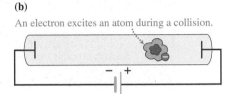

An electron excites an atom during a collision.

(c)

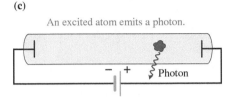

An excited atom emits a photon.

Photon

Figure 28.15 shows what you see if you observe a mercury discharge tube through a grating that has 500 lines per millimeter. The blue central maximum shows the color of the tube that you see with the naked eye. Different color lines on both sides are the 1st order maxima of the brightest lines that together produce the color of the central maximum.

FIGURE 28.15 Mercury spectrum as seen through a grating.

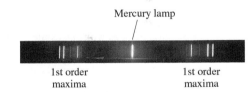

Mercury lamp

1st order maxima   1st order maxima

EXAMPLE 28.3   **Discharge tube excitation of hydrogen**

Use the results of Example 28.2 to estimate the potential difference needed between the cathode and anode in a discharge tube filled with hydrogen to produce hydrogen's visible light spectral lines.

Sketch and translate An illustration of the gas in the discharge tube is shown in Figure 28.14a–c. For the atom to emit a visible photon, it must transition from the $n = 3, 4, 5,$ or 6 state to the $n = 2$ state. Therefore, the colliding electrons must have enough kinetic energy to induce the transition of the electrons in the hydrogen atoms from the $n = 1$ ground state to the $n = 3, 4, 5,$ or 6 state. Because the transition from $n = 1$ to $n = 6$ requires the most energy, the potential difference between cathode and anode must be high enough to induce that transition. The system is the colliding electron, the electrodes and the electric field they produce, the hydrogen atoms, and the emitted photons.

The overall process has three parts.

I. The electric potential energy of the system is converted into the kinetic energy of the electron, as shown in part (a) of the figure on the top of the right column.
II. The electron collides with and excites the atom [part (b)].
III. The excited atom returns to the ground state, emitting a photon [part (c)].

(a)

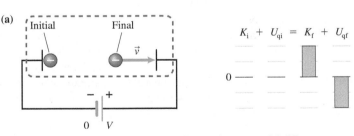

$$K_i + U_{qi} = K_f + U_{qf}$$

The electron gains kinetic energy while crossing a potential difference.

(b)

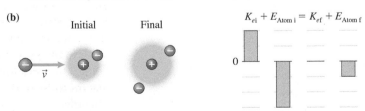

$$K_{ei} + E_{Atom\ i} = K_{ef} + E_{Atom\ f}$$

The electron excites an atom when it collides with it.

(c)

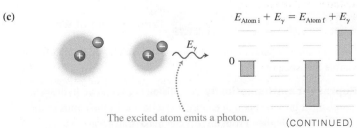

$$E_{Atom\ i} + E_\gamma = E_{Atom\ f} + E_\gamma$$

The excited atom emits a photon.

(CONTINUED)

Simplify and diagram Assume first that the electron's initial kinetic energy is zero and that it increases only as a result of the decrease in electric potential energy of the system. Assume also that the electric potential at the initial position of the electron is 0 V. Before the colliding electron encounters the hydrogen atom, assume the atom is in the ground state and that the Bohr model is a reasonable way to model the atom. Lastly, assume the electron is not moving relativistically. Energy bar charts shown earlier represent the three stages of this process.

Represent mathematically We can use the bar charts to help us represent these three processes mathematically.

Part I: In the initial state, the electron is at rest at zero electric potential. In the final state, the electron has positive kinetic energy and the system's electric potential energy has decreased.

$$K_i + U_{qi} = K_f + U_{qf}$$

Using the expressions for nonrelativistic kinetic energy and electric potential energy:

$$0 + q_e V_i = \tfrac{1}{2} m_e v^2 + q_e V_f$$

Subtracting $q_e V_i$ from both sides, we have

$$0 = \tfrac{1}{2} m_e v^2 + q_e V_f - q_e V_i$$

$$= \tfrac{1}{2} m_e v^2 + q_e \Delta V = \tfrac{1}{2} m_e v^2 - e \Delta V$$

Part II: The electron collides with and excites an atom from the $n = 1$ to the $n = 6$ state. In the initial state, the colliding electron has considerable kinetic energy, and the atom is in the $n = 1$ ground state. In the final state, the colliding electron stops, and the atom is in the $n = 6$ excited state.

$$K_i + E_{\text{Atom i}} = K_f + E_{\text{Atom f}}$$

Inserting the expressions for kinetic energy and the total energy of the hydrogen atom:

$$\tfrac{1}{2} m_e v^2 + \left(\frac{-13.6 \text{ eV}}{n_i^2} \right) = 0 + \left(\frac{-13.6 \text{ eV}}{n_f^2} \right)$$

Part III: The excited atom emits a photon and transitions to the $n = 2$ state. Initially, the atom is in the $n = 6$ state. In the final state, the atom is in the $n = 2$ state and a photon has been emitted.

$$E_{\text{Atom i}} = E_{\text{Atom f}} + E_\gamma$$

Insert the expressions for the total energy of the atom and the energy of the photon:

$$\left(\frac{-13.6 \text{ eV}}{n_i^2} \right) = \left(\frac{-13.6 \text{ eV}}{n_f^2} \right) + \frac{hc}{\lambda}$$

Solve and evaluate We are interested in finding ΔV. Using our mathematical representation of Part I and solving for ΔV:

$$\Delta V = \frac{1}{e} \left(\tfrac{1}{2} m_e v^2 \right)$$

We can determine the kinetic energy from Part II. Solving for $(1/2) m_e v^2$ gives

$$\tfrac{1}{2} m_e v^2 = \left(\frac{-13.6 \text{ eV}}{n_f^2} \right) - \left(\frac{-13.6 \text{ eV}}{n_i^2} \right)$$

$$= (-13.6 \text{ eV}) \left(\frac{1}{n_f^2} - \frac{1}{n_i^2} \right)$$

Substituting this into the expression for ΔV:

$$\Delta V = \frac{1}{e} (-13.6 \text{ eV}) \left(\frac{1}{n_f^2} - \frac{1}{n_i^2} \right)$$

Inserting the appropriate values and converting to joules:

$$\Delta V = \frac{1}{1.6 \times 10^{-19} \text{ C}} (-13.6 \text{ eV}) \left(\frac{1}{6^2} - \frac{1}{1^2} \right) \left(1.6 \times 10^{-19} \frac{\text{J}}{\text{eV}} \right)$$

$$= 13.2 \text{ V}$$

This value is the smallest potential difference that the electrons need to cross in order to acquire enough kinetic energy to excite hydrogen atoms to the maximum $n = 6$ state so that they can emit photons in the visible spectrum. The potential difference we found is not the potential difference across the discharge tube, but rather the potential difference across the average distance that the colliding electrons travel between two consecutive collisions. The required potential difference across the cathode and anode of the tube is much greater than this.

- -

Try it yourself Estimate the minimum potential difference required to allow the hydrogen in the tube to emit photons from the $n = 4$ to the $n = 2$ transition.

Answer   12.8 V.

Another way to produce visible light is by the collisions of hot gas atoms with each other. In order for visible light to be produced, the collisions must cause one of the atoms to become excited to the $n = 3$ state or higher. How hot does the gas need to be in order for this process to occur?

| QUANTITATIVE EXERCISE 28.4 | Temperature at which an H atom emits visible photons |

Estimate the temperature of hydrogen gas at which you might observe visible line spectra.

Represent mathematically A collision between two hydrogen atoms must provide enough energy to excite one of the atoms to the $n = 3$ state or higher. From there it can make transitions to the $n = 2$

state and emit visible photons. Recall that the average kinetic energy (thermal energy) of a gas particle at absolute temperature T is given by this expression (from Section 12.4):

$$\overline{K} = \tfrac{3}{2} k_B T$$

Here $k_B = 1.38 \times 10^{-23}$ J/K, Boltzmann's constant. A hydrogen atom in the ground state ($n = 1$) with energy E_1 needs additional energy ΔE for it to be excited to the state with energy E_3 or higher. Choosing the system of interest to be one of the colliding hydrogen atoms, we can use the idea of energy conservation to represent this process mathematically:

$$E_1 + \Delta E = E_3$$

Assume that ΔE equals the average kinetic energy of a single hydrogen atom, meaning one of the colliding particles converts all of its kinetic energy to excite the other. E_1 and E_3 are the energies of the atom in the indicated states:

$$E_n = \frac{-13.6 \text{ eV}}{n^2}$$

Solve and evaluate Putting the above equations together, we have

$$\frac{-13.6 \text{ eV}}{1^2} + \tfrac{3}{2} k_B T = \frac{-13.6 \text{ eV}}{3^2}$$

Solving for T, inserting the appropriate values, and converting to joules, we get

$$T = \frac{2(-13.6 \text{ eV})}{3(1.38 \times 10^{-23} \text{ J/K})} \left(\frac{1.6 \times 10^{-19} \text{ J}}{1 \text{ eV}} \right) \left(\frac{1}{3^2} - \frac{1}{1^2} \right)$$

$$= 93{,}000 \text{ K}$$

At this temperature, collisions will excite almost all of the atoms into the $n = 3$ state. However, even at lower temperatures, some will be excited to the $n = 3$ state (remember the Maxwell distribution). It is clearly easier to excite atoms using an electric field to accelerate free electrons in a cathode ray tube (which then collide with the atoms) than it is to raise the temperature of all of the atoms.

Try it yourself Estimate the temperature at which an average hydrogen atom is ionized.

Answer

At a temperature of 10^5 K or less, the collisions between atoms are violent enough to easily transfer the 13.6 eV of energy needed to ionize an atom.

The "Try it yourself" question of the last exercise indicates that the temperature at which hydrogen ionizes as a result of collisions with other atoms is almost the same as the temperature needed for hydrogen atoms to become excited to higher energy states so they can emit visible light. Ionized hydrogen gas no longer emits spectral lines, instead emitting a continuous spectrum. This is because the energy states of free electrons can have any positive value of energy, so when they combine with protons to form neutral hydrogen again, the emitted photons do not have discrete wavelengths. A gas with most of its atoms ionized due to collisions is called a *high-temperature plasma*. The cores of stars are at millions of kelvins and consist mostly of high-temperature plasma. Gas can also be ionized if placed in a strong electric field, producing a *low-temperature plasma* such as exists in neon lights or discharge tubes.

Continuous spectra from stars

The Sun consists almost entirely of hydrogen and helium atoms, so we would expect its light to consist primarily of hydrogen and helium emission lines. Instead, light from the Sun and many other high-temperature objects is close to a continuous black body spectrum that depends only on the object's surface temperature (**Figure 28.16**).

The continuous spectrum from these objects must depend on the distribution of the random kinetic energy (thermal energy) and temperature of the surface particles and not on their atomic compositions. On the hot Sun (5700 K on the surface), the material is dense and the atoms are continually colliding with each other. Evidently, these collisions distort the shapes of atoms during the collisions and excite them in unpredictable ways so that the light emitted when they de-excite is characteristic of the temperature of the gas and not of the types of atoms in the gas. These continuous (black body) spectra have the same shape for all objects at a particular temperature, but they differ significantly for objects at different temperatures (Figure 28.16).

Other stars emit similar black body radiation with the peak intensity at a wavelength that depends on their temperatures. However, if you look closely at radiation from the Sun and other stars, you discover important details. Observational Experiment Table 28.2 (on the next page) helps us analyze them.

FIGURE 28.16 As the temperature increases, the black body radiation increases in intensity and the peak moves toward shorter wavelength.

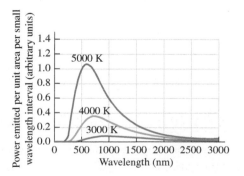

OBSERVATIONAL
EXPERIMENT TABLE 28.2

Stellar spectra

Observational experiment	Analysis
We observed a spectrum of the Sun using a handheld spectrometer and then recorded the same spectrum using a computer-controlled light detector.	The figure shows a photo of the spectrum from the first experiment and the spectral curve produced by the computer in the second experiment. In the visible region, we see a continuous spectrum but with several dark lines. The lines look dark in the photo, but the matching spectral curve (intensity versus wavelength) shows they are relative reductions in brightness: less light is emitted at particular wavelengths compared to the continuous spectrum. 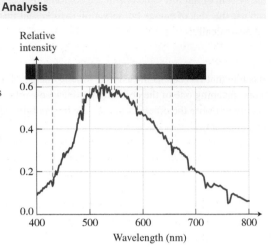
	If we look at a narrow part of the visible spectrum of the Sun, we see dips in the intensity at particular wavelengths of the intensity-versus-wavelength graph. The wavelengths of the dips in the solar spectrum match the wavelengths of the line spectra emitted by a sodium vapor lamp in a lab. Intensity is shown in relative units. You can see that light from the lamp is much brighter than light from the Sun. 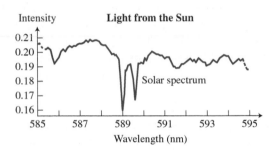 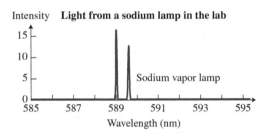

Pattern

If we observe a narrow region (516.6–517.4 nm) of the solar spectrum, we see dips that match the wavelengths of the line spectra of specific elements (iron, magnesium, etc.). The spectra of other stars are similar—continuous emission spectra with specific dark lines.

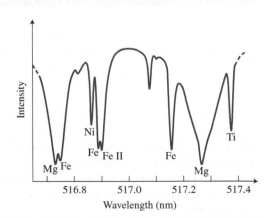

Why would photons be missing at particular wavelengths? The first explanation that might come to mind is that sodium is not present on the surface of the Sun. However, this explanation does not work, because we find dark lines matching the emission lines for almost all elements known on Earth.

Another explanation is that the dark lines are not the result of missing elements, but of the absorption of photons traveling from the Sun to Earth. Imagine that some sodium atoms in the ground state exist in the region between the Sun and the observer on Earth. Sodium atoms can absorb those photons coming from the Sun's surface whose energies match possible energy transitions in the atoms (see Figure 28.10). Soon after absorption, the atoms reemit these photons but in random directions. Thus the original photons that were absorbed would be "missing" from the continuous spectrum arriving at Earth. If we take a picture of the arriving spectrum, it should have dark lines at the locations of the missing photons. This is known as an **absorption spectrum**—a continuous spectrum with missing photons of specific wavelengths. This explanation is currently accepted as the correct one for the absorption spectrum of the Sun and other stars.

Satellite spectrometers above Earth's atmosphere show the same dark lines in the Sun's spectrum that we observe on Earth. Thus, the absorption must occur just outside the Sun's hot surface, indicating the presence of different types of atoms in the gas surrounding the Sun.

Spectral analysis is used in chemistry, engineering, transportation, medicine, and other technological fields. The main principles for these applications are the same: capture and analyze the wavelengths of the photons, and compare their wavelengths to the wavelengths of the known elements emitted or absorbed by a particular object to determine its composition.

Why are plants green?

Light is needed for the process of photosynthesis in plants. The process starts when pigments absorb energy delivered by light photons. Chlorophyll extracted from soft leaves absorbs light most strongly in the blue portion of the electromagnetic spectrum, followed by the red portion. It is a poor absorber of green and near-green portions of the spectrum, which it reflects, producing the green color of chlorophyll-containing tissues. **Figure 28.17** shows the percentage of light energy absorbed by chlorophyll molecules at the wavelengths of red and blue light. The peaks are essentially inverted dips on an absorption spectrum that you are familiar with. Notice two plots, marked chlorophyll a and chlorophyll b. These represent two chemical variants of chlorophyll that are universal constituents of wild vascular plants. Both chlorophylls show absorption maxima at wavelengths corresponding to blue and red.

FIGURE 28.17 Absorption of light by chlorophyll.

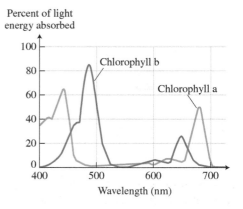

REVIEW QUESTION 28.3 What materials and measuring instruments do you need in order to observe the absorption spectrum of a gas?

28.4 Lasers

When excited gas atoms emit light, they produce photons at different times and in different directions; thus the corresponding waves are incoherent (have different phases) and spread out in many directions. This process is called **spontaneous emission**. Lasers, on the contrary, emit very narrow, almost coherent beams of light. How is this accomplished?

In a laser, many excited atoms return from the same excited state to the same ground state simultaneously. The result is emission of in-phase photons traveling in the same direction. This process, called **stimulated emission**, was first suggested by Einstein. The word *laser* is an acronym for "light amplification by stimulated emission of radiation."

How lasers work

Imagine that you have an atom in an excited state. You also have a photon whose energy equals exactly the energy difference between the atom's excited state and the ground state (**Figure 28.18a**). This photon can cause the excited atom to "de-excite" and emit a photon of exactly the same frequency (Figure 28.18b)—the stimulated emission mentioned above. The passing photon does not lose any energy; it only triggers the process. The result is two photons of the same frequency traveling in the same direction.

FIGURE 28.18 Stimulated emission and production of laser light. (a) In the initial state, a photon whose energy exactly equals the difference in energy between the excited and ground states passes through, stimulating the excited atom. (b) In the final state, the excited atom's electron drops to the lower energy state, emitting a photon. Now two synchronized photons have the same energy.

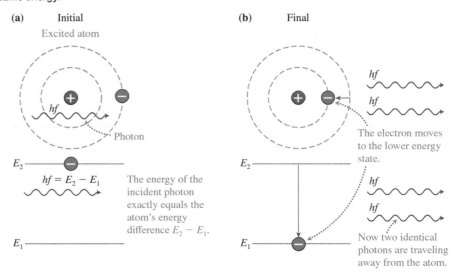

(a)   Initial

Excited atom

hf

Photon

E_2

$hf = E_2 - E_1$ The energy of the incident photon exactly equals the atom's energy difference $E_2 - E_1$.

E_1

(b)   Final

hf

hf

The electron moves to the lower energy state.

E_2

hf

hf

E_1

Now two identical photons are traveling away from the atom.

FIGURE 28.19 Steps leading to the production of a laser beam.

(a) Excitation

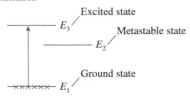

Excited state

E_3

Metastable state

E_2

Ground state

E_1

(b) Transition to metastable state

E_3

E_2

E_1

(c) An inverted population

E_3

E_2

Transition from state 2 to 1 is unlikely.

E_1

(d) An unlikely spontaneous transition from 2 to 1 starts the laser beam.

E_3

E_2

E_1

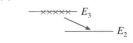

(e) A stimulated transition adds another photon to the beam.

E_3

E_2

E_1

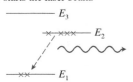

To make stimulated emission effective, many more atoms must be in the same excited state than in the ground state, a situation known as **population inversion**. It is generally difficult to create population inversion because the average time that an atom spends in an excited state is very short (about 10^{-8} s). However, some atoms enter excited states called **metastable states**, in which they can remain for a significantly longer time (about 10^{-3} s). These atoms are good candidates for population inversion.

Suppose that the atoms in a material make the transition from ground state 1 to excited state 3 (**Figure 28.19a**) by absorbing photons. Instead of quickly returning to state 1, they make a more probable spontaneous transition to a metastable state 2 (Figure 28.19b). These atoms will remain in that state for a relatively long time since transitions from state 2 to state 1 are comparatively unlikely to occur. Thus, more atoms will be in state 2 than in the ground state 1 (Figure 28.19c). This is an example of population inversion.

One of these excited atoms eventually transitions to ground state 1 and emits a photon of energy $E_2 - E_1$ (Figure 28.19d). That photon passes through the material, many of whose atoms are in excited metastable energy state 2. This photon causes another atom in the metastable state 2 to de-excite via stimulated emission, resulting in two synchronized photons moving through the material (Figure 28.19e). Each of these photons can then stimulate another transition, resulting in a total of four photons. This process continues and creates an avalanche of new photons by stimulated emission—an intense laser beam (**Figure 28.20**).

The *lasing* material is placed between two mirrors that reflect the emitted photons back and forth through the material. In continuous lasers, one mirror is only partially reflecting and allows photons to leave the lasing region at a nearly constant rate.

FIGURE 28.20 Stimulated emission produces an increasing number of in-phase photons in the laser cavity.

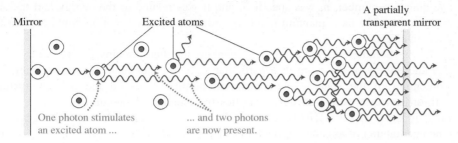

Continual excitation of the atoms back to the metastable state provides a continuous source of stimulated emission photons.

Lasers can be used to transmit thousands of simultaneous telephone calls along thin glass fibers, read DVDs, scan bar codes, and remove certain kinds of tumors. Lasers are also used surgically to correct nearsightedness and farsightedness. All such procedures, including LASIK (laser-assisted in situ keratomileusis), PRK (photorefractive keratectomy), and other corneal ablation methods of vision correction, employ lasers to remove corneal tissue to reshape the eye so that it can focus an image directly on the retina.

Another important application is lidar (light detection and ranging). Lidar can be thought of as radar in which, instead of pulses of radio waves, pulses of laser light are used to measure the distance to an object. The differences in time of the return of the pulses and the changes in their wavelengths can be used to make three-dimensional images of the object. Lidar is used to make high-resolution maps.

REVIEW QUESTION 28.4 What is the main difference between spontaneous and stimulated emission of light?

28.5 Quantum numbers and Pauli's exclusion principle

Bohr's model was successful in explaining certain features of hydrogen's spectrum and structure, but it could not explain the structure of multi-electron atoms or some of the fine details of the hydrogen spectrum. We will describe one of these details next.

The Zeeman effect

In 1896 Dutch physicist Pieter Zeeman performed an experiment in which light from a known source passed through a region with sodium atoms in a strong magnetic field and then through a grating. He observed that when the magnetic field was on, a particular sodium spectral line was split into three closely spaced lines. The result of a similar experiment using mercury vapor is shown in **Figure 28.21**. The figure shows the split for the 404.7-nm spectral line in a 2.9-T magnetic field. The split is very small—about 1 nm. This phenomenon, which became known as the Zeeman effect, could only be explained after Bohr's model was extended.

Sommerfeld's extension of the Bohr model and additional quantum numbers

In 1915 Arnold Sommerfeld, a German theoretical physicist, modified Bohr's model to account for the Zeeman lines. Later, his modifications helped explain the spectral lines of more complex atoms. One of his innovations was the suggestion that electrons

FIGURE 28.21 Zeeman split of the 404.7-nm line of mercury in a 2.9-T magnetic field.

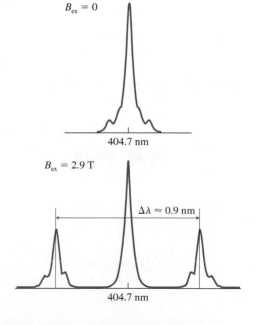

could have elliptical orbits. He proposed this through the introduction of two new quantum numbers related to the motion of the electron in the atom. (Prior to this, only one quantum number, n, was specified, and it was related to the energy and the size of the atom.) The new quantum numbers and the associated motion could explain the Zeeman effect.

In the Sommerfeld model, the n quantum number remains an indicator of the size and energy of the atom in a particular state, but not of the rotational (angular) momentum of the state. A new quantum number l, called the **orbital angular momentum quantum number**, identifies the angular momentum of a state whose value is $L = \sqrt{l(l+1)}(h/2\pi)$, where $l = 0, 1, 2, \ldots, n-1$ and $h/2\pi$ acts as a fundamental unit of angular momentum (see **Figure 28.22**).

FIGURE 28.22 The allowed electron $n = 1$ to 4 orbits in the Sommerfeld model.

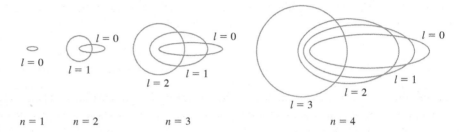

For the $n = 1$ energy state, l can only equal zero. Thus the angular momentum of that state could only be $L = \sqrt{0(0+1)}(h/2\pi) = 0$. For the $n = 2$ state, l can be zero or 1, resulting in two possible states $L = 0$ or $L = \sqrt{2}(h/2\pi)$. These two states correspond to electron orbits that look very different (see the shapes of the different n and l orbits in Figure 28.22). The orbit of the $l = 0$ state is that of an elongated, flat ellipse. Recall that the angular momentum of an object moving in an orbit is $L = mrv \sin \theta$, where θ is the angle between the r and v vectors. When the orbit is elongated, r and v are nearly parallel or antiparallel; therefore, θ is zero. Thus, $L = 0$. For the more circular higher L orbit, r and v are perpendicular and $\sin \theta$ is approximately 1.

According to the Sommerfeld model, in an $n = 2$ state, an electron zipping around the $l = 1$ orbit is the equivalent of an electric current loop, which therefore produces a magnetic dipole field. The elongated $l = 0$ orbit should not produce any magnetic field. This differs from Bohr's model, in which an electron in the $n = 1$ ground state has angular momentum and moves in a circular orbit, thus producing a circular current and a magnetic field. Sommerfeld's prediction is consistent with experimental results, which show that the ground state of hydrogen does not have this magnetic field. But how does the Sommerfeld model help explain the Zeeman effect?

Consider hydrogen. The electrons in most hydrogen atoms in a low-temperature gas are in the $n = 1$ ground state, the orbit nearest the nucleus. Suppose this cool gas is irradiated by photons of just the right frequency to excite an atom from the $n = 1$ state to one of the two possible $n = 2$ states of the atom. If the gas is placed in a strong magnetic field, the atoms in states with nonzero angular momentum would behave like tiny magnets (with their own dipole moments) and interact with the magnetic field. This would result in the slightly different energies of the two $n = 2$ states. The stronger the magnetic field, the greater the difference in energies of the states.

Zeeman's experiments on hydrogen gas using sensitive equipment and strong magnetic fields $(\vec{B}_{ex})$ produced three distinct spectral lines for transitions from the $n = 2$, $l = 1$ state to the $n = 1$, $l = 0$ state. This result could be explained if the $\vec{B}$ field due to the electron's motion in the $n = 2$, $l = 1$ state could be oriented in three discrete directions (see **Figure 28.23a**). This caused the $n = 2$, $l = 1$ state to split into three separate states (Figure 28.23b) due to different orientations of the electron orbit relative to the external magnetic field. The stronger the external magnetic field, the greater the splitting.

This reasoning led to the introduction of a third quantum number, the **magnetic quantum number** m_l. If this quantum number is allowed to have values

FIGURE 28.23 Explanation for the Zeeman effect.

(a)

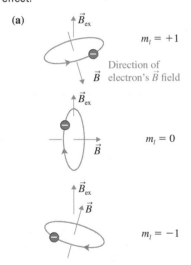

(b) Energy shifts due to the orientation of the magnetic dipole in the magnetic field

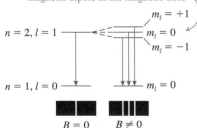

(c)

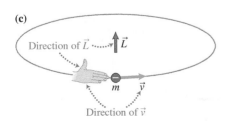

$m_l = 0, \pm 1, \pm 2, \ldots, \pm l$, then the Zeeman effect for the $n = 2$, $l = 1$ to $n = 1$, $l = 0$ transition is explained: three slightly different energy photons would be emitted, as shown in Figure 28.23b. Apparently, m_l is related to the orientation of the angular momentum vector of the electron in its orbit (the method for determining the orientation of the orbit's angular momentum is shown in Figure 28.23c). The component L_z of that angular momentum in the direction of $\vec{B}_{ex}$ is typically defined as the z-axis. L_z is then equal to m_l times the fundamental unit of angular momentum $h/2\pi$, or

$$L_z = m_l \frac{h}{2\pi} \tag{28.10}$$

Understanding the splitting of spectral lines in a magnetic field led to many applications, specifically to studying different magnetic fields. By analyzing the wavelengths of the spectral lines in a magnetic field (like the split shown in Figure 28.21), scientists determined the magnitude of the $\vec{B}$ field of the Sun: about 4 T, 100,000 times stronger than the magnetic field of Earth.

Another difficulty with observed spectral lines

At about the same time that Zeeman made his observations, Thomas Preston performed Zeeman effect experiments with an even better magnet. He observed that some of the spectral lines for many elements did not split into three lines when a magnetic field was applied, but instead split into four, six, or more lines. This became known as the "anomalous" Zeeman effect. Sommerfeld's model could not explain it.

Around 1920 Sommerfeld and Alfred Landé suggested that the nucleus was affecting the angular momentum of the electron, thus producing this anomalous Zeeman effect. They incorporated this idea by assigning a quantum number to the nucleus, an ad hoc mathematical proposal with little physical model in mind.

Otto Stern and Walther Gerlach decided to test this idea by passing moving silver atoms through a nonhomogeneous magnetic field (**Figure 28.24**). A nonhomogeneous field was needed because the force that the magnetic field exerts on a stationary magnetic dipole differs from the force the field exerts on a single moving charged particle. This force is proportional to the rate of change per unit distance of the $\vec{B}$ field as opposed to just the magnitude of the $\vec{B}$ field. The combination of a pointed north pole of a magnet and a flat south pole produces a large rate of change in the vertical direction (Figure 28.24). Moving atoms were needed to make the split on the screen visible: the magnetic field would exert forces on two dipoles in opposite directions. If the Sommerfeld and Landé models were correct, then moving silver atoms should split into two streams, corresponding to the two values for the quantum number of the nucleus, rather than scatter randomly (the classical physics prediction). The experimental outcome matched the prediction based on the Sommerfeld and Landé model. However, even when a testing experiment matches the prediction, the model can still be wrong. Read on.

FIGURE 28.24 Stern-Gerlach experiment.

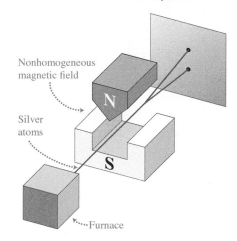

Nonhomogeneous magnetic field

N

Silver atoms

S

Furnace

A new fourth quantum number

In 1923–1924, Wolfgang Pauli disproved the Sommerfeld and Landé model using the theory of relativity. With relativistic corrections, the Landé and Sommerfeld model consequently made predictions that were contradicted by the Stern-Gerlach experiments.

In trying to explain the anomalous Zeeman effect, Pauli uncovered many unexplained patterns in the periodic table of the elements. Why did elements in the same column (group) have similar properties when they had very different numbers of electrons but very different properties from adjacent columns? Why didn't the properties of atoms change gradually with additional electrons? Why were atoms with an even number of electrons less chemically active than those with odd numbers? Why were elements in the periodic table with 2, 8, or 18 electrons almost chemically inert?

Finally, and perhaps most importantly, why at low temperatures weren't nearly all electrons in multi-electron atoms in the $n = 1$ ground state?

Pauli realized that these patterns in the periodic table and the anomalous Zeeman effect could be explained by a single, simple rule, namely, that each available state in the atom could hold at most *one* electron, an idea called the **Pauli exclusion principle**. This principle required the introduction of a fourth quantum number, but a different one from that of the Sommerfeld and Landé model. The fourth quantum number concerned the electron itself and was called m_s, the **spin quantum number**, because it was related to the electron's intrinsic angular momentum (called **spin**). Pauli reasoned that in a magnetic field, an electron's spin could orient along the direction of the external field $\vec{B}_{\text{ex}}$ or opposite to the external field $\vec{B}_{\text{ex}}$, but not in any other direction—like a compass needle that could point only precisely north or south. These two electron spin orientations correspond to values of $\pm 1/2$ for the quantum number m_s.

By suggesting this additional electron property, Pauli could explain the outcome of the Stern-Gerlach experiment and account for the anomalous Zeeman effect. Together with the exclusion principle, which states that the allowed states of an atom could be completely characterized by the four quantum numbers n, l, m_l, and m_s, and that only one electron could exist in each state, Pauli's ideas could account for many of the patterns observed in the periodic table (we will discuss the patterns in the periodic table in Section 28.7).

A failed attempt at a classical explanation for the spin quantum number

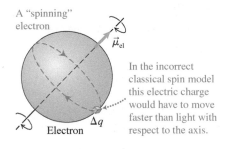

FIGURE 28.25 An incorrect classical model of electron spin.

A "spinning" electron

$\vec{\mu}_{\text{el}}$

In the incorrect classical spin model this electric charge would have to move faster than light with respect to the axis.

Δq

Electron

In 1925 Samuel Goudsmit and George Uhlenbeck attempted to explain the physical meaning of the m_s quantum number. For the electron to have angular momentum, it would have to be like a tiny spinning sphere, similar to Earth spinning about its axis (**Figure 28.25**). However, small charged parts Δq at the edge of this small spinning spherical shell of charged material would have to move faster than light speed to have the necessary angular momentum. This assumption contradicts special relativity. The spin of the electron seemed to be some sort of new property that could not be understood using classical physics.

Summary of quantum numbers

By 1925 scientists had four quantum numbers that could describe the state of an electron in an atom. Also, Pauli's exclusion principle accounted for the distribution of electrons in multiple-electron atoms. Let's summarize the four quantum numbers:

1. *Principal quantum number n* can be any positive integer:

$$n = 1, 2, 3, \dots \tag{28.11}$$

and it indicates the energy of the electron in the atom. The larger the value of n, the larger the energy of the atom.

2. *Orbital angular momentum quantum number l* characterizes the orbital angular momentum of the electron. For a particular value of n, the allowed l values are

$$l = 0, 1, 2, \dots, n - 1 \tag{28.12a}$$

The magnitude of the angular momentum of the electron is

$$L = \sqrt{l(l + 1)}\, \frac{h}{2\pi} \tag{28.12b}$$

3. *Magnetic quantum number m_l* describes the component of the electron's orbital angular momentum along an external magnetic field (Figure 28.23a). The energy of each l state in an external magnetic field splits according to the state's m_l value, which results in the additional lines in the spectra observed by Zeeman

(Figure 28.23b). The magnetic quantum number m_l can have values that depend on the quantum number l:

$$m_l = 0, \pm 1, \pm 2, \ldots, \pm l \tag{28.13a}$$

The component of the angular momentum in the z-direction is

$$L_z = m_l(h/2\pi) \tag{28.13b}$$

4. *Spin magnetic quantum number m_s* Each electron has an intrinsic spin and corresponding magnetic dipole. The spin magnetic quantum number m_s indicates the two directions that this spin dipole can be oriented relative to an external magnetic field:

$$m_s = \pm \tfrac{1}{2} \tag{28.14}$$

The spin quantum number doubles the number of states available for an electron in an atom.

Pauli's exclusion principle states that there is a relationship between these quantum numbers and the electrons in an atom, regardless of the type of atom.

Pauli exclusion principle Each electron in an atom must have a unique set of quantum numbers n, l, m_l, and m_s.

Consider the case of an $n = 2$ state for electrons in an atom. The possible values of l are 0 and 1. For $l = 0$, the only possible value of m_l is 0. For $l = 1$, the possible values of m_l are -1, 0, and 1. For each of these possibilities, m_s can be $\pm 1/2$. Thus, there are eight possible $n = 2$ states, as listed in **Table 28.3**.

An electron with $n = 2$ is said to be in the $n = 2$ shell of the atom. If a particular type of atom has eight $n = 2$ electrons, the atom is said to have a filled $n = 2$ shell—the shell can hold no more electrons.

You can use similar reasoning for other electron shells of atoms. You should find that the $n = 3$, $l = 2$ subshell of an atom can hold 10 electrons. The $n = 3$, $l = 1$ subshell can hold 6 electrons, and the $n = 3$, $l = 0$ subshell can hold 2 electrons. Thus, the $n = 3$ shell can hold a maximum of 18 electrons.

Pauli's ideas of atomic structure explain the possible states of electrons in atoms, and also the reason that not all electrons in multi-electron atoms are in the lowest energy state (the $n = 1$ state). However, these ideas do not explain how to calculate the values of the quantities that are based on these quantum numbers (energy, orbital angular momentum, etc.). Physicists needed to construct an entirely new theory for describing this subatomic world, the subject of the next section.

TABLE 28.3 The possible $n = 2$ states of an atom

$l = 0, m_l = 0, m_s = +1/2$
$l = 0, m_l = 0, m_s = -1/2$
$l = 1, m_l = -1, m_s = +1/2$
$l = 1, m_l = -1, m_s = -1/2$
$l = 1, m_l = 0, m_s = +1/2$
$l = 1, m_l = 0, m_s = -1/2$
$l = 1, m_l = +1, m_s = +1/2$
$l = 1, m_l = +1, m_s = -1/2$

REVIEW QUESTION 28.5 In what energy states are the 10 electrons in the atom neon? Label each by its set of four quantum numbers.

28.6 Particles are not just particles

While the work of Bohr, Sommerfeld, and Pauli explained many observed properties of atoms, it did not explain why quantum numbers existed. Why were the energies of atoms and the orbital and spin angular momenta of electrons quantized in the first place? Physicists began arriving at answers in the mid-1920s.

A proposal for the wave nature of particles

The theoretical foundation for this effort began in 1924 with a young French physicist, Louis de Broglie. Knowing that light had both wave and particle behaviors, he thought it seemed reasonable that the elementary constituents of matter might also have these behaviors. Given that the wavelength λ of a photon is $\lambda = h/p_{\text{photon}}$, de Broglie suggested that elementary particles also had a wavelength that could be determined in a similar way:

$$\lambda = \frac{h}{p} \tag{28.15}$$

where h is Planck's constant and p is the particle's momentum (which could potentially be relativistic).

If matter has wave-like properties, then why don't macroscopic objects such as human beings have them as well? Using Eq. (28.15), the wavelength of a 50-kg person walking at speed 1 m/s would be 1.33×10^{-35} m. This wavelength is a hundred billion billion times smaller than the size of the nucleus of an atom. Thus, there is no real-world situation that would cause a macroscopic object to exhibit wave-like behavior. However, the smaller the mass of the object, the larger its wavelength. For example, for an electron (mass $m = 9.11 \times 10^{-31}$ kg) moving at a speed of 1000 m/s (a typical speed for electrons in cathode ray tubes), its wavelength would be

$$\lambda = \frac{6.63 \times 10^{-34}\ \text{J} \cdot \text{s}}{(9.11 \times 10^{-31}\ \text{kg})(1000\ \text{m/s})} = 7.3 \times 10^{-7}\ \text{m} = 730\ \text{nm}$$

This is just a little longer than the wavelength of red light. Thus, the wave-like properties of atomic-sized particles might be observable, and might be very important for understanding their behavior.

De Broglie waves and Bohr's third postulate

De Broglie's radical idea contributed to building a new model of subatomic physics. Consider an electron moving around the nucleus of an atom in a stable Bohr orbit. Instead of modeling the electron as a point-like particle, de Broglie's ideas allow us to model it as a matter wave vibrating around the nucleus.

As the wave returns to the starting point in its orbit around the nucleus, it can interfere constructively with itself (**Figure 28.26a**); it can interfere destructively with itself; or some intermediate interference could occur (Figure 28.26b). If anything other than perfect constructive interference occurs, the wave will ultimately dampen and vanish as it circles many times around the nucleus. Thus, for the electron wave to be stable, an exact integer number n of electron wavelengths must be wrapped around the nucleus (**Figure 28.27**), with a circumference of $2\pi r$. Mathematically (and assuming the electron is not moving relativistically):

$$2\pi r = n\lambda = n\frac{h}{p} = n\frac{h}{mv}, \quad n = 1, 2, 3, \ldots$$

Because n must be a positive integer, this equation represents a quantization condition for the hydrogen atom. If we take this equation, multiply both sides by mv, and divide both sides by 2π, we get

$$mvr = n\frac{h}{2\pi}$$

This is precisely Bohr's third postulate [Eq. (28.2)]! Therefore, de Broglie's idea of the electron behaving like a standing wave explains Bohr's third postulate, which originally seemed rather arbitrary except that it led to the correct states of the hydrogen atom.

FIGURE 28.26 (a) Constructive and (b) destructive electron waves.

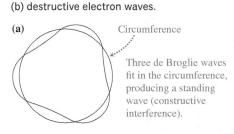

(a)   Circumference

Three de Broglie waves fit in the circumference, producing a standing wave (constructive interference).

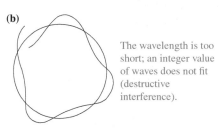

(b)

The wavelength is too short; an integer value of waves does not fit (destructive interference).

FIGURE 28.27 Electron standing waves.

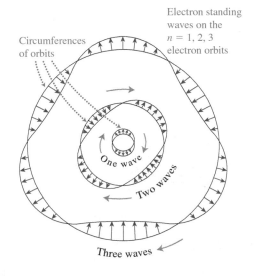

Electron standing waves on the $n = 1, 2, 3$ electron orbits

Circumferences of orbits

One wave

Two waves

Three waves

Testing de Broglie's matter wave idea

Recall our earlier experiment with a low-intensity photon beam irradiating a double slit (Section 27.4). The individual flashes of single photons on the screen gradually built up an interference pattern, suggesting that even individual photons had wave-like behavior. In Testing Experiment **Table 28.4**, we replicate that experiment using electrons instead of photons.

  28.4  **Do electrons behave like waves?**

Testing experiment	Prediction	Outcome
We shoot electrons one by one toward a screen with two closely spaced narrow slits (similar to Young's double-slit experiment). The electrons that pass through the slits reach a screen that lights up (flashes) when hit. The electron speed is about 1000 m/s. 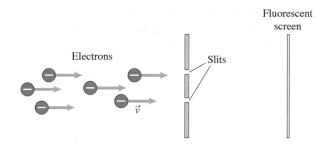	If the electrons behave only like particles, there should be two images of the slits on the screen. Image of slits If the electrons possess wave-like properties, they should hit the screen like particles but gradually build up an alternating bright and dark interference pattern. Electron interference image	An alternating bright and dark interference pattern gradually appears.

Conclusion

Electrons do indeed exhibit wave-like behavior. The experiment supports (that is, it does not disprove) de Broglie's hypothesis.

Another test of de Broglie waves

De Broglie's hypothesis was tested again a few years later. Shortly after de Broglie proposed his matter wave hypothesis, two scientists in New York, Clinton Davisson and his young colleague Lester Germer, were probing the structure of the atom by firing low-speed electrons at a nickel target made of small crystals and observing the resulting electron scattering (**Figures 28.28a** and **b**). The experiments did not give them any interesting results until they had a small accident in the laboratory in 1925: their

FIGURE 28.28 Davisson and Germer's experiment and accidental finding. (a) An electron beam excites a sample, and scattering is detected. (b) The scattering from a sample made up of small crystals. (c) The scattering from a sample accidentally turned into a large crystal.

(a)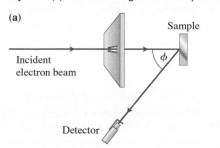

(b)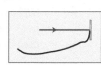
Scattering from small crystals: a smooth curve

(c)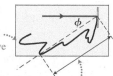
Scattering from a large crystal: peaks produced by electron-wave interference

The distance to the curve is proportional to the number of electrons scattered in a narrow cone at angle θ.

equipment malfunctioned, and the nickel target previously consisting of many small crystals suddenly melted and resolidified into one large crystal.

When Davisson and Germer resumed the experiments with the large crystal, they saw a new pattern (compare Figure 28.28c with b). As they changed the position of the detector, they found alternating directions at which electrons were scattered and not scattered. The results made no sense to them until they heard about de Broglie's hypothesis. The large crystal was behaving like a three-dimensional interference grating, and they had observed the wave-like behavior of the electrons. This story is a perfect example of how accidental observations can turn into fundamental testing experiments, or how important serendipity is in science. Later, similar experiments performed with hydrogen nuclei and alpha particles supported the idea that all subatomic particles exhibit wave-like properties.

EXAMPLE 28.5 **Electron's wavelength**

An electron moves across a 1000-V potential difference in a cathode ray tube. What is its wavelength after crossing this potential difference?

Sketch and translate We've sketched the process below. The electron starts at the cathode (where the potential is 0 V) and accelerates to the anode (where the potential is +1000 V). We know the mass and the charge of the electron; thus we can determine its momentum at the end, and from its momentum, its wavelength.

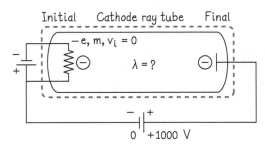

Simplify and diagram We assume that the kinetic energy of the electron is zero at the cathode where it begins and that it moves nonrelativistically. We can represent the process with an energy bar chart. The system of interest is the electron and the electric field produced by the cathode ray tube. In the initial state, the system has zero kinetic and electric potential energy (since $V = 0$ at the cathode). In the final state, the system has positive kinetic energy and negative electric potential energy (since the electron's electric charge is negative).

$$K_i + U_{qi} = K_f + U_{qf}$$

Represent mathematically We can use the bar chart to apply energy constancy to the system ($-e$ represents the charge of the electron):

$$K_i + U_{qi} = K_f + U_{qf}$$

Inserting the appropriate energy expressions:

$$0 + 0 = \tfrac{1}{2} m v_f^2 + (-e)V_f$$

or

$$\tfrac{1}{2} m v_f^2 = eV_f$$

The electron's momentum when it reaches the anode is $p = mv_f$, and its wavelength, from Eq. (28.15), is $\lambda = h/p$.

Solve and evaluate Use the preceding energy equation to solve for the electron's final speed, v_f:

$$v_f = \sqrt{\frac{2eV_f}{m}}$$

We should check to see if the electron is nonrelativistic:

$$v_f = \sqrt{\frac{2(1.6 \times 10^{-19}\ \text{C})(1000\ \text{V})}{9.11 \times 10^{-31}\ \text{kg}}} = 1.9 \times 10^7\ \text{m/s}$$

This speed is extremely fast, but only about 6% of light speed, so using the low-speed expression for momentum was justified. Combining this with the expression for the momentum and wavelength of the electron, we get

$$\lambda = \frac{h}{p} = \frac{h}{mv_f} = \frac{h}{m\sqrt{\dfrac{2eV_f}{m}}} = \frac{h}{\sqrt{2meV_f}}$$

Inserting the appropriate values:

$$\lambda = \frac{6.63 \times 10^{-34}\ \text{J} \cdot \text{s}}{\sqrt{2(9.11 \times 10^{-31}\ \text{kg})(1.6 \times 10^{-19}\ \text{C})(1000\ \text{V})}}$$

$$= 3.9 \times 10^{-11}\ \text{m}$$

This is a tiny wavelength, the same order of magnitude as the wavelength of X-rays and the size of an atom! This knowledge of the value of the electron's wavelength led to the development of microscopes that use electrons to image individual atoms, a task that would be impossible with visible light.

Try it yourself Imagine that the electrons in this example were used to investigate the structure of a crystal. When the electrons reflect off the crystal, they form an interference pattern. What would happen to the distances between the peaks in the pattern if the electrons were accelerated by a potential difference that is half the value used in this example?

Answer

The electron wavelength would be longer, and the peaks in the pattern would be spread farther apart.

Wave functions

Louis de Broglie's revolutionary hypothesis changed the way scientists viewed elementary particles. However, it was not clear exactly what was "waving" in a de Broglie matter wave. In a water wave, the water "waves." In a light wave the $\vec{E}$ and $\vec{B}$ fields are "waving." But what is "waving" in an electron wave?

De Broglie himself thought of the wave as some sort of vibrating shroud surrounding the true electron (which he still thought of as a point-like particle). This idea is known as de Broglie's **pilot wave model**; the wave guides the actual particle within it. Others thought the electron wasn't point-like and was "smeared" over a certain volume; it was this smeared electron that was vibrating as it moved through space.

In 1926, German scientist Max Born suggested that a particle's matter wave is a mathematical distribution related to the likelihood of measuring the particle to be at various locations at a specific time. If the value (technically, the square of the value) of this matter wave at a particular location is near zero, then there is almost zero chance of detecting the particle at that location at that time. If the value is large, then there is a large chance of finding the electron there.

Erwin Schrödinger coined the term **wave function** for these matter waves and developed an equation for determining them. The field of **quantum mechanics** is based on understanding the particles of the micro-world in terms of wave functions and the Schrödinger equation (the mathematics of which is beyond the scope of this textbook). Based on all the above, now we can see electrons not as particles with a specific location, but as clouds smeared in three-dimensional space.

TIP In the previous paragraphs (and throughout the rest of this section) we use the word "particle" to refer to a fundamental constituent of matter (an electron, for example), even though we have learned that these constituents have both wave- and particle-like behaviors.

Atomic wave functions

Understanding matter in terms of wave functions lets us construct a better model of the hydrogen atom. When the Schrödinger equation is used to determine the electron wave functions, three of the quantum numbers (n, l, and m_l) appear naturally, without needing to be put in by hand. The fourth number, the spin magnetic quantum number m_s, appears naturally when quantum mechanics is extended so that it incorporates Einstein's theory of special relativity. This model resulted in a detailed understanding of the periodic table of the elements based on fundamental physics ideas. We will learn later in the book (Chapter 30) that this theory of relativistic quantum mechanics predicts the existence of antimatter and ultimately evolves into the standard model of particle physics, a theory that describes all the known elementary particles and their interactions.

REVIEW QUESTION 28.6 How did the Davisson and Germer experiment serve as a testing experiment for the idea that subatomic particles possess wave-like properties?

28.7 Multi-electron atoms and the periodic table

Together, the four quantum numbers n, l, m_l, and m_s specify the wave functions of possible electron states in atoms. Examples of the square of the $n = 2$ wave functions (that is, the values that generate the probability waves) are shown in **Figure 28.29**. The more darkly shaded regions indicate where the electron's probability wave has larger values; therefore, there is a greater probability of finding the electron at those locations. The $l = 0$ state actually has a greater probability than the $l = 1$ and 2 states of the electron's being at the nucleus—the dark region at the center of its probability distribution. This depiction provides an explanation for the lower energy in multi-electron atoms of the $l = 0$ electron states.

FIGURE 28.29 Probability distributions (wave functions squared) for $n = 2$ states.

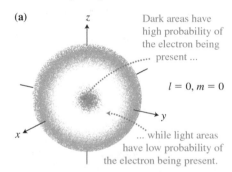

(a) Dark areas have high probability of the electron being present ...

$l = 0, m = 0$

... while light areas have low probability of the electron being present.

(b) $l = 1, m = \pm 1$

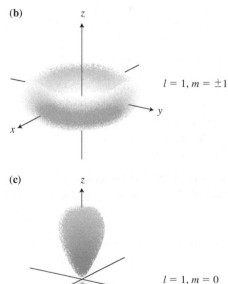

(c) $l = 1, m = 0$

Subshells and number of electrons in them

Early in the study of the emission and absorption of light by atoms, scientists used letters to designate the different states of orbital angular momentum quantum number l. This notation is still widely used today. Table 28.5 summarizes the letter designations. A state with $n = 1$ and $l = 0$ is called a $1s$ state; an $n = 4$ and $l = 2$ state is called a $4d$ state.

TABLE 28.5 Letter designations of different l states and maximum number of electrons in that subshell

l value	0	1	2	3	4	5	6
Letter	s	p	d	f	g	h	i
Maximum number of electrons	2	6	10	14	18	22	26

Specifying the n and l quantum numbers specifies a group of states called a **subshell**. Specifying just the n quantum number specifies a group of subshells that is called a **shell**. The l quantum number also determines the maximum number of electrons that can populate that subshell.

How is the maximum number of electrons in the third row of Table 28.5 determined? Let's look at the $n = 1$, $l = 0$ s subshell. In this subshell, m_l can only be zero and m_s can be $+1/2$ or $-1/2$. Thus, there are two different states with $n = 1$ and $l = 0$. According to Pauli's exclusion principle, each state can hold just one electron; so the $1s$ shell can hold at most two electrons. This reasoning holds for any s state.

A subshell is said to be filled if each state in the subshell is occupied by an electron. The number of states in a subshell depends on the number of different allowed combinations of m_l and m_s for the given value of l. In the following exercise we consider the number of states in an $l = 2$ (or d) subshell.

CONCEPTUAL EXERCISE 28.6   **States in $3d$ subshell**

Determine the number of states and the quantum number designation of each state for the $3d$ subshell.

Sketch and translate For a $3d$ subshell, $n = 3$ and $l = 2$ (see Table 28.5). For an $l = 2$ state, the m_l quantum number can have the values $m_l = -2, -1, 0, 1,$ or 2. For each of these five values of m_l, the m_s quantum number can have values $m_s = -1/2$ or $+1/2$. Thus, there are 10 unique quantum states for the $3d$ subshell (5 m_l values times 2 m_s values).

Simplify and diagram The 10 states and their corresponding quantum numbers are summarized in Table 28.6. If one electron resides in each of these states, such as occurs in a copper atom, the atom is said to have a filled $3d$ subshell.

- -

Try it yourself How many electrons can be in a $4f$ subshell?

TABLE 28.6 The number of states and the quantum number designation of each state for the $3d$ subshell

n	l	m_l	m_s
3	2	-2	$+1/2$
3	2	-1	$+1/2$
3	2	0	$+1/2$
3	2	1	$+1/2$
3	2	2	$+1/2$
3	2	-2	$-1/2$
3	2	-1	$-1/2$
3	2	0	$-1/2$
3	2	1	$-1/2$
3	2	2	$-1/2$

Answer

14 because $l = 3$, $m_l = -3, -2, -1, 0, 1, 2,$ or 3, and $m_s = -1/2$ or $+1/2$.

Electron configuration for the ground state of an atom

The electrons in an atom normally occupy the states with lowest energy. In this case the atom is said to be in its ground state. An electron configuration for the ground state of an atom indicates the subshells occupied by its electrons and the number of electrons in each subshell. For example, a neutral sodium atom has 11 electrons distributed as follows: $1s^2 2s^2 2p^6 3s^1$. The superscript represents the number of electrons in each subshell: $1s^2$ indicates that two electrons occupy the $1s$ subshell, $2p^6$ indicates that six electrons occupy the $2p$ subshell, and so forth. The subshell $3s$ has only a single electron. The maximum numbers of electrons that can be in a subshell are listed in Table 28.7. The order of filling is shown in Figure 28.30. Note, for example, that electrons in the $4s$ subshell have lower energy than electrons in the $3d$ subshell. We see that for sodium, 10 electrons are needed to fill the $1s$, $2s$, and $2p$ subshells. The 11th electron in sodium occupies a state with the next available lowest energy subshell, the $3s$ subshell. When sodium is ionized to form Na^+, it loses the electron from the $3s$ subshell since that electron is most weakly bound to the nucleus.

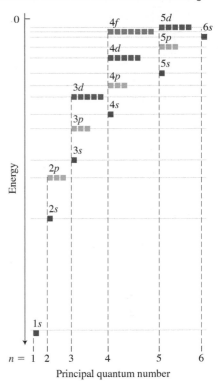

FIGURE 28.30 The order of subshell filling.

TABLE 28.7 Atomic subshells from lowest to highest energy (approximate)

	Quantum Numbers		Number of Quantum States	
n	l	m_l	In the Subshell	Total
1	0 (s)	0	2	2
2	0 (s)	0	2	8
	1 (p)	−1, 0, +1	6	
3	0 (s)	0	2	18
	1 (p)	−1, 0, +1	6	
	2 (d)	−2, −1, 0, +1, +2	10	
4	0 (s)	0	2	32
	1 (p)	−1, 0, +1	6	
	2 (d)	−2, −1, 0, +1, +2	10	
	3 (f)	−3, −2, −1, 0, +1, +2, +3	14	

The periodic table

Many of the properties of an atom, such as its tendency to form positive or negative ions and its tendency to combine with certain other atoms to form molecules, depend on the number of electrons in its outermost subshell and on the type of subshell it is. In addition, groups of atoms with the same number of outer electrons in subshells with the same letter designation (s, p, d, etc.) usually exhibit similar properties. This leads naturally to the classification of the atoms as presented in the **periodic table of the elements** (Table 28.8, on the next page). In the table, elements in the same column have similar electron configurations for their outer electrons and therefore similar chemical properties.

REVIEW QUESTION 28.7 What is the difference between a shell and a subshell?

TABLE 28.8 Periodic table of the elements*

I	II											III	IV	V	VI	VII	0
1 **H** 1.0080																	2 **He** 4.0026
3 **Li** 6.941	4 **Be** 9.0122											5 **B** 10.81	6 **C** 12.011	7 **N** 14.0067	8 **O** 15.9994	9 **F** 18.9984	10 **Ne** 20.179
11 **Na** 22.9898	12 **Mg** 24.305	Transition elements										13 **Al** 26.9815	14 **Si** 28.086	15 **P** 30.9738	16 **S** 32.06	17 **Cl** 35.453	18 **Ar** 39.948
19 **K** 39.102	20 **Ca** 40.08	21 **Sc** 44.956	22 **Ti** 47.90	23 **V** 50.941	24 **Cr** 51.996	25 **Mn** 54.9380	26 **Fe** 55.847	27 **Co** 58.9332	28 **Ni** 58.71	29 **Cu** 63.54	30 **Zn** 65.37	31 **Ga** 69.72	32 **Ge** 72.59	33 **As** 74.9216	34 **Se** 78.96	35 **Br** 79.909	36 **Kr** 83.80
37 **Rb** 85.467	38 **Sr** 87.62	39 **Y** 88.906	40 **Zr** 91.22	41 **Nb** 92.906	42 **Mo** 95.94	43 **Tc** (99)	44 **Ru** 101.07	45 **Rh** 102.906	46 **Pd** 106.4	47 **Ag** 107.870	48 **Cd** 112.40	49 **In** 114.82	50 **Sn** 118.69	51 **Sb** 121.75	52 **Te** 127.60	53 **I** 126.9045	54 **Xe** 131.30
55 **Cs** 132.906	56 **Ba** 137.34	57 **La** 138.906	72 **Hf** 178.49	73 **Ta** 180.948	74 **W** 183.85	75 **Re** 186.2	76 **Os** 190.2	77 **Ir** 192.2	78 **Pt** 195.09	79 **Au** 196.967	80 **Hg** 200.59	81 **Tl** 204.37	82 **Pb** 207.2	83 **Bi** 208.981	84 **Po** (210)	85 **At** (210)	86 **Rn** (222)
87 **Fr** (223)	88 **Ra** 226.03	89 **Ac** 227.028	104 **Rf** (261)	105 **Db** (262)	106 **Sg** (266)	107 **Bh** (264)	108 **Hs** (269)	109 **Mt** (268)	110 **Ds** (271)	111 **Rg** (272)	112 **Cn** (285)	113 **Nh** (286)	114 **Fl** (289)	115 **Mc** (288)	116 **Lv** (293)	117 **Ts** (294)	118 **Og** (294)

Lanthanide series

58 **Ce** 140.12	59 **Pr** 140.908	60 **Nd** 144.24	61 **Pm** (147)	62 **Sm** 150.4	63 **Eu** 151.96	64 **Gd** 157.25	65 **Tb** 158.925	66 **Dy** 162.50	67 **Ho** 164.930	68 **Er** 167.26	69 **Tm** 168.934	70 **Yb** 173.04	71 **Lu** 174.97

Actinide series

90 **Th** 232.038	91 **Pa** 231.036	92 **U** 238.029	93 **Np** 237.048	94 **Pu** (242)	95 **Am** (243)	96 **Cm** (248)	97 **Bk** (249)	98 **Cf** (249)	99 **Es** (254)	100 **Fm** (257)	101 **Md** (258)	102 **No** (259)	103 **Lr** (260)

*Atomic masses of stable elements are those adopted in 1969 by the International Union of Pure and Applied Chemistry. For those elements having no stable isotope, the mass number of the "most stable" isotope is given in parentheses.

28.8 The uncertainty principle

Quantum mechanics provided a theoretical basis for observations of spectral lines. It also explained how different chemical elements could have similar chemical properties. Another profound prediction of quantum mechanics concerned the determinability of certain physical quantities.

Think of a classical particle. The ability to determine its velocity and its position would appear to be limited only by the precision of the measuring instruments. However, quantum mechanics predicts something quite different for subatomic particles.

Recall that if we send a laser beam through a single slit, it forms a pattern of bright and dark fringes on the screen behind the slit. What if you send a beam of electrons though the same horizontally oriented single slit and use a scintillating screen to register the electrons after they pass through the slit (**Figure 28.31**)? Electrons hit the screen at different locations. Instead of a sharp horizontal image of the slit as shown in Figure 28.31a, we see a set of "bright" and "dark" bands in the y-direction, parallel to the slit, as in Figure 28.31b. The pattern looks similar to the diffraction pattern that light produced when passing through a slit. Electron waves are being diffracted by the slit. The angular position θ_1 of the first dark region (where no electrons are detected) above or below the central maximum is determined by the light diffraction equation [Eq. (24.6) from Section 24.5]:

$$\sin \theta_1 = \frac{\lambda}{w}$$

Here w is the slit width and λ is the electron's wavelength (**Figure 28.32**).

Now, let's consider a single electron passing through the slit. Because the slit has width w, the electron's y-coordinate position can be determined only to within w. However, in order to hit the screen in any location other than directly across from the slit, the electron must have some y-component of momentum once it has passed through the slit. (We're considering the particle's momentum rather than velocity because the momentum determines its wavelength.)

TIP Remember that a significant diffraction pattern is being produced because of the wave-like properties of the electron combined with the narrowness of the slit.

A connection exists between the narrowness of the slit and the apparent y-component of the electron's momentum once it passes through. The narrower the slit, the larger the y-component of the electron's momentum might be. To put it another way, the more precisely we attempt to determine the y-coordinate of the electron's position y as it passes through the slit (by using a narrower slit), the less precisely we can determine the y-component of the electron's momentum p_y. It seems there is a limit on how precisely one can simultaneously determine the position y and momentum component p_y of the electron.

Let's see if we can make this idea more quantitative. The angular location of the first dark fringe is

$$\sin \theta_1 = \frac{\lambda}{w}$$

Inserting the expression for the wavelength of the electron, we have

$$\sin \theta_1 = \frac{(h/p)}{w}$$

Use Figure 28.32 to write $\sin \theta_1$ in terms of momentum:

$$\sin \theta_1 = \frac{p_y}{p} = \frac{(h/p)}{w} = \frac{h}{wp}$$

Combining and rearranging the second and fourth terms above, we get

$$w p_y = h$$

Note that w is the width of the slit and quantifies how determinable the y-position of the electron is (all we know is that the electron went through the slit). Assuming that nearly all of the electrons are detected between the first dark fringes on either side of the central maximum, p_y quantifies how determinable the y-component of the momentum of an electron is once it has passed through the slit (all we know is that it reaches the screen between the two dark fringes). We indicate this by saying $w = \Delta y$ and $p_y = \Delta p_y$. The relationship then becomes

$$\Delta y \, \Delta p_y \approx h$$

FIGURE 28.31 Electron diffraction patterns.

(a)

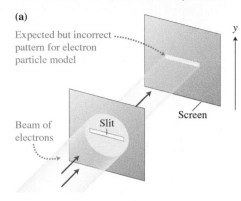

Expected but incorrect pattern for electron particle model

Beam of electrons

Slit

Screen

y

(b)

Instead of a sharp horizontal image of the slit as shown in part (a), we see a set of bright and dark bands above and below the central slit maximum, as shown here.

Beam of electrons

FIGURE 28.32 Single-slit diffraction with electrons.

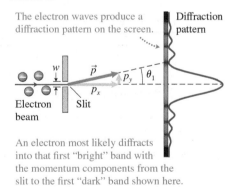

The electron waves produce a diffraction pattern on the screen.

Diffraction pattern

w

$\vec{p}$

p_y

θ_1

p_x

Electron beam

Slit

An electron most likely diffracts into that first "bright" band with the momentum components from the slit to the first "dark" band shown here.

Since the right-hand side of this equation is not zero, neither Δy nor Δp_y is exactly determinable. These limitations did not come from using a particular measuring device. They result solely from the electron having wave-like properties. Quantum mechanics sets a fundamental limit on how well the position and momentum of a particle in a particular direction can be determined simultaneously.

A more detailed analysis of the situation leads to an improved relationship (we switched to the x-direction, as it is the typical way to write this expression, but the same relations can be derived for the y- and z-directions):

$$\Delta x \, \Delta p_x \geq \frac{h}{4\pi} \tag{28.16}$$

The factor 4π is a result of a more careful statistical analysis. Equation (28.16) is one example of what is known as the **uncertainty principle** of quantum mechanics, formulated by German physicist Werner Heisenberg in 1925.

Uncertainty principle for energy

Imagine that we have an atom that can potentially emit a photon of energy E. Remember that $E = pc$ for a photon. But because a fundamental limit exists on the determinability of the momentum of a particle, a fundamental limit should also exist on the determinability of the energy of the photon:

$$\Delta E = (\Delta p)c$$

Combining this with the uncertainty principle, we have

$$\Delta x \left(\frac{\Delta E}{c}\right) \geq \frac{h}{4\pi}$$

Photons travel at the speed of light, so they travel a distance Δx in a time interval $\Delta t = (\Delta x/c)$. Therefore, we can express Δx equivalently as $c\Delta t$. The equation then becomes

$$\Delta E \, \Delta t \geq \frac{h}{4\pi} \tag{28.17}$$

In words, if a system (such as an atom in an excited state) exists for a certain time interval Δt before it emits a photon, then the energy of the state is only determinable within a range of ΔE. Thus, the energy states of an atom should be represented as "bands" whose thickness is determined by how long the atom can be in this state— the shorter-lived the state, the greater the energy uncertainty and the wider its energy spread (**Figure 28.33**).

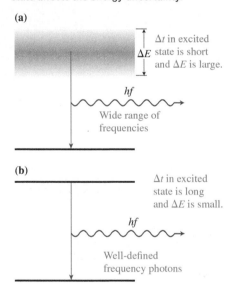

FIGURE 28.33 The lifetime Δt in an excited state affects the energy uncertainty.

(a)

ΔE Δt in excited state is short and ΔE is large.

hf

Wide range of frequencies

(b)

Δt in excited state is long and ΔE is small.

hf

Well-defined frequency photons

QUANTITATIVE EXERCISE 28.7 **Width of an energy state**

The lifetime of a hydrogen atom in the first excited state ($n = 2$) is about 10^{-9} s. Estimate the *width* of this energy state (the determinability of the energy of the state).

Represent mathematically Using Eq. (28.17) we can relate the uncertainty in the energy of the state to its lifetime:

$$\Delta E \, \Delta t \geq \frac{h}{4\pi}$$

Solve and evaluate Solving for ΔE, we have

$$\Delta E = \frac{h}{4\pi \, \Delta t}$$

We have changed the inequality to an equal sign, so we are calculating the minimum determinability of the energy of the state. Inserting the appropriate values gives

$$\Delta E = \frac{6.63 \times 10^{-34} \text{ J} \cdot \text{s}}{4\pi(10^{-9} \text{ s})} = 5.28 \times 10^{-26} \text{ J} \approx 5 \times 10^{-26} \text{ J}$$

Comparing this uncertainty to the value of the energy when in that energy state, 5.4×10^{-19} J, we see that the determinability of the energy is less than one-millionth of the value of the energy. This means that the energy of the $n = 2$ state is sharply defined, though not exactly determinable.

Try it yourself Using the uncertainty principle, determine the minimum determinability of the position of a bowling ball whose momentum is 50 ± 2 kg · m/s in the direction of motion.

Answer

of the physical quantities of a macroscopic system.

principle does not place any practical limits on the determinability

of an atomic nucleus. From this we can conclude that the uncertainty

$\approx 3 \times 10^{-35}$ m. This is 100 billion billion times smaller than the size

When we perform any measurements in physics we always encounter experimental uncertainty, although classical physics places no fundamental limits on how precise a measurement can be. Moreover, the uncertainty in one measurement generally is not related to the uncertainty in some other measurement. In the subatomic world the situation is different. The wavelike properties of particles make many pairs of physical quantities not simultaneously exactly determinable, for example, the components of position and momentum of a particle in a particular direction, or the energy and lifetime of a quantum state.

Tunneling

The energy-time interval uncertainty relationship, $\Delta E \, \Delta t \geq (h/4\pi)$, predicts that for a short time interval Δt the energy of a quantum system is only determinable to within $\Delta E \approx h/(4\pi \, \Delta t)$. In other words, energy conservation can be violated by an amount ΔE for a short time $\Delta t < h/(4\pi \, \Delta E)$. This means that the energy even of an isolated system is not constant in time but fluctuates. This result leads to an important phenomenon called **quantum tunneling**, which allows particles to have nonzero chances of existing in locations where classical physics forbids their existence.

In classical mechanics, the energy of a system is exactly determinable for all practical purposes because it is a macroscopic system. (Imagine a skier not having enough speed to go over a hill—he stops on the side of the hill before getting over the top.) However, a subatomic system's energy is not exactly determinable, so a particle that seems to be totally trapped actually isn't.

Consider a particle interacting with other particles so that the potential energy U of the particle-surroundings system as a function of position x looks as shown in **Figure 28.34** (the rectangular hump is the energy barrier). The horizontal line represents the system's energy E. The particle's kinetic energy at each position is then $K = E - U$. A classical particle on the left side of the barrier bounces back and forth between positions 0 and x_1. It cannot escape that region because its kinetic energy would be negative between positions x_1 and x_2 (the region where $U > E$), and that's impossible. In quantum mechanics, however, the square of a wave function determines the probability of finding the particle at different locations. Solving the Schrödinger equation for this situation reveals that the amplitude of the wave function decreases exponentially into the barrier region, rather than immediately becoming zero, meaning that after the barrier there is some nonzero wave function. This result means there is a nonzero probability that the particle will be detected on the right side of the barrier. It will have "tunneled through" the barrier.

FIGURE 28.34 A particle can pass through an energy barrier.

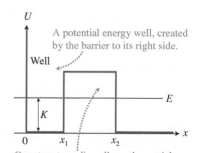

A potential energy well, created by the barrier to its right side.

Quantum tunneling allows the particle to cross this barrier, where the kinetic energy is negative.

Hydrogen bonds and proton tunneling in DNA

A DNA molecule, which stores the genetic code, is made up of two complementary strands that coil about each other in a helical form. The two strands pair with each other through hydrogen bonding between adenine (A) and thymine (T) molecules and between cytosine (C) and guanine (G) molecules. A hydrogen bond forms when a hydrogen atom bonds with a more electronegative atom (such as oxygen or fluorine).

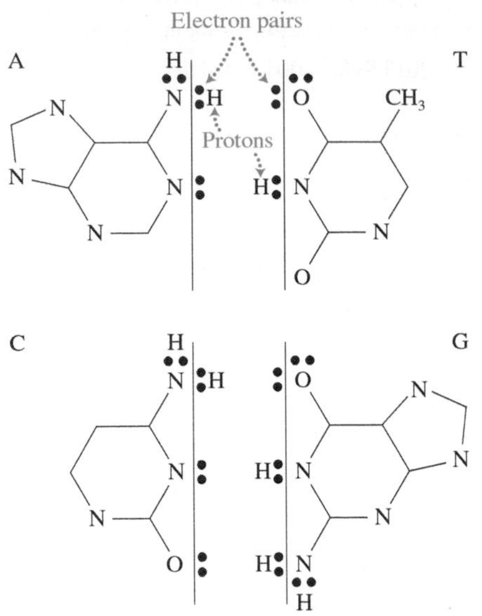

FIGURE 28.35 Hydrogen bonding between the two strands of a DNA molecule.

In each strand of the DNA molecule, an electron pair extends from either a nitrogen or an oxygen atom (consider the top bond in the A-T pair of molecules in **Figure 28.35**). The proton nucleus of a hydrogen atom, H in the figure, resides in a potential well created by the electron pair, represented by the two dots (⁚) on the A side of the figure. On the T side is another electron pair where the electric potential well is not as deep. The proton usually resides on the A side, but it can tunnel through the barrier between the wells to briefly reside on the T side (see **Figure 28.36**). In the next bond down in Figure 28.35, the well is deeper on the T side, and the proton spends more time there than on the A side.

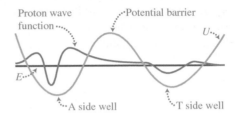

FIGURE 28.36 Energy wells for a hydrogen nucleus associated with DNA strands. In a pairing between A and T, some small probability exists that a hydrogen nucleus can cross the kinetic energy barrier between the wells.

A new DNA molecule is formed when these H bonds are unzipped and new A, T, C, or G molecules bond with the complementary molecule on each side of the unzipped DNA to form two identical DNA molecules. However, there is a slight chance that a proton will have tunneled to the wrong side during the unzipping process. In that case, a different code is produced in the new DNA—a mutation occurs.

Let's consider the implications of this situation. Because the proton is a tiny particle that must be described using the principles of quantum mechanics, it is *impossible* for the genetic code to be 100% stable. A finite probability exists that because of proton tunneling, the proton will be on the wrong side of its bond when DNA is replicated. This approximately 1-nm displacement of protons through classically forbidden barriers plays a small part in the evolution of living organisms.

REVIEW QUESTION 28.8 What experimental evidence supports the uncertainty principle?

Summary

Nuclear model of atom The scattering of alpha particles at large angles from a thin foil indicated that the mass of an atom must be localized in a tiny nucleus (about 10^{-15} m in diameter) at the center of the atom. Low-mass electrons circle the nucleus. (Section 28.1)

Bohr's outdated model of one-electron atom
- A negatively charged electron moves in a stable circular orbit about a positively charged nucleus.
- In the stable orbits, the angular momentum of the electron is quantized and equal to a positive integer n times $h/2\pi$.
- When an electron transitions from one stable orbit to another, the atom's energy changes and a photon is emitted or absorbed. (Section 28.2)

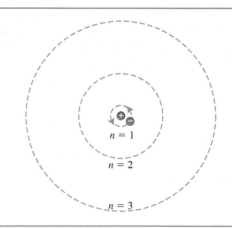

$$hf = E_i - E_f$$

Wave-particle duality EM radiation has a dual wave-particle (photon) behavior. Subatomic particles also have dual wave-particle behaviors. The wavelength $\lambda_{particle}$ depends on the momentum p of the particle. (Section 28.6)

$$\lambda = h/p$$

$$\lambda_{particle} = h/p \qquad \text{Eq. (28.15)}$$

Quantum numbers Electron states in atoms are described by four quantum numbers:

Principal quantum number n indicates the energy of the state, but *not* the angular momentum as in Bohr's atom.
Orbital angular momentum quantum number l indicates the orbital angular momentum L of the electron wave and its shape.
Magnetic quantum number m_l indicates the projection L_z of the electron's angular momentum on the direction of an external $\vec{B}_{ex}$ field.
Spin magnetic quantum number m_s indicates the two directions that the electron's spin can be oriented relative to an external $\vec{B}_{ex}$ field. (Section 28.5)

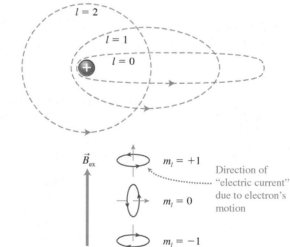

Direction of "electric current" due to electron's motion

$$n = 1, 2, 3, \ldots \qquad \text{Eq. (28.11)}$$

$$l = 0, 1, 2, \ldots, n-1 \qquad \text{Eq. (28.12a)}$$

$$L = \sqrt{l(l+1)}\,\frac{h}{2\pi} \qquad \text{Eq. (28.12b)}$$

$$m_l = 0, \pm 1, \pm 2, \ldots \pm l \qquad \text{Eq. (28.13a)}$$

$$L_z = m_l\,(h/2\pi) \qquad \text{Eq. (28.13b)}$$

$$m_s = \pm 1/2 \qquad \text{Eq. (28.14)}$$

Pauli exclusion principle Each available state in the atom can hold at most one electron. Each electron in an atom must have a unique set of quantum numbers n, l, m_l, and m_s. (Section 28.5)

Uncertainty principle The products of the uncertainties of position and momentum and of energy and lifetime of a state must be greater than or equal to $h/4\pi$. (Section 28.8)

$$\Delta x\,\Delta p_x \geq \frac{h}{4\pi} \qquad \text{Eq. (28.16)}$$

$$\Delta E\,\Delta t \geq \frac{h}{4\pi} \qquad \text{Eq. (28.17)}$$

Questions

Multiple Choice Questions

1. What could the plum pudding model of the atom account for (choose all answers that apply)?
   (a) The mass of the atom
   (b) The electric charge of the atom
   (c) Line spectra
   (d) The stability of the atom
   (e) Scattering experiments

2. What could the planetary model account for (choose all answers that apply)?
   (a) The mass of the atom
   (b) The electric charge of the atom
   (c) Line spectra
   (d) The stability of the atom
   (e) Scattering experiments

3. What was Bohr's model?
   (a) A revolutionary physics model that established modern physics
   (b) A compromise between classical and modern physics to account for experimental evidence
   (c) A classical model, as it relied on Newton's laws and Coulomb's law.

4. If you consider the energy of an electron-nucleus system to be zero when the electron is infinitely far from the nucleus and at rest, then
   (a) the energy of a hydrogen atom is positive.
   (b) the energy of a hydrogen atom is negative.
   (c) the sign of the energy depends on where in the atom the electron is located.

5. If you choose the electron to be the system (as opposed to an electron-nucleus system), its potential energy in the atom
   (a) is positive.   (b) is negative.
   (c) is zero.   (d) a single particle does not have potential energy.

6. Which of the spectra in Figure Q28.6 correspond to each of the following three processes?
   (I) Light emitted by a gas placed in a strong electric field
   (II) Light emitted by a tungsten wire at a very high temperature
   (III) Light from the Sun reaching Earth

FIGURE Q28.6

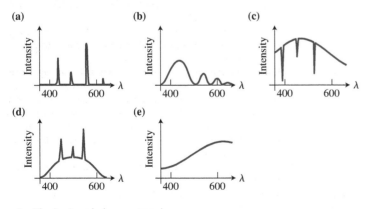

7. The Sun's emission spectrum is
   (a) continuous.   (b) line.
   (c) a combination of continuous and line. (d) none of those.

8. The Sun's spectrum has
   (a) a continuous emission component and an emission line component.
   (b) a continuous emission component and an absorption line component.
   (c) a continuous absorption component and an emission line component.

9. What do the dark lines in the Sun's spectrum indicate?
   (a) Elements that are missing from the Sun
   (b) Elements that are present in the Sun's atmosphere
   (c) Elements that are present in Earth's atmosphere

10. What does wave-particle duality mean?
    (a) Waves and particles have the same properties.
    (b) Particles are simultaneously waves.
    (c) The behavior of elementary particles can be described simultaneously using physical quantities that describe the behavior of waves and of particles.

11. What does the uncertainty principle define?
    (a) The experimental uncertainty of our measurement
    (b) The uncertainty of our knowledge of some physical quantities that we try to determine simultaneously
    (c) The limits of our knowledge about nature

12. Why does the term "electron orbit" have no meaning in quantum mechanics?
    (a) The electron is considered to be at rest.
    (b) The electron does not follow a particular orbit.
    (c) The electron is a cloud smeared in a three-dimensional space.

Conceptual Questions

13. How do we know that the plum pudding model is not an accurate model of an atom?

14. How do we know that the planetary model is not an accurate model of the atom?

15. How do we know that Bohr's model is not an accurate model of the atom?

16. Why study atomic models that we do not think are accurate?

17. How is the contemporary model of the atom different from the planetary model? From Bohr's model?

18. Explain carefully why the scattering at large angles of alpha particles from atoms is inconsistent with the plum pudding model of the atom but is consistent with the nuclear model.

19. Describe how an energy state diagram helps explain the emission and absorption of photons by an atom.

20. Use available resources to learn what the Lyman and Paschen series are. Draw an energy diagram describing the series.

21. How can one determine if the Sun contains iron?

22. If you do not see helium absorption lines in the Sun's spectrum, does it mean that helium is missing from the Sun? Explain your answer.

23. Explain how an atom with only one electron can produce so many spectral lines.

24. A group of hydrogen atoms is contained in a tube at room temperature. Infrared radiation, visible light, and ultraviolet rays pass through the tube, but only the ultraviolet rays are absorbed at the wavelengths of the Lyman series. Explain.

25. If the principal quantum number n is 5, what are the values allowed for the orbital quantum number l? Explain how the maximum possible l quantum number is related to the n quantum number.

26. If the orbital quantum number is 3, what are possible values for the magnetic quantum number?

27. Explain why an electron cannot move in an orbit whose circumference is 1.5 times the electron's de Broglie wavelength.

28. If Planck's constant were approximately 50% bigger, would atoms be larger or smaller? Explain your answer.

29. Why do fluorine and chlorine exhibit similar properties?

30. Why do atoms in the first column of the periodic table exhibit similar properties?

31. How do astronomers know that atoms of particular elements are present in a star?

32. Explain how some elements can go undetected when they are actually present in a star.

33. What are the differences between spontaneous emission of light and stimulated emission?

34. What property of laser light makes it useful for (a) cutting metal and (b) carrying information?

Problems

Below, BIO indicates a problem with a biological or medical focus. Problems labeled EST ask you to estimate the answer to a quantitative problem rather than derive a specific answer. Asterisks indicate the level of difficulty of the problem. Problems with no * are considered to be the least difficult. A single * marks moderately difficult problems. Two ** indicate more difficult problems.

28.1 and 28.2 Early atomic models and Bohr's model of the atom: quantized orbits

1. * The electron in a hydrogen atom spends most of its time 0.53×10^{-10} m from the nucleus, whose radius is about 0.88×10^{-15} m. If each dimension of this atom was increased by the same factor and the radius of the nucleus was increased to the size of a tennis ball, how far from the nucleus would the electron be?

2. * EST A single layer of gold atoms lies on a table. The radius of each gold atom is about 1.5×10^{-10} m, and the radius of each gold nucleus is about 7×10^{-15} m. A particle much smaller than the nucleus is shot at the layer of gold atoms. Roughly, what is its chance of hitting a nucleus and being scattered? (The electrons around the atom have no effect.)

3. ** (a) Determine the mass of a gold foil that is 0.010 cm thick and whose area is 1 cm × 1 cm. The density of gold is 19,300 kg/m³. (b) Determine the number of gold atoms in the foil if the mass of each atom is 3.27×10^{-25} kg. (c) The radius of a gold nucleus is 7×10^{-15} m. Determine the area of a circle with this radius. (d) Determine the chance that an alpha particle passing through the gold foil will hit a gold nucleus. Ignore the alpha particle's size and assume that all gold nuclei are exposed to it; that is, no gold nuclei are hidden behind other nuclei.

4. * An object of mass M moving at speed v_0 has a direct elastic collision with a second object of mass m that is at rest. Using the energy and momentum conservation principles, it can be shown that the final velocity of the object of mass M is $v = (M - m)v_0/(M + m)$. Using this result, determine the final velocity of an alpha particle following a head-on collision with (a) an electron at rest and (b) a gold nucleus at rest. The alpha particle's velocity before the collision is $0.010c$; $m_{alpha} = 6.6 \times 10^{-27}$ kg; $m_{electron} = 9.11 \times 10^{-31}$ kg; and $m_{gold\ nucleus} = 3.3 \times 10^{-25}$ kg. (c) Based on your answers, could an alpha particle be deflected backward by hitting an electron in a gold atom?

5. * Describe what happens to the energy of the atom and represent your reasoning with an energy bar chart when (a) a hydrogen atom emits a photon and (b) a hydrogen atom absorbs a photon.

6. * How do we know that the energy of the hydrogen atom in the ground state is -13.6 eV?

7. * Determine the wavelength, frequency, and photon energies of the line with $n = 5$ in the Balmer series.

8. * Determine the wavelengths, frequencies, and photon energies (in electron volts) of the first two lines in the Balmer series. In what part of the electromagnetic spectrum do the lines appear?

9. * **Invent an equation** An imaginary atom is observed to emit electromagnetic radiation at the following wavelengths: 250 nm, 2250 nm, 6250 nm, 12,250 nm, ... Invent an empirical equation for calculating these wavelengths; that is, determine $\lambda = f(n)$, where f is an unknown function of an integer n, which can have the values 1, 2, 3, 4, ...

10. * Write three basic equations that are needed to derive the expressions for the allowed radii and the allowed energy of electron states in the Bohr model of the atom.

11. * If we know the value of n for the orbit of an electron in Bohr's model of a hydrogen atom, we can determine the values of three other quantities related to either the electron's motion or the atom as a whole. Briefly describe these quantities.

12. * EST Is it possible for a hydrogen atom to emit an X-ray? If so, describe the process and estimate the n value for the energy state. If not, indicate why not.

13. * A gas of hydrogen atoms in a tube is excited by collisions with free electrons. If the maximum excitation energy gained by an atom is 12.09 eV, determine all of the wavelengths of light emitted from the tube as atoms return to the ground state.

14. * Some of the energy states of a hypothetical atom, in units of electron volts, are $E_1 = -31.50$, $E_2 = -12.10$, $E_3 = -5.20$, and $E_4 = -3.60$. (a) Draw an energy diagram for this atom. (b) Determine the energy and wavelength of the least energetic photon that can be absorbed by these atoms when initially in their ground state.

15. ** Show that the frequency of revolution of an electron around the nucleus of a hydrogen atom is $f = (4\pi^2 k_C^2 e^4 m/h^3)(1/n^3)$.

16. * Are we justified in using nonrelativistic energy equations in the Bohr theory for hydrogen? (That is, is the electron's speed smaller than $0.1c$?)

17. * Determine the ratio of the electric force between the nucleus and an electron in the ground state of the hydrogen atom and the gravitational force between the two particles. Based on your answer, are we justified in ignoring the gravitational force in the Bohr theory?

28.3 Spectral analysis

18. * A group of hydrogen atoms in a discharge tube emit violet light of wavelength 410 nm. Determine the quantum numbers of the atom's initial and final states when undergoing this transition.

19. * Draw an energy state diagram for a hydrogen atom and explain (a) how an emission spectrum is formed and (b) how an absorption spectrum is formed.

20. * Explain what spectral lines could be emitted by hydrogen gas in a gas discharge tube with an 11.5-V potential difference across it. What assumptions did you make?

21. * The fractional population of an excited state of energy E_n compared to the population of the ground state of energy E_0 is $N_n/N_0 = e^{-(E_n - E_0)/k_B T}$. At approximately what temperature T are 20 % of hydrogen atoms in the first excited state?

22. * Draw an energy bar chart that describes the ionization process for a hydrogen atom due to collisions with other atoms.

28.4 Lasers

23. (a) A laser pulse emits 2.0 J of energy during 1.0×10^{-9} s. Determine the average power emitted during that short time interval. (b) Determine the average light intensity (power per unit area) in the laser beam if its cross-sectional area is 8.0×10^{-9} m².

24. A pulsed laser used for welding produces 100 W of power during 10 ms. Determine the energy delivered to the weld.

25. * BIO **Welding the retina** A pulsed argon laser of 476.5-nm wavelength emits 3.0×10^{-3} J of energy to produce a tiny weld to repair a detached retina. How many photons are in the laser pulse?

26. ** BIO **More welding the retina** A laser used to weld the damaged retina of an eye emits 20 mW of power for 100 ms. The light is focused on a spot 0.10 mm in diameter. Assume that the laser's energy is deposited in a small sheet of water of 0.10-mm diameter and 0.30-mm thickness. (a) Determine the energy deposited. (b) Determine the mass of this water. (c) Determine the increase in temperature of the water (assume that it does not boil and that its heat capacity is 4180 J/kg·C°).

27. ** Compare and contrast the energy state diagrams for a laser and for an atom of a hot gas that emits light.

28. * Explain why hot gases do not emit coherent light, but lasers do.

28.5 and 28.6 Quantum numbers and Pauli's exclusion principle and Particles are not just particles

29. * (a) An electron moves counterclockwise around a nucleus in a horizontal plane. Assuming that it behaves as a classical particle, what is the direction of the magnetic field produced by the electron? (b) The same one-electron atom is placed in an external magnetic field whose $\vec{B}$ field points horizontally from right to left. Draw a picture showing what happens to the atom.

30. * EST Estimate the wavelength of a tennis ball after a good serve.

31. ** What is the wavelength of the electron in a hydrogen atom at a distance of one Bohr radius from the nucleus?

32. * EST Estimate the average wavelength of hydrogen atoms at room temperature.

33. * Describe how you will determine the wavelength of an electron in a cathode ray tube if you know the potential difference between the electrodes.

34. * An electron first has an infinite wavelength and then after it travels through a potential difference has a de Broglie wavelength of 1.0×10^{-10} m. What is the potential difference that it traversed? Draw a picture of the situation and an energy bar chart, and explain why the electron's wavelength decreased.

35. * Describe two experiments whose outcomes can be explained using the concept of the de Broglie wavelength.

36. * How does the wavelength of an electron relate to the radius of its orbit in a hydrogen atom? Why?

37. * Discuss the similarities and differences in the way a hydrogen atom is pictured in Bohr's model and in quantum mechanics.

38. * A high-energy particle scattered from the nucleus of an atom helps determine the size and shape of the nucleus. For best results, the de Broglie wavelength of the particle should be the same size as the nucleus (approximately 10^{-14} m) or smaller. If the mass of the particle is 6.6×10^{-27} kg, at what speed must it travel to produce a wavelength of 10^{-14} m?

39. * (a) Use de Broglie's hypothesis to determine the speed of the electron in a hydrogen atom when in the $n = 1$ orbit. The radius of the orbit is 0.53×10^{-10} m. (b) Determine the electron's de Broglie wavelength. (c) Confirm that the circumference of the orbit equals one de Broglie wavelength.

40. * Repeat Problem 28.39 for the $n = 3$ orbit, whose radius is 4.77×10^{-10} m. Three de Broglie wavelengths should fit around the $n = 3$ orbit.

41. * A beam of electrons accelerated in an electric field is passing through two slits separated by a very small distance d and then hits a screen that glows when an electron hits it. What do you need to know about the electrons to be able to predict where on the screen, which is L meters from the slits, you will see the brightest and the darkest spots?

42. (a) An electron is in an $n = 4$ state. List the possible values of its l quantum number. (b) If the electron happens to be in an $l = 3$ state, list the possible values of its m_l quantum number. (c) List the possible values of its m_s quantum number.

43. (a) An electron in an atom is in a state with $m_l = 3$ and $m_s = +1/2$. What are the smallest possible values of l and n for that state? (b) Repeat for an $m_l = 2$ and $m_s = -1/2$ state.

44. (a) An atom is in the $n = 7$ state. List the possible values of the electron's l quantum number. (b) Of these different states, the electron occupies the $l = 4$ state. List the possible values of the m_l quantum number.

45. ** Draw schematic orbits and arrows representing the different m_l quantum states for an $l = 2$ atomic electron in a magnetic field.

28.7 Multi-electron atoms and the periodic table

46. List the n, l, m_l, and m_s states available for an electron in a $4p$ subshell.

47. List the n, l, m_l, and m_s states available for an electron in a $4f$ subshell.

48. Identify the values of n and l for each of the following subshell designations: $3s$, $2p$, $4d$, $5f$, and $6s$.

49. (a) Determine the electron configuration of sulfur (its atomic number is 16) and explain how you did it. (b) Why are sulfur and oxygen (atomic number 8) in the same group on the periodic table?

50. (a) Determine the electron configuration of silicon (atomic number 14) and explain how you did it. (b) Why are carbon and silicon in the same group of the periodic table?

51. Determine the electron configuration for iron (atomic number 26) and explain how you did it.

52. Manganese (atomic number 25) has two $4s$ electrons. How many $3d$ electrons does it have? Explain your answer.

53. * Determine the electron configurations of four elements of group I in the periodic table. Explain why these elements are likely to have similar properties. Note that a higher s shell fills before the next lower d shell—the electron in the s shell spends more time on average closer to the nucleus.

54. * Determine the electron configurations of three elements in group VI of the periodic table. Explain why these elements are likely to have similar properties.

28.8 The uncertainty principle

55. The uncertainty in the x-position of a 3×10^{-8}-g dust particle is about 1×10^{-6} m. Determine the uncertainty in the particle's velocity component v_x.

56. * Describe the experimental evidence that supports the concept of the uncertainty principle.

57. * The lifetime of the hydrogen atom in the $n = 3$ second excited state is 10^{-9} s. What is the uncertainty of that energy state of the atom? Compare this uncertainty with the magnitude of the -1.51-eV energy of the atom in that state.

58. * Use the uncertainty principle to discuss whether lasers can emit 100% monochromatic light. (Note that monochromatic light means light with a precisely defined frequency.)

General Problems

59. * (a) Determine the radii and energies of the $n = 1$, 2, and 3 states in the He^+ ion (it has two protons in its nucleus and one electron). (b) Construct an energy state diagram for this ion. Indicate any assumptions that you made.

60. * (a) Determine the radii and energies of the $n = 1$, 2, and 3 states of a sodium ion in which 10 of its electrons have been removed. (b) Construct an energy state diagram for the ion. Indicate any assumptions that you made.

61. * **EST** A uranium atom with $Z = 92$ has 92 protons in its nucleus. It has two electrons in an $n = 1$ orbit. Estimate the radius of this orbit. Indicate any assumptions that you made.

62. * **EST** Estimate the energy needed to remove an electron from (a) the $n = 1$ state of iron ($Z = 26$ is the number of protons in the nucleus) and (b) the $n = 1$ state of hydrogen ($Z = 1$).

63. An electron in a hydrogen atom changes from the $n = 4$ to the $n = 3$ state. Determine the wavelength of the emitted photon.

64. An electron in a He^+ ion changes its energy from the $n = 3$ to the $n = 1$ state. Determine the wavelength, frequency, and energy of the emitted photon.

65. Determine the energy, frequency, and wavelength of a photon whose absorption changes a He^+ ion from the $n = 1$ to the $n = 6$ state.

66. * A helium ion He^+ emits an ultraviolet photon of wavelength 164 nm. Determine the quantum numbers of the ion's initial and final states.

67. ** The average thermal energy due to the random translational motion of a hydrogen atom at room temperature is $(3/2)k_B T$. Here k_B is the Boltzmann constant. Would a typical collision between two hydrogen atoms be likely to transfer enough energy to one of the atoms to raise its energy from the $n = 1$ to the $n = 2$ energy state? Explain your answer. [Note: Earth's free hydrogen is in the molecular form H_2. However, the above reasoning is similar for atomic and molecular hydrogen.]

Reading Passage Problems

BIO **Electron microscope**   The *scanning electron microscope* (*SEM*) uses high-energy electrons to form three-dimensional images of the surfaces of biological, geological, and integrated circuit samples. The electrons in the SEM have much shorter wavelengths than the light used in visible wavelength microscopes and consequently can produce high-resolution images of a surface, revealing details of about 1 to 5 nm in size with magnification ranging from 25 to 250,000.

A color-enhanced image of the inner surface of a lung is shown in **Figure 28.37**. The cavities are alveoli where inhaled oxygen is exchanged into the blood for carbon dioxide, a waste product removed in the lungs during exhalation.

A schematic diagram shows how a SEM works (see **Figure 28.38**). Electrons are emitted from a hot tungsten cathode (the electron gun) and accelerated

FIGURE 28.37 A scanning electron microscope photograph of lung tissue showing alveoli.

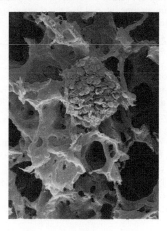

FIGURE 28.38 A schematic of a scanning electron microscope. Secondary electrons from the sample are recorded by the detector, producing a three-dimensional surface image.

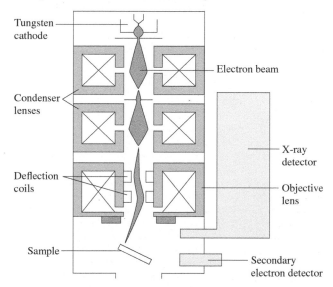

Tungsten cathode

Electron beam

Condenser lenses

X-ray detector

Deflection coils

Objective lens

Sample

Secondary electron detector

across a potential difference of a few hundred volts to 40 kV. Two or more condenser lenses, each consisting of a coil of wire through which current flows, produce a magnetic field that bends and focuses the electron beam so that it irradiates a small 0.4-nm to 5-nm spot on the sample. The focused beam is then deflected so that it scans over a rectangular area of the sample surface.

The focused electron beam knocks electrons out of surface atoms. These so-called secondary electrons have low energy (< 50 eV) and are caught by a secondary electron detector. The number of secondary electrons produced depends on the orientation and nature of the exposed surface, thus forming a vivid three-dimensional image. The electron beam can also knock inner electrons out of the sample atoms. When outer electrons move into these vacant inner electron orbits, X-ray photons are emitted. The wavelengths of these photons are characteristic of the particular type of atom and thus provide information about the types of atoms in the sample.

68. Which answer below is closest to the speed of an electron accelerated from rest across a 100-V potential difference?
    (a) 600 m/s
    (b) 6×10^4 m/s
    (c) 6×10^6 m/s
    (d) 6×10^7 m/s

69. Which answer below is closest to the wavelength of an electron accelerated from rest across a 100-V potential difference?
    (a) 1×10^{-11} m
    (b) 1×10^{-10} m
    (c) 1×10^{-9} m
    (d) 1×10^{-8} m

70. Compared to the electron in Problem 28.69, the wavelength of an electron accelerated from rest across a 10,000-V potential difference would be
    (a) the same length.
    (b) 10 times longer.
    (c) 1000 times longer.
    (d) 1/10 times as long.
    (e) 1/1000 times as long.

71. An electron in the SEM electron beam is moving parallel to the paper and downward toward the bottom of the page. In which direction should a magnetic field point to deflect the electron toward the right as seen while looking at the page?
    (a) Right
    (b) Left
    (c) Toward the top of the page
    (d) Toward the bottom of the page
    (e) Into the paper
    (f) Out of the paper

72. The SEM can also detect the types of atoms in the sample by measuring which of the following?
    (a) Energy of the secondary electrons
    (b) The wavelengths of X-rays emitted from the sample
    (c) Both (a) and (b)

73. The secondary electrons detected by the SEM detector are
    (a) outer electrons knocked out of atoms on or near the surface of the sample.
    (b) X-rays produced when outer electrons fall into vacant inner electron orbits.
    (c) electrons in the electron beam slowed by passing through the top layer of the sample.
    (d) none of the above.
    (e) (a)–(c) are correct.

BIO **Electron transport chains in photosynthesis and metabolism** Electron transport chains (ETCs), large proteins through which electrons move, play important roles in two of nature's fundamental processes: (1) the conversion of electromagnetic energy from the Sun into the energy in the chemical bonds in glucose molecules and (2) the conversion of the energy in these glucose molecules into useful forms for metabolic processes, such as muscle contraction, building proteins, and respiration.

Respiratory ETCs are located in the inner membranes of mitochondria, the power plants of eukaryotic cells. ETCs have molecular mass of 800,000 atomic mass units (u), are about 20 nm long, and vary in width from 3 to 6 nm. Each ETC has about 20 sites at which an energetic "free" electron traveling through the chain can make temporary stops. Most of the resting sites have metal ions at their centers. At three places along the chain the electron-protein system goes into a significantly lower energy state (≈ 0.4 eV). The released energy catalyzes a reaction that converts two other molecules into ATP, which carries the energy gained to other parts of the cell for different processes that require chemical energy.

How does a lone energetic electron traverse such a long distance along the bumpy electrical potential hills and valleys of a large protein and not have its energy transformed into thermal energy or light? The answer is still uncertain, but several possibilities have been proposed: (1) the ground state electron wave functions at resting sites are spread out (delocalized) over neighboring bonded atoms, which in turn are spread over and overlap with the next resting site; (2) an electron-atom system is excited into a higher energy, very delocalized state, after which the electron comes back down from this conduction state into a new location; or (3) the electron tunnels through the potential energy barrier between the resting sites. We'll examine briefly possibilities (2) and (3).

Explanation (2): Consider the visible spectra of metal atoms at the resting sites. If the excited state is delocalized over many atoms, as occurs in a semiconductor, this excited so-called conduction band is much broader than the excited state of a single atom. Thus, the visible spectra of a metal ion with a spread out excited conduction band should be somewhat broader than the excited state of a localized resting site with no conduction band. However, the observed visible spectra of the metal ion resting sites in ETCs look similar to metal ion sites found in other proteins not involved in electron conduction, with no apparent broader energy conduction band. Another problem with this explanation is the difficulty of exciting an electron at one site into the higher energy conduction band. The random kinetic energy at room temperature is about 0.025 eV, whereas visible spectra indicate that higher energy bands are 2–3 eV above the ground state. At room temperature, it is almost impossible to "promote" the electron to the conduction band.

Explanation (3): Tunneling can occur between the ground state of the atom at one site and an energy state at a nearby site—a state that is empty and slightly above the ground state. After the electron transfer, the electron-local atom system at the new site moves down to the slightly lower energy ground state, which prevents the electron from tunneling back. The system can also transition into a somewhat lower energy state and catalyze a reaction that converts the site's extra energy into the stable bond of some other molecule. Tunneling times seem consistent with measured electron transfer rates. But the actual transfer mechanism is still an open question.

74. If all of the electron resting sites were in a line along the length of an ETC, what would be the approximate distance from one site to the next?
    (a) 400 nm
    (b) 20 nm
    (c) 1 nm
    (d) (1/20) nm

75. Electron tunneling involves electrons
    (a) jumping to a conduction band and then falling down at a different place.
    (b) diffusing with a small protein from one place to another.
    (c) converting to a photon and moving at light speed to another site.
    (d) passing through a potential barrier to a different location.
    (e) having an uncertain energy for a short time interval.
    (f) (d) and (e)

76. Suppose the barrier height above the electron energy $(U_{\text{barrier}} - E_{\text{electron}})$ is 1 eV and the barrier is 2 nm wide. According to the uncertainty principle, what is the minimum time interval that the electron's energy is in this classically forbidden region?
    (a) 3×10^{-16} s
    (b) 3×10^{-15} s
    (c) 3×10^{-14} s
    (d) 3×10^{-13} s
    (e) 3×10^{-12} s

77. Electron transport chains are fundamental parts of
    (a) the conversion of glucose into energetic molecules used in the body.
    (b) molecules used for fuel in trains and other vehicles.
    (c) the photosynthetic conversion of light into stable chemical compounds.
    (d) the passing of electric current in nerve cells.
    (e) (a) and (c)
    (f) (a), (c), and (d)

78. The conduction band model of electron transport in electron transport chains may not work because
    (a) the electron transport chain is not entirely metal ions.
    (b) the conduction band is too localized.
    (c) the electron has little chance of being "promoted" into the conduction band.
    (d) the visible spectrum of the resting site is not a broad band as is expected.
    (e) (c) and (d)
    (f) (a), (b), (c), and (d)

Nuclear Physics

In the late 1800s and early 1900s, Pierre and Marie Curie were investigating the radiation produced by salts of uranium. While extracting uranium from uranium ore, Marie made an accidental observation: the residual material produced more radiation than even the pure uranium. Could that material be a new element? Testing this hypothesis required processing several tons of the uranium ore. After 4 years of laborious, painstaking work, Curie isolated two new elements. She called them polonium (named after her native Poland) and radium. Marie Curie was awarded two Nobel Prizes for this work, in physics (1903) and in chemistry (1911), the former shared with her husband Pierre and Henri Becquerel. In this chapter, we will learn the mechanism by which uranium and other elements produce radiation.

- Who is Marie Curie, and why is her statue holding a model of the element polonium?
- What are the natural sources of ionizing radiation?
- How does carbon dating work?

BY 1911, the atomic nucleus was known to scientists from experiments with alpha particles scattering on thin gold foil. These scattering experiments indicated that the nucleus occupied about $10^{-12}\%$ of the atom's volume and yet contained 99.97% of the atom's mass. In this chapter we will investigate several questions about the nucleus: what is its structure, and what processes do nuclei undergo?

BE SURE YOU KNOW HOW TO:

- Use the right-hand rule for magnetic force to determine the direction of the force exerted by a magnetic field on a moving charged particle (Section 20.3).
- Relate mass to energy using the special theory of relativity (Section 26.9).

29.1 Radioactivity and an early nuclear model

Wilhelm Roentgen's discovery in 1895 of what became known as X-rays left open the question of what X-rays are. In this section we will review the findings that followed this discovery.

Becquerel and the emissions from uranyl crystals

This image of the uranium-laden cross that Becquerel used for his investigations appeared on a photographic plate.

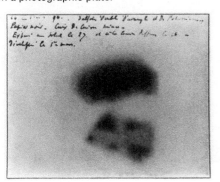

Henri Becquerel, the third generation of a family of French scientists, was very interested in Roentgen's X-rays. In 1896, Becquerel was working with potassium uranyl sulfate crystals (crystals containing the element uranium). These crystals would glow in the dark after exposure to sunlight. In addition, Becquerel found that the crystals would produce images on photographic plates. At that time, scientists thought that photographic plates could only be exposed by light, ultraviolet rays, and X-rays. Consequently, Becquerel hypothesized that the uranium crystals absorbed the Sun's energy, which then caused them to emit X-rays. However, Becquerel also discovered that crystals that had not been exposed to sunlight still formed images on photographic plates. This meant that the uranium emitted radiation without an external source of energy.

Did the crystals emit electrically neutral high-energy X-rays—electromagnetic waves? If they did, then an external magnetic field would not deflect them. However, when Becquerel passed the rays through a region with a magnetic field, he found that the field deflected the rays. Therefore, they were electrically charged particles, not X-rays. Strangely, the magnetic field deflected the rays in two opposite directions, as if both negatively and positively charged particles were being emitted. He also found that the rays could discharge an electroscope independently of the sign of its initial charge, also consistent with the idea that the rays contained both positively and negatively charged particles.

Pierre and Marie Curie and the particles responsible for Becquerel's rays

In 1896—1898, Marie and Pierre Curie continued the work of Becquerel. They found a way to measure the rate of emission of "Becquerel rays" by using an electrometer, a device similar to an electroscope invented by Pierre. A charged electrometer maintains its charge in a dry room for a relatively long time (minutes) because dry air normally contains mostly electrically neutral particles. However, an electrometer discharges much more quickly when placed near a sample containing uranium salts because uranium rays ionize air molecules. By recording the time it takes the electrometer to discharge, one can infer the ion concentration in the air.

Marie Curie conducted several observational experiments on uranium salts using an electrometer.

Experiment 1: Curie measured the time of discharge of the electrometer for samples of uranium salt of different masses. Doubling the mass of the same salt caused a discharge in half the time.

Experiment 2: Curie measured the time of discharge of the electrometer for different uranium salts for which the mass of uranium was known. As long as the total uranium mass in the salt stayed the same, the discharge time did not change.

Experiment 3: Curie measured the time of discharge of the electrometer using the same sample of uranium salt, but varying the amount of light incident on it, its temperature, and its wetness. Despite all these variations, the discharge time did not change.

Based on her observations, Marie Curie came up with the following conclusions. Assuming that the ion concentration measured by the electrometer was due to the radiation from the uranium salts, then the amount of radiation depended only on

Uranium salts.

the amount of the uranium used and not on the chemical composition of the salt, its temperature, the presence of water, or the amount of light shining on it. These findings suggested that the electrons in the atoms were not responsible for the rays. Marie and Pierre Curie concluded that the Becquerel rays must come from the nuclei of atoms.

Rutherford and experiments investigating the charge of emitted particles

In 1899 Ernest Rutherford and his colleagues in England investigated the ability of uranium salts to ionize air. He set up two parallel metal plates with a potential difference between them. Because the air between the plates was a dielectric, there was no current in the circuit. However, when a uranium sample was placed between the plates, a considerable current was detected. The current was due to the ions and free electrons created by the uranium's radiation, which were then pulled to the plates by the electric field (**Figure 29.1**).

Rutherford then covered the uranium sample with thin aluminum sheets to investigate how metal layers affected the amount of radiation. He observed that the amount of current decreased as he added more sheets, but only up to a point. After this point, no further decrease in radioactivity was observed, even with the addition of more aluminum sheets. He proposed that the radiation consisted of at least two components, one of which was not significantly absorbed by the aluminum sheets. He reasoned that the component of radiation that was absorbed by the aluminum sheets consisted of charged particles. If true, then these particles should be deflected by a magnetic field. In contrast, if there were no charged particles in the rays, the magnetic field should not affect them.

In 1903, Rutherford tested his reasoning by passing rays from radium through a magnetic field. In Testing Experiment **Table 29.1**, we describe the experiment.

FIGURE 29.1 A schematic of Rutherford's uranium radiation detection device.

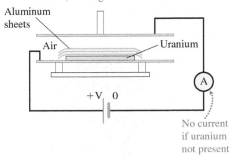

Radiation from uranium caused air molecules between the plates to ionize, causing an electric current.

No current if uranium not present

TESTING
EXPERIMENT TABLE 29.1   ☢   **Testing the deflection of radiation by magnetic fields**

Testing experiment	Prediction	Outcome
Radiation emanating from a radioactive sample passes through a $\vec{B}$ field pointing into the page. A scintillating screen (which glows when hit with radiation) is in the plane perpendicular to the beam.	If the radiation contains positively charged particles, then according to the right-hand rule, the magnetic field should deflect them upward with respect to the original direction.	The screen glowed in three places: straight ahead, deflected up, and deflected down. The location of the downward-deflected radiation was much farther from the original beam than the location of the upward-deflected radiation.

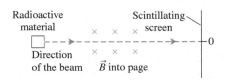

If the radiation contains negatively charged particles, then the magnetic field should deflect them downward with respect to the original direction.

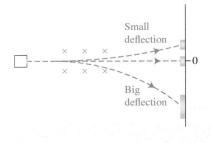

Conclusion

The radiation emanating from the radioactive sample consisted of positively charged particles, negatively charged particles, and particles with no electric charge.

Rutherford reasoned that the greater downward deflection could be the result of the downward-deflected negatively charged particles having a smaller mass-to-charge ratio than the upward-deflected positively charged particles. In the same year, using more powerful magnets, Rutherford found that the upward-bending particles were positively charged, with a mass-to-charge ratio twice that of a hydrogen ion. These positively charged massive particles were called **alpha particles**. These are the same alpha particles that Rutherford and his colleagues used later to probe the structure of the atom (Chapter 28). The downward-bending negatively charged particles had the same mass-to-charge ratio as that of electrons and were called **beta particles**. The radiation that was not deflected by the magnetic field was thought to be high-energy electromagnetic waves, called **gamma rays**.

Subsequent studies revealed that many elements with high atomic numbers were radioactive; all of them emitted alpha particles, beta particles, and/or gamma rays. Some elements with low atomic numbers were also radioactive, but generally these emitted only beta particles and gamma rays.

The early model of the nucleus

Based on these experiments, scientists developed a provisional explanation for radioactivity: the nucleus of an atom is made of positively charged alpha particles and negatively charged electrons. Their electrostatic attraction holds them together. When a nucleus contains a large number of alpha particles, they start repelling each other more strongly than the electrons can attract them, and the alpha particles leave the nucleus, a process called **alpha decay**. Alpha decay leaves behind a large number of electrons that also repel each other; thus beta particles are emitted (**beta decay**). After each transformation, the nucleus is left in an excited state and emits a high-energy photon, a gamma ray (**gamma decay**). This model provided a start for nuclear physics and was later significantly modified when new findings emerged.

A hydrogen atom, however, is lighter than an alpha particle. What, then, is the composition of its nucleus? In 1918, Rutherford noticed that when alpha particles were shot into nitrogen gas, particles that were knocked out of the nitrogen nuclei moved in curved paths that indicated they were positively charged. Further testing indicated that the particles had the same magnitude charge as an electron (only positive) and a mass that equaled that of a hydrogen nucleus. The particle was called a **proton**. Protons must be the nuclei of hydrogen atoms. The proton became an important part of a new model of the nucleus that began to emerge.

REVIEW QUESTION 29.1 How do we know that radioactive emission consists of three components: positively charged particles, negatively charged particles, and radiation with no electric charge?

29.2 A new particle and a new nuclear model

Werner Heisenberg's uncertainty principle (discussed in Chapter 28) had a profound effect on the nuclear model, in which protons and electrons were thought to be the primary nuclear constituents. However, it turned out that electrons were too light to reside inside the nucleus. Read on to follow the story.

Size of the nucleus: too small for an electron

Consider an electron confined to a carbon nucleus whose radius is approximately 2.7×10^{-15} m. The position determinability Δx of such an electron can be no greater than the size of the nucleus (**Figure 29.2**). According to the uncertainty principle, this electron must have a momentum determinability of

$$\Delta p \geq \frac{h}{4\pi \Delta x} = \frac{6.63 \times 10^{-34} \, \text{J} \cdot \text{s}}{4\pi (2.7 \times 10^{-15} \, \text{m})} = 1.95 \times 10^{-20} \, \text{kg} \cdot \text{m/s}$$

We can use this value as an estimate for the momentum of the electron confined to the nucleus. Using the equation $p = mv$, we would find that the speed of the electron is hundreds of times the speed of light, so we need to treat the electron relativistically. Using energy ideas, we find that the kinetic energy of the electron is the total energy of the electron minus its rest energy:

$$K = E_{\text{total}} - E_{\text{rest}}$$

Inserting the appropriate expressions:

$$K = \sqrt{(pc)^2 + (mc^2)^2} - mc^2$$

We can evaluate this expression using our estimate for the momentum of the electron, the mass of the electron, and the speed of light:

$$K = \sqrt{[(1.95 \times 10^{-20} \, \text{kg} \cdot \text{m/s})(3.00 \times 10^8 \, \text{m/s})]^2 + [(9.11 \times 10^{-31} \, \text{kg})(3.00 \times 10^8 \, \text{m/s})^2]^2}$$
$$- (9.11 \times 10^{-31} \, \text{kg})(3.00 \times 10^8 \, \text{m/s})^2$$
$$= 5.77 \times 10^{-12} \, \text{J}$$

How does this result compare to the electric potential energy between the electron and the nucleus? When studying the Bohr model of the hydrogen atom, we found that the total energy of a system (potential plus kinetic) must be negative for the system to be bound (Chapter 28). Let's think, for example, about a carbon nucleus. Carbon is assigned atomic number 6, so its nucleus has a charge of $+6e$. If there are both protons and electrons in the nucleus, there must be 6 more protons than electrons. We'll make a rough estimate of the electric potential energy between an electron in a carbon nucleus and the rest of the nucleus by assuming that the electron is an average distance from the protons equal to half the diameter of the nucleus, or about 2.7×10^{-15} m. The electric potential energy between the electron and rest of the nucleus (total charge $+7$) is approximately

$$U_q = k_C \frac{q_{\text{nucleus}} q_e}{r} = \left(9 \times 10^9 \, \frac{\text{N} \cdot \text{m}^2}{\text{C}^2}\right) \frac{(7 \cdot 1.6 \times 10^{-19} \, \text{C})(-1.6 \times 10^{-19} \, \text{C})}{2.7 \times 10^{-15} \, \text{m}}$$
$$\approx -6.0 \times 10^{-13} \, \text{J}$$

This value is slightly more than just one-tenth the kinetic energy of the electron. In other words, the total energy of the system (kinetic plus potential) is positive! This result means that an electron in the carbon nucleus would very rapidly escape the nucleus. This form of carbon is stable, however, and does not emit electrons. Similar reasoning applies to other nuclei.

Thus we have determined that, because of the uncertainty principle and their tiny mass, electrons cannot be components of nuclei.

The search for a neutral particle

Yet another mystery existed: an alpha particle had the charge of two protons ($+2e$) but had four times the mass of a proton. If there are no electrons to balance the two positive charges of the protons, what else is contributing to the alpha particle's mass? In 1920, Rutherford proposed that a neutral particle with the approximate mass of a proton, produced by the capture of an electron by a proton, should exist. This hypothesis stimulated a search for the particle. However, its electrical neutrality complicated the search because almost all experimental techniques of the time could detect only charged particles.

FIGURE 29.2 According to the uncertainty principle, if an electron is confined within a nucleus, its speed would be greater than light speed. Thus, it cannot be within the nucleus.

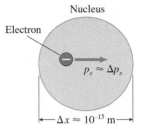

If an electron is in a nucleus, its position uncertainty is the size of the nucleus and its momentum can be estimated by the uncertainty principle.

TIP   Notice that if you substitute the mass of a proton into the equation for kinetic energy, the kinetic energy goes to zero, making the proton a possible constituent of a nucleus due to its larger mass.

In 1928, German physicist Walther Bothe and his student Herbert Becker took an initial step in the search. They bombarded beryllium atoms with alpha particles and found that neutral radiation left the beryllium atoms. This radiation was initially thought to be high-energy gamma ray photons.

Detecting neutral radiation with paraffin

In 1932, Irène Joliot-Curie, one of Marie Curie's daughters, and her husband, Frédéric Joliot-Curie, used a stronger source of alpha particles to repeat and extend Bothe's experiment. They bombarded beryllium atoms with alpha particles, as Bothe had done. However, in their experiment they placed a block of paraffin (a wax-like substance) beyond the beryllium and a particle detector beyond the paraffin. They found that the neutral radiation leaving the beryllium ejected protons from the paraffin. The Joliot-Curies knew that ultraviolet and high-frequency visible photons could knock electrons from a metal surface (the photoelectric effect discussed in Chapter 27) and made the reasonable assumption that high-frequency gamma ray photons could knock protons out of paraffin.

The neutral radiation is not a gamma ray photon

In England, James Chadwick repeated the Joliot-Curie experiments. He placed a variety of materials, not just paraffin, beyond the beryllium (**Figure 29.3**). By comparing the energies and momenta of the particles knocked out of these different materials by the rays from the beryllium atoms, he found that the rays were not gamma ray photons but instead uncharged particles with a mass approximately equal to that of the proton. He called the new particle the **neutron**. In 1935, Chadwick received the Nobel Prize in physics for this work.

Revising ideas of the structure of the nucleus

The discovery of the neutron was a major factor in revising ideas of nuclear structure, as was the realization that electrons were definitely not constituents of the nucleus. A new model for a nucleus consisting of protons and neutrons evolved. The protons, each of charge $+e$, accounted for the electric charge of the nucleus, while the uncharged neutrons accounted for the extra mass. For example, two protons and two neutrons make a helium nucleus; five protons and five neutrons make a boron nucleus (**Figure 29.4**). The most common form of uranium is composed of 92 protons and 146 neutrons, with electric charge $+92e$, and 238 protons and neutrons.

Describing atomic nuclei

Nuclei are now represented by the notation $^A_Z X$, where X is the letter abbreviation for the element (e.g., Fe for iron, Na for sodium), and Z is the number of protons in the nucleus, called the **atomic number**. N (not shown in the notation) is the number of neutrons in the nucleus. $A = Z + N$ is the **mass number**, the total number of **nucleons** (protons plus neutrons). Thus, helium with two protons and two neutrons is shown as $^4_2 He$, and the most common form of uranium, with 92 protons and 146 neutrons (a total number of protons and neutrons of $A = 92 + 146 = 238$), is $^{238}_{92} U$.

The masses of atoms and nuclei are very small if given in kilograms. Thus, another useful mass unit, called the **atomic mass unit (u)**, was developed.

FIGURE 29.3 Chadwick's apparatus for detecting neutrons.

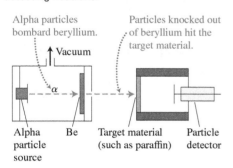

Alpha particles bombard beryllium.

Particles knocked out of beryllium hit the target material.

↑ Vacuum

α

Alpha particle source   Be   Target material (such as paraffin)   Particle detector

---

FIGURE 29.4 The structure of a boron nucleus.

 = neutron

 = proton

One atomic mass unit (u) equals one-twelfth of the mass of a carbon atom with six protons and six neutrons in the nucleus ($^{12}_6 C$), including the six electrons in the atom. In terms of kilograms,

$$1\ u = 1.660539 \times 10^{-27}\ kg$$

Isotopes

In an earlier chapter we described the mass spectrometer—a device that allows scientists to measure the mass of an elementary particle by observing its motion in a magnetic field (Chapter 20). Scientists use mass spectrometers to measure the mass of ionized atoms, as described in Observational Experiment **Table 29.2**.

OBSERVATIONAL EXPERIMENT TABLE 29.2 Measurement of atomic mass

Observational experiment	Analysis
We accelerate carbon ions to high speeds by letting them pass through a region with high potential difference and then through a magnetic field, observing the paths they take. Ions of a single chemical element take different paths. 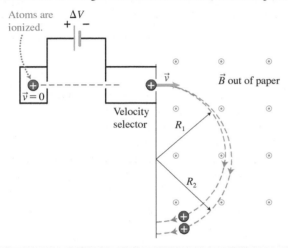	The force exerted by the magnetic field on the positive ions is perpendicular to their velocity. Thus, ions move in a semicircle. We assume that the force exerted on the ions by the magnetic field is the only force exerted on them. Newton's second law in radial component form becomes $(mv^2/R) = qvB$ or $R = (mv/qB)$. Ions entering the region with magnetic field have the same kinetic energy: $K = (mv^2/2) = q\,\Delta V$. Thus their speeds are $$v = \sqrt{\frac{2q\,\Delta V}{m}}$$ Therefore, the radius of curvature for the ions of different mass is determined by $$R = \frac{mv}{qB} = \frac{m}{qB}\sqrt{\frac{2q\,\Delta V}{m}} = \sqrt{\frac{m^2 2q\,\Delta V}{q^2 B^2 m}} = \sqrt{\frac{2m\,\Delta V}{qB^2}}$$

Pattern

Atoms of a single chemical element with the same atomic number have different masses.

Because atoms of a particular element have the same number of protons, the differences in mass among them can be explained only if we assume that the number of neutrons is different. Atoms with a different number of neutrons are called **isotopes** of that particular element. For example, carbon comes in three naturally occurring isotopic forms: $^{12}_{6}\text{C}$, $^{13}_{6}\text{C}$, and $^{14}_{6}\text{C}$, with six, seven, and eight neutrons, respectively. Each isotope has six protons and six electrons surrounding the nucleus.

The electron structure of an element's isotopes is the same, which means that their chemical behaviors are almost identical. However, the nuclei behave quite differently. Generally, only a small number (usually one or two) of the possible isotopes of an element are stable. The rest undergo radioactive decay, which we will consider shortly.

CONCEPTUAL EXERCISE 29.1 **Reading symbolic representations of atomic nuclei**

Determine the number of protons and neutrons in each of the following nuclei: $^{6}_{3}\text{Li}$, $^{16}_{8}\text{O}$, $^{27}_{13}\text{Al}$, $^{56}_{26}\text{Fe}$, $^{64}_{30}\text{Zn}$, and $^{107}_{47}\text{Ag}$.

Sketch and translate For lithium, $Z = 3$, so it has three protons in the nucleus. Lithium has $A = 6$ nucleons; thus there are three neutrons ($N = 3$).

Simplify and diagram We use the same strategy to determine the number of protons (Z) and the number of neutrons (N) in the other elements and summarize the results in the table shown at right.

$^{6}_{3}\text{Li}$		$^{16}_{8}\text{O}$		$^{27}_{13}\text{Al}$		$^{56}_{26}\text{Fe}$		$^{64}_{30}\text{Zn}$		$^{107}_{47}\text{Ag}$	
Z	N	Z	N	Z	N	Z	N	Z	N	Z	N
3	3	8	8	13	14	26	30	30	34	47	60

Try it yourself Determine Z and N for potassium-39, $^{39}_{19}\text{K}$.

Answer $Z = 19$ and $N = 20$.

CONCEPTUAL EXERCISE 29.2 **Reading and understanding the symbols in the periodic table**

Determine the chemical elements represented by the symbol X for each of the following nuclei: $^{20}_{10}$X, $^{52}_{24}$X, $^{59}_{27}$X, and $^{93}_{41}$X.

Sketch and translate To determine the element, we use the periodic table (Table 28.8) and the lower left number (Z) in the symbol.

Simplify and diagram The results for the four symbols are shown in the table at right.

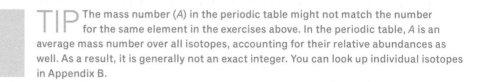

$^{20}_{10}$X	$^{52}_{24}$X	$^{59}_{27}$X	$^{93}_{41}$X
Ne (neon)	Cr (chromium)	Co (cobalt)	Nb (niobium)

Try it yourself Determine the chemical element with the following Z and A: $^{63}_{29}$X.

Answer

An isotope of copper, that is, copper-63.

> TIP The mass number (A) in the periodic table might not match the number for the same element in the exercises above. In the periodic table, A is an average mass number over all isotopes, accounting for their relative abundances as well. As a result, it is generally not an exact integer. You can look up individual isotopes in Appendix B.

REVIEW QUESTION 29.2 Explain why the nucleus cannot contain any electrons.

29.3 Nuclear force and binding energy

So far, we have found that the nucleus consists of positively charged protons and neutral neutrons. The positively charged protons repel each other (**Figure 29.5a**). There are no electrons in the nucleus to cancel this repulsive force.

How can the protons stay bound together when they repel each other?

Nuclear force

Some attractive force must balance this electrical repulsive force and must attract neutrons as well—it has to be an attractive force for both protons and neutrons. (Figure 29.5b). We call this attractive force a **nuclear force**.

The nuclear force must also weaken to nearly zero extremely rapidly with increasing distance between nucleons. If it didn't, then nuclei of nearby atoms would be attracted to each other, clumping together into ever-larger nuclei. Since this does not occur, the nuclear force must be very short range—no greater than about 2×10^{-15} m. When nucleons are farther apart, the nuclear force they exert on each other is essentially zero (Figure 29.5c).

This 2×10^{-15}-m distance limit is important since 2×10^{-15} m is smaller than the size of many of the higher-Z nuclei. Therefore, not every nucleon is attracted to every other nucleon because some of them are farther away than this maximum range. If they were attracted equally strongly, then a massive uranium nucleus with 238 protons and neutrons should be exceedingly more stable than a helium nucleus with only 4 protons and neutrons simply because of the greater number. However, experiments show that these two nuclei are approximately equally stable. Thus, a proton in a small nucleus must be bound by the nuclear force to about the same number of protons and neutrons as a proton in a large nucleus. Evidently, the nuclear force involves only nearest neighbor nucleons (Figure 29.5b and c).

However, in a nucleus with many protons, the repulsive force exerted on a particular proton by all the other protons must be strong; this repulsion might exceed the strong nuclear attraction between a proton and its neighbors. If this model of the nuclear force is correct, then we predict that such nuclei might be unstable. Could it be that we have found a reason for radioactive decay? Before we explore this possibility, let's consider another prediction based on the hypothesis of this attractive nuclear force.

FIGURE 29.5 The nuclear force. (a) Protons exert an electric repulsion on each other. (b) This repulsion is balanced by the nuclear force. (c) Nearby nucleons exert a nuclear force; more distant nucleons do not.

(a)

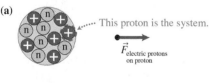

This proton is the system.

$\vec{F}_{\text{electric protons on proton}}$

(b)

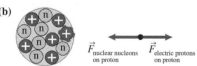

$\vec{F}_{\text{nuclear nucleons on proton}}$ $\vec{F}_{\text{electric protons on proton}}$

(c)

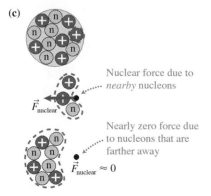

Nuclear force due to *nearby* nucleons

$\vec{F}_{\text{nuclear}}$

Nearly zero force due to nucleons that are farther away

$\vec{F}_{\text{nuclear}} \approx 0$

Mass defect

Recall from our study of atoms (Chapter 28) that we must add energy to a system in order to separate the electrons from the nucleus, ionizing the atom. When the sum of added energy, electrical potential energy, and kinetic energy reaches zero, the electron becomes unbound from the nucleus. The magnitude of the energy that must be added to ionize the atom is known as the **binding energy** of the atom (also called the ionizing energy).

We could make similar statements about the nucleus itself, composed of protons and neutrons. The nucleus is a bound system, and therefore its **nuclear potential energy** plus electric potential energy (positive since the protons repel each other) plus kinetic energy of the protons and neutrons must be negative. We neglect the energy of the interaction between the nucleus and the electrons because electrons are 100,000 times farther from protons in the nucleus than protons are from neutrons. The binding energy of the nucleus is the energy that must be added to the nucleus to separate it into its component protons and neutrons. The energy can be added through a heating mechanism, by using a photon of light, or through a collision with a fast-moving particle.

Recall that all objects with mass have a corresponding rest energy given by $E_0 = mc^2$ (Chapter 26). According to Einstein's equation, the mass m of an object equals E_0/c^2, where E_0 is the corresponding rest energy of the object. Because the other three forms of energy of the nucleus add to a negative value, the rest energy of the nucleus should be less than the rest energy of the separated protons and neutrons. Consequently, the mass of the nucleus should be less than the total mass of its constituents. To check whether this prediction matches experimental evidence, we need to collect data on the masses of the nuclei and their constituents.

The data for the masses are below (see Appendix B). Note that we use the masses of atoms, not the nuclei, because in practice they are much easier to determine. The fact that the mass of the atom includes the mass of the electrons will not change the result, as you will see later.

$$m_{\text{helium}} = 4.002602 \text{ u}; \; m_{\text{hydrogen}} = 1.007825 \text{ u}; \; m_{\text{neutron}} = 1.008665 \text{ u}$$

Mass of helium constituents:

$$2m_{\text{hydrogen}} + 2m_{\text{neutron}} = 2(1.007825 \text{ u}) + 2(1.008665 \text{ u}) = 4.032980 \text{ u}$$

Missing mass:

$$(2m_{\text{hydrogen}} + 2m_{\text{neutron}}) - m_{\text{helium}} = 4.032980 \text{ u} - 4.002602 \text{ u} = 0.030378 \text{ u}$$

The missing mass is on the order of 1% of the mass of the nucleus. Our prediction matched the outcome of the experiment. The mass of the helium nucleus is less than the sum of the masses of the protons and neutrons that comprise it (**Figure 29.6**).

In Chapter 6 we learned that mass is constant in an isolated system. Now we have found that this is not true for an atomic nucleus because it is a bound system. This means that the mass of any bound system is less than the mass of its constituents, but at a macroscopic level the difference is tiny. Now we understand that mass is not a conserved quantity. However, energy is. We will use the idea of energy conservation to understand what happens to the energy of a system when a nucleus is formed.

Let us represent a hypothetical process of making helium nuclei from their constituents using a bar chart (**Figure 29.7a** on the next page). The system is an even and equal number of protons and neutrons. The initial state is when the particles are far apart, and the final state is when they form groups of two protons and two neutrons, helium nuclei. Thus the protons and neutrons initially have zero potential energy and zero kinetic energy. The protons repel each other, so to bring them closer we need to add energy to the system (E_{Add}), for example, by doing work on the protons. Once the particles are so close that the attractive nuclear forces are larger than the repulsive electric forces, the particles start accelerating toward each other, forming helium nuclei. In the final state, each nucleus (two protons and two neutrons) has negative total potential energy (some positive electric potential energy U_{qf} and large negative nuclear potential energy U_{nf}).

FIGURE 29.6 The mass of a helium nucleus is less than that of its constituents.

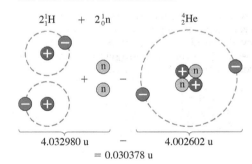

$2{}^1_1\text{H} \quad + \quad 2{}^1_0\text{n} \qquad\qquad {}^4_2\text{He}$

$\underbrace{\qquad\qquad}_{4.032980 \text{ u}} \quad - \quad \underbrace{\qquad\qquad}_{4.002602 \text{ u}}$

$= 0.030378 \text{ u}$

FIGURE 29.7 Energy bar charts representing the process of making a helium nucleus out of several constituents.

(a)

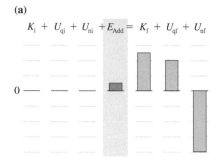

$$K_i + U_{qi} + U_{ni} + E_{Add} = K_f + U_{qf} + U_{nf}$$

(b)

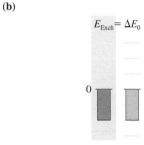

$$E_{Exch} = \Delta E_0$$

Because the final negative nuclear potential energy is much larger in magnitude than the sum of the positive electric potential energy and added energy, some energy converts into the kinetic energy K_f of the chaotic motion of the nuclei, which manifests as thermal energy of the system. We therefore see that to form helium nuclei, some energy initially needs to be added to the system, but at the end of the process, a larger amount of energy is converted into thermal energy, which can leave the system through the process of heating.

If the description above is correct, it should be consistent with the description in terms of masses. Refer back to the calculation on the previous page; instead of masses, we will use corresponding rest energies. The initial state is when the particles are at rest and far apart. In the final state, we have helium nuclei and we will assume that the thermal energy has already left the system. Initially the system has large total rest energy. In the final state, the rest energy of the system is smaller than the initial rest energy. This is consistent with the previous energy analysis: the initial total energy of the system is zero and the final total energy of the helium nuclei (after the thermal energy produced in the process has left the system) is negative. The change in the total rest energy of the system, $\Delta E_0 = E_{0f} - E_{0i}$, is therefore equal to the total energy that the system exchanged with the environment ($E_{Exch} = E_{Add} - K_f$). That is, $\Delta E_0 = E_{Exch}$. We can represent this relation using a new energy bar chart (Figure 29.7b), which is similar to the bar charts we used to represent the first law of thermodynamics in Chapter 15.

We will see in Section 29.5 how this prediction of thermal energy release leads to the production of energy through fusion. As we will also see then, the rest energies in typical processes that involve nuclear particles are much larger in magnitude than the released thermal energies.

Helium is not unique in having a smaller mass than its constituents. For example, the nucleus of lithium $_3^7\text{Li}$ is made of three protons and four neutrons. The atom also has three electrons. Thus we can compare it to three hydrogen atoms $_1^1\text{H}$ and four neutrons. The total mass of the three hydrogen atoms and four neutrons is 7.058135 u, whereas the mass of a lithium atom is 7.016003 u, or 0.042132 u less than the mass of its constituents. On the basis of such findings, we can now define a new physical quantity called **mass defect**.

Mass defect The mass defect of a nucleus is the difference in mass between the constituents of the atom (as hydrogen atoms and neutrons) and the mass of the atom itself. The mass defect Δm is

$$\Delta m = \left[Z m_{\text{hydrogen atom}} + (A - Z) m_{\text{neutron}} \right] - m_{\text{atom}} \qquad (29.1)$$

As we mentioned above, the reason the mass of the hydrogen atom is used rather than the mass of the proton is to account for the mass of the electrons. When the mass of the atom is subtracted, the mass of the electrons cancels. In practice, it is also much easier to measure the masses of atoms than the masses of just their nuclei, which is why it's more useful to define mass defect in terms of the masses of atoms.

Binding energy *BE* of the nucleus

Earlier, we found the mass defect of a helium nucleus. This defect is related to the binding energy of the atomic nucleus.

Binding energy The binding energy of a nucleus is the rest energy equivalent of its mass defect:

$$BE = \Delta m \cdot c^2 \qquad (29.2)$$

The binding energy represents the total energy needed to separate the nucleus into its component nucleons.

The binding energy of a nucleus is most easily expressed in units of million electron volts (MeV), where

$$1 \text{ MeV} = 10^6 \text{ eV} = 1.602 \times 10^{-13} \text{ J}$$

Recall that mass defect is easily expressed in atomic mass units (u). It is convenient to be able to quickly convert the mass defect of a nucleus into the corresponding binding energy. Because 1 u has units of mass, $1 \text{ u} \cdot c^2$ has units of energy. The goal is to convert $1 \text{ u} \cdot c^2$ into MeV:

$$1 \text{ u} \cdot c^2 = 1(1.660539 \times 10^{-27} \text{ kg})(2.9979 \times 10^8 \text{ m/s})^2$$
$$= 1.4924 \times 10^{-10} \text{ J}$$
$$= (1.4924 \times 10^{-10} \text{ J})\left(\frac{1 \text{ MeV}}{1.6022 \times 10^{-13} \text{ J}}\right) = 931.5 \text{ MeV}$$

Rearranging the last equation, we get the following conversion factor:

$$1 \text{ u} = 931.5 \text{ MeV}/c^2 \tag{29.3}$$

As you can see, the mass defect allows us to calculate the binding energy of any nucleus and consequently the potential energy of attraction of the particles inside it.

Binding energy per nucleon

The binding energy of the nucleus of sodium-23 ($^{23}_{11}\text{Na}$) is 187 MeV; for lithium-7 ($^{7}_{3}\text{Li}$), the binding energy is 39.2 MeV; and earlier we found that the binding energy of helium-4 was 28.3 MeV. These differences means that more energy is needed to separate a sodium-23 nucleus into protons and neutrons than to separate a lithium-7 nucleus into protons and neutrons, and both require more energy than is needed to separate a helium-4 nucleus into protons and neutrons.

Does this mean that sodium is more stable than lithium, and that lithium is more stable than helium? Not necessarily. Sodium has an "unfair" advantage in total binding energy because it has more nucleons. Instead, a better measure of stability is the binding energy of a nucleus divided by its number of nucleons, A. For sodium-23, its binding energy per nucleon is $(187 \text{ MeV})/(23 \text{ nucleons}) = 8.1 \text{ MeV/nucleon}$. For lithium-7, it is $(39.2 \text{ MeV})/(7 \text{ nucleons}) = 5.6 \text{ MeV/nucleon}$, and for helium-4, it is $(28.3 \text{ MeV})/(4 \text{ nucleons}) = 7.1 \text{ MeV/nucleon}$. With this comparison, we can see that sodium is a more stable nucleus than helium, and that helium is more stable than lithium. Put another way, more energy is needed per nucleon to separate the sodium nucleus into its constituents than is needed for the helium nucleus, and the least stable is the lithium nucleus. Binding energy per nucleon is a much better indicator of nuclear stability than total binding energy.

Binding energy per nucleon Binding energy per nucleon is the binding energy BE of the atom divided by the number of nucleons A in the atom's nucleus:

$$\text{Binding energy per nucleon} = BE/A \tag{29.4}$$

Binding energy per nucleon is a good indicator of the stability of a nucleus; the larger the binding energy per nucleon, the more stable the nucleus.

REVIEW QUESTIONS 29.3 What is the difference between nuclear binding energy and the energy needed to ionize an atom when in the ground state (atomic binding energy)?

29.4 Nuclear reactions

In 1919 Ernest Rutherford became the first person to transmute one element into another when he shot helium nuclei (alpha particles) at nitrogen and produced some oxygen-17 atoms (an isotope of oxygen that has nine neutrons and eight protons). This process, called a **nuclear reaction**, can be represented as follows:

$$\,^{4}_{2}\text{He} + \,^{14}_{7}\text{N} \rightarrow \,^{17}_{8}\text{O} + \,^{1}_{1}\text{H}$$

Rutherford's work was followed by many remarkable advances in the study of nuclear physics.

Representing nuclear reactions

Nuclear reactions, like chemical reactions, involve the transformation of reactants into different products. In a chemical reaction, molecules are transformed into other molecules, but the number of each type of atom remains the same. In a nuclear reaction, nuclei are transformed into different nuclei, resulting in different elements. In these reactions two nuclei may interact to form one or more new nuclei (e.g., $\,^{1}_{1}\text{H} + \,^{7}_{3}\text{Li} \rightarrow \,^{4}_{2}\text{He} + \,^{4}_{2}\text{He}$). Or a single nucleus may divide into two or more new nuclei, or perhaps emit a small particle, thus leaving behind a different nucleus (e.g., the radioactive decay experiments observed by Curie and others).

The advantage of writing nuclear reactions as shown above is that atomic masses found in atomic mass tables can be used to analyze the energy transformations that occur during the reactions. Even though atomic masses are used, the energy transformations are associated almost entirely with the rest energies of the reactant and product nuclei.

Rules for nuclear reactions

In balancing chemical reactions, the number of atoms of each type must be the same both before the reaction and after. Similar rules apply to nuclear reactions. Consider Observational Experiment **Table 29.3**, which describes reactions that have been observed to take place and reactions that have never been observed to occur.

OBSERVATIONAL
EXPERIMENT TABLE 29.3   Deducing rules for allowed nuclear reactions

Observational experiment	Analysis
Nuclear reactions that have been observed:	*Nuclear reactions that have been observed:*
(a) $\,^{1}_{1}\text{H} + \,^{7}_{3}\text{Li} \rightarrow \,^{4}_{2}\text{He} + \,^{4}_{2}\text{He}$	(a) $A: 1 + 7 = 4 + 4$ $Z: 1 + 3 = 2 + 2$
(b) $\,^{226}_{88}\text{Ra} \rightarrow \,^{222}_{86}\text{Rn} + \,^{4}_{2}\text{He}$	(b) $A: 226 = 222 + 4$ $Z: 88 = 86 + 2$
(c) $\,^{14}_{6}\text{C} \rightarrow \,^{14}_{7}\text{N} + \,^{0}_{-1}e$	(c) $A: 14 = 14 + 0$ $Z: 6 = 7 - 1$
In the above reactions, the products are observed to have significantly more kinetic energy than the reactants.	Both A and Z numbers are equal on both sides of the equations.
Nuclear reactions that have never been observed:	*Nuclear reactions that have never been observed:*
(d) $\,^{4}_{2}\text{He} + \,^{27}_{13}\text{Al} \rightarrow \,^{32}_{15}\text{P} + \,^{1}_{0}\text{n}$	(d) $A: 4 + 27 \neq 32 + 1$ $Z: 2 + 13 = 15 + 0$
(e) $\,^{2}_{1}\text{H} + \,^{3}_{1}\text{H} \rightarrow \,^{4}_{2}\text{He} + \,^{1}_{1}\text{H}$	(e) $A: 2 + 3 = 4 + 1$ $Z: 1 + 1 \neq 2 + 1$
	Either A or Z numbers are not equal on both sides of the equations.

Patterns

Two patterns appear in the observed nuclear reactions:

- The total numbers of nucleons A of the reactants and of the products are equal.
- The total Z numbers of the reactants and of the products are equal.
  (Z is the electric charge of the charged particles involved in the reaction.)

The second pattern, in which Z of the reactants is equal to Z of the products, is the result of electric charge conservation: in an isolated system, the total electric charge of the reactants equals that of the products; therefore, the charge is constant.

TIP The symbol for an electron is $_{-1}^{0}e$. It has charge $Z = -1$ and has zero nucleons, $A = 0$. It can be included in the reaction rules using this notation—see Reaction (c) in Table 29.3.

Why is the number of nucleons (protons plus neutrons) on both sides of the reactions equal? If a neutron vanished, it wouldn't violate charge conservation, but neutrons don't seem to do that. Evidently, they can transform into protons (see Reaction (c) in Table 29.3), but they cannot vanish. At the moment, we don't have a reason for why this pattern exists (but we will return to it in Chapter 30). For now, we will remember that in nuclear reactions the total charge of all participants and the total number of nucleons remain constant before and after the reaction. We summarize these two rules below.

Rule 1: Constant electric charge The total electric charge of the reacting nuclei and particles equals the total charge of the nuclei and particles produced by the reaction. This condition is satisfied if the total atomic number Z of the reactants equals the total atomic number of the products, including $Z = -1$ for free electrons.

Rule 2: Constant number of nucleons The total number of nucleons A (protons plus neutrons) of the reactants always equals the total number of nucleons of the products.

QUANTITATIVE EXERCISE 29.3 **Determine the missing product in the following reactions**

(a) $_{2}^{4}\text{He} + _{6}^{12}\text{C} \rightarrow _{7}^{15}\text{N} + ?$

(b) $_{1}^{2}\text{H} + _{1}^{3}\text{H} \rightarrow _{2}^{4}\text{He} + ?$

(c) $_{0}^{1}\text{n} + _{92}^{235}\text{U} \rightarrow _{54}^{140}\text{Xe} + ? + 2_{0}^{1}\text{n}$

(d) $_{55}^{137}\text{Cs} \rightarrow ? + _{-1}^{0}e$

Represent mathematically To determine the products, we need to use the rules stating that the A and Z numbers remain constant throughout the reaction.

(a) $4 + 12 = 15 + A$
     $2 + 6 = 7 + Z$

(b) $2 + 3 = 4 + A$
     $1 + 1 = 2 + Z$

(c) $1 + 235 = 140 + A + 2 \cdot 1$
     $0 + 92 = 54 + Z + 2 \cdot 0$

(d) $137 = A + 0$
     $55 = Z + (-1)$

Solve and evaluate Solving for A and Z in each case, we find the following:

(a) $A = 1, Z = 1$. The unknown product must be a hydrogen nucleus, $_{1}^{1}\text{H}$.

(b) $A = 1, Z = 0$. The unknown product must be a neutron, $_{0}^{1}\text{n}$.

(c) $A = 94, Z = 38$. The unknown product must be a strontium nucleus, $_{38}^{94}\text{Sr}$.

(d) $A = 137, Z = 56$. The unknown product must be a barium nucleus, $_{56}^{137}\text{Ba}$.

- -

Try it yourself Determine the missing product in this reaction: $_{6}^{14}\text{C} \rightarrow ? + _{-1}^{0}e$.

Answer $_{7}^{14}\text{N}$.

Energy conversions in nuclear reactions

In the three observed reactions in Table 29.3, the products had significantly more kinetic energy than the reactants. How did the products get this kinetic energy? One hypothesis is that some of the rest mass energy of the reactants was converted to kinetic energy. If this is correct, then the mass of the reactants should be greater than the mass of the products. Let's test this hypothesis using the reaction described in Testing Experiment **Table 29.4**.

TESTING
EXPERIMENT TABLE 29.4

Accounting for the extra kinetic energy

Testing experiment	Prediction	Outcome
In the following reaction, the products have more kinetic energy than the reactants: $${}^1_1\text{H} + {}^7_3\text{Li} \rightarrow {}^4_2\text{He} + {}^4_2\text{He} + \text{kinetic energy}$$ Is this consistent with the idea that some rest mass energy of the reactants was transformed into kinetic energy of the products? The atomic masses of the atoms involved are $$m({}^1_1\text{H}) = 1.007825\ \text{u}$$ $$m({}^7_3\text{Li}) = 7.016003\ \text{u}$$ $$m({}^4_2\text{He}) = 4.002602\ \text{u}$$	Mass of reactants: $$m({}^1_1\text{H}) + m({}^7_3\text{Li}) = 1.007825\ \text{u} + 7.016003\ \text{u}$$ $$= 8.023828\ \text{u}$$ Mass of products: $$m({}^4_2\text{He}) + m({}^4_2\text{He}) = 4.002602\ \text{u} + 4.002602\ \text{u}$$ $$= 8.005204\ \text{u}$$ The energy equivalent of this mass difference is $$\Delta mc^2 = (8.023828\ \text{u} - 8.005204\ \text{u})c^2\left(\frac{931.5\ \text{MeV}}{\text{u} \cdot c^2}\right)$$ $$= 17.3\ \text{MeV}$$ We predict that the products will have 17.3 MeV more kinetic energy than the reactants.	Through a complex process, the products are found to have 17.3 MeV more kinetic energy than the reactants, as predicted.

Conclusion

This experiment supports the idea that some of the rest energy of the reactants is converted into kinetic energy of the products.

These kinds of reactions, where rest energy is converted into kinetic energy, are called **exothermic**. The opposite can occur as well. In certain nuclear reactions the kinetic energy of the reactants is greater than the kinetic energy of the products. In this case, the rest energy of the products is greater than that of the reactants. In these **endothermic** reactions, some of the kinetic energy of the reactants is converted into rest energy of the products. Both types of reactions can be understood in terms of energy constancy in an isolated system.

Rule 3: Constant energy The rest energy of the reactants equals the rest energy of the products plus any change in kinetic energy of the system:

$$\sum_{\text{reactants}} mc^2 = \sum_{\text{products}} mc^2 + \Delta K$$

ΔK can be either positive (if the products' rest energy is less that of the reactants) or negative (if the products' rest energy is greater). ΔK is also known as the reaction energy Q. Rearranging, we find an expression for Q:

$$Q = \left(\sum_{\text{reactants}} m - \sum_{\text{products}} m\right)c^2 \tag{29.5}$$

We will have several opportunities to use this idea in the following sections.

REVIEW QUESTION 29.4 Is it possible for the nuclear reaction ${}^4_2\text{He} + {}^6_3\text{Li} \rightarrow {}^{11}_6\text{C} + 2\,{}^0_{-1}e$ to occur? Explain your answer.

29.5 Nuclear sources of energy

Ernest Rutherford once said, "Anyone who expects a source of power from the transformation of the atom is talking moonshine."

Rutherford had an enormous impact on early atomic and nuclear physics, but he missed the mark with this statement. Let's consider more carefully why nuclear reactions can be a source of power.

Binding energy and energy release

Figure 29.8 is a graph of the binding energy per nucleon (BE/A) versus mass number A for atomic nuclei. Notice that nuclei with $A \approx 60$ have the greatest binding energy per nucleon, while very small and very large nuclei have less binding energy per nucleon. The higher the binding energy per nucleon, the more energy is needed to split the nucleus into its constituent protons and neutrons; therefore, the nuclei with the highest binding energy per nucleon—nuclei around $A \approx 60$—should be the most stable (in fact, iron, with $A = 56$, is the most stable element).

The graph in Figure 29.8 allows us to predict two new phenomena. First, note that the binding energy per nucleon for nuclei with small atomic numbers is less than that for nuclei with larger atomic numbers (up to about 60). Therefore, if two smaller nuclei were to combine to make a heavier one, the difference in their total binding energy should be released. Second, the graph shows that nuclei with very large atomic numbers have less binding energy per nucleon than those with atomic numbers near 60. Thus if there was a way to break one of these high-Z nuclei into two or more smaller-Z nuclei, the difference in binding energy could be released.

Now we have two predictions that are based on the patterns in the graph:

- When two small nuclei combine, energy should be released.
- When a large nucleus breaks apart into elements with atomic numbers below iron, energy should be released.

It turns out that both of these predicted processes do occur in nature. Note also that the first prediction is consistent with the analysis based on the bar chart in Figure 29.7a.

Fusion and chemical elements

Small nuclei $(A < 60)$ do not spontaneously join to form heavier ones because the nuclei repel each other due to their positive electric charges. For the attractive nuclear force to be exerted, the nuclei must be very close to each other—less than 10^{-14} m apart. In solids and liquids the nuclei of neighboring atoms are about 10^{-10} m apart, far too distant from one another to have a chance to fuse. However, if a material is so hot that it ionizes completely (that is, it becomes a plasma), and the pressure is high enough, a small percentage of nuclei will come close enough so that the nuclear force pulls them together (**Figure 29.9a**) and a huge amount of energy is converted into the kinetic energy of the products (Figure 29.9b). This process, called **fusion**, occurs naturally in stars, which begin their lives with mostly hydrogen and a little helium. Fusion could potentially be a source of clean energy on Earth; however, the temperatures and pressures needed test the limits of current technology.

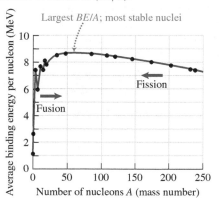

FIGURE 29.8 A graph of binding energy versus number of nucleons (BE/A).

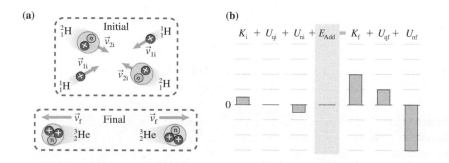

FIGURE 29.9 (a) A greatly simplified example of a fusion reaction. In plasma, ions are moving so fast that they can get close enough to each other during a collision that the nuclear force binds them together. (b) An energy bar chart representing the process.

The combining of light nuclei into heavier ones happens in the cores of stars where the temperature and pressure are very high. Stars with masses comparable to that of the Sun produce elements from helium to carbon through fusion. More massive stars produce elements up to iron. These stars finish their lives by cooling down, and the heavy elements produced in their cores remain there. In contrast, stars that are much more massive than the Sun explode at the end of their lives. These explosions, known as supernovae, produce elements heavier than iron.

Supernovae contribute to the chemical composition of the universe in two ways. First, the elements lighter than iron that are produced in the star's core before the explosion are ejected into space. (Lighter stars that do not explode as supernovae produce those elements, too, but those elements remain in their cores.) Second, elements heavier than iron that are produced during the explosion are also ejected into space. The Sun and Earth have a high abundance of elements heavier than helium; thus the Sun and Earth are made of elements produced long ago inside stars that exploded as supernovae. This means that many of the atoms in our bodies came from these supernovae.

QUANTITATIVE EXERCISE 29.4 ▸ **Fusion energy in the Sun**

The energy released by the Sun comes from several sources, including the so-called proton-proton chain of fusion reactions. This chain of reactions can be summarized as follows:

$$2{}_1^1\text{H} + 2{}_0^1\text{n} \rightarrow {}_2^4\text{He} + \text{energy}$$

Determine the amount of energy (in MeV) released in this chain of reactions. Use the masses ${}_1^1\text{H}$, 1.007825 u; ${}_0^1\text{n}$, 1.008665 u; and ${}_2^4\text{He}$, 4.002602 u to determine the rest energy converted to other forms.

Represent mathematically The energy released, the reaction energy Q, is determined using Eq. (29.5):

$$Q = \left(\sum_{\text{reactants}} m - \sum_{\text{products}} m \right) c^2$$

Solve and evaluate Inserting the appropriate values and converting to MeV gives

$$Q = [\, 2(1.007825\text{ u}) + 2(1.008665\text{ u}) - 4.002602\text{ u} \,]$$
$$\times c^2 \left(\frac{931.5 \text{ MeV}}{\text{u} \cdot c^2} \right) = 28 \text{ MeV}$$

Compare this energy to the energy needed to remove an electron from a hydrogen atom, 13.6 eV. This energy has an order of magnitude typical of chemical reactions. Thus the energy released in nuclear chain reactions is about a million times greater than that released in chemical reactions.

- -

Try it yourself What fraction of the total reactant rest energy in the proton-proton chain reactions is converted to other forms of energy during the above fusion reaction?

Answer   About (0.03 u)/(4.0 u) = 0.008, or 0.8%.

Just as in any reaction, once the reactants have been used up, the reaction stops. This means that eventually the Sun will run out of hydrogen and start burning helium into carbon. After helium is exhausted, fusion will cease (as the Sun is not massive enough to be able to fuse carbon), and our star will slowly go dark. How long will the Sun be able to shine?

EXAMPLE 29.5 ▸ **How long will the Sun shine?**

According to the present models of stellar structure, the central parts of stars, which contain about 10% of the star's mass, have conditions favorable for fusion. Assuming the Sun is made of pure hydrogen, how much time will it take the Sun to radiate the energy released when 10% of its hydrogen has fused into helium? The mass of the Sun is about 2.00×10^{30} kg. The total luminosity L (also power, or energy/second) emitted by the Sun is $L = 3.85 \times 10^{26}$ W.

Sketch and translate Imagine the Sun as an object that simultaneously releases energy through fusion reactions and then radiates it as electromagnetic radiation.

Simplify and diagram Assume that during a fusion reaction in the Sun's interior, two protons and two neutrons join together to form a helium nucleus. The amount of energy released is equal to

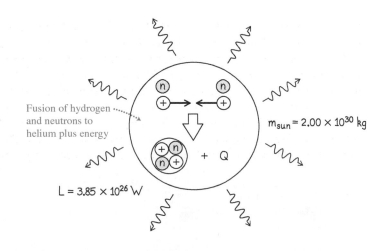

Fusion of hydrogen and neutrons to helium plus energy

$m_{sun} = 2.00 \times 10^{30}$ kg

$L = 3.85 \times 10^{26}$ W

approximately 0.008 of the rest energy of the participating particles (see Quantitative Exercise 29.4). Further, assume that there are no other sources of energy, so that the rate at which energy is produced inside the Sun by fusion reactions is the same as the rate at which the energy is radiated. Finally, assume that the Sun emits the same amount of energy every second during its lifetime.

Represent mathematically The total energy released by fusion reactions throughout the lifetime of the Sun is equal to 0.8% of the rest energy of the 10% of the Sun's mass energy that is available for fusion:

$$E_{\text{released}} = (0.008)(0.10)m_{\text{Sun}}c^2$$

This equals the luminosity L of the Sun times the time interval Δt during which the Sun will shine, or

$$E_{\text{released}} = L\,\Delta t$$

Solve and evaluate Setting the above two expressions equal to each other, solving for Δt, and inserting the appropriate values, we estimate the expected lifetime of the Sun:

$$\Delta t = \frac{E_{\text{released}}}{L}$$

$$= \frac{(0.008)(0.10)(2.00 \times 10^{30}\,\text{kg})(3.00 \times 10^8\,\text{m/s})^2}{3.85 \times 10^{26}\,\text{W}}$$

$$= 3.7 \times 10^{17}\,\text{s}$$

$$= 3.7 \times 10^{17}\,\text{s}\left(\frac{1\,\text{year}}{3.16 \times 10^7\text{s}}\right) = 1.2 \times 10^{10}\,\text{years}$$

This result is about 2.5 times the age of Earth. Nuclear fusion is a possible mechanism for powering the Sun for billions of years.

Try it yourself Studying the energy emitted by stars of different masses, astronomers found that the energy that a typical star emits every second is proportional to its mass raised to the power of 3.5, that is, $L \propto m^{3.5}$. Will a 10-solar-mass star take more or less time than the Sun to exhaust its nuclear fuel, assuming that both the star and the Sun have 10% of their mass in hydrogen available for fusion?

Answer

$$= \frac{L_{\text{Star}}}{10L_{\text{Sun}}} = \frac{L_{\text{Star}}}{m_{\text{Star}}^{3.5}} = 10\left(\frac{m_{\text{Sun}}}{10m_{\text{Sun}}}\right)^{3.5} = 0.0032$$

$$\frac{\Delta t_{\text{Star}}}{\Delta t_{\text{Sun}}} = \frac{\left(\dfrac{E_{\text{released}}}{L}\right)_{\text{star}}}{\left(\dfrac{E_{\text{released}}}{L}\right)_{\text{Sun}}} = \frac{\dfrac{(0.008)(0.10)(10m_{\text{Sun}})c^2}{L_{\text{star}}}}{\dfrac{(0.008)(0.10)m_{\text{Sun}}c^2}{L_{\text{Sun}}}}$$

The star will take significantly less time—about 0.003 times the time for the Sun:

FIGURE 29.10 Nuclear fission.

(a) Original heavy nucleus and a moving neutron

^{235}U

(b) The neutron enters the nucleus, increasing its energy.

^{236}U*

(c) The nucleus becomes unstable.

^{236}U*

(d) The excited nucleus divides into two smaller nuclei and several neutrons.

(e) The new neutrons can hit other heavy nuclei, producing a chain reaction.

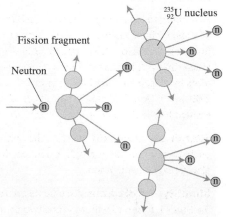

$^{235}_{92}$U nucleus

Fission fragment

Neutron

Fission and nuclear energy

We saw from the graph in Figure 29.8 that energy is released when very heavy nuclei split into nuclei of atoms that reside near the middle of the periodic table. However, the heavy nuclei do not just spontaneously split. One way the process can be initiated is the collision of a moving neutron with a heavy nucleus, which disturbs the equilibrium of the nucleus and causes it to split into two smaller nuclei. This process is called **fission** (**Figure 29.10a–d**). Heavy nuclei have a high neutron-to-proton ratio; however, nuclei near the middle of the periodic table (which are the typical product nuclei of these kinds of reactions) usually have a somewhat lower ratio. Thus splitting a heavy nucleus usually releases several neutrons, which can then catalyze the splitting of other

FIGURE 29.11 Energy bar chart representing fission.

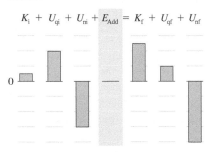

$$K_i + U_{qi} + U_{ni} + E_{Add} = K_f + U_{qf} + U_{nf}$$

nuclei, causing a chain reaction (Figure 29.10e). This reaction results in a huge release of energy, the kinetic energy of the products (see the bar chart in **Figure 29.11**).

A common fission reaction is

$$^1_0n + ^{235}_{92}U \rightarrow ^{141}_{56}Ba + ^{92}_{36}Kr + 3^1_0n + energy$$

Notice that one neutron absorbed by one uranium nucleus produces two smaller nuclei plus three new neutrons that can cause three new fission reactions. Let's calculate the energy released by the above fission reaction.

QUANTITATIVE EXERCISE 29.6   **Energy from a fission reaction**

Determine the energy released in the above fission reaction. Use the following masses: ^1_0n, 1.008665 u; $^{235}_{92}U$, 235.043924 u; $^{141}_{56}Ba$, 140.914411 u; $^{92}_{36}Kr$, 91.926156 u. One neutron collides with and momentarily joins the U-235 atomic nucleus to form an excited U-236, which quickly splits into Ba-141, Kr-92, and three neutrons.

Represent mathematically The energy Q released in the reaction is the energy equivalent to the mass difference between the reactants and products.

$$Q = \left[m(^1_0n) + m(^{235}_{92}U) \right]c^2$$
$$- \left[m(^{141}_{56}Ba) + m(^{92}_{36}Kr) + 3m(^1_0n) \right]c^2$$
$$= \left[m(^{235}_{92}U) - m(^{141}_{56}Ba) - m(^{92}_{36}Kr) - 2m(^1_0n) \right]c^2$$

Solve and evaluate Inserting the appropriate values and converting to MeV, we get

$$Q = \left[235.043924\ u - 140.914411\ u - 91.926156\ u - 2(1.008665\ u) \right]$$
$$\times c^2 \left(\frac{931.5\ MeV}{u \cdot c^2} \right) = 173\ MeV$$

This is a typical energy release during fission reactions. This reaction produces over 300 million times more energy than burning one octane molecule.

Try it yourself Your friend says that nuclear power plants could release energy by the fission of iron-56 (very abundant) instead of the much less abundant uranium-235. How do you respond?

Answer Iron-56 has about the highest binding energy per nucleon of any nucleus. This means energy must be added to the nucleus to cause it to break into smaller nuclei with lower BE/A—exactly the opposite of what is desired in a nuclear power plant.

Lise Meitner.

Bohr's liquid drop model of the nucleus

Quantitative Exercise 29.6 represents an experiment performed by German scientists Lise Meitner, Otto Hahn, and Fritz Strassmann in the 1930s. Their goal was not to split uranium, but instead to produce elements heavier than uranium. They thought that bombarding uranium with neutrons would lead to neutron capture and subsequent beta decay that in turn would increase the number of protons in the nucleus, thus creating heavier elements. (You will find the explanation of how neutron capture leads to the production of heavier elements in Section 29.6.) To their surprise, the nuclei produced in the reaction behaved like isotopes of barium and other nuclei with about half the mass of uranium. They were at a complete loss to explain this result. At this point, Meitner, who was Jewish, had to flee Germany under Adolf Hitler's regime. She immigrated to Sweden. There, she asked her nephew Otto Robert Frisch (who was working in Denmark with Niels Bohr) to help interpret these results.

Meitner and Frisch decided that Bohr's **liquid drop model** of a nucleus could explain the strange experimental results (see Figure 29.10). In this model, the nucleus was compared to a drop of water: just as surface tension holds the water drop together, nuclear forces hold the nucleons together. However, in heavy nuclei, such as uranium, there are too many electrically charged protons present. The protons repel each other and can overwhelm the effect of the "surface tension," especially if the nucleus is not spherical. The model suggested that such a nucleus could stretch and divide into two smaller pieces. This meant that the uranium nucleus is very unstable, ready to split with the slightest provocation—such as being hit by a neutron. Meitner and Frisch had come up with a model for fission. Unfortunately, the Nobel Prize committee did not think that the work of Lise Meitner contributed enough to the discovery of fission, and so she did not share the Nobel Prize given to Otto Hahn in 1944.

REVIEW QUESTION 29.5 Explain how fusion can lead to a release of energy. Does this release mean that energy is not conserved in nuclear fusion reactions?

29.6 Mechanisms of radioactive decay

Now that we understand the stability of nuclei in terms of their binding energy per nucleon and the energy released in nuclear reactions (the Q of the reactions), we can explain why some nuclei undergo radioactive decay while others do not.

Alpha decay

Of the small nuclei, helium $\left(^{4}_{2}\text{He}\right)$ is one of the most stable. Because of this, in a larger nucleus, groups of two protons and two neutrons tend to form helium nuclei within the larger nucleus (like a small group of friends joined together in a large crowd). In addition to the attractive forces that the two protons and two neutrons in a helium nucleus exert on each other, a repulsive electric force is exerted on the helium nuclei by the other protons in the larger nucleus. When the nucleus is large enough, the electric repulsion becomes significant compared to the nuclear attraction. Then there is a chance that one of the helium nuclei will leave the larger nucleus. This is the emission of an alpha particle.

When one of the alpha particles leaves, this emission reduces the number of protons in the original nucleus and also reduces the electric repulsion between the protons that remain. Because a helium nucleus also contains two neutrons, the total number of nucleons A in the original nucleus decreases by four. **Figure 29.12** illustrates the alpha decay of radium-226. When radium undergoes this decay, it becomes radon-222, and an alpha particle is released. The mass of the new radon nucleus (called the **daughter nucleus**) plus the alpha particle is less than the original radium nucleus (called the **parent nucleus**). Some of the rest energy of the radium nucleus is converted into the kinetic energy of the products.

$$^{226}_{88}\text{Ra} \rightarrow {}^{222}_{86}\text{Rn} + {}^{4}_{2}\text{He} + 4.9\ \text{MeV}$$

FIGURE 29.12 Alpha decay of radium-226. The result is a new element plus release of an alpha particle.

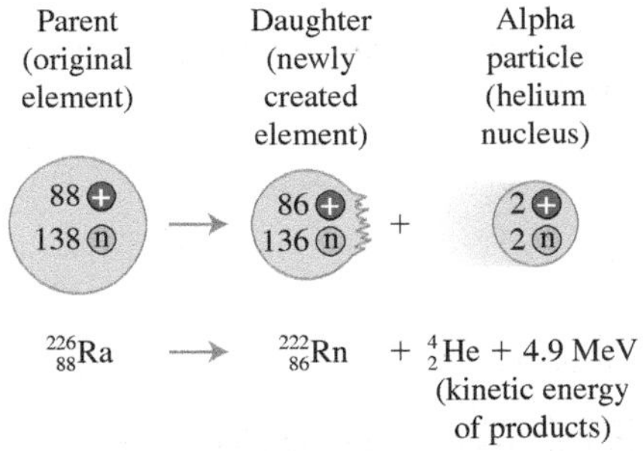

$$^{226}_{88}\text{Ra} \longrightarrow {}^{222}_{86}\text{Rn} + {}^{4}_{2}\text{He} + 4.9\ \text{MeV}$$
(kinetic energy of products)

QUANTITATIVE EXERCISE 29.7 **Kinetic energy produced during alpha decay**

Determine the kinetic energy of the product nuclei when polonium-212 undergoes alpha decay. The masses of the nuclei involved in the decay are $m\left(^{212}_{84}\text{Po}\right)$, 211.9889 u; $m\left(^{208}_{82}\text{Pb}\right)$, 207.9766 u; and $m\left(^{4}_{2}\text{He}\right)$, 4.0026 u.

Represent mathematically The alpha decay of polonium-212 is

$$^{212}_{84}\text{Po} \rightarrow {}^{208}_{82}\text{Pb} + {}^{4}_{2}\text{He} + \text{energy}$$

Use Eq. (29.5) to find the energy Q released during this reaction:

$$Q = \left[m\left(^{212}_{84}\text{Po}\right) - m\left(^{208}_{82}\text{Pb}\right) - m\left(^{4}_{2}\text{He}\right)\right]c^2$$

Solve and evaluate Insert the appropriate values and convert to MeV:

$$Q = (211.9889\ \text{u} - 207.9766\ \text{u} - 4.0026\ \text{u})c^2$$

$$= (0.0097\ \text{u})c^2 \left(\frac{931.5\ \text{MeV}}{\text{u}\cdot c^2}\right)$$

$$= 9.0\ \text{MeV}$$

Of this 9.0 MeV of released energy, 8.8 MeV is converted to the kinetic energy of the alpha particle $\left(^{4}_{2}\text{He}\right)$. Most of the remaining 0.2 MeV is converted to the kinetic energy of the recoiling lead-208 nucleus. In addition, a small fraction of the energy may be released as an additional product in the reaction: a gamma ray (a high-energy photon).

Try it yourself Determine the product nucleus when thorium-232 $\left(^{232}_{90}\text{Th}\right)$ undergoes alpha decay.

Answer   Radium-228 $\left(^{228}_{88}\text{Ra}\right)$.

Beta decay

The particle emitted from a nucleus during beta decay is an electron. But we know from the uncertainty principle that electrons cannot reside in the nucleus. How can we explain this apparent contradiction? One explanation is that a neutron inside a nucleus can spontaneously decay into a proton and an electron. The proton stays inside the nucleus, while the electron created during the decay leaves. This neutron decay reaction could be written as follows:

$$\,^1_0\text{n} \rightarrow \,^1_1\text{p} + \,^{\ 0}_{-1}e$$

where p represents a proton. Notice that both the total electric charge and the total nucleon number on both sides of the equation are equal, but, as we will see later, this equation is incomplete.

If this explanation is correct, then during beta decay, a particular element should transform to an element with a Z number that is larger by one. This is exactly what Ernest Rutherford and his colleague Frederick Soddy predicted while studying radioactive decay. Another prediction that follows from this explanation is that the heavier isotopes of a particular element should produce more beta radiation than the lighter isotopes of that element. This is exactly what is observed. For example, carbon-14 is a heavy isotope of carbon that has two more neutrons than protons, and it is unstable, undergoing beta decay, whereas carbon-12 and carbon-13 are stable. However, when scientists studied beta decay in greater detail, they encountered two difficulties.

The problem with conservation of angular momentum arising from spin You learned in Chapter 28 that electrons have an intrinsic spin quantum number. This spin quantum number was observed in many other experiments to be a conserved quantity. However, in beta decay experiments, it seemed not to be. For example, during the beta decay of a neutron, all three participating particles have spin values equal to $\pm 1/2$. But because there is only one particle on one side of the reaction but two particles on the other side, it is not possible that the total spin on both sides could be equal. Either spin is not a conserved quantity, or our understanding of beta decay is flawed.

The problem with energy conservation The second problem involved energy conservation. It, too, seemed to not be conserved in beta decay. The total energy of the products was always observed to be less than the total energy of the reactants, but it was not possible to detect where that missing energy went. This pattern was so persistent that originally Niels Bohr suggested that maybe energy was not always conserved in the subatomic world. However, this idea was too radical for most physicists to take seriously.

In 1930, Wolfgang Pauli proposed an explanation for beta decay that did not require abandoning energy conservation or spin number conservation. He hypothesized that some unknown particle carried away the missing energy and accounted for the discrepancy in spin number. This particle had zero electric charge, zero mass, and a spin number of either $+1/2$ or $-1/2$. Enrico Fermi called the particle a **neutrino**, meaning "little neutral one." Because the neutrino was thought to have zero mass, it was expected to travel at the speed of light.

Although this sounds like the description of a photon, there is a subtle difference between a photon and a neutrino. An atom can absorb a photon and as a result change its energy state. Allowed energy transitions occur only between states that differ by 1 in the angular momentum quantum number l. The photon's intrinsic spin must then be equal to the change in angular momentum of the atom, in this case, 1. A neutrino's intrinsic spin, however, must be $1/2$.

The existence of neutrinos was confirmed 25 years after Pauli's proposal. Large numbers of neutrinos produced by the Sun and other astronomical objects continually pass through our bodies. Yet they cause no damage and leave almost no trail because of the extremely small likelihood of them interacting with the atoms in their path.

We can now correct the equation for the decay of a neutron using our knowledge of neutrinos. The correct equation is

$$_0^1n \rightarrow {}_1^1p + {}_{-1}^0e + \bar{\nu}$$

where $\bar{\nu}$ is the antineutrino, the particle that is responsible for the conservation of angular momentum and energy in the decay. We will learn more about antineutrinos and why they are needed in this reaction as opposed to neutrinos in Chapter 30. Examples of other reactions involving beta decay, shown in **Figure 29.13** and represented mathematically below, always result in the production of a neutrino ν or an antineutrino $\bar{\nu}$ (we will discuss neutrinos and antimatter in more detail in Chapter 30). The first reaction is called **beta-minus decay**, as it produces a negative electron:

$$_6^{14}C \rightarrow {}_7^{14}N + {}_{-1}^0e + \bar{\nu}$$

$$_{11}^{22}Na \rightarrow {}_{10}^{22}Ne + {}_1^0e + \nu$$

The second beta decay is an example of a somewhat less common decay known as **beta-plus decay**, which produces a positron $\left({}_1^0e\right)$, which is otherwise identical to an electron but has positive charge e instead of the negative charge $-e$ of the electron. The positron is the antiparticle to the electron (we will discuss positrons further in Chapter 30).

What is the mechanism behind beta decay? Because neutrinos interact with atoms only extremely rarely, it's unlikely that the mechanism is electromagnetic in nature. For this and other technical reasons, Enrico Fermi proposed a new interaction in 1934 called the **weak nuclear interaction**, which explained beta decay. Like the nuclear interaction, the weak interaction is very short range and is much weaker (hence its name).

A free neutron (a neutron that is not inside a nucleus) is not a stable particle. It spontaneously decays rather quickly (in about 15 min on average), producing a proton, an electron, and an antineutrino. The weak interaction is responsible for this decay. When neutrons are bound inside nuclei they are very stable.

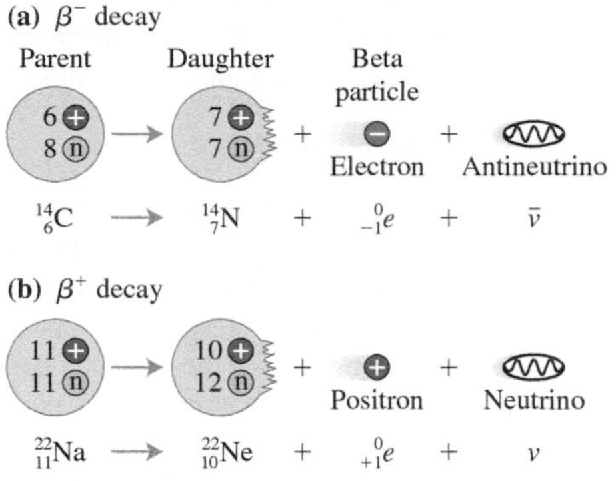

Gamma decay

We already know that alpha and beta decays are usually accompanied by rays that are not deflected by a magnetic field but that are easily detected by their interactions with atoms. These rays are gamma (γ) rays—photons that have higher energy and therefore shorter wavelength than X-rays.

One explanation for why gamma rays accompany alpha and beta decay is that the energy of the nucleus is quantized in a similar way to the energy in atoms. If this is true, then after alpha or beta decay the nucleus could be left in an excited state from which it then emits one or more photons to return to its ground state. Because the binding energy of a nucleus is much higher than that of the atom's electrons, the energy differences between states are also likely to be much larger. As a result, the energy of the emitted photons is much larger. Gamma decay happens, for example, when boron-12 undergoes beta decay to form carbon-12. The carbon-12 that is produced is in an excited state $\left({}_6^{12}C*\right)$ (the asterisk indicates that the nucleus is in an excited state). When it returns to its lowest energy state, a gamma ray photon is emitted (**Figure 29.14**):

$$_5^{12}B \rightarrow {}_6^{12}C* + {}_{-1}^0e$$

$$_6^{12}C* \rightarrow {}_6^{12}C + \gamma$$

FIGURE 29.13 Two types of beta decay. (a) Beta-minus decay, releasing an electron and an antineutrino. (b) Beta-plus decay, releasing a positron and a neutrino.

(a) β^- decay

Parent   Daughter   Beta particle

$6 \oplus$ $7 \oplus$
$8 \, ⓝ$ $7 \, ⓝ$ $+$ $\ominus$ Electron $+$ Antineutrino

$_6^{14}C \rightarrow {}_7^{14}N + {}_{-1}^0e + \bar{\nu}$

(b) β^+ decay

$11 \oplus$ $10 \oplus$
$11 \, ⓝ$ $12 \, ⓝ$ $+$ $\oplus$ Positron $+$ Neutrino

$_{11}^{22}Na \rightarrow {}_{10}^{22}Ne + {}_{+1}^0e + \nu$

FIGURE 29.14 Gamma decay. (a) After beta decay, the daughter nucleus is left in an excited state and drops to the ground state, emitting a gamma ray photon. (b) An energy level diagram for the gamma ray emission process.

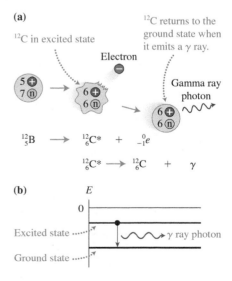

(a)

^{12}C in excited state

^{12}C returns to the ground state when it emits a γ ray.

Electron

$5 \oplus$
$7 \, ⓝ$ $\rightarrow$ $6 \oplus$
$6 \, ⓝ$ $\rightarrow$ $6 \oplus$
$6 \, ⓝ$ Gamma ray photon

$_5^{12}B \rightarrow {}_6^{12}C* + {}_{-1}^0e$

$_6^{12}C* \rightarrow {}_6^{12}C + \gamma$

(b) E

0

Excited state

Ground state γ ray photon

QUANTITATIVE EXERCISE 29.8 | **Alpha and beta decay**

The following nuclei undergo different types of radioactive decay. Determine the daughter nucleus for each, and write an equation representing each decay reaction: (a) $^{239}_{94}$Pu alpha decay, (b) $^{144}_{58}$Ce beta-minus decay, (c) and $^{65}_{30}$Zn beta-plus decay. The latter produces a positron.

Represent mathematically

(a) We have to subtract $A = 4$ and $Z = 2$ (the $^4_2$He alpha particle) from the $^{239}_{94}$Pu. Thus, the daughter nucleus has $A = 239 - 4 = 235$ and $Z = 94 - 2 = 92$, which is uranium-235.

(b) We have to subtract $A = 0$ and $Z = -1$ (the $^0_{-1}e$ beta-minus particle, an electron) from the $^{144}_{58}$Ce. Thus, the daughter nucleus has $A = 144 - 0 = 144$ and $Z = 58 - (-1) = 59$, which is praseodymium-144.

(c) We have to subtract $A = 0$ and $Z = +1$ (the ^0_1e beta-plus particle, a positron) from the $^{65}_{30}$Zn. Thus, the daughter nucleus has $A = 65 - 0 = 65$ and $Z = 30 - 1 = 29$, which is copper-65.

Solve and evaluate The reactions are then

(a) $^{239}_{94}$Pu $\rightarrow$ $^{235}_{92}$U + $^4_2$He + energy

(b) $^{144}_{58}$Ce $\rightarrow$ $^{144}_{59}$Pr + $^0_{-1}e$ + $\bar{\nu}$ + energy

(c) $^{65}_{30}$Zn $\rightarrow$ $^{65}_{29}$Cu + ^0_1e + ν + energy

"Energy" here means the kinetic energy of the products and possibly one or more gamma rays.

- -

Try it yourself Identify the daughter nucleus and write a reaction equation for the beta-minus decay of $^{131}_{53}$I.

Answer $^{131}_{53}$I $\rightarrow$ $^{131}_{54}$Xe + $^0_{-1}e$ + $\bar{\nu}$ + energy.

EXAMPLE 29.9 | **Beta decay in our bodies**

The body normally contains about 7 mg of radioactive potassium-40 $^{40}_{19}$K (it is present in some of the foods we eat, such as bananas). Each second, about 2×10^3 of these potassium nuclei undergo either beta-plus or beta-minus decay. Assume that all decays are beta-plus and that 40% of the energy released is absorbed by the body (the percent depends on the ratio of beta-minus decay to beta-plus decay, body thickness, the energy of the neutrinos produced in the decays, and other factors). Determine the energy in MeV transferred to the body each second by $^{40}_{19}$K beta-plus decay. What is the rate of energy transfer in watts?

Sketch and translate A single beta-plus decay is sketched below. If we determine the energy released by a single decay, we can then determine the rate of energy transfer to the body by multiplying the energy of one decay by 2000 (the number of decays that occur every second) and then taking 40% of that number.

Simplify and diagram The system of interest is the potassium atom and its decay products.

Represent mathematically The Q of a single beta-plus decay is

$$Q = \left[m\left(^{40}_{19}K\right) - m\left(^{40}_{18}Ar\right) - m\left(^0_1e\right) \right]c^2$$

The rate of energy transferred to the body ($\Delta U/\Delta t$) is 40% of Q times the number ($\Delta N/\Delta t$) of decays per second, or

$$\frac{\Delta U}{\Delta t} = \frac{\Delta N}{\Delta t}(0.4Q)$$

Solve and evaluate From Appendix B, we find the masses of the atoms in atomic mass units (u):

$$Q = (39.964000 \text{ u} - 39.962384 \text{ u} - 5.4858 \times 10^{-4} \text{ u})c^2$$

$$= (0.001068 \text{ u})c^2\left(\frac{931.5 \text{ MeV}}{\text{u} \cdot c^2}\right) = 0.995 \text{ MeV}$$

The energy transferred to the body per second is then

$$\frac{\Delta U}{\Delta t} = (2000 \text{ decays/s})(0.995 \text{ MeV})(0.4) = 800 \text{ MeV/s}$$

In watts this is

$$\frac{\Delta U}{\Delta t} = (800 \text{ MeV/s})\left(\frac{1.6 \times 10^{-13} \text{ J}}{\text{MeV}}\right)$$

$$= 1.3 \times 10^{-10} \text{ W} \approx 1 \times 10^{-10} \text{ W}$$

This is extremely small compared to the average human metabolic rate of 100 W. We'll see in a later section whether these beta particles cause any negative health effects.

- -

Try it yourself Is more or less energy released by potassium-40 beta-minus decay than beta-plus decay?

Answer The product of the beta-minus decay is $^{40}_{20}$Ca, which has mass 39.962591 u. This is slightly greater than the 39.962383-u $^{40}_{18}$Ar product mass of the beta-plus decay. The beta-plus decay releases more energy.

REVIEW QUESTION 29.6 What was the observational evidence related to beta decay that made scientists think that a new particle was involved?

29.7 Half-life, decay rate, and exponential decay

Radioactive materials are used for medical diagnoses, and they are part of the world in which we live. Radioactive isotopes can also help determine the age of bones and other archeological artifacts through radioactive dating. For isotopes to serve these purposes, we need a quantitative description of how the number of radioactive nuclei in a sample changes and the rate at which radioactive decay occurs. These quantitative measures include half-life, the decay rate of nuclei, and the application of the exponential function to decay.

Half-life

Suppose a radioactive sample has a number N_0 of radioactive nuclei at time $t_0 = 0$. We can determine this number by dividing the mass of the sample by the mass of one atom. Using a particle detector such as a Geiger counter, we can estimate the number of nuclei that decay in a short time interval. By continually subtracting the number of decays in a short time interval from the initial number, we can determine the number N of radioactive nuclei that remain in the sample as a function of time. In Observational Experiment Table 29.5 we look for a pattern in the variation of N with time t.

OBSERVATIONAL EXPERIMENT TABLE 29.5 **The number N of radioactive nuclei versus time t**

Observational experiment	Analysis
We measure the decay rate in a radioactive sample and determine the number of radioactive nuclei in the sample as a function of time:	• The number decreased from N_0 to $0.50N_0$ from time 0 to 1 min.
	• The number decreased from $0.50N_0$ to $0.25N_0$ from time 1 min to 2 min.

Number N of radioactive nuclei	t (min)
N_0	0
$0.50N_0$	1
$0.25N_0$	2
$0.13N_0$	3
$0.06N_0$	4
$0.03N_0$	5
undetectable	10

• The number decreased from $0.25N_0$ to $0.13N_0$ from time 2 min to 3 min.

• The number decreased from $0.13N_0$ to $0.06N_0$ from time 3 min to 4 min.

• The number decreased from $0.06N_0$ to $0.03N_0$ from time 4 min to 5 min.

Pattern

For this sample, the number N of radioactive nuclei in the sample decreased by half during each 1-min time interval.

The data in Table 29.5 are plotted in **Figure 29.15**. As noted in the table, the number N of radioactive nuclei present in this particular radioactive sample decreased by half during each 1-min time interval. The reduction by one-half in a particular time interval, which varies depending on the type of nuclei, is a general property of radioactive samples called the **half-life** of the sample.

Half-life The half-life T of a particular type of radioactive nucleus is the time interval during which the number of nuclei in a given sample is reduced by one-half. After n half-lives, the fraction of radioactive nuclei that remain is

$$\frac{N}{N_0} = \frac{1}{2^n} \text{ at } t = nT \qquad (29.6)$$

FIGURE 29.15 A graph of radioactive decay.

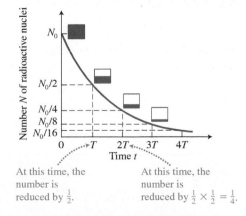

At this time, the number is reduced by $\frac{1}{2}$.

At this time, the number is reduced by $\frac{1}{2} \times \frac{1}{2} = \frac{1}{4}$.

TABLE 29.6 Half-lives and decay constants of some common nuclei

Isotope	Half-life	Decay constant (s^{-1})
$^{87}_{37}Rb$	4.75×10^{10} years	4.62×10^{-19}
$^{238}_{92}U$	4.47×10^{9} years	4.91×10^{-18}
$^{40}_{19}K$	1.28×10^{9} years	1.72×10^{-17}
$^{239}_{94}Pu$	2.41×10^{4} years	9.11×10^{-13}
$^{14}_{6}C$	5730 years	3.83×10^{-12}
$^{226}_{88}Ra$	1600 years	1.37×10^{-11}
$^{137}_{55}Cs$	30.0 years	7.32×10^{-10}
$^{90}_{38}Sr$	28.9 years	7.60×10^{-10}
$^{3}_{1}H$	12.3 years	1.79×10^{-9}
$^{60}_{27}Co$	5.27 years	4.17×10^{-9}
$^{131}_{53}I$	8.03 day	9.99×10^{-7}
$^{11}_{6}C$	20.4 min	5.66×10^{-4}

Note that after one half-life, the ratio $N/N_0 = 1/2$. During the next half-life, there is another 50% reduction in the number of nuclei. Thus, after two half-lives, $N/N_0 = 1/2^2 = 1/4$, and so forth. For the radioactive sample in Table 29.5, the half-life is 1 min. According to Eq. (29.6), the fraction of radioactive nuclei that remains at $t = 4$ min $= 4T$ is $N/N_0 = 1/2^4 = 1/16 = 0.06$, in agreement with the data in the table.

This method can be used for fractional half-lives. For example, after 4.5 min (4.5 half-lives), the fraction of nuclei that remain is $N/N_0 = 1/2^{4.5} = 1/22.6 = 0.044$. The half-lives for a variety of nuclei are listed in Table 29.6.

Determining the source of carbon in plants

The complex photosynthesis process in plant growth can be summarized by the following chemical reaction:

$$6CO_2 + 6H_2O + sunlight \rightarrow C_6H_{12}O_6 + 6O_2$$

What is the source of the CO_2 in this process—is it part of the minerals found in the ground in which the plants grow, such as calcium carbonate $(CaCO_3)$, or is it the gaseous CO_2 in the air? A radioactive decay experiment was originally used to answer this question.

The carbon isotope carbon-11 $\left(^{11}_{6}C\right)$ is radioactive, with a half-life of 20 min. Carbon-11 was incorporated into the CO_2 in the air surrounding growing barley plants in a controlled environment. Investigators found that radioactive carbon-11 became part of the carbohydrate molecules produced by photosynthesis in the barley. This result is evidence that carbon in plants comes from CO_2 in the atmosphere. More evidence could be obtained by repeating the experiment with carbon-11 incorporated into soil minerals.

Decay rate

When working with radioactive samples, we seldom know or measure directly the number of radioactive nuclei in a sample. Instead, we measure the change in the number of nuclei that decay, ΔN, during a certain time interval Δt. From this measurement we can calculate the number of radioactive nuclei that still remain in the sample because the change in the number ΔN of nuclei that decay during time interval Δt is proportional to the starting number N of radioactive nuclei in the sample, and is proportional to Δt:

$$\Delta N = -\lambda N \Delta t \qquad (29.7)$$

This relationship makes sense if we assume that the absolute value of ΔN is the number of decayed nuclei. The number that decay is greater if the number N in the sample is greater. The number that decay during a certain time interval should also be proportional to that time interval Δt as long as the time interval is short compared to the half-life of the radioactive sample. λ is a proportionality constant, called the **decay constant**, that depends on the type of radioactive nucleus (see Table 29.6). The greater the value of λ, the greater the rate at which the radioactive nuclei decay. The ratio of ΔN and Δt is called the **decay rate**, or the **activity**, of a sample of radioactive material.

Decay rate (activity) A The ratio of the number of nuclei that decay ΔN during a certain time interval Δt and that time interval is called the decay rate, or activity, of a sample of that type of radioactive nucleus:

$$\text{Decay rate (activity) } A = -\frac{\Delta N}{\Delta t} = \lambda N \qquad (29.8)$$

where λ is the decay constant for that particular type of nucleus and ΔN is a negative number due to the decrease in the number of nuclei.

The unit of decay rate is the becquerel (Bq) and equals 1 decay/s. An older unit of decay rate is the curie (Ci), where $1 \text{ Ci} = 3.70 \times 10^{10} \text{ Bq}$. This unusual number for the curie was chosen because it represents roughly the activity of 1 g of pure radium, the radioactive element that Marie Curie isolated from tons of uranium ore. Radioactive tracers used in medicine usually have activities of tens of kilobecquerels or microcuries (μCi; $1 \mu\text{Ci} = 10^{-6} \text{ Ci}$). Radiation detection devices such as Geiger counters are used to measure the activity of radioactive samples.

Exponential function and decay

The N-versus-t data in Table 29.5 and in Figure 29.15 are characteristic of a variety of phenomena that can be represented mathematically by what is called the exponential function. We can use Eq. (29.8) to derive an expression for N as a function of time t. To start, we turn the Δt in Eq. (29.8) into a very short time interval $\Delta t'$ during which the number N changes by $\Delta N'$. We then rearrange the equation to

$$\frac{\Delta N'}{N} = -\lambda \Delta t'$$

These short time changes can be summed up for the time interval starting from time 0 when N_0 nuclei are present to some later time t when N nuclei remain. The result of this summation is the exponential function

$$N = N_0 e^{-\lambda t}$$

Because the activity A of a sample is proportional to N, the activity is also described by an exponential function:

$$A = -\frac{\Delta N}{\Delta t} = \lambda N = \lambda N_0 e^{-\lambda t}$$

The number of nuclei present in the sample decreases exponentially with time, as does the activity of the sample.

Exponential decay The number N of radioactive nuclei that remain at time t if the number at time zero was N_0 is given by an exponential function:

$$N = N_0 e^{-\lambda t} \qquad (29.9)$$

where λ is the decay constant of the nucleus, and $e = 2.718\ldots$ is the base of the natural logarithms. The activity A of the sample also decreases exponentially:

$$A = -\frac{\Delta N}{\Delta t} = \lambda N_0 e^{-\lambda t}$$

N is an exponentially decreasing function of time and is plotted in the graph shown in Figure 29.15. We can now determine the number N of radioactive nuclei that remain in a sample at time t compared to the number N_0 at time zero by using either Eq. (29.6) or Eq. (29.9). The former equation, $(N/N_0) = (1/2^n)$, is often easier to use. However, rearrangement of the exponential function ($N = N_0 e^{-\lambda t}$) leads to an important equation for calculating the unknown age of a sample, which we will use when we discuss radioactive dating in the next section.

Equations for radioactive decay have numerous applications. For example, they can be used in medicine to measure the volume of blood in a patient. The process for this measurement is illustrated in the following example.

EXAMPLE 29.10 Estimating blood volume

When a small amount of radioactive material is placed in 1.0 cm^3 of water, it has an activity of 75,000 decays/min. Imagine that an identical amount of radioactive material is injected into an individual's bloodstream. Later, after the material has spread throughout the blood, a 1.0-cm^3 sample of blood taken from the individual has an activity of 16 decays/min. Determine the individual's total blood volume.

Sketch and translate The total blood volume can be thought of as a large container of unknown volume and the 1.0-cm^3 sample as a small portion of that. We know that a direct relationship exists between the decays/min in the sample and the decays/min in the total volume of blood. Therefore, we can use the value of one to determine the value of the other.

Simplify and diagram Assume that during this process the person does not drink any water (so that their blood volume doesn't change), that all the radioactive material is absorbed by the blood, and that none is filtered out by the kidneys. Assume also that the material disperses evenly throughout the blood. Lastly, assume that the half-life of that type of material is much longer than the time interval needed to complete the experiment so that the activity of the radioactive material can be considered constant.

Represent mathematically We use a ratio method to determine the blood volume by comparing the activity A and volume V of the two amounts of blood (the whole body and the sample):

$$\frac{V_{blood}}{V_{sample}} = \frac{A_{blood}}{A_{sample}}$$

Solve and evaluate Solve for V_{blood} and substitute the known quantities into the above to determine the blood volume:

$$V_{blood} = \frac{A_{blood}}{A_{sample}}V_{sample} = \left(\frac{75,000 \text{ decays/min}}{16 \text{ decays/min}}\right)(1.0 \text{ cm}^3)$$
$$= 4.7 \times 10^3 \text{ cm}^3$$

where $(4.7 \times 10^3 \text{ cm}^3)(1 \text{ L}/10^3 \text{ cm}^3) = 4.7 \text{ L}$. This value is within the 4- to 6-L range that is considered a normal human blood volume.

Try it yourself Suppose the radioactive sample had a half-life of 1.0 h and you waited 2 h for the injected sample to distribute uniformly before measuring the activity of 1 cm^3 of blood. What would the activity of that blood sample be?

Answer

In two half-lives, the activity would be reduced to $(1/2^2)(16 \text{ decays/min}) = 4 \text{ decays/min}$.

Decay rate and half-life

The values of the half-life and decay constant for several different radioactive nuclei were reported in Table 29.6. Because these two quantities are both related to how quickly a sample of radioactive material decays, we expect that a relationship exists between them. What is that relationship?

At time $t = T$ (one half-life), the number N of radioactive nuclei remaining is one-half the number N_0 at time zero. Using the equation $N = N_0 e^{-\lambda t}$, we can write

$$\tfrac{1}{2} N_0 = N_0 e^{-\lambda T}$$

or

$$\tfrac{1}{2} = e^{-\lambda T}$$

Take the natural logarithm of both sides of this equation and solve for the half-life T:

$$\ln\left(\tfrac{1}{2}\right) = -\lambda T \Rightarrow T = -\frac{1}{\lambda}\ln\left(\tfrac{1}{2}\right)$$

or

$$T = \frac{0.693}{\lambda} \tag{29.10}$$

This relationship between the half-life of the material and its decay constant makes sense. If the decay constant λ is large, then the material decays rapidly and consequently has a short half-life T.

REVIEW QUESTION 29.7 What does it mean that the half-life of a particular isotope is 200 years?

29.8 Radioactive dating

In the previous section, you saw how to calculate the fraction N/N_0 of radioactive nuclei that remain in a sample after a known time interval. Archeologists and geologists, in contrast, are interested in the reverse calculation—determining the age (that is, time interval that has passed) of a radioactive sample from the known fraction N/N_0 of radioactive nuclei that remain in a sample. Their method is based on a rearrangement of Eq. (29.9), $N = N_0 e^{-\lambda t}$.

First, divide both sides of the equation by N_0 and then take the natural logarithm of each side to get

$$\ln\left(\frac{N}{N_0}\right) = -\lambda t$$

Now substitute $\lambda = 0.693/T$ [Eq. (29.10)] and rearrange to get an expression for the age of a radioactive sample:

$$t = -\frac{\ln(N/N_0)}{0.693}T \qquad (29.11)$$

In this equation T is the half-life of the radioactive material and t is the sample's age when N radioactive nuclei remain. The sample started at time zero with N_0 radioactive nuclei.

TIP Notice that in Eq. (29.11) you can use whatever units you want for t and T (seconds, years, thousands of years) provided that you are using the same units for t and T.

Carbon dating

An important type of radioactive dating, called carbon dating, is used to determine the age of objects that are less than 40,000 years old. Most of the carbon in our environment consists of the isotope carbon-12, but about 1 in 10^{12} carbon atoms has a radioactive carbon-14 nucleus with a half-life of about 5700 years. Carbon-14 is continuously produced in our atmosphere by collisions of neutrons from the solar wind with nitrogen nuclei in the atmosphere, resulting in the reaction

$$^{14}_{7}\text{N} + ^{1}_{0}\text{n} \rightarrow ^{14}_{6}\text{C} + ^{1}_{1}\text{H}$$

Atmospheric carbon-14 radioactively decays at the same rate as it is created by the above reaction: a stable equilibrium exists.

Any plant or animal that metabolizes carbon incorporates about one carbon-14 atom into its structure for every 10^{12} carbon-12 atoms it metabolizes. Because carbon is no longer absorbed and metabolized by the organism after death, the carbon-14 in the bones starts to transform by negative beta decay into nitrogen-14. After 5700 years the carbon-14 concentration decreases by one-half. After two half-lives, or 11,400 years after death, the concentration has dwindled to one-fourth what it was when the organism died. A measurement of the current carbon-14 concentration indicates the age of the remains.

EXAMPLE 29.11 Age of bone

A bone found by an archeologist contains a small amount of radioactive carbon-14. The radioactive emissions from the bone produce a measured decay rate of 3.3 decays/s. The same mass of fresh cow bone produces 30.8 decays/s. Estimate the age of the sample.

Sketch and translate The situation is sketched below.

Fresh bone   Old bone

$\dfrac{\Delta N}{\Delta t}$ $(t_i = 0)$ $\dfrac{\Delta N}{\Delta t}$ $(t_f = ?)$

$= 30.8 \dfrac{\text{decays}}{\text{s}}$ $= 3.3 \dfrac{\text{decays}}{\text{s}}$

Simplify and diagram We assume that the concentration of radioactive carbon-14 in the atmosphere is the same today as it was at the time of death of the animal being studied. We assume the half-life of this isotope of carbon is 5730 years.

Represent mathematically The decay rate and number of radioactive nuclei are proportional to each other $(\Delta N/\Delta t = -\lambda N)$. Therefore, we can determine the ratio of carbon-14 in the old bone and what it was when the animal died.

$$\frac{(\Delta N/\Delta t)_{\text{now}}}{(\Delta N/\Delta t)_{\text{at death}}} = \frac{-\lambda N_{\text{now}}}{-\lambda N_{\text{at death}}} = \frac{N_{\text{now}}}{N_{\text{at death}}} = \frac{N}{N_0}$$

(CONTINUED)

From here we can use Eq. (29.11) to determine the age of the bone.

$$t = -\frac{\ln(N/N_0)}{0.693} T$$

Solve and evaluate The fraction of carbon-14 atoms remaining in the old bone is

$$\frac{N}{N_0} = \frac{3.3}{30.8} = 0.107$$

Substituting this value into Eq. (29.11), we find the age of the bone:

$$t = -\frac{\ln(0.107)}{0.693}(5730 \text{ years}) = 18,500 \text{ years}$$

The unit is correct and the order of magnitude is reasonable.

Try it yourself The ratio of carbon-14 in an old bone compared to the number in a fresh bone of the same mass is 0.050. Determine the age of the bone.

Answer   25,000 years.

Radioactive decay series

Many of the nuclei produced in radioactive decays are themselves radioactive. In fact, the decay of the original parent nucleus may start a chain of successive decays called a **decay series**. An example of a decay series is illustrated in **Figure 29.16**. First, the alpha decay of $^{238}_{92}\text{U}$ (top right) leads to the formation of $^{234}_{90}\text{Th}$. Thorium-234 then undergoes beta decay to form $^{234}_{91}\text{Pa}$. The series continues until $^{218}_{84}\text{Po}$ is formed. The series branches at this point, as polonium-218 can decay by either alpha or beta emission. Branching occurs at several other points as well. Eventually, the stable lead isotope $^{206}_{82}\text{Pb}$ is formed (lower left) and the series ends.

Radioactive series replenish our environment with nuclei that would normally have disappeared long ago. For example, radium ($^{226}_{88}\text{Ra}$) has a half-life of 1600 years.

FIGURE 29.16 A radioactive decay series. The sequence begins in the upper right with uranium-238.

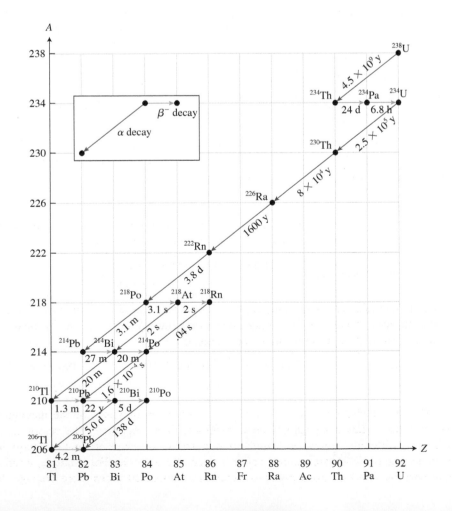

During the approximately 5×10^9 years of our solar system's existence, the original abundance of radium-226 would have long since been depleted by radioactive decay. However, the supply is continually replenished as uranium-238, with an extremely long 4.5×10^9-year half-life, decays and leads to the production of radium-226 via a series of reactions.

REVIEW QUESTION 29.8 How does the presence of radioactive carbon-14 help us determine the age of the bone of an animal that died thousands of years ago?

29.9 Ionizing radiation and its measurement

Ionizing radiation consists of photons and moving particles whose energy exceeds the 10 to 30 eV needed to ionize atoms and molecules. Ionizing radiation exists in many forms, such as ultraviolet, X-ray, and gamma ray photons, alpha and beta particles emitted during the radioactive decay of nuclei, and cosmic rays (high-energy particles) that reach Earth from space.

Life has evolved on Earth in the presence of a steady background of radiation, most of it coming from emissions of radioactive nuclei in Earth's crust and from cosmic rays and their collision products passing down through Earth's atmosphere. Using ionizing radiation for the purpose of diagnosing and treating health problems began more than 100 years ago. Today, our exposure from human-made sources of ionizing radiation accounts for about 40% of our total exposure.

Ionizing radiation's effects on living organisms are divided into two categories: genetic damage and somatic damage. Genetic damage occurs when DNA molecules in reproductive cells (those leading to eggs or sperm) are altered (mutated) by the radiation. These genetic changes are passed on to future generations. Somatic damage involves cellular changes to all other parts of the body except the reproductive cells.

Four different physical quantities are typically used to describe ionizing radiation and its effect on the matter that absorbs it. The first quantity, the decay rate or activity of a radioactive source, was defined in Section 29.7. The remaining three quantities are the absorbed dose, the relative biological effectiveness (RBE), and the dose equivalent.

The **absorbed dose** indicates the amount of ionizing radiation absorbed by biological material.

Absorbed dose The absorbed dose is the energy E of ionizing radiation absorbed by the exposed part of the body divided by the mass m of that exposed part:

$$\text{Absorbed dose} = \frac{\text{Energy absorbed by exposed material}}{\text{Mass of absorbing material}} \qquad (29.12)$$

The SI unit of absorbed dose is the gray (Gy), named after British physicist Louis Harold Gray, where 1 Gy = 1 J/kg. Another commonly used unit for absorbed dose is called the rad, short for radiation absorbed dose, where

$$1 \text{ rad} = 10^{-2} \text{ J/kg} = 10^{-2} \text{ Gy} \qquad (29.13)$$

The absorbed dose is not a completely satisfactory indicator of the damage we might expect when a living organism absorbs ionizing radiation. The reason is that when two different forms of ionizing radiation deposit the same amount of energy into organic material, the damage they cause is generally different. For example, a sample that absorbs 1 rad of alpha particles experiences about 20 times more damage than

TABLE 29.7 **The approximate relative biological effectiveness (RBE) of different types of radiation**

Type	RBE
X-rays and gamma rays	1
Electrons	1–2
Neutrons	4–20
Protons	10
Alpha particles	20
Heavy ions	2–5

when it absorbs 1 rad of X-rays. Alpha particles move through matter more slowly and slow down more gradually than other forms of radiation, and as a result, they interact with a larger number of atoms. X-rays and gamma rays, by contrast, deposit a large fraction of their energy within a small number of atoms.

Because of the variation in damage caused by different types of radiation, it is useful to define a quantity called **relative biological effectiveness (RBE)**. This measure indicates approximately the relative damage caused by different types of ionizing radiation. The RBE of several different types of radiation is listed in Table 29.7.

Relative biological effectiveness (RBE) Relative biological effectiveness is a dimensionless number that indicates the relative damage caused by a particular type of radiation compared with 2×10^5-eV X-rays. The approximate RBE for a particular type of radiation can be found in tables.

The last quantity is the **dose equivalent**, or simply **dose**—an indicator of the net biological effect of ionizing radiation.

Dose (dose equivalent) This quantity is a measure of the net biological effect (damage) of a particular exposure to ionizing radiation. The dose is the product of the absorbed dose and the relative biological effectiveness:

$$\text{Dose (dose equivalent)} = (\text{Absorbed dose})(\text{RBE})$$

$$= \left(\frac{\text{Energy absorbed}}{\text{Mass of absorbing material}} \right)(\text{RBE}) \quad (29.14)$$

The SI unit of dose is the sievert (Sv), named after Swedish medical physicist Rolf Sievert, where $1\ \text{Sv} = (1\ \text{Gy})(\text{RBE})$. Another commonly used unit is the rem (roentgen equivalent in man or in mammal), where $1\ \text{rem} = (1\ \text{rad})(\text{RBE})$.

QUANTITATIVE EXERCISE 29.12   **Ions produced by a chest X-ray**

In a typical chest X-ray, about 10 mrem (10×10^{-3} rem) of radiation is absorbed by about 5 kg of body tissue. The X-ray photons used in such exams each have energy of about 50,000 eV. Determine about how many ions are produced by this X-ray exam. Assume that each photon will produce approximately 100 ions because its energy is much larger than the ionization energy of atoms and molecules.

Represent mathematically We can estimate first the number of X-ray photons absorbed and then try to use that number to estimate the number of ions formed. We can accomplish this using Eq. (29.14):

$$\text{Dose} = (\text{Absorbed dose})(\text{RBE})$$

$$= \left(\frac{\text{Energy absorbed}}{\text{Mass of absorbing material}} \right)(\text{RBE})$$

Solve and evaluate The RBE of X-rays is 1. Thus, the dose of X-rays in rem equals the absorbed dose in rad. For an absorbed dose of 10 mrad, we can insert the known values and solve the equation to determine the energy absorbed per kilogram of exposed tissue:

$$10\ \text{mrad} = (10 \times 10^{-3}\ \text{rad}) \left(\frac{10^{-2}\ \text{J/kg}}{1\ \text{rad}} \right) = 10^{-4}\ \text{J/kg}$$

Because 5 kg of tissue receives this dose, the total absorbed energy is

$$(5\ \text{kg})(10^{-4}\ \text{J/kg}) = 5 \times 10^{-4}\ \text{J}$$

We can determine the number of X-ray photons by dividing this result by the energy of a single photon:

$$\frac{5 \times 10^{-4}\ \text{J}}{\left(50{,}000\ \dfrac{\text{eV}}{\text{photon}} \right)\left(\dfrac{1.6 \times 10^{-19}\ \text{J}}{\text{eV}} \right)} = 6 \times 10^{10}\ \text{photons}$$

or 60 billion photons. Because each photon produces 100 ions, the number of ions is approximately

$$6 \times 10^{10}\ \text{photons} \times 100\ \text{ions/photon} = 6 \times 10^{12}\ \text{ions}$$

- -

Try it yourself How many ions may be produced by a bilateral mammogram, assuming that each breast has a mass of 0.5 kg, the absorbed dose for both breasts is 0.2 rad, and 50,000-eV X-ray photons are used?

Answer

$\approx 3 \times 10^{11}$ photons, or 3×10^{13} if each photon produces 100 ions.

Sources of ionizing radiation

The U.S. Environmental Protection Agency estimates that the average dose of ionizing radiation received by a person in the United States or in Canada is about 300–400 mrem/year. The sources of this radiation can be divided into two broad categories: natural and human-made.

Natural sources The three major natural sources of radiation are the following.

1. Radioactive elements in the Earth's crust, such as uranium-238, potassium-40, and radon-226. Small amounts of radioactive radon, an inert gaseous atom, diffuse out of the soil into buildings, exposing the occupants.
2. Foods that naturally contain radioactive isotopes, such as Brazil nuts, lima beans, and bananas.
3. Cosmic rays. A cosmic ray is an elementary particle moving at almost the speed of light. The original source of cosmic rays is primarily supernova explosions of stars in our galaxy.

Human-made sources Human-made radiation comes predominantly from medical applications such as X-rays used in diagnostic procedures, the medical use of radioactive nuclei as tracers, and gamma rays used in cancer treatment. Together, these sources account for a dose of about 60 mrem/year (see Table 29.8). Smoking and exposure to radon significantly increase the dose/year.

TABLE 29.8 Ionizing radiation sources and doses

Exposure mechanism	Typical dose
Having a smoke detector	0.008 mrem/year[1]
Living near nuclear plant	0.009 mrem/year[2]
Living near coal-fired plant	0.03 mrem/year[1]
5-hour airplane flight	2.5 mrem[1]
Natural gas stoves and heaters	6–9 mrem/year[3]
Dental X-ray	0.5 mrem/X-ray[1]
Chest X-ray	10 mrem/X-ray[1]
Smoking (one half-pack per day)	18 mrem/year[1]
Radon	200 mrem/year[2]

Sources: [1]American Nuclear Society;
[2]PBS Frontline; [3]Lawrence Berkeley Lab.

REVIEW QUESTION 29.9 Why are X-rays, gamma rays, and fast-moving electrically charged particles potentially harmful to our bodies?

Summary

Terminology Nuclei are represented by the notation $^A_Z X$, where X is the letter abbreviation for that element, the atomic number Z is the number of protons in the nucleus, N is the number of neutrons in the nucleus, and the mass number $A = Z + N$ is the number of nucleons (protons plus neutrons) in the nucleus. (Section 29.2)

$^7_3$Li

Isotopes Atoms with a common number of protons but differing numbers of neutrons are called isotopes. (Section 29.2)

Mass defect is the difference between the total mass of the proton and neutron components of a nucleus and its mass when the components are bound together.

Binding energy of a nucleus is the rest energy equivalent of its mass defect. The binding energy represents the total energy needed to separate a nucleus into protons and neutrons. (Section 29.3)

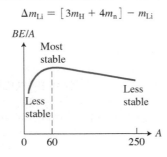

$\Delta m_{Li} = [3m_H + 4m_n] - m_{Li}$

mass defect $= \Delta m$
$= [Zm_{\text{hydrogen atom}}$
$\qquad + (A - Z)m_{\text{neutron}}] - m_{\text{atom}}$ Eq. (29.1)

$BE = \Delta m \cdot c^2$ Eq. (29.2)

Binding energy per
nucleon $= BE/A$ Eq. (28.4)

Types of radioactive decay
Alpha decay A heavy nucleus emits an alpha particle (α), the nucleus of a helium atom $^4_2$He.

Parent   Daughter   Alpha particle

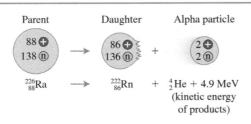

$^{226}_{88}$Ra $\longrightarrow$ $^{222}_{86}$Rn $+$ $^4_2$He $+$ 4.9 MeV
(kinetic energy of products)

α decay:
$^A_Z X \rightarrow ^{A-4}_{Z-2}X + ^4_2He +$ energy

Beta decay The nucleus ejects a beta particle (β^- electron or β^+ positron) and an antineutrino $\bar{\nu}$ or neutrino ν.

Parent   Daughter   Beta particle

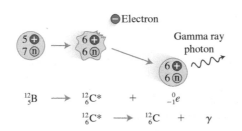

$^{14}_6$C $\longrightarrow$ $^{14}_7$N $+$ $^0_{-1}e$ $+$ $\bar{\nu}$

β^- decay:
$^1_0n \rightarrow ^1_1p + ^0_{-1}e + \bar{\nu}$

Gamma decay A nucleus in an excited state ($^A_Z X^*$) emits a high-energy photon called a gamma ray. (Sections 29.1 and 29.6)

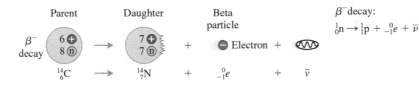

$^{12}_5$B $\longrightarrow$ $^{12}_6$C* $+$ $^0_{-1}e$
$^{12}_6$C* $\longrightarrow$ $^{12}_6$C $+$ γ

γ decay:
$^A_Z X^* \rightarrow ^A_Z X + \gamma$

Half-life The half-life T of a particular type of radioactive nucleus is the time interval during which the number of these nuclei in a given sample is reduced by one-half. (Section 29.7)

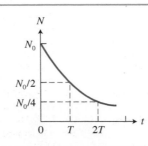

$N/N_0 = 1/2^n$ at $t = nT$ Eq. (29.6)
where N_0 is the number at time $t = 0$, and N is the number at time t and n is the number of half-lives that have passed

Questions

Multiple Choice Questions

1. What were Becquerel rays?
   (a) Positively charged particles
   (b) Negatively charged particles
   (c) High-energy photons
   (d) All of the above

2. To decide whether Becquerel rays contained electrically charged particles, scientists could
   (a) let them pass through a magnetic field.
   (b) let them pass through an electric field.
   (c) study the existing literature.
   (d) both a and b

3. Why are there are no electrons in the nucleus?
   (a) The nucleus consists of protons and neutrons only.
   (b) Electrons do not exert a strong nuclear force.
   (c) Their kinetic energy is greater than the magnitude of the potential energy of their interaction with the nucleus.
   (d) Electrons are too big to fit into a nucleus.

4. What is radioactive decay?
   (a) A physical quantity
   (b) An observable physical phenomenon
   (c) A law of physics
   (d) A unit

5. Elements with higher decay constants decay
   (a) slower.
   (b) faster.
   (c) The decay constant does not relate to how fast the substance decays.

6. One can speed up the rate of natural radioactive decay of a substance by
   (a) increasing its temperature.
   (b) illuminating it with high energy photons.
   (c) turning it into a powder.
   (d) No physical or chemical changes affect the rate of decay.

7. If the original number of radioactive nuclei in a sample is N_0, then after three half-life time intervals how many nuclei are left?
   (a) $N_0/3$ nuclei
   (b) $N_0/6$ nuclei
   (c) $N_0/8$ nuclei
   (d) $N_0/9$ nuclei

8. During alpha decay,
   (a) a neutron decays into an electron, a proton, and an antineutrino.
   (b) a photon is emitted by the nucleus changing its energy state.
   (c) a helium nucleus is emitted from the nucleus.
   (d) a hydrogen nucleus is emitted from the nucleus.

9. What happens during beta decay?
   (a) Electrons originally in the nucleus escape.
   (b) A neutron originally in the nucleus converts into a proton, an electron, and an antineutrino.
   (c) An alpha particle originally in the nucleus escapes.
   (d) Photons are emitted by the excited nucleus.

10. A sample of cesium-137 is placed in a lead container with a hole through which beta particles and gamma rays are emerging. A Geiger counter is placed next to the lead container as shown in Figure Q29.10. Placing the setup into which of the following *uniform* fields can result in an increase in the number of counts per unit time?
    (a) A magnetic field pointing perpendicularly into the page
    (b) A magnetic field pointing perpendicularly out of the page
    (c) An electric field pointing perpendicularly out of the page
    (d) An electric field pointing perpendicularly into the page
    (e) None of the above.

FIGURE Q29.10

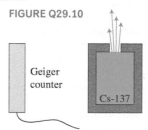

Conceptual Questions

11. How did scientists figure out that radioactive materials emit radiation from the nucleus rather than from the electron structure of atoms and molecules?

12. Describe the experiments that helped scientists understand that uranium salts produce both fast-moving charged particles and high-energy photons.

13. Why did Becquerel have burns on his chest near the vest pocket in which he carried a test tube containing radium?

14. How did Marie Curie discover radium?

15. How did Rutherford determine that radioactivity consists of three types of particles, including a high-energy photon?

16. How did scientists devise the concept of half-life?

17. Explain why heavier isotopes of atoms with a small number of protons are unstable.

18. The half-life of radium-226 is 1600 years. Earth is about 4 billion years old, so essentially all of the radium-226 present originally should have decayed. How can you explain the fact that radium-226 still remains in our environment in moderate abundance?

19. Why is a high temperature usually needed for fusion to occur?

20. Compare and contrast isotopes and ions.

21. Compare and contrast fission and fusion.

22. Compare and contrast the binding energy of a nucleus and the ionization energy of an atom.

Problems

Below, BIO indicates a problem with a biological or medical focus. Problems labeled EST ask you to estimate the answer to a quantitative problem rather than derive a specific answer. Asterisks indicate the level of difficulty of the problem. Problems with no * are considered to be the least difficult. A single * marks moderately difficult problems. Two ** indicate more difficult problems.

29.1 and 29.2 Radioactivity and an early nuclear model and A new particle and a new nuclear model

1. EST Estimate the magnitude of the repulsive electrical force between two protons in a helium nucleus. How are they held together in the nucleus?

2. * EST Estimate the magnitude of the repulsive electrical force that the rest of a gold nucleus ($^{197}_{79}$Au) exerts on a proton near the edge of the nucleus. State any assumptions used in your estimation. This is an order-of-magnitude estimate, so do not become bogged down in details.

3. Determine the number of protons, neutrons, and nucleons in the following nuclei: $^9_4$Be, $^{16}_8$O, $^{27}_{13}$Al, $^{56}_{26}$Fe, ^{64}Zn, and ^{107}Ag.

4. * EST Suppose the radius of a copper nucleus is 10^{-14} m and the radius of the copper atom is about 0.1 nm. Estimate the atom's size if the nucleus were the size of a pin head.

5. ** EST **Electrons and nucleons in body** Estimate the total number of (a) nucleons and (b) electrons in your body. (c) Indicate roughly the volume in cubic centimeters occupied by these nucleons.

29.3 Nuclear force and binding energy

6. * Measurements with a mass spectrometer indicate the following particle masses: $^7_3$Li, 7.016003 u; $^1_1$H, 1.007825 u; and $^1_0$n, 1.008665 u. Compare the mass of the lithium atom to the mass of the particles of which it is made. What do you conclude? Note: 1 u = 1.660539×10^{-27} kg.

7. * A 5.5-g marble initially at rest is dropped in the Earth's gravitational field. How far must it fall before its decrease in gravitational potential energy is 938 MeV, the same as the rest mass energy of a proton?

8. * Determine the rest mass energies of an electron, a proton, and a neutron in units of mega-electron volts.

9. EST Use Figure 29.8 to estimate the total binding energy of $^{197}_{79}$Au.

10. * Determine the total binding energy and the binding energy per nucleon of $^{12}_6$C.

11. * Determine the binding energies per nucleon for $^{238}_{92}$U and $^{120}_{50}$Sn. Based on these numbers, which nucleus is more stable? Explain.

12. * Determine the energy that is needed to remove a neutron from ^7_3Li to produce ^6_3Li plus a free neutron. (Hint: Compare the mass of ^7_3Li with that of $^6_3\text{Li} + ^1_0\text{n}$.)

29.4 and 29.5 Nuclear reactions and Nuclear sources of energy

13. Insert the missing symbol in the following reactions.
    (a) $^4_2\text{He} + ^{12}_6\text{C} \rightarrow ^{15}_7\text{N} + ?$
    (b) $^2_1\text{H} + ^3_1\text{H} \rightarrow ^4_2\text{He} + ?$
    (c) $^1_0\text{n} + ^{235}_{92}\text{U} \rightarrow ^{140}_{54}\text{Xe} + ? + 2^1_0\text{n}$
    (d) $^3_1\text{H} \rightarrow ? + ^0_{-1}e$

14. Explain why the following reactions violate one or more of the rules for nuclear reactions.
    (a) $^4_2\text{He} + ^{27}_{13}\text{Al} \rightarrow ^{32}_{15}\text{P} + ^1_0\text{n}$
    (b) $^2_1\text{H} + ^3_1\text{H} \rightarrow ^4_2\text{He} + ^1_1\text{H}$
    (c) $^1_0\text{n} + ^{238}_{94}\text{Pu} \rightarrow ^{140}_{54}\text{Xe} + ^{96}_{40}\text{Zr} + 2^1_0\text{n}$
    (d) $^{14}_6\text{C} \rightarrow ^{14}_7\text{N} + ^0_1e$

15. * Explain why the following reaction does not occur spontaneously: $^4_2\text{He} \rightarrow ^3_1\text{H} + ^1_1\text{H}$.

16. ** The following reaction occurs in the Sun: $^3_2\text{He} + ^4_2\text{He} \rightarrow ^7_4\text{Be}$. How much energy is released?

17. * **Nuclear reaction in the Sun** Oxygen is produced in stars by the following reaction: $^{12}_6\text{C} + ^4_2\text{He} \rightarrow ^{16}_8\text{O}$. How much energy is absorbed or released by the reaction in units of mega-electron volts?

18. * **Another reaction in the Sun** One part of the carbon-nitrogen cycle that provides energy for the Sun is the reaction $^{12}_6\text{C} + ^1_1\text{H} \rightarrow ^{13}_7\text{N} + 1.943 \text{ MeV}$. Using the known masses of ^{12}C and ^1H and the results of this reaction, determine the mass of ^{13}N.

19. * Determine the missing nucleus in the following reaction and calculate how much energy is released: $^{232}_{92}\text{U} \rightarrow ? + ^4_2\text{He} + \text{energy}$.

20. * Determine the missing nucleus in the following reaction and calculate its mass: $? \rightarrow ^{211}_{83}\text{Bi} + ^4_2\text{He} + 8.20 \text{ MeV}$.

21. * Determine (a) the number of protons and neutrons in the missing fragment of the reaction shown below and (b) the mass of that fragment:
    $$^1_0\text{n} + ^{235}_{92}\text{U} \rightarrow ? + ^{136}_{54}\text{Xe} + 12^1_0\text{n} + 126.5 \text{ MeV}$$

22. * **More energy for the Sun** A series of reactions in the Sun leads to the fusion of three helium nuclei (^4_2He) to form one carbon nucleus ($^{12}_6\text{C}$). (a) Determine the net energy released by the reactions. (b) What fraction of the total mass of the three helium nuclei is converted to energy?

23. * **Another Sun process** A series of reactions that provides energy for the Sun and stars is summarized by the following equation: $6^2_1\text{H} \rightarrow 2^1_1\text{H} + 2^1_0\text{n} + 2^4_2\text{He}$. (a) Determine the net energy released by the reaction. (b) Convert this answer to units of joules per kilogram of deuterium (^2_1H).

24. * **Equation Jeopardy** Equations for determining the mass defect for two nuclear reactions are shown below. Represent each reaction in symbolic form (as in the previous problems). Decide whether each reaction results in energy release or energy absorption.
    (a) $235.0439 \text{ u} + 1.0087 \text{ u} \rightarrow 95.9343 \text{ u} + 137.9110 \text{ u} + 2(1.0087 \text{ u}) + \text{energy}$
    (b) $3(4.002602 \text{ u}) \rightarrow 12.000000 \text{ u} + \text{energy}$

29.6 Mechanisms of radioactive decay

25. * In 1913, Frederick Soddy collected the following data related to the radioactive transformation of uranium (Figure P29.25). The first product that appears in the sample is thorium, then protactinium, then another isotope of uranium, and so on. Examine the series of the transformation found by Soddy and explain using your knowledge of alpha and beta decays. Discuss what quantities are constant in each process.

FIGURE P29.25

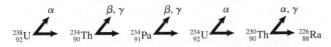

26. * In the 1930s, Meitner, Hahn, and Strassmann did experiments irradiating uranium with neutrons. They predicted three possible outcomes:
    - Production of a new element (if the neutron undergoes beta decay in the nucleus).
    - Production of a heavier isotope of uranium (if the neutron stays in the nucleus).
    - Production of a slightly lighter nucleus (if the neutron captured by the nucleus causes one or two alpha particles to leave).

    Instead, the nuclei produced in the reaction included isotopes of barium and other nuclei with about half the mass of uranium. How could they explain their findings?

27. * The following nuclei produced in a nuclear reactor each undergo radioactive decay. Determine the daughter nucleus formed by each decay reaction: (a) $^{239}_{94}\text{Pu}$ alpha decay; (b) $^{144}_{58}\text{Ce}$ beta-minus decay; (c) $^{129}_{53}\text{I}$ beta-minus decay; and (d) $^{60}_{30}\text{Zn}$ beta-plus decay.

28. * Potassium-40 ($^{40}_{19}\text{K}$) can undergo both beta-plus and beta-minus decay. Determine the daughter nucleus in each case.

29. * Radon-222 ($^{222}_{86}\text{Rn}$) is released into the air during uranium mining and undergoes alpha decay to form $^{218}_{84}\text{Po}$ of mass 218.0090 u. Determine the energy released by the decay reaction. Most of this energy is in the form of alpha particle kinetic energy.

30. * Carbon-11 ($^{11}_6\text{C}$) undergoes beta-plus decay. Determine the product of the decay and the energy released.

31. * (a) Determine the total binding energy of radium-226 ($^{226}_{88}\text{Ra}$). (b) Determine and add together the binding energies of a radon-222 ($^{222}_{86}\text{Rn}$) and an alpha particle. (c) Determine the difference of these numbers, which equals the energy released during alpha decay of ^{226}Ra.

32. ** BIO **Potassium decay in body** The body contains about 7 mg of radioactive $^{40}_{19}\text{K}$ that is absorbed with the foods we eat. Each second, about 2.0×10^3 of these potassium nuclei undergo beta decay (either beta-minus or beta-plus). About how many potassium nuclei are in the average body and what fraction decay each second?

33. * A radioactive ^{60}Co nucleus emits a gamma ray of wavelength 0.93×10^{-12} m. If the cobalt was initially at rest, use constancy of momentum to determine its speed following the gamma ray emission.

34. * A radioactive series different from those shown in Figures 29.16 and P29.25 begins with $^{232}_{90}\text{Th}$ and undergoes the following series of decays: $\alpha\beta^-\beta^-\alpha\alpha\alpha\alpha\beta^-\alpha\beta^-$. Determine each nucleus in the series.

29.7 Half-life, decay rate, and exponential decay

35. * BIO **O_2 emitted by plants** Propose a hypothetical experiment to determine whether O_2 emitted from plants comes from H_2O or from CO_2, the basic input molecules that lead to plant growth.

36. * A radioactive sample initially undergoes 4.8×10^4 decays/s. Twenty-four hours later, its activity is 1.2×10^4 decays/s. Determine the half-life of the radioactive species in the sample.

37. * Cesium-137, a waste product of nuclear reactors, has a half-life of 30 years. Determine the fraction of ^{137}Cs remaining in a reactor fuel rod (a) 120 years after it is removed from the reactor, (b) 240 years after, and (c) 1000 years after.

38. * A sample of radioactive technetium-99 of half-life 6 h is to be used in a clinical examination. The sample is delayed 15 h before arriving at the lab for use. Determine the fraction of radioactive technetium that remains.

39. ** BIO **Radiation therapy** If 120 mg of radioactive gold-198 with half-life 2.7 days is administered to a patient for radiation therapy, what is the gold-198 activity 3 weeks later if none is eliminated from the body by biological means?

40. **Fuel rod decay** How many years are required for the amount of krypton-85 ($^{85}_{36}\text{Kr}$) in a spent nuclear reactor fuel rod to be reduced to 1/8 of its original value? 1/32? 1/128? The half-life of ^{85}Kr is 10.8 years.

41. * BIO **Wisconsin glacier tree sample** A tree sample was uprooted and buried 60,000 years ago during part of the Wisconsin glaciation. How many years after it was buried was the radioactive carbon-14 (^{14}C) in the root reduced to (a) 1/2 of its original value? (b) 1/4? (c) 1/8? (d) What fraction remained after 60,000 years? The carbon-14 was not replenished after the tree stopped growing.

42. * How many years are required for the amount of strontium-90 (^{90}Sr) released from a nuclear explosion in the atmosphere to be reduced to (a) 1/16 of its original value? (b) 1/64? (c) 1/100?

43. ** BIO **Student swallows radioactive iodine** A student accidentally swallows 0.10 μg of iodine-131 while pipetting the radioactive material. (a) Determine the number of ^{131}I atoms swallowed (the atomic mass of ^{131}I is approximately 131 u). (b) Determine the activity of this material. The half-life of ^{131}I is 8.02 days. (c) What is the mass of radioactive iodine-131 that remains 21 days later if none leaves the body?

44. * An unlabeled container of radioactive material has an activity of 90 decays/min. Four days later the activity is 72 decays/min. Determine the half-life of the material. When will its activity reach 9 decays/min?

45. * One gram of pure, radioactive radium produces 130 J of energy per hour due to radium decay only and has an activity of 1.0 Ci. Determine the average energy in electron volts released by each radioactive decay.

29.8 Radioactive dating

46. **Age of mallet** A mallet found at an archeological excavation site has 1/16 the normal carbon-14 decay rate. Determine the mallet's age.

47. * The ^{235}U in a rock decays with a half-life of 7.04×10^8 years. A geologist finds that for each ^{235}U now remaining in the rock, 2.6 ^{235}U have already decayed to form daughter nuclei. Determine the age of the rock.

48. A sample of water from a deep, isolated well contains only 30% as much tritium as fresh rainwater. How old is the water in the well?

49. * The decay rate of ^{14}C from a bone uncovered at a burial site is 12.6 decays/min, whereas the decay rate from a fresh bone of the same mass is 1610 decays/min. Approximately how old is the uncovered bone?

50. ** BIO The tree sample described in Problem 29.41 contains 50 g of carbon when it is discovered. (a) If 1 in 10^{12} carbon atoms in a fresh tree sample are carbon-14, how many carbon-14 atoms would be in 50 g of carbon from a fresh tree? (b) Calculate the carbon-14 activity of the sample. (c) Determine the age of the buried tree if its 50 g of carbon has an activity of -2.2 s^{-1}.

29.9 Ionizing radiation and its measurement

51. Design an experiment in which radioactive nuclei are used to test the ability of different detergents to remove dirt from clothes.

52. * BIO **Radiation dose** A 70-kg person receives a 250-mrad whole-body absorbed dose of radiation. (a) How much energy does the person absorb? (b) Is it better to absorb 250 mrad of X-rays or 250 mrad of beta particles? Explain. (c) What is the dose in each case?

53. BIO **Dose due to different types of radiation** Determine the dose equivalent of a 70-mrad absorbed dose of the following types of radiation: (a) X-rays, (b) beta particles, (c) protons, (d) alpha particles, and (e) heavy ions.

54. * BIO **Potassium decay in body** The yearly whole-body dose caused by radioactive ^{40}K absorbed in our tissues is 17 mrem. (a) Assuming that ^{40}K undergoes beta decay with an RBE of 1.4, determine the absorbed dose in rads. (b) How much beta particle energy does an 80-kg person absorb in one year? Note: ^{40}K also emits gamma rays, many of which leave the body before being absorbed. Because fatty tissue has low potassium concentration and muscle has a higher concentration, gamma ray emissions indicate indirectly a person's fat content.

55. ** BIO **More potassium** Determine the number of ^{40}K nuclei in the body of an 80-kg person using the information provided in the previous problem and the fact that ^{40}K has a radioactive half-life of 1.28×10^9 years. Assume that each ^{40}K beta decay results in 1.4 MeV of energy that is deposited in the person's tissue.

56. * BIO **X-ray exam** During an X-ray examination a person receives a dose of 80 mrem in 4.0 kg of tissue. The RBE of X-rays is 1.0. (a) Determine the total energy absorbed by the 4.0 kg of body tissue. (b) Determine the energy in joules of each 40,000-eV X-ray photon. (c) Determine the number of photons absorbed by that tissue.

General Problems

57. Estimate the density of an atomic nucleus.
58. Estimate the radius of a sphere that would hold all of your mass if its density equaled the density of a nucleus.
59. Determine the chemical elements represented by the symbol X in each of the following: (a) $^{23}_{11}$X; (b) $^{56}_{26}$X; (c) $^{64}_{30}$X; and (d) $^{107}_{47}$X.

60. * EST Estimate the temperature at which two protons can come close enough together to form an isotope of a helium nucleus.

61. ** EST **Building a fusion reactor** The mass of a helium nucleus is less than the mass of the nucleons inside it. (a) Explain how this observation led scientists to the idea that it is possible to convert hydrogen into helium to produce thermal energy. (b) Will this process mean that energy is not conserved? Explain. (c) Why do you think scientists need very high temperatures and high pressures for this reaction? (d) Estimate the temperature at which two protons will join together due to their nuclear attraction. Remember that nuclear forces are effective at distances less than or equal to 10^{-15} m. (Hint: Use an energy approach, not a force approach.) (e) Suggest possible ways of containing ionized hydrogen to make the reaction possible (note that all solid containers will melt at this temperature).

62. ** **World energy use** World energy consumption in 2005 was about 4×10^{20} J. (a) Determine the number of deuterium nuclei ($^2_1$H) that would be needed to produce this energy. The fusion of two deuterium nuclei releases about 4 MeV of energy. (b) Determine the volume of water of density 1000 kg/m^3 needed to supply the energy. One mole of water has a mass of 18 g, and about one in every 6700 hydrogen atoms in water is a deuterium $^2_1$H.

63. * Convert the 200 MeV per nucleus energy that is released by ^{235}U fission to units of joules per kilogram. By comparison, the energy released by burning coal is 3.3×10^7 J/kg.

64. ** (a) Determine the energy used by a 1200-W hair dryer in 10 min. (b) Approximately how many ^{235}U fissions must occur in a nuclear power plant to provide this energy if 35% of the energy released by fission produces electrical energy? The energy released per fission is about 200 MeV.

65. * EST Estimate the number of ^{235}U nuclei that must undergo fission to provide the energy to lift your body 1 m. Indicate any assumptions made.

66. ** **Comparing nuclear and coal power plants** Suppose that a nuclear fission power plant and a coal-fired power plant both operate at 40% efficiency. Determine the ratio of the mass of coal that must be burned in 1 day of operating a 1000-MW plant compared to the mass of ^{235}U that must undergo fission in the same plant. The energy released by burning coal is 3.3×10^7 J/kg. The energy released per ^{235}U fission is 200 MeV.

67. ** **Uranium energy potential** The world's uranium supply is approximately 10^9 kg (10^6 tons), 0.7% of which is ^{235}U. (a) How much energy is available from the fission of this ^{235}U? (b) The world's energy consumption rate for production of electricity is about 10^{20} J/year. At this rate, how many years would the uranium last if it were used to provide all our electrical energy?

68. * A radioactive sample contains two different types of radioactive nuclei: A, with half-life 5.0 days, and B, with half-life 30.0 days. Initially, the decay rate of the A-type nucleus is 64 times that of the B-type nucleus. When will their decay rates be equal?

69. ** After a series of alpha and beta-minus decays, plutonium-239 ($^{239}_{94}$Pu) becomes lead-207 ($^{207}_{82}$Pb). Determine the number of alpha and beta-minus particles emitted in the complete decay process. Explain your method for determining these numbers.

70. * BIO EST **Estimate number of ants in nest** To estimate the number of ants in a nest, 100 ants are removed and fed sugar made from radioactive carbon of a long half-life. The ants are returned to the nest. Several days later, of 200 ants taken from the nest, only 5 are radioactive. Roughly how many ants are in the nest? Explain your calculation technique.

71. ** BIO **Ion production due to ionizing radiation** A body cell of 1.0×10^{-5}-m radius and density 1000 kg/m^3 receives a radiation absorbed dose of 1 rad. (a) Determine the mass of the cell (assume it is shaped like a sphere). (b) Determine the energy absorbed by the cell. (c) If the average energy needed to produce one positively charged ion is 100 eV, determine the number of positive ions produced in the cell. (d) Repeat the procedure for the 3.0×10^{-6}-m-radius nucleus of the cell, where its genetic information is stored. Indicate any assumptions that you made.

Reading Passage Problems

BIO **Nuclear accident and cancer risk** If 10,000 people are exposed to an average radiation dose of 1 rem, it is estimated that six cancer deaths will result. In March, 2011, a 9.0-magnitude earthquake and subsequent tsunami rocked Japan, causing hydrogen explosions in four reactors at the Fukushima Daiichi

Nuclear Power Station. As a result, radioactive materials such as cesium-137 and iodine-131 were released into the atmosphere. About 2 million people living within 80 km of the Fukushima reactors were each exposed to an average dose of approximately 1 rem.

72. Which answer below is closest to the Fukushima total person-rem exposure?
    (a) 2×10^4 person·rem   (b) 2×10^6 person·rem
    (c) 2×10^8 person·rem   (d) 2×10^{10} person·rem

73. Which answer below is closest to the number of statistically likely cancer deaths to be caused by this event?
    (a) 10   (b) 100   (c) 1000
    (d) 10,000   (e) 100,000

74. About 4000 of the 2 million residents die each year from cancer caused by other factors, such as smoking. If those Fukushima accident cancer deaths occurred during the 5 years following the accident, which number below is closest to the ratio of Fukushima-caused cancer deaths to normal cancer deaths for the exposed population?
    (a) 1/2000   (b) 1/20   (c) 1   (d) 20/1   (e) 200/1

75. Which answer below is closest to the number of cancer deaths in the United States due to an estimated 20 mrad/year average radon exposure to each person?
    (a) 50   (b) 500   (c) 5000   (d) 50,000   (e) 500,000

76. Which is the most important reason why it is difficult statistically to identify cancer deaths due to radiation exposure from the Fukushima accident?
    (a) An unknown number of people stayed in their homes following the accident.
    (b) The radiation detection devices were inaccurate.
    (c) The fraction of people who left the area was unknown.
    (d) The normal cancer rate significantly exceeded the cancer caused by the accident.
    (e) Authorities often distort information they provide about radiation exposure.

Was the sculpture of David from Notre Dame? Neutron activation analysis (NAA) is a sensitive analytical technique used in environmental studies, semiconductor quality control, forensic science, and archeological studies to identify trace and rare elements and determine the effect of rare elements on chronic diseases. In NAA a tiny specimen under investigation is irradiated with neutrons. The neutrons are absorbed by nuclei in the sample, thus creating artificially radioactive nuclei. These nuclei decay via the emission of alpha particles, beta particles, and gamma rays. The energy and wavelength of the gamma rays are characteristic of the element from which they were emitted.

NAA has been used to help identify the sources of art and sculpture by an elemental analysis of the paint or stone used. For example, the Metropolitan Museum of Art recently determined that stone used in the mid-13th-century *Head of King David* sculpture that they acquired in 1938 matched stone from

the cathedral of Notre Dame in Paris. Previously, it had been assumed that the sculpture came from Amiens Cathedral north of Paris. Figure 29.17 shows the elemental content of stone from Notre Dame (the red line) and from Amiens Cathedral (the blue line). Although the patterns on this logarithmic scale look similar, they are significantly different for some of the 16 elements tested.

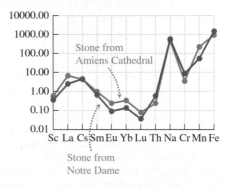

FIGURE 29.17 NAA comparison showing that the relative amounts of different elements in stone from Cathedral of Notre Dame and from Amiens Cathedral are distinctively different.

77. In neutron activation analysis (NAA) a particular elemental concentration in a sample is determined by measuring which of the following?
    (a) Alpha particles   (b) Beta particles
    (c) Gamma rays   (d) Neutrons
    (e) Two of these choices

78. Which answer below is closest to the ratio of the relative amount of lutetium (Lu) in stone from Notre Dame compared to stone from Amiens Cathedral?
    (a) 0.5   (b) 1   (c) 2   (d) 10   (e) 1000

79. What does neutron absorption by sodium ($^{23}_{11}$Na) produce?
    (a) $^{22}_{11}$Na   (b) $^{24}_{11}$Na   (c) $^{24}_{12}$Mg   (d) $^{23}_{12}$Mg   (e) $^{22}_{10}$Ne

80. Neutron absorption by potassium produces an excited $^{40}_{19}$K nucleus, which emits a beta-minus and a gamma ray. Which nucleus below is a product of this decay?
    (a) $^{40}_{19}$K   (b) $^{40}_{18}$Ar   (c) $^{40}_{20}$Ca   (d) $^{41}_{19}$K   (e) $^{39}_{19}$K

81. Rank the elements Cr, Mn, and Na in stone samples from Notre Dame compared to stone from Amiens Cathedral as the best indicators for distinguishing the two types of stone (rated from best to worst).
    (a) Cr, Mn, Na   (b) Mn, Cr, Na
    (c) Na, Mn, Cr   (d) Mn, Na, Cr

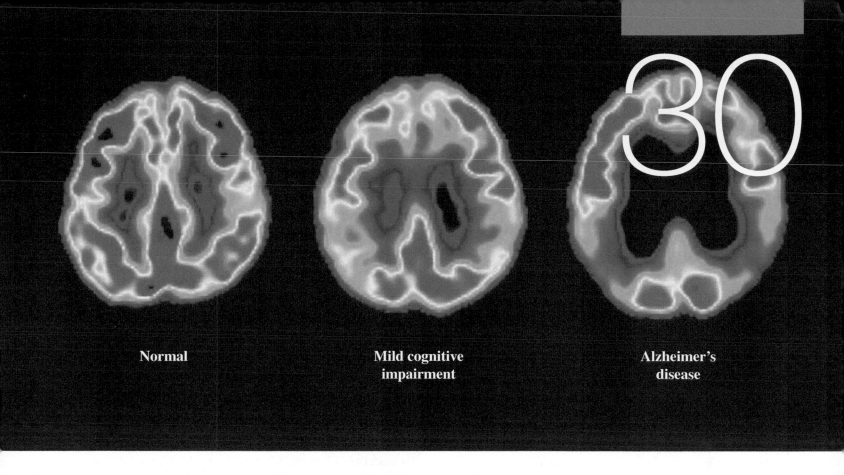

Normal

Mild cognitive
impairment

Alzheimer's
disease

Particle Physics

To obtain a three-dimensional image of the brain of a person who is suspected to have Alzheimer's disease, doctors use positron emission tomography (PET). Before the PET scan, the patient is given an injection of a glucose-like molecule containing fluorine-18, a radioactive isotope of fluorine. After an hour, the substance is absorbed into the patient's brain. Radioactive decay of the fluorine nuclei produces positrons, and by observing the photons that originate in the reactions of these positrons with nearby electrons, doctors can locate the active parts of the brain. A comparison of the active and inactive regions in the completed scan with scans of normal brains helps physicians determine whether Alzheimer's disease or other forms of dementia are present.

- What is antimatter?
- What are the fundamental particles and interactions in nature?
- What was the Big Bang, and how has the universe evolved since?

Beta decay produces antineutrinos, a form of antimatter. Every known particle has a corresponding antiparticle. In this chapter we investigate elementary particles such as the positron, which is an antielectron, and their fundamental interactions. By the end of this chapter, you should be able to understand the physics behind PET as well as the basic components of the universe, how they interact, and how they were discovered. This area of physics is called **particle physics**.

BE SURE YOU KNOW HOW TO:

- Use the right-hand rule to determine the direction of the magnetic force exerted by a magnetic field on a moving charged particle (Section 20.4).
- Explain beta decay (Section 29.6).
- Write an expression for the rest energy of a particle (Section 26.9).

30.1 Antiparticles

As you may recall, the word "atom" means indivisible. Physicists thought at one time that atoms were the smallest constituents of matter. However, in the late 19th century, physicists learned that atoms had an internal structure consisting of negatively charged electrons and a positively charged nucleus. The discovery of radioactivity and subsequent investigations indicated that the nucleus itself has a complex structure. In addition, the investigation of black body radiation and the photoelectric effect led scientists to conclude that light can be modeled as a photon—an object that has particle-like properties. By 1930, physicists had identified four particles—the electron, the proton, the neutron, and the photon. At that time, these were the only known truly **elementary particles**, a term used to indicate the simplest and most basic particles. This changed with the proposal and discovery of so-called antiparticles.

Antielectrons predicted

At the end of the 1920s, physicists solving the Schrödinger equation for the possible electron wave functions in the hydrogen atom produced solutions that were labeled by three quantum numbers: n, l, and m_l. However, the Schrödinger equation did not explain the existence of the fourth quantum number of the electron in the atom, its spin m_s.

In 1928 Paul Dirac extended quantum mechanics to incorporate Einstein's theory of special relativity. Dirac's model of relativistic quantum mechanics successfully predicted the existence of the spin quantum number for the electron. However, along with this prediction came an unexpected complication. Dirac's model also predicted that free electrons (electrons that are not interacting with other objects or fields) had an infinite number of possible quantum states with negative total energy. As a result, a free electron in a positive energy state should be able to transition to one of these negative energy states by emitting a photon whose energy equals the difference between those two states. All free electrons in the universe would transition to increasingly negative energy states, continually releasing electromagnetic energy. This model certainly was inconsistent with the observed behavior of electrons. How could a free electron have negative total energy? (Remember that only certain types of potential energies of systems, such as electric potential energy, can be negative.)

Dirac suggested that these negative energy states were occupied by an infinite number of so-called virtual electrons. Because only one electron can occupy an electron state according to the Pauli exclusion principle (Chapter 28), free electrons with positive energy could not then transition into these states. Over the next three years Dirac modified his model and eventually proposed that one of these virtual electrons in a negative energy state could be lifted out of its negative energy state to become a positive energy free electron. The empty state left behind would behave like a positively charged electron—like a particle with an electric charge of $+1.6 \times 10^{-19}$ C. Many quantum physicists disagreed with many aspects of his model. But Dirac argued that such particles should exist. He called them *antielectrons*, a new type of particle that had not yet been observed.

Antielectrons detected

To check Dirac's prediction, physicists had to search for a particle that behaved exactly like an electron but had a positive electric charge. At that time, to determine the sign of elementary particles, scientists used cloud chambers. A cloud chamber contained a gas that was supersaturated with vaporized water or alcohol. A charged particle like an electron or alpha particle passing through the vapor caused the vapor to condense into visible droplets, leaving a trail marking the particle's trajectory.

In 1932, American physicist Carl Anderson was studying cosmic rays, protons, and other particles coming to Earth from the Sun and from supernova explosions. He used a cloud chamber placed in a magnetic field. One of his photographs (**Figure 30.1a**) included the curved path of a particle passing through the chamber. The magnetic field pointed into the plane of the page.

The direction of travel of the particle and its charge were not immediately apparent from its path. If it entered the chamber from the top, then using the right-hand rule for magnetic force, it must be a negatively charged particle in order to curve as shown (Figure 30.1b). However, if it entered the chamber from the bottom, then it must be a positively charged particle (Figure 30.1c). How could Anderson distinguish between these two possibilities?

Notice that the lower part of the path is less curved than the upper part. Recall that a uniform magnetic field causes a charged particle to travel in a circular path (Chapter 20). The radius of its path can be expressed as

$$r = \frac{mv}{|q|B}$$

This means that the radius of the curved path is largest when the speed is largest—on the bottom half of the path. The particle passed through a thin lead plate in the middle of the chamber. This caused the particle's speed to decrease. Thus, the particle must have been traveling from the bottom of the chamber to the top, meaning it was a positively charged particle.

Anderson was able to determine the charge-to-mass ratio of the particle and found that it was the same as that for an electron. But it was positively charged! Anderson realized that the cloud chamber trace must have been produced by one of Dirac's antielectrons. Anderson found many more such tracks in other cloud chamber photographs. He called the new particle a **positron**.

Pair production

Electrons are abundant in our world, but there are few positrons. Under most circumstances, positrons exist for very short time periods. How are they produced? One way is during the radioactive decay of certain nuclear isotopes—for example, $^{11}_{6}C$, $^{13}_{7}N$, $^{15}_{8}O$, $^{22}_{11}Na$, and $^{40}_{19}K$ (we discussed this topic in Chapter 29).

Positrons can also be produced during the interaction of a high-energy gamma ray photon γ with matter. The photon has zero rest mass—you can't measure a photon's mass on a scale because the photon cannot exist at rest. However, under the right conditions this photon can simultaneously produce an electron $^{0}_{-1}e$ and a positron $^{0}_{+1}e$ (particles with nonzero rest mass). The process is called **pair production** and is represented by the reaction below:

$$\gamma \rightarrow {}^{0}_{-1}e + {}^{0}_{+1}e$$

For the photon to simultaneously produce both an electron and a positron, its energy must at least equal the combined rest energies of the electron and the positron. For the minimal energy calculation, we assume the kinetic energies of the electron and positron after their production are zero. The rest energy of the electron (and of the positron) is $E_0 = mc^2$. Thus, the photon must have double this energy:

$$E_{\text{photon}} = 2mc^2 = 2(9.11 \times 10^{-31}\,\text{kg})(3.00 \times 10^8\,\text{m/s})^2$$

$$= 1.64 \times 10^{-13}\,\text{J}$$

In electron volts, this is

$$E_{\text{photon}} = (1.64 \times 10^{-13}\,\text{J})\left(\frac{1\,\text{eV}}{1.6 \times 10^{-19}\,\text{J}}\right) = 1.02 \times 10^6\,\text{eV}$$

$$= 1.02\,\text{MeV}$$

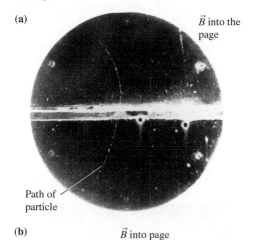

FIGURE 30.1 (a) The cloud chamber photograph taken by Anderson. (b) The direction of the force that the magnetic field would exert on an electron if it were moving from the top to the bottom. (c) The direction of the force that the magnetic field would exert on a positron if it were moving from the bottom to the top.

(a)

$\vec{B}$ into the page

Path of particle

(b)

$\vec{B}$ into page

$\vec{F}_{\vec{B}\text{ on } \ominus}$ $\vec{v}$

(c)

$\vec{B}$ into page

$\vec{F}_{\vec{B}\text{ on } \oplus}$ $\vec{v}$

FIGURE 30.2 (a) A gamma ray photon converts to an electron and a positron. (b) If in a magnetic field, the magnetic force causes the particles to move apart in spiral circles. (c) A cloud chamber photograph of this pair production.

(a)

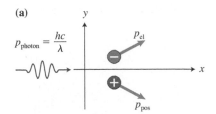

(b)

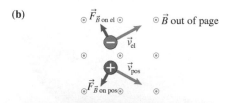

(c)

Positron   Electron

The frequency of such a photon is

$$f = \frac{E_{photon}}{h} = \frac{1.64 \times 10^{-13}\,\text{J}}{6.63 \times 10^{-34}\,\text{J}\cdot\text{s}} = 2.47 \times 10^{20}\,\text{Hz}$$

and its wavelength is

$$\lambda = \frac{c}{f} = \frac{3.00 \times 10^8\,\text{m/s}}{2.47 \times 10^{20}\,\text{Hz}} = 1.21 \times 10^{-12}\,\text{m}$$

This wavelength is about $1/100$ times the wavelength of an X-ray photon, making it a gamma ray photon.

If the positron and electron produced are to possess some kinetic energy, then a higher energy gamma ray is needed. If this pair production occurs in a magnetic field, the two particles should spiral in opposite directions.

To examine this idea, consider the photon as the system—it has momentum in the x-direction before producing the pair (**Figure 30.2a**). Immediately after pair production, the photon no longer exists, so the momentum of the two particles should be the same as the momentum of the original photon. The electron and positron should fly apart at an angle relative to each other, so that the y-components of momentum perpendicular to the original momentum of the photon cancel each other.

If we place the system in a magnetic field, the field exerts a force on the electron and on the positron in opposite directions because they have opposite electric charge (Figure 30.2b). This force will be perpendicular to their velocities, so they should curve in opposite directions. In addition, the radius of their circular paths should decrease as their speeds decrease, resulting in an inward spiraling motion. The photographs of pair production by gamma ray photons obtained by Anderson are consistent with this reasoning—see Figure 30.2c.

Pair annihilation

We have just learned that high-energy gamma ray photons can be used to create electron-positron pairs. Is the reverse process possible? If an electron and positron meet, will one or more photons be produced? The answer is yes, and the process is called **pair annihilation**.

CONCEPTUAL EXERCISE 30.1 ▶ **Pair annihilation**

Imagine that an electron and a positron meet and annihilate each other. Assume that they are moving directly toward each other at the same constant speed. Will one or two photons be produced? Write a reaction equation for this process. In what direction does the photon or photons move after the process?

Sketch and translate If one photon is produced, the reaction for this process is

$$_{-1}^{\,0}e + _{+1}^{\,0}e \rightarrow \gamma$$

If two photons are produced, the reaction is

$$_{-1}^{\,0}e + _{+1}^{\,0}e \rightarrow \gamma + \gamma$$

Simplify and diagram Assume that the process occurs in isolation, which means that the total electric charge and momentum of the system must remain constant. Electric charge is constant in both versions of the process. Because the two particles were moving directly toward each other at the same constant speed, the total initial momentum of the

system is zero. In the first version of the process, however, momentum is not conserved. If only one photon (with momentum $p = E/c$) is produced, the system's final momentum after the annihilation cannot be zero. Therefore, for the momentum of the system to be constant in this process, two photons must be produced, and they must travel in opposite directions, as shown in the figure.

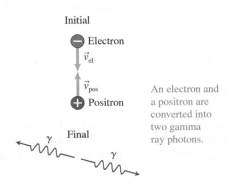

An electron and a positron are converted into two gamma ray photons.

(The details of the reaction are much more complicated than discussed here and are beyond the scope of this book.)

Try it yourself In this pair annihilation, two gamma rays were produced. But in pair production, an electron-positron pair is produced by only one gamma ray photon. How is this possible?

Answer

The momentum of the single gamma ray photon can be converted to the combined momentum of the two particles (see Figure 30.2).

The previous exercise involved a positron. We have already discussed the creation of a positron and an electron by a high-energy gamma ray photon as well as the production of positrons by radioactive decay. Let's consider another possibility. Recall the explanation of the beta decay of a neutron in nuclear physics (Chapter 29). A free neutron is an unstable particle with half-life of about 10 min. A free neutron decays into a proton, an electron, and an antineutrino (the antiparticle of a neutrino):

$$\,_0^1 n \rightarrow \,_1^1 p + \,_{-1}^0 e + \,_0^0 \bar{\nu}$$

The electric charge of the system during this process is constant $(0 = +1 - 1 + 0)$, as is its nucleon number $(+1 = +1 + 0 + 0)$. The rest energy of the neutron is greater than the combined rest energies of the proton and electron; therefore, the system's energy can possibly be constant as well, provided the proton, electron, and neutrino have sufficient kinetic energy. Finally, an appropriate combination of the momenta of the three particles in the final state can equal the initial momentum of the neutron—the momentum of the system can be constant. This process could happen without interaction with the environment. Is there a similar process by which a positron is produced instead of an electron?

Beta-plus decay: transforming a proton into a neutron

Since the rest energy of the neutron is greater than the rest energy of the proton, it seems that transforming a proton into a neutron cannot occur. However, if a proton captures a gamma ray photon, it is possible for a neutron and the other particles to be produced:

$$\gamma + \,_1^1 p \rightarrow \,_0^1 n + \,_{+1}^0 e + \,_0^0 \nu$$

The proton absorbs the photon and then decays into a neutron, a positron, and a neutrino. This process is called **beta-plus decay** to indicate that a positron (not an electron) is produced as the result of the process.

Beta-plus decay can also occur without the proton absorbing a photon. This can happen inside a nucleus if some of the energy of the nucleus converts into the additional rest energy of the products. Nuclei that can undergo beta-plus decay include carbon-11, nitrogen-13, oxygen-15, fluorine-18, and potassium-40. In fact, about 20,000 positrons are produced each minute in your body from the beta-plus decay of potassium-40:

$$\,_{19}^{40}K \rightarrow \,_{18}^{40}Ar + \,_{+1}^0 e + \,_0^0 \nu$$

The potassium nucleus transforms into an argon nucleus and in the process emits a positron and a neutrino. The potassium-40 nuclei in your body come from the potassium in foods you eat, such as bananas.

Note that the positively charged positrons produced in your body by this radioactive decay travel infinitesimal distances. They are attracted to negatively charged electrons and abruptly undergo pair annihilation, producing high-energy gamma rays, most of which leave your body. This process is what makes positron emission tomography possible.

FIGURE 30.3 Positron emission tomography (PET).

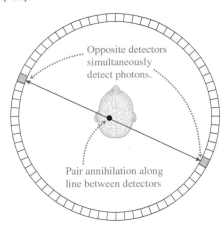

Opposite detectors simultaneously detect photons.

Pair annihilation along line between detectors

Positron emission tomography (PET)

In the chapter opening story, we described positron emission tomography (PET), a process for imaging the brain using radioactive fluorine-18 injected into a person's body. The isotope undergoes beta-plus decay continually, producing positrons. The positrons immediately meet electrons and annihilate, producing a pair of gamma ray photons that move in opposite directions. The PET chamber surrounding the patient's head detects the pairs of simultaneously appearing photons (**Figure 30.3**) and determines where in the brain they were emitted. Combining many pairs of gamma rays produces a three-dimensional image of the active parts of the brain—the places where more photons are produced.

This technique not only allows doctors to find the active and inactive regions of the brain in sick patients, but also helps brain researchers identify parts of the brain that are active when a person performs a particular task, such as listening to a poem being read or thinking about a mathematical problem.

Other antiparticles

Other particles have antiparticles as well. The positively charged proton that is part of all nuclei has a negatively charged antiproton of the same mass but opposite electric charge. Even though the neutron has zero electric charge, it too has an antimatter counterpart; other properties besides charge differentiate the neutron from the antineutron (see Section 30.3 on quarks). Some particles are their own antiparticle; the photon is an example, and possibly the neutrino. Later in this chapter we will consider why our universe is composed almost entirely of ordinary particles and almost no antiparticles.

REVIEW QUESTION 30.1 Why does proton decay occur only inside nuclei and not when a proton is free?

30.2 Fundamental interactions

In our studies of physics we have learned about many different forces: gravitational, electric, elastic, buoyant, etc. Some of these forces characterize *fundamental* interactions, and some characterize *nonfundamental* interactions. Fundamental interactions are the most basic interactions known, such as the electromagnetic interaction between charged particles. Nonfundamental interactions, such as friction, can be understood in terms of fundamental interactions. Friction is a macroscopic manifestation of the electromagnetic interaction between the electrons of the two surfaces that are in contact. We use the term "interaction" rather than "force" when speaking in general terms because interactions can also be represented using energy ideas. The four fundamental interactions are the gravitational interaction, the electromagnetic interaction, the strong interaction, and the weak interaction.

Gravitational interaction All objects in the universe participate in **gravitational interactions** due to their mass (and, according to general relativity, also due to their energy content through $m = E/c^2$). The gravitational force that two objects exert on each other is proportional to the product of their masses and the inverse square of their separation $1/r^2$. This interaction is important for mega-objects (planets, stars, galaxies, etc.), much less important for objects in our daily lives (e.g., the gravitational force that one human exerts on another), and extremely insignificant for microscopic objects (atoms, electrons, etc.).

Electromagnetic interaction Electrically charged objects participate in **electromagnetic interactions**. The interaction is electric if the objects are at rest or in motion with respect to each other. The interaction is magnetic only if the objects are moving with respect to each other. Similar to the gravitational interaction, the electromagnetic interaction between two charged particles also decreases as the inverse square of the distance between them. The electromagnetic interaction between nuclei and electrons is important in understanding the structure of atoms.

The electromagnetic interaction is tremendously stronger than the gravitational interaction. For example, the electrostatic force that an electron and a proton exert on each other in an atom is about 10^{39} times greater than the gravitational force that they exert on each other (**Figure 30.4**).

Atoms are electrically neutral, but because they are composed of distributions of charged particles (protons and electrons), they do participate in electromagnetic interactions with each other. These interactions contribute to the formation of molecules and to holding liquids and solids together. Because these interactions occur between overall electrically neutral objects, they are called **residual interactions**. Interactions such as friction, tension, and buoyancy are nonfundamental interactions and also are macroscopic manifestations of residual electromagnetic interactions.

Strong interaction Atomic nuclei are made up of protons that repel each other via electromagnetic interactions and neutrons that do not significantly interact electromagnetically. The interaction that binds protons and neutrons together into a nucleus is a residual interaction of the **strong interaction**. The strong interaction is a very short range interaction, exerted by protons and neutrons only on their nearest neighbors within the nucleus (a range of $\sim 3 \times 10^{-15}$ m). This residual interaction of the strong interaction is called the nuclear interaction, and the force characterizing it is called the **nuclear force**.

Weak interaction The weak interaction is responsible for beta decay. Protons, neutrons, electrons, and neutrinos all participate in it. The **weak interaction** is (as the name implies) significantly weaker than the strong interaction and has a significantly shorter range, about 10^{-18} m or less.

Mechanisms of fundamental interactions

How do elementary particles interact with each other? Consider an analogy. Humans interact using speech. A vibration in a person's larynx produces a sound wave that travels through the air to another person. That sound wave vibrates the eardrums of the other person. This vibration is then converted into an electrical signal that travels to the brain, where it is interpreted. How can we use this idea to understand the microscopic mechanism behind the four fundamental interactions?

Earlier in this text we developed a field model for the interactions between electrically charged objects (Chapter 18). The idea behind this model was that electrically charged particles produce a disturbance in the region around them called an electric field. When a charged particle moves, the "signal" produced by that movement ripples through the electric field at a finite speed (light speed), and only when that signal reaches other charged particles does the force exerted on them by the field change.

We later used the field model to explain the mechanism behind the magnetic interaction, ultimately predicting the existence of electromagnetic waves, of which visible light is just one example. We also discussed how electromagnetic waves could be modeled as streams of photons and showed that this model was necessary to explain microscopic phenomena such as the photoelectric effect.

During the first half of the 20th century, physicists realized that the photon actually played a more central role in electromagnetic phenomena. Researchers learned that the mechanism behind all microscopic electromagnetic phenomena was the exchange of photons between electrically charged particles.

FIGURE 30.4 The electric force that atomic size objects exert on each other is much greater than the gravitational force they exert on each other.

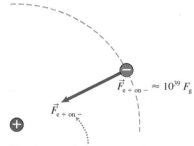

The electrostatic force exerted by a proton on an electron in a hydrogen atom is about 10^{39} times greater than the gravitational force.

This particle exchange mechanism has since been successful in describing the weak and strong interactions, and it has had some success in describing the gravitational interaction (although there are still many unresolved difficulties). In each of the four fundamental interactions, when two objects interact, one of the two objects emits a particle that is then absorbed by the other. This emitted and absorbed particle is called a **mediator**. In the electromagnetic interaction, the mediator is the photon. Two electrons repel each other through the exchange of photons.

But the particle exchange mechanism of particle interactions appears to have a serious problem. Where does the energy required to create the mediating particle come from? To answer this question, we need to think about the uncertainty principle (Chapter 28):

$$\Delta E \Delta t \geq \frac{h}{4\pi}$$

The energy of an atomic-scale system is only determinable to within the range ΔE. When the system is involved in a process, the smaller the time interval Δt in which the process occurs, the larger the energy uncertainty of the system during the process:

$$\Delta E \approx \frac{h}{4\pi \Delta t}$$

This helps us understand how the mediator particle can exist. The energy of an atomic-scale system is not exactly determinable; it fluctuates from instant to instant. In the case of two electrons interacting, these fluctuations manifest as what are known as **virtual photons** being exchanged between the electrons. These mediator photons are called "virtual" because they exist only as extremely small energy fluctuations of the system. Virtual photons do not have independent energy of their own that could initiate a chemical change in your retina or in an electronic detector. As a result, virtual photons (and other virtual particles in general) cannot be experimentally detected in any direct way.

We see from the uncertainty principle that for interactions that are mediated over long distances and therefore require the mediator particles to exist for long time intervals Δt to travel between the particles, the energy uncertainty ΔE of the system must be small (large Δt, small ΔE). This means that photons mediating long-range electromagnetic interactions should have low total energy. This is possible because photons have zero rest energy and can have very low frequencies (recall $E_{\text{photon}} = hf$).

Interaction mediators

We can now think of the four fundamental interactions as exchange processes of four different mediators. The mediators of the electromagnetic, strong, and weak interactions have all been discovered. The hypothetical mediator for the gravitational interaction, the so-called **graviton**, has not. Below we summarize the four types of interaction mediators.

Photons (electromagnetic interaction) The electromagnetic interaction can be modeled as an exchange of photons. Photons are massless particles and always travel at the speed of light. They have zero electric charge and are their own antiparticles (the photon and antiphoton are the same particle).

Gluons (strong interaction) Gluons are massless and have zero electric charge but interact strongly with each other. As a result, the strong interaction has a very short range. There are eight different kinds of gluons. The exchange of gluons mediates the interaction of quarks and allows them to be bound together into protons, neutrons, and more exotic particles. (More on quarks below.)

Gluons were first hypothesized as the mediators of quark interactions in the 1960s, but over a decade later there was still no direct experimental evidence for them. In 1976,

Mary Gaillard, Graham Ross, and John Ellis proposed a search for gluons using specific collision events. The gluon was subsequently discovered at DESY (Deutsches Elektronen-Synchrotron) in 1979.

$W^{\pm}$ **and Z^0 (weak interaction)** Three particles mediate the weak interaction: the positively charged W^+, its antiparticle, the negatively charged W^-, and the neutral Z^0 (which is its own antiparticle). The existence of these particles was predicted in the 1960s by a theory that unified the electromagnetic and weak interactions. This electroweak theory of Sheldon Glashow, Abdus Salam, and Steven Weinberg remained largely theoretical until 1983, when the $W^{\pm}$ and Z^0 particles were discovered at the European Organization for Nuclear Research (CERN) in Geneva, Switzerland.

Electroweak theory predicted (correctly) that these mediators have a very large rest energy. Producing them, therefore, required the colliding particles to have extremely high total energy. When the $W^{\pm}$ and Z^0 particles were first predicted, the accelerator at CERN could accelerate protons and antiprotons only to energies of tens of giga-electron volts (1 GeV is one billion electron volts), which was well below the rest energy of the weak interaction mediators. This explains why it took so many years to confirm the mediators' existence—technology had to catch up.

Gravitons (gravitational interaction) Based on fairly widely accepted assumptions about what a quantum theory of gravity should be, the graviton is predicted to be electrically neutral and have zero mass and therefore to travel exclusively at the speed of light.

Our current understanding of the four fundamental interactions is summarized in Table 30.1. For reference, recall that the mass of a proton is 1.67×10^{-27} kg.

TIP Although the properties of the graviton seem identical to those of the photon, they are very different particles. In this text we won't discuss the properties that distinguish them.

TABLE 30.1 Fundamental interactions

Interaction (Force)	Range	Mediating particle		Relative strength
Electromagnetic	Infinity $\left(F \sim \dfrac{1}{r^2} \right)$	Photon		10^{-2}
		Mass 0 kg	Electric charge 0	
Strong	Short (about 10^{-15} m)	Gluon		1
		Mass 0 kg	Electric charge 0	
Weak	Short (about 10^{-18} m)	$W^{\pm}$		10^{-13}
		Mass 1.43×10^{-25} kg	Electric charge $+e$ for W^+ $-e$ for W^-	
		Z^0		
		Mass 1.62×10^{-25} kg	Electric charge 0	
Gravitational	Infinity $\left(F \sim \dfrac{1}{r^2} \right)$	Graviton		10^{-38}
		Mass 0 kg	Electric charge 0	

As you can see, the mediator particles are massless except for the $W^{\pm}$ and Z^0, which have masses of about 100 times that of a proton. This huge mass explains why the weak interaction has such a short range. When produced as virtual particles, the $W^{\pm}$ and Z^0 can only last very briefly, so briefly that even if they are traveling at near light speed they can travel only an average of 10^{-18} m.

In Table 30.1 we listed the masses of the interaction mediators in kilograms. This is not the common practice in particle physics because the values are so small in those units. In fact, mass is not generally used at all. Instead, the "mass" of a particle is given in terms of its rest energy $E_0 = mc^2$ measured in electron volts (eV). Typical particle "masses" are in the mega-electron volt (1 MeV = 10^6 eV) or giga-electron volt (1 GeV = 10^9 eV) range.

Units in particle physics

Convert the masses of the $W^{\pm}$ and Z^0 particles into electron volts.

Represent mathematically First we need to determine the rest energies of the particles in joules, then convert them into electron volts, where $1 \text{ eV} = 1.6 \times 10^{-19}$ J.

Solve and evaluate The rest energy of the $W^{\pm}$ particles is

$$E_0 = mc^2 = (1.43 \times 10^{-25} \text{ kg})(3.00 \times 10^8 \text{ m/s})^2$$
$$= 1.29 \times 10^{-8} \text{ J}$$

Converting to eV, we get

$$(1.29 \times 10^{-8} \text{ J})\left(\frac{1 \text{ eV}}{1.6 \times 10^{-19} \text{ J}}\right) = 80.6 \times 10^9 \text{ eV} = 80.6 \text{ GeV}$$

Using the same method, the rest energy of the Z^0 particle is 91.3 GeV.

Try it yourself Determine the mass in mega-electron volts of the 1.67×10^{-27}-kg proton and of the 9.11×10^{-31}-kg electron.

Answer   938 MeV and 0.51 MeV.

REVIEW QUESTION 30.2 Why do two protons repel each other—for example, in an ionized gas of hydrogen—but attract each other when they are inside a nucleus?

FIGURE 30.5 Inside the SLAC National Accelerator Laboratory.

FIGURE 30.6 Inside the Large Hadron Collider (LHC).

30.3 Elementary particles and the Standard Model

Since the early 20th century, physicists have been discovering new elementary particles using particle accelerators such as the electron-positron collider at the SLAC National Accelerator Laboratory (**Figure 30.5**), the proton-antiproton collider at Fermilab in Illinois, and the Large Hadron Collider (LHC; see **Figure 30.6**) at CERN in Switzerland. Because the total energy of accelerated particles is significantly greater than the electron and proton rest energies, it is possible for additional particles to be produced in the collisions. The properties of these additional particles can be determined using elaborate detectors. Most of the particles produced are not stable; they quickly decay into other particles until eventually only stable particles remain.

The interaction mediators we discussed in the previous section are the mechanism behind the four fundamental interactions. These form one category of elementary particles. The building blocks of matter form two more categories, **leptons** and **hadrons**, which are the topic of this section.

Leptons

Leptons interact only through the weak, electromagnetic (if electrically charged), and gravitational interactions (though very weakly), not through the strong interaction. The electron is an example of a lepton, as is the neutrino ν_e, which is produced in beta-plus decay (Chapter 29). Note the symbol e as the subscript for this neutrino, which we did not use before. The symbol indicates that it is an "electron neutrino," and you will see that there other types of neutrinos. The electron neutrino is electrically neutral. These two particles form what is called a generation (or family) of leptons.

In 1936 the first member of a second lepton generation was discovered by American physicists Carl Anderson and Seth Neddermeyer. The **muon** μ is similar to the electron in that it is negatively charged, but it is approximately 200 times more massive. It also is unstable, with a half-life of 1.5×10^{-6} s. The corresponding **muon neutrino** ν_μ was discovered in 1962. A third generation of leptons was later discovered: the **tau** τ (by Martin Lewis Perl and his collaborators in a series of experiments between 1974 and 1977 at the SLAC) and the **tau neutrino** ν_τ (in 2000 at Fermilab). The tau is 17 times more massive than the muon and has an even shorter half-life of 2.0×10^{-13} s.

Electrons, muons, and taus have their own distinct antiparticles, but it is not yet clear whether the antiparticles of neutrinos are distinct or whether neutrinos are their own antiparticles. The electron and the three neutrinos are stable particles and do not decay. Some of the properties of leptons are shown in **Table 30.2.**

TABLE 30.2 Lepton generations (families)

	Electron generation		Muon generation		Tau generation	
	Mass (GeV)	**Electric charge in e**	**Mass (GeV)**	**Charge 1.6×10^{-19} C**	**Mass (GeV)**	**Electric charge in e**
Particle	Electron e		Muon μ		Tau τ	
	5.1×10^{-4}	-1	0.106	-1	1.777	-1
Antiparticle	Positron e^+		Antimuon $\overline{\mu}$		Antitau $\overline{\tau}$	
	5.1×10^{-4}	$+1$	0.106	$+1$	1.777	$+1$
Paired neutrino	Electron neutrino ν_e		Muon neutrino ν_μ		Tau neutrino ν_τ	
	$<2 \times 10^{-9}$	0	$<1.9 \times 10^{-4}$	0	$<1.8 \times 10^{-2}$	0
Paired antineutrino	Electron antineutrino $\overline{\nu}_e$		Muon antineutrino $\overline{\nu}_\mu$		Tau antineutrino $\overline{\nu}_\tau$	
	$<2 \times 10^{-9}$	0	$<1.9 \times 10^{-4}$	0	$<1.8 \times 10^{-2}$	0

Until 1998, physicists had no evidence to suggest that the masses of the neutrinos were anything other than zero (though the possibility was considered as early as the 1950s). In 1998 the Super-Kamiokande neutrino detector in Japan (see **Figure 30.7**) established that the weak interaction allows for processes that can convert one type of neutrino into another, a process called **neutrino oscillation.** This process is only possible if one or more of the neutrinos have nonzero mass. Since then, additional evidence from both particle physics and astrophysics experiments suggests that all three neutrinos have a rest energy of about 1 eV. Experiments are under way to measure this more precisely. Because neutrinos do not interact via the electromagnetic or strong interaction, they are extremely difficult to detect.

FIGURE 30.7 Inside the Super-Kamiokande neutrino detector.

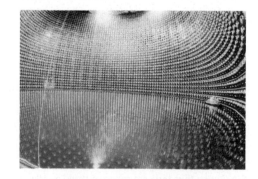

Hadrons

Two different types of hadrons can be distinguished—**baryons** and **mesons.** You already know two examples of baryons, the proton and the neutron, but there are many more. In 1949–1952, new higher mass baryons were discovered in interactions between cosmic rays and the atmosphere: the lambda particle Λ^0 and a set of four similar baryons known as the delta particles Δ^-, Δ^0, Δ^+, and Δ^{++}. Additional baryons were discovered in later years.

The first example of a meson was Hideki Yukawa's suggestion in 1935 of the existence of new particles that mediated the strong interaction. He called these particles mesons. The motivation for Yukawa's ideas came from combining relativity theory with quantum theory. The masses of Yukawa's mesons were predicted to be about 200 times the electron mass. In 1947 physicists discovered a meson in cosmic rays that participated in strong interactions and that had the correct properties to be Yukawa's meson (they called it a pi-meson, or pion for short). Later, additional mesons were discovered, such as the K^+ and K^0. Mesons and pions, however, were not the mediators of the strong interaction, as we will learn later.

The difference between baryons and mesons lies in their internal structure and is discussed in the next subsection. With the exception of the proton, all hadrons are unstable. Their half-lives range from 610 s for the neutron to 10^{-24} s for the shortest lived. A small sample of some hadrons and their properties is provided in **Table 30.3**. Hundreds of hadrons have been discovered since 1950.

TABLE 30.3 **Properties of hadrons**

	Particle, antiparticle	Mass (GeV)	Electric charge (in e)	Half-life (s)
Baryons				
Proton	$p, \bar{p}$	0.938	1, −1	Infinity
Neutron	$n, \bar{n}$	0.940	0, 0	610
Lambda	$\Lambda^0, \overline{\Lambda}^0$	1.116	0, 0	1.8×10^{-10}
Delta	$\Delta^{++}, \overline{\Delta}^{++}$	1.232	+2	3.9×10^{-24}
	$\Delta^+, \overline{\Delta}^+$	1.232	+1	3.9×10^{-24}
	$\Delta^0, \overline{\Delta}^0$	1.232	0	3.9×10^{-24}
	$\Delta^-, \overline{\Delta}^-$	1.232	−1	3.9×10^{-24}
Omega	Ω^-, Ω^+	1.672	−1, 1	5.7×10^{-11}
Mesons				
Pion	π^+, π^-	0.140	1, −1	1.8×10^{-8}
	π^0, self	0.135	0, 0	5.9×10^{-17}
Eta	η^0, self	0.548	0, 0	$<10^{-18}$
Kaon	K^+, K^-	0.498	1, −1	8.6×10^{-9}
	$K^0, \overline{K}^0$	0.498	0	No definite lifetime

Quarks

The large number of hadrons and the differences in their properties led to the belief that they were not truly elementary particles. In 1964 Murray Gell-Mann and his collaborators in the United States and Kazuhiko Nishijima in Japan independently proposed a new model of hadrons. In this model hadrons are complex objects made of a small number of more fundamental particles that combine in different combinations to make all of the existing hadrons. The idea is similar to combining protons, neutrons, and electrons in different combinations to make the elements of the periodic table.

The concept of hadrons having an internal structure helped explain the results of experimental investigations in which electron beams with energies of about 25 GeV (25 billion eV) were shot into a sample of liquid hydrogen. Electrons were scattered at various angles, similar to the alpha particles shot by Rutherford's colleagues at gold foil atoms. Electrons that were scattered at large angles had energy changes as if they had collided with tiny particles within the protons (**Figure 30.8**) rather than with the protons themselves.

These particles had to be electrically charged for the proton to be charged. Surprisingly, the experiments showed that the particles had *fractional* electric charge. These particles became known as **quarks**. When the quark model was first invented, only three quarks were needed to build all the hadrons that were known. These quarks were given the whimsical names **up, down,** and **strange**. Since then, three more quarks have been proposed (called **charm, top,** and **bottom**) to explain the structure of the ever more massive hadrons being produced in increasingly powerful accelerators. This brings the total number of quarks to six, all of which have since been discovered experimentally.

These different quark types are known in the physics community as **flavors** (having nothing to do with the sense of taste). Quarks interact through the strong (via gluons), weak (via the $W^\pm$ and Z^0), and electromagnetic (via photons) interactions. Because the quarks have nonzero mass, they also interact gravitationally, but only extremely

FIGURE 30.8 Electron scattering from protons indicates that they are made of three quarks.

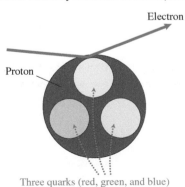

Three quarks (red, green, and blue)

weakly. The difference between baryons and mesons comes from their quark content: baryons are bound states of three quarks, while mesons are bound states of one quark and one antiquark.

Recall that leptons come in three generations (electron, muon, or tau), each with a negatively charged member and a neutrino for a total of six leptons. The quarks also fall into three generations (up/down, charm/strange, top/bottom) for a total of six quarks. Is this a coincidence? In other words, is there a deep connection between leptons and quarks? Particle physicists hope to find a connection, and that it motivates investigations into grand unification, but no connection has yet been established.

Several properties of particles determine whether they participate in particular interactions. For example, objects with nonzero electric charge participate in the electromagnetic interaction. The property that determines whether a particle participates in the strong interaction is known as **color charge** (it is not related to the colors we can see, but rather is a technical term used for this property). Because quarks participate in the strong interaction, they have color charge. A particle with color charge can interact with another color-charged particle by exchanging gluons.

Unlike electric charge, color charge comes in three varieties, known as **red, green, and blue** (again, these do not correspond to the colors we see). Just as neutral atoms have a net electric charge of zero, composite particles such as the proton and neutron are made of three quarks with complementary colors, which results in an overall color-neutral object. In other words, a proton has one red quark, one blue quark, and one green quark. This neutrality is analogous to the effect of shining complementary beams of red, blue, and green light on a surface. That surface glows white, which corresponds to color neutral (sometimes called colorless).

We mentioned earlier that quarks have fractional electric charge. The up quark has charge $(2/3)e$, where e is the magnitude of the charge of the electron. The down quark has charge $(-1/3)e$. It is reasonable to suggest, then, that the proton is composed of two up quarks and one down quark, or uud for short (**Figure 30.9**). This results in a total electric charge of $(2/3)e + (2/3)e + (-1/3)e = e$, consistent with the charge of the proton. The three quarks also have color charge, one red, one green, and one blue, so that the proton is color neutral. In the following exercise we determine the quark content of a neutron.

FIGURE 30.9 A proton is made of three charged quarks, with colors red, green, and blue.

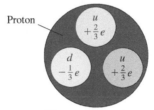

The proton's net charge is $+e$, and the proton is color neutral.

CONCEPTUAL EXERCISE 30.3   **Making a neutron**

What combination of quarks will have the correct properties to be a neutron?

Sketch and translate Because the electric charge of the neutron is zero, the charges of its component quarks must add to zero. The neutron is a baryon, so it is composed of three quarks (and no antiquarks).

Simplify and diagram This is accomplished if we include one up quark and two down quarks (udd).

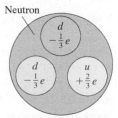

The neutron's net charge is zero, and the neutron is color neutral.

$$2(-1/3) + (2/3) = 0$$

Two d quarks   One u quark

- -

Try it yourself How does the proton composition differ from the neutron composition? What would have to happen for a neutron to be converted into a proton?

Answer

The proton has two up quarks and one down quark. A neutron could be converted into a proton if one down quark were converted to an up quark. This sounds very similar to what occurs in beta decay.

We stated earlier that an antineutron has properties besides charge that differentiate it from a neutron. An antineutron is composed of three antiquarks, namely, one up antiquark and two down antiquarks.

Figure 30.10 summarizes our understanding of the structure of matter and the role of quarks in that structure. All of the matter that makes up our local world is composed of electrons and two types of quarks (up and down).

FIGURE 30.10 A summary of particles (matter) and their interactions.

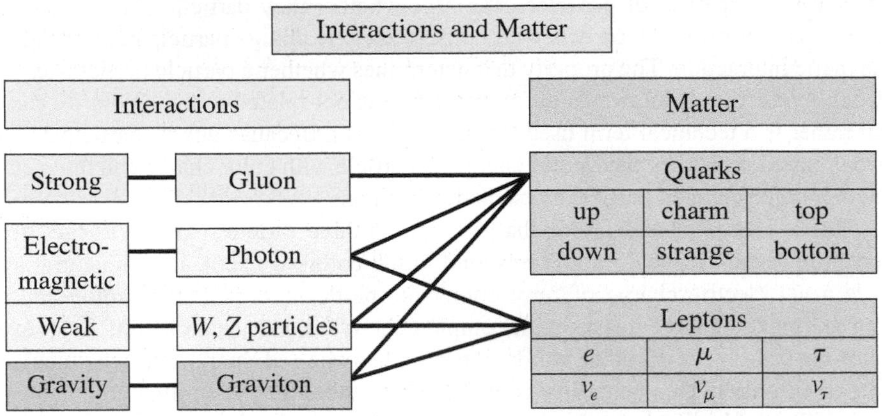

Confinement

No experiment has ever produced a quark in isolation. Every quark and antiquark ever produced has always been part of a hadron (either a three-quark baryon or a quark-antiquark meson). Trying to split a proton into quarks by shooting a high-energy particle into it produces additional quark-antiquark pairs rather than separate quarks. These pairs then combine with the original proton's quarks to form new baryons and mesons. This phenomenon is called **confinement,** and it is an indication of a feature of the strong interaction that is very different from the other three interactions. The strong interaction between quarks is weakest when they are close together and gets stronger the farther apart the quarks are. The reasons why only the strong interaction exhibits confinement are beyond the scope of this book, but the feature is crucial to explaining the structure and stability of the protons and neutrons in every atomic nucleus in your body.

Development of the Standard Model

With this basic understanding of elementary particles (interaction mediators, leptons, and hadrons) we now can discuss the **Standard Model**, the name given to the combined theory of the building blocks of matter and their interactions. Its first versions were constructed during the first half of the 20th century, and it has continued to evolve ever since.

In the late 1940s physicists Richard Feynman, Julian Schwinger, and Sin-Itiro Tomonaga independently combined the ideas of special relativity and quantum mechanics into a single model of the electromagnetic interaction that explains all electromagnetic phenomena. (The three of them were jointly awarded the Nobel Prize in physics in 1965 for this work.) Their model is known as **quantum electrodynamics** (QED) and is a cornerstone of the Standard Model. In 1957–1959 Julian Schwinger, Sidney Bludman, and Sheldon Glashow, working separately, proposed a model in which the weak interaction was mediated by two massive charged particles, which were later called the W^+ and W^- particles. Chen-Ning Yang and Robert Mills developed the mathematical framework needed to describe this model.

In 1967 Sheldon Glashow, Abdus Salam, and Steven Weinberg independently put forth a model that unified the electromagnetic and weak interactions into a single interaction, which they called the **electroweak model**. This model made several striking predictions. First, it predicted that in the very distant past, when the universe was much smaller and very much hotter, all particles were massless. Second, it predicted the existence of a particle, which became known as the **Higgs particle** after physicist Peter Higgs. Third, it predicted that as the universe cooled, the Higgs particle began interacting significantly with other elementary particles, reconfiguring them into the familiar forms they have today (most having nonzero mass). This is known as the **Higgs mechanism**.

In 1973, using Yang's and Mills' mathematical framework, Harald Fritzsch and Murray Gell-Mann formulated **quantum chromodynamics** (QCD), a mathematical model of the strong interaction in terms of the exchange of gluons between quarks. By the end of 1974, results from the first experiments testing QCD began arriving. Among them was the discovery of the so-called J/ψ (J-psi) particle, a meson composed of one charm and one anticharm quark. The existence of the J-psi particle was a successful prediction of QCD. Between 1976 and 1979 scientists discovered the tau lepton and bottom quark and found direct evidence for gluons.

The 1980s brought the discoveries of the predicted weak interaction mediators $W^{\pm}$ and Z^0. The 1990s gave us the top quark, and in 2000 the tau neutrino was discovered. In July 2012 CERN announced the discovery of the long-sought-after Higgs particle, which was a triumph of the Standard Model (see **Figure 30.11**).

Although the predictions of the Standard Model of particle physics are very consistent with the outcomes of particle physics experiments, several unanswered questions remain and are the focus of active research:

1. Can the strong interaction be unified with the electroweak interaction, just as the weak and electromagnetic interactions were unified?

2. Why are there only three families of quarks/leptons?

3. Are the Standard Model particles truly fundamental, or do they have internal structure?

4. Are there additional particles beyond those predicted by the Standard Model?

FIGURE 30.11 A look inside ATLAS, one of the detectors of the Large Hadron Collider (LHC) at CERN, which was instrumental in the discovery of the Higgs particle. (a) The vertical tunnel leading underground to the accelerator part of the detector. (b) The front side of the heart of the ATLAS detector is 25 m in diameter and 46 m long and weighs about 7000 tons.

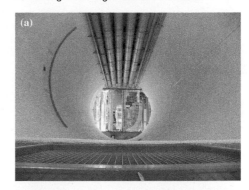

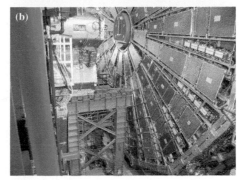

Summary of the Standard Model

The Standard Model is the currently accepted model of fundamental particles and their interactions (see Figure 30.10). It includes the following:

1. Quarks and leptons make up the matter of the universe. Quarks participate in all four fundamental interactions. Leptons participate in all but the strong interaction.

2. The model of strong interactions (QCD) mediated by gluons.

3. The model of electromagnetic (QED) and weak interactions mediated by photons and the $W^{\pm}$ and Z^0 particles.

4. The Higgs particle, which explains, through the Higgs mechanism, why some of the fundamental particles have nonzero mass.

The Standard Model does not include the gravitational interaction. How to combine this interaction with the Standard Model is a challenging, unsolved problem in physics. One framework for potentially doing this is *string theory*, and while much progress has been made in the past 40 years, no one knows yet whether string theory can achieve this goal.[1]

---

[1]The details of the discovery of the Higgs particle are beyond the scope of this book. If you wish to learn more about it, you can look up the following article: Lincoln, D. (2012).The Higgs boson: Is the end in sight? *The Physics Teacher* 50, 332.

REVIEW QUESTION 30.3 Using what you have learned about particle physics, describe as many differences as you can between a proton and an electron.

30.4 Cosmology

As we have seen, almost all elementary particles have distinct antiparticles. Furthermore, when particles are produced, they are always produced in matter-antimatter pairs. Why, then, is our universe not filled with equal numbers of particles and antiparticles? Why is there an imbalance? These questions are answered in part by particle physics and by **cosmology**—a branch of physics that studies the composition and evolution of the universe as a whole.

Recall that independently of which direction we look, distant galaxies are moving away from us—the universe is expanding (Chapter 26). If we could run this expansion in reverse, the universe would get smaller, denser, and hotter. About 13.7 billion years ago, the entire universe would have been in an unimaginably hot and dense state of nearly zero size, which then rapidly expanded. This initial expansion is called the **Big Bang** (see the timeline from the Big Bang to the present in **Figure 30.12**). The Big Bang model of cosmology was first proposed in 1927 by a Belgian Catholic priest, Georges Lemaître, and more fully developed by George Gamow, Ralph Alpher, and

FIGURE 30.12 The development of the universe since the Big Bang 13.7 billion years ago.

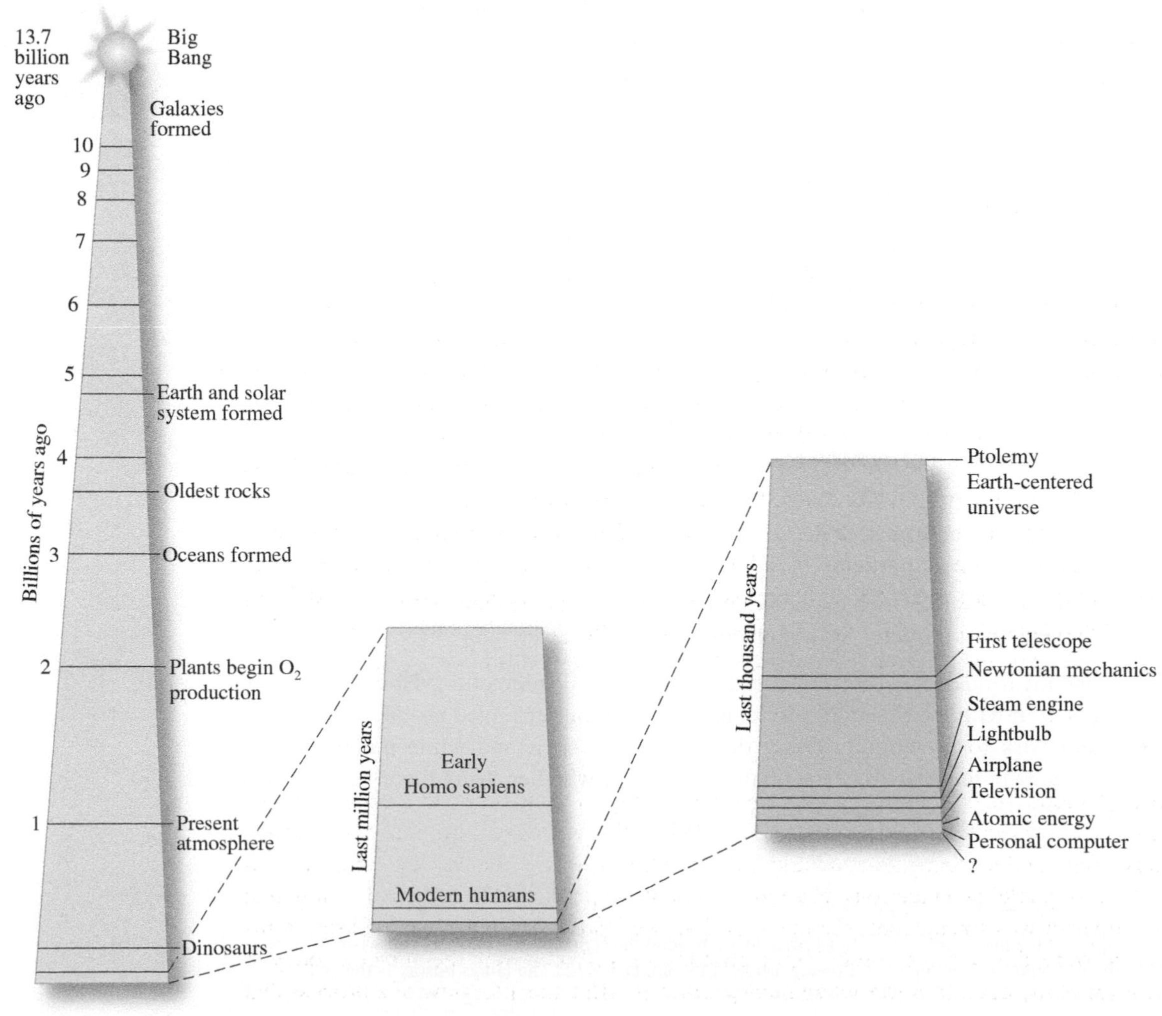

Robert Herman in the late 1940s. The model explained the observed red shifts of spiral nebulae, now called galaxies. It also accounted for the fact that the red shifts were greater for more distant nebulae than for closer nebulae. Below is a brief summary of what we know about the history of the universe.

Inflation

From $t = 0$ s to $t = 10^{-43}$ s, the average temperature of the universe was around 10^{32} K. At that time it was so hot that quarks and leptons could easily be converted into each other. At about $t = 10^{-36}$ s, the temperature of the universe had fallen to about 10^{28} K. For the first time, the universe became "cold" enough so that quarks and leptons became distinguishable particles. This caused a fundamental change in the structure of the universe, resulting in an extremely rapid exponential expansion known as cosmic **inflation**. During the time interval $t = 10^{-36}$ s to $t = 10^{-32}$ s, the linear size of the universe increased by a factor of 10^{26} and then settled into a more gradual expansion. During inflation, small fluctuations in the density of the universe decreased. Areas where the density was slightly above average would later act as the seeds of galaxy formation. Even with these fluctuations, the density of the universe at a particular point differed from the average by only $1/1000$ of 1%. The universe was an extremely hot plasma of quarks, leptons, and interaction mediator particles.

From $t = 10^{-32}$ s to $t = 10^{-12}$ s, the average temperature was still so high that the random thermal motion of particles was at ultra-relativistic speeds (very close to light speed), and particles and antiparticles were continuously created and annihilated in collisions. However, for some reason (there are multiple explanations for this), the annihilations led to a small excess of quarks over antiquarks and leptons over antileptons. Understanding the details of these processes is an important goal of current research in physics.

Nucleosynthesis

At about $t = 10^{-6}$ s, the universe had cooled enough that quarks and gluons were able to combine to form baryons such as protons and neutrons. Because there was an excess of quarks over antiquarks, an excess of baryons over antibaryons was produced. The temperature was no longer high enough for particle collisions to create proton and antiproton pairs or neutron and antineutron pairs. Therefore, the baryons and antibaryons annihilated each other, destroying all antibaryons and leaving a small number of baryons. Only one in 10 billion protons survived this annihilation. These are the protons that we find in the universe today. After this annihilation, temperatures became low enough that the thermal motion of particles was no longer relativistic.

By a few minutes after the Big Bang, the temperature had dropped to about a billion degrees K, and the average density of the universe was close to the density of air at sea level on Earth today. For the first time, protons and neutrons were able to combine to form the simplest nuclei: deuterium (a heavy isotope of hydrogen with one proton and one neutron), helium, and trace amounts of lithium. This process is known as **Big Bang nucleosynthesis**. Some protons combined into helium nuclei, but most remained free as hydrogen nuclei. The ratio of hydrogen to helium nuclei was roughly 12 to 1.

Atoms, stars, galaxies

After 370,000 years the universe became cold enough for electrons to finally combine with nuclei to form the first neutral atoms. When this happened, the universe became transparent to electromagnetic radiation. Prior to this, the universe was a plasma, an ionized gas of nuclei and electrons. Plasmas do not allow photons to travel freely. Now that the universe was transparent, the photons that were produced when the neutral atoms formed were able to travel freely. These photons are still present in the universe today in the form of the **cosmic microwave background** (CMB), the afterglow of a process that took place more than 13 billion years ago. Due to the expansion of the universe, these photons have been red-shifted so that they have an effective temperature today of about 2.7 K. This can be thought of as the ambient temperature of the present universe.

The discovery of the CMB by Arno Penzias and Robert Wilson in 1965 was one of the most significant pieces of supporting evidence for the Big Bang model, as it had predicted this radiation. The model also made predictions about the relative abundances of hydrogen, helium, and lithium that resulted from nucleosynthesis. These predictions are consistent with experiments as well.

When the universe had cooled enough, the gravitational interaction became the dominant driver of its further evolution. Density fluctuations caused some regions to begin contracting. This process led to the formation of the first galaxies and stars just 500,000 years after the Big Bang. These early stars went through their life cycles, some ending in a violent collapse and explosion known as a **supernova**. Through nuclear fusion processes, heavier elements such as carbon, oxygen, iron, and gold were produced in supernovae and released into space to become parts of planets such as our own.

The present matter on Earth consists of quarks that were part of the early universe. These quarks combined to form protons and neutrons in the early universe, which then formed complex nuclei and atoms during the life cycles of stars. Thus, our bodies really are composed of matter that was created near the dawn of time and processed in stellar explosions before becoming part of us.

In the latter part of the 20th century two serious problems with this picture of the universe arose that have yet to be resolved. First, galaxies aren't nearly massive enough to explain the way stars move within them. Second, the universe is speeding up in its expansion rather than slowing down, as the gravitational interaction would predict. What possible explanations have physicists devised to explain these unexpected experimental results?

REVIEW QUESTION 30.4 Explain how the Big Bang model predicts the existence of the cosmic microwave background (CMB).

30.5 Dark matter and dark energy

Earlier (in Chapter 5), we learned that Earth's speed around the Sun is determined by the radius of Earth's orbit and the mass of the Sun. In a similar way, the solar system's speed around the Milky Way galaxy is determined by the radius of the solar system's orbit and the total mass of everything that lies between that orbit and the center of the galaxy (**Figure 30.13**).

When astronomers measure the mass of all the stars and gas that they can see, however, they find that the total mass is only about 5% of the mass needed to account for the speed of the solar system around the center of the galaxy. The pioneer of this work was American astronomer Vera Cooper Rubin. In the early 1980s, she found the discrepancy between the predicted rotational motion of galaxies and the observed motion by studying graphs of the orbital speeds of stars in a galaxy versus their distance to the galactic center. The same pattern was found for the motion of galaxies relative to each other; there is far too little visible mass to account for the galaxies' motion. Apparently, the universe is "missing" about 95% of the mass needed to account for the observed motion of stars and galaxies. How can this contradiction be resolved?

Dark matter

The first evidence of the problem came in 1933 when astrophysicist Fritz Zwicky looked at the galaxies in the Coma cluster. He found that the galaxies at the edge of the cluster were traveling much too fast to remain part of the cluster. He speculated

FIGURE 30.13 Our galaxy, with the position of the solar system marked.

Solar system

that there must be some unseen **dark matter** present in the Coma cluster, but for about 40 years his observation was the only evidence for its existence.

In the 1970s Rubin presented further evidence. She looked at individual galaxies and found that stars near the edge of a galaxy were also traveling too fast to remain part of the galaxy. It was at this point that the dark matter explanation started to become more widely accepted.

Scientists suggested a new form of unseen matter and devised experiments to detect it as directly as possible. But as of 2017, direct detection of dark matter has not been accomplished. Astronomers are actually more certain about what this unseen dark matter is not than about what it is. First, it does not emit photons or otherwise participate in the electromagnetic interaction (this is why it is called "dark"). As a result, dark matter cannot be some sort of dark cloud of protons or gaseous atoms, because these could be detected by the scattering of radiation passing through them. Several hypotheses have been proposed, however, for what this dark matter might be. Two of these go by the names MACHOs (massive compact halo objects) and WIMPs (weakly interacting massive particles).

MACHOs (massive compact halo objects) MACHOs could be black holes, neutron stars, or brown dwarfs (objects that were not massive enough to achieve nuclear fusion and become stars). Astronomers have been trying to detect MACHOs using their gravitational effects on the light from distant objects (a phenomenon called gravitational microlensing). When one of these MACHOs passes in front of a distant object, such as a star or another galaxy, the light bends around the MACHO and is "focused" for a short time interval, making the light appear brighter. The MACHO Project has observed about 15 such lensing events over a span of 6 years of observations. This small number of events translates into MACHOs accounting for at most 20% of the dark matter in our galaxy. There must be another (or an additional) explanation.

WIMPs (weakly interacting massive particles) WIMPs are more exotic in nature. They are elementary particles, but not the quarks and leptons that make up ordinary matter. They are "weakly interacting" because they can pass through ordinary matter with almost no interaction, and light is not absorbed or emitted by them. They are "massive" in that their mass is not zero. Prime candidates for WIMPs include neutrinos, axions, and neutralinos, described next (the latter two are not Standard Model particles and hence require the Standard Model to be extended to accommodate their existence).

Cosmologists have a good idea of how many neutrinos there are in the universe. The Standard Model gives a zero mass to the neutrinos, but the model can be extended to include nonzero masses. Recent experiments have strongly suggested that neutrinos have a rest energy in the range of 0.1 to 1.0 eV, a million times smaller than that of the electron. For many years astronomers hoped that neutrinos could be the dark matter. However, it appears that neutrinos are too light for this. Some **grand unified theories** (theories that combine the strong, weak, and electromagnetic interactions into a single interaction) predict the existence of a so-called "sterile neutrino" that could be even less interactive than Standard Model neutrinos and also far more massive. Physicists do not know how to detect such a particle, but if it exists in sufficient abundance it could account for dark matter.

Physicists originally proposed the existence of **axions** to explain a mystery about the properties of the neutron. These proposed axions have very small mass (less than that of the electron), no electric charge, and very little interaction with Standard Model particles. They would have been produced abundantly in the Big Bang. Current searches for axions include Earth-based laboratory experiments and astronomical searches in the halo of our galaxy and in the Sun. However, axions have never been experimentally observed.

The **neutralino** is an example of another type of hypothetical particle known as a **superpartner**. Several outstanding problems in theoretical physics are understood better by suggesting the existence of what is known as **supersymmetry**. Supersymmetry is an extension of the Standard Model that effectively doubles the number of elementary particles, gives insight into the cosmological constant problem (to be described shortly), allows for a more precise understanding of the unification of interactions in grand unified theories, and gives a potential candidate for dark matter. None of the superpartners have ever been detected, but their detection is one of the primary design goals of the Large Hadron Collider.

Supersymmetry predicts that the lightest superpartner is stable and very weakly interacting. There are many versions of supersymmetry, but in some of them the lightest superpartner is the neutralino, predicted to have a mass of 100 to 1000 times the mass of a proton. Astronomers and physicists hope to detect neutralinos by using underground detectors, searching the universe for signs of their interactions, or producing them in particle accelerators.

Because none of the particles suggested as a solution to the dark matter problem have been detected, the mystery of the missing matter of the universe is still largely unsolved.

Although the dark matter problem has been around since the 1930s, one idea that seemed irrefutable through the 20th century was that the gravitational interaction between all the massive objects in the universe would gradually slow the expansion rate of the universe. Then in 1998, two independent experiments using the Hubble Space Telescope produced observations of very distant supernovae that showed that the universe is currently expanding more rapidly than it was 5–10 billion years ago. The expansion of the universe has not been slowing down; it has been speeding up! No one expected this, and no one at the time knew how to explain it. The 2011 Nobel Prize in physics was awarded to Saul Perlmutter, Brian P. Schmidt, and Adam G. Riess for the discovery of the accelerating universe. This led to extensions of the Big Bang model that could potentially explain this accelerated expansion. The most widely accepted of these is known as **dark energy**.

Dark energy

Physicists came up with many ideas to explain the accelerating expansion of the universe. Here are a few:

- A discarded feature of Einstein's general theory of relativity (our current best model of the gravitational interaction) known as the **cosmological constant**
- The existence of a strange kind of energy-fluid that fills space and has a repulsive gravitational effect
- A modified version of general relativity that includes a new kind of field that creates this cosmic acceleration

Let's look at each of these ideas.

The cosmological constant model An idea introduced and discarded by Einstein 80 years ago has been resurrected to explain recent observations. The cosmological constant is a term that Einstein included in general relativity to allow for a static universe. At the time he constructed general relativity, there was no experimental evidence for the Big Bang or the expansion of the universe. The dominant model of the universe was known as the **steady state model**, which asserted that the universe essentially didn't change in any major way as time passed. But general relativity predicted that a static universe was unstable. Einstein introduced the cosmological constant into general relativity in an attempt to allow the theory to accommodate a steady state universe.

Einstein could have predicted the expansion of the universe 10 years before it was discovered, but even he couldn't accept such a radical idea. It is said that he considered this his greatest blunder. However, because the cosmological constant introduces a repulsive effect into the equation to balance the attractive effect of gravity, it can also be used to describe the accelerated expansion.

The dark energy model The second idea is actually a suggestion about what the cosmological constant term means physically. It seems to represent a type of dark energy that is present at every point in space with equal density. Even as the universe expands, this density does not decrease because it is a property of space itself. What's even stranger is that this energy has a negative pressure. In general relativity this produces a gravitationally repulsive effect on space.

As the universe expands, the matter density decreases, but the dark energy density remains constant. This means that as time passes, the attractive gravitational effect of the matter decreases while the repulsive gravitational effect of the dark energy remains the same. In other words, as time goes on, the repulsion gets stronger relative to the attraction. This is precisely what astronomers have observed—the expansion is more rapid today than it was in the past.

Modified general relativity The third idea is in some sense the least radical. Just as general relativity is an improvement on Newton's view and is able to make more accurate predictions, perhaps the gravitational interaction theory can be revised so that it would make even better predictions than general relativity. The challenge has been to modify the theory in such a way that it doesn't contradict experiments that have already been done. Thus far physicists have been unsuccessful in doing this.

Mysteries Of these three ideas, dark energy has become the favored one. Various versions have been set forth, and the simplest of them represent the dark energy mathematically in general relativity as a cosmological constant. Dark energy models have a problem, though, and it comes from a basic feature of the combination of relativity and quantum mechanics.

This combined model predicts that each elementary particle in the Standard Model is actually an excitation of an associated quantum field, similar to how the photon is one quantum of excitation of the electromagnetic field. Each of these quantum fields has a certain minimum energy, called its **zero point energy**. In many dark energy models, the dark energy is the sum of the zero point energies of all the quantum fields in the universe. But when physicists make estimates of what value these models predict for the cosmological constant, they get a result that is 10^{120} times the observed value. It's been said that this is the largest disagreement between prediction and experiment in all of science. This so-called **cosmological constant problem** is a major unsolved problem in physics.

Supersymmetry, which predicts a doubling of all the types of elementary particles, suggests a possible solution. Each of these superpartner particles also has an associated quantum field. But the zero point energies of each of these additional quantum fields contribute oppositely to the dark energy compared to their nonsuper counterparts. This means the total dark energy density would be zero. That is not consistent with observations—we do observe an accelerated expansion—but the dark energy density is calculated to be very small. That's a much better prediction than 10^{120} times the observed value! None of the superpartner particles have been observed, however, which means that supersymmetry is not present in its full form in this universe. One of the goals of the Large Hadron Collider is to produce some of these superpartner particles and see whether supersymmetry can resolve the cosmological constant problem.

Explaining the accelerated expansion of the universe is an even greater puzzle than explaining dark matter. We know how much dark energy there is because we know how it affects the universe's expansion. Other than that, it is a complete mystery. It turns out that roughly 73% of the total energy of the universe is dark energy. The dark matter's rest energy makes up about 23%. The rest—everything on Earth, everything ever observed with all of our instruments, all normal matter—adds up to ~4% of the energy in the universe (see **Figure 30.14**). We understand only a few percent of what comprises the universe we live in. That is a humbling but highly motivating realization.

FIGURE 30.14 The proportion of matter, dark matter, and dark energy in the universe.

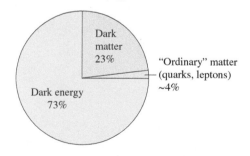

TIP Dark matter and dark energy are ideas invented by physicists to explain patterns observed in nature. Because neither of these has yet been detected, it is possible that one or both of these ideas could be disproven.

REVIEW QUESTION 30.5 How do we know that much more mass exists in the Milky Way galaxy than is present in the galaxy's stars and interstellar gas?

30.6 Is our pursuit of knowledge worthwhile?

In this textbook, we have been building models that describe the behavior of only 4% of the contents of our universe. The nature of the remaining 96% of our universe currently remains an unsolved problem. Is it worthwhile trying to learn about these other mysterious components of our universe?

In the late 1800s, the prime minister of England asked Michael Faraday what use there was for his idea of electromagnetic induction. Faraday could not say. But today we have electric power generators, microphones, credit card readers, electric guitar pickups, and electromagnetic braking systems for hybrid vehicles. All are based on electromagnetic induction.

In 1897, J. J. Thomson discovered and characterized the electron—he could not have imagined how this understanding would lead to the computing, communication, and entertainment devices that pervade our everyday lives.

In the 1930s, physicists at MIT were studying microwaves. What use would they have? Many historians believe that microwave radar saved England in World War II. Today, microwaves, a form of nonionizing radiation, are involved in satellite communications and GPS, and we use microwaves to cook our food.

Will our eventual knowledge of the other 96% of the universe someday make people's lives better? It's impossible to say for sure, but history suggests that it very likely will. Perhaps it will lead to new sustainable energy sources, the ability to easily travel to other planets, or ways to protect us from cosmic events that potentially threaten life on Earth. The most amazing future applications of understanding our universe are the ones no one has yet devised.

Summary

Antiparticles Each particle has an antiparticle with identical mass but opposite electric charge, color, etc. Some particles (such as the photon) are their own antiparticles. (Section 30.1)

Fundamental interactions All interactions between objects are gravitational, electromagnetic, weak, or strong. All other interactions are nonfundamental and can be understood in terms of these. (Section 30.2)

Elementary particles Every known particle is a quark (up, down, etc.), a lepton (electron, neutrino, etc.), or an interaction mediator (photon, gluon, etc.)

Standard Model The elementary particles together with the four fundamental interactions and the Higgs particle comprise the Standard Model of particle physics. (Section 30.3)

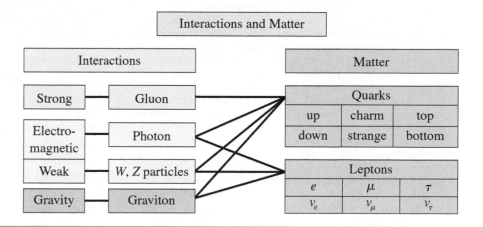

Cosmology The universe rapidly expanded from a hot, dense state 13.7 billion years ago, an event called the Big Bang. As it expanded and cooled, nuclei formed, followed by atoms, stars, and galaxies. (Section 30.4)

Dark matter and dark energy Most of the energy in the universe is thought to consist of two mysterious forms. Dark matter interacts only gravitationally and clumps around galaxies. Most of the mass of galaxies is proposed to be dark matter. Dark energy is thought to be spread uniformly throughout the universe and is responsible for its accelerated expansion. (Section 30.5)

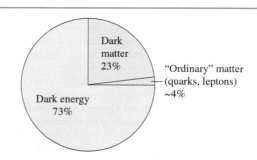

Questions

Multiple Choice Questions

1. What is the composition of all baryons? (a) Three different color quarks (b) Three quarks of any color (c) Two quarks of any color (d) Three gluons (e) Two quarks of complementary color
2. What is the composition of all mesons? (a) Three different color quarks (b) Three quarks of any color (c) Two quarks of any color (d) Three gluons (e) Two quarks of complementary color
3. What is the weak interaction mediated by? (a) Gluons (b) The W^+, W^-, or Z^0 particles (c) Photons (d) Gravitons (e) Mesons
4. What is the strong interaction mediated by? (a) Gluons (b) The W^+, W^-, or Z^0 particles (c) Photons (d) Gravitons (e) Mesons

Conceptual Questions

5. Give three examples of particles that are elementary (as far as is currently known) and three examples of those that are not.
6. What differs between a particle and its antiparticle?

7. Billions of neutrinos continually pass through each square centimeter of your body each second. Why don't you notice them?
8. How do we know that protons are very stable particles?
9. Free neutrons (those that are not part of a nucleus) decay into protons plus other particles with a half-life of about 10 min. Free protons, however, do not decay into neutrons plus other particles. Explain why.
10. What particle(s) could potentially not have an antiparticle or could be said to be its own antiparticle? Explain your reasoning.
11. Give an example of a long-range interaction and a short-range interaction. What is the difference between the mechanisms for each?
12. What are the components of the Standard Model?
13. Why do scientists need particle accelerators to discover new particles and to study the properties of known particles?
14. In 1930 the proton and neutron were thought to be elementary particles. We now know this is not true. How could physicists have been mistaken?

Problems

Below, **BIO** indicates a problem with a biological or medical focus. Asterisks indicate the level of difficulty of the problem. Problems with no * are considered to be the least difficult. A single * marks moderately difficult problems. Two ** indicate more difficult problems.

30.1 Antiparticles

1. Explain Dirac's reasoning for the existence of antielectrons.
2. How do scientists know that antiparticles exist in nature?
3. **BIO** Explain how positron emission tomography (PET) works.
4. A proton and an antiproton, both with negligible kinetic energy, annihilate each other to produce two photons. Determine the energy in electron volts of each photon and the frequency and wavelength of each photon.
5. An isolated slow-moving electron and positron annihilate each other. Why are two photons produced instead of just one?
6. Use Newtonian circular motion concepts to show that the radius r of the circle in which a charged particle spirals while moving perpendicular to a magnetic field is proportional to the particle's speed v.
7. A cosmic ray photon enters a cloud chamber from above. While inside the chamber the photon converts into an electron-positron pair. Draw a sketch of this situation. Label the photon, electron, and positron tracks and indicate the direction of the magnetic field within the cloud chamber.
8. A particle enters a cloud chamber from above traveling at $0.50c$ and leaves a curving track. The magnetic field within the cloud chamber has magnitude 0.120 T and points out of the page. The track curves to the right and has a radius of 0.020 m. Determine everything you can about this particle.

30.2 Fundamental interactions

9. Compare and contrast the fundamental interactions.
10. Even though the electromagnetic interaction is so much stronger than the gravitational interaction, the gravitational interaction clearly is much more relevant in everyday life. Do you agree with this statement? Explain either way.
11. Explain the mechanism behind elementary particle interactions.
12. * Explain how the energy to produce interaction mediators is provided.
13. What is the difference between a real particle and a virtual particle?
14. * Make an analogy between the interactions of elementary particles and some real-life interactions. Indicate which object in one system corresponds to each object in the other. Describe the limitations of your analogy.

30.3 Elementary particles and the Standard Model

15. Why are neutrinos difficult to detect?
16. Describe the evidence for neutrinos having nonzero mass.
17. What criteria do scientists use to classify elementary particles?
18. Compare and contrast leptons and hadrons.
19. How do we know that baryons are made of quarks? What is the experimental evidence that supports this idea?
20. Compare and contrast mesons and baryons.
21. How do quarks interact with each other inside a baryon?
22. * In what way is the interaction of electrically neutral atoms similar to the interactions of color-neutral protons and neutrons?
23. Describe the phenomenon of confinement.
24. What were four important steps in the building of the Standard Model?
25. ** What is the significance of the Higgs particle and the Higgs mechanism?
26. * What major piece of physics is not part of the Standard Model?
27. * Describe several open questions of the Standard Model.

30.4 Cosmology

28. * What is inflation, and what eventually happened as a result of the density fluctuations in the universe?
29. * What is the origin of the elementary particles?
30. * How would the universe be different if the number of particles in it had equaled the number of antiparticles?
31. * What process produced the cosmic microwave background?

32. * Our bodies contain significant amounts of carbon, oxygen, nitrogen, and other "heavy" elements. If only "light" elements such as hydrogen and helium were produced during Big Bang nucleosynthesis, how were the heavier elements produced?

30.5 Dark matter and dark energy

33. * What is the evidence that a large proportion of the mass of the universe is in the form of dark matter? Explain carefully.
34. * Describe as many hypotheses as you can about the nature of dark matter.
35. * What is the experimental evidence for dark energy? Explain carefully.
36. ** Describe as many hypotheses as you can that explain the accelerated expansion of the universe.
37. ** What is supersymmetry, and why is it a useful idea in physics?
38. ** What is the cosmological constant problem? Describe a possible resolution for it.

30.6 Is our pursuit of knowledge worthwhile?

39. What are the potential benefits of continued research in physics?

General Problems

40. * An electron and a positron are traveling directly toward each other at a speed of $0.90c$ (90% of light speed) with respect to the lab in which the experiment is being performed. The electron and positron collide, annihilate, and produce a pair of photons. Determine the wavelength of each photon. If you think they will have the same wavelength, explain why.
41. * In order to discover the W^+ and W^- particles, a large particle accelerator at CERN collided protons and antiprotons together at extremely high speeds. What is the minimum speed the proton and antiproton would need to produce a $W^\pm$ pair?

Reading Passage Problems

The solar neutrino problem Our Sun continually converts hydrogen into helium through nuclear fusion. The chain of reactions that occurs can be combined into the single reaction shown here:

$$4{}_1^1\text{H} \rightarrow {}_2^4\text{He} + 2e^+ + 2\nu_e$$

Four hydrogen nuclei fuse into one helium nucleus. In the process two positrons and two electron neutrinos are produced. These positrons interact immediately with electrons present in the Sun and annihilate with them, producing gamma ray photons. Because the Sun is a plasma, these photons interact strongly with it and push outward, supporting the Sun against gravitational collapse. The neutrinos, however, have only an extremely small chance of interacting with matter and therefore stream out of the Sun's core unhindered, making their way to Earth and beyond. So many neutrinos are produced in the fusion reactions that ~ 100 billion of them pass through each square centimeter of your skin each second.

Solar neutrinos act as a "window" into the fusion reactions occurring in the Sun that normally would be obscured from our view. For example, if we can measure the rate at which solar neutrinos reach Earth we can estimate the nuclear reaction rate in the Sun. We also have an independent way to measure the nuclear reaction rate—by measuring the intensity of the Sun's radiation. The higher the intensity of the Sun's radiation, the greater the nuclear reaction rate at its core. These two methods should produce consistent results. But they don't. The amount of electron neutrinos detected is less than half the amount expected. This is the **solar neutrino problem**.

The resolution of the problem comes from suggesting that some of the electron neutrinos convert into other types of neutrinos (muon or tau) on their way to Earth. This phenomenon is called **neutrino oscillation**. For this to be possible, the neutrinos cannot have zero mass. The experiment that first detected solar neutrinos could not detect muon or tau neutrinos. In 2001 a new detector was used that could detect all three types of neutrinos. About two-thirds of the electron neutrinos had in fact oscillated into muon and tau neutrinos. This resolved the solar neutrino problem.

42. Why don't you notice the solar neutrinos passing through your body?
   (a) The Sun is very far away so the neutrinos have very little energy by the time they get to Earth.
   (b) Nearly all the neutrinos oscillate into other types that can pass easily through your body.
   (c) Solar neutrinos only participate in nuclear fusion reactions and since nuclear fusion doesn't happen in your body you don't notice the neutrinos.
   (d) Neutrinos only interact very weakly with atoms.
   (e) Only a very small number of them pass through your body each second.

43. In 1987 a supernova was detected in the Large Magellanic Cloud. Neutrinos coming from the supernova were detected, but there were fewer than expected. What might be the reason for that?
   (a) The supernova actually occurred farther from Earth than astronomers thought.
   (b) Some of the neutrinos were converted into other types of neutrinos as they traveled to Earth.
   (c) The model used to predict the number of neutrinos produced in the supernova makes some unreasonable assumptions.
   (d) The detector used was not as efficient as the designers thought.
   (e) All of these are possible explanations.

44. Originally a different way to resolve the solar neutrino problem was suggested—the nuclear reaction rate wasn't as high as astrophysicists thought. What would have to be true for this suggestion to be reasonable?
   (a) The mass of the Sun would have to be significantly different than thought.
   (b) The brightness of the Sun would have to be significantly different than thought.
   (c) The size of the Sun would have to be significantly different than thought.
   (d) The orbits of the planets would have to be significantly different than thought.
   (e) Both (b) and (c) are correct.

Mathematics Review

A study of physics at the level of this textbook requires some basic math skills. The relevant math topics are summarized in this appendix. We strongly recommend that you review this material and become comfortable with it as quickly as possible so that, during your physics course, you can focus on the physics concepts and procedures that are being introduced without being distracted by unfamiliarity with the math that is being used.

A.1 Exponents

Exponents are used frequently in physics. When we write 3^4, the superscript 4 is called an **exponent** and the **base number** 3 is said to be raised to the fourth power. The quantity 3^4 is equal to $3 \times 3 \times 3 \times 3 = 81$. Algebraic symbols can also be raised to a power—for example, x^4. There are special names for the operation when the exponent is 2 or 3. When the exponent is 2, we say that the quantity is **squared**; thus, x^2 is x squared. When the exponent is 3, the quantity is **cubed**; hence, x^3 is x cubed.

Note that $x^1 = x$, and the exponent is typically not written. Any quantity raised to the zero power is defined to be unity (that is, 1). Negative exponents are used for reciprocals: $x^{-4} = 1/x^4$. The exponent can also be a fraction, as in $x^{1/4}$. The exponent $\frac{1}{2}$ is called a **square root**, and the exponent $\frac{1}{3}$ is called a **cube root**.

For example, $\sqrt{6}$ can also be written as $6^{1/2}$. Most calculators have special keys for calculating numbers raised to a power—for example, a key labeled y^x or one labeled x^2.

Exponents obey several simple rules, which follow directly from the meaning of raising a quantity to a power:

1. When two powers of the same quantity are multiplied, the exponents are added:

$$(x^n)(x^m) = x^{n+m}$$

   For example, $(3^2)(3^3) = 3^5 = 243$. To verify this result, note that $3^2 = 9$, $3^3 = 27$, and $(9)(27) = 243$.
       A special case of this rule is $(x^n)(x^{-n}) = x^{n+(-n)} = x^0 = 1$.

2. The product of two different base numbers raised to the same power is the product of the base numbers, raised to that power:

$$(x^n)(y^n) = (xy)^n$$

   For example, $(2^4)(3^4) = 6^4 = 1296$. To verify this result, note that $2^4 = 16$, $3^4 = 81$, and $(16)(81) = 1296$.

3. When a power is raised to another power, the exponents are multiplied:

$$(x^n)^m = x^{nm}$$

   For example, $(2^2)^3 = 2^6 = 64$. To verify this result, note that $2^2 = 4$, so $(2^2)^3 = (4)^3 = 64$.

If the base number is negative, it is helpful to know that $(-x)^n = (-1)^n x^n$, and $(-1)^n$ is $+1$ if n is even and -1 if n is odd.

QUANTITATIVE EXERCISE A.1 Simplifying an exponential expression

Simplify the expression $\dfrac{x^3y^{-3}xy^{4/3}}{x^{-4}y^{1/3}(x^2)^3}$ and calculate its numerical value when $x = 6$ and $y = 3$.

Represent mathematically, solve, and evaluate We simplify the expression as follows:

$$\frac{x^3x}{x^{-4}(x^2)^3} = x^3x^1x^4x^{-6} = x^{3+1+4-6} = x^2; \quad \frac{y^{-3}y^{4/3}}{y^{1/3}} = y^{-3+\frac{4}{3}-\frac{1}{3}} = y^{-2}$$

$$\frac{x^3y^{-3}xy^{4/3}}{x^{-4}y^{1/3}(x^2)^3} = x^2y^{-2} = x^2\left(\frac{1}{y}\right)^2 = \left(\frac{x}{y}\right)^2$$

For $x = 6$ and $y = 3$, $\left(\dfrac{x}{y}\right)^2 = \left(\dfrac{6}{3}\right)^2 = 4.$

If we evaluate the original expression directly, we obtain

$$\frac{x^3y^{-3}xy^{4/3}}{x^{-4}y^{1/3}(x^2)^3} = \frac{(6^3)(3^{-3})(6)(3^{4/3})}{(6^{-4})(3^{1/3})([6^2]^3)}$$

$$= \frac{(216)(1/27)(6)(4.33)}{(1/1296)(1.44)(46{,}656)} = 4.00$$

which checks.

This example demonstrates the usefulness of the rules for manipulating exponents.

QUANTITATIVE EXERCISE A.2 Solving an exponential expression for the base number

If $x^4 = 81$, what is x?

Represent mathematically, solve, and evaluate We raise each side of the equation to the $\frac{1}{4}$ power: $(x^4)^{1/4} = (81)^{1/4}$. $(x^4)^{1/4} = x^1 = x$, so $x = (81)^{1/4}$ and $x = +3$ or $x = -3$. Either of these values of x gives $x^4 = 81$.

Notice that we raised *both sides* of the equation to the $\frac{1}{4}$ power. As will be explained in Section A.3, an operation performed on both sides of an equation does not affect the equation's validity.

A.2 Scientific notation and powers of 10

In physics, we frequently encounter very large and very small numbers, and it is important to use the proper number of significant digits when expressing a quantity. We can deal with both these issues by using **scientific notation**, in which a quantity is expressed as a decimal number with one digit to the left of the decimal point, multiplied by the appropriate power of 10. If the power of 10 is positive, it is the number of places the decimal point is moved to the right to obtain the fully written-out number. For example, $6.3 \times 10^4 = 63{,}000$. If the power of 10 is negative, it is the number of places the decimal point is moved to the left to obtain the fully written-out number. For example, $6.56 \times 10^{-3} = 0.00656$. In going from 6.56 to 0.00656, the decimal point is moved three places to the left, so 10^{-3} is the correct power of 10 to use when the number is written in scientific notation. Most calculators have keys for expressing a number in either decimal (floating-point) or scientific notation.

When two numbers written in scientific notation are multiplied (or divided), multiply (or divide) the decimal parts to get the decimal part of the result, and add (or subtract) the powers of 10 to get the power-of-10 portion of the result. You may have to adjust the location of the decimal point in the answer to express it in scientific notation. For example,

$$(8.43 \times 10^8)(2.21 \times 10^{-5}) = (8.43 \times 2.21)(10^8 \times 10^{-5})$$
$$= (18.6) \times (10^{8-5}) = 18.6 \times 10^3$$
$$= 1.86 \times 10^4$$

Similarly,

$$\frac{5.6 \times 10^{-3}}{2.8 \times 10^{-6}} = \left(\frac{5.6}{2.8}\right) \times \left(\frac{10^{-3}}{10^{-6}}\right) = 2.0 \times 10^{-3-(-6)} = 2.0 \times 10^3$$

Your calculator can handle these operations for you automatically, but it is important for you to develop good "number sense" for scientific notation manipulations.

A.3 Algebra

Solving equations

Equations written in terms of symbols that represent quantities are frequently used in physics. An **equation** consists of an equals sign and quantities to its left and to its right. Every equation tells us that the combination of quantities on the left of the equals sign has the same value as (is equal to) the combination on the right of the equals sign. For example, the equation $y + 4 = x^2 + 8$ tells us that $y + 4$ has the same value as $x^2 + 8$. If $x = 3$, then the equation $y + 4 = x^2 + 8$ says that $y = 13$.

Often, one of the symbols in an equation is considered to be the *unknown*, and we wish to solve for the unknown in terms of the other symbols or quantities. For example, we might wish to solve the equation $2x^2 + 4 = 22$ for the value of x. Or we might wish to solve the equation $x = v_0 t + \frac{1}{2} at^2$ for the unknown a in terms of x, t, and v_0.

An equation can be solved by using the following rule:

> **An equation remains true if any operation performed on one side of the equation is also performed on the other side.** The operations could be (a) adding or subtracting a number or symbol, (b) multiplying or dividing by a number or symbol, or (c) raising each side of the equation to the same power.

QUANTITATIVE EXERCISE A.3 Solving a numerical equation

Solve the equation $2x^2 + 4 = 22$ for x.

Represent mathematically, solve, and evaluate First we subtract 4 from both sides. This gives $2x^2 = 18$. Then we divide both sides by 2 to get $x^2 = 9$. Finally, we raise both sides of the equation to the $\frac{1}{2}$ power. (In other words, we take the square root of both sides of the equation.) This gives $x = \pm\sqrt{9} = \pm 3$. That is, $x = +3$ or $x = -3$. We can verify our answers by substituting our result back into the original equation: $2x^2 + 4 = 2(\pm 3)^2 + 4 = 2(9) + 4 = 18 + 4 = 22$, so $x = \pm 3$ does satisfy the equation.

Notice that a square root always has *two* possible values, one positive and one negative. For instance, $\sqrt{4} = \pm 2$, because $(2)(2) = 4$ and $(-2)(-2) = 4$. Your calculator will give you only a positive root; it's up to you to remember that there are actually two. Both roots are correct mathematically, but in a physics problem only one may represent the answer. For instance, if you can get dressed in $\sqrt{4}$ minutes, the only physically meaningful root is 2 minutes!

QUANTITATIVE EXERCISE A.4 Solving a symbolic equation

Solve the equation $x = v_0 t + \frac{1}{2} at^2$ for a.

Represent mathematically, solve, and evaluate We subtract $v_0 t$ from both sides. This gives $x - v_0 t = \frac{1}{2} at^2$. Now we multiply both sides by 2 and divide both sides by t^2, giving $a = \dfrac{2(x - v_0 t)}{t^2}$.

As we've indicated, it makes no difference whether the quantities in an equation are represented by variables (such as x, v, and t) or by numerical values.

The quadratic formula

Using the methods of the previous subsection, we can easily solve the equation $ax^2 + c = 0$ for x:

$$x = \pm\sqrt{\frac{-c}{a}}$$

For example, if $a = 2$ and $c = -8$, the equation is $2x^2 - 8 = 0$ and the solution is

$$x = \pm \sqrt{\frac{-(-8)}{2}} = \pm \sqrt{4} = \pm 2$$

The equation $ax^2 + bx = 0$ is also easily solved by factoring out an x on the left side of the equation, giving $x(ax + b) = 0$. (To *factor out* a quantity means to isolate it so that the rest of the expression is either multiplied or divided by that quantity.) The equation $x(ax + b) = 0$ is true (that is, the left side equals zero) if either $x = 0$ or $x = -\frac{b}{a}$. Those are the equation's two solutions. For example, if $a = 2$ and $b = 8$, the equation is $2x^2 + 8x = 0$ and the solutions are $x = 0$ and $x = -\frac{8}{2} = -4$.

But if the equation is in the form $ax^2 + bx + c = 0$, with a, b, and c all nonzero, we cannot use our standard methods to solve for x. Such an equation is called a **quadratic equation**, and its solutions are expressed by the **quadratic formula**:

Quadratic formula For a quadratic equation in the form $ax^2 + bx + c = 0$, where a, b, and c are real numbers and $a \neq 0$, the solutions are given by the quadratic formula:

$$x = \frac{-b \pm \sqrt{b^2 - 4ac}}{2a}$$

In general, a quadratic equation has two roots (solutions). But if $b^2 - 4ac = 0$, then the two roots are equal. By contrast, if $b^2 - 4ac$ is negative, then both roots are complex numbers and cannot represent physical quantities. In such a case, the original quadratic equation has mathematical solutions, but no physical solutions.

QUANTITATIVE EXERCISE A.5 Solving a quadratic equation

Find the values of x that satisfy the equation $2x^2 - 2x = 24$.

Represent mathematically, solve, and evaluate First we write the equation in the standard form $ax^2 + bx + c = 0$: $2x^2 - 2x - 24 = 0$. Then $a = 2$, $b = -2$, and $c = -24$. Next, the quadratic formula gives the two roots as

$$x = \frac{-(-2) \pm \sqrt{(-2)^2 - 4(2)(-24)}}{(2)(2)}$$

$$= \frac{+2 \pm \sqrt{4 + 192}}{4} = \frac{2 \pm 14}{4}$$

so $x = 4$ or $x = -3$. If x represents a physical quantity that takes only nonnegative values, then the negative root $x = -3$ is nonphysical and is discarded.

As we've mentioned, when an equation has more than one mathematical solution or root, it's up to *you* to decide whether one or the other or both represent the true physical answer. (If neither solution seems physically plausible, you should review your work.)

Simultaneous equations

If a problem has two unknowns—for example, x and y—then it takes two independent equations in x and y (that is, two equations for x and y, where one equation is not simply a multiple of the other) to determine their values uniquely. Such equations are called **simultaneous equations** (because you solve them together). A typical procedure is to solve one equation for x in terms of y and then substitute the result into the second equation to obtain an equation in which y is the only unknown. You then solve this equation for y and use the value of y in either of the original equations in order to solve for x. In general, to solve for n unknowns, we must have n independent equations.

QUANTITATIVE EXERCISE A.6   **Solving two equations in two unknowns**

Solve the following pair of equations for x and y:

$$x + 4y = 14$$
$$3x - 5y = -9$$

Represent mathematically, solve, and evaluate The first equation gives $x = 14 - 4y$. Substituting this for x in the second equation yields, successively, $3(14 - 4y) - 5y = -9$, $42 - 12y - 5y = -9$, and $-17y = -51$. Thus, $y = \frac{-51}{-17} = 3$. Then $x = 14 - 4y = 14 - 12 = 2$. We can verify that $x = 2$, $y = 3$ satisfies both equations.

An alternative approach is to multiply the first equation by -3, yielding $-3x - 12y = -42$. Adding this to the second equation gives, successively, $3x - 5y + (-3x) + (-12y) = -9 + (-42)$, $-17y = -51$, and $y = 3$, which agrees with our previous result.

As shown by the alternative approach, simultaneous equations can be solved in more than one way. The basic method we describe is easy to keep straight; other methods may be quicker, but may require more insight or forethought. Use the method you're comfortable with.

A pair of equations in which all quantities are symbols can be combined to eliminate one of the common unknowns.

QUANTITATIVE EXERCISE A.7   **Solving two symbolic equations in two unknowns**

Use the equations $v = v_0 + at$ and $x = v_0 t + \frac{1}{2} at^2$ to obtain an equation for x that does not contain a.

Represent mathematically, solve, and evaluate We solve the first equation for a:

$$a = \frac{v - v_0}{t}$$

We substitute this expression into the second equation:

$$x = v_0 t + \frac{1}{2}\left(\frac{v - v_0}{t}\right)t^2 = v_0 t + \frac{1}{2} vt - \frac{1}{2} v_0 t$$

$$= \frac{1}{2} v_0 t + \frac{1}{2} vt = \left(\frac{v_0 + v}{2}\right)t$$

When you solve a physics problem, it's often best to work with symbols for all but the final step of the problem. Once you've arrived at the final equation, you can plug in numerical values and solve for an answer.

A.4 Logarithmic and exponential functions

The base-10 logarithm, or **common logarithm** (log), of a number y is the power to which 10 must be raised to obtain y: $y = 10^{\log y}$. For example, $1000 = 10^3$, so $\log(1000) = 3$; you must raise 10 to the power 3 to obtain 1000. Most calculators have a key for calculating the log of a number.

Sometimes we are given the log of a number and are asked to find the number. That is, if $\log y = x$ and x is given, what is y? To solve for y, write an equation in which 10 is raised to the power equal to either side of the original equation: $10^{\log y} = 10^x$. But $10^{\log y} = y$, so $y = 10^x$. In this case, y is called the **antilog** of x. For example, if $\log y = -2.0$, then $y = 10^{-2.0} = 1.0 \times 10^{-2.0} = 0.010$.

The log of a number is positive if the number is greater than 1. The log of a number is negative if the number is less than 1, but greater than zero. The log of zero or of a negative number is not defined, and $\log 1 = 0$.

Another base that occurs frequently in physics is the quantity $e = 2.718\ldots$. The **natural logarithm** (ln) of a number y is the power to which e must be raised to obtain y: $y = e^{\ln y}$. If $x = \ln y$, then $y = e^x$. Most calculators have keys for ln x and for e^x. For example, $\ln 10.0 = 2.30$, and if $\ln x = 3.00$, then $x = e^{3.00} = 20.1$. Note that $\ln 1 = 0$.

Logarithms with any choice of base, including base 10 or base e, obey several simple and useful rules:

1. $\log(ab) = \log a + \log b$

2. $\log\left(\dfrac{a}{b}\right) = \log a - \log b$

3. $\log(a^n) = n \log a$

A particular example of the second rule is

$$\log\left(\frac{1}{a}\right) = \log 1 - \log a = -\log a$$

since $\log 1 = 0$.

QUANTITATIVE EXERCISE A.8 ⟩ **Solving a logarithmic equation**

If $\frac{1}{2} = e^{-\alpha T}$, solve for T in terms of α.

Represent mathematically, solve, and evaluate We take the natural logarithm of both sides of the equation: $\ln\left(\frac{1}{2}\right) = -\ln 2$ and $\ln\left(e^{-\alpha T}\right) = -\alpha T$. The equation thus becomes $-\alpha T = -\ln 2$, and it follows that $T = \dfrac{\ln 2}{\alpha}$.

The equation $y = e^{\alpha x}$ expresses y in terms of the exponential function $e^{\alpha x}$. The general rules for exponents in Appendix A.1 apply when the base is e, so $e^x e^y = e^{x+y}$, $e^x e^{-x} = e^{x+(-x)} = e^0 = 1$, and $(e^x)^2 = e^{2x}$.

FIGURE A.1

Area
$A = ab$

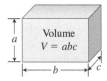

Volume
$V = abc$

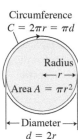

Circumference
$C = 2\pi r = \pi d$

Radius
$\leftarrow r \rightarrow$

Area $A = \pi r^2$

$\leftarrow$ Diameter $\rightarrow$
$d = 2r$

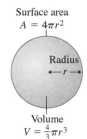

Surface area
$A = 4\pi r^2$

Radius
$\leftarrow r \rightarrow$

Volume
$V = \frac{4}{3}\pi r^3$

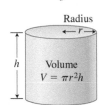

Radius
$\leftarrow r \rightarrow$

h Volume
$V = \pi r^2 h$

A.5 Areas and volumes

Figure A.1 illustrates the formulas for the areas and volumes of common geometric shapes:

- A rectangle with length a and width b has area $A = ab$.
- A rectangular solid (a box) with length a, width b, and height c has volume $V = abc$.
- A circle with radius r has diameter $d = 2r$, circumference $C = 2\pi r = \pi d$, and area $A = \pi r^2 = \pi d^2/4$.
- A sphere with radius r has surface area $A = 4\pi r^2$ and volume $V = \frac{4}{3}\pi r^3$.
- A cylinder with radius r and height h has volume $V = \pi r^2 h$.

A.6 Plane geometry and trigonometry

Following are some useful facts about angles:

1. Interior angles formed when two straight lines intersect are equal. For example, in **Figure A.2**, the two angles θ and ϕ are equal.
2. When two parallel lines are intersected by a diagonal straight line, the alternate interior angles are equal. For example, in **Figure A.3**, the two angles θ and ϕ are equal.
3. When the sides of one angle are each perpendicular to the corresponding sides of a second angle, then the two angles are equal. For example, in **Figure A.4**, the two angles θ and ϕ are equal.
4. The sum of the angles on one side of a straight line is 180°. In **Figure A.5**, $\theta + \phi = 180°$.
5. The sum of the angles in any triangle is 180°.

FIGURE A.2

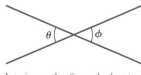

Interior angles formed when two straight lines intersect are equal:
$\theta = \phi$

FIGURE A.3

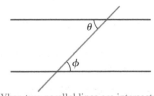

When two parallel lines are intersected by a diagonal straight line, the alternate interior angles are equal:
$\theta = \phi$

FIGURE A.4

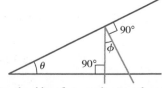

When the sides of one angle are each perpendicular to the corresponding sides of a second angle, then the two angles are equal:
$\theta = \phi$

FIGURE A.5

The sum of the angles on one side of a straight line is 180°:
$\theta + \phi = 180°$

Similar triangles

Triangles are **similar** if they have the same shape, but different sizes or orientations. Similar triangles have equal angles and equal ratios of corresponding sides. If the two triangles in **Figure A.6** are similar, then $\theta_1 = \theta_2$, $\phi_1 = \phi_2$, $\gamma_1 = \gamma_2$, and $\dfrac{a_1}{a_2} = \dfrac{b_1}{b_2} = \dfrac{c_1}{c_2}$.

If two similar triangles have the same size, they are said to be **congruent**. If triangles are congruent, one can be rotated to where it can be placed precisely on top of the other.

Right triangles and trig functions

In a right triangle, one angle is 90°. Therefore, the other two acute angles (*acute* means less than 90°) have a sum of 90°. In **Figure A.7**, $\theta + \phi = 90°$. The side opposite the right angle is called the **hypotenuse** (side c in the figure). In a right triangle, the square of the length of the hypotenuse equals the sum of the squares of the lengths of the other two sides. For the triangle in Figure A.7, $c^2 = a^2 + b^2$. This formula is called the **Pythagorean theorem**.

If two right triangles have the same value for one acute angle, then the two triangles are similar and have the same ratio of corresponding sides. This true statement allows us to define the functions **sine**, **cosine**, and **tangent**, which are ratios of a pair of sides. These functions, called **trigonometric functions** or **trig functions**, depend only on one of the angles in the right triangle. For an angle θ, these functions are written $\sin\theta$, $\cos\theta$, and $\tan\theta$.

In terms of the triangle in Figure A.7, the sine, cosine, and tangent of the angle θ are as follows:

$$\sin\theta = \frac{\text{opposite side}}{\text{hypotenuse}} = \frac{a}{c}$$

$$\cos\theta = \frac{\text{adjacent side}}{\text{hypotenuse}} = \frac{b}{c}$$

$$\tan\theta = \frac{\text{opposite side}}{\text{adjacent side}} = \frac{a}{b}$$

Note that $\tan\theta = \dfrac{\sin\theta}{\cos\theta}$. For angle ϕ, $\sin\phi = \dfrac{b}{c}$, $\cos\phi = \dfrac{a}{c}$, and $\tan\phi = \dfrac{b}{a}$.

In physics, angles are expressed in either degrees or radians, where π radians $= 180°$. Most calculators have a key for switching between degrees and radians. Always be sure that your calculator is set to the appropriate angular measure.

Inverse trig functions, denoted, for example, by $\sin^{-1} x$ (or $\arcsin x$), have a value equal to the angle that has the value x for the trig function. For example, $\sin 30° = 0.500$, so $\sin^{-1}(0.500) = \arcsin(0.500) = 30°$. Note that $\sin^{-1} x$ does *not* mean $\dfrac{1}{\sin x}$.

FIGURE A.6

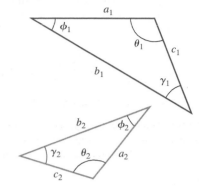

Two similar triangles: Same shape but not necessarily the same size

FIGURE A.7

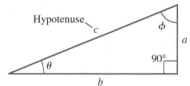

For a right triangle:
$\theta + \phi = 90°$
$c^2 = a^2 + b^2$ (Pythagorean theorem)

QUANTITATIVE EXERCISE A.9   **Using trigonometry I**

A right triangle has one angle of 30° and one side with length 8.0 cm, as shown in **Figure A.8**. What is the angle ϕ, and what are the lengths x and y of the other two sides of the triangle?

FIGURE A.8

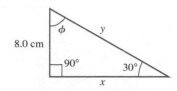

Represent mathematically, solve, and evaluate $\phi + 30° = 90°$, so $\phi = 60°$.

$$\tan 30° = \frac{8.0\ \text{cm}}{x}, \text{ so } x = \frac{8.0\ \text{cm}}{\tan 30°} = 13.9\ \text{cm}$$

To find y, we use the Pythagorean theorem: $y^2 = (8.0\ \text{cm})^2 + (13.9\ \text{cm})^2$, so $y = 16.0\ \text{cm}$.

Or we can say $\sin 30° = 8.0\ \text{cm}/y$, so $y = 8.0\ \text{cm}/\sin 30° = 16.0\ \text{cm}$, which agrees with the previous result.

Notice how we used the Pythagorean theorem in combination with a trig function. You will use these tools constantly in physics, so make sure that you can employ them with confidence.

QUANTITATIVE EXERCISE A.10 ⟩ **Using trigonometry II**

A right triangle has two sides with lengths as specified in **Figure A.9**. What is the length x of the third side of the triangle, and what is the angle θ, in degrees?

FIGURE A.9

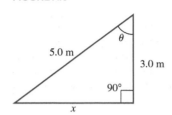

Represent mathematically, solve, and evaluate The Pythagorean theorem applied to this right triangle gives $(3.0 \text{ m})^2 + x^2 = (5.0 \text{ m})^2$, so $x = \sqrt{(5.0 \text{ m})^2 - (3.0 \text{ m})^2} = 4.0$ m. (Since x is a length, we take the positive root of the equation.) We also have

$$\cos \theta = \frac{3.0 \text{ m}}{5.0 \text{ m}} = 0.600, \text{ so } \theta = \cos^{-1}(0.600) = 53.1°$$

In this case, we knew the lengths of two sides but none of the acute angles, so we used the Pythagorean theorem first and then an appropriate trig function.

FIGURE A.10

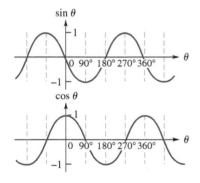

FIGURE A.11

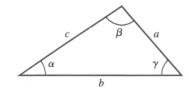

In a right triangle, all angles are in the range from 0° to 90°, and the sine, cosine, and tangent of the angles are all positive. This must be the case, since the trig functions are ratios of lengths. But for other applications, such as finding the components of vectors, calculating the oscillatory motion of a mass on a spring, or describing wave motion, it is useful to define the sine, cosine, and tangent for angles outside that range. Graphs of $\sin \theta$ and $\cos \theta$ are given in **Figure A.10**. The values of $\sin \theta$ and $\cos \theta$ vary between $+1$ and -1. Each function is periodic, with a period of 360°. Note the range of angles between 0° and 360° for which each function is positive and negative. The two functions $\sin \theta$ and $\cos \theta$ are 90° out of phase (that is, out of step). When one is zero, the other has its maximum magnitude.

For any triangle (see **Figure A.11**)—in other words, not necessarily a right triangle—the following two relations apply:

1. $\dfrac{\sin \alpha}{a} = \dfrac{\sin \beta}{b} = \dfrac{\sin \gamma}{c}$ (law of sines)

2. $c^2 = a^2 + b^2 - 2ab \cos \gamma$ (law of cosines)

Some of the relations among trig functions are called trig identities. The following table lists only a few, those most useful in introductory physics:

Useful trigonometric identities

$\sin(-\theta) = -\sin(\theta)$ ($\sin \theta$ is an odd function)

$\cos(-\theta) = \cos(\theta)$ ($\cos \theta$ is an even function)

$\sin 2\theta = 2 \sin \theta \cos \theta$

$\cos 2\theta = \cos^2 \theta - \sin^2 \theta = 2 \cos^2 \theta - 1 = 1 - 2 \sin^2 \theta$

$\sin(\theta \pm \phi) = \sin \theta \cos \phi \pm \cos \theta \sin \phi$

$\cos(\theta \pm \phi) = \cos \theta \cos \phi \mp \sin \theta \sin \phi$

$\sin(180° - \theta) = \sin \theta$

$\cos(180° - \theta) = -\cos \theta$

$\sin(90° - \theta) = \cos \theta$

$\cos(90° - \theta) = \sin \theta$

A.7 Examples of operations with vectors

EXAMPLE A.11 **Graphical addition of vectors**

A car travels 200 km west, 100 km south, and finally 150 km at an angle 60° south of east. Determine the net displacement of the car.

Reasoning Use the graphical addition technique with the three displacements drawn tail to head, as shown in **Figure A.12a**. To find the resultant displacement $\vec{R}$, draw an arrow from the tail of the first

displacement to the head of the last (Figure A.12b). We measure the magnitude of the resultant with a ruler and find that its length is 5.2 cm. Since each centimeter represents 50 km, the magnitude of the resultant displacement is $(5.2\ \text{cm})(50\ \text{km/cm}) = 260\ \text{km}$. Using a protractor, we confirm that the direction of the resultant is 60° south of west.

FIGURE A.12 (a) Three displacement vectors placed tail to head. (b) The resultant displacement.

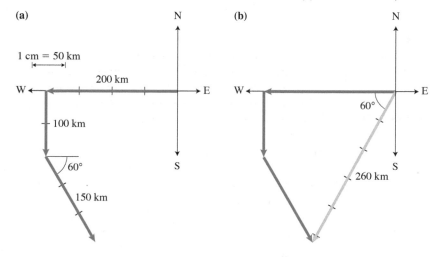

EXAMPLE A.12 **Components of vectors**

Suppose we have a vector $\vec{N}$ of the magnitude of 10 units that is directed at 20° above the horizontal direction as shown in **Figure A.13a**. The hypotenuse and the x- and y-scalar components of $\vec{N}$ form a triangle for which:

$$\cos\theta = \frac{\text{adjacent side}}{\text{hypotenuse}} = \frac{N_x}{N} \quad \text{or} \quad N_x = N\cos\theta$$

$$\sin\theta = \frac{\text{opposite side}}{\text{hypotenuse}} = \frac{N_y}{N} \quad \text{or} \quad N_y = N\sin\theta$$

where N is the magnitude of the vector $\vec{N}$. Using the magnitude N and the angle, we can calculate the values of the scalar components of $\vec{N}$:

$$N_x = N\cos\theta = (10\ \text{units})\cos 20° = 9.4\ \text{units}$$
$$N_y = N\sin\theta = (10\ \text{units})\sin 20° = 3.4\ \text{units}$$

If the vector pointed in the opposite direction, as in Figure A.13b, the magnitude of the scalar stays the same but the signs are negative:

$$N_x = -N\cos\theta = -9.4\ \text{units}$$
$$N_y = -N\sin\theta = -3.4\ \text{units}$$

The signs depend on the orientation of the scalar components relative to the x- and y-axes.

FIGURE A.13 N_x and N_y can be calculated from known N and θ.

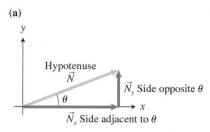

EXAMPLE A.13   Components of vectors

Determine the x- and y-components of each of the force vectors shown in Figure A.14.

Reasoning $F_{1x} = +24$ N   and $F_{1y} = +32$ N; $F_{2x} = -35$ N   and $F_{2y} = +35$ N. Note that the angle between $\vec{F}_3$ and the negative x-axis is $30°$ and not $60°$. Thus, $F_{3x} = -26$ N and $F_{3y} = -15$ N.

FIGURE A.14

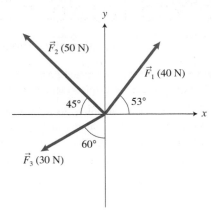

EXAMPLE A.14   Addition using components

Determine the sum of the three forces shown in Figure A.14. You can use the results of the component calculation in Example A.13.

Reasoning The x- and y-components of the resultant force are:

$$R_x = F_{1x} + F_{2x} + F_{3x} = +24 \text{ N} + (-35 \text{ N}) + (-26 \text{ N})$$
$$= -37 \text{ N}$$

$$R_y = F_{1y} + F_{2y} + F_{3y} = +32 \text{ N} + 35 \text{ N} + (-15 \text{ N})$$
$$= +52 \text{ N}$$

The magnitude of the resultant force is:

$$R = \sqrt{(-37 \text{ N})^2 + (52 \text{ N})^2} = 64 \text{ N}$$

The resultant vector is in the second quadrant (it has a negative x-component and a positive y-component) and makes the following angle above the negative x-axis:

$$\theta = \arctan \frac{52 \text{ N}}{37 \text{ N}} = \arctan 1.41 = 55°$$

Thus, the sum of the three forces has a magnitude of 64 N and points $55°$ above the negative x-axis.

Atomic and Nuclear Data

Atomic number (Z)	Element	Symbol	Mass number (A)	Atomic mass (u)	Percent abundance	Decay mode	Half-life $t_{1/2}$
0	(Neutron)	n	1	1.008 665		β^-	10.4 min
1	Hydrogen	H	1	1.007 825	99.985	stable	
	Deuterium	D	2	2.014 102	0.015	stable	
	Tritium	T	3	3.016 049		β^-	12.33 yr
2	Helium	He	3	3.016 029	0.000 1	stable	
			4	4.002 602	99.999 9	stable	
			6	6.018 886		β^-	0.81 s
3	Lithium	Li	6	6.015 121	7.50	stable	
			7	7.016 003	92.50	stable	
			8	8.022 486		β^-	0.84 s
4	Beryllium	Be	7	7.016 928		electron capture (EC)	53.3 days
			9	9.012 174	100	stable	
			10	10.013 534		β^-	1.5×10^6 yr
5	Boron	B	10	10.012 936	19.90	stable	
			11	11.009 305	80.10	stable	
			12	12.014 352		β^-	0.020 2 s
6	Carbon	C	10	10.016 854		β^+	19.3 s
			11	11.011 433		β^+	20.4 min
			12	12.000 000	98.90	stable	
			13	13.003 355	1.10	stable	
			14	14.003 242		β^-	5 730 yr
			15	15.010 599		β^-	2.45 s
7	Nitrogen	N	12	12.018 613		β^+	0.011 0 s
			13	13.005 738		β^+	9.96 min
			14	14.003 074	99.63	stable	
			15	15.000 108	0.37	stable	
			16	16.006 100		β^-	7.13 s
			17	17.008 450		β^-	4.17 s
8	Oxygen	O	14	14.008 595		EC	70.6 s
			15	15.003 065		β^+	122 s
			16	15.994 915	99.76	stable	
			17	16.999 132	0.04	stable	
			18	17.999 160	0.20	stable	
			19	19.003 577		β^-	26.9 s
9	Fluorine	F	17	17.002 094		EC	64.5 s
			18	18.000 937		β^+	109.8 min
			19	18.998 404	100	stable	
			20	19.999 982		β^-	11.0 s
10	Neon	Ne	19	19.001 880		β^+	17.2 s
			20	19.992 435	90.48	stable	
			21	20.993 841	0.27	stable	
			22	21.991 383	9.25	stable	

Atomic number (Z)	Element	Symbol	Mass number (A)	Atomic mass (u)	Percent abundance	Decay mode	Half-life $t_{1/2}$
11	Sodium	Na	22	21.994 434		β^+	2.61 yr
			23	22.989 770	100	stable	
			24	23.990 961		β^-	14.96 hr
12	Magnesium	Mg	24	23.985 042	78.99	stable	
			25	24.985 838	10.00	stable	
			26	25.982 594	11.01	stable	
13	Aluminum	Al	27	26.981 538	100	stable	
			28	27.981 910		β^-	2.24 min
14	Silicon	Si	28	27.976 927	92.23	stable	
			29	28.976 495	4.67	stable	
			30	29.973 770	3.10	stable	
			31	30.975 362		β^-	2.62 hr
15	Phosphorus	P	30	29.978 307		β^+	2.50 min
			31	30.973 762	100	stable	
			32	31.973 908		β^-	14.26 days
16	Sulfur	S	32	31.972 071	95.02	stable	
			33	32.971 459	0.75	stable	
			34	33.967 867	4.21	stable	
			35	34.969 033		β^-	87.5 days
			36	35.967 081	0.02	stable	
17	Chlorine	Cl	35	34.968 853	75.77	stable	
			36	35.968 307		β^-	3.0×10^5 yr
			37	36.965 903	24.23	stable	
18	Argon	Ar	36	35.967 547	0.34	stable	
			38	37.962 732	0.06	stable	
			39	38.964 314		β^-	269 yr
			40	39.962 384	99.60	stable	
			42	41.963 049		β^-	33 yr
19	Potassium	K	39	38.963 708	93.26	stable	
			40	39.964 000	0.01	β^+	1.28×10^9 yr
			41	40.961 827	6.73	stable	
20	Calcium	Ca	40	39.962 591	96.94	stable	
			42	41.958 618	0.64	stable	
			43	42.958 767	0.13	stable	
			44	43.955 481	2.08	stable	
			47	46.954 547		β^-	4.5 days
			48	47.952 534	0.18	stable	
24	Chromium	Cr	50	49.946 047	4.34	stable	
			52	51.940 511	83.79	stable	
			53	52.940 652	9.50	stable	
			54	53.938 883	2.36	stable	
26	Iron	Fe	54	53.939 613	5.9	stable	
			55	54.938 297		EC	2.7 yr
			56	55.934 940	91.72	stable	
			57	56.935 396	2.1	stable	
			58	57.933 278	0.28	stable	
27	Cobalt	Co	59	58.933 198	100	stable	
			60	59.933 820		β^-	5.27 yr

Atomic number (Z)	Element	Symbol	Mass number (A)	Atomic mass (u)	Percent abundance	Decay mode	Half-life $t_{1/2}$
28	Nickel	Ni	58	57.935 346	68.08	stable	
			60	59.930 789	26.22	stable	
			61	60.931 058	1.14	stable	
			62	61.928 346	3.63	stable	
			64	63.927 967	0.92	stable	
29	Copper	Cu	63	62.929 599	69.17	stable	
			65	64.927 791	30.83	stable	
37	Rubidium	Rb	96	95.934 27		stable	
38	Strontium	Sr	90	89.907 320		α	28.9 yr
47	Silver	Ag	107	106.905 091	51.84	stable	
			109	108.904 754	48.16	stable	
48	Cadmium	Cd	106	105.906 457	1.25	stable	
			109	108.904 984		EC	462 days
			110	109.903 004	12.49	stable	
			111	110.904 182	12.80	stable	
			112	111.902 760	24.13	stable	
			113	112.904 401	12.22	stable	
			114	113.903 359	28.73	stable	
			116	115.904 755	7.49	stable	
50	Tin	Sn	120	119.902 197	32.4		
53	Iodine	I	127	126.904 474	100	stable	
			129	128.904 984		β^-	1.6×10^7 yr
			131	130.906 124		β^-	8.03 days
54	Xenon	Xe	128	127.903 531	1.9	stable	
			129	128.904 779	26.4	stable	
			130	129.903 509	4.1	stable	
			131	130.905 069	21.2	stable	
			132	131.904 141	26.9	stable	
			133	132.905 906		β^-	5.4 days
			134	133.905 394	10.4	stable	
			136	135.907 215	8.9	stable	
55	Cesium	Cs	133	132.905 436	100	stable	
			137	136.907 078		β^-	30 yr
			138	137.911 017		β^-	32.2 min
56	Barium	Ba	131	130.906 931		EC	12 days
			133	132.905 990		EC	10.5 yr
			134	133.904 492	2.42	stable	
			135	134.905 671	6.59	stable	
			136	135.904 559	7.85	stable	
			137	136.905 816	11.23	stable	
			138	137.905 236	71.70	stable	
79	Gold	Au	197	196.966 543	100	stable	
			198	197.968 242		β^-	2.7 days
81	Thallium	Tl	203	202.972 320	29.524	stable	
			205	204.974 400	70.476	stable	
			207	206.977 403		β^-	4.77 min

Atomic number (Z)	Element	Symbol	Mass number (A)	Atomic mass (u)	Percent abundance	Decay mode	Half-life $t_{1/2}$
82	Lead	Pb	204	203.973 020	1.4	stable	
			205	204.974 457		EC	1.5×10^7 yr
			206	205.974 440	24.1	stable	
			207	206.975 871	22.1	stable	
			208	207.976 627	52.4	stable	
			210	209.984 163		α, β^-	22.3 yr
			211	210.988 734		β^-	36.1 min
83	Bismuth	Bi	208	207.979 717		EC	3.7×10^5 yr
			209	208.980 374	100	stable	
			211	210.987 254		α	2.14 min
			215	215.001 836		β^-	7.4 min
84	Polonium	Po	209	208.982 405		α	102 yr
			210	209.982 848		α	138.38 days
			215	214.999 418		α	0.001 8 s
			218	218.008 965		α, β^-	3.10 min
85	Astatine	At	218	218.008 685		α, β^-	1.6 s
			219	219.011 294		α, β^-	0.9 min
86	Radon	Rn	219	219.009 477		α	3.96 s
			220	220.011 369		α	55.6 s
			222	222.017 571		α, β^-	3.823 days
87	Francium	Fr	223	223.019 733		α, β^-	22 min
88	Radium	Ra	223	223.018 499		α	11.43 days
			224	224.020 187		α	3.66 days
			226	226.025 402		α	1 600 yr
			228	228.031 064		β^-	5.75 yr
89	Actinium	Ac	227	227.027 749		α, β^-	21.77 yr
			228	228.031 015		β^-	6.15 hr
90	Thorium	Th	227	227.027 701		α	18.72 days
			228	228.028 716		α	1.913 yr
			229	229.031 757		α	7 300 yr
			230	230.033 127		α	75.000 yr
			231	231.036 299		α, β^-	25.52 hr
			232	232.038 051	100	α	1.40×10^{10} yr
			234	234.043 593		β^-	24.1 days
91	Protactinium	Pa	231	231.035 880		α	32.760 yr
			234	234.043 300		β^-	6.7 hr
92	Uranium	U	232	232.03713		α	72 yr
			233	233.039 630		α	1.59×10^5 yr
			234	234.040 946		α	2.45×10^5 yr
			235	235.043 924	0.72	α	7.04×10^8 yr
			236	236.045 562		α	2.34×10^7 yr
			238	238.050 784	99.28	α	4.47×10^9 yr
93	Neptunium	Np	236	236.046 560		EC	1.15×10^5 yr
			237	237.048 168		α	2.14×10^6 yr
94	Plutonium	Pu	238	238.049 555		α	87.7 yr
			239	239.052 157		α	2.412×10^4 yr
			240	240.053 808		α	6 560 yr
			242	242.058 737		α	3.73×10^5 yr

Answers to Select Odd-Numbered Problems

Chapter 17

Multiple-Choice Questions

1. (d)   3. (f)   5. (a), (c), (d), (f)   7. (b)   9. (c)   11. (c)

Problems

1. 1.25×10^{18}
3.

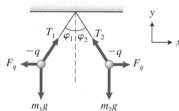

$m_1 \tan \varphi_1 = m_2 \tan \varphi_2$

5. 6.1×10^{-7} kg
7. 57 N, repulsive
9. 3.6×10^{-10} N
11. (a) 0   (b) 0   (c) 2×10^{-3} N, right
13. 2.2×10^{-9} N to the left
21. (a) Positive   (b) Increase their separation.

(c)
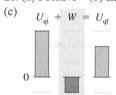
$U_{qi} + W = U_{qf}$

23. (a) Negative   (b) Decrease their separation.

(c)
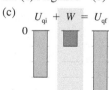
$U_{qi} + W = U_{qf}$

25. (a) 4.3×10^{-9} N   (b) 3.6×10^{-9} N   (c) 3.9×10^{-13} J
(d) The work done is equal to the change in electrical potential energy.
31. $U_D < U_A < U_C < U_B$
43. (a) 8.5×10^{-3} m/s   (b) 7.4×10^{-3} m/s
45. 4×10^{-8} C
47. 9×10^{-6} C
49. (a) 1.9×10^6 m/s   (b) Initial kinetic energy is 2.9×10^{-15} J
51. 2.4×10^7 m/s
53. 3.2 C
57. (b)   59. (b)   61. (b)   63. (a)   65. (b)   67. (a)

Chapter 18

Multiple-Choice Questions

1. (c)   3. (e)   5. (d)   7. (a)

Problems

1.
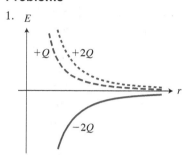

7. 0.27 m to the right of the 4.0-nC charge
9. 2.4×10^4 N/C at 242° counterclockwise from the x-axis

11. (a) $E_1 = E_0 + \dfrac{kq}{d^2}$ (b) $E_2 = E_0 - \dfrac{kq}{d^2}$

13. (a) 0.55 s   (b) 0.87 m
17. 9.8×10^5 N/C upward
21. -1.2×10^9 J

23. $\dfrac{2kq}{d}$

25. $E = 0$, $V = -\dfrac{4\sqrt{2}kq}{d}$

27. (a) -5.4×10^2 V   (b) -2.0×10^2 V
29. (a) -1.1×10^{-11} J   (b) 7.8 J   (c) 7.8%
33. (a) $E_F = E_G > E_A = E_B = E_C = E_D$
(b) $V_A > V_C = V_D > V_B > V_F > V_G$
(c) IV, I, III, II
35. 0
37.

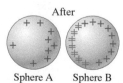

39. (a) No   (b) Yes
41. (a) 9.3×10^6 V/m   (b) 1.5×10^{-12} N
43. (a) 2.4×10^4 V/m   (b) 4.0×10^{-5} m
45. (a) $q = C|\Delta V|$, $U_q = \frac{1}{2}C|\Delta V|^2$ (b) Same charge, U_q/κ
47. (a) 1.7×10^{-7} F   (b) 1.2×10^{-8} C   (c) 4.1×10^{-10} J
51. 9×10^3 V
55. (a) 7.5×10^6 J   (b) 1.7 h
57. (a) All charges are positive.   (b) $K_B > K_A = K_D > K_C$
(c) $v_A > v_B = v_C > v_D$
59. 9.5 g, 4.4×10^{-6} C
63. (b)   65. (b)   67. (b)   69. (b)   71. (d)   73. (a)

Chapter 19

Multiple-Choice Questions

1. (c), (d)   3. (f)   5. (c)   7. (a)   9. (a)   11. (a), (d), (g)   13. (a)

Problems

3. 3.4×10^{22}

11.

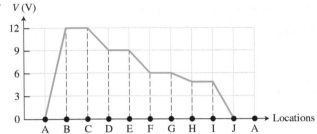

13. $R = 12\ \Omega, P = 12\ \text{W}$
15. 1.2 A
17. (a) $I_{50\Omega} > I_{100\Omega} > I_{150\Omega}$ (b) $\Delta V_{50\Omega} = \Delta V_{100\Omega} = \Delta V_{150\Omega}$
19. (b) 2.7 kΩ
21. (a) 1.6 W   (b) More
23. 0.12 kW
29. (a) $\mathcal{E}_1 - IR_3 - \mathcal{E}_2 - IR_2 - IR_1 = 0$ (b) 0.50 A clockwise
(c) 11.5 V
31. $I_2 = 1.6\ \text{A}, I_3 = 0.40\ \text{A}, 24\ \text{V}$
33. (a) $I_1 + I_3 = I_2, \mathcal{E}_2 - \mathcal{E}_1 - I_2R_2 - I_1R_1 = 0, \mathcal{E}_2 - I_3R_3 - I_2R_2 = 0,$
$I_1 = \dfrac{3}{110}\ \text{A}, I_2 = \dfrac{15}{110}\ \text{A}, I_3 = \dfrac{12}{110}\ \text{A}$
(b) 3.3 V
35. (a) 40 Ω   (b) 0.25 A
37. 20 Ω
41. $q_1 = 0.0040\ \text{C}, q_2 = 0.0060\ \text{C}$
43. (b) $I_1 = 0.15\ \text{A}, I_2 = 0.023\ \text{A}, I_3 = 0.13\ \text{A}$ (c) 2.5 V
45. 15.3 A
47. 13 W, 6.7 W
49. (a) $P = \dfrac{\mathcal{E}^2}{R(1 + r/R)^2}$
51. $I_1 > I_3 > I_4 > I_0 = I_2$
53. $R = 38\ \Omega, I = 0.040\ \text{A}, P_{\text{batt}} = 0.060\ \text{W}$
55. 2.0 m
57. 1.2
59. (a) 2   (b) $\frac{1}{4}$ (c) 0.61
61. (a) 30 V, 15 Ω
63. 2.2 W
65. (a) 3.3 V   (b) 0.13 W, 0.20 W
69.

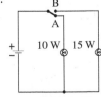

73. (a) $\sim 600\ \Omega$ (b) $\sim 2.0 \times 10^5\ \Omega$ (c) No
77. (a) Circuit 1   (b) Right terminal at higher potential   (c) 70 Ω
79. 10^9
81. (a) $1.1 \times 10^{-5}\text{A}$ (b) 7.16×10^{13} (c) New current is $5.7 \times 10^{-6}\text{A}$
83. (e)   85. (e)   87. (d)   89. (d)   91. (d)   93. (b)

Chapter 20

Multiple-Choice Questions

1. (d)   3. (d)   5. (b)   7. (a), (e), (f), (h)   9. (d)

Problems

3. No
7. 3.4×10^{-3} N west
15. (a) Segment a: 4.5×10^{-3} N out of page; segment b: 4.5×10^{-2} N upward; segment c: 4.5×10^{-3} N into page
(b) 0, 7.8×10^{-4} N·m, causing loop to rotate counterclockwise
19.

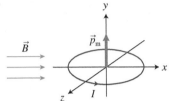

0.12 T
21. 2.5×10^{-11} N west
23. 1.4×10^7 m/s
29. 0.22 A
37. (a) 1.22×10^{-15} s   (b) 1.31×10^{-4} A   (c) 0.388 T, 1.85×10^{-23} A·m²
39. (a) 5.0×10^5 A   (b) Use coils with multiple loops.
41. (a) 1.2×10^6 m/s   (b) 1.7 m
45. (a) Into the page
47. (d)   49. (e)   51. (d)   53. (b)   55. (c)   57. (a)

Chapter 21

Multiple-Choice Questions

1. (b), (c), (d), (f), (g), (i)   3. (c)   5. (b)   7. (b)   9. (b)   11. (d)   13. (a)

Problems

13. (a) Clockwise   (b) Counterclockwise
19. 4.5×10^{-5} V
23. (b) a to b
27. 0.089 V
35. (a) 0.02 s
37. 7.6 cm²
39. (a) 25 cm²   (b) $(38\ \text{V})\sin\left[(2\pi)(80\ \text{s}^{-1})t\right]$ (c) 26 V
43. (a) 70.7 V   (b) 0.020 s   (c) 0.500 A   (d) 25.0 W
45. (a) 0.074 H   (b) 1.4 ms   (c) 0.011 A
47. (a) 2×10^{-9} F   (b) 2×10^7 Hz
49. 100
53. (a) 0.38 J   (b) 0.38 V   (c) 3.8 Ω
55. 4.1×10^{-5} V
65. 8 V
67. (a) Yes   (b) 4.8 s   (c) −23 m/s²
69. (d)   71. (e)   73. (c)   75. (b)

Chapter 22

Multiple-Choice Questions

1. (d)   3. (a), (b)   5. (b)   7. (c)   9. (b)   11. (c)

Problems

1. 17 m
3.

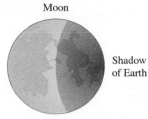

Moon

Shadow
of Earth

5. Inverted
9. 65° below the horizontal

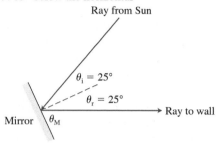

Ray from Sun

$\theta_i = 25°$

$\theta_r = 25°$

Ray to wall

Mirror θ_M

11. (b) 120°   (c) 90° or 45°
13. 25°, 25°
15. 17°
17. $\theta_i = \theta_r = 45.6°$
19. 13 cm
21. 1.4
23. 64°
25. 58°
27. No
29. 24°
33. 1.25, yes
35. (a) 39.6°   (b) 1.6, 38°
37. (a) 39.3°   (b) No
43. (a) 33.9°   (b) 1.43
45. $1.44 \leq n_g \leq 1.88$
49. $\theta_1 = 60°, \theta_2 = 33°, \theta_3 = 27°, \theta_4 = 47°$

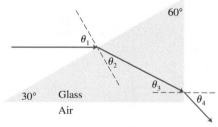

60°

θ_1

θ_2

θ_3

θ_4

30°   Glass

Air

53. $x = \dfrac{Hl}{h + H}$
55. 1.1 m

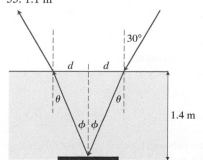

30°

d d

θ θ

ϕ ϕ

1.4 m

57. (a)   59. (e)   61. (e)   63. (c)   65. (c)   67. (d)

Chapter 23

Multiple-Choice Questions

1. (c)   3. (b)   5. (d)   7. (b)   9. (d)   11. (c)   13. (d)

Problems

17. (a) 0.86 m behind mirror   (b) 0.49 m
19. 6.7 cm in front of the mirror on the principal axis
27. (a) $s' \approx 15$ cm, real, inverted

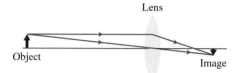

Lens

Object

Image

(b) $s' \approx 38$ cm, real, inverted

Lens

Object

Image

(c) $s' \approx -10$ cm, virtual, upright

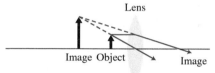

Lens

Image Object   Image

29. (a) $s' \approx -23$ cm, virtual, upright

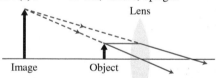

Lens

Image   Object

(b) $s' \approx -4$ cm, virtual, upright

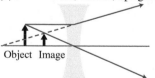

Object Image

33. (a) $s' \approx 12$ cm, real, inverted   (b) $s' \approx -2.7$ cm, virtual, upright

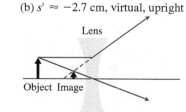

Lens   Lens

Image

Object   Object Image

37. 9.3 cm
39. ~3
41. 6.1 cm from the lens on the side opposite the object
43. (a) 252 cm   (b) −40 cm
45. 25 cm

47.

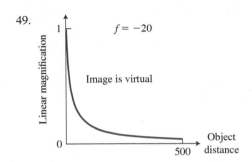

49.

51. (a) 2.10 cm   (b) 2.11 cm   (c) 2.29 cm
53. (a) −2.4 m   (b) 0.36 m
55. (a) Hyperopia, 0.50 m, infinity   (b) Myopia, 25 cm, 3.5 m
57. 5.8, increase, decrease
59. (a) 4.8 cm   (b) 31
61. (a) 20 cm to the right of the second lens   (b) Upright   (c) Real
63. (a)

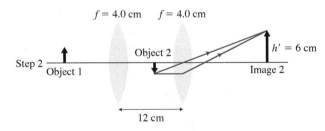

(b) $m = 2$
65. −17 cm
67. (a) 1.1 m to the left of the second lens   (b) 1.5 mm   (c) 270
69. Distance between lenses is 1.24 m, 4.4
71. (a) 8.1 cm   (b) −20
73. 46 cm, 3.8 cm from eyepiece
79. 24 cm to 18 m
83. (a)   85. (a)   87. (a)   89. (b)   91. (d)   93. (b)

Chapter 24

Multiple-Choice Questions

1. (d)   3. (e)   5. (b)   7. (a), (b)   9. (b), (d), (g)

Problems

3. (a) 0.25°   (b) 13 mm
5. 39°

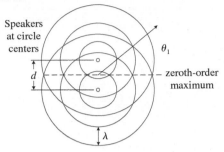

7.

11. For $\lambda = 656$ nm, $\theta_1 = 20.3°$, $\theta_2 = 44.1°$; for $\lambda = 410$ nm, $\theta_1 = 12.6°$, $\theta_2 = 25.8°$
13. No change
15. 530 nm
17. (a) 1.5×10^{-6} m   (b) 4 Gb
23. (a) Yes   (b) No   (c) 480 nm   (d) 240 nm
25. (a) 600 nm   (b) 300 nm
33. 2.5°
35. 0.13 mm
37. 1.1×10^{-7} rad
41. 1.8 cm
45. (a) The magenta band appears between bands 2 and 3 for all gratings illuminated by white light.   (b) 4 m

47.

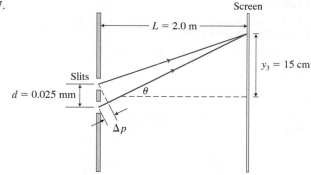

(b) Coherent light source, $\theta_3 = 4.29°$, $\lambda = 620$ nm
51. (a) 0.84°   (b) 3.1°
53. 130 nm, 389 nm
55. 50 μm
57. 0.30°
59. (a) 13 μm   (b) 9.1 μm
61. 5.2 mm
63. 0.65 cm
65. (a) 657 nm, 4.57×10^{14} Hz   (b) 993 nm, 3.02×10^{14} Hz
(c) The light may be Doppler shifted because the galaxy Hydra is receding from us.
67. ~0.04 mm
69. ~10 km
71. ~0.6 mm
73. 0.18 mm
75. 1×10^5 Hz
77. (d)   79. (d)   81. (d)   83. (b)   85. (d)

Chapter 25

Multiple-Choice Questions

1. (a)   3. (b)   5. (a)   7. (c)   9. (a), (c), (e)

Problems

1. 10 cm
3. (a) 0.40 W/m^2 (b) 0.17 W/m^2
9. 6300 years ago, yes
11. 1.8 to 10 km
13. 100 m
15. 80 km
17. (a) 6.67×10^{-8} T   (b) 9.00×10^5 N/C
19. 500 m
21. (a) 3.0×10^6 N/C

(b) $B_z = (1.0 \times 10^{-2} \text{ T}) \cos\left[2\pi\left(\dfrac{t}{1.0 \times 10^{-16} \text{ s}} - \dfrac{x}{3.0 \times 10^{-8} \text{ m}}\right)\right]$

(c) $E_y = (3.0 \times 10^6 \text{ N/C}) \cos\left[2\pi\left(\dfrac{t}{1.0 \times 10^{-16} \text{ s}} - \dfrac{x}{3.0 \times 10^{-8} \text{ m}}\right)\right]$

(d) Far field
23. UV-A: 7.5×10^{14} Hz $< f < 9.4 \times 10^{14}$ Hz;
UV-B: 9.4×10^{14} Hz $< f < 1.1 \times 10^{15}$ Hz
25. (a) 6.3×10^7 W   (b) 1.4×10^3 W/m^2
29. (a) 4.4×10^{13} W   (b) 3×10^{14} N/C
33. (a) 45°   (b) 72°
35. (a) 0.13   (b) 0.19
37. 39°
41. 6×10^{-23} kg/m^3
43. 4×10^{-18} W

47. $B_{max} = 6.7 \times 10^{-8}$ T, $\bar{u} = 1.8 \times 10^{-9}$ J/m^3,
$u_{max} = 3.6 \times 10^{-9}$ J/m^3, $I = 0.53$ W/m^2
49. $\dfrac{I_0}{8} \sin^2(2\theta)$
51. (c)   53. (d)   55. (e)   57. (b)   59. (d)

Chapter 26

Multiple-Choice Questions

1. (c)   3. (b)   5. (a)   7. (b)   9. (d)   11. (d)

Problems

1. 5 m/s
7. 1.3×10^{-10} s
9. $\Delta t_0 = 12$ s, $v_{spaceship} = 2.5 \times 10^8$ m/s
11. (a) 2.2×10^4 s   (b) Yes
13. 2.993×10^8 m/s
17. 12 cm
19. 2.9×10^8 m/s
21. Yes
23. (a) 4.5 y   (b) 1.1 y
(c)

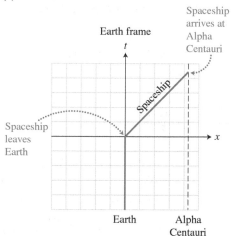

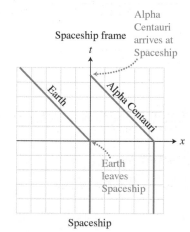

27. 50 m/s
33. (a) $1 + \frac{1}{2} \times 10^{-12}$ (b) 1.00005   (c) 1. 00504   (d) 1.1
(e) 1.3   (f) 3.9
35. 4.2×10^7 m/s, 1.3×10^8 m/s, 2.6×10^8 m/s
37. 299,791,858 m/s
39. 1.7×10^{19} J, 5.4×10^{18} J, 1.2×10^{19} J

41. 59%
43. (a) 7.3 (b) 2.5
45. (a) 1.96×10^{-35} kg   (b) 4.2×10^{-10}
47. 7.4×10^{-5} kg
49. (a) 3.3×10^{-8} kg/s   (b) 1.1 kg
51. (a) 3.6×10^{26} J   (b) 4.5×10^{-10}
53. 11.8 km/s
57. No
59. (a) 4.5 y   (b) 1.1 y
61. (a) 10 s   (b) 4.4 s   (c) 390 m   (d) Yes
63. (a) 2.8×10^8 m/s   (b) 2.9×10^{20} J, 9.0×10^{19} J
65.

Person making measurement	Time interval for Bob's travel	Time interval for Charlie's travel
Alice	14 y	12 y
Bob	10 y	6.4 y
Charlie	11 y	5.8 y
Darien	14 y	12 y

67. (a) $E_{\text{therm}} = 1 \times 10^6$ J, $E_0/E_{\text{therm}} = 4 \times 10^{11}$ (b) 4×10^{-12}
69. (c)   71. (c)   73. (a)   75. (c)   77. (e)   79. (b)

Chapter 27

Multiple-Choice Questions

1. (c)   3. (e)   5. (e)   7. (c)   9. (c)   11. (d)

Problems

1. 9.35 μm
3. (a) 72.5 nm   (b) 480 nm   (c) 9.67 μm
5. 1.3×10^{-5} m^2
7. $P_{\text{Earth}}^{\text{received}} = 2 \times 10^{17}$ W, $P_{\text{Earth}}^{\text{emitted}} = 2 \times 10^{17}$ W
9. 0.0379 eV
11.

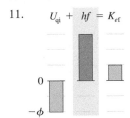

$U_{\text{qi}} + hf = K_{\text{ef}}$

13. (a) 5.1×10^{14} Hz   (b) 1.0 eV
17. 1.2×10^{15} Hz, 250 nm
19. 4.1 eV
21. (a) 4×10^{-3} W   (b) 30 nm
23. $(5.0 \times 10^{15})P\lambda$, where the λ must be given in nm and P in W
25. 1.4×10^{18}, 2 W
27. 2.6×10^{19}
29. (a) 7.0×10^{17} (b) 6.7×10^{-10} kg·m/s
33. (b) 9.7×10^{18} Hz, 3.1×10^{-11} m
35. 5.0×10^{11}
37. 2.0×10^{-12} W/m^2
39. (a) 8.0×10^{19} (b) 13 A
45. (a) 0.69 V   (b) 1.4 μA
47. 1×10^2
49. (a) 9.1×10^{-22} W   (b) 3.9×10^2 s
51. 1.0×10^{28} J, about 26 times greater
53. (a) 2.1×10^{11} W   (b) 7.0×10^5 K
55. (c)   57. (d)   59. (c)   61. (d)   63. (e)

Chapter 28

Multiple-Choice Questions

1. (a), (b), (d)   3. (b)   5. (d)   7. (a)   9. (b)   11. (b)

Problems

1. 2 km
3. (a) 2×10^{-4} kg   (b) 6×10^{20} (c) 1.5×10^{-28} m   (d) 9×10^{-4}
7. 435 nm, 6.90×10^{14} Hz, 2.86 eV
9. $\lambda = 250(2n - 1)^2$ nm $= [1000(n - 1)n + 250]$ nm
11. $r_n = \dfrac{h^2}{4\pi k_C e^2 m_e} n^2$, $E_n = \dfrac{-13.6 \text{ eV}}{n^2}$, $L_n = n\dfrac{h}{2\pi}$
13. 658 nm, 122 nm, 103 nm
17. 2.3×10^{39}, yes
21. 4.6×10^{23} K
23. (a) 2.0×10^9 W   (b) 2.5×10^{17} W/m^2
25. 7.2×10^{15}
29. (a) Upward

(b)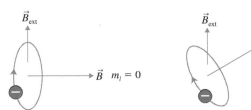

31. (a) 2.2×10^6 m/s   (b) 3.3×10^{-10} m
41. The potential difference V through which the electrons are accelerated
43. (a) 3, 4   (b) 2, 3
45.

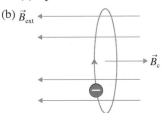

47. $(n, l, m_l, m_s) = (4, 3, \pm 3, \pm\frac{1}{2}), (4, 3, \pm 2, \pm\frac{1}{2}), (4, 3, \pm 1, \pm\frac{1}{2}), (4, 3, 0, \pm\frac{1}{2})$
49. (a) $1s^2, 2s^2, 2p^6, 3s^2, 3p^4$
(b) Because their outermost subshell is a p shell and they have the same number of electrons (4) in this subshell
51. $1s^2, 2s^2, 2p^6, 3s^2, 3p^6, 4s^2, 3d^6$

53. H: $1s^1$
Li: $1s^2, 2s^1$
Na: $1s^2, 2s^2, 2p^6, 3s^1$
K: $1s^2, 2s^2, 2p^6, 3s^2, 3p^6, 4s^1$
55. 2×10^{-18} m/s
57. 3.3×10^{-7} eV, $\Delta E_{min}/(1.51 \text{ eV}) = 2.2 \times 10^{-7}$
59. (a) 0.027 nm, 0.106 nm, 0.239 nm

(b)

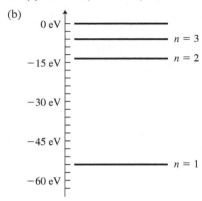

61. 5.8×10^{-13} m
63. 1.88 μm
65. 52.9 eV, 1.28×10^{16} Hz, 23.5 nm
67. No   69. (b)   71. (f)   73. (a)   75. (f)   77. (e)

Chapter 29

Multiple-Choice Questions

1. (d)   3. (c)   5. (b)   7. (c)   9. (b)

Problems

1. 58 N. The protons are held together by the strong nuclear interaction.
3.

	Protons	Neutrons	Nucleons
$^9_4$Be	4	5	9
$^{16}_8$O	8	8	16
$^{27}_{13}$Al	13	14	27
$^{56}_{26}$Fe	26	30	56
^{64}Zn	30	34	64
^{107}Ag	47	60	107

5. (a) 4×10^{28} (b) 2×10^{28} (c) 2×10^{-10} cm^3
7. 2.8 nm
9. 1.5 GeV
11. 7.57 MeV, 8.50 MeV, $^{120}_{50}$Sn
13. (a) $^1_1$H   (b) $^1_0 n$ (c) $^{94}_{38}$Sr   (d) $^3_2$He
17. 7.16 MeV released

19. $^{228}_{90}$Th, 5.41 MeV
21. (a) 38, 50   (b) 87.905586 u
23. (a) 43.2 MeV   (b) 3.45×10^{14} J/kg
27. (a) $^{235}_{92}$U   (b) $^{144}_{59}$Pr   (c) $^{129}_{54}$Xe   (d) $^{60}_{29}$Cu
29. 5.65 MeV
31. (a) 1732 MeV   (b) 1708 MeV, 28.30 MeV, 1737 MeV
(c) 4.87 MeV
33. 7.2×10^3 m/s
37. (a) 0.063   (b) 0.00391   (c) 9.24×10^{-11}
39. 4.94×10^{12} s^{-1}
41. (a) 5700 y   (b) 11,400 y   (c) 17,100 y   (d) 6.8×10^{-4}
43. (a) 4.6×10^{14} (b) 4.6×10^8 s^{-1} (c) 6.1×10^{-7} g
45. 6.09 MeV
47. 1.3×10^9 y
49. 40,000 y
53. (a) 7.0×10^{-4} Sv   (b) 1.2×10^{-3} Sv   (c) 7.0×10^{-3} Sv
(d) 7.0×10^{-3} (e) 2×10^{28}
55. 8.0×10^{19}
57. 2.4×10^{17} kg/m^3
59. (a) Na   (b) Fe   (c) Zn   (d) Ag
63. 8.2×10^{13} J/kg
65. 2×10^{13}
67. (a) 5.7×10^{20} J   (b) 5.7 y
69. 8, 4
71. (a) 4.2×10^{-12} kg   (b) 4.2×10^{-14} J   (c) 2.6×10^3
(d) 1.1×10^{-13} kg, 1.1×10^{-15} J, 71
73. (c)   75. (d)   77. (c)   79. (b)   81. (b)

Chapter 30

Multiple-Choice Questions

1. (a)   3. (b)

Problems

7.

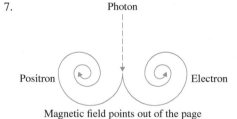

Magnetic field points out of the page

17. Their fundamental interactions
21. Through all four fundamental interactions (gravity, electromagnetic, strong, and weak)
25. Higgs gives mass to the other particles.
31. The formation of neutral atoms 370,000 years after the Big Bang
41. 2.9977×10^8 m/s
43. (b)

Credits

Chapter 1

Opener: Muratart/Shutterstock; p. 2: John Reader/Science Source; p. 3: Gary Yim/Shutterstock; p. 5: Frank Polich/Reuters; p. 8 bottom: Passion Images/Shutterstock; p. 9: Vladischern/Fotolia.

Chapter 2

Opener: MadCircles/E+/Getty Images; Fig. 2.26: Jim Sugar/Corbis Documentary/Getty Images.

Chapter 3

Opener: FStop Images GmbH/Alamy Stock Photo; p. 61: Derek Bayes/Lebrecht Music and Arts Photo Library/Alamy Stock Photo; Fig. 3.7: Vereshchagin Dmitry/Shutterstock; p. 81: US Navy Permissions Dept./Academx Publishing; p. 82: Dotshock/Shutterstock.

Chapter 4

Opener: Jerry Wolford/Polaris/Newscom; p. 92: Henn Photography/Cultura/Getty Images; p. 710: Adam Rowley/Alamy Stock Photo; p. 922 bottom: Stefan Wermuth/Reuters.

Chapter 5

Opener: AM Corporation/Aflo Co. Ltd./Alamy Stock Photo.

Chapter 6

Opener: Joel Kowsky/NASA; Fig. 6.5: Ted Kinsman/Science Source; Fig. 6.7: Joel Kowsky/NASA; P6.52: Reinhard Dirscherl/Alamy Stock Photo.

Chapter 7

Opener: Christian Charisius/dpa picture alliance archive/Alamy Stock Photo; Fig. 7.3: Ted Foxx/Alamy Stock Photo; P7.77: Wikipedia User:BrandonR.

Chapter 8

Opener: Damir Sagolj/REUTERS/Alamy Stock Photo; Fig. 8.1: Alexander Y/Shutterstock; Fig. 8.13: Dylan Martinez/Reuters; Fig. 8.15: GIPhotoStock/Science Source; Fig. 8.22: Nmiskovic/Shutterstock; Fig. 8.24: Stringer Shanghai/Reuters; p. 241: Eric Schrader.

Chapter 9

Opener: Matt Tilghman/Shutterstock; p. 255: NASA.

Chapter 10

Opener: Afonskaya/iStock Editorial/Getty Images; Fig. 10.9: NASA; p. 299: David Hancock/Alamy Stock Photo; Fig.10.14: Cathy Keifer/Shutterstock; Fig.10.13: Joerg Reuther/imageBROKER/Alamy Stock Photo; Fig.10.16: AP Images; Fig.10.17: Eric Schrader/Fundamental Photographs, NYC.

Chapter 11

Opener: Merlin D. Tuttle/Science Source.

Chapter 12

Opener: MaxWebb/Shutterstock; Fig. 12.1: Richard Megna/Fundamental Photographs; Fig. 12.2: Egmont Strigl/ImageBROKER/Alamy Stock Photo.

Chapter 13

Opener: Topseller/Shutterstock; Fig.13.14: Akg-images/Superstock; Fig.13.15: Richard Megna/Fundamental Photographs, NYC.

Chapter 14

Opener: Biophoto Associates/Science Source; Fig.14.5: Thomas Otto/Fotolia; Fig.14.6: Dmitry Naumov/Fotolia; Fig.14.7: Ustyujanin/Shutterstock.

Chapter 15

Opener, left: William Osgood Field. 1941. Muir Glacier: From the Glacier Photograph Collection. Boulder, Colorado USA: National Snow and Ice Data Center. Digital media; right: Bruce F. Molnia. 2004. Muir Glacier: From the Glacier Photograph Collection. Boulder, Colorado USA: National Snow and Ice Data Center. Digital media; p. 464: Kirk Geisler/Shutterstock; Fig.15.17: Ted Kinsman/Science Source.

Chapter 16

Opener: Richard Megna/Fundamental Photographs, NYC; Fig.16.1a: iPics/Fotolia; p. 486: Michael Dalton/Fundamental Photographs, NYC; Fig.16.6: Misu/Fotolia; p. 491: Kenneth Sponsler/Shutterstock.

Chapter 17

Opener: Suksawad/Shutterstock; Fig. 17.2: GIPhotoStock/Science Source; Fig. 17.19: Sciencephotos/Alamy Stock Photo; p. 525: Ted Kinsman/Science Source; Fig. 17.21: Clive Streeter/Dorling Kindersley Ltd.; P17.15: Blickwinkel/Hartl/Alamy Stock Photo.

Chapter 18

Opener: Lyle Leduc/Photographer's Choice RF/Getty images; Fig.18.28: Matzsoca/Shutterstock.

Chapter 19

Opener: Bacho12345/123RF.

Chapter 20

Opener: CoolKengzz/Shutterstock; Fig. 20.7: Stephen Oliver/Dorling Kindersley Ltd.; p. 647: Thomas Dobner 2007/Alamy Stock Photo.

Chapter 21

Opener: Phanie/Alamy Stock Photo; Fig. 21.10: Sergei Kozak/Alamy Stock Photo; P21.50: Erik Charlton, Hammerhead, https://www.flickr.com/photos/erikcharlton/2935553971/in/photostream, © CC BY 2.0.

Chapter 22

Opener: Richard Megna/Fundamental Photographs, NYC; Fig. 22.1: Atith Jiranaphawiboon/123RF; Fig. 22.8: Artem Mazunov/Shutterstock; Fig. 22.9: Van Heuvelen, Alan; Fig. 22.14: Silkfactory/iStock/Getty Images; Fig. 22.17: Jeremy Woodhouse/Photodisc/Getty Images; P22.34: Dcwcreations/Shutterstock; p. 710: Pedrosala/Shutterstock.

Chapter 23

Opener: Juice Images242/Juice Images/Alamy Stock Photo; p. 713: I.Dr/Shutterstock; Fig. 23.3: Major Pix/Alamy Stock Photo; Fig. 23.20: Vladislav Lebedinski/Fotolia; Fig. 23.22: David Paul Morris/Bloomberg/Getty Images; P23.36: Anion/Fotolia.

Chapter 24

Opener: Richard Megna/Fundamental Photographs, NYC.; Fig. 24.4: Dietrich Zawischa, https://www.itp.unihannover.de/fileadmin/arbeitsgruppen/zawischa/static_html/multibeam.html; Fig. 24.7: Sciencephotos/Alamy Stock Photo; Fig. 24.9: Berenice Abbott/Science Source; Fig. 24.12a: GIPhotoStock/Science Source; Fig. 24.13: Langdu/Shutterstock; Fig. 24.15: Kelley/Fotolia; Fig. 24.16a: Andrew Lambert Photography/Science Source; Fig. 24.16b: Eric Schrader/Fundamental Photographs, NYC.; Fig. 24.19: Adam Filipowicz/Shutterstock; Fig. 24.20: Doug Schnurr/Fotolia; Fig. 24.21: Richard Megna/Fundamental Photographs, NYC.; Fig. 24.27: Images courtesy of Nick Strobel at www.astronomynotes.com; Fig. 24.29: Cphoto/Fotolia.

Chapter 25

Opener: Gorazd Planinsic; Fig. 25.9: National Oceanic and Atmospheric Administration (NOAA); P25.8: JPL-Caltech/R. Hurt/NASA; P25.41: P. Challis/R. Kirshner (CfA)/B. Su/NASA.

Chapter 26

Opener: Rafal Olechowski/123RF.

Chapter 27

Opener: NASA; Fig. 27.1: Dr. Arthur Tucker/Science Source; Fig. 27.10: Richard Megna/Fundamental Photographs, NYC; Fig. 27.11: Wilhelm Röntgen/Library of Congress; Fig. 27.18: The Physics Teacher 52, 212 (2014); doi: 10.1119/1.4868933; P27.62: NASA.

Chapter 28

Opener: Oxford Designers/Illustrators Ltd/Pearson Education, Inc.; Fig. 28.12: Wabash Instrument Corp/Fundamental Photographs, NYC; Fig. 28.37: Stefan Diller/Science Source.

Chapter 29

Opener: Joaquin Ossorio Castillo/Shutterstock; p. 922 bottom: Marcel Clemens/Shutterstock; p. 922 top: Science Source; p. 932: Chonrawit boonprakob/Shutterstock; p. 934: Chonrawit boonprakob/Shutterstock; p. 938: Photo Researchers/Science History Images/Alamy Stock Photo.

Chapter 30

Opener: Photo Researchers/Science History Images/Alamy Stock Photo; Fig. 30.1: Lawrence Livermore National Laboratory; Fig. 30.2: Lawrence Livermore National Laboratory; Fig. 30.5: Peter Ginter/RGB Ventures/SuperStock/Alamy Stock Photo; Fig. 30.6: Xenotar/iStock/Getty Images Plus; Fig. 30.7: Super Kamiokande Collaboration Japan/NASA; Fig. 30.13: JPL-Caltech/NASA.

Index of physics terms

For users of the two-volume edition, pages 1–499 are in Volume 1 and pages 500–981 are in Volume 2.

A

Absolute temperature, 364, 365–367
Absorbed dose, 949–950
Absorption, 691, 897
Absorption process, 888–889
Absorption spectrum, 897
AC circuits, 668–674, 678
    capacitors in, 670–671, 674, 678
    resistors in, 668–670, 678
    solenoids (inductors) in, 672–674, 678
Acceleration, 30–32, 51
    average, 32
    constant, 30–37, 39
    defined, 31, 32, 42
    force and, 61–66
    free fall, 33–34, 66
    as function of time, 291–292
    instantaneous, 32
    Newton's second law, 61–66
    radial, 122, 123–127, 139
    rotational, 254, 259
    sign of, 33
    simple harmonic motion, 292–293, 308
    tangential, 261
    units of, 32
    velocity change determined from, 32–33
    velocity change method to find, 121–122, 139
Activity, of radioactive material, 944
Adiabatic processes, 453
Age of the universe, 837
Air bags, 51, 75–76
Air conditioners. *See* Thermodynamic pumps
Air pressure. *See* Atmospheric pressure
Airplanes
    how they fly, 427
    test pilot blackouts, 118, 126
    plastic bottle on, 371, 375–376
Alpha decay, 924, 939, 942, 952
Alpha particles, 924
Alternating current (AC), 599, 667, 668
    AC circuits, 668–674, 678
Altitude, pressure and, 405–406
Ammeters, 580
Ampere (unit, A), 576, 624
Amplitude, 287
    vibrational motion, 287, 308
    waves, 319, 324–325
Aneroid barometer, 358
Angle of incidence, 690, 706
Angle of reflection, 690, 706
Angle of refraction, 693
Angular acceleration, 254
Angular magnification, 739–740, 744

Angular momentum. *See* Rotational momentum
    orbital, 900, 902
Angular position, 252–253
Angular size, 739
Angular velocity, 253–254
Anode, 628, 855
Antennas, 791–792, 797, 802
Antielectrons, 958–960
Antimatter, 962
Antinodes, 338
Antiparticles, 958–962, 979
    beta-plus decay, 961–962
    pair annihilation, 960–961
    pair production, 959–960
Antireflection coatings, *See* Lens coatings
Arc length, 252
Archimedes, 387
Archimedes' principle, 400
Area of support, 238
Assumptions, 4
Atmospheric pressure, 395–398
Atomic mass, 360, 927
Atomic mass unit (u), 926
Atomic nucleus, 881–883, 952
    binding energy of, 930–931, 935, 952
    description of, 926
    liquid drop model of, 938–939
    model of, 881–884, 915, 924–928
    nuclear force, 928–929
    size of, 924–925
    structure of, 881–883, 926
    symbolic representations, 927, 952
Atomic number, 926
Atomic physics, 880–920
    de Broglie waves, 904–907
    early atomic models, 881–885
    lasers, 897–899
    multi-electron atoms, 907–909
    particle nature, 903–907
    periodic table, 909–910, 928
    quantized orbits, 885–891
    quantum numbers, 887, 899–903, 915
    spectral analysis, 892–897
    tunneling, 913
    uncertainty principle, 910–914, 915, 924
    wave functions, 907
    wave-particle duality, 864–866, 915
    Zeeman effect, 899, 900–901
Atomic structure
    Balmer's equation, 891
    Bohr radius, 887
    Bohr's model, 885–891, 915
    Dalton model, 881
    electrons, 506
    energy states (ground, excited), 888–889

    multi-electron atoms, 907–909
    nucleus, 881–883
    periodic table, 909–910, 926
    quantized orbits, 885–891
    Rutherford (planetary) model, 881–884
    shells and subshells, 908–909
    thermal excitation, 892
    Thomson ("plum pudding") model, 881
Atomic wave functions, 907
Atoms, 355, 506, 973–974
    ground state, 909
    magnetic properties of, 640
    metastable states, 898
    nucleus of, 881–883
    thermal excitation, 892
Atwood machine, 101–102, 264–265
Auroras, 616, 632
Average acceleration, 32
Average velocity, 30
Avogadro's number, 360, 379
Axions, 975
Axis of rotation, 220–221
Axis of polarizer, 786

B

$\vec{B}$ field. *See* Magnetic field
Balancing, 240
Balloons, 378, 405
    electrostatic charge, 501–502, 507
    hot-air, 386, 405, 453–454
Balmer's equation, 891
Balmer series, 890
Bands (interference), 752
Bar charts, 149
    Bernoulli, 422–423
    impulse-momentum, 157–166
    nuclear reactions and, 930, 935–938
    photoelectric effect and, 853, 859
    rotational momentum, 269
    work-energy, 182–184, 271–273
    work-heating-energy, 449–454
Barometers, 358, 397
Baryons, 967–968
Batteries, 575, 576–578, 587
    emf, 677
    internal resistance, 594–595
Beat frequencies, 335
Beats, 334–335
Becquerel (unit, Bq), 945
Becquerel, Henri, 922
Bernoulli bar charts, 422–423
Bernoulli's effect, 416–418
Bernoulli's equation, 420–423, 435
    applications of, 424–427
Beta decay, 924, 940–941, 942, 952
Beta-minus decay, 941

Beta particles, 924
Beta-plus decay, 941, 961–962
Big Bang, 972–973
Binding energy, 929–931, 935, 952
Binding energy per nucleon, 931
Biologically equivalent dose. *See* Dose
    equivalent
Black body, 848
Black body radiation, 848–853
    Stefan-Boltzmann's law, 849, 875
    Wien's law, 849, 875
Black holes, 206–207, 255–256, 839–840
Blood pressure, 420
Blood flow
    speed of, 343–344, 419
Blue quarks, 969
Bohr radius, 887
Bohr's liquid drop model of nucleus,
    938–939
Bohr's model of the atom, 885–891, 915
Bohr's postulates, 886, 904
Boiling, 460–461
    boiling point, 364
Boltzmann's constant, 366, 482
Bottom quarks, 968
Boyle's law, 370
Brewster's law, 804
Brownian motion, 354–355
Buoyant force, 398–401, 407
    applications and examples, 401–406
    Archimedes' principle, 400
    defined, 400, 407

C

Calorie, 446
Cameras, 734–736
    camera obscura, 689
    pinhole, 688–689
    ray diagram, 728
    red eye effect, 692
Capacitance, 559–561
Capacitive reactance, 670–671
Capacitor circuits, 598–599
Capacitors, 558–563, 565
    in AC circuits, 670–671, 674, 678
    charging, 559
    components of, 559
    energy of, 561–562
    parallel-plate, 560
    RC circuits, 599
Carbon, isotopes of, 944
Carbon dating, 947–948
Carbon dioxide, 303–304
    climate change and, 441, 468–469
Carnot's principle, 480
Cars
    momentum and collisions, 155, 165
    drag force, 432
    engine cooling, 466
    engine heating, 457
    lightning and, 535
    power of, 203

seat belts and air bags, 51, 75–76
    friction and, 107–108
    stopping distance, 13, 160–162
Cathode, 628, 855
Cathode ray tube (CRT), 628, 868
Cathode rays, 868
Cause-effect relationship, 64
CD as reflection grating, 762
Celsius scale, 364
Center of gravity, 220
Center of mass, 218–219, 230–233
    calculating, 231
    defined, 219, 242
    mass distribution and, 233
    quantitative definition, 232
Centripetal acceleration. *See* Radial
    acceleration
CERN, 965, 966, 971
Chain reaction, 937
Charge. *See* Electric charge
Charged objects, 503–504, 506–507,
    514–515, 528
Charging,
    by rubbing, 506, 511
    by touching, 509
    by induction, 510
    capacitors and, 559
Charles's law, 370
Charm quarks, 968
Chemical energy, 178
Chromatic aberrations, 759–760
Circuit breakers, 572, 599
Circuit diagrams, 580
Circuits. *See* Electric circuits
Circular motion, 118–146
    conceptual difficulties with, 133
    at constant speed, 119–121, 128–132,
      139
    in magnetic field, 630–631
    Newton's second law, 127–128, 133
    period, 126–127, 139
    planetary motion, 133–137
    qualitative dynamics of, 119–121
    radial acceleration, 122, 123–127, 139
Classical velocity transformation equation,
    827
Climate change, 441, 468–469
Coefficient of kinetic friction, 93
Coefficient of performance. *See* Performance
    coefficient
Coefficient of static friction, 93, 108
Coherent monochromatic waves, 759–760,
    777
Collisions, 199–202, 208
    elastic and inelastic, 201, 208
    momentum and, 149–150, 160–162,
      164–167, 199–200
    particles in ideal gas, 361–364
    in two dimensions, 164–167
Color
    complementary, 767
    of light, 758

of sky, 703, 805
    spectrum, 762
Color charge (quarks), 969
Commutator, 627
Compass, 617, 618
Complementary colors, 767
Compound microscope, 741–743
Concave lenses, 726–727, 744
Concave mirrors, 715–719, 744
Conductance, 582
Conduction, 463–465, 470
Conductive heating/cooling, 464, 470
Conductivity, 603–604
    semiconductors, 606–607
    superconductivity, 605
Conductors, 507–511, 528
    defined, 510, 528
    in electric fields, 552–555
    grounding, 511, 553–554
    human body as, 510
Confinement, 970
Constancy of rotational momentum, 267
Conservation of charge, 511
Conservation of energy, 181–186
Conservation of mass, 149, 168
Conservation of momentum, 156
Conserved quantities, 149, 156, 168, 181
Constancy of momentum, 151, 155–156
Constant pressure processes. *See* Isobaric
    processes
Constant temperature processes. *See*
    Isothermal processes
Constant velocity linear motion, 24–30
Constant volume processes. *See* Isochoric
    processes
Constructive interference, 330
Contact forces, 43–44
Continuity equation, 419, 435
Convection, 465–467, 470
Converging lenses. *See* Convex lenses
Convex lenses, 725–726, 738, 744
Convex mirrors, 715–716, 719–721, 744
Cooling
    conductive heating/cooling, 464, 470
    evaporation, 468, 470
    phase change, 458–459
    radiative heating/cooling, 467, 470
Coordinate system, 14, 21
Copernicus, 2
Cornea, 736
Cosmic inflation, 973
Cosmic microwave background (CMB),
    973–974
Cosmic rays, 631–632, 958–959
Cosmological constant, 976
Cosmological constant model, 976
Cosmological constant problem, 977
Cosmology, 972–974, 979
    Big Bang, 972–973
    cosmic microwave background, 973–974
    inflation, 973
    mysteries in, 977

nucleosynthesis, 973
    stars and galaxies, 973–974
Coulomb (unit, C), 511, 528, 547, 576
Coulomb's constant, 513
Coulomb's law, 512–516, 528
Critical angle of incidence, 696
Crystal lattice, 575
Crystals, ionic, 557
Curie (unit, Ci), 945
Curie, Marie and Pierre, 922–923
Current. *See* Electric current
Current-carrying wire,
    magnetic field of, 619
    magnetic force on, 621–626
Current loop,
    as magnetic dipole, 628
    torque exerted on, 628
Curvature of space, 838–839
Cutoff frequency, 856, 861, 864, 875

D
Dalton atomic model, 881
Damping, 304–305
Dark energy, 976–977, 979
Dark energy model, 977
Dark matter, 974–976, 979
Data, linearizing, 74–75
Daughter nucleus, 939
DC circuits, 572–615. *See also* Electric
    circuits
DC (direct current), 575, 599, 668. *See also*
    Electric current
de Broglie, Louis, 904, 907
de Broglie waves, 904–907
Decay constant, 944
Decay rate, 944–945, 946
Decay series, 948–949
Decibels (unit, dB), 333
Density, 387–389
    defined, 359, 379, 407
    energy density, 422, 563, 800–801
    floating and, 388–389
    of gases, 359, 387
    of solids and liquids, 387–389
Depolarization, 615
Destructive interference, 330
Diamagnetic materials, 640
Dielectric breakdown, 558
Dielectric constant, 556, 565
Dielectric heating, 795
Dielectrics, 509–510, 528
    in an electric field, 555–558
Diffraction, 768–772, 777
    circular opening, 772
    double-slit interference, 752–757
    gratings, 762, 911
    single-slit, 769–771, 777
Diffuse reflection, 690, 691, 706
Diffusion, 378
Diopters, 738
Dipoles
    electric, 509–510, 511

human heart, 540, 547, 563–564
    magnetic dipole moment, 627–628
Dirac, Paul, 958
Direct current (DC), 572, 575, 599. *See also*
    Electric circuits; Electric current
Direct current electric motor, 626–627
Discharge tube excitation, 893–894
Discharging, 511, 553
Dispersion of light. *See* Refractive index;
    Color, of light
Displacement, 21, 22, 42, 292
    at constant acceleration, 34–37
    defined, 21, 29, 42
    for waves, 317–321
    scalar component of, 22
    from velocity graphs, 29–30
    work and, 180
Distance, 21, 42, 133–134
Divergent lenses. *See* Concave lenses
Domains (magnetic), 641
Doppler effect, 340–344
    for electromagnetic waves, 834–837, 842
    for sound, 343, 345
Doppler radar, 835
Dose (radiation), 950
Dose equivalent, 950
Double-slit interference, 752–757, 777
Down quarks, 968
Drag force, 95, 420, 431–434, 435
    Stokes's law, 431–432
Drift (electron), 575
Driven vibrations. *See* Forced vibrations
Drude model, 605
Dynamics, 65

E
$\vec{E}$ field. *See* Electric field
Earth
    auroras, 632
    climate change, 441, 468–469
    magnetic field, 631, 632, 645
    magnetic poles of, 617
    planetary motion and, 2–3, 133–137
    tides, 274–275
Eddy currents, 658–659
Eddy current waste separator, 659
Efficiency
    quantum, 867
    thermodynamic engine, 490
Einstein, Albert, 355, 817–818, 838,
    859–862, 976
Elastic collisions, 201, 208
Elastic potential energy, 178, 190–192, 208
Electric charge, 500–534
    charge separation, 524–527
    conductors and insulators, 507–511, 528
    contemporary model of, 506–507
    Coulomb's law, 512–516, 528
    defined, 511, 528
    electrostatic interactions, 501–507,
        536–542
    flow of, 574–575

fluid models of, 505–506
    negative, 502, 506, 528
    particle model of, 506
    positive, 502, 506, 528
    unit of (coulomb), 511, 528, 576
    Van de Graaff generators, 524–526
    Wimshurst machine, 526
Electric circuits, 572–615
    AC circuits, 668–674, 678
    capacitor circuits, 598–599
    circuit breakers, 572, 599
    circuit diagrams, 580
    DC circuits, 572, 575
    household wiring, 589, 599
    Joule's law, 589–592, 608
    Kirchhoff's junction rule, 595, 600, 608
    Kirchhoff's loop rule, 592–594, 600, 608
    LEDs (light-emitting diodes), 583–584,
        591, 593
    lightbulbs, 578–579
    measuring current and potential difference,
        580
    Ohm's law, 581–586, 608
    qualitative analysis of, 586–589
    quantitative reasoning on, 597–598
    RC circuits, 599
    resistors, 580, 596–600, 602–607
    short circuit, 595–596
    symbols in, 580, 584
Electric current, 573–576. *See also* Electric
    circuits; Electromagnetic induction
    alternating current (AC), 599, 667
    ammeters, 580
    batteries, 575, 576–578, 587
    defined, 576, 608
    direct current (DC), 575, 599, 668
    direction of, 575–576
    induction of, 650–652
    magnetic field produced by, 619–620,
        624–625, 632–634
    magnetic force and, 621–628
    Ohm's law, 581–586, 608
    unit of (ampere), 576, 624
Electric dipoles, 509–510, 511
Electric field, 535–571, 963
    capacitors, 558–563, 565
    conductors in, 552–555
    defined, 538, 565
    dielectric materials in, 555–558
    due to multiple charged objects, 539–540,
        543
    due to single point-like charged object,
        537–538
    energy density of, 563
    model, 536–542
    relating to electric potential, 550–552, 565
    shielding, 554–555
    source charge, 538, 543
    superposition principle for, 539–540, 565
    test charges, 538
    unit of (newtons/coulomb), 538
    unit of (volts/meter), 798

Electric field lines, 540–542
Electric forces, 502, 512–516
  compared with gravitational, magnetic forces, 513–514, 626
  Coulomb's law, 512–516
Electric generators, 666–668, 678
Electric motors, 626–627
Electric potential, 546–550, 565
  batteries and, 575, 576–577
  defined, 547, 565
  due to multiple charges, 547
  due to single charged object, 546–547
  equipotential surfaces, 549–550, 565
  graphing, 577–578
  potential difference (voltage), 548–549, 573–575
  relating to electric field, 550–552, 565
  units of (joules/coulomb, volt), 547
  voltmeters, 580–581
Electric potential energy, 516–521, 528
  defined, 517, 519, 528
  graphing vs. distance, 520
  Joule's law, 590, 608
  multiple charge systems, 520–521
  unit of (joule), 519
Electric power, 589–591
  kilowatt-hours, 591
  unit of (watt), 590
Electrical resistance, 604, 608. *See also* Resistance; Resistors
Electricity. *See* Electric circuits; Electric current; Electric field
  magnetism compared with, 677
Electrocardiography, 563–564
Electrodes, 578
Electrolytes, 607
Electromagnetic induction, 649–684
  defined, 653, 678
  eddy currents, 658–659
  Faraday's law of, 659–662
  Lenz's law, 657, 658–659, 678
  magnetic flux, 654–656, 678
  motional emf, 666
  transformers, 674–676, 678
Electromagnetic interactions, 963, 964, 965
Electromagnetic spectrum, 796–797
Electromagnetic waves, 784–812, 963
  applications, 793–795
  defined, 808
  Maxwell's equations, 677, 789
  Doppler effect for, 834–837, 842
  energy densities, 799–801
  light as, 790–791, 852, 857
  mathematical description, 798–799
  polarization, 802–807
  production of, 789–790
  speed of, 789–790
Electromagnets, 621, 641, 662–664
Electromotive force. *See* Emf
Electron microscope, 918–919
Electron particle accelerator, 833
Electron volt (unit, eV), 832, 853

Electrons, 506–507
  in atomic models, 883
  drift, 575
  electric charge and, 506, 508, 511
  electric field and, 545–546, 551–552
  conduction, 508, 575
  kinetic energy of, 859–860
  interference of, 905
  magnetic field and, 628–629, 630–634, 640
  in semiconductors, 606–607
  quantized orbits, 885–891
  in shells and subshells, 908
  spin, 902
  wavelength, 906
Electroscope, 508–509, 854–855
Electrostatic force, 502
Electrostatic interactions, 501–507, 536–542
  charged objects, 503–504, 506–507, 514–515, 528
  explanations for, 504–507
  and electric field, 536–542
  uncharged objects, 503–504
Electroweak model, 971
Elementary particles, 958, 966–970, 979
  fundamental interactions, 970
  hadrons, 967–968
  leptons, 966–967
  quarks, 968–970
  Standard Model, 970–971, 975, 979
  supersymmetry, 975–976
Elements
  isotopes, 927–928
  nuclear fusion, 935–936
  nucleosynthesis, 973
  periodic table of, 909–910, 928
Emf, 577, 608, 791
  of battery, 677
  of generator, 667–668, 678
  magnetic flux and, 661–662
  motional, 666
  of transformer, 675, 676
Emission process, 888
Emission spectra, 892
  spectral analysis, 892–897
Emitter (EM waves), 793
Endothermic reactions, 934
Energy, 176–216. *See also* Internal energy; Kinetic energy; Mechanical energy; Potential energy; Thermal energy; Work
  binding, 929–931, 935, 952
  conservation of, 181–186
  dark, 976–977, 979
  of electromagnetic waves, 799–800
  from fission reaction, 938
  generalized work-energy principle, 184–186, 208, 448
  nuclear sources of, 935–939
  power and, 202–204, 208
  quantifying, 186–192
  relativistic, 830–833, 842

  rest, 832, 842
  total, 181, 184, 208
  types of, 178, 184
  uncertainty principle for, 912–913
  work and, 177–181, 422–423
  work-energy bar charts, 182–184, 449
  zero point, 977
Energy conservation, 181–186
Energy conversion, 192–194
Energy density, 422, 563, 800–801
Energy quantization, 851–852
Energy states
  Bohr atomic model, 888–889
  excited state, 888
  ground state, 888
  width of, 912–913
Energy transfer, 447, 470
  conduction, 463–465, 470
  convection, 465–467, 470
  direction of, 478, 480
  evaporation, 468, 470
  heating, 447, 451, 455–458
  radiation, 467, 470
  second law of thermodynamics, 480
  through resonance, 307, 308
  work and, 447, 470
Engines
  Carnot's principle, 480
  efficiency, 490–492
  horsepower, 203
  thermodynamic, 478–480, 488–492
Entropy, 480–488, 495
  defined, 482, 495
  irreversible processes and, 477–480
  second law of thermodynamics, 483, 486, 495
Environment, 52, 77
Equilibrium, 227–230, 242, 482
  center of mass and, 230–233, 240
  conditions for, 228
  rotating objects and, 239–240, 242
  stability of, 237–241, 242
  static equilibrium, 227–230, 242
  thermal, 457
Equilibrium position, 286, 308
Equilibrium state, 482
Equipotential surfaces, 549–550, 565
Equivalence, principle of, 838
Escape speed, 205–206
Estimates, 8, 162–163
Ether, 753, 814–817, 852
Evaporation, 468, 470
Excited state, 888
Exclusion principle, 902, 915
Exothermic reactions, 934
Expansion of universe, 836–837, 976
Experiments, role of, 3–4
Exponential decay, 945
Exponential function, 945
Extended bodies at rest, 217–250
  center of gravity, 220

center of mass, 218–219, 230–233, 242
equilibrium, 227–230, 233–241, 242
rigid bodies, 218
torque, 220–226, 242
Extended light sources, 687, 706
External interactions, 52
Eye, human, 736
Eyepiece (microscope), 742

F

Fahrenheit scale, 364
Far point, 736
Farad (unit, F), 559
Faraday, Michael, 653, 978
Faraday's law of electromagnetic induction, 659–662
Farsightedness, 736, 737–738
Fermi, Enrico, 940
Ferromagnetic materials, 640–641
Feynman, Richard, 970
Fiber optics, 701–702
Field lines
electric, 540–542
magnetic, 618–620
First law of thermodynamics, 441–475.
See also Heating
derived, 448–450
gas processes, 451–455
stated, 449, 470
1st order maximum, 754
Fission, 937–938
Flavors (quarks), 968
Floating, 388–389, 403–405. *See also*
Buoyant force
stability of ships, 404–405
Flow rate, 418–419, 435
Fluid dynamics. *See* Fluids in motion
Fluids in motion, 415–440
applications and process analysis, 424–427
Bernoulli bar charts, 422–423
Bernoulli's effect, 416–418
Bernoulli's equation, 420–423, 435
continuity equation, 419, 435
drag force, 420, 431–434, 435
flow rate, 418–419, 435
laminar flow, 420
moving across surfaces, 416–418
Poiseuille's law, 429–431, 435
pressure and, 416, 417, 428–429
speed and, 417
Stokes's law, 431–432
streamlines, 420
terminal speed, 432–433
turbulent flow, 420
viscous fluid flow, 428–431
work and energy, 421–422
Fluids, 355. *See also* Fluids in motion;
Static fluids
Archimedes' principle, 400
buoyant force, 398–401, 407
density of, 387–389, 407

Pascal's first law, 389–390, 407
Pascal's second law, 393–394, 407
pressure inside, 389–391
viscous fluids, 428–431
Flywheels, 273
Focal length, 717, 731, 744
Focal plane, 717, 719, 728
Focal point, 717, 726
lenses, 726
mirrors, 717, 719, 744
virtual, 719, 726
Force, 52–59, 77. *See also* Electric forces;
Magnetic forces; Pressure; Torque
acceleration and, 61–66, 70–75
adding force vectors, 55
contact and noncontact, 54–55
defined, 52, 77
equal and opposite, 70–75
friction, 89–95
gravitational, 66–67, 133–138
interactions, 52–55
measuring, 55–56
net, 55
Newton's second law, 61–66
Newton's third law, 70–75
normal, 90
nuclear, 928–929
restoring, 286
scalar components, 75–77
tension, 55, 87
unit of (newton), 52, 64
vector components, 85–87, 109
Force diagram, 53–54, 77
Forced convection, 466
Forced vibration, 305–306
Franklin, Benjamin, 505
Free-body diagram. *See* Force diagram
Free electrons, 508, 575
Free fall, 33–34
Free-fall acceleration, 66
Freezing, 459–460
freezing point, 364
Frequency, 287–288
beats, 335
cutoff, 856, 861, 864, 875
Doppler effect, 340–344, 345
electromagnetic waves, 795–796
frequency spectrum, 334
natural, 308
sound, 331–332, 333–335
vibrational motion, 287, 308
waves, 319, 331
Frequency spectrum, 334
Friction, 89–95, 109, 192–194
energy, work and, 192–194
inclines and, 99–100
kinetic, 94–95, 109
rolling, 95
static, 89–93, 109
in vibrational motion, 304–305
Fringes (interference), 752
Fundamental (sound), 334

Fundamental frequency, 334, 336–337
Fundamental interactions, 962–966, 970, 979.
See also Interactions
mechanisms of, 963–964
Fuses, 599
Fusion
heat of, 459–460
nuclear, 935–937

G

g field, 537
Galaxies, 974
Galileo, 2, 5, 33, 741
Galilean velocity transformation equation, 827
Galvanometer, 650–651
Gamma decay, 924, 941, 952
Gamma ray photons, 959–960
Gamma rays, 797, 924
Gases, 352–385. *See also* Ideal gases
Boyle's law, 370
characteristics of, 355–356
Charles's law, 370
density, 359, 379, 387
diffusion, 378
entropy of expanding, 483, 484–485
first law of thermodynamics, 451–455
Gay-Lussac's law, 370
heating of, 446, 451
mass of, 360
moles, 360, 379
phase change, 458, 460–461
pressure, 356–359, 379
spectra of, 884, 897
speed of particles, 361–363, 373–374
temperature and, 352, 364–368, 379
thermal energy, 376, 379, 442–443
universal gas constant, 366
work, 443–445, 470
Gauge pressure, 358
Gay-Lussac's law, 370
General relativity, 838–840. *See also* Special
relativity
black holes, 839–840
curvature of space, 838–839
gravitational red shift, 839
gravitational time dilation, 839
gravitational waves, 839–840
modified general relativity, 977
precession, 839
spacetime, 839
Generalized impulse-momentum principle, 156–159
Generalized work-energy principle, 184–186, 208
Geodesic lines, 838
Global positioning system (GPS), 136, 794–795, 813, 840–841
Global warming, 441
Gluons, 964–965
GPS. *See* Global positioning system
Grand unified theories, 975

Graphical representation, 22–24. *See also*
    Bar charts
    bar charts, 149
    field lines, 541, 618
    force diagrams, 53–54, 55
    motion diagrams, 15–18, 42
    position-versus-time graphs, 22–23
    ray diagrams, 687, 728
    slope, 25
    spacetime diagrams, 824–826
    trendline, 24
    velocity-versus-time graphs, 29–30
    vectors, 18–20
Gratings, 760–764, 777
Gravitational constant, 135
Gravitational field, 536, 537
Gravitational force, 66–67, 77, 133–138, 139
    compared with electric, magnetic forces,
      513–514, 626
    Kepler's laws and, 136–137
    law of, 66–67
    Newton's third law and, 135–136
Gravitational interactions, 962, 965
Gravitational potential energy, 178, 204–207,
    208
    Bernoulli's equation, 422
    defined, 205
    quantifying, 186–187
    zero level of, 186
Gravitational red shift, 839
Gravitational time dilation, 839
Gravitational waves, 839–840
Gravitons, 964, 965
Gravity, 52–53, 77, 133–138. *See also*
    Gravitational force
    escape speed, 205–206
    gravitational interactions, 962, 965
    law of universal gravitation, 133–138, 139,
      204
    planetary motion and, 133–137
Gray (unit, Gy), 949
Green quarks, 969
Greenhouse effect, 303–304, 468–469
Ground state, 888, 909
Grounding, 511, 553–554
Gyroscopes, 271

H
Hadrons, 966–968
Half-life, 821, 943–944, 946, 952
Half-wave dipole antenna, 791–792
Harmonic motion, simple, 290, 292–295, 308
Harmonics (sound), 334
Heat engines, 478–480
Heat of fusion, 459–460
Heat of vaporization, 460, 461
Heat pumps. *See* Thermodynamic pumps
Heating, 446–451, 455–470, 470. *See*
    also First law of thermodynamics;
    Temperature
    of a gas, 446, 451
    conduction, 463–465, 470

convection, 465–467, 470
defined, 446, 470
dielectric, 795
energy transfer, 447, 451, 455–458, 470
evaporative, 468, 470
mechanisms, 463–469
microwaves, 795
passive solar, 467
phase change, 458–463, 470
quantitative analysis of, 450
radiative heating/cooling, 467, 470
specific heat, 455–458
units (joule, calorie), 446
work and, 447
Heating rate, 463
Heisenberg, Werner, 912, 924
Henry (unit, H),, 672
Hertz, Heinrich, 790
Hertz (unit, Hz), 288
Higgs mechanism, 971
Higgs particle, 971
High-temperature plasma, 895
Hole (semiconductor), 606–607, 872–873
Hooke's law, 190–191, 292
Horsepower (unit, hp), 203
House wiring, 589, 599
Hubble, Edwin, 836–837
Hubble constant, 836
Hubble's law, 836–837
Hubble Space Telescope, 976
Huygens, Christiaan, model of light, 704
Huygens' principle, 331, 345
Hydraulic lift, 390–391
Hydrogen atom
    energy states of, 889–890
    photons emitted by, 889–890
    size of, 886
Hydrogen bonds, 913–914
Hydrogen spectrum, 763
Hyperopia, 736, 737–738
Hypothesis, 3–4
    testing, 53

I
Ice, 388–389
    phase change, 458–459
Ideal battery. *See* Internal resistance
Ideal gas law, 365–367, 379
    absolute temperature scale, 365–367
    applications of, 370–371
    limitations of, 374
    processes, 368–370
    testing, 368–372
Ideal gas model, 356, 379
Ideal gases, 356
    Boyle's law, 370
    Charles's law, 370
    Gay-Lussac's law, 370
    isobaric processes, 368, 369, 370
    isochoric processes, 368, 369, 370
    isothermal processes, 368, 369, 370
    Maxwell speed distribution, 373

quantitative analysis, 361–364
thermal energy, 376, 379, 442–443
Ideal polarizer, 788
Image
    inverted, 718
    lenses, 728–730
    mirrors, 713–714, 717–720
    virtual, 714, 718
Image distance, 721–723, 744
Impedance, 327
Impulse, 147, 153–159, 168
    defined, 154
    generalized impulse-momentum
      principle, 156–159, 168
    impulse-momentum bar charts, 157–159
    momentum and, 153–156
Impulse-momentum bar charts, 157–159
Impulse-momentum equation, 154,
    155–156
Incandescent lightbulbs, 578–579,
    583, 606
    power of, 591
Incident light, 690, 696
Inclines, 99–100
Index of refraction. *See* Refractive index
Induced current. *See* Electromagnetic
    induction
Inductance, 672
Induction. *See* Electromagnetic induction
Inductive reactance, 672, 673–674
Inertia
    moment of, 259
    rotational, 259, 262–266
Inertial reference frames, 60–61, 133,
    818–819
Inflation, 973
Infrared radiation, 848
Infrared waves, 797
Infrasound, 331
Instantaneous acceleration, 32
Instantaneous velocity, 30
Insulators, 507, 509–510, 528
Intensity, 326
    of electromagnetic waves, 801
    of light, 808
    of sound, 332–333
    of waves, 326, 345
Intensity level, 332–333, 345
Interactions, 52–55, 962–966, 970
    electromagnetic, 963, 964, 965
    electrostatic, 501–507, 536–542
    fundamental, 962–966, 970, 979
    gravitational, 962, 965
    mediators, 964–965
    nonfundamental, 962
    residual, 963
    strong, 963, 964, 965
    weak, 963, 965
Interference, 752–768
    bands/fringes, 752, 759
    constructive vs. destructive, 330, 753
    double-slit, 752–757, 777

electrons, 905
1st order maximum, 754
gratings, 760–764
mechanical waves, 328-330
*m*th bright band, 754–755
thin-film, 764–768, 777
electromagnetic waves (light), 752–757, 760–771
X-ray, 871–872
0th order maximum, 754
Interference pattern, 753
Interferometers, 815–816, 840
Internal energy, 178, 193, 208
Internal interactions, 52
Internal resistance, 594–595
Interrogation, 9
Invariance, 817, 838
Inverted image, 718
Ionizing radiation, 949–951
absorbed dose, 949–950
sources of, 951
Iris, 736
Irreversible processes, 477–480
Isobaric processes, 368, 369, 370
Charles's law, 370
first law of thermodynamics, 452
Isochoric processes, 368, 369, 370, 452
Isolated system, 148, 168
Isoprocesses, 368
Isothermal processes, 368, 369, 370
Boyle's law, 370
first law of thermodynamics, 451–452
gas expansion, 484
Isotopes, 927–928, 952
half-life, 943–944

J
Jeopardy-style problems, 38–39, 46
Jet propulsion, 163–164
Joliot-Curie, Irène and Frédéric, 926
Joule (unit, J), 180, 186, 191, 446, 519, 547
Joule's law, 589–592, 608

K
Kelvin temperature scale, 364, 365–367
Kepler's laws, 136–137
Kilogram (kg), 63
Kilowatt-hour, 591
Kinematics, 13–50
constant acceleration, 30–37
constant velocity linear motion, 24–30
defined, 22
one-dimensional motion, 13–50
rotational, 252–256
vibrational motion, 288–292
Kinetic energy, 178, 187–189
defined, 187, 208
relativistic, 831–832, 842
rotational, 271–274, 276
vibrational systems, 295
Kinetic friction force, 90, 94–95, 109
Kirchhoff's junction rule, 595, 600, 608

Kirchhoff's loop rule, 592–594, 600, 608, 675
Knowledge, construction/pursuit of, 369–370, 978

L
Laminar drag force, 431, 435
Laminar flow, 420
Large Hadron Collider (LHC), 966, 971, 976
Lasers, 897–899
Latent heat of fusion, 459–460
Lavoisier, Antoine, 148
Law of constancy of mass, 148–149
Law of reflection, 690
Law of universal gravitation, 133–138, 139
Laws, 2–3, 818
LCDs (liquid crystal displays), 806–807
LEDs (light-emitting diodes), 583–585, 593–594, 606–607, 776, 872–874
blinking (in mechanics), 16, 58–59
opening voltage, 584
power of, 591
structure of, 874
Length contraction, 822–824, 842
Lenses, 725–738
angular magnification, 739–740, 744
chromatic aberration, 759–760
coatings, 767
concave, 726–727, 744
convex, 725–726, 738, 744
focal length, 726, 731, 744
focal point, 726
glasses, 712, 728
image distance, 744
image location, 728–730
linear magnification, 732–733
in microscopes, 742
nearsightedness and farsightedness, 736–738
optics of human eye, 736
qualitative analysis, 725–730
quantitative analysis, 730–733
ray diagrams, 727–728
resolving ability, 773–774
single-lens optical systems, 735–738
thin lenses, 730–731
Lenz's law, 657, 658–659, 678
Leptons, 966–967
Light. *See also* Color; Light rays; Light waves; Optics; Quantum optics
absorption, 691, 888–889
diffraction, 768–772, 777
electromagnetic wave model, 790–791, 852, 857
emission of, 888–889
ether model, 852
fiber optics, 701–702
incident, 690, 696
intensity, 808
interference, 752–768

mirages, 702–703
particle model of, 704, 752–757, 852
photons, 864–867
polarization, 802–807
prisms, 702, 758
propagation, 686–689
quantum model of, 852
ray diagrams, 687, 696
reflection, 685, 689–692, 698–701, 706
refraction, 685, 692–696, 698–701, 706
shadows and semi-shadows, 688
sources, 686–688, 706
speed of, 818
visible, 797, 894
wave model of, 704–705, 771, 776, 852
Light-emitting diodes. *See* LEDs
Light field photography, 735–736
Light rays, 687–688, 706
ray diagrams, 687, 696
Light waves. *See also* Wave optics
ether and, 753, 814–817, 852
polarization, 802–807
spectrometers, 762–763
Light-year (unit, ly), 824
Lightbulbs, 578–579, 583, 606, 687
Lightning, 535
Limit of resolution, 773
Line spectrum, 884
Linear magnification, 723–724, 732–733, 744
Linear momentum, 147–175, 168. *See also* Momentum
defined, 151
impulse and, 153–156
Linear motion, 14–15
at constant acceleration, 35–37
at constant velocity, 24–30, 37–38
in one dimension, 13–50
quantities for describing, 21–22
Linearizing data, 74–75
Linearization-style problem, 74–75
Linearly polarized waves, 786
Liquid drop model of nucleus, 938–939
Liquids. *See also* Fluids
density of, 387
phase change, 458–461
specific heats, 456
viscosity, 429–430
Longitudinal waves, 317
Lorentz velocity transformation. *See* Velocity transformation
Low-temperature plasma, 895
Lyman series, 890

M
MACHOs (massive compact halo objects), 975
Macrostate, 481
Magnetic dipole moment, 627–628
Magnetic domains, 641
Magnetic field lines, 618–620

Magnetic fields, 618–621, 642. *See also* Magnetic forces
   circular motion in, 630–631
   cosmic rays, 631–632
   of bar magnet, 618
   of current-carrying wire, 620, 633
   of circular loop, 620, 633
   defined, 618
   deflection of radiation by, 923–924
   direction of, 617, 618, 634, 642
   electric current induced by, 650–654
   electric currents and, 619–621, 624–625, 632–634
   field lines, 618, 642
   force on moving charged particle, 629–630
   of horseshoe magnet, 618, 621
   right-hand rule for, 620, 642
   of solenoid, 620, 633
Magnetic flux, 654–656, 678
Magnetic forces, 621–632, 642. *See also* Magnetic fields
   compared with gravitational, electric forces, 626
   on current-carrying wire, 621–628
   direction of, 621–622, 626, 628–629
   magnitude of, 642
   on moving charged particle, 628–632
   right-hand rule for, 622, 630, 642
   on single moving charged particle, 628–629
   strength of, 625–626
   torque and, 627, 642
Magnetic induction tomography, 684
Magnetic permeability, 633, 672–673, 789
Magnetic quantum number, 900–901, 902–903
Magnetic resonance imaging (MRI), 616, 647
Magnetic torque, 627, 642
Magnetism, 616–648, 677, 963. *See also* Electromagnetic induction; Magnetic fields; Magnetic forces
   compared with electricity, 677
   diamagnetic materials, 640
   electromagnets, 621, 641, 662–664
   ferromagnetic materials, 641
   magnetic interactions, 617
   paramagnetic materials, 641
   poles of magnet, 617, 642
   properties of atoms, 640
   properties of various materials, 639–641
Magnification
   angular, 739–740, 744
   compound microscope, 742–743
   linear, 723–724, 732–733, 744
   telescopes, 740–741
Magnifying glasses, 728, 739–740
Malus's law, 788
Mass, 62–63, 148–149. *See also* Center of mass
   acceleration and, 62–65
   as conserved quantity, 149
   constancy of mass in isolated system, 148–149

defined, 63
   energy and, 832, 867
   gravitational force and, 134–135
   mass defect, 929–930, 952
   molar mass, 360
Mass defect, 929–930, 952
Mass number, 926
Mass spectrometer, 636–638, 927
Matter, 353–355, 979. *See also* Atomic structure; Elementary particles; Particle physics
   antimatter (antiparticles), 958–962, 979
   atoms, 355
   dark matter, 974–976, 979
   gases, liquids, and solids, 355–356
   particle model of, 355
   phase change, 458–463
   Standard Model, 970–971, 975, 979
   structure of, 353–355
Maximum static friction force, 90, 92
Maxwell speed distribution, 373
Maxwell's equations, 677, 795
Mechanical energy, 184
Mechanical waves, 315–351. *See also* Sound; Waves
   defined, 316
   types of, 317
Mechanics, Newtonian, 51–83
   interactions, 52–55
Mediators, 964–965
Meitner, Lise, 938–939
Melting, 459–460
Mesons, 967–968
Metals, 853
   cutoff frequencies, 864
   structure, 853
   work function, 853
Metastable states, 898
Michelson-Morley experiment, 813, 815–817, 818
Microscopes, 740, 741–743
Microstate, 481
Microwaves, 794
Millikan, Robert, 861–862
Mirages, 702–703
Mirror equation, 721–723
Mirrors, 689–690, 712–724
   center of curvature, 716
   concave, 715–719, 744
   converging, 717
   convex, 715–716, 719–721, 744
   focal length, 717, 744
   focal plane, 717, 719
   focal point, 717, 744
   image distance, 721–723, 744
   image location, 714–715, 718, 720
   magnification, 723–724
   mirror equation, 721–723
   plane, 713–715, 744
   principal axis, 716
   real inverted image, 718
   reflection, 716–717

rotating, 698–699
   virtual focal point, 719
   virtual image, 714, 718, 744
Models, 5
Modeling, 5–6
Molar mass, 360
Mole, 360, 379
Molecule, 355
Moment of inertia, 259
Momentum, 147–175
   collisions, 149–150, 160–162, 164–167, 199–200
   as conserved quantity, 156, 166
   constancy of isolated system, 151, 155–156
   generalized impulse-momentum principle, 156–159, 168
   impulse and, 153–159, 168
   impulse-momentum bar charts, 157–159
   jet propulsion, 163–164
   linear, 147–175, 168
   of photons, 866–867, 875
   radioactive decay, 166–167
   relativistic, 828–830, 842
   rotational, 266–271, 276
Monochromatic waves, 759–760, 777
Moon, orbital motion, 134
Motion, 14–15. *See also* Circular motion; Kinematics; Linear motion; Newton's laws; Rotational motion; Vibrational motion
   analyzing, 37–41
   conceptual description, 15–18
   at constant acceleration, 30–34
   at constant velocity, 24–30, 37–39
   defined, 14
   force and, 56–59
   free fall, 33–34
   modeling, 14–15
   Newton's laws of, 60–75, 77
   one-dimensional, 13–50
   projectile motion, 102–107, 109
   reference frames, 14, 23–25, 42, 60–61
   relativity of, 14
   representing with data tables and graphs, 22–24
   rolling, 142
   translational, 218
Motion diagrams, 15–18, 23, 42, 288–289
Motional emf, 666
Motors, 626–627
Multi-electron atoms, 907–909
Multimeter, 581
Muon decay experiment, 821–822
Muon neutrino, 966
Muons, 966–967
Musical instruments, 334, 340
Myopia, 736, 737

N

n (quantum number), 887
n-type semiconductor, 873

Natural convective heating, 466
Natural frequency, 308
Near point, 736
Nearsightedness, 736, 737
Negative electric charge, 502, 506, 528
Negative ion, 506
Neon bulbs, 578–579
Net force, 55
Neutralinos, 975
Neutral object, 506
Neutral particle, 925–926
Neutral radiation (neutron), 926
Neutrinos, 940–941
Neutrino oscillation, 967
Neutrons, 926–927
Newton, Isaac, 61
Newton (N), 52, 64
Newtonian mechanics, 51–83
Newton's law of universal gravitation,
    133–138, 139, 204
Newton's laws of motion, 60–75, 77
    applications of, 84–117
    inertial reference frames, 60–61
Newton's first law of motion, 60–61, 77
    stated, 61
Newton's second law of motion,
    61–66, 77
    applications of, 85–87, 101–102
    circular motion component form,
        127–128, 133
    in component form, 97–98, 109
    inertial reference frame and, 133
    for one-dimensional processes,
        67–70
    for rotational motion, 260–264
    stated, 64
    vector components, 75–77
Newton's second law for rotational
    motion, 260–266
Newton's third law of motion, 70–75, 77
    electric force and, 514
    equal-magnitude, opposite forces, 72,
        77, 623
    gravitational force and, 135–136
    stated, 72
Newton's model of light, 704
Nodes, 338
Non-ohmic devices, 582–584
Noncontact forces, 43–44
Nonfundamental interactions, 962
Normal forces, 90
Normal line, 690
Normal vector, 627
North pole, 617
Nuclear binding energy. See Binding
    energy
Nuclear decay. See Radioactive decay
Nuclear fission, 937–938
Nuclear force, 928–929, 963
Nuclear fusion, 935–937
Nuclear model of the atom, 881–884,
    915, 924–928

Nuclear physics, 921–956. See also
    Radioactive decay; Radioactivity
    binding energy, 929–931, 935, 952
    fission, 937–938
    fusion, 935–937
    ionizing radiation, 949–951
    mass defect, 929–930, 952
    nuclear force, 928–929
    nuclear reactions, 932–934
    nuclear sources of energy, 935–939
    radioactive dating, 947–949
    radioactive decay, 939–942, 952
    radioactivity, 922–924, 952
    terminology, 952
    uncertainty principle, 924
Nuclear potential energy, 929
Nuclear reactions, 932–934
Nucleons, 926, 931
Nucleosynthesis, 973
Nucleus. See Atomic nucleus

O
Object of interest, 14
Objective (microscope), 742
Observational experiment, 3, 10
Observer, 14
Ohm (unit), 582
Ohmic devices, 582–583
Ohm's law, 581–586, 608
    stated, 582, 608
One-dimensional motion, 13–50
Open-closed pipe, 339–340
Open-open pipe, 338–339
Opening voltage, 584
Optical axis. See Principal axis
Optical boundary, 696
Optical fiber, 701–702
Optical instruments
    cameras, 734–736
    chromatic aberration, 759–760
    human eye, 736
    magnifying glass, 728, 732, 739–740
    microscopes, 741–743
    Rayleigh criterion, 773–774, 777
    spectrometers, 762–763
    telescopes, 740–741
Optical power, 738
Optics, 751–783. See also Quantum optics;
    Wave optics
    human eye, 736
    lenses, 725–738
    mirrors, 689–690, 712–724
Orbital angular momentum quantum number,
    900, 902
Oscillations. See Vibrational motion
Outcome of testing experiment, 3–4

P
p-n junction, 873
p-type semiconductor, 873
Pair annihilation, 960–961
Pair production, 959–960

Parallel circuits, 587–588, 598, 602
Parallel plate capacitor, 560
Parallel resistors, 596–597, 608
Paramagnetic materials, 640–641
Parent nucleus, 939
Particle accelerators, 833, 966, 971
Particle model
    of electric charge, 506
    of light, 704, 752–757, 852, 865
    of matter, 355
Particle physics, 957–981
    antiparticles, 958–962, 979
    axions, 975
    confinement, 970
    cosmology, 972–974, 979
    dark energy, 976–977, 979
    dark matter, 974–976, 979
    elementary particles, 958, 966–970,
        979
    fundamental interactions, 962–966,
        970, 979
    Standard Model, 970–971, 975, 979
Particles, 354
    Maxwell speed distribution, 373
    particle motion, temperature and,
        367–368
    rest energy of, 832
    thermal energy and, 446
    wave nature of, 904
    wave-particle duality, 864–866, 915
Pascal (unit, Pa), 358
Pascal's first law, 389–390, 407
Pascal's second law, 393–394, 395, 407
Paschen series, 890
Path length, 21, 42
Pauli, Wolfgang, 901–902, 940
Pauli exclusion principle, 902, 903, 915
Pendulum, 297–299, 308
Penumbra, 688
Performance coefficient, 493
Period, 126–127, 139, 287
    of simple pendulum, 298–299
    of vibration (spring system), 293–294
    vibrational motion, 287, 308
    of wave, 319
Periodic table of the elements, 909–910,
    928
Permanent magnet, 641
Permeability. See Magnetic permeability
Permittivity. See Vacuum permittivity
PET scans, 957, 962
Phase, of waves, 320
Phase change, 458–463, 470, 764
Photocells, 872
Photocopier, 500, 526–527
Photocurrent, 856
Photoelectric effect, 853–864, 875, 963
    Einstein's hypothesis, 859–861
    electromagnetic wave model, 857–858
    quantum model explanation, 859–864
Photography, 735–736. See also Cameras
    purple fringing, 759

Photons, 864–867, 875
  dual nature of, 864–866, 915
  electromagnetic interactions, 964
  emission of, 888–889, 894
  energy of, 885
  gamma ray photons, 959–960
  momentum of, 866–867, 875
  particle-like properties, 864
  temperature and, 894–895
  virtual, 964
  wave-like properties, 864
  X-ray, 869–870
Phototube, 855
Physical quantities, 6–8
Pilot wave model, 907
Pinhole camera, 688–689
Pitch (sound), 333–334
Planck's constant, 851, 861
Planck's hypothesis, 851–853
Plane mirrors, 713–715, 744
Planetary (atomic) model, 881–884
Planetary motion, 2–3, 133–137
  Kepler's laws of, 136–137
  precession, 839
Plasma, 895
Point-like objects, 5, 513
Point source of light, 688
Poiseuille's law, 429–431, 435
Poisson spot, 771
Polar molecules, 556
Polarization, 509, 785–788, 802–807, 808
  Brewster's law, 803
  light polarizers, 802–803
  Malus's law, 788
  polarizing angle, 803
  by reflection, 803–804, 808
  by scattering, 805–806
  unpolarized light, 802
  unpolarized waves, 786
Polarizers, 786, 802–803
Poles (of magnet), 617, 642
Population inversion, 898
Position, 21, 42, 292
  during constant acceleration, 34–37
  as function of time, 289–291
  rotation and, 221–222
Position equation, 26–28
Position-versus-time graphs, 22–23
Positive electric charge, 502, 506, 528
Positive ion, 506
Positron emission tomography (PET), 957, 962
Positrons, 959
Potential difference, 548–549. *See also* Electric potential
Potential energy, 178
  elastic, 178, 190–192, 208
  electric, 516–521, 528
  gravitational, 178, 186, 204–207, 208
  nuclear, 929
Power, 202–204, 208
  of waves, 326

Powers of 10, 6, 8
Precession, 839
Prediction, 3
Prefixes, 6
Pressure
  altitude and, 405–406
  atmospheric pressure, 395–398
  buoyant force, 398–401, 407
  defined, 357, 379
  depth and, 391–395
  gases, 356–359
  in gas processes, 451–455
  gauge pressure, 358
  inside a fluid, 389–391
  measuring, 358
  Pascal's first law, 389–390, 407
  Pascal's second law, 393–394, 395, 407
  P-versus-*V* graphs, 443–445
  static fluids, 289–291
  unit of (pascal), 358
Pressure waves, 332
Primary coil, 675–676
Principal axis, 716
Principal quantum number, 887, 902
Principle of equivalence, 838
Prisms, 702, 758
Problem solving strategy, 11–12
Problem types
  estimation, 8
  equation jeopardy problems, 38–39, 46
  linearization problems, 74–75
  multiple-possibility problems 300–301
Projectile motion, 102–107, 109
Projectiles, 102
Proper length, 823, 842
Proper reference frame, 820–821, 822, 842
Proper time interval, 820–821, 822, 842
Proportions, 83, 515
Proton tunneling, 913–914
Protons, 924
Pulleys, 101–102
Pulsars, 251, 269–270
Pulses, wave propagation, 316–318, 321–323, 326

Q
Quanta, 851, 859
Quantization, 511, 851
  quantized orbits (electron), 885–891
Quantum chromodynamics, 971
Quantum efficiency, 867
Quantum electrodynamics, 970
Quantum mechanics, 907
Quantum numbers, 887, 899–903, 915
  additional, 899–901
  magnetic, 900–901, 902–903
  orbital angular momentum, 900, 902
  Pauli exclusion principle, 902, 903, 915
  principal (*n*), 887, 902
  spin magnetic, 901–902, 903
Quantum optics, 847–879
  black body radiation, 848–853, 875

light-emitting diodes (LEDs), 874–875
  photocells, 872
  photoelectric effect, 853–864, 875
  photons, 864–867, 875
  solar cells, 872–873
  X-rays, 867–872
Quantum tunneling, 913
Quarks, 968–970
  charm, top, bottom, 968
  color charge, 969
  flavors, 968–969
  up, down, strange, 968

R
Rad (unit), 949
Radar, 793–794, 835
Radial acceleration, 122, 123–127, 139
Radian, 252
Radiation, 467, 470
  climate change and, 468–469
  deflection by magnetic fields, 923
Radiation absorption, 303–304
Radiative heating/cooling, 467, 470
Radio waves, 793, 797
  FM, 796
Radioactive dating, 947–949
Radioactive decay, 166–167, 924, 939–942, 952
  alpha, 939, 942, 952
  beta, 940–941, 942, 952
  beta-minus, 941
  beta-plus, 941
  decay rate, 944–945, 946
  decay series, 948–949
  exponential, 945
  gamma, 941, 952
  mechanisms for, 939–942
Radioactivity, 922–924, 952. *See also* Radioactive decay
  activity of material, 944
  decay series, 948–949
  half-life, 943–944, 946, 952
  isotopes, 927–928, 952
Radium, 923–924, 939
Radius, acceleration and, 124–126
Radius vector, 289
Radon, 166–167
Rainbows, 710
Ray diagrams, 687, 696
  concave mirrors, 717–718
  convex mirrors, 720
  lenses, 727–728
Rayleigh criterion, 773–774, 777
RC circuit, 599
Real image, 718
Real inverted image, 718
Receiver (EM waves), 793
Red (quarks), 969
Red shift, 834, 839
Reference frames, 14, 23–25, 42
  inertial, 60–61, 133, 818–819
  noninertial, 60

proper, 820–821, 822, 842
relativity and, 818
simultaneity and, 818–819
time dilation, 820–821, 842
Reflection, 685, 689–692, 706
angle of, 690, 706
angle of incidence, 690, 706
critical angle of incidence, 696
diffuse, 690, 691, 706
fiber optics, 701–702
gratings, 762
and interference 764–766
law of, 690, 706
mirrors, 689–690, 716–717
polarization by, 803–804, 808
sky color, 703
specular, 690, 691, 706
total internal reflection, 696–698, 706
of mechanical waves, 317–318, 326–327, 338
Refraction, 685, 692–701, 706
angle of, 693
and Huygens principle, 704–705
mirages, 702–703
prisms, 702
Snell's law, 694, 706
wave model and, 705
Refractive index, 694, 757–759, 777
and color of light, 758
Refrigerators. See Thermodynamic pumps
Relative biological effectiveness (RBE), 949, 950
Relative magnetic permeability, 672–673
Relativistic effects, 819, 830
Relativistic energy, 830–833
Relativistic kinetic energy, 831–832, 842
Relativistic momentum, 828–830
Relativistic velocity transformation, 827
Relativity. See General relativity; Special relativity
Rem (unit), 950
Residual interactions, 963
Resistance, 582, 604, 608
batteries, 594–595
capacitive, 670–671
defined, 604, 608
Ohm's law, 581–586, 608
Resistivity, 603–605
Resistors, 580, 596–600, 602–607
in AC circuits, 668–670, 678
equivalent resistors, 601–602
in parallel, 596–597, 608
properties of, 602–607
RC circuits, 599
in series, 596, 608
Resolve, 773
Resolving power, 772–774
Resonance, 306–307
Resonant energy transfer, 307, 308
Rest energy, 831, 832, 842
Retina, 736
Reversible processes, 478

Revolution, 254
Reynolds number, 430–431
Right-hand rule, 270
for magnetic field, 620, 642
for magnetic force, 622, 630, 642
for rotational velocity, momentum, and torque, 270
Rigid bodies, 218
rotational motion of, 262–263, 276
Rockets, 163–164
Roentgen, Wilhelm, 868–869
Rolling friction, 95
Rolling motion, 142
Root-mean-square (rms),
current and voltage, 669
speed 363, 368
Rotational acceleration, 254, 259
Rotational inertia, 259
Rotational kinetic energy, 271–274, 276
Rotational momentum, 266–271, 276
isolated system, 267–268
right-hand rule for, 270
of shrinking object, 269–270
Rotational motion, 251–283
axis of rotation, 220–221
inertia and, 259, 262–266, 276
kinematics of, 252–256, 276
Newton's second law for, 260–266
physical quantities affecting, 257–259
rigid bodies, 262–263, 276
torque, 220–226, 257–259
vector nature of, 270–271
Rotational position, 252–253
Rotational velocity, 253–254, 270–271
changes in, 259
right-hand rule for, 270
Rough estimates, 8
Rutherford, Ernest, 881, 923, 932
Rutherford atomic model, 881–884
Rydberg constant, 884

S
Satellites, 137–138
Scalar components
of a vector, 85–87
of displacement, 21–22
Scalar quantities, 8–9
multiplication of a vector by a scalar, 20
Scattering, of light, 805–806
Schrödinger, Erwin, 958
Science process, 3–4
Scientific notation, 8
Seat belts, 51, 75–76
Second law of thermodynamics, 476–499
direction of processes, 478, 480
entropy, 480–488, 495
entropy change, 484–486
irreversible processes, 477–480
reversible processes, 478
stated using entropy, 483, 486, 495

thermodynamic engines, 478–480, 488–492, 495
thermodynamic pumps, 492–494, 495
Secondary coil, 675–676
Seismic waves, 318, 327
Seismometers, 654
Semi-shadow, 688
Semiconductors, 606–607, 873
Series circuits, 586–587, 598
Series resistors, 596, 608
Shadows, 688
Shells, 908
Shielding, 554–555
Short circuit, 595–596
SI system/units, 6
Sievert (unit, Sv), 950
Significant digits, 7–8, 22
Silicon, 606, 872–873
Simple harmonic motion (SHM), 290, 292–295, 308
Simple pendulum, 297–299, 308
Simultaneity, 818–819
Single-slit diffraction, 769–771, 777
Sky color, 703, 805
Slope, 25
Snell's law, 694, 706
Sodium, 607
Solar cells, 872–873
Solar constant, 801–802
Solar heating, 467
passive, 467
Solenoid, 620, 672–674
Solids, 355
density of, 387
heats of fusion and vaporization, 459–460, 461
phase change, 458–460, 461
specific heats, 456
Sommerfeld's extension of Bohr model, 899–901
Sound, 331–335
beats and beat frequencies, 334–335
Doppler effect, 340–344, 345
frequencies, 331–332, 333–335
frequency spectrum, 334
fundamental frequency, 334
harmonics, 334
infrasound, 331
intensity level, 332–333, 345
loudness, 332
pitch, 333–334
pressure waves, 332
ultrasound, 327, 331
waveform, 334
wind instruments, 334, 340
Sound waves, 318, 331–335, 705
Source (of waves), 316
Source charge, 538
South pole (of magnet), 617
Space, curvature of, 838–839
Spacetime, 839
Spacetime diagrams, 824–826, 842

Spark gap transmitter, 790
Special relativity, 813–846. *See also* General
    relativity
    Doppler effect for EM waves, 834–837,
        842
    expansion of the universe, 836–837
    Hubble's law, 836–837
    length contraction, 822–824, 842
    Michelson-Morley experiment, 813,
        815–817, 818
    postulates of, 817–818, 842
    relativistic effects, 819, 830
    relativistic energy, 830–833, 842
    relativistic kinetic energy, 831–832, 842
    relativistic momentum, 828–830, 842
    simultaneity, 818–819
    spacetime diagrams, 824–826, 842
    special notations, 830
    time dilation, 819–822, 842
    velocity transformations, 827–828
Specific heat, 455–458
Spectra, 762–763
    absorption spectrum, 897
    electromagnetic, 796–797
    lines, 884, 901
    spectrometers, 762–763
    stellar, 895–897
Spectral analysis, 892–897
    continuous spectra from stars, 895–897
    discharge tube excitation, 893–894
    line spectra from hot gases, 884
    thermal excitation, 892
Spectral curve, 848
Spectrometers, 636–638, 762–763
Specular reflection, 690, 691, 706
Speed, 26, 42
    defined, 26, 42
    of fluid flow, 417
    of gas particles, 361–363
    root mean square speed, 363
    terminal speed, 432–433
    of waves, 319, 321–323, 345
Speed of light
    in a vacuum, 818
    inertial reference frames, 818
    refractive index and, 757–758, 777
Spin, 902
Spin (magnetic) quantum number, 901–902,
    903
Spontaneous emission, 897
Spring constant, 190, 191
Springs
    cart-spring system, 292–295, 308
    elastic potential energy, 191–192, 295
    Hooke's law, 190–191
    period of vibration, 293–294
    spring constant, 190, 191
    vertical, 300–302
    vibrational motion, 285–287, 292–295,
        300–302
Stable equilibrium, 239
Stable orbits, 886, 887

Standard Model, 970–971, 975, 979
Standing waves, 335–340
    in air columns, 338–340
    antinodes, 338
    defined, 335, 345
    frequencies, 336–337, 339, 340
    fundamental frequency, 336–337
    nodes, 338
    open-closed pipe, 339–340
    open-open pipes, 338–339
    on strings, 335–338
    wind instruments, 340
Stars, 973–974. *See also* Black holes; Sun
    fusion in, 936
    pulsars, 251, 269–270
    supernovae, 936, 974
State function, 442
Static equilibrium, 221–222, 227–230,
    242
Static fluids, 386–414
    buoyant force, 398–402, 407
    Pascal's first law, 389–390, 407
    Pascal's second law, 393–394, 407
    pressure inside, 289–291
    pressure variation by depth, 391–395
Static friction force, 89–93, 109
    car acceleration and, 107–108
Steady state model, 976
Stefan-Boltzmann's law, 849, 875
Stellar spectra, 895–897
Step-down transformer, 675
Stern-Gerlach experiment, 901, 902
Stimulated emission, 897
Stokes's law, 431–432
Stopping distance, 160–162
Stopping potential, 858–859, 860, 875
Strange quarks, 968
Streamlines, 420
String theory, 971
Strong interactions, 963, 964, 965
Strong nuclear force. *See* Nuclear force
Subshells, 908–909
Sun
    gravitational field, 536
    how long will it shine?, 376–378,
        936–937
    mass of, 376, 867
    power of surface radiation, 850
    surface temperature, 376–377, 849
    thermal energy of, 376–378
Superconductivity, 605, 640
Supernovae, 936, 974
Superpartners, 975
Superposition principle, 327–329, 330,
    345, 752
    for electric fields, 539–540, 565
    electric potential and, 547
Supersymmetry, 975–976
System, 52, 77, 445
    defined, 52, 77
    entropy change of, 486
    equilibrium state, 482

T
Tangential velocity, 254
Tau, 966–967
Tau neutrinos, 966
Telescopes, 739
Temperature, 9, 364–368, 379. *See also*
    Climate change; Heating; Thermal
    energy
    absolute (Kelvin), 364, 365–367
    of gases, 352, 364–368
    in gas processes, 451–455
    particle motion and, 367–368
    phase change, 458–463, 470
    radiation and, 467, 468–469, 470
    specific heat, 455–458
    superconductivity and, 605
    thermometers, 364
Temperature change, 470
Temperature scales, 364, 365–367, 379
    conversions, 366
Terminal speed, 432–433
Tesla (unit, T). 625
Test charges, 538
Testing experiment, 3–4, 11
Thermal conductivity, 464
Thermal energy, 178, 449, 470. *See also* First
    law of thermodynamics
    defined, 470
    of ideal gas, 376, 379, 442–443
    particle motion and, 446
    transfer, 447, 451, 455–458, 470
Thermal equilibrium, 457
Thermal excitation, 892
Thermodynamic engines, 478–480, 488–492,
    495
    defined, 488
    efficiency, 490–492
    quantitative analysis, 488–491
Thermodynamic pumps, 492–494, 495
Thermodynamics, 441–499. *See also* First
    law of thermodynamics; Second law of
    thermodynamics
    adiabatic processes, 453
    entropy, 480–488, 495
    first law of, 448–450, 470
    heating, 446, 448–450
    irreversible processes, 477–480
    reversible processes, 478
    isobaric processes, 452
    isochoric processes, 452
    isothermal processes, 451–452
    second law of, 476–499
    specific heat, 455–458
Thermography, 467
Thermometers, 364
Thin-film interference, 764–768, 777
Thin lens equation, 730–731
Thin lenses, 730–731
Thomson, J. J., 868, 978
Thomson atomic model, 881
Tides, 274–275
Time, 21, 42

position as function of, 289–291
position-versus-time graphs, 22–23
velocity-versus-time graphs, 29–30
Time dilation, 819–822, 839, 842
Time interval, 21, 42
Top quarks, 968
Torque, 220–226, 257–259. *See also*
Rotational motion
axis of rotation, 220–221
current-carrying coil and, 627
defined, 224, 242
magnetic, 627, 642
net, 228, 240, 257–258, 259
sign of, 225
static equilibrium and, 227–230
vector nature of, 270–271
zero, 222, 225
Torricelli's experiments, 395–397
Total energy, 181, 184, 208
Total internal reflection, 696–698, 706
Transcranial magnetic stimulation (TMS),
649
Transformers, 674–676, 678
Translational motion, 218
Transverse waves, 317, 319
Traveling sinusoidal wave, 319–321, 345
Trendline, 23
Trilateration, 794
Tunneling, 913
Turbulent drag force, 431–432, 435
Turbulent flow, 420

U

Ultrasound, 327, 331
Ultraviolet catastrophe, 850
Umbra, 688
Uncertainty principle, 910–914, 915, 924
Uniform electric field, 542
inside a capacitor, 559
Uniform motion. *See* Constant velocity
linear motion
Unit, defined, 6
Units of measurement, 6–7
Universal gas constant, 366
Universal gravitation, law of, 133–138,
139
Universe
age of, 837
Big Bang, 972–973
cosmological constant model, 976
cosmology, 972–974, 979
dark energy model, 977
electroweak model, 971
expansion of, 836–837, 976
inflation, 973
modified general relativity, 977
nucleosynthesis, 973
proportion of matter in, 977
stars and galaxies, 973–974
steady state model, 976
Up quarks, 968
Uranium, 923, 938–939

V

V field. *See* Electric potential
Vacuum permeability, 789
Vacuum permittivity, 789
Van de Graaff generators, 524–526
Vector components, 85–87
Vector quantities, 9, 151
Vectors, 18–20, 85–87
addition of, 19
multiplication by a scalar, 20
normal, 627
radius vector, 289
subtraction of, 19–20
in two dimensions, 85–87
vector components, 85–87, 109
Vehicles. *See* Cars
Velocity, 42. *See also* Acceleration
average, 30
change, determined from acceleration,
32–33
constant velocity linear motion, 24–30,
37–39
defined, 26, 42
as function of time, 291–292
graphing, 29–30
instantaneous, 30
momentum and, 151
rotational, 252–253, 270–271
speed and, 26
tangential, 254
vibrational motion, 291–292
Velocity arrows, 16, 26
Velocity change arrows, 16–17
Velocity change method, 121–122, 139
Velocity transformation, 827–828
Velocity-versus-time graphs, 29–30
Vibrational motion, 284–314. *See also*
Mechanical waves
amplitude, 287, 308
damping, 304–305
defined, 308
energy and, 295–296
equilibrium position, 286, 308
external driving force, 305–307
frequency, 287–288, 308
friction and, 304–305
kinematics of, 288–292
period, 287, 298–299, 308
position as function of time, 289–291
resonance, 306–307
restoring force, 286
simple harmonic motion, 290, 292–295,
308
simple pendulum, 297–299, 308
springs, 285–287, 292–295, 300–302, 308
velocity and acceleration as function of
time, 291–292
Virtual focal point, 719, 726
Virtual image, 714, 718, 744
Virtual photons, 964
Viscosity, 429–430
Viscous, 428

Viscous fluid flow, 428–431
Vision
human eye, 736, 749
20/20, 782
Voltage, 548. *See* Potential difference
opening voltage of LEDs, 584
Voltmeters, 580–581
Volume, 355
in gas processes, 451–455
von Lenard, Philipp, 855–856, 859

W

Waste separator. *See* Eddy current waste
separator
Water. *See also* Ice
density-temperature dependence, 414
molecular structure, 511, 556
objects floating on, 388–389, 403–405
phase change, 458–459, 460–461
polar molecules, 556
Watt (W), 202, 590
Wave front, 317
Wave functions, 907
Wave model of light, 704–705, 771, 776, 852,
865
applications, 774–776
Wave motion, 316, 321–324
interference, 330
Wave optics, 751–783
coherent monochromatic waves, 759–760,
777
diffraction, 768–772
gratings, 760–764, 777
interference, 752–768
monochromatic light, 764–767, 777
Poisson spot, 771
resolving power, 772–774
spectra, 762–763
thin-film interference, 764–768, 777
Wave-particle duality, 864–866, 915
Wave speed, 319, 321–323, 345
Waveforms, 334
Wavelength, 320, 762, 795–796
Wavelets, 331, 704, 705
Waves, 315–351. *See also* Electromagnetic
waves; Light waves; Sound; Standing
waves
amplitude, 319, 324–325
coherent, 759–760
compression waves, 327, 331
Doppler effect, 340–344, 834–837
energy of, 324–325
frequency, 319, 331
gravitational, 839–840
Huygens' principle, 331, 345
impedance, 327
intensity, 326, 345
interference, 330
longitudinal, 317
mathematical descriptions of, 318–321
monochromatic waves, 759–760, 777
period, 319

Waves (*Continued*)
  phase, 320
  polarization, 785–788
  power, 326
  pressure waves, 332
  pulse propagation, 316–318, 321–323, 326
  reflection of, 317–318, 326–327, 764–765
  seismic, 318, 327
  sound, 318, 331–335, 705
  speed of, 319, 321–323, 345
  standing, 335–340, 345
  superposition of, 327–329, 330, 345,
      764–766
  three-dimensional, 325
  transverse, 317, 319
  two-dimensional, 324–325
  types of, 317
  wave front, 317
  wavelength, 320
Weak interactions, 963, 965
Weak nuclear interaction, 941
Weakly interacting massive particles.
      See WIMPs
Weight, 67
Weightlessness, 137–138
White light, 757, 759, 762
Wien's law, 849, 875

WIMPs (weakly interacting massive
      particles), 975–976
Wimshurst machine, 526, 573
Wind instruments, 334, 340
Work, 176–216
  Bernoulli's equation, 422–423
  defined, 180, 208, 470
  energy and, 177–181
  external forces and, 177–178
  friction and, 192–194
  in gas processes, 443–445, 470
  heating and, 447
  kinetic energy and, 178, 187–189
  negative work, 178–180, 184
  as physical quantity, 180–181
  positive work, 180, 184
  potential energy and, 178, 208
  power and, 202–204
  thermal energy and, 178
  unit of (Joule), 180, 186
  work-energy bar charts, 182–184,
      449
  work-energy principle, 184–186,
      208, 421–422, 445–447
  work-heating-energy equation, 445
  zero work, 178–180
Work-energy bar charts, 182–184, 449

Work-energy principle, 184–186, 208, 448
  fluid flow, 421–422
Work function, 853
World line, 824–825, 842

X

x-component of the displacement, 22
X-ray interference, 871–872
X-ray photons, 869–870
X-rays, 867–872, 922
  danger/safety of, 870–871, 950
  discovery of, 868–869
  ions produced by, 950
  properties, 797
  X-ray machine, 548–549

Y

Young's double-slit experiment, 752–757

Z

Zeeman effect, 899, 900–901, 902
Zero level of gravitational potential energy,
      186
Zero point energy, 977
0th order maximum, 754
Zero torque, 222, 225
Zero work, 178–180

Base Units of the SI System

The SI system of units is built on the seven base units listed below. All other derived units are combinations of two or more of these base units.

BASE UNITS OF THE SI SYSTEM

Base Quantity	Name	Abbreviation
Length	meter	m
Mass	kilogram	kg
Time	second	s
Electric current	ampere	A
Temperature	kelvin	K
Amount of substance	mole	mol
Luminous intensity	candela	cd

An example of a derived unit is the unit for kinetic energy, the joule (J), which is expressed in terms of the kg, m, and s:

$$1 \text{ J} = 1 \text{ kg} \left(\frac{1 \text{ m}}{1 \text{ s}} \right)^2$$

Definitions of the base units

meter (m) The distance that light travels in a vacuum in a time interval of $\left(\dfrac{1}{299,792,458} \right)$ s.

kilogram (kg) The mass of an international standard cylindrical platinum-iridium alloy stored in a vault in Sèvres, France by the International Bureau of Weights and Measures.

second (s) The duration of 9,192,631,770 vibrations of radiation emitted by a particular transition of a cesium-133 atom.

ampere (A) The constant electric current that, when flowing in two very long parallel straight wires placed 1 m apart in a vacuum, causes them to exert a force on each other of 2×10^{-7} newton per meter length of wire.

kelvin (K) $1/273.16$ of the thermodynamic temperature of the triple point of water.

mole (mol) The amount of a substance that contains as many elementary entities as there are carbon atoms in 0.012 kg of the carbon isotope carbon-12. The elementary entities could be atoms, molecules, ions, electrons, other particles, or specified groups of such particles.

candela (cd) The luminous intensity in a given direction of a light source that emits monochromatic radiation of frequency 540×10^{12} Hz and that has a radiant intensity in that direction of $1/683$ W per steradian.

Problem Solving in Physics

Physics problem solving starts with analyzing information given in a representation of a physical situation, such as a written problem statement (words), a video, or a photo. It ends with a numerical answer, another representation (graphical or algebraic), an argument, an evaluation, or an experimental design. For all problems, you need to start by translating the initial representation of the situation/process/task into physical quantities or processes. Visualizing the situation is extremely helpful, and thus making a sketch or a drawing of the situation/process is the first step that is necessary to solve any problem.

If the problem asks for a numerical answer, you obtain it from a mathematical representation. Both words and math are abstract representations of physical processes, and it is challenging to directly translate from one into the other. To bridge this gap, in this text you'll learn how to represent a described physical process with a drawing—a sketch, diagram, bar chart, etc., depending on the situation. These visual representations will help you identify what quantities you know and what you don't know, and how to find the right equations to use to solve the problem. You can also use them to evaluate your answer: is the answer consistent with the representation you used to write the equation?

For easy reference, the following tables list the Physics Tool Boxes and the type of representation you will learn in each, outline the general four-step problem-solving strategy used throughout the book, and list the Problem-Solving Strategies, which show how the strategy can be modified for broad classes of problems for different topics.

Physics Tool Box		Page	Example
2.1	Constructing a motion diagram	17	
3.1	Constructing a force diagram	54	
4.1	Determining the scalar components of a vector	86	
4.2	Using Newton's second law in component form	88	
5.1	Estimating the direction of acceleration during two-dimensional motion	121	
6.1	Constructing a qualitative impulse-momentum bar chart	157	
7.1	Constructing a qualitative work-energy bar chart	184	
14.1	Constructing a bar chart for a moving fluid	423	
18.1	Estimating the $\vec{E}$ field at a position of interest	540	
21.1	Determining the direction of an induced current	657	
23.1	Constructing a ray diagram to locate the image of an object produced by a concave mirror	718	
23.2	Constructing a ray diagram to locate the image produced by a convex mirror	720	
23.3	Constructing a ray diagram for single-lens situations	727	

Problem-Solving Strategy

The following general four-step problem-solving strategy is best used for problems asking for a numerical answer. If a problem is qualitative, the first two steps apply; if the problem is about a simple calculation, the last two steps apply. For problems that ask for experimental design, evaluation of an argument, analysis of video or photo data, etc., you might need a combination of the steps.

Sketch and translate

- Convert the problem's everyday language into the language of physical quantities.
- Draw a sketch of the situation and label known and unknown quantities. Be sure to label the coordinate axes if necessary.

Simplify and diagram

- Make necessary simplifications. (What interactions can you neglect? What model can you choose for the object?)
- Draw an appropriate physics representation (from your Physics Tool Box).

Represent mathematically

- Use the physics representation to construct a mathematical representation of the same situation.
- Isolate the variable(s) that you need to solve for.

Solve and evaluate

- Plug the values of the physical quantities into the mathematical representation. (Do not forget the units!)
- Solve for the unknown variable(s).
- Evaluate your answer: is it reasonable? Is it consistent with the physics representation in the Simplify and diagram step? Are the units correct? Does it lead to the correct predictions in extreme cases?

Problem-Solving Strategy		Page
2.1	Kinematics	40
3.1	Applying Newton's laws for one-dimensional processes	67
4.1	Analyzing processes involving forces in two dimensions	96
5.1	Processes involving constant speed circular motion	128
6.1	Applying the generalized impulse-momentum principle	159
7.1	Applying the work-energy principle	195
8.1	Applying static equilibrium conditions	234
10.1	Analyzing vibrational motion	300
11.1	Analyzing wave processes	329
12.1	Applying the ideal gas law	374
13.1	Analyzing situations involving static fluids	401
14.1	Applying Bernoulli's equation	424
15.1	Using thermodynamics for gas law processes	453
17.1	Analyzing processes involving electric charges	521
19.1	Applying Kirchhoff's rules	600
21.1	Problems involving electromagnetic induction	662
24.1	Analyzing processes using the wave model of light	774

College Physics: Explore and Apply is more than just a book. It is your learning companion. As a companion, it will not just tell you about physics, it will serve as a guide to help you build your own physics ideas. The ideas will be yours, not somebody else's. As a result, these ideas will be much easier for you to use when you need them: to succeed in your physics course, to obtain a good score on exams, and to apply to your future career and everyday life.

On the cover of this book, a young woman with a backpack rides a bicycle toward a big city. This image symbolizes the process you will use to learn physics: from observing and exploring nature, to creating your own explanations, to the application of new ideas and tools to practical problems of everyday life. This is exactly what physicists do when they "do" physics. The physics you learn in this book will help you learn to think differently about everything else, to base your ideas on evidence not on authority.

At this point you should turn to the first chapter and begin reading. That's where you'll learn the details of the approach that the book uses, what physics is, and how to be successful in the physics course you are taking.

Mastering Physics

Mastering™ is the most effective and widely used online homework, tutorial, and assessment system for the sciences. Please visit pearson.com/mastering/physics to access the below items and much much more!

- The Pearson eText allows you to access your textbook any time anywhere. You can take notes, highlight, bookmark, and watch embedded videos. Available as an app for your smartphone—just visit your app store to learn more.

- Dynamic Study Modules (DSMs) help you study effectively on your own by continuously assessing your activity and performance in real time and adapting to your level of understanding. Also available as an app for your smartphone.

INTERCEPTAG
Authentic P Pearson

Please visit us at **www.pearson.com** for more information. To order any of our products, contact our customer service